# Chemical Analysis

# Chemical Analysis

Kenneth A. Rubinson

**Little, Brown and Company**
Boston    Toronto

Library of Congress Cataloging-in-Publication Data

Rubinson, Kenneth A.
  Chemical analysis.

  Includes index.
  1. Chemistry, Analytic.  I. Title.
QD75.2.R83    1986        543        86-18528
ISBN 0-316-76087-0

Library of Congress Catalog Card No. 86-18528


ISBN 0-316-76087-0


9  8  7  6  5  4  3  2  1

DON

Published simultaneously in Canada
by Little, Brown & Company (Canada) Limited

Printed in the United States of America


Produced by R. David Newcomer Associates
Interior Design by Brenn Lea Pearson
Copyedited by Yvonne Howell
Illustrations by Irene Imfeld Design

### Credits

  Fig. 7.1. Adapted from E. B. Sandall, *Colormetric Determi-
nation of Trace Metals,* 3rd ed. (New York: Wiley–
Interscience, 1959). Reprinted by permission of John Wiley &
Sons.
  Excerpts pp. 268–269. Reprinted with permission from
P. A. Siskos et al., *Talanta* 30 (1983): 980–982. Copyright
1983, Pergamon Press, Ltd.

*(continued on page 911)*

# Preface

*It is the mark of an educated mind to expect that amount of exactness which the nature of the particular subject admits. It is equally unreasonable to accept merely probable conclusions from a mathematician and to demand scientific proofs from an orator.*

Aristotle
Ethics, *Book 1, Chapter 3*

## Text Organization

The logical connections of the text chapters are shown in the figure on page vi. Each chapter is written so that the most general topics are presented early in the chapter and the presentation of the details progresses smoothly from simple to more complex and complete. It is hoped that this organization allows the reader to attain progressive depths of coverage of each topic simply by stopping further along in a given chapter. Topics that would interfere in this progression are separated into Supplementary Sections that appear at the ends of chapters. These include such topics as the details of complexation equilibrium calculations, more specific nomenclature of spectrometry, the general principles of operation of parts of various instruments, and mathematical proofs and demonstrations.

The general principles of analysis (e.g., sampling, the need for a measurement, the presence of upper and lower limits to the instrumental or wet chemical method, the inevitability of interferences, and the methods used to overcome difficulties in carrying out the operations) are covered in the four chapters of Part II (Chapters 4 – 7). Included are the techniques used with a number of different analytical methods, such as statistical tests, standardization, and sample preparation.

In order to give some sense of how typical problems involving chemical analysis are solved, where appropriate, readings are quoted that describe how the techniques of the chapter are used in conjunction with others in finding a solution. Among the topics included are automatic analysis methods, the question of whether there is blood on the Shroud of Turin, identification of paint samples in forensic analysis, and electrochemical analysis *in situ* in the brain. Some advanced topics are presented, though less rigorously, in this way as well.

Logical Connections of the Text Chapters

The aim of *Chemical Analysis* is to bring readers to the point where they can read specialized texts and compendia (e.g., Kolthoff and Elving) and allow an understanding of how the details fit into the whole problem-solving framework and the types of problems they or their co-workers might face.

## Acknowledgments

I am grateful to the following individuals who have given me the benefit of their deep expertise in suggesting ways to improve the presentation of the topics and correcting errors. Any errors and obfuscation that remain are my own.

*Charles Boss*, North Carolina State University; *Joseph A. Caruso*, University of Cincinnati; *J. A. Cox*, Southern Illinois University; *Johannes Coetzee*, University of Pittsburgh; *Alice J. Cunningham*, Emory University; *Richard A. Day*, University of Cincinnati; *Thomas Fogg*, Science Applications International; *William T. Gilbert*, University of Cincinnati; *Ted Haupert*, Sacramento State University; *Harvey Herman*, University of North Carolina; *John Lanning*, University of Colorado; *Chui Liu*, Arizona State University; *Charles Lochmuller*, Duke University; *Harry B. Mark, Jr.*, University of Cincinnati; *Hershel Markovitz*, Carnegie-Mellon University; *Linda B. McGown*, Oklahoma State University; *John Oakes*, Gonzaga University; *Thomas C. Pinkerton*, Upjohn Company; *Michael Seymour*, Hope College; *Peter Uden*, University of Massachusetts at Amherst; *Wesley W. Wendlandt*, University of Houston; *Allen West*, Lawrence University; *James Winefordner*, University of Florida; *Bernhard Wunderlich*, Rensselaer Polytechnic Institute; *Alexander Yacynych*, Rutgers University; *John Zimmerman*, Wabash College

## Case Studies Quoted in the Text

# Contents

**Part IV   SEPARATIONS**                                       451

# Chemical Analysis

# Part I
# BACKGROUND AND REVIEW

# 1

# Preliminaries

## 1.1 Introduction

What do archaeology, anthropology, botany, chemistry, engineering, forensic science, genetics, geology, materials science, medicine, pharmacology, and toxicology have in common? The contemporary practice of these areas of endeavor depends on chemical analyses.

Everyone who works with materials and their transformations is dealing with chemistry, and we find out about these materials using methods of chemical analysis. Analytical chemistry is, then, really the study of how we learn about materials: their elemental and molecular composition. The compositions of solids, liquids, gases, solutions, glasses, flames, and other forms of matter determine their characteristics and their best uses.

Chemical analysis involves problem solving, and most of the chapters of this text include a case study that shows how the methods of chemical analysis are applied to solve them. These problems vary from identifying the cause of sludge buildup in a chemical plant, to verifying the origins of the Shroud of Turin, to identifying an illicit drug.

Analytical chemists are interested in the specific methods employed for such useful analyses. And those analytical chemists who are involved in research are usually trying to find better methods for learning about the constitution of matter.

The general principles and language of analytical **methodology,** that is, how an analysis is carried out, are worth knowing. One important reason is the necessity for discussing your work with others and learning about the results of their efforts from them.

Inevitably, many technical words, phrases, abbreviations, and acronyms are used. These are useful since they are often more precise and can be used to communicate ideas more efficiently and correctly than less specialized language. These terms are, then, worth knowing. In addition, they will enable you to learn more easily about the subjects from sources more advanced than this textbook. This understanding will open the literature of chemical analysis to you.

As you learn the language step by step, it is likely that you will find that all the methods are straightforward in conception. However, actually carrying out chemical analyses requires significantly more practice and knowledge of the details. There is no way to learn how to do specific analyses well without working with your own hands. Learning the tricks of the art of analysis, in addition to the science, is a study of many years. In other words, learning how to do an analysis takes much longer than learning how each is done in general and the ideas behind it.

## 1.2 Some Definitions

### Identification, Determination, Analysis, and Assay

Let us be as precise as we can about the use of the words **identification** and **determination** versus **analysis** and **assay.** Elements, ions, and compounds are identified or determined in a sample. We may say that the analyst has identified penicillin in the preparation or has determined that penicillin is present. It is incorrect to indicate that a sample is "determined."

However, only samples are analyzed; an element is not analyzed. We may say that the sample was analyzed for penicillin. To state that copper was analyzed is incorrect unless it means that a sample of copper (ore or wire, for instance) was analyzed to determine the other elements present.

On the other hand, if we intend to determine what fraction of some sample is a *named* material, the word *assay* should be used. For instance, if the amount of a specific element, ion, or compound, say carbon in iron, is to be determined, we can say that an assay for carbon is to be run.

Some confusion among these terms may occur when the sample is an element or compound of some unknown purity. Saying an analysis of iron will be made means that an identification (or determination) of the impurities will be made on a sample of iron. However, if the iron content itself is to be determined, the correct word for the procedure is *assay.* Thus an *analysis of iron* conveys a different meaning from an *assay for iron.*

## Verification

Most chemical analyses are developed only after much trial and error. For each analytical procedure, an extremely important requirement is to demonstrate that the procedure measures what the analyst says it measures on a specific type of sample. This is called **verification** of the analytical method. For instance, an analyst asserts that a new procedure may be used to measure the amount of cobalt in samples of animal livers. By comparing her results with results of older, accepted methodologies, the analyst must show that they agree.

However, it may be that when some methodology is used with a different type of sample, some unexpected problems might become obvious. (If the problems are not obvious, you do not even know about the difficulty. But more of that later.) What may have happened is that the sample contains something that causes an **interference** with the method. In developing chemical analyses, almost always there are problems with interfering species. These interfering species cause the result to be greater or less than what it would be if they were absent. The part of a material that causes the interference consists of one or more **interferents.** For example, when one uses the new methodology, which works for specimens of liver, to find the cobalt content of bone, high levels of calcium interfere in the assay. Calcium is an interferent in the analysis. As a result, the new analytical procedure for cobalt that is verified for use with animal liver specimens is not verified for bone.

## Methods and Methodology

The problems with interferents is one of the reasons that the term *analytical chemical methodology* appears rather than the term you might expect, *analytical chemical method.* The methodology of an analysis consists of a series of **methods.** A method is a fixed sequence of actions to be carried out. It is a fixed procedure. These procedures are the methods that will be discussed in this book. However, the analytical methodology consists of a choice of the best combinations of methods to achieve the end desired.[1] The steps of chemical analysis methodology are outlined in Figure 1.1.

## 1.3 Solving a Problem

To solve a problem employing the methodology of chemical analysis, we must first decide on a method for taking a sample. To be more specific, say there are 120 tons of some ore sitting outside the laboratory window. How can a representative sample be obtained that will fit inside the door?

Next, a decision must be made about what method can be used to prepare the sample for the assay. For instance, assume that the instruments that are available can

---

[1] The term **technique** is somewhat similar to the word *method,* although it also is used to refer to a way of doing a single operation in a method. An example of this use is "She has a good titration technique." Also, the series of operations making up either a specific method or making up the entire methodology can also be labeled as a **protocol.** This term is perhaps used more often in biological analysis.

6

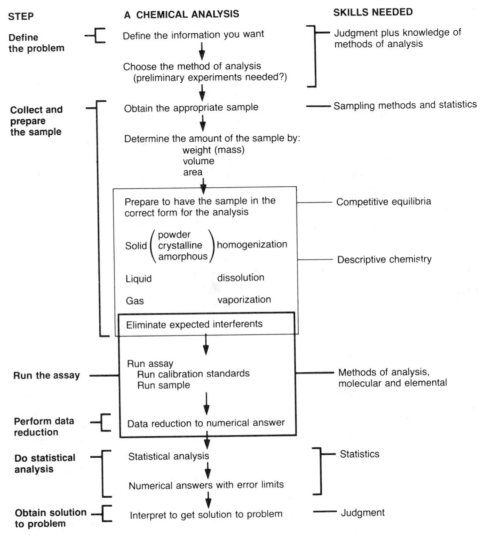

**STEP**　　　　　　**A CHEMICAL ANALYSIS**　　　　　**SKILLS NEEDED**

**Define the problem**

Define the information you want

Choose the method of analysis
(preliminary experiments needed?)

Judgment plus knowledge of methods of analysis

**Collect and prepare the sample**

Obtain the appropriate sample

Sampling methods and statistics

Determine the amount of the sample by:
weight (mass)
volume
area

Prepare to have the sample in the correct form for the analysis

Competitive equilibria

Solid $\left(\begin{array}{l}\text{powder}\\\text{crystalline}\\\text{amorphous}\end{array}\right)$ homogenization

Descriptive chemistry

Liquid　　　　　　dissolution

Gas　　　　　　　vaporization

Eliminate expected interferents

**Run the assay**

Run assay
　Run calibration standards
　Run sample

Methods of analysis, molecular and elemental

**Perform data reduction**

Data reduction to numerical answer

**Do statistical analysis**

Statistical analysis

Statistics

Numerical answers with error limits

**Obtain solution to problem**

Interpret to get solution to problem

Judgment

**Figure 1.1**　Diagram of the process of solving a problem using the methodology of chemical analysis. In the left column are the six general steps toward the solution. In the middle column is a more detailed list of the stages involved. The top box delineates the steps of sample preparation. Notice that the last part of this process is also part of the assay step. The lower box contains the steps involved in the assay. The right column lists some of the less tangible skills involved in the process. Contemporary analysis is a combination of chemical principles, instruments and electronics, and practiced judgment. The "hardware" is an essential but relatively small part.

only measure some property of an aqueous solution. But so far there is only 10 kg of solid chunks of ore in a bucket in the lab. What shall we do with it?

Then, we must choose the assay method(s) for the analysis. These are now usually done using some instrumental method. If some interferents present in the prepared sample would be especially detrimental to performing the given assay, these should be removed if possible. The sample preparation method must be chosen with the characteristics of the assay method in mind.

Finally, we must decide how to treat the data obtained from the instrument. Usually we use statistical tests. These tests help us decide how well we really know the numerical answer to the analytical question. In addition, if some interferents cannot be removed, the effects of the interferents must be accounted for by a correction to the assay results. In the optimum case, we must obtain knowledge of how each (and every!) interferent interacts with the desired assay method and then apply a correction to the result to account for the effects of the interference.

This problem-solving process can be illustrated with a specific problem and its solution, as found in an example from the literature of analytical chemistry. It is the first of a number of cases that are used to illustrate how solutions to problems are actually found. This one comes from the oil industry and is adapted from an article in the journal *Analytical Chemistry*. [*Anal. Chem. 48* **1976**, 972A.] The steps follow those of Figure 1.1.

Some solids were building up in a reactor of a petrochemical processing plant, and the condition threatened to require a complete shutdown. This buildup was accompanied by relatively high corrosion rates of the metal reactors. The problem was, thus, *defined:* Can the solids be eliminated? From previous experience, the analysts believed the problem might be due to the presence of chlorine, but the presence of sulfur was also a possibility. The question was how much chlorine was in the system during the process.

The *samples* of the feed material that reacted in the synthetic process were obtained for an analysis of chlorine. In the laboratory of the company, the chlorine was to be determined by two different methods. Through one, the *total* chlorine in a sample could be determined. This is done by a thermal method of analysis in which are measured changes in the temperature of a sample as it is heated. Through the second method, the chlorine that originated in *organic* chlorine compounds could be determined. It is based on an electrochemical method — one that depends on the oxidation or reduction of a substance.

The crude feed material was a liquid. Samples of this liquid needed to be *prepared* differently for the two assay methods.

After the measurements of chlorine were made, the *numerical results* from the two methods were found to differ. The assay for organic chlorine compounds showed that there were no chlorinated organic compounds in the sample. The assay using the thermal method to determine the total chlorine content showed apparent high levels of chlorine.

By *statistical analysis* of results from a number of samples, the analysts determined that the different results from the two methods could not arise due to random experimental errors made by the technicians who did the analyses.

A check for interferences was done. The assay using the thermal method was found to be influenced by high levels of sulfur. The sulfur acted as an interferent in the assay.

As a result, the analysts modified the analytical methodology. They changed the technique for the sample preparation; any sulfur present in the sample was removed before the thermal-method assay was run. After that modification, the results from both methods agreed. The chemists concluded that the problem did not arise from excessive chlorine. Their worries about chlorine were due to an unanticipated interference, and the original analysis was in error.

The *answer* to the original question was that the problem of the sludge and corrosion lay in the high sulfur levels of the feedstock. There is more to the story, but this introduction serves to illustrate the sometimes circuitous steps carried out in solving a problem using chemical analyses. The path was hardly straightforward.

## More Definitions

There are some other terms that appear in Figure 1.1 that need definition. *Data reduction* is part of the assay. A machine or method does *not* give a direct measure of the amount of a given material in a sample. The quantitative result is found from calculations using the response of the instrument as it depends on the analyte content. This calculation may be done by a computer inside the instrument with the output displayed directly as the content of the material. On the other hand, this same calculation may be done by a separate procedure. For instance, using an instrument we might find that a certain amount of yellow light is absorbed or that the sample solution conducts electricity or that the sample's weight changes a given amount when it is heated. Each assay method will have different calculations done, using the experimental data to find the desired information. You will learn about these different calculations mostly by doing the exercises at the ends of the chapters. Thus a computer will make only a small part of the analysis easier: the data reduction and statistical analysis. The rest still requires judgment and personal skill.

A statistical analysis of the results is required since each analysis and each sample analyzed is not exactly the same every time. To evaluate this variation, three or more samples of the same material are usually treated and assayed together. Then a quantitative measure can be found of how trustworthy the results are. The result is called the *error limits*. You will learn how to calculate these with the statistical methods of Chapter 4.

Two more terms occur in the "skills needed" column in Figure 1.1. One is **competitive equilibrium.** An example is when two bases such as acetate ($CH_3COO^-$) and chloride ($Cl^-$) ions compete for the same proton in an acid–base reaction. The fundamentals of competitive equilibria *must* be mastered in order to understand chemical analysis. Not only do the problems of preparing a sample and eliminating interferents depend on it, but understanding the assay method itself may rely on it. The ideas of competitive equilibria even apply to such apparently unlikely objects as the flames used in assays.

The other term, **descriptive chemistry,** is, in fact, one of the strongest thinking-tools required for solving problems in chemical analysis. Descriptive chemistry is the body of knowledge of the reactions of elements and compounds with each other under specific conditions and the properties of these reaction products, such as their volatilities or solubilities. Not all of the necessary descriptive chemistry may be fascinating and easy to unearth from numerous possible sources, but the knowledge comprises an extremely potent contribution to solving many problems involving chemical analysis as well as to developing safe laboratory procedures.

The descriptive chemistry often cannot be codified in mathematical terms but is understood as numerous rules with numerous, specific exceptions. Some of these rules are followed regularly, whereas others apply only occasionally. The organization of the periodic table is an example of descriptive chemistry. For instance, the elements that appear in a column of the table show many similar chemical traits, such as volatility, formation of stable oxides, or ease of oxidation. Knowledge of this descriptive chemistry with its rules and exceptions is one major characteristic that distinguishes an experienced analyst from a novice.

## 1.4 Basic Measures

A **measure** labels the extent or size of anything, especially as determined in comparison to a standard. For instance, the gram is defined as exactly 1/1000th of a standard mass kept at the International Bureau of Weights and Measures near Paris. It was selected to be identical with the mass of 1.0000 mL of water at the temperature of its maximum density, 3.98 °C. Since originally defined, this value of the mass of water has been more accurately measured and differs from the original by about 3 parts in 10,000. However, the standard gram has been kept the same and is close enough for all but the most exacting measurements. The centimeter similarly is defined by a standard. Finally, the size of a degree of temperature on the Celsius and Kelvin scales is defined as 1/100th of the difference between the normal freezing and boiling points of water. It is interesting to note that these three fundamental measures are arbitrarily defined.

Different measures are used for different physical states and for different ranges of concentration. These measures have been developed as the need arose over more than a century of endeavor. As with any human activity, the measuring units evolved for good reasons for their various practical uses. However, overall they might seem to be quite a messy system. Attempts to standardize the units have been made: the Système International (SI). These are only slowly being adopted.

Depending on the disciplinary areas and the age of publications, the SI units may or may not be used. As a result, when you read the results of others, you will have to recognize the units used and should be able to interconvert various units of measure to those most useful to you.

### Atomic Mass

The **atomic masses** are a set of relative numbers whose absolute values depend on a defined standard. The standard that has been chosen is the mass of the isotope $^{12}C$. This is defined to be mass 12 exactly, and 1/12th of this mass is called an **atomic mass**

**Table 1.1**   Prefix Notation

| Exponential | Prefix Name | Symbol |
|---|---|---|
| $10^{12}$ | tera- | T- |
| $10^9$ | giga- | G- |
| $10^6$ | mega- | M- |
| $10^3$ | kilo- | k- |
| $10^{-1}$ | deci- | d- |
| $10^{-2}$ | centi | c- |
| $10^{-3}$ | milli- | m- |
| $10^{-6}$ | micro- | $\mu$- |
| $10^{-9}$ | nano- | n- |
| $10^{-12}$ | pico- | p- |
| $10^{-15}$ | femto- | f- |

**unit** (AMU). An AMU is also named the **dalton** after one of the key chemists in history. Atomic masses are commonly called the **atomic weights.** It can be argued that the name *atomic weight* is somewhat less correct than the name *atomic mass.* However, on a day-to-day basis for chemical analysis there is no essential difference.

## Mole

Using one of the basic ideas of chemistry, we can define the **mole** as a specific number of units *or* a weight containing that number of units. A mole as a number contains $6.022 \times 10^{23}$ particles of any kind: an Avogadro's number of things. A mole as a number can be used when measuring atoms, ions, or molecules or units such as a chemical formula group.

As a weight, the mole is the mass formula weight (in AMU) expressed in grams. The **formula weight** is its common name. One mole is the mass of $6.02205 \times 10^{23}$ chemical-formula units. This is the definition used on a day-to-day basis. In addition, there are two other commonly used specific definitions with which you probably are familiar. If the particles (or formula units) are atoms, the mole weight is called the **atomic weight** or **gram-atomic weight.** If the particle (or formula) is a molecule, the weight is called the **molecular weight** or **gram-molecular weight.** The three different names may be abbreviated f.w., a.w., and m.w., respectively.

## Prefix Notation

The range of concentrations that are analyzed and the size of the samples that can be analyzed range widely and benefit from a language to express this range. Prefix names have been given to each factor of 1000 in size. The prefixes that name the exponential powers of 10 are shown in Table 1.1. For example, the prefix *kilo* signifies 1000 times the measurement unit, as in kilogram. Analyses can now be done on small amounts of materials in the picogram range. Similarly, the average concentration of an element in a planet's mass might be desired. The mass of such a "material" is in the range of $10^{15}$ Tg.

## 1.5 Measures of Composition: Units of Content

Three basic *types* of expressions are employed to characterize the compositions of materials that are not composed of a single pure element or a single pure compound. These are shown in Table 1.2 and described below.

### Weight-to-Weight Measures

**Weight-to-weight** measures express the ratio of the weight of one component to the weight of the whole. **Weight percent** (abbreviated wt% or % w/w) is the ratio of the weight of a component as a part of the whole mixture expressed as a percentage. Occasionally it is called **parts per hundred.**

---

**EXAMPLE**

If a sample weighing 1.2304 grams contains 0.1012 g of iron, its content of iron is

$$\frac{0.1012}{1.2304} = 0.0822$$

of the whole, and it has

$$0.0822 \times 100 = 8.22 \text{ wt\% iron in it.}$$

---

If a sample contains a smaller amount of some component than is conveniently expressed as wt%, the next smaller commonly used unit is **parts per thousand** (abbreviated ppt or ‰). This is analogous to parts per hundred.

---

**EXAMPLE**

If a sample weighing 1.2304 grams contains 0.0101 g of iron, its content of iron is

$$\frac{0.0101 \text{ g}}{1.2304 \text{ g}} = 0.0082$$

of the whole, and it has

$$0.0082 \times 1000 = 8.2 \text{ ppt iron in it.}$$

---

Smaller fractions are described using **parts per million** (ppm, p.p.m.), **parts per billion** (ppb, p.p.b.),[2] and **parts per trillion** (ppt, pptr). Note that there can be confu-

---

[2] Some confusion may occur because the use of *billion* in Europe often means $10^{12}$, whereas *billion* in the United States is $10^9$.

**Table 1.2**   Units of Content

| Component Measured by | Total Sample Measured by | Nomenclature | Abbreviation |
|---|---|---|---|
| Weight* | Weight | Weight to weight | w/w |
| Weight | Volume | Weight to volume | w/v |
| Volume | Volume | Volume to volume | v/v |

\* Mass is more correct. But this is practically always expressed as, for example, weight to weight.

sion between the abbreviations for parts per thousand and parts per trillion when both are abbreviated ppt. No confusion exists when pptr is used.

It is worth noting that in common usage, if no other abbreviation such as w/v or v/v is listed, then %, ppt, ppm, and ppb all refer to the weight-to-weight measure. Weight-to-weight measures are used for solid and liquid samples.

A weight-to-weight measure also can be expressed as a specific fraction. For example: The sample contains 33 $\mu$g/mg of ingredient X. The following equivalencies also hold:

$$\text{parts per million} = \mu\text{g/g}$$
$$\text{parts per billion} = \text{ng/g}$$
$$\text{parts per trillion} = \text{pg/g}$$

To get an intuitive idea of the size of these measures, 3 ppb of one second of time is the time it takes light to travel one meter. One ppm of 50 L (13 gallons) of water is one drop.

### Weight-to-Volume Measures

In a **weight-to-volume** measure, a component's weight is related to the total volume of the substance. The volumes in weight-to-volume measures are related to the weight of water: one $cm^3$ of water at 3.98 °C weighs 1.000 g. Thus, to express a weight-to-volume value if the weight measure is grams, the measure of the solution volume is mL. If the weight measure is kg, the volume measure is liters. A more complete list is shown in the appropriate column of Table 1.3. If a w/v measure is expressed as, for instance, parts per million, it is usually written ppm(w/v). Only seldom does a weight-to-volume measure not refer to a liquid sample.

---

**EXAMPLE**

If 2.354 g of $KNO_3$ is dissolved in exactly 250 mL of total solution, the w/v measure is

$$\left(\frac{2.354}{250}\right) \times 1000 = 9.42 \text{ ppt(w/v)}$$

---

**Molarity** and **normality** are two other common w/v measures which are probably familiar to you. However, they are also **number-to-volume** measures.

**Table 1.3**  More Common Units Used to Express Content

| Name | Abbreviation | Units Used w/w | w/v | v/v |
|------|-------------|-----|-----|-----|
| parts per thousand | ‰ or ppt | mg/g | mg/mL<br>g/L | mL/L |
| parts per million | ppm | μg/g<br>mg/kg | μg/mL<br>mg/L | nL/mL<br>μL/L |
| parts per billion | ppb | ng/g<br>μg/kg | ng/mL<br>μg/L | nL/L |

## Number-to-Volume Measures of Content

Molarity (M) is defined as the number of moles of solute in 1 liter of solution. This is a number-to-volume measure. (Since a mole can be a weight of material, it is a weight-to-volume measure as well.)

---

**EXAMPLE**

If 2.354 g of $KNO_3$ is dissolved in exactly 250 mL of total solution, the molar concentration will be

$$\frac{(2.354 \text{ g}/101.1 \text{ g mol}^{-1})}{0.250 \text{ L}} = 0.0931 \text{ molar}$$

since the molecular weight of $KNO_3$ is 101.1 g mol$^{-1}$.

---

It is, perhaps, more common to find ppt, ppm, and so forth used in analyses rather than molar quantities because of the relative directness of the calculations. Compare the two examples above, for example.

Normality *(N)* is used in two general areas: analyses depending on acid–base reactions and those depending on oxidation–reduction reactions. The normality of a solution can be related to the molarity by multiplying the molarity value by the number of **equivalents** in a mole.

The number of equivalents in a mole of an acid or base is the number of protons that can actually be donated by a molecule. For example, acetic acid has the formula $H_3CCOOH$ (or $H_4C_2O_2$). Each molecule has four protons. However, the descriptive chemistry of experiments involving acetic acid in water shows that only one of them is donated. That is,

$$CH_3COOH = H^+ + CH_3COO^-$$

in water. Thus it has one equivalent to donate in water, so the normality is the same as the molarity. However, sulfuric acid ($H_2SO_4$) can easily donate two protons to a base. A one-molar sulfuric acid solution contains two equivalents of protons and thus is two normal. The general equation for normality is

normality = molarity × number of equivalents per mole

In oxidation – reduction reactions, the number of equivalents equals the number of electrons that are actually donated or accepted in an oxidation – reduction reaction.

---

**EXAMPLE**

Permanganate is used as a reagent in a number of analyses. It reacts by donating five electrons to form manganous ion:

$$5\ e^- + 8\ H^+ + MnO_4^- = Mn^{++} + 4\ H_2O$$

Since five electrons are donated per permanganate ion, the numerical value of the normality will be five times the value of the molarity. Thus a 0.1-M solution of permanganate will be equal to a 0.5-$N$ solution under these conditions.

---

### Volume-to-Volume Measures

A third type of content measure expresses the volume of a component as a part of the total volume of material. This **volume-to-volume** measure is expressed in a similar way as the other two types: volume percent (% v/v), as well as ppt v/v, ppm v/v, and ppb v/v. It is mostly used for liquid components of a liquid sample or gaseous components in a gaseous sample. Some of the expressions for volume-to-volume measures are collected in Table 1.3.

There is another volume/volume measure commonly used in conjunction with chemical analyses. It is written in the notation of ratios — for example, 1 : 2 methanol – water. This means that one volume of methanol is mixed with two times that volume of water to make a solution. Such measures are used in instructions for mixed-solvent preparations.

---

## 1.6 The pH and Other Logarithmic Scales

---

In 1909 the Danish biochemist Sven P. L. Sorensen invented a scale to make it easier to manipulate, calculate, and discuss the wide range of concentrations of hydrogen ions in solution. Instead of writing the concentration of protons, $[H^+]$, he decided to use the negative logarithm of the concentration, $-\log[H^+]$. This value is called the pH. (A brief review of logarithms appears in Supplement 1A.)

Consider the following examples. The hydrogen ion concentration of a 0.1-M solution of acetic acid in water at 25 °C is 0.0013 M. But the concentration of $H^+$ in pure water is 0.0000001 M. And a 0.1-M solution of NaOH has $[H^+] =$ 0.00000000000011 M. It is a lot easier to write or discuss the equivalent pH values of 2.89, 7.00, and 12.9,[3] respectively!

---

[3] This is not exactly pH 13 since concentrated base solutions do not behave as simply as the theory of dilute solutions would predict.

Since chemical concentrations are usually molar or less, a common practice is to use the negative logarithm of a number to represent concentrations. The operation of taking the negative log of a number is signified by the letter $p$ placed before the number.

---

**EXAMPLES**

$$-\log K = pK$$
$$-\log[Ca^{++}] = pCa = pCa^{++}$$
$$-\log[H^+] = pH$$

---

## 1.7 Reading Equations

In any science, quantitative relationships are expressed mathematically in equations. Equations are succinct expressions of the ideas. They are really shorthand for sentences or paragraphs that express in a quantitative manner the general knowledge based on experience.

Take the case of the ideal gas equation. It describes relationships that have been found from experimental data. In words: The pressure of a gas multiplied by its volume is a constant which is dependent on temperature.

Written algebraically,

$$PV = nRT \tag{1-1}$$

where

> $P$ is the pressure of the gas,
> $V$ is the volume of the gas,
> $T$ is the temperature in K,
> $R$ is a constant of proportionality, and
> $n$ equals the number of moles of gas present.

This equation is a concise description of experimentally determined behavior. Mathematics is a powerful, symbolic way of describing it.

Equation 1-2 is more formidable. The terms are probably unfamiliar, but the mathematical relationships should be clear. Let us work our way from the algebraic expression back through to the characteristics of the experiments that it describes. The equation, called the Ilkovic equation, applies when using an electrochemical method.

At time $t$, a quantity $i_L$, the limiting current, is described by

$$i_L = (7.082 \times 10^4)nv^{2/3}\, t^{1/6}D_A^{1/2}C^0 \tag{1-2}$$

where

> $n$ is the number of electrons transferred per molecule,
> $i_L$ is the current in amperes,

$v$ is grams of mercury flowing per second,
$t$ is the time in seconds,
$D_A$ is the diffusion coefficient in $cm^2 s^{-1}$, and
$C^0$ is the concentration of reduced species in $mol/cm^3$.

If we insert word-definitions into the equation and ignore units (such as cm or g), the equation says: The limiting current at time $t$ equals $7.082 \times 10^4$ times the product of the number of electrons transferred per molecule, the number of grams of mercury flowing per second raised to the 2/3rd power, the time to the 1/6th power, the square root of the diffusion constant, and the concentration of something in $mol/cm^3$. But that's not a lot of help!

One way to approach the meaning of the equation is to look at one term at a time, starting with the familiar ones. The following is an example of such an approach.

The term $i_L$ represents the limiting current because that is what the introductory clause says. As yet, you may not know what a limiting current is. (However, you could look it up in the index of this book to find out in more detail.) For now, you may simply be satisfied that this is an electrical current that is at an extreme (that is, limiting) value. This limiting current is a function of a number of **variables** and a constant. The constant, $7.082 \times 10^4$, should present no problem.

In Eq. 1-2, the factor that is probably the most familiar to you is $t$, the time. Next you might focus on $C^0$, a concentration related to the more familiar molar units, moles/liter.[4] $C^0$ will be 1000 times smaller, though, since there are 1000 $cm^3$ in a liter.

The number $n$ is the number of electrons per molecule of substance. You may be somewhat rusty in your electrochemistry. But even so, electrons per atom or per molecule at least sounds like it has something to do with oxidation and reduction. If necessary, you can always go and look up more information in your introductory chemistry textbooks under the topic of electrochemistry.

All that are left are the parameters $v$ and $D_A$. You most likely have not seen either of these before in your studies of chemistry. Let us begin with $v$ and reason out what it means. The description of $v$ states something about flowing mercury in units of grams/second. This suggests that the experiment uses flowing mercury somehow. However, the units are strange. The amount of mercury is not written as a volume but is stated in g/s. We normally express flow as volume/time — say, $cm^3/s$. We can get to g/s through the density, which is expressed in $g/cm^3$, that is, $(cm^3/s)(g/cm^3) = g/s$. Perhaps the value of the constant includes the density of mercury.

The only term for which we have not figured out a reasonable or partial explanation is that written $D_A$. If you do not know exactly what a diffusion constant is, at least you can guess from its name that the value is treated as a *constant* under the experimental conditions. Does it apply to mercury? That seems unlikely. For if it applied to mercury, its value (to whatever power) would be the same in all the experiments. It seems likely that such a constant value would have been included in the numerical

---

[4] Mole is defined in Section 1.4. Liter is the same as $(decimeter)^3$. As yet, the use of $dm^3$ has not entered the analytical literature to any great extent.

constant of the equation, $7.082 \times 10^4$. This conclusion is not certain; it is a reasonable guess. However, if $D_A$ does not apply to mercury, it probably applies to the molecule (element, ion?) oxidized or reduced in the experiment.

Let us see what further information we can infer from considering the evidence from the equation. In an experiment, we usually can measure values such as time, rate of flow of mercury, concentration of the compound in solution, and magnitude of an electrical current such as the limiting current $i_L$. Thus the only quantities that will not be measurable independently are the diffusion constant $D_A$ and the number of electrons transferred $n$. These variables must refer to the type of information that can be found from the method. Alternately—and this is quite important—if we have some way of knowing $D_A$ and $n$, then the method could be used to measure the concentration. Thus you might guess that the method to which the equation applies could be used to measure concentrations of compounds that can be oxidized or reduced. You would be correct in that guess.

To review about reading the equation, first we recognized the terms that were familiar to us. Then we tackled the unfamiliar parameters and developed some suggestions of how and where to find out more about them. Finally, the information in the equation and some practical guesses led to a suggestion of the use of the equation and method. The equation will probably be used to determine $n$ and $D_A$ or, alternatively, the concentration of some chemical species that can transfer electrons.

Thus, by working one term at a time through an equation that appears complicated, you can relate it to your past experience. In this way, it can be understood at least partially. Further, you can find which if any terms are unclear and need further study.

## 1.8 Reading Graphs

A number of different types of graphs are employed in association with chemical analyses. Selected aspects of these are noted in this section.

### Cartesian Graphs

Cartesian graphs used in chemical analysis usually consist of plots of the result of an assay as it depends on the content of analyte in the sample, such as illustrated in Figure 1.2. Each axis consists of a linear scale, which should be labeled with the quantities that are plotted. In this figure, one labeling convention is shown, namely, labels with the units inside parentheses. In this case, the instrument output is a linear scale, but the units are arbitrary. For instance, the output may simply be the voltage measured by an attached voltmeter. In such a graph, the slope of the line (the ratio of change in instrument response/change in analyte content) is the sensitivity of the instrument. The straight-line relationship between the data points (the points inside the circles) indicates that the instrument produces an output linear with the analyte content.

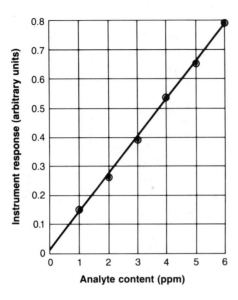

**Figure 1.2** Cartesian graph with two linear scales. Placing the units of the measure in parentheses is one of the common conventions. This graph is representative of numerous calibration plots found in instrumental chemical analysis.

## Log – Log Graphs

A number of analytical methods can produce linear responses over a far larger range of concentrations of analyte than can be plotted clearly on a Cartesian graph. In such situations, it may be convenient to plot the results on a log – log graph, which consists of logarithmic scales[5] on the perpendicular axes of the graph, as shown in Figure 1.3. Here the response is again linear with analyte content — for example, a ten-fold increase in analyte results in a ten-fold increase in counts per second. Note that the parenthetical units label on the vertical axis is $(s^{-1})$ since counts is a unitless value.

The data points at the left in Figure 1.3 have vertical bars extending from them. These are called **error bars** and indicate the precision of the data. The lengths become too small to show as the number of counts increases. When such bars are shown, the caption of the graph should describe what statistical measurement they indicate. One common length of extension is out to the values at plus and minus one standard deviation, a quantity described in detail in Sections 4.3 – 4.7.

## Semilog Graphs

On a semilogarithmic graph, one axis (usually the vertical one) has a logarithmic scale, and the axis perpendicular to it has a linear scale. Figure 1.4 shows an example of a plot of concentration versus time. Semilog graphs are usually made to linearize the relationship between the plotted values. A straight-line relationship on a semilog plot means that there is a constant percentage change in the quantity noted on the logarithmic axis.

---

[5] A brief review of logarithms appears in Supplement 1A.

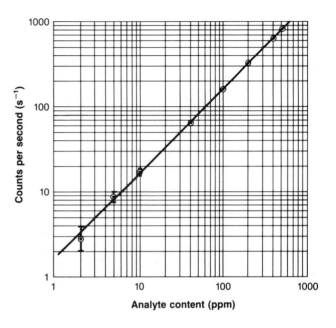

**Figure 1.3** Three-cycle by three-cycle log–log graph with error bars. This graph is representative of a calibration plot for a number of instrumental methods which have a wide range of linear response.

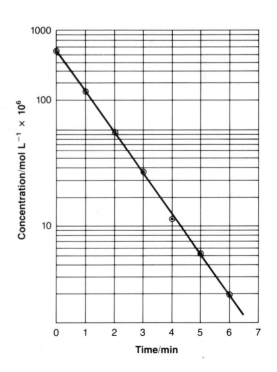

**Figure 1.4** Semilogarithmic graph with three log cycles on the vertical axis. The manner of labeling the units of the measure illustrates another of the common conventions. In this, each number label is unitless. This plot is representative of data from kinetic methods of analysis.

The labels on the axes of Figure 1.4 illustrate another labeling convention. The numerical values are to be unitless, so the quantities on each axis are quotients. The horizontal axis has the label (time/min), which means that the numerical values are the experiment times divided by one minute.

The denominator of the label on the vertical axis is (mol L$^{-1}$ × 10$^6$). Using the negative exponent avoids the confusion possible if the label were written (mol/L × 10$^6$) and is recommended. The factor 10$^6$ means that the units of concentration that are written have been *multiplied* by 10$^6$. Thus the value of 1000 on the scale is 1 × 10$^{-3}$ molar.

### Linearized Graphs

Sometimes it is useful to plot some mathematical function of a measurable quantity on a graph axis. A simple example is shown in Figure 1.5. Here a quantity called $\Lambda$ is plotted against the square root of the normality of a solution of HCl in water. The units along the horizontal axis then are (equivalents L$^{-1}$)$^{1/2}$ or $N^{1/2}$. When such manipulated units appear, you can be moderately certain that the purpose is to produce a linear relationship between the values plotted. In this case, at the lower concentrations the line connecting the data points is straight and extrapolated to zero. It is useful to use some change in the appearance of the line to indicate the extrapolation—a dashed line segment serves the purpose here.

If you want to find the normality of the solution that corresponds to the value at the arrow, a simple calculation is needed. Since the factor of 100 multiplies the scale, the value of 4 corresponds to 0.04 $N^{1/2}$. The solution normality is then 0.0016 $N$ at the arrow. Similarly, the data point at the right corresponds to a 0.005-$N$ solution (0.071 $N^{1/2}$).

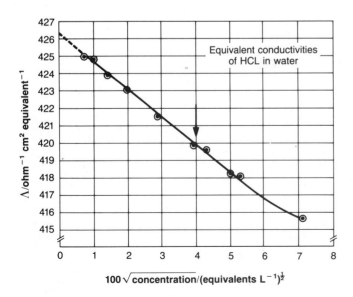

**Figure 1.5**   Plot on Cartesian axes but with a linearizing function plotted on one axis. This plot shows the equivalent conductivities of hydrochloric acid in water at 25 °C. This type of graph illustrates a fundamental theory in the chemistry of electrolyte solutions.

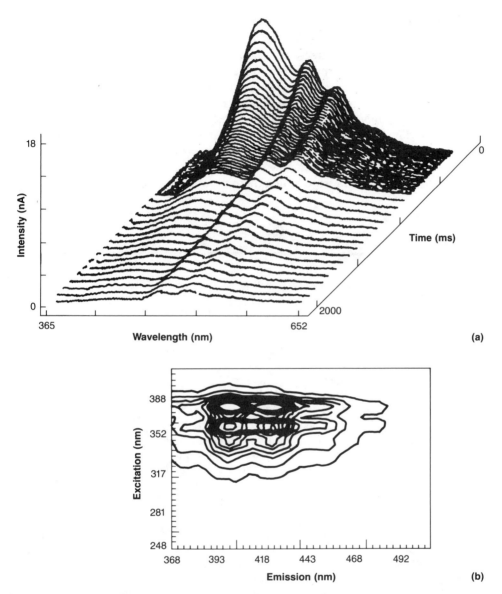

**Figure 1.6** Two examples of three-dimensional plots. (a) Shows instrument response as a current in nA versus both light wavelength and time. The experiment involves the measurement of light emitted by a solution of 2-naphthoic acid and *p*-aminobenzoic acid. (b) Shows a contour plot of the amount of light emitted by a $8.3 \times 10^{-5}$-M solution of anthracene in cyclohexane. As the contours indicate, the amount of light measured depends on both the wavelength of light impinging on the solution (excitation) and the wavelength of light given off (emission). This is a fluorescence measurement. Fluorescence methods are described in Chapter 17. [Figures reprinted with permission from (a) Goeringer, D. E.; Pardue, H. L. *Anal. Chem.* **1981,** *53,* 272–276 and (b) Warner, I. M.; Patonay, G. P.; Thomas, M. P. *Anal. Chem.* **1985,** *57,* 463A–483A. Copyright 1981, 1985 American Chemical Society.]

### Three-dimensional Graphs

In some cases, it is advantageous to plot data as a function of two different quantities. Two different displays of a three-dimensional format are shown in Figure 1.6. Figure 1.6a shows a plot of an instrument output versus wavelength and time. It is drawn so as to have perspective. Figure 1.6b is a contour plot with each contour representing a different specific level of instrument response.

Because they can be so useful, such plots are becoming more common along with computerized instrumentation. Such graphs can illustrate trends that might not be seen as easily with a series of two-dimensional graphs such as those discussed above. The detailed calculations associated with the analysis are done by the computer using the stored numerical data from which the plots are constructed.

## Suggestions for Further Reading

Readings of case studies in problem solving using chemical analytical techniques. A few are the same as in this text.
Grasselli, J. G. "The Analytical Approach"; American Chemical Society: Washington, D.C., 1983.

An important and complete treatise on the methods and methodology of chemical analysis. The treatise consists of three parts:
       Part I: Theory and Practice
       Part II: Analytical Chemistry of the Elements
       Part III: Analysis of Industrial Products
Part I of the first edition is complete but now is significantly out of date (the first came out in 1959, the last in 1976). The second edition is incomplete but began appearing in 1978. Part II is a list of the methods (with references) used for each element and is still useful. However, Chemical Abstracts contains references that are more up to date. The level of writing varies, but you should be able to make your way through it after your analytical course.
Kolthoff, I. M.; Elving, P. J., Eds. "Treatise on Analytical Chemistry," 2nd ed.; Wiley: New York, 1978.

A book of definitions of the terms used in chemical analysis. Despite the title, it is *not* the final word as to how the terms are now used. IUPAC (The International Union of Pure and Applied Chemistry) published it, but the reports in *Pure & Applied Chemistry,* which appear from time to time, are more up to date.
Irving, H.; West, T. S.; Freiser, H. "Compendium of Analytical Nomenclature: Definitive Rules 1977"; Pergamon Press: Oxford and New York, 1978.

A multivolume set of descriptions of instrumental methods of analysis. These are at an intermediate level but are not "how to" oriented.
Kuwana, T., Ed. "Physical Methods in Modern Chemical Analysis"; Academic Press: New York, 1978.

## Problems

1.1    What is the molarity of $K^+$ in a solution that contains 62.4 ppm (w/v) of $K_3Fe(CN)_6$?

1.2    A 200-mg tablet of the drug ibuprofen (m.w. 306.3) contains 20 mg of the drug with the rest filler.
    a.   What is the w/w percentage of the tablet that is the drug?
    b.   What is the w/w percentage filler?
    c.   How many ppm w/w of the tablet is active material?

1.3    Average human blood serum contains 2.2 mM $K^+$ and 77.5 mM $Cl^-$.
    a.   What are their concentrations (w/v) in ppm?
    b.   What are the w/w concentrations of potassium and chloride per mL? Assume that at body temperature, the density of serum is 1.005 g/mL.
    c.   What are the w/w concentrations of potassium and chloride per liter?

1.4    One-half container—0.5 L—of summer coolant is added to a car's radiator. What is the percentage (v/v) of coolant in the system if the total volume is 4.5 L?

1.5    Water is considered "hard" when it contains 100 ppm of $CaCO_3$.
    a.   How many mg of $Ca^{++}$ and how many mg of $CO_3^-$ are in 0.5 L of a 100-ppm calcium carbonate solution?
    b.   What is the molarity of the solution?

1.6    An impurity found in motor oil will form a sludge on the spark plugs when at or above 100 ppm (w/v) levels. How many mg of impurity would this be in a quart of oil? (1.06 quarts = 1 L)

1.7    In many states, a person is considered legally intoxicated if his or her blood contains 0.1% ethanol (v/v). What volume of ethanol in the blood is enough to produce legal intoxication in a 70-kg person with a total blood volume of 5.45 L?

1.8    A new metal composite has a density of 2.543 $g/cm^3$. Part of its composition consists of 14.4 ppt (w/w) graphite fibers. What weight of an airfoil weighing 35.00 kg and made of the composite is graphite?

1.9    Two shakes from a salt shaker dispenses 150 mg of NaCl. This is added to a container holding 0.75 L of hot water on a stove.
    a.   What is the content of $Na^+$ and $Cl^-$ in the water in g/L and in mg/mL?
    b.   What are the molar concentrations of $Na^+$, $Cl^-$, and NaCl?

1.10   If 40 rat hairs weighing 0.15 mg each were found in a 1.00-kg sample of grain, what is the w/w content of rat hairs in ppm and mg/kg?

1.11   10.0 mg of $K_3Fe(CN)_6$ were added to 0.250 L of water.
    a.   What is the molarity of the solution?
    b.   What is the solution's salt content in ppm (w/v)?

    c.   If a 10.0-mL sample of the solution is taken, what is the salt content in ppm (w/v) of the sample?

    d.   In a oxidation–reduction reaction, $K_3Fe(CN)_6$ takes up one electron per molecule. What is the normality of the salt solution for an oxidation–reduction reaction?

**1.12**   If the hydrogen ion concentration of a solution is 0.010 M, what is the pH of the solution?

**1.13**   If a solution's pH is 9.05, what is the value of $[H^+]$?

**1.14**   The compound 2,6-dichlorophenol indophenol (m.w. 256), abbreviated DCIP, is used in an analysis to determine the amount of vitamin C. Two electrons per mole of DCIP are transferred in the reaction. A 250-mL solution containing 15.0 mg of DCIP has what normality?

**1.15**   How many grams of $H_2SO_4$ should be added to water to make 500.0 mL of 0.030-$N$ solution? How many grams of HCl should be added to a 50:50 water:methanol mixture to make 500.0 mL of 0.030-$N$ solution?

**1.16**   Iodate ion can be reduced as well as accept a proton in a reaction. The general reactions for these processes are

$$12\ H^+ + 2\ IO_3^- + 10\ e^- \rightarrow I_2 + 6\ H_2O \qquad \text{redox}$$
$$HIO_3 \rightleftharpoons H^+ + IO_3^- \qquad \text{acid–base}$$

If 5.2 g of $KIO_3$ is contained in 1.00 L of solution,

    a.   What is the molarity of the iodate solution?

    b.   For an oxidation–reduction reaction, what is the solution normality?

    c.   For an acid–base reaction, what is the normality?

**1.17**   An acid is 0.620 $N$. To what volume must 1.000 L be diluted to make it 0.500 $N$?

**1.18**   Concentrated reagent $H_2SO_4$ has a specific gravity of 1.84 and contains 96.0% of $H_2SO_4$ by weight. What is the molarity of the solution?

**1.19**   How many mL of $HNO_3$ solution whose density is 1.3393 g mL$^{-1}$ and which contains 55.00% $HNO_3$ by weight are required to prepare 1800 mL of 0.3100-$N$ solution?

**1.20**   If 25.00 mL of an HCl solution exactly reacts with 0.2178 g of pure $Na_2CO_3 \cdot H_2O$ replacing both sodiums, what is the normality of the acid?

**1.21**   What is the molarity of a 6.0% (w/w) NaCl solution if the specific gravity of the resulting solution is 1.0413?

**1.22**   A nitric acid solution containing 7.20% (w/v) of $HNO_3$ is needed. To what volume must 50.00 mL of concentrated $HNO_3$ solution—75.00% $HNO_3$, specific gravity 1.4337—be diluted to obtain the 7.20% (w/v) solution?

**1.23**   When 50.02 mL of an aqueous solution containing only $H_2SO_4$ was treated with an excess $BaCl_2$ solution, a precipitate of $BaSO_4$ was obtained. This was filtered off and dried and found to weigh 1.2930 g. Assuming that the small amount of dissolved $BaSO_4$ can be neglected, what is the molarity of the original $H_2SO_4$ solution?

1.24   To make a special alkaline solution, an experimenter dissolves 7.932 g of BaO, 3.976 g of NaOH, and 1.682 g of $Na_2O$ in enough water to make 1000.0 mL of the solution.
   a.   Calculate the normality of the solution.
   b.   If someone wants to know the molarity of the solution, what problems will be encountered?

## SUPPLEMENT 1A

# Mathematical Review

## 1A.1 Logarithms

The logarithm of a number is the power to which 10 must be raised to give the number. Thus, for the number $M$ and its logarithm $a$, the following relationship is true:

$$M = 10^a; \qquad a = \log M \qquad \textbf{(1A-1)}$$

The antilogarithm or antilog is the number itself (here $M$) which corresponds to the logarithm (here $a$).

With the advent of hand calculators, the use of logarithms for calculation has almost disappeared. However, logarithms are still used as a shorthand for noting a wide range of numbers. They are used to enumerate equilibrium constants as well as concentrations of species in a solution. For this reason, it is necessary to be able to calculate the logs of numbers and vice versa.

Since the logarithm is an exponent, it follows the rules of exponential mathematics. The rules for manipulating logs are simple. The log of a product is the sum of the logs of the multiplied numbers.

$$\log(ab) = \log a + \log b$$

The log of the quotient of two numbers is given by

$$\log(a/b) = \log a + \log(1/b) = \log a - \log b$$

As for other exponents, multiplying two exponential numbers gives the same result as adding the exponents.

$$10^a \cdot 10^b = 10^{a+b}$$

and

$$\log(10^a \cdot 10^b) = \log(10^{a+b}) = a + b$$

---

**EXAMPLE**

$$\text{The value of } \log(3.45 \times 10^{-4}) = \log 3.45 + \log 10^{-4}$$
$$= 0.5378 + (-4)$$
$$= -3.4622$$

---

In addition, you will see the abbreviation of logarithm to the base 10 as lg and $\log_{10}$. This notation differentiates it from the logarithm to the base $e(= 2.71828)$, which is written ln or $\log_e$. For all numbers,

$$\log_e M = 2.3026 \log_{10} M \qquad \text{(1A-2)}$$

---

**EXAMPLE**

$$\ln 100 = 4.6052 = 2.3026 \times 2$$

---

## 1A.2 Quadratic Equations

In calculations that involve equilibria, you will inevitably encounter quadratic equations. These are equations in which the highest power of the unknown is the square. A general quadratic equation has the form

$$ax^2 + bx + c = 0 \qquad \text{(1A-3)}$$

Here $a$, $b$, and $c$ are constants. This can be solved in closed form.

$$x = \frac{-b \pm (b^2 - 4ac)^{1/2}}{2a} \qquad \text{(1A-4)}$$

For the calculations relating to chemical equilibria, approximate methods can often be used, although it is easy to calculate the answers with the simplest hand calculator. It is still useful to learn about the approximation methods because the resulting simplified equations can help understand important behavior in a complicated equilibrium system. Approximations help us see an overview. These are discussed further in Section 2.5.

## 1A.3 Equations of Higher Degree

Equations that contain an unknown to a power higher than two (for example, $x^3$) are classed in the category of equations of higher degree. With a simple hand calculator, a successive approximation to answers is quite easy. With the programmable calculators, even quite complicated equations can be solved in a few minutes by successive approximation. The method requires making an estimate of the value of the variable that will fit the function, calculating the value of the function, and making a closer approximation until you are satisfied with the closeness of the true and the approximate result.

**EXAMPLE**

Find the solution to the equation

$$(2 + x)^2 + 5x^3 = 8.33 = f(x)$$

If this is a description of a real physical system, then we may expect that the value of $x$ we want to find will be a positive number. We begin with $1 = x$. The trials and results are shown in the table below.

| $x$ Chosen | $f(x)$ Calculated | Comments |
|---|---|---|
| 1 | 14 | Too high. The equation is monotonic, so try a lower value. |
| 0.5 | 6.875 | Somewhat too low. |
| 0.7 | 9.00 | Somewhat too high; $x$ is between 0.5 and 0.7. |
| 0.65 | 8.396 | Nearly right. |
| 0.63 | 8.167 | |
| 0.64 | 8.28 | |
| 0.645 | 8.338 | |
| ⋮ | ⋮ | |

If higher accuracy is needed and meaningful, the approximation can be made more exact.

## SUPPLEMENT 1B

# Mathematics of Rates of Reactions, Decay Times, and Half-lives

A number of phenomena occur in nature that can be described with the following algebraic equation:

$$\frac{\Delta[X]}{\Delta t} = -k[X] \qquad \textbf{(1B-1)}$$

when $\Delta[X]$ is the change in $[X]$ that occurs in the time interval $\Delta t$ (the value of $[X]$ is a function of time and may also be written $[X(t)]$), and $k$ is a proportionality constant. Its value denotes the *fraction* of $[X]$ that changes over the interval $\Delta t$ (the value of $k$ has the units time$^{-1}$).

The equation states that the change in concentration of some substance X is proportional to the amount of X present. The minus sign in the equation indicates that $[X]$ decreases with time. This type of behavior is called first order **exponential decay.** By inspecting Eq. 1B-1, we can infer that if $k$ is large, then $[X]$ changes relatively rapidly. If $k$ is small, then $[X]$ changes relatively slowly. Thus, by decreasing or increasing the value of $k$, the equation can be made to describe a slower or faster exponential decay.

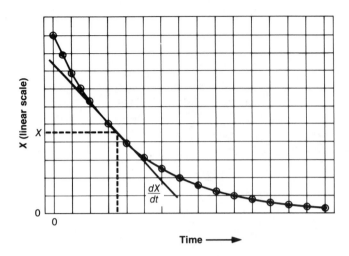

**Figure 1B.1**   One way to determine *k* of the exponential decay is to measure *dX/dt* together with *X* at a point on the data curve. This technique tends to be less accurate since little of the data is used.

In Eq. 1B-1, the quantity [X] is a chemical concentration. Such changes in concentration occur in chemistry in simple chemical reactions and in radioactive decay. However, the same type of behavior is also seen in the responses of parts of electronic circuits. In that case, $X$ represents electrical current or voltage. Since chemical reactions and electronic circuits are both part of contemporary analysis, Eq. 1B-1 can be generalized. Instead of [X], the letter $X$ alone will be used with the understanding that it can signify a chemical concentration as well as other quantities.

It is easier mathematically to manipulate Eq. 1B-1 with time intervals that are infinitesimally small. This is signified by putting $dt$ in place of $\Delta t$ and $dX$ in place of $\Delta X$. The information contained in the equation is the same. With the substitution, the equation becomes

$$\frac{dX}{dt} = -kX \tag{1B-2a}$$

The above equation is a description of experimentally measurable behavior. The data consist of measurements of $X$ at various times. From the data, the characteristic value $k$ can be calculated in two different ways.

The first procedure is illustrated in Figure 1B.1. It requires measuring $X$ as a function of time—$dX/dt$—together with the value of $X$. Algebraically, we can rearrange Eq. 1B-2a to

$$\frac{dX/dt}{X} = -k \tag{1B-2b}$$

The value of $dX/dt$ in the equation is the slope of a tangent at the point $X(t)$ on the graph. However, even though $X(t)$ can be found relatively exactly, a major problem ensues because it is often difficult to measure the slope as precisely.

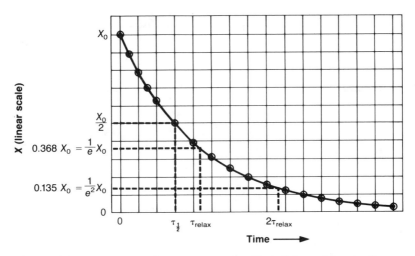

**Figure 1B.2**   Graph of $X$, in units of $X_0$, versus time. The half-time and the relaxation time are shown.

An alternate, and inherently far more accurate, method requires measuring $X(t)$ at a series of times and calculating $k$ from all the data together. To do so we need an equation that relates $X$ alone directly to $k$. Equation 1B-3 satisfies these criteria and is written without proof:

$$X(t) = X_0\, e^{-kt} \qquad\qquad \textbf{(1B-3)}$$

Equation 1B-3 describes the results of an exponential decay in which the *first* measurement of $X(t)$ is made at a time that we shall call $t = 0$. Call the value of $X(t)$ at this first measurement $X_0$. The time in the term $X(t)$ is, then, a time $t$ after $t = 0$. As you know, $e$ is the natural logarithm base ($= 2.71828$). A graph of the function $X(t)$ versus $t$ is plotted in Figure 1B.2.

However, the value of $k$ determined from the graph in Figure 1B.2 is no more accurate than the slope determined under these conditions; an alternate form of the equation is far superior. This alternate form is obtained by taking the logarithm of both sides of Eq. 1B-2.

$$\ln X(t) = \ln(X_0\, e^{-kt})$$

And, from the properties of logarithms,

$$\ln \frac{X(t)}{X_0} = -kt \qquad\qquad \textbf{(1B-4)}$$

If the logarithm of the ratio $(X(t)/X_0)$ is plotted versus time, the slope will be $-k$. A simple way to obtain such a plot is to graph $(X(t)/X_0)$ on semilogarithmic graph paper. Such a plot appears in Figure 1B.3. The ordinate value is $\log(X(t)/X_0)$ or $\ln(X(t)/X_0)$,

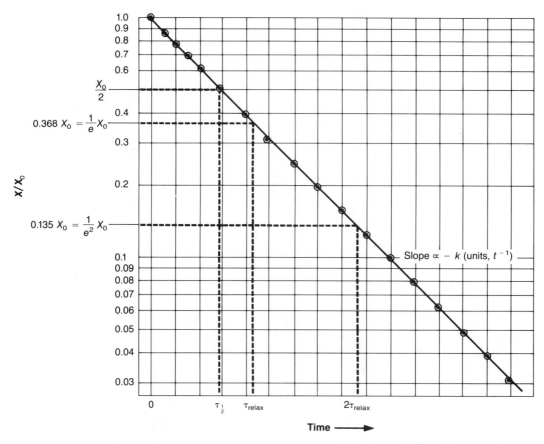

**Figure 1B.3**    Graph of $(X/X_o)$ on a logarithmic scale versus time. In this case, $X$ has the value $X_o$ at $t = 0$. The value of $X/X_o$ is on the left scale. The slope of the plot is proportional to $-k$. The value of $k$ is the characteristic rate constant for the process of exponential decay. The same value of $k$ would be obtained from a plot of $X$ alone. The locations of the half-time, the relaxation time, and twice the relaxation time are shown.

with time on the abscissa. Further, if the base-10 logarithm is used, the slope differs by a factor of 2.303 compared to a plot based on the natural (base $e$) logarithm.

---

**EXAMPLE**

The value of $k'$ from a plot of $\log[X(t)]$ versus time is 0.283 s$^{-1}$. What is the value of $k$ that would be used in Eq. 1B-4 to produce the same plot for $\ln[X(t)]$ versus time?

**SOLUTION**

The equation that produces the value of $k'$ is

$$\frac{\Delta \log[X(t)]}{\Delta t} = -k't$$

It is necessary to convert the base-10 logarithm to a base-$e$ logarithm. This is done by multiplying both sides of the equation by $2.303 = \ln 10$. Thus

$$\frac{2.303 \cdot \Delta\log[X(t)]}{\Delta t} = \frac{\Delta\ln[X(t)]}{\Delta t} = -2.303 \, k't$$

Therefore $k = 2.303 \, k'$ and $k = 2.303 \times 0.283 = 0.652 \text{ s}^{-1}$.

---

As the size of $X$ decreases from $X_0$, there will be a time when

$$X(t) = \tfrac{1}{2}X_0 \tag{1B-5}$$

This special time is called the **half-time, decay time,** or **half-life,** abbreviated $t_{1/2}$ or $\tau_{1/2}$. Which of the three names is used depends on whether the equation is applied to chemical kinetics, electronics, or radioactivity, respectively.

The value of $t_{1/2}$ is related to $k$ in the following way. At the half-time, the following equality is true *by definition:*

$$X(t_{1/2}) = \frac{X_0}{2} = X_0 \, e^{-kt_{1/2}}$$

or

$$\ln \tfrac{1}{2} = \ln(e^{-kt_{1/2}}) \tag{1B-6a}$$

After the mathematical operations, Eq. 1B-6a becomes

$$-\ln 2 = -kt_{1/2} \tag{1B-6b}$$

This can be rearranged to solve for $k$.

$$k = \frac{\ln 2}{t_{1/2}} = \frac{0.693}{t_{1/2}} \tag{1B-7}$$

## Relaxation Time

Another characteristic time of the exponential decay is when

$$X(t) = (1/e)X_0 = 0.368 \, X_0$$

This time is called the **relaxation time** or **response time** and is a quantity most often used in electronics. The longer the relaxation time, the slower is the decay. The relaxation time of an exponential decay is also illustrated in Figures 1B.2 and 1B.3.

## Supplement Problems

---

S1.1   The first-order rate constant for the decomposition of cyclobutane ($C_4H_8$) at 500 °C is $9.2 \times 10^{-3} \text{ s}^{-1}$.
     a.   How long will it take for 90.0% of a sample of 0.100-M $C_4H_8$ to decompose at 500 °C?

b. Under the same conditions, how long will it take for 90.0% of a sample of 0.05-M $C_4H_8$ to decompose?

S1.2   The radioactive isotope strontium-90 decays with a half-life of 28 years through the reaction

$$^{90}_{38}Sr \rightarrow ^{90}_{38}Y + \beta^-$$

What is the rate constant for the decay in the units of $yr^{-1}$?

S1.3   A measured voltage decreases exponentially with a half-time of $1 \times 10^{-4}$ s. What is the relaxation time of the decay?

S1.4   Requires calculus. Assume that Eq. 1B-3 describes the behavior of a chemical reaction. Show algebraically that the slope of $\ln[X(t)]$ versus time is the same as a plot of $\ln([X(t)]/[X(0)])$ versus time.

S1.5   The following data were determined by experiment on the kinetics of the special compound rarium chloride (RmCl), which reacts in solution to form rarerium chloride (RrCl).

| Time(s) | [RmCl] (g/L) |
|---------|--------------|
| 0 | 2.13 |
| 5 | 1.71 |
| 10 | 1.37 |
| 20 | 0.88 |
| 40 | 0.36 |
| 60 | 0.15 |

a. Plot on Cartesian graph paper $\ln[X(t)]$ and $\log[X(t)]$. What are the slopes of the two plots?
b. What are the half-times from each of the graphs?
c. Plot on semilogarithmic graph paper $X(t)$ versus time. Is the rate constant $k$ the slope?

S1.6   An aqueous solution of hydrogen peroxide ($H_2O_2$) was held at 40.0 °C. The solution initially was found to contain 0.530-M $H_2O_2$. The concentration of $H_2O_2$ was determined at a number of subsequent times. The results are shown in the table.

| Time(s) | [$H_2O_2$] (M) |
|---------|----------------|
| 0 | 0.530 |
| 5 000 | 0.472 |
| 10 000 | 0.420 |
| 20 000 | 0.335 |
| 40 000 | 0.208 |
| 60 000 | 0.133 |

a. What is the half-time for the reaction under these conditions?
b. If the original concentration of $H_2O_2$ were 0.832 M, how much would be left in solution after 4.00 days?
c. What is the rate constant for the reaction?

S1.7   The radioactivity of a sample of cobalt-61 was measured. At an initial time, 31,422 disintegrations $s^{-1}$ were counted: 0.50 h later, under the same conditions, 23,613 disintegrations $s^{-1}$ were counted. What is the half-life of cobalt-61?

S1.8   Instead of measuring the disappearance of a reactant over time, we can observe the appearance of its product.

$$X \xrightarrow{k} P$$

However, Eq. 1B-3 will not describe the result obtained by measuring the product. If the amount of product reaches a limiting value of $P_\infty$, the exponential equation takes the form

$$P(t) = P_\infty(1 - e^{-kt})$$

For a first-order reaction, the rate constant is the same as for the disappearance of $X$.

    a.   In terms of $P(t)$ and $P_\infty$, what is the equation to use in place of Eq. 1B-4 to produce a linear semilogarithmic plot with slope $-k$?

    b.   What function of $P(t)$ and $P_\infty$ is plotted versus time on a semilogarithmic graph to find the rate constant?

# 2

# Simple and Competitive Chemical Equilibria

## 2.1 Properties of Chemical Equilibria

A knowledge of the principles of chemical equilibria and the ability to make calculations about competitive equilibria are essential in understanding the fundamentals of chemical analysis. These principles underlie a number of different stages of an analysis:

1. In sample preparation — making the sample homogeneous and removing parts of the sample that will interfere with the analysis.
2. In assay methods such as titration (Chapters 9 and 10).
3. In most instrumental assay methods, to explain interferences.

You will find it far harder to understand the methods presented in this text if you cannot first work easily with the ideas and calculations of chemical equilibria.

## The Concept of Chemical Equilibrium

If nitrogen gas, $N_2$, and hydrogen gas, $H_2$, are placed together in a container at one atmosphere pressure and 400 °C (a dull-red heat) and left for a week, an analysis of the contents of the vessel should show the presence of hydrogen, nitrogen, and ammonia. The reaction that occurs can be written

$$N_2(g) + 3 H_2(g) \rightleftharpoons 2 NH_3(g) \tag{2-1}$$

Since all three chemical species are present, the reaction did not go to completion. If it had, either $N_2$ or $H_2$ would be used up and would be absent.[1]

If ammonia gas alone is placed in a container under the same conditions, and the contents are analyzed a week later, the analysis again should show the presence of the same three components: hydrogen, nitrogen, and ammonia. The reaction can be written

$$2 NH_3(g) \rightleftharpoons N_2(g) + 3 H_2(g) \tag{2-2}$$

The conclusion drawn from this second experiment is that reaction 2-2 does not go to completion either.

One possible explanation of the data is that neither reaction has had time to progress all the way to its ultimate state; perhaps the reaction will go to completion if it is allowed to react further. This hypothesis can be tested by running both reactions for a few more weeks, followed by another analysis of the contents. If the analytical results remain the same, we can say that the system of ammonia, nitrogen, and hydrogen has reached its equilibrium condition. At equilibrium, the identities of reactants and products are the same regardless of the side (that is, in this case, nitrogen and hydrogen alone or ammonia alone) from which the equilibrium is started.

It has been found *from experiments* that the reactions of nitrogen, hydrogen, and ammonia can be described by a constant, the **equilibrium constant,** which is dependent on temperature. For the ammonia reaction

$$2 NH_3(g) \rightleftharpoons N_2(g) + 3 H_2(g) \tag{2-2}$$

the equilibrium constant $K$ is

$$K = \frac{[H_2]^3[N_2]}{[NH_3]^2} \tag{2-3}$$

where the square brackets indicate that molar concentration units are to be used. Each concentration is raised to the power of its stoichiometric coefficient.

An alternate means of expressing the equilibrium constants of gases uses the pressures of the species. For reaction 2-2,

$$K_p = \frac{P_{H_2}^3 P_{N_2}}{P_{NH_3}^2} \tag{2-4}$$

---

[1] The term completion must be defined arbitrarily. If it means that at least one reactant is absent, one must state the concentration below which the presence of that reactant cannot be measured.

The subscript $p$ in $K_p$ is used to denote this type of equilibrium constant, since the value of $K_p$ differs from $K$ with molar concentration units.

For the general reaction

$$aA + bB \rightleftharpoons cC + dD \tag{2-5}$$

the concentrations of the reactants and products at equilibrium are related by the equation

$$K = \frac{[C]^c[D]^d}{[A]^a[B]^b} \tag{2-6}$$

$K$ is a constant, the value of which depends on the temperature and the identities of the products and reactants. Equation 2-6 expresses the **law of mass action.** In other words, a quantitative description of the equilibrium follows the law of mass action. Equation 2-6 also can be called a **mass-action expression.**

As can be seen in Eqs. 2-3 and 2-6, the equilibrium constant expressions are written with the products of the reaction (the species on the right side) in the numerator and reactants (the species on the left side) in the denominator. In addition, the concentration of each species is raised to a power equal to its stoichiometric coefficient in the chemical equation.

If the mass-action expressions as they are written describe the behavior exactly, the system is said to follow *ideal* behavior. In real equilibrium mixtures, this ideal behavior is nearly true for gases at less than a few atmospheres pressure and for solutions that are dilute.[2] For the present, we shall continue to use the molar concentrations in the equilibrium expressions.

## Conventions

The chemical equations and equilibrium constant expressions only *represent* the reactions that occur. This may seem too trivial to state overtly; however, the equations allow the use of relatively simple concepts to represent reactions that we know are highly complicated in their chemical details. For instance, consider the equilibrium constant for the dissociation of acetic acid. The chemical equilibrium can be represented by the equation

$$CH_3COOH + H_2O \rightleftharpoons CH_3COO^- + H_3O^+ \tag{2-7a}$$

The exact formula of the species formed by water interacting with the proton from the acid is uncertain. We can write $H_3O^+$ as above. However, some workers maintain that three waters associate with the proton to form $H_7O_3^+$, and others suggest that $H_9O_4^+$ is formed by association with four waters. Equation 2-7a might then be written

$$CH_3COOH + 3\,H_2O \rightleftharpoons CH_3COO^- + H_7O_3^+ \tag{2-7b}$$

---

[2] This imprecise term is used because the upper range of ideal behavior depends on the chemical system. Ions in solution are dilute up to about 1 mM. Depending on the materials involved, organic solutes in organic solvents can behave as ideal systems over a wide range: Some exhibit ideal behavior only up to 1% solute, whereas others may do so from 0–100%.

In other words, the number of associated water molecules to be specified in the equation is questionable. Recognizing this, sufficient information is conveyed when the equation is written

$$CH_3COOH \rightleftharpoons CH_3COO^- + H^+ \qquad \text{(2-7c)}$$

Actually, both the acetic acid molecules and the acetate ions are themselves strongly associated with waters of hydration. These waters have been ignored as well in Eqs. 2-7. Also, for the most part, the molecular structures of the species are ignored in these simple equations (for example, $H_9O_4^+$ is thought to be tetrahedral).

Besides the selective simplification of the chemistry, less important abbreviations make the chemical equations easier to write. As long as the significance is understood, there is no confusion. For instance, acetate ion,

$$H_3C - C \overset{\displaystyle O}{\underset{\displaystyle O^-}{\diagup\diagdown}}$$

can be abbreviated as $^-Ac$, $^-OAc$, Ac, or OAc. The form $^-OAc$ is used in this text because it is short, indicates the oxygen to which the proton is bound, and retains a reminder of the charge to help avoid errors in balancing the charges in stoichiometric equations. Therefore the simple equation representing ionization of acetic acid in water can be written

$$HOAc \rightleftharpoons H^+ + {}^-OAc \qquad \text{(2-7d)}$$

The equilibrium constant will then be written

$$K = \frac{[H^+][^-OAc]}{[HOAc]} = 1.85 \times 10^{-5} \text{ at } 25°: \quad pK = 4.74 \qquad \text{(2-8)}$$

Often, for acid dissociation reactions such as for acetic acid, the equilibrium constant is written with a subscript as $K_a$. The subscript $a$ is for acid. You may also see the equilibrium constant written $K_d$ in certain circumstances, with $d$ for dissociation.

### Properties of Equilibrium Constants

The equilibrium reaction 2-7d can be reversed and written

$$H^+ + {}^-OAc \rightleftharpoons HOAc \qquad \text{(2-9)}$$

Then the mass-action expression becomes

$$K' = \frac{[HOAc]}{[H^+][^-OAc]} \qquad \text{(2-10)}$$

The equilibrium constant written with a prime $K'$ differs from $K$. Comparison of the two equilibrium expressions, Eqs. 2-8 and 2-10, shows that

$$K' = \frac{1}{K} \qquad \text{(2-11)}$$

$K'$ is called an **association equilibrium constant.**

The algebraic relation 2-11 illustrates a general rule:

*When a chemical equilibrium equation is reversed, the equilibrium constant must be changed to its reciprocal.*

If we add Eq. 2-7d to itself

$$\begin{aligned}
&\text{HOAc} \rightleftharpoons \text{ H}^+ + \text{ }^-\text{OAc} \\
+ \quad &\underline{\text{HOAc} \rightleftharpoons \text{ H}^+ + \text{ }^-\text{OAc}} \\
&2\,\text{HOAc} \rightleftharpoons 2\,\text{H}^+ + 2\,^-\text{OAc}
\end{aligned} \qquad \textbf{(2-12)}$$

then, according to the rules for writing mass-action expressions, the equilibrium constant for reaction 2-12 is

$$K = \frac{[\text{H}^+]^2[^-\text{OAc}]^2}{[\text{HOAc}]^2} = K_a(\text{HOAc}) \cdot K_a(\text{HOAc}) = (1.85 \times 10^{-5})^2 \qquad \text{at } 25\,^\circ\text{C}$$

$$\textbf{(2-13)}$$

This example of adding the chemical equations illustrates another general rule:

*When chemical equilibrium equations are added, the associated equilibrium constants must be multiplied together to obtain the equilibrium constant for the sum equation.*

Most equilibrium calculations relating to real systems can be carried out with the operations of addition and reversal of chemical equations and the appropriate manipulation of the mass-action expressions. To obtain numerical results, the data, including the stoichiometry of the chemical reaction and the value of the equilibrium constant (at the temperature of the experiment), must be available. In the next section begins a short discussion of the calculations regarding equilibria of chemical reactions that are not ideal.

## 2.2 Activities and Activity Coefficients

Equations of the type

$$K = \frac{[\text{C}]^c[\text{D}]^d}{[\text{A}]^a[\text{B}]^b} \qquad \textbf{(2-6)}$$

accurately describe the equilibrium concentrations of ideal systems, with a value of $K$ that is constant. In reality, equilibria in solutions, especially ionic solutions, do not behave so simply.

Why do the concentrated solutions not behave as ideal solutions? The reason is that there are interactions between molecules and especially between ions. These interactions are exceedingly difficult to quantify. Consider a simple salt solution such as 0.1 M sodium chloride in water. Assume that the only other species present is water with a pH of 7. This allows us to ignore the very low concentrations of protons and hydroxide ions. Even with this simplification, there are interactions of

$Na^+$ with $Na^+$
$Cl^-$ with $Cl^-$
$Na^+$ with $H_2O$
$Cl^-$ with $H_2O$
$H_2O$ with $H_2O$
$Na^+$ with $Cl^-$

The situation quickly becomes extremely complicated!

Suppose an equilibrium constant was determined experimentally by measuring the concentrations of the species involved. In the case of acetic acid dissociation, $[H^+]$, $[^-OAc]$, and $[HOAc]$ would be measured and the equilibrium constant calculated from the concentrations in mol $L^{-1}$. Further, suppose that this experiment was repeated for various concentrations of acid. As the solution is made more concentrated, the measured equilibrium "constant" changes! However, rather than throw out the very useful concept of mass action, an *effective* concentration can be substituted in place of the measured ones. This allows the idea of an equilibrium constant to be useful over a wider range of conditions.

These effective concentrations are called **activities** (abbreviated $a$), and the activity of a salt tends to be lower than the concentration of the salt. The activities of various species under *specific* solution conditions are found through experimental measurements. This information is conveniently stored as **activity coefficients,** which relate the effective concentration (that is, the activity) to the molar concentration.

Activities can be related to the concentrations of solutes on three scales:

the molal scale, for which $\qquad a_A(m) = \gamma_A m_A$ $\qquad$ **(2-14a)**

the molar scale, for which $\qquad a_A(c) = y_A c_A$ $\qquad$ **(2-14b)**

the mole-fraction[3] scale for which $\quad a_A(N) = f_A N_A$ $\qquad$ **(2-14c)**

Here $m_A$ is the molality, $c_A$ the molarity, and $N_A$ the mole fraction of solute A; $\gamma_A$, $y_A$, and $f_A$ are called, respectively, the molal, molar, and rational activity coefficients of species A. As can be seen in the representative example in Table 2.1, activity coefficients are concentration dependent, but at concentrations less than 0.2 $m$, the three types of activity coefficients differ by less than 1% for sodium chloride solutions. In the table the subscript $\pm$ means that the activity coefficient reflects the concentrations of both ions of the salt.

---

**EXAMPLE**

What is the activity of sodium chloride in an aqueous solution in which the concentration of added NaCl is 0.10 M, and the temperature is 25 °C?

---

[3] A mole fraction is the fraction that a component is of the total number of moles in a mixture. For a two-component system composed of $x$ moles of X and $w$ moles of W, the mole fraction of X $= x/(x + w)$.

**Table 2.1**   Experimental Activity Coefficients of Solute NaCl and Water in Solutions at 25 °C*†

| Molality | Molarity | $\gamma_\pm$ | $y_\pm$ | $f_\pm$ | $a_{H_2O}$ |
|----------|----------|--------------|---------|---------|------------|
| 0 | 0 | 1.000 | 1.000 | 1.000 | 1.000 |
| 0.001 | 0.000997 | 0.965 | 0.965 | 0.965 | 0.9999 |
| 0.01 | 0.009969 | 0.902 | 0.902 | 0.902 | 0.9997 |
| 0.1 | 0.09953 | 0.778 | 0.779 | 0.781 | 0.9966 |
| 0.2 | 0.1987 | 0.735 | 0.738 | 0.740 | 0.9933 |
| 1.0 | 0.9788 | 0.656 | 0.668 | 0.681 | 0.9669 |
| 2.0 | 1.921 | 0.668 | 0.693 | 0.716 | 0.9316 |
| 4.0 | 3.696 | 0.783 | 0.845 | 0.896 | 0.8515 |
| 6.0 | 5.305 | 0.986 | 1.111 | 1.199 | 0.7598 |

* The activity can be related to the concentration of an ion on three scales:

$$\begin{aligned} \text{molal scale:} \quad & a(m) = \gamma m \\ \text{molar scale:} \quad & a(c) = yc \\ \text{mole-fraction scale:} \quad & a(N) = fN \end{aligned}$$

Here $m$ is the molality, $c$ the molarity, and $N$ the mole fraction; $\gamma$, $y$, and $f$ are, respectively, the molal, molar, and rational activity coefficients. At less than 0.2 $m$, the three activity coefficients differ by less than 1%. In the table the subscript $\pm$ means that the activity coefficient relates to both ions.

† *Source:* Data from R. A. Robinson and R. H. Stokes, "Electrolyte Solutions," 2nd ed. rev.; Butterworths: London, 1965.

**SOLUTION**

From the information in Table 2.1 and Eq. 2-14b,

$$a_{NaCl}(c) = 0.779 \times 0.10 = 0.0779 \text{ M}$$

The effective concentration of sodium chloride is less than the concentration measured by, say, adding a known weight of NaCl to make a known volume of solution.

## Ionic Strength

The activity of an ion-forming solute depends on the concentrations of *all* the different ions in a solution. As a result, activity coefficients have been measured and tabulated only for relatively simple solutions and mostly only for water as solvent. For more complex solutions, such as seawater, approximations can be made with the aid of a computer, a list of the molar concentrations of all the ionic species, and one of a number of different mathematical models that have been found to describe the activities reasonably well.

In addition, the activity coefficients depend on the magnitude of the *charges* on the ions in the solution. For instance, a doubly charged ion has a greater effect on the activities than the same concentration of a singly charged ion. A single number that is used to characterize the total ionic concentration, and which accounts for the charges on them, is the solution **ionic strength.** Ionic strength is defined with molar concentrations $c_j$ of each ionic species $j$ and the associated ionic charges $z_j$. The defining equation is

$$\mu = \tfrac{1}{2} \sum_j c_j z_j^2 \qquad (2\text{-}15)$$

Each type of negative and positive ion is included in the sum. Note that neutral molecules do not contribute to the ionic strength.

---

**EXAMPLE**

What is the ionic strength of a 0.01 M $CaCl_2$ solution? The salt is completely dissociated under these conditions.

**SOLUTION**

In the solution, the calcium concentration is 0.01 M, and the chloride ions are present at 0.02 M. Therefore

$$\mu = \tfrac{1}{2}(0.01 \times 2^2 + 0.02 \times 1^2) = \tfrac{1}{2}(0.04 + 0.02)$$
$$= 0.03 \text{ M}$$

---

**EXAMPLE**

What is the ionic strength of a 0.01-M solution of wishful sulfate $WsSO_4$, which under the conditions of the experiment is exactly 50% dissociated?

**SOLUTION**

In the solution, both the $Ws^{++}$ and $SO_4^-$ concentrations are 0.005 M. The undissociated, uncharged $WsSO_4$ makes no contribution to $\mu$. Therefore

$$\mu = \tfrac{1}{2}(0.005 \times 2^2 + 0.005 \times 2^2) = \tfrac{1}{2}(0.04) = 0.02 \text{ M}$$

---

## Single-ion Activities

A theoretical interpretation of the interactions of ions in solution provides an equation that can be used to assign activity coefficients to each species of ion in a solution. For aqueous solutions at 25 °C,

$$-\log \gamma_i = 0.5\, z_i^2 \sqrt{\mu} \qquad (2\text{-}16)$$

where $z_i$ is the charge on the ion species $i$, $\gamma_i$ the molal activity coefficient, and $\mu$ the ionic strength of the solution. The numerical value 0.5 is only slightly temperature-dependent. This equation applies within about 3% accuracy to solutions with ionic strengths less than about 1 mM. At these concentrations the molal and molar activity coefficients are essentially equal. As a result, $y_i$ can be substituted for $\gamma_i$ in Eq. 2-16 in the equation's applicable range.

---

**EXAMPLE**

Calculate the single-ion activity coefficients you would expect to find for $Na^+$ and $Cl^-$ in an aqueous solution containing 0.001 M NaCl and 0.002 M $H_2SO_4$ at 25 °C. Under these conditions, both electrolytes are completely dissociated.

**SOLUTION**

The ionic strength of the electrolyte solution is calculated from Eq. 2-15.

$$\mu = \tfrac{1}{2}([Na^+] + [Cl^-] + [H^+] + [SO_4^-] \cdot 4)$$

Substituting the appropriate concentrations of the ions, this becomes

$$\mu = \tfrac{1}{2}([0.001] + [0.001] + [\,0.004] + [0.002] \cdot 4) = 0.007 \text{ M}$$

The activity coefficients for both $Na^+$ and $Cl^-$ are expected to be equal since both possess unit charge:

$$-\log \gamma_{Cl} = -\log \gamma_{Na} = 0.5 \cdot 1^2\sqrt{0.007}$$

Thus

$$\gamma_{Cl} = \gamma_{Na} = 10^{-0.0418} = 0.908$$

---

From this calculation we find that the activity coefficients for both sodium and chloride ions are expected to be the same and to equal the activity coefficient for the salt $\gamma_\pm$. This is a general result for so-called symmetric electrolytes, in which the anion(s) and cation(s) are in the ratio 1:1 (e.g., KCl, CaSO$_4$, and La[Fe(CN)$_6$]). The electrolyte must be completely dissociated.

When the activity coefficient is measured for a solution of electrolyte $A_xB_y$ with $x \neq y$, this simple result is not true. A **mean[4] activity coefficient** $\gamma_\pm$ is measured, and the calculated activity coefficients for the individual ions are not expected to be equal to it. The mean activity coefficient is defined in relation to the individual-ion activity coefficients by

$$\gamma_\pm^{(x+y)} = \gamma_A^x \cdot \gamma_B^y \qquad \text{for } A_xB_y \qquad\qquad \textbf{(2-17a)}$$

or, taking the logarithm of both sides,

$$(x+y) \log \gamma_\pm = x \log \gamma_A + y \log \gamma_B \qquad \text{for } A_xB_y \qquad \textbf{(2-17b)}$$

---

**EXAMPLE**

What is the mean activity coefficient expected for NaCl in a solution 0.001 M NaCl and 0.002 M H$_2$SO$_4$?

**SOLUTION**

Using the results of the previous example,

$$\gamma_\pm^{(1+1)} = \gamma_{Na}^1 \cdot \gamma_{Cl}^1 = 0.908 \cdot 0.908$$
$$\gamma_\pm^2 = (0.908)^2$$
$$\gamma_\pm = 0.908$$

This is a specific case of a symmetric electrolyte.

---

[4] As indicated in Eq. 2-17a, this is a geometric mean, not an algebraic mean.

## Equilibrium Constants and Activity Coefficients

Using activities in place of concentrations allows us to extend the range in which an equilibrium is found to obey the law of mass action. For example, the complicated behavior of the equilibrium of acetic acid with concentration can be described by Eq. 2-18. The complexity is incorporated into the behavior of the factor consisting of the ratio of activity coefficients:

$$K = \frac{[H^+][^-OAc]}{[HOAc]} \cdot \frac{y_{H^+} \cdot y_{-OAc}}{y_{HOAc}} \tag{2-18}$$

In the practice of chemical analysis, the ability to calculate activities and activity coefficients is less important than understanding that ionic equilibria change with ionic strength. When ionic equilibria are involved in an analysis, it is necessary to keep the ionic strength as constant as possible to obtain the most precise results.

## Standard States

The activity coefficients $y$ and $y$ are usually regarded as dimensionless quantities. The definitions in Eqs. 2-14 therefore imply that $a(m)$ and $a(c)$ have dimensions of molal and molar, respectively. The mole-fraction activity $a(N)$ is a more fundamental quantity, mainly because it is independent of volume. Both the mole-fraction activity and its associated activity coefficient $f$ are dimensionless quantities. The reason it is unitless and the important consequences are described next.

The mole-fraction activity is defined as a ratio of the effective concentration of a species relative to its **standard state.** The standard state is *defined* in a different way for different parts of an equilibrium system. The standard state for

*solids is a pure solid*
*solvents is a pure solvent*
*solutes is a 1 molar solution of that solute*
*gases is a concentration of pure gas at standard temperature (273 K) and*
  *pressure (1 atm)*

The definition of the mole-fraction activity incorporating the applicable standard state of substance A, $f_A^\circ$, is

$$\frac{f_A}{f_A^\circ} = a_A(N) \tag{2-19}$$

The effective concentration is written as $f_A$ in order not to confuse it with the molar concentration [A]. The value of $f_A^\circ$ is chosen as appropriate to the system: solids, solutes, and so forth. For instance, if the system is an aqueous NaCl solution, $f^\circ$ is 1 molar NaCl. (The correct term for the effective concentration is **fugacity** and is closely related to the mole-fraction activity coefficient.)

## The Activities of Solids

The concept of an activity as a dimensionless ratio unifies a great deal of equilibrium chemistry. We *should* use activities in all equilibrium constant expressions. Thus, instead of writing

$$K_{\text{concentration}} = \frac{[H^+][^-OAc]}{[HOAc]} \qquad (2\text{-}8)$$

it is more proper to write

$$K_{\text{thermodynamic}} = \frac{a_{H^+} \cdot a_{-OAc}}{a_{HOAc}}$$

Although this type of equilibrium constant is not often used in day-to-day analyses, it serves to illustrate the origins of the rules for including or excluding a species in the mass-action expression. For instance, let us consider the equilibrium that includes a pure solid dissolving in a solvent. No matter how much solid dissolves, any solid that remains continues to have the same concentration — that of the pure solid. So the activity of the remaining solid *at all times* is

$$a_{\text{solid}} = f^\circ / f^\circ = 1$$

---

**EXAMPLE**

What is the thermodynamic equilibrium constant for a saturated solution of sodium chloride in contact with solid sodium chloride?

**SOLUTION**

$$K_{\text{thermodynamic}} = \frac{a_{Na^+} a_{Cl^-}}{a_{NaCl(s)}} = (a_{Na^+})(a_{Cl^-})$$

Such an equilibrium constant is called the **solubility product** and is designed $K_{sp}$. More about solubilities and the related solubility products is presented in Chapter 8.

---

It is most important to understand that there is nothing inherently different between solubility products and the mass-action expressions that were described earlier. As a result, for chemical equilibria that include solids, the rules for the addition and reversal of the stoichiometric equations and allied mass-action expressions are the same as for other equilibria.

## The Activity of Water in Solutions

When we write the equilibrium constant for a reaction such as acetic acid dissociation (Eq. 2-8), the water that is present does not appear in the equilibrium expression. Why not? And is there any time when the water activity must be left in?

Let us begin answering the questions by writing the equation for the reaction and include the water:

$$HOAc + n\,H_2O = H(H_2O)_n^+ + {}^-OAc \qquad (2\text{-}20)$$

The mass-action expression is then

$$K_{\text{thermodynamic}} = \frac{a_{\text{H(H}_2\text{O)}_n^+} \cdot a_{-\text{OAc}}}{a_{\text{HOAc}} a_{\text{H}_2\text{O}}^n} \qquad (2\text{-}21a)$$

The key to answer the questions is to find the value of $a_{\text{H}_2\text{O}}$. Representative values for NaCl solutions can be seen in the right column in Table 2.1.

But what is the value of $a_{\text{H}_2\text{O}}$ for a 0.1 M solution of acetic acid? Since $a_{\text{H}_2\text{O}} = f/f^\circ$, we can make an approximate calculation. The concentration of pure water is around 55.5 M (1000 g/L and 18 g/mol). One liter of a solution that is 0.1 M in acetic acid requires adding 6.0 g (5.7 cm$^3$) of pure acetic acid to approximately 995 cm$^3$ of water: 1000 mL of solution contains 995 mL of water. The concentration of water itself is lower by only about 0.5% in the acid solution compared with pure water. The activity of water is within a few parts-per-thousand of unity; that is, $a_{\text{H}_2\text{O}} \approx 995/1000$.

Leaving the water activity out of the mass-action expression is really an approximation:

$$K_{\text{thermodynamic}} = \frac{a_{\text{H(H}_2\text{O)}_n^+} \cdot a_{-\text{OAc}}}{a_{\text{HOAc}} a_{\text{H}_2\text{O}}^n} \approx \frac{a_{\text{H(H}_2\text{O)}_n^+} \cdot a_{-\text{OAc}}}{a_{\text{HOAc}}} \qquad (2\text{-}21b)$$

In other words, Eq. 2-18 is not exact, but it is an approximation that is adequate for all but the most exacting work. If the solution is very concentrated, however, this approximation may no longer be reasonable.

### Mixed Equilibrium Constants

Ideally, the equilibrium constants we use always should be thermodynamic equilibrium constants that utilize the activities of each of the species involved. In reality, analytical chemical problems (among others) involve extremely complicated solutions and mixtures, and it simply is not clear what values should be used for the activity coefficients in these complicated solutions. In fact, usually only a few of the components in the solution are known. Thus concentration equilibrium constants are used for calculations most of the time.

However, with certain methods, notably the electrochemical ones discussed in Chapter 12, the activities of certain species are actually measured. Perhaps the most common case is with the measurement of the activity of protons, $a_{\text{H}^+}$, made by measuring the solution pH. This proton *activity* value $a_{\text{H}^+}$ is often used in the equilibrium expression along with the *concentrations* of other species. When activities and concentrations occur in equilibrium expressions together, the equilibrium constant is said to be a **mixed equilibrium constant.** For instance, the mixed equilibrium constant for acetic acid is

$$K_{\text{mixed}} = \frac{a_{\text{H}^+} \cdot [^-\text{OAc}]}{[\text{HOAc}]} \qquad (2\text{-}22)$$

The only problem with such mixed constants is that they cannot be manipulated using all the known rules of chemistry. However, they are adequate for calculations relating to a series of experiments in which the conditions remain the same for every experiment. This is one reason that standardized and reproducible conditions are so important in chemical analyses.

## 2.3 Introduction to Simultaneous Equilibria

Sometimes a chemical system can be described by a single mass-action expression and a single stoichiometric equation. It appears as if only one chemical reaction occurs. However, competitive equilibria are more typical, that is, when two or more different reactions occur *that have at least one species in common.* For example, if two bases, $B^-$ and $N$, compete for protons in solution by the reactions

$$H^+ + B^- \rightleftharpoons HB \qquad (2\text{-}23a)$$
$$H^+ + N \rightleftharpoons HN^+ \qquad (2\text{-}23b)$$

then $H^+$ is the species in common. All solution equilibria are competitive equilibria. Simple (single) equilibria might *appear* to hold because of approximations made in modeling the real system.

In the competitive equilibria represented by Eqs. 2-23, a proton may associate with only one of the two competing bases. Or it may associate with both by spending part of the time in association with one and part of the time with the other. This seems straightforward enough. However, some problems arise in quantitative calculations of concentrations of the species present: here $[HB]$, $[HN^+]$, $[B^-]$, $[N]$, and $[H^+]$. The algebra often becomes quite complicated and involved. Even professional chemists begin to run to their computers when there are more than three chemical components mixed in solution.

The essential algebra for even the most complicated equilibrium calculations has been demonstrated in Section 2.1. No matter how complex the algebra of a problem appears, you will only be adding and reversing chemical equations and manipulating the equilibrium constants using the two general rules listed on page 38. However, until you work through a number of problems of each type, you should not expect to have a clear understanding of the techniques.

The rules of manipulating equilibrium constants and chemical equilibrium equations also allow us to calculate quantitatively the concentrations of chemical species at equilibrium under conditions where the same information might be difficult to measure (or even unobtainable) by direct experiment. However, you should understand that these extrapolations are dependent on the quality of the calculations, and many imperfections are possible. Some of these difficulties were noted in the discussion of the nature of activities and activity coefficients. For calculations in this text, such difficulties are ignored.

### Weak and Strong Acids and Bases

Let us write the general formula HB for an acid consisting of a proton $H^+$ bound to an anionic base $B^-$. The dissociation of this general acid is written

$$HB \rightleftharpoons H^+ + B^- \qquad (2\text{-}23a)$$

**Figure 2.1** Graph of the fraction of molecules dissociated as a function of concentration for a strong acid (HCl) and a weak acid (acetic) in water. The strong acid is fully dissociated at low concentrations, and the fraction dissociated slowly decreases with concentration as a result of activity effects. The weak acid is less dissociated, but the extent of dissociation depends strongly on concentration in the low range. The activity effects are slight for acetic acid, since at the higher concentrations the majority of the acid is undissociated and not ionic.

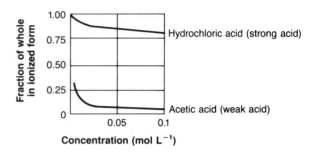

The acid HB dissociates to produce a proton and the acid's **conjugate base, B⁻.**

As represented by their $pK_a$ values, in most solvents, including water, most acids fall into one of two categories: **weak acids** and **strong acids.** As illustrated in Figure 2.1, when a weak acid is added to water, only a small fraction of it dissociates. The equilibrium of the dissociation reaction, represented by Eq. 2-23, lies far to the left. Acetic acid is an example of a weak acid. On the other hand, a strong acid is one for which the equilibrium lies far to the right side. Only a small fraction remains undissociated. Hydrochloric acid (HCl) is an example of a strong acid.

If HB is a *weak* acid, then B⁻ is classifiable as a *strong* base; weak acids result from the proton dissociation from strong conjugate bases. If HB is a *strong* acid, then B⁻ is classified as a *weak* base; strong acids result from the proton dissociation from weak conjugate bases. This is the nomenclature of elementary Brønsted acid–base theory.

Because few intermediate-strength acids exist, by study of the calculations for both weak and strong acids most of the characteristics of simultaneous equilibrium calculations can be demonstrated and understood. The following discussion of simultaneous equilibria and the associated calculations is ordered as follows:

1. The simple equilibrium of acetic acid dissociation: first exact solutions and then approximations.
2. The competitive equilibrium between acetate ion and hydroxide ion.
3. The competitive equilibrium between hydrochloric acid and water.
4. The competitive equilibrium between water and more than one acid or base.

All the calculations are done ignoring the effects that require activities to be used in place of concentrations. In other words, the complicated effects engendered by changes in the ionic strength are ignored, and concentrations are used in the mass-action expressions.

## 2.4 Weak Acids: Simple Equilibria

What is the pH of the solution at 25 °C when acetic acid (a weak acid) is added to water to a concentration of 0.1 M?

From tables such as those in Appendix II, the equilibrium constant for acetic acid dissociation is found to be

$$K_a = 1.8 \times 10^{-5} \text{ M}; \qquad pK_a = 4.74 \tag{2-24}$$

The stoichiometric equation for the acid dissociation is

$$\text{HOAc} \rightleftharpoons \text{H}^+ + \text{OAc}^- \tag{2-25a}$$

$$[0.1 - X] \quad [X] \quad [X] \tag{2-25b}$$

The chemical Eq. 2-25a says that for each molecule of acid that dissociates, one proton and one acetate ion is produced. In Eq. 2-25b, the value $[X]$ represents the molar concentration of the acetic acid that dissociates. The concentration $[X]$ must be in molar units because the equilibrium constant is in molar units (i.e., $K_a = 1.8 \times 10^{-5}$ M). Note that the algebraic concentrations written in Eq. 2-25b are derived from the stoichiometric information in Eq. 2-25a.

The information in the chemical Eq. 2-25a, the associated mass-action equation, and the equilibrium constant value are combined in the expression below.

$$K_a = \frac{[\text{H}^+][^-\text{OAc}]}{[\text{HOAc}]} = \frac{[X][X]}{[0.1 - X]} = 1.8 \times 10^{-5} \tag{2-26a}$$

Once this equation is set up, we can solve it exactly, or we can simplify it by approximation. Do not confuse the method of solving the equation, which is strictly algebra, with the chemical behavior being described. The stoichiometry as well as the $K_a$ value were *found through experimentation*. This chemistry is expressed in Eqs. 2-24 and 2-25.

In the two subsections following, Eq. 2-26a will be solved both ways: first exactly and then after making an approximation. As you get more familiar with the approximation methods, you will get more confident in their use. Unfortunately there are no set rules that can always be applied.

### Exact Solution

We write Eq. 2-26a as a quadratic function:

$$X^2 = (1.8 \times 10^{-5})(0.1 - X)$$

After multiplication and rearrangement, the equation becomes

$$X^2 + (1.8 \times 10^{-5})X - 1.8 \times 10^{-6} = 0 \tag{2-26b}$$

We then solve for $X$ using the quadratic formula (Supplement 2A).

$$X = \frac{-1.8 \times 10^{-5} \pm \sqrt{3.24 \times 10^{-10} + 7.20 \times 10^{-6}}}{2}$$

$$X = \frac{-1.8 \times 10^{-5} \pm 2.683 \times 10^{-3}}{2} = \frac{2.67 \times 10^{-3}}{2} \approx 1.33 \times 10^{-3}$$

Note that the negative value solution, $X = -1.35 \times 10^{-3}$, was not used because a negative concentration has no meaning.

Since $X = [H^+]$, the pH of the solution is

$$-\log(1.33 \times 10^{-3}) = 2.87.$$

Notice that even though an "exact" solution was calculated, there was an approximation made in setting up the equation to solve. The protons that were contributed to the acid solution by the water were not accounted for. The water alone contributes about $10^{-7}$ M $H^+$. However, in comparison, the water contributes only about 1/10,000th the $H^+$ of that present from the acetic acid dissociation ($10^{-4} = 10^{-7}/10^{-3}$). It is reasonable to ignore this small extra $H^+$ contribution. Experimentally, a one-part-in-10,000 change is extremely difficult to detect. You will see later in this section an example in which the water protons cannot be ignored.

### An Approximation

Now an approximate solution is calculated. Let us simplify Eq. 2-26a by ignoring the value of $X$ in the denominator. This means that we *assume* that $X$ will be so small relative to 0.1 that the answer to this approximate calculation will still be within 1% of the value that was calculated exactly. This limit of 1% error is set arbitrarily. (Note that no other approximation makes sense; we cannot ignore the value of $X$ in the numerator because there then would be no equation to solve.)

The simplified equation that needs to be solved is

$$K_a = \frac{X^2}{0.10} \qquad\qquad (2\text{-}27)$$

If the value for $K_a$ is substituted, then

$$X^2 = 1.8 \times 10^{-5} \times 0.1 = 1.8 \times 10^{-6}$$

The solution is obtained by finding the square root, so

$$X = 1.34 \times 10^{-3} \text{ M} = 1.34 \text{ mM}$$

The answer, $1.34 \times 10^{-3}$ M, *is* within 1% of $1.33 \times 10^{-3}$ M, which was the answer calculated without mathematical approximations.

How can you tell whether an approximation will yield an accurate result? Sometimes you can decide simply by inspecting the equation being solved. In the case of acetic acid, you might compare

$$(1.8 \times 10^{-5})^{1/2} = (18 \times 10^{-6})^{1/2} \approx 4 \times 10^{-3}$$

with 0.10. The value of 0.004 is small enough relative to 0.10 that an approximation might work. However, there are no regular, general rules to help us.

You might ask why we should solve such a simple equation approximately at all. There is a good reason. Simplifications such as this often can make our perception of a problem clearer. It might be easier to see what trends of behavior are expected without going through an entire calculation. This kind of judgment is useful in stopping us from pursuing the wrong directions when attempting to solve a problem.

Another reason for employing such simplifications occurs in more complicated equilibrium problems. Then, obtaining the solution itself may become far easier mathematically. In addition, as the calculations become messier, the need to perceive a trend may be much more profound—a need that a computer may not fulfill. In other words, these simplifications help us avoid getting lost along the way to the solution.

### Percent Dissociation

Another term used to describe the dissociation of weak acids is **percent dissociation.** The number describes the percentage of the total acid in the solution that is in the dissociated (ionized) form. The total acid is the sum of both the dissociated and undissociated acid.

The percent dissociation is, algebraically,

$$\% \text{ dissociation} = 100 \times \frac{[\text{dissociated acid}]}{[\text{total acid added}]} \quad (2\text{-}28)$$
$$= 100 \times \frac{\text{moles dissociated acid}}{\text{moles acid added}}$$

The percent dissociation is simply 100 times the fraction dissociated, which is the vertical axis in Figure 2.1.

---

**EXAMPLE**

What is the percent dissociation of acetic acid in a 0.100-M solution?

**SOLUTION**

We use the results from in the solved problem above. For a 0.100 M acetic acid solution, the resulting proton (and acetate) concentration is $1.34 \times 10^{-3}$. Then

$$\% \text{ dissociation} = \frac{1.34 \times 10^{-3}}{0.100} \times 100 = 1.34\%$$

of the acid originally added.

---

### A Strong Base in Water: Hydrolysis

What happens when, instead of adding acetic acid to the water, we add the conjugate strong base, acetate ion, instead? This can be done by making a solution of a salt such as potassium acetate. The known chemistry of this compound shows that it is completely dissociated in the solution to yield hydrated potassium ions and acetate ions. Some of the acetate ions react with the water with the following chemistry

$$^-OAc + H_2O \rightleftharpoons HOAc + OH^- \quad (2\text{-}29)$$

This type of reaction is called **hydrolysis** (meaning splitting water).

Can we calculate the pH of this acetate salt solution? We have nearly enough information to solve the problem, using the rules for manipulating chemical equations. The reaction 2-29 describes a competition for protons between acetate and hydroxide. Two equations contain the information necessary to describe this competition. The first involves the acetate–acetic-acid equilibrium.

$$HOAc \rightleftharpoons H^+ + {}^-OAc; \qquad K_a = 1.8 \times 10^{-5} \text{ M} \qquad \text{at 25 °C} \quad \textbf{(2-7)} \text{ and } \textbf{(2-8)}$$

The equation describing the hydroxide reaction is the expression for water dissociation. This last piece of necessary information is found in Appendix II.

$$H_2O \rightleftharpoons H^+ + OH^-; \qquad K_w = 1.0 \times 10^{-14} \qquad \text{at 25 °C} \qquad \textbf{(2-30)}$$

We can use this information to describe the hydrolysis reaction of Eq. 2-29. Equation 2-29 can be written as the sum of these two chemical equations. However, acetate occurs on the left, so the first of the two equations must be reversed. The equilibrium constants must be treated by the usual rules (Section 2.1).

Thus

$$
\begin{array}{ll}
H^+ + {}^-OAc \rightleftharpoons HOAc & K' = 1/K_a = 5.56 \times 10^4 \\
+ \qquad H_2O \rightleftharpoons H^+ + OH^- & K_w = 10^{-14} \\
\hline
\cancel{H^+} + {}^-OAc + H_2O \rightleftharpoons \cancel{H^+} + OH^- + HOAc & K_h = K'K_w = K_w(1/K_a) = K_w/K_a
\end{array}
$$

The sum equation is the same as Eq. 2-29 after the $H^+$ is canceled from both sides. The equilibrium constant of the sum $K_h$ is often called the **hydrolysis constant** — in this case for the hydrolysis of acetate. However, it is not a special kind of equilibrium constant. On the contrary, it is like any other equilibrium constant but has been labeled with a special name alluding to the type of chemical reaction to which it applies.

The numerical value of $K_h$ can be calculated from the values of $K_a$ and $K_w$ for the acids of the two bases ${}^-OAc$ and $OH^-$, respectively. From the value of $K_h$, the pH of a solution of sodium acetate can be calculated in the same way as for acetic acid.

This example illustrates some of the power of the methods of calculation involving simultaneous equilibria. For instance, the calculation shows the relationship between the properties of an acid and its conjugate base. As a result, if we measure the properties of the acid in water, we do not have to do the experiment of adding the base to water in order to measure its properties quantitatively. On the other hand, if the base does not yield the "expected" value of pH in an experiment, you might expect that either the experiments measuring the properties of the acid or base were not done correctly, or perhaps the material in one of them might have been impure. In other words, the separate experiments on an acid and its conjugate base can be used to check each other through the use of these calculational methods.

## 2.5 Strong Acids in Water

The reaction of a strong acid, abbreviated in a general form as HB, placed in a water solution is represented by

$$HB + H_2O \rightleftharpoons H_3O^+ + B^- \qquad \textbf{(2-31)}$$

For strong acids, the equilibrium lies far to the right; the concentration of the undissociated form of the acids is essentially zero. Examples of strong acids in water are $HCl$ and $HNO_3$. The reaction 2-31 is an example of a competitive equilibrium.

This system is considered to be a competitive equilibrium because there are two equilibria present that involve the proton. The acid, for example $HCl$, has given up its proton to form its conjugate base,

$$HCl \rightleftharpoons H^+ + Cl^- \qquad (2\text{-}32)$$

while the water equilibrium is shifting simultaneously:

$$H_2O \rightleftharpoons H^+ + OH^- \qquad (2\text{-}30)$$

This example of a strong acid is a competive equilibrium because the water itself, $H_2O$, and the base of the acid, $B^-$, compete for protons. In this case, the water dominates, forming $H_3O^+$ (or more correctly, $H(H_2O)_n^+$). (Contrast this to the case of a *weak* acid in which the conjugate base of the acid dominates water for the proton.)

Because these equilibria occur simultaneously, equations representing both equilibria must be solved simultaneously. Algebraically, we must solve these in a manner similar to the methods used in the previous section for the examples of hydrolysis. We might try to add the two stoichiometric equations and manipulate the equilibrium constants to get an equation such as reaction 2-31:

$$\begin{array}{r} H_2O \rightleftharpoons H_2O \\ HCl \rightleftharpoons H^+ + Cl^- \\ \hline H_2O + HCl \rightleftharpoons H_3O^+ + Cl^- \end{array}$$

We rapidly run into trouble since $K_a$ ($HCl$) is so large that no meaningful calculation can be done. The problem cannot be solved in the same way as before, and a different approach is necessary. In the end, we want to find the total concentration of protons $[H^+]_{tot}$. This is the species that links the equilibria, and finding its concentration will solve the problem. The procedure to find $[H^+]_{tot}$ is described next.

We know that

$$K_w = [H^+][OH^-] = 10^{-14} \qquad \text{at 25 °C} \qquad (2\text{-}30)$$

or

$$[H^+]_{water} = K_w/[OH^-] = 10^{-14}/[OH^-] \qquad (2\text{-}33)$$

In addition, there are protons from the acid, and a strong acid is completely dissociated in the solution:

$$HCl \rightleftharpoons H^+ + Cl^-; \qquad K \gg 1$$

We require a quantitative relationship for a completely dissociated species that serves the same purpose in the calculation as an equilibrium constant, namely, a numerical relationship between concentrations. If $C_{HCl}$ is the concentration of the strong acid solution, then from the stoichiometric equation, Eq. 2-32, and the idea of complete dissociation,

$$[H^+]_{acid} = C_{HCl} \qquad (C_{HCl} = [Cl^-]) \qquad (2\text{-}34)$$

An equation of this type is needed for any calculation involving a completely disso-ciated species.

How do we calculate $[H^+]_{tot}$? If we could separately label the protons from the acid and those from the water, it would be true that

$$[H^+]_{tot} = [H^+]_{acid} + [H^+]_{water} \tag{2-35}$$

This kind of equation is called a **conservation equation,** or **conservation relation.** It reflects the fact that mass is conserved.

The solution to the algebraic problem lies in finding the values of the two terms on the right side of Eq. 2-35. The answers have already been written in Eqs. 2-33 for $[H^+]_{water}$ and 2-34 for the acid. Thus we now have a value for both $[H^+]_{water}$ and for $[H^+]_{acid}$. These concentrations have each been written in terms of quantities we can measure.

After substitution into the conservation relation, Eq. 2-35, it can be written in two equivalent ways:

$$[H^+]_{total} = C_{HCl} + K_w/[OH^-] \tag{2-36a}$$

or

$$[H^+]_{total} = C_{HCl} + [H^+]_{water} \tag{2-36b}$$

Here the equilibria for water and the added acid are linked through their common product $[H^+]$. The concentration $C_{HCl}$ appears because the strong acid dissociates completely. Also, a mass-action expression appears for water either explicitly, Eq. 2-36a, or implicitly, Eq. 2-36b, since the water only slightly dissociates.

### A Relatively High-Acid Concentration: $10^{-3}$ M

To find the concentration $[H^+]_{tot}$, we need the value of $C_{HCl}$, the concentration of added acid. Let us calculate the value of $[H^+]$ in a $10^{-3}$-M HCl solution in water. To allow approximations to be used, the required accuracy of the calculations is limited to $\pm 1\%$.

We substitute $10^{-3}$ M into Eq. 2-36b and write

$$[H^+]_{tot} = 0.001 + [H^+]_{water} \tag{2-37}$$

But how do we find the value of the second term, the proton concentration from the water? The information in Eqs. 2-30 and 2-33 says

$$H_2O \rightleftharpoons H^+ + OH^-; \qquad K_w = [H^+][OH^-] = 10^{-14}$$

The water dissociation is like any simple dissociation of a weak acid. From the stoichiometry of the reaction, we know that *before* any HCl is added,

$$[H^+]_{water} = [OH^-] \qquad \text{and} \qquad X^2 = 10^{-14}$$

So

$$[H^+]_{water} = [OH^-] = 10^{-7}$$

But this is the *maximum* value of $[H^+]_{water}$. It is important to realize that the water *alone* cannot produce a concentration of protons greater than $10^{-7}$ M at 25 °C.

But because $10^{-7}$ is small relative to 0.001, we can make an algebraic approximation to solve the problem: Ignore the second term in Eq. 2-37. Thus, in a $10^{-3}$-M HCl solution,

$$[H^+] = 10^{-3} \text{ M} + \sim 0$$

**Review**    To solve this problem, we required two stoichiometry equations,

$$HCl \rightleftharpoons H^+ + Cl^- \qquad (2\text{-}32)$$

and

$$H_2O \rightleftharpoons H^+ + OH^- \qquad (2\text{-}30)$$

and the information from the mass-action expression from the weak acid, water,

$$K_w = [H^+][OH^-] = 10^{-14} \qquad (2\text{-}33)$$

and the concentration of the strong acid, HCl, and its stoichiometry,

$$[H^+]_{acid} = C_{HCl} \qquad (2\text{-}34)$$

An approximation was made. It was found that the water equilibrium was unimportant and that the value of $[H^+]_{tot} = [H^+]_{acid}$ was within $\pm 1\%$.

This method of solution reflects some general rules for solving simultaneous equilibria problems:

1. *Include all the* applicable *equilibria.*
2. *For the applicable equilibria,* all *the stoichiometry equations are required.*
3. *For the applicable equilibria,* all *the numerical values associated with the mass-action expressions (or equivalents, such as $C_A$) are needed.*

No matter how the problem is stated, the stoichiometric information and equilibrium constants *must* be present or derivable in order to solve the simultaneous equilibrium problem.

In the example of HCl, the applicable equilibria are HCl and $H_2O$. The stoichiometric equations are Eqs. 2-30 and 2-32. The numerical values associated with the equilibria are $K_w = 10^{-14}$ and $C_{HCl} = 10^{-3}$ M. If the acid were a weak acid, the values of $K_w$ and the appropriate $K_a$ would be used.

After this information is available, we then decide if we can ignore some of it within the level of approximation we choose. In the example above, a 1% numerical accuracy was specified. Thus the small contribution of $[H^+]_{water}$ was ignored.

## A Strong Acid at Low Concentration

If a $10^{-10}$ M solution of HCl is made up in water, what is the value of $[H^+]$ expected in the solution? Let us calculate the result to within 1% numerical accuracy ignoring activity effects.

For this very dilute acid solution, Eq. 2-36 becomes

$$[H^+]_{tot} = 10^{-10} \text{ M} + [H^+]_{water}$$

Since the protons donated from water alone are $10^{-7}$ M, we can ignore the contribution from the acid, so

$$[H^+] = 10^{-7} \text{ M}$$

Approximations such as this cannot be made when we calculate the expected pH of a $10^{-7}$-M HCl solution.

### A Strong Acid at an Intermediate Concentration: $10^{-7}$ M

For the relatively concentrated $10^{-3}$-M acid, we found that $[H^+]$ due to added acid was so large that the algebra could be simplified by ignoring the protons donated by the water. When the acid contributes significantly fewer protons than the water alone, we can ignore the acid. Let us now calculate $[H^+]$ in a $10^{-7}$-M HCl solution in water at 25 °C. When only enough strong acid is added to produce initially $[H^+]_{acid} = 10^{-7}$ M, both $[H^+]_{acid}$ and $[H^+]_{water}$ in Eq. 2-36 are about the same. Neither can be ignored.

What happens when this small amount of a strong acid is added can be described in words. The pure water at 25 °C initially contains $[H^+]_{water} = 10^{-7}$ M. Then enough strong acid is added to contribute $[H^+]_{acid} = 10^{-7}$ M. Addition of the acid does not change the equilibrium *constant* for water,

$$K_w = 10^{-14} = [H^+][OH^-] \qquad (2\text{-}33)$$

but it does change the position of the equilibrium. The water equilibrium is shifted back slightly toward undissociated water as the solution $[OH^-]$ decreases in response to an increase in $[H^+]$. But the decrease of $[OH^-]$ to form water is accompanied by a *decrease* in $[H^+]$:

$$OH^- + H^+ \rightleftharpoons H_2O$$

So the final proton concentration is less than the sum of the initial values $[H^+]_{acid}$ and $[H^+]_{water}$. This chemical reaction occurs in less than one microsecond.

What happens to the chloride ions? Since $Cl^-$ is such a weak base, it does not protonate; the $Cl^-$ remains completely dissociated. Changes in the proton concentration do *not* shift the acid equilibrium back toward undissociated HCl.

### A Calculation Trick: The Charge Balance

We now take a short digression from our calculation. All the stoichiometric information that is needed in this problem is included in the two stoichiometry equations 2-30 and 2-32. However, when a calculation cannot be simplified by approximation, it is convenient to transform the stoichiometric information into an equation of **charge balance,** or **charge conservation.** The equation is a statement that the solution is electrically neutral: The number of positive ions equals the number of negative ions. Let us take an example from a different solution, that of potassium phosphate.

Since the only ions in solution are $H^+$, $OH^-$, $K^+$, $H_2PO_4^-$, $HPO_4^=$, and $PO_4^{3-}$, the charge-balance equation is written

$$[K^+] + [H^+] = [\,OH^-] + [H_2PO_4^-] + 2\,[HPO_4^=] + 3\,[PO_4^{3-}]$$

The coefficient in front of each concentration term equals the absolute value of the number of elementary charges on the ionic species. The general algebraic form for the charge-balance equation is

$$\sum_{\text{anions}} |z_i|\, c_i = \sum_{\text{cations}} |z_j|\, c_j \qquad \qquad \textbf{(2-38)}$$

---

**EXAMPLE**

Write the charge balance equation for a solution that contains $Fe^{2+}$, $Fe^{3+}$, $H_2O$, $H^+$, $^-OH$, $Fe(CN)_6^{4-}$, and $(CH_3CH_2)_2O$.

**SOLUTION**

Since neutral species do not appear,

$$2[Fe^{2+}] + 3[Fe^{3+}] + [H^+] = [OH^-] + 4\,[Fe(CN)_6^{4-}]$$

---

The charge-balance equation for the solution of HCl in water is

$$[H^+] = [Cl^-] + [OH^-] \qquad \qquad \textbf{(2-39)}$$

as derived from the stoichiometric equations. If $C_{HCl}$ is relatively large, such as when $[HCl] = 10^{-3}$ M, then Eq. 2-39 becomes

$$[H^+] = [Cl^-]$$

which is the same information as we obtained from the HCl dissociation stoichiometry. The calculation is that of a single equilibrium. If $C_{HCl}$ is small relative to $[H^+]_{water}$, such as when $[HCl] = 10^{-10}$ M, then Eq. 2-39 becomes

$$[H^+] = [OH^-]$$

which is the water dissociation stoichiometry; again a single equilibrium is enough. Such an analysis indicates which is the important simple equilibrium to consider. In the intermediate range, we shall find the charge-balance equation more directly useful than when one system or the other dominates the equilibrium.

## The Algebra Using Charge Balance

We return now to our calculation concerning a $10^{-7}$-M acid solution. Once the stoichiometric information has been included in the charge-balance equation, all the information is contained in three equations:

$$K_w = 10^{-14} = [H^+][OH^-] \qquad \qquad \textbf{(2-30)}$$

$$C_{HCl} = [Cl^-] \qquad \qquad \textbf{(2-34)}$$

and

$$[H^+] = [Cl^-] + [OH^-] \qquad \text{(2-39)}$$

Substituting Eqs. 2-30 and 2-34 into Eq. 2-39, we find

$$[H^+] = C_{\text{HCl}} + K_w/[H^+]$$

which is a quadratic equation in $[H^+]$. Upon multiplying both sides by $[H^+]$ and rearranging, we get

$$[H^+]^2 - C_{\text{HCl}}[H^+] - K_w = 0$$

Upon substituting numerical values for $C_{\text{HCl}}$ ($= 10^{-7}$ M) and $K_w (10^{-14})$, we can use the quadratic formula to solve for $[H^+]$. The formula gives

$$[H^+] = 1.62 \times 10^{-7} \text{ M} \qquad \text{and} \qquad [H^+] = -0.62 \times 10^{-7} \text{ M}$$

We reject the negative value because it is unrealistic. The final pH of the solution is

$$-\log[H^+] = -\log(1.62 \times 10^{-7}) = 6.79$$

ignoring ionic strength effects on the activities.

Relevant to the chemistry of the solution, consider the change in $[H^+]_{\text{tot}}$ upon adding $10^{-7}$-M acid if no shift in the water's equilibrium had occurred. Then

$$[H^+]_{\text{acid}} + [H^+]_{\text{water}} = 2 \times 10^{-7} \text{ M}; \qquad \text{pH} = 6.70$$

The true pH, 6.79, is higher because the added acid shifts the water equilibrium slightly toward water and thus the pH toward neutrality.

## 2.6 Strong Bases in Water

The calculation for predicting the pH of an aqueous solution of a strong base is similar to that for a strong acid. In an aqueous solution of a general base B, the following reaction occurs,

$$B + H_2O \rightleftharpoons OH^- + HB^+ \qquad \text{(2-40)}$$

If B is a strong base, the equilibrium lies far to the right. In a solution of a strong base, after the reaction with the solvent has occurred, the only base in solution is the hydroxide ion contributed by water. The hydroxide ion concentration can be linked to value of $[H^+]$ by

$$K_w = [H^+][OH^-]$$

Equation 2-41 describes the hydroxide concentration in a solution that has a concentration $C_B$ of a strong base,

$$[OH^-]_{\text{tot}} = C_B + [OH^-]_{\text{water}} \qquad \text{(2-41)}$$

Compare this equation with Eq. 2-36b. As you might guess, when the concentration of strong base is greater than about $10^{-5}$ M, the second term can be ignored for $\pm 1\%$ arithmetic accuracy. Calculations at low and intermediate concentrations also parallel those for strong acids. Examples of strong bases are hydroxide, $OH^-$; amide, $NH_2^-$; and oxide, $O^-$. Compounds composed of these bases with sodium and/or potassium are completely dissociated in water.

## Summary

In making the calculations as illustrated in this chapter, the following general rules were followed.

1. *Include all the possible* applicable *equilibria.*
2. *Write all the stoichiometry equations of the applicable equilibria.*
3. *Write the mass-action expressions including equilibrium constants or the concentrations of fully dissociated species.*
4. *If necessary, use the data to write a conservation relation. See whether the problem can be simplified to a single equilibrium.*
5. *If the calculation applies to an intermediate range, where two or more equilibria are involved approximately equally, use the stoichiometric information to write a charge-balance equation. This, together with the mass-action expressions and numerical values, should be sufficient to make the calculation.*
6. *Carry out the algebra.*

## Reminder

Do not lose sight of the fact that all the equilibrium calculations described are made to match and explain experiments. The properties of the equilibrium constant (law of mass action) and its value merely codify the experimental information. In addition, the rules of addition and reversal of chemical-equilibrium equations allow us to calculate concentrations under conditions that may not be obtainable experimentally. But there are many imperfections in the calculations. Some of these difficulties were reflected in the discussion of activity coefficients.

Nevertheless, equilibrium calculations can be used to determine, if only approximately, equilibrium properties that might be too costly to determine experimentally for every case. More importantly, though, they help to explain the chemistry that is encountered in analyses and thus can help in optimizing the analytical conditions.

## 2.7 Solutions of Polybasic Acids: More Simultaneous Equilibria

Polybasic acids are those that contain a base that can bind to more than one dissociable proton. These include $H_3PO_4$, $H_2CO_3$, and $H_2SO_4$. Although the use of the appropriate equilibrium-constant expressions and stoichiometric equations remains similar to their use in simpler problems, the insight needed to solve equations or to simplify them by approximations is greater. (Some people coming upon these more difficult problems for the first time say that it is not insight that is needed but cunning.) The problems are solved mathematically by finding the solutions for the appropriate simultaneous equations.

As before, we do the calculations to help understand (and mimic) the behavior of the chemical species present. This is where the importance of such calculations lie. A clear description of the equilibrium is useful and in some cases essential in a number of analytical chemical methods.

### Dibasic Acid: Carbonate

A chemical system that is commonly used to illustrate the principles of polybasic acids is that of carbonate ion. Its behavior is one of the simpler ones for dibasic acids. Nevertheless, the techniques that are used in the carbonate equilibrium calculations are general. So it is worthwhile to understand the approach and details. Extra interest in the carbonate equilibrium comes from its importance. For instance, the acid–base system of carbonate/bicarbonate ($CO_3^=/HCO_3^-$) is the main determinant of the pH of your blood. The equilibrium system is also important in the following diverse phenomena: the formation of sea shells, which are calcium carbonate; the formation of carbonate scale in boilers; and the carbon dioxide content of the atmosphere and its changes due to the burning of fossil fuels.

From experiments, the following stoichiometries and equilibrium constants have been determined.

1st ionization     $H_2CO_3 \rightleftharpoons H^+ + HCO_3^-;$     $K_1 = 4.2 \times 10^{-7}$ M;     $pK_1 = 6.38$
$$(2\text{-}42)$$

2nd ionization     $HCO_3^- \rightleftharpoons H^+ + CO_3^=;$     $K_2 = 4.8 \times 10^{-11}$ M;     $pK_2 = 10.32$
$$(2\text{-}43)$$

The subscript nomenclature of Eqs. 2-42 and 2-43 is general. The steps of the ionization — labeled 1st, 2nd, 3rd, and so on — start from the fully protonated form of the acid. Connected with each ionization is a $pK_a$ value with the same subscript number as ionization label: $pK_1$, $pK_2$, $pK_3$, and so on.

Given the information included in Eqs. 2-42 and 2-43, we shall be calculating the expected values of such quantities as $[H_2CO_3]$, $[HCO_3^-]$, and $[CO_3^=]$, and how they are related to the solution pH.

A primary question is whether there is enough information to make such calculations. One way to view this question is strictly mathematically. There are four *possible* unknown values in the equations: $[H^+]$, $[H_2CO_3]$, $[HCO_3^-]$, and $[CO_3^=]$. There also are four equations interrelating the four quantities: the two stoichiometric equations and the two mass-action expressions (not written out above). With four equations and four unknowns, in principle the *algebraic* problem must be solvable. To obtain numerical values, the $pK_a$ values and at least one concentration, such as that of $NaHCO_3$, are needed. An exact solution of the equations can be awkward, and approximations can simplify the mathematics significantly.

One way of simplifying the mathematics is to use chemical principles — specifically, a conservation equation and sometimes a charge-balance equation — on the system. For the carbonate equilibrium, a conservation equation can be derived from the stoichiometric equations 2-42 and 2-43. As expressed in Eq. 2-44 below, the total carbonate concentration (the sum of all three carbonate species in the solution) remains constant. Call the total concentration of all carbonate species $C_{tot}$. Then

$$C_{tot} = [H_2CO_3] + [HCO_3^-] + [CO_3^=] \qquad (2\text{-}44)$$

You might recognize this equation as being similar to Eq. 2-35 for the conservation of protons in the calculations involving strong acids. It is important to recognize that the conservation equation is equivalent to *two* equations, namely, both stoichiometry equations. Thus the information needed to solve any problem concerning carbonate solutions is included in the conservation relation, the two mass-action expressions, and the numerical values of $C_{tot}$, $pK_1$, and $pK_2$.

## Calculations without Approximations

The carbonate equilibrium equations can be solved as simultaneous equations without finding simplifying approximations or finding the principal equilibrium under the conditions at hand. What is necessary is to solve an equation like

$$[H^+]^4 + (K_1 + [Na^+])[H^+]^3 + (K_1K_2 + [Na^+]K_1 - K_w - K_1C_{tot})[H^+]^2 +$$
$$+ ([Na^+]K_2 - K_w - 2K_2C_{tot})K_1[H^+] - K_1K_2K_w = 0 \qquad (2\text{-}45)$$

This equation is specifically for a mixture of three stock solutions with the same concentration $C_{tot}$, containing either $Na_2CO_3$, $NaHCO_3$, or $H_2CO_3$. The value of $[Na^+]$ is determined by the proportions of each of the stock solutions in the final mix. The derivation of Eq. 2-45 is described in Supplement 2A. Perhaps it is now more obvious why it is worthwhile to simplify and use approximations in calculating the expected chemical properties of such solutions.

## Carbonate at pH 6.38

The complete equation resulting from all the chemical information will not be solved here. Instead, we shall investigate how to simplify the calculations. As done earlier, simplifications can be made when one or another term in the sum in Eq. 2-44 can be ignored. The following calculation illustrates such a simplification.

Let us calculate the concentrations of $H_2CO_3$, $HCO_3^-$, and $CO_3^=$ expected in a solution with 0.10 M total carbonate when the pH is 6.38.

(This aqueous solution might be made by adding 0.10 mol of $NaHCO_3$ to 950 mL of water. Then the pH can be brought to 6.38 with concentrated HCl and the solution diluted to 1.000 liter. In this way, the addition of needed HCl will not produce a diluted solution.)

Notice that pH 6.38 is equal to $pK_1$ for carbonate. This special pH value is used here to show how approximations can simplify the calculations.

The statement of the problem has given us two different pieces of information. First, the total carbonate concentration $C_{tot}$ is 0.10 M. Second, having the solution pH = 6.38 means that $[H^+] = 4.2 \times 10^{-7}$. Let us substitute what is known into the mass-action expressions:

$$K_1 = \frac{[H^+][HCO_3^-]}{[H_2CO_3]} = \frac{(4.2 \times 10^{-7})[HCO_3^-]}{[H_2CO_3]} = 4.2 \times 10^{-7} \qquad \textbf{(2-46)}$$

$$K_2 = \frac{[H^+][CO_3^=]}{[HCO_3^-]} = \frac{(4.2 \times 10^{-7})[CO_3^=]}{[HCO_3^-]} = 4.8 \times 10^{-11} \qquad \textbf{(2-47)}$$

The value of $[H^+]$ has been substituted into the equations. Upon rearranging the equations, *ratios* of the concentrations of the carbonate species can be calculated.

From $K_1$:    $4.2 \times 10^{-7}/4.2 \times 10^{-7} = 1 = [HCO_3^-]/[H_2CO_3]$    **(2-46)**

From $K_2$:    $4.8 \times 10^{-11}/4.2 \times 10^{-7} = 1.14 \times 10^{-4} = [CO_3^=]/[HCO_3^-]$    **(2-47)**

The third equation, the conservation equation for a solution with a total carbonate concentration of 0.10 M, is

$$0.10 \text{ M} = [H_2CO_3] + [HCO_3^-] + [CO_3^=] \qquad \textbf{(2-48)}$$

All the information we have is now included in the three equations 2-46, 2-47, and 2-48.

The rest of the calculation is straightforward but messy algebra. However, there are a number of choices as to the approach taken. For instance, we could substitute the ratios of the carbonate species Eqs. 2-46 and 2-47 into the conservation equation, Eq. 2-48, and solve for one of the three carbonate species ($HCO_3^-$ would be convenient). Then, by substituting $[HCO_3^-]$ into the ion-concentration ratios, we could calculate the concentrations of all three carbonate species. However, let us see if an approximation might make the calculation even easier.

### An Approximation

Look at the carbonate species ratios that were calculated above.

$$1 = [HCO_3^-]/[H_2CO_3] \qquad \textbf{(2-46)}$$

$$1.14 \times 10^{-4} = [CO_3^=]/[HCO_3^-] \qquad \textbf{(2-47)}$$

The carbonic acid concentration $[H_2CO_3]$ and the bicarbonate ion concentration $[HCO_3^-]$ are the same. Their ratio is one. On the other hand, compared with the bicarbonate ion concentration $[HCO_3^-]$, the carbonate ion concentration $[CO_3^=]$ is quite small — only about 1/10,000th the amount. In the conservation equation, Eq. 2-48, the value of $[CO_3^=]$ hardly contributes to the sum. We shall lose little accuracy if we ignore $[CO_3^=]$.

First, we can simplify the conservation equation. It becomes

$$0.10 \text{ M} = [H_2CO_3] + [HCO_3^-] \qquad \textbf{(2-49)}$$

Being able to ignore the presence of $CO_3^=$ causes a second, significant simplification in the algebra. Since we assume that $CO_3^=$ is not present, then the mass-action expression containing $[CO_3^=]$ *is no longer needed or applicable*. It is *most* important that you understand this point. We ignore its concentration just as with the strong acid calculation in Section 2.5; we ignored either $[H^+]_{water}$ or $[H^+]_{acid}$. There the proton concentration from the added acid was either far larger or smaller than the concentration due to the water present. Here the concentration of other carbonate species are far larger than $[CO_3^-]$.

As a result there are now only two equations with two unknowns to solve. These are the mass-action expression for $K_1$ at pH 6.38

$$1 = [HCO_3^-]/[H_2CO_3] \tag{2-46}$$

and the modified conservation relation

$$0.10 \text{ M} = [H_2CO_3] + [HCO_3^-] \tag{2-49}$$

Equation 2-46 says

$$[H_2CO_3] = [HCO_3^-] \tag{2-46}$$

And by using the conservation relation, we obtain

$$[H_2CO_3] = [HCO_3^-] = 0.05 \text{ M}$$

Thus the answer to the original question is that at pH 6.38

$$[H_2CO_3] = 0.05 \text{ M}$$
$$[HCO_3^-] = 0.05 \text{ M}$$

and, from Eq. 2-47,

$$[CO_3^-] = [HCO_3^-] \times 1.14 \times 10^{-4} = 1.14 \times 10^{-4} \times 0.05 \text{ M} = 5.7 \times 10^{-6} \text{ M}$$

The procedures used in solving this problem are the same as those that can be used for all such simultaneous equilibria.

## Overview

Carrying out calculations to mimic and explain simultaneous equilibrium systems more complicated than carbonate use other algebraic methods. These allow calculations to be made systematically so as to account for the important chemical effects. One of the most useful of these involves manipulating ratios such as $[HCO_3^-]/[CO_3^-]$ and $[H_2CO_3]/[HCO_3^-]$ in a methodical way. Primarily used to describe complexation equilibria, the method is described in Supplement 2B.

For the equilibrium systems described in this chapter, given the specific conditions involved (for instance, pH), there is either a single principal equilibrium or two approximately equally influential principal equilibria. If only a single equilibrium is involved, the algebra is straightforward. If two principal equilibria are involved, it is useful to use a conservation equation and the charge-balance equation together to solve the algebra.

Note that it is generally possible to ignore the reactions having $pK_a$ values that are three or more pH units away from the pH value of the solution. In that case, one of the species will be less than 1/1000th of the other. In the carbonate equilibrium algebra, the relationship is seen in Eqs. 2-46 and 2-47. Since most organic and inorganic acids have $pK_1$ and $pK_2$ values that differ by three pH units ($K_1$ and $K_2$ differing by three orders of magnitude), this simplification is *commonly done.* You have seen the simplification carried out both in the calculation involving carbonate and in the calculation involving hydrolysis.

The general procedure for carrying out the algebra is as follows.

1. *Write the mass-action expressions and stoichiometric equations for all the* applicable *equilibria.*
2. *Using the stoichiometric equations, derive the conservation equation.*
3. *Use the equilibrium constant (mass-action) expressions to find the* ratios *of the species needed.*
4. *Decide if approximations can be made and do so if possible.*
5. *Solve the equations for the desired concentrations using the charge-balance equation if useful.*

Do not lose sight of the fact that these calculations are mathematical models made to match the chemical system in the solution. For example, rejecting the need for the $K_2$ equilibrium constant and $[CO_3^=]$ in the conservation equation for carbonate (Eqs. 2-47 and 2-48), merely says that at the specific pH the reaction

$$HCO_3^- \rightleftharpoons H^+ + CO_3^=$$

plays no *significant* part in determining the solution pH. More about this important point will be presented in the chapter on titrations (Chapter 9).

## 2.8 Oxidation–Reduction Equilibria: Balancing Equations

**Oxidation–reduction** reactions are those in which at least one electron is transferred between two reactants. Often the name is shortened to **redox** reactions. **Electrochemical** methods of analysis (Chapter 12) involve the properties of redox reactions: The voltages and/or currents of oxidation–reduction reactions are measured. A preliminary requirement in understanding the techniques is the facile balancing of electrochemical equations. A number of ways are used to do this balancing. But chemical analyses usually do not require knowing the formal oxidation state of some specific atom in a molecule. Thus a single equation-balancing method suffices, as in the following procedure. Another method, in which oxidation states are assigned, is reviewed in Supplement 2C.

Oxidation–reduction reactions can be divided into two **half-reactions:** One half-reaction represents an oxidation, and the other represents a reduction.

**EXAMPLE**

Write the half-reactions for

$$6\,I^- + O_2 + 4\,H^+ \rightleftharpoons 2\,I_3^- + 2\,H_2O$$

**SOLUTION**

| | |
|---|---|
| Oxidation half-reaction: | $3\,I^- \rightleftharpoons I_3^- + 2\,e^-$ |
| Reduction half-reaction: | $4\,e^- + 4\,H^+ + O_2 \rightleftharpoons 2\,H_2O$ |

Once the stoichiometry of the oxidation–reduction reaction is found, the individual half-reactions are balanced independently and then added together to balance the overall reaction. The steps for balancing a redox equation are

1. *Write the unbalanced chemical equation for the half-reactions involved. This requires knowing both the reacting species and the product species of the reaction.*
2. *Balance the material, beginning with the species other than water. Then, if the reaction is in aqueous solution, add water and protons or water and hydroxide. Two different situations are possible:*
   a. *In acid, only water and protons are allowed.*
   b. *In basic solution, water and hydroxide ions are allowed.*
3. *Make the charge balance. The half-reactions will require adding electrons to balance the charge.*
4. *Obtain the full reaction by multiplying (where necessary) and adding the half-reactions so that the electrons cancel.*

Let me emphasize that the first step, that of writing the oxidized and reduced species, is not done by magic. This is descriptive chemistry; someone has done experiments to show what species are present in the reaction.

---

**EXAMPLE**

Balancing a half-reaction

Balance the half-reaction for the reduction of permanganate, $MnO_4^-$, in an aqueous acid solution.

**SOLUTION**

Step 1:  The descriptive chemistry of manganese in acid indicates that permanganate will be reduced to divalent manganese. The reactant is $MnO_4^-$, and the product is $Mn^{++}$. This chemical equation (not balanced) is written

$$MnO_4^- \rightleftharpoons Mn^{++}$$

Step 2:  In acid, $H_2O$ and $H^+$ are allowed to balance material.

$$MnO_4^- \rightleftharpoons Mn^{++} + 4\,H_2O$$

to balance the oxygens of permanganate. To match the protons that are now on the right, we can add protons to the left.

$$8\,H^+ + MnO_4^- \rightleftharpoons Mn^{++} + 4\,H_2O$$

Step 3:  The left side has net 7+ charges, the right has net 2+. We need 5 $e^-$ on the left.

$$5\,e^- + 8\,H^+ + MnO_4^- \rightleftharpoons Mn^{++} + 4\,H_2O \qquad \text{(balanced)}$$

The equation has both charge and material balance.

---

**EXAMPLE**

Balancing a full redox reaction

An organic compound is oxidized with the inorganic oxidant permanganate. (The reaction is the basis for one method of analysis for alcohols.) The chemistry of the reaction of methanol with *basic* permanganate has been thoroughly investigated. Following from experimental data, the reaction is described by

$$MnO_4^- + CH_3OH \rightleftharpoons CO_2(g) + MnO_2(s) \qquad \text{(unbalanced)}$$

**SOLUTION**

We split this into its two half-reactions.
For permanganate:

Step 1. $\qquad\qquad\qquad\qquad MnO_4^- \rightleftharpoons MnO_2(s)$

Step 2. $\qquad\qquad\qquad\qquad MnO_4^- \rightleftharpoons MnO_2(s) + 2\ OH^-$

But since $OH^-$ and water are the relevant species in a basic, aqueous solution,

$$2\ H_2O + MnO_4^- \rightleftharpoons MnO_2(s) + 4\ OH^-$$

Step 3. $\qquad\qquad 3\ e^- + 2\ H_2O + MnO_4^- \rightleftharpoons MnO_2(s) + 4\ OH^-$

For methanol:

Step 1. $\qquad\qquad\qquad\qquad\qquad CH_3OH \rightleftharpoons CO_2$

Next, one *more* oxygen is needed on the left and more hydrogens on the right to complete the material balance. We can add hydroxide(s) on the left as long as we add one *less* water on the right to balance the oxygens. In addition, *at least* 4 $OH^-$ are required to take up the hydrogens from the methanol since no free $H^+$ can show up in a solution of base. Therefore

Step 2. $\qquad\qquad\qquad 6\ OH^- + CH_3OH \rightleftharpoons CO_2 + 5\ H_2O$

(Another way to think of this step is that on the left side, there are 4 $OH^-$ to take up the protons from the methanol, 1 $OH^-$ to balance the oxygen, and 1 $OH^-$ to pick up the proton from the oxygen-balancing $OH^-$, for the total of 6 $OH^-$. Alternately, the number of $OH^-$ and $H_2O$ needed may be solved by trial and error.)

Step 3. $\qquad\qquad\qquad 6\ OH^- + CH_3OH \rightleftharpoons CO_2 + 5\ H_2O + 6\ e^-$

For permanganate plus methanol:

Step 4.   To eliminate the electrons between the two half-reactions, the permanganate equation must be multiplied by 2 and the half-reactions added.

$$2 \times [3\ e^- + 2\ H_2O + MnO_4^- \rightleftharpoons MnO_2(s) + 4\ OH^-]$$

$$+ \qquad\qquad 6\ OH^- + CH_3OH \rightleftharpoons CO_2 + 5\ H_2O + 6\ e^-$$

$$\overline{6\ e^- + 4\ H_2O + 2\ MnO_4^- + 6\ OH^- + CH_3OH \rightleftharpoons 2\ MnO_2 + 8\ OH^- + CO_2 + 5\ H_2O + 6\ e^-}$$

After canceling species that appear on both sides of the equality, the answer is

$$2\ MnO_4^- + CH_3OH \rightleftharpoons 2\ MnO_2(s) + 2\ OH^- + CO_2(g) + H_2O \qquad \text{(balanced)}$$

A number of problems for review on balancing redox equations appear at the end of the chapter.

## 2.9 The Nernst Equation

In this section we review the relationships between electrical potentials and the chemical reactions that can cause them. Other names for the electrical potential are the *electromotive force* (EMF), electrochemical potential, and voltage. A chemical transformation *must involve charged species* for a voltage to be generated by it.

The electrical potential, which is associated with an equilibrium involving charge transfers, is described quantitatively by the **Nernst equation.**

$$E = E° - \frac{RT}{n\mathscr{F}} \ln K_{act} \qquad (2\text{-}50)$$

$E$ is the electrochemical potential.

$E°$ is the electrochemical potential for the reaction under standard state conditions (described in Section 2.2).

$R$ is the gas constant (8.3144 joules $K^{-1}$ $mol^{-1}$).

$T$ is the temperature in K. (K is the same as °K.)

$n$ is the number of moles of charges transferred in the reaction per mole of reactant.

$\mathscr{F}$ is the *Faraday,* which represents the number of coulombs in a mole of electrons ($\mathscr{F} = 96{,}486$ coulombs $mol^{-1}$ = 96,486 joules $V^{-1}$ $mol^{-1}$).

$K_{act}$ is the thermodynamic equilibrium constant. To provide a correct description of the equilibrium and its related potential, all concentrations must be expressed as activities.

The equation applies to two types of chemical processes that involve the motion of electrical charge. One type is the oxidation–reduction reactions discussed earlier; in these, an electron is transferred. The other process involves transfer of ionic charge by **diffusion** of ions from a region where they are more concentrated to a region where they are more dilute. Diffusion is the process in which, by random motion, molecules eventually become equally distributed in the volume accessible to them. As will be reviewed below, equilibria associated with each of the processes — both redox reactions (transfer of $e^-$) and concentration changes (transfer of ionic charges) — can *always* be treated separately with the Nernst equation. When *both* redox and concentration changes occur simultaneously, the EMF of the total reaction equals the sum of the EMFs of each type. Such calculations will be demonstrated later in this chapter.

Two special situations involving equilibria described by the Nernst equation are worth noting. First, when the concentration of each of the species present is equal to its respective standard state, then each activity in the mass-action expression equals 1. Under this special condition, all mass-action expressions have the same equilibrium constant, namely,

$$K_{act} = 1; \qquad \text{all species in standard states}$$

And thus,

$$\log K_{act} = 0; \quad \text{all species in standard states}$$

The Nernst equation then becomes

$$E = E^\circ; \quad \text{all species in standard states}$$

$E^\circ$ is called the **standard potential.**

The second special condition occurs when a chemical reaction has come to equilibrium. At that point, there can be no further *net* reaction. There can be no further *net* transfer of charge. This condition occurs only if there is no voltage to drive the charges, which means that

$$E = 0; \quad \text{at equilibrium}$$

Next consider the following point: Electrochemical potentials can be either positive or negative. What does this mean in terms of the chemistry? The answer, in a general sense, can be found by analyzing some of the properties of the Nernst equation,

$$E = E^\circ - \frac{RT}{n\mathscr{F}} \ln K_{act} \tag{2-50}$$

Let us investigate this question through the use of a general reduction reaction,

$$A^{++} + e^- \rightleftharpoons B^+$$

Let $K_{act}$ represent the equilibrium constant for the simple reaction:

$$K_{act} = \frac{[B^+]}{[A^{++}]}$$

Further, let us assume that the standard potential $E^\circ = 0$ volts. Since there is one charge transferred, $n = 1$ in the Nernst equation. Since the equilibrium condition is reached when $E = 0$,

$$E = 0 = 0 - \frac{RT}{n\mathscr{F}} \ln K_{act}$$

This is true only when $K_{act} = 1$, which means that

$$[A^{++}] = [B^+]; \quad \text{equilibrium}$$

Now let us assume that we mix two more solutions of the same species $A^{++}$ and $B^+$ but with different concentration ratios.

The first solution has    $[A^{++}] = 10\,[B^+]$.
The second solution has $[A^{++}] = 0.1\,[B^+]$.

The potentials can be calculated from the Nernst equation by substituting the appropriate $K_{act}$.

The first,    $[A^{++}] = 10\,[B^+]$, has $K_{act} = 0.1$ and $E = +(RT/n\mathscr{F})$ volts.
The second, $[A^{++}] = 0.1\,[B^+]$, has $K_{act} = 10$ and $E = -(RT/n\mathscr{F})$ volts.

The arithmetic sign of the potential changes when the equilibrium is shifted from left to right. When the reaction "goes" to the right, the potential is positive. When the reaction "goes" to the reactant side, the potential is negative.

This also is a general result.

> *A chemical reaction proceeds* spontaneously as written *if the voltage associated with it is positive.*

This statement means that the products are formed in preference to the reactants if the potential for the cell is greater than zero. The reaction does not proceed toward forming the products if the potential is negative.

There is one other term associated with electrochemical half-reactions that you should know. The pair of oxidized and reduced species that can interconvert, such as $Fe^{+++}/Fe^{++}$ and $Tl^{0}/Tl^{+}$, is called a **redox couple.** The symbol $Fe^{+++}/Fe^{++}$ (or $Fe^{3+}/Fe^{2+}$) is read as *the iron three – iron two redox couple.*

### Electrochemical Cell Construction

An electrochemical cell is constructed of two half-cells as represented by the experimental arrangement shown in Figure 2.2. This example consists of two half-cells: One forms a Tl(I)/Tl(0) couple and the other a Fe(III)/Fe(II) couple. The Tl(I)/Tl(0) half-cell consists of a thallium metal wire or foil partly immersed in a 1-M $Tl^{+}$ solution. The metallic thallium and the thallous ion in solution comprise the electrochemical couple. The Fe(III)/Fe(II) half-cell contains a platinum wire or foil immersed in a solution with 1-M $Fe^{+++}$ and 1-M $Fe^{++}$ present. Through the platinum wire, we can measure the potential of this electrochemical couple. These two half-cells are connected between their electrodes through a voltage-measuring device. One other connection is needed to complete the circuit. Shown here is a porous glass or porous ceramic disk which prevents mixing of the two solutions while allowing electrical contact between them. This is a common form of a **salt bridge.** A number of different forms of salt bridges are described in Supplement 12A.

These types of electrochemical cells are so common that a shorthand has been developed to show the same information as is contained in the figure of the apparatus. In this shorthand, a salt bridge is drawn as two parallel vertical lines ∥. An interface between a solid electrode and a solution (and any boundary between phases — solids, liquids, or gases) is a single vertical line |. If more than one relevant soluble species is present, the components' labels are separated by a comma. For instance,

$$\|Fe^{++}(1 \text{ M}), Fe^{+++}(1 \text{ M})|$$

Convention now requires that for an electrochemical cell, the reaction involving oxidation be written on the left. However, for some time previously the reverse convention was assumed. With the contemporary convention, the cell shown in the figure can be represented compactly by

$$Tl|Tl^{+}(1 \text{ M})\|Fe^{++}(1 \text{ M}), Fe^{+++}(1 \text{ M})|Pt$$

This is the nomenclature for electrochemical cells used from now on in this text.

**Figure 2.2**   Diagram of an electrochemical cell composed of two half-cells.
Each solution with its solid electrode can
be described by a redox half-reaction. The
content of each solution is indicated in the
figure. However, no counter ions are
shown; the counter ions in this case might
be $NO_3^-$, $SO_4^-$, $Cl^-$, and so forth. The
circuit is completed through a porous plug
that allows current to flow through it but
prevents the two solutions from mixing
and reacting. Often this porous plug will
contain a concentrated solution of KCl or
similar simple salts. The connection is
then called a salt bridge. The voltage of
the electrochemical cell is easily
measured using a digital voltmeter, as
illustrated. Using such an instrument, we
can measure the voltage without
appreciable current flowing. As a result,
no appreciable oxidation or reduction
occurs. This means that the measured
voltage is an *equilibrium* potential.

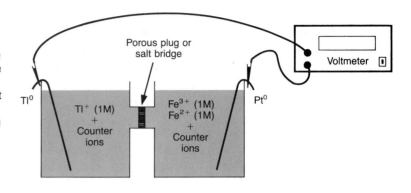

Consistent with this convention, the cell potential is then

$$E_{\text{cell}} = E_{\text{right}} - E_{\text{left}}; \qquad E_i \text{ for reduction} \qquad (2\text{-}51)$$

when the potentials $E_{\text{right}}$ and $E_{\text{left}}$ are reduction potentials.

### Redox Equilibria, Half-Reactions, and the Nernst Equation

In the same way that a redox reaction can be split into half-reactions, the Nernst
equation can be used to express the concentration dependence of a half-reaction
potential. In that case, the sign in front of the logarithmic term depends on the way
the half-reaction is written. It is positive $(+)$ if written as an oxidation and negative $(-)$
if written as a reduction. The electrons that appear in the half-reactions do not appear
in the equilibrium constants for the half-reactions.[5]

---

[5]The basis for this practice is that each half-reaction potential is thought of as being measured against a
reference-compound half-reaction with its reactants and products both in their standard states (having
unit activity). For instance,

$$Fe^{3+} + e^- \rightleftharpoons Fe^{2+}$$
$$\underline{+ \quad [\text{reference(reduced form, } a = 1) \rightleftharpoons \text{reference(oxidized form, } a = 1) + e^-]}$$
$$Fe^{3+} + \text{reference(reduced form, } a = 1) \rightleftharpoons Fe^{2+} + \text{reference(oxidized form, } a = 1)$$

The electrons for the implicit reference reaction cancel those of the half-reaction while both chemical
forms in the reference half-reaction have unit activity in the mass-action expression. Thus the equilibrium
is described by

$$E_{\text{Fe}} = E_{\text{Fe}}^\circ - (RT/n\mathscr{F}) \log([Fe^{3+}][1]/[Fe^{2+}][1])$$

Numerical substitutions for $R$, $T = 298$ K (25 °C), and $\mathscr{F}$ can be made. Along with a conversion from natural to base-10 logarithms, the potential (in volts) for an electrochemical *reduction* is expressed as

$$E = E° - \frac{0.059}{n\mathscr{F}} \log K_{act} \qquad \text{at 298 K} \qquad (2\text{-}52)$$

---

**EXAMPLE**

Write the equilibrium constants for the half-reactions associated with the redox reaction between thallium and iron.

**SOLUTION**

The half-reactions and their sum are

$$Tl^+ + e^- \rightleftharpoons Tl^0 \qquad\qquad K_{Tl} = 1/[Tl^+]$$
$$\underline{+ \qquad Fe^{2+} \rightleftharpoons Fe^{3+} + e^-} \qquad K_{Fe} = [Fe^{3+}]/[Fe^{2+}]$$
$$Fe^{2+} + Tl^+ \rightleftharpoons Tl + Fe^{3+} \qquad K_{eq} = \frac{[Fe^{3+}]}{[Tl^+][Fe^{2+}]}$$

The associated equilibrium constants are written at the right. No factor such as $[e^-]$ appears. The usual rules for addition of stoichiometric chemical reactions apply. As a result, the thermodynamic equilibrium constant is the product of the two half-reaction constants,

$$K_{eq} = K_{Fe} \cdot K_{Tl}$$

The mass-action expression for the sum reaction is the product of those for the two half-reactions. Because the expression for $K_{eq}$ is a thermodynamic equilibrium constant, the activity of $Tl^0$, metallic thallium, is unity.

---

## Cataloging Cell Potentials

For the ions and metals, the **reduction potentials** for oxidation–reduction reactions are conveniently stored as the potentials for half-reactions with all reactants in their standard states. A selected list of these appear in Appendix III. Two examples of these reduction potentials are

$$Tl^+ + e^- \rightleftharpoons Tl^0 \qquad E° = -0.34 \text{ V}$$

and

$$O_2 + 2\,H_2O + 4\,e^- \rightleftharpoons 4\,OH^- \qquad E° = 0.401 \text{ V}$$

The equations are a shorthand for the more complete expressions

$$Tl^+(1 \text{ M}) + e^- \rightleftharpoons Tl^0 \qquad\qquad E° = -0.34 \text{ V}$$

and

$$O_2(1 \text{ atm}, 273°) + 2\,H_2O(\text{pure}) + 4\,e^- \rightleftharpoons 4\,OH^-(1 \text{ M}) \qquad E° = 0.401 \text{ V}$$

Since the potentials are standard potentials $E°$, the species are in their respective standard states.

By experiment, only the voltages associated with full cells can be measured. One half-cell can be the **reference** against which all other half-cells are measured. From the measurements of each cell against the reference, the differences in potential between *all* the others can be found. For instance, Table 2.2 lists the results of a set of ten hypothetical experiments and the calculated half-cell potentials. Ten cells require five half-cell potentials.

Although the half-cell potentials are fixed relative to one another, there are an infinite number of absolute half-cell potentials that can still be assigned as indicated by the three different sets of half-cell potentials at the bottom of Table 2.2. An *arbitrary* fixed value must be given to one half-cell in order to assign a specific voltage to each of the other half-cells. In electrochemistry, the half-cell that has been chosen to be the reference and assigned a potential of zero is

$$2\ H^+(1\ M) + 2\ e^- \rightleftharpoons H_2(1\ atm); \qquad E° = 0.00\ V\ \text{exactly, any}\ T$$

Two names have been given to this half-cell: the *Standard Hydrogen Electrode* (SHE) and the *Normal Hydrogen Electrode* (NHE).

**Table 2.2**  Hypothetical Cell Potentials and Possible Half-cell Potentials

The electrochemical half-cells are labeled A, B, C, D, and E. Ten cells composed of sets of the five half-cells produce these potentials.

There are an *infinite* number of *sets* of half-cell potentials that will produce the measured cell potentials. For instance:

Set I: A $+1.22$; B $-0.34$; C 0.98; D $-1.06$; E 0.51.
Set II: A 0.86; B $-0.70$; C 0.62; D $-1.42$; E 0.15.
In Set III, cell E is set arbitrarily to zero volts.
So Set III: A 0.71; B $-0.85$; C 0.47; D $-1.57$; E 0.0 exactly.

| Cell Composed of | Measured Cell Potential (V) |
|---|---|
| A against B | 1.56 |
| A against C | 0.24 |
| A against D | 2.28 |
| A against E | 0.71 |
| B against C | $-1.32$ |
| B against D | 0.72 |
| B against E | $-0.85$ |
| C against D | 2.04 |
| C against E | 0.47 |
| D against E | $-1.57$ |

**EXAMPLE**

On the scale based on NHE, the following half-cell potentials are found.

$$Fe^{3+} + e^- \rightleftharpoons Fe^{2+} \quad E° = +0.771 \text{ V}$$
$$Pb^{2+} + 2 e^- \rightleftharpoons Pb^0 \quad E° = -0.126 \text{ V}$$

What would the potential of the iron half-cell be if based on a standard lead electrode?

$$Pb^{2+} + 2 e^- \rightleftharpoons Pb^0 \quad E° = 0.00 \text{ V exactly?}$$

**SOLUTION**

To visualize the answer, a simple figure with the three potentials involved can be drawn.

$$
\begin{array}{ll}
+0.771 & Fe^{3+}/Fe^{2+} \\
\\
0.00 & H^+/H_2 \\
-0.126 & Pb^{2+}/Pb
\end{array}
$$

If lead is now assigned zero, the diagram would appear as

$$
\begin{array}{ll}
+0.897 & Fe^{3+}/Fe^{2+} \\
\\
0.126 & H^+/H_2 \\
0.000 & Pb^{2+}/Pb
\end{array}
$$

To summarize, half-cell potentials have the following properties:

*The Standard Hydrogen Electrode (SHE) is assigned a potential of exactly zero volts at all temperatures.*
*When all reactions are written as reductions,*

$$oxidized\ form\ +\ n\ e^- = reduced\ form$$

*the reactions proceeding to the right* more *readily than $H^+/H_2$ are assigned positive voltages. Those proceeding with a smaller driving force are assigned negative half-cell voltages.*
*The half-cell potential is a quantitative measure of the tendency of the half-reaction to proceed from left to right.*

## Calculating Full-cell Potentials

Since any two half-cells can be combined to form an electrochemical cell, the data for a large number of cells can be stored efficiently as half-cell potentials. For example, from 50 half-cell potentials it is possible to calculate the expected potentials for more than 1200 different pairs.

The cell potentials are calculated by adding two half-reactions and finding the expected potential from the sum. However, the voltages associated with half-reactions are handled differently from the equilibrium constants that are associated with adding and reversing stoichiometric equations. The following rules are needed to calculate full-cell potentials from the tables of half-reactions.

a.  *If the direction in which a half-reaction is written is reversed, the sign of its half-cell potential is reversed.*
b.  *If the half-reaction is multiplied by a positive number, its voltage remains unchanged.*

The basis for these rules is described in Supplement 2D.

For review, two examples of regenerating electrochemical cell potentials from half-reaction potentials are shown.

---

**EXAMPLE**

Calculate the potential (EMF) of the electrochemical cell illustrated in Figure 2.2 from the half-reactions. The cell is at the standard state. If current is allowed to flow, in which half-cell do you expect oxidation to occur?

**SOLUTION**

The cell reaction can be written

$$Fe^{2+} + Tl^+ \rightleftharpoons Tl + Fe^{3+}$$

From the table in Appendix III, the half-cell reactions and reduction potentials are

$$Tl^+ + e^- \rightleftharpoons Tl^0 \qquad E° = -0.34 \text{ V}$$
$$Fe^{3+} + e^- \rightleftharpoons Fe^{2+} \qquad E° = 0.77 \text{ V}$$

For the reaction in which thallium is reduced and iron is oxidized, the following sum is written. Note that the reaction for iron is reversed from the sense above. $E_1$ and $E_2$ are merely convenient labels.

$$
\begin{array}{ll}
Tl^+ + e^- \rightleftharpoons Tl & E_1 = E°_{red} = -0.34 \text{ V} \\
\underline{Fe^{2+} \rightleftharpoons Fe^{3+} + e^-} & \underline{E_2 = -E°_{red} = -0.77 \text{ V}} \\
Fe^{2+} + Tl^+ \rightleftharpoons Tl + Fe^{3+} & E_{tot} = E_1 + E_2 = -1.11 \text{ V}
\end{array}
$$

With a negative potential, is this reaction spontaneous as written? With iron being oxidized and thallium reduced, the reaction is not spontaneous (to the right) as written. The reaction will proceed spontaneously to the left. We expect the oxidation to occur in the Tl half-cell.

---

**EXAMPLE**

Find the potential of the electrochemical cell with the reaction

$$Cu^0 + 2 Fe^{3+} \rightleftharpoons 2 Fe^{2+} + Cu^{2+}$$

with all solutions 1 M in the ionic species.

**SOLUTION**

This is a problem with the addition of half-cell reactions, which have different $n$-values. The reaction can be run in an electrochemical cell,

$$Cu|Cu^{2+}(1 \text{ M})\|Fe^{2+}(1 \text{ M}), Fe^{3+}(1 \text{ M})|Pt$$

The two half-reactions and their reduction potentials are

$$Cu^{2+} + 2 e^- \rightleftharpoons Cu^0 \qquad E° = 0.34 \text{ V}$$
$$Fe^{3+} + e^- \rightleftharpoons Fe^{2+} \qquad E° = 0.77 \text{ V}$$

The algebraic equations that are added to obtain the required full-cell equation are

$$Cu^0 \rightleftharpoons Cu^{2+} + 2\ e^- \qquad E_1 = -E° = -0.34\ V$$

$$\underline{2 \times [Fe^{3+} + e^- \rightleftharpoons Fe^{2+}]} \qquad E_2 = E° = 0.77\ V$$

$$Cu^0 + 2\ Fe^{3+} \rightleftharpoons 2\ Fe^{2+} + Cu^{2+} \qquad E_{cell} = E_1 + E_2 = 0.43\ V$$

By the rules, when the equation for copper is reversed, the potential changes sign. When the equation for iron is multiplied, the potential remains the same. This reaction will be spontaneous to the right as written since the voltage is positive.

## Concentrations and the Nernst Equation

An electric potential is generated by changes in concentrations of ionic species. The difference in concentration is equivalent to making a solution more concentrated or more dilute. Let us consider a reaction in which the direction of the process is dilution.

The Nernst equation includes, essentially, two steps,

$$E = \underbrace{E°}_{\substack{\text{reaction under} \\ \text{standard conditions}}} \underbrace{-\frac{RT}{n\mathscr{F}} \ln K_{act}}_{\text{dilution step}} \tag{2-50}$$

The value of $E°$ derives from a reaction done under standard-state conditions and the second, logarithmic, term results from any changes (diluting or concentrating) from the standard-state conditions.

The voltage associated with a dilution alone can be measured using an electrochemical cell, such as the one for copper:

$$Cu|Cu^{2+}(0.1\ M)\|Cu^{2+}(1\ M)|Cu$$

The only difference between the two half-cells is the concentration of $Cu^{++}$ in the solution phase. The cell is called a **concentration cell.** The reaction in this cell is equivalent to diluting a 1-M solution of $Cu^{++}$ ions to produce a 0.1-M solution. The equation that describes the full-cell reaction is

$$Cu^{2+}(0.1\ M) \rightleftharpoons Cu^{2+}(1\ M) \tag{2-53}$$

In this case, the value of $n$ is equal to the charge of the $Cu^{++}$ ion, so the Nernst potential for the cell is

$$E_{cell} = E°_{cell} - \frac{0.059}{2} \log \frac{[1]}{[0.1]} \tag{2-54a}$$

As you can calculate, the logarithmic term has a value of $-0.029$ V. Thus

$$E_{cell} = (E°_{cell} - 0.029)\ V \tag{2-54b}$$

However, to find the cell potential, the value of $E°_{cell}$ is required.

To find the origin of $E°_{cell}$, let us divide the "dilution" reaction into two separate steps as shown in Figure 2.3. The horizontal axis represents the electrochemical

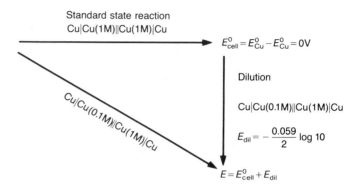

**Figure 2.3**    Diagram of the division of a copper concentration cell into a cell in the standard state and a cell equivalent to dilution. The horizontal and vertical arrows indicate the two processes. The total reaction cell is represented along the diagonal arrow.

reaction under the standard conditions for a copper versus copper cell. The vertical axis represents a copper concentration cell. At this time, it may seem unnecessary to include a cell with identical half-cells. However, this step is helpful in generalizing this type of mathematical analysis. For both individual steps a cell is diagrammed; such cells can be used to measure the potential experimentally. As can be seen from the figure, the value of $E^{\circ}_{cell}$ in Eq. 2-54 is for a cell

$$Cu|Cu(1\ M)\|Cu(1\ M)|Cu \qquad \text{with } E^{\circ}_{cell} = E^{\circ}_{Cu} - E^{\circ}_{Cu} = 0\ V$$

In this case, the standard cell is copper versus copper. Thus Eq. 2-54b becomes

$$E_{cell} = (0 - 0.029)\ V = -0.029\ V \qquad \textbf{(2-54b)}$$

The entire potential results from the concentration difference alone. The concentration cell is drawn with the lower concentration on the left since oxidation occurs there during a spontaneous reaction.

The separation of the Nernst equation into two separate steps presents a useful outline for calculating the potentials expected for a system that is not at standard conditions. The steps for this calculation are

1. *Calculate the value of* $E^{\circ}_{cell}$ *from the applicable tables.*
2. *Determine the value of* n.
3. *Calculate the contribution due to concentration changes away from standard conditions in both half-cells.*

## How to Find the *n*-Value

It is not always easy to decide what *n*-value is correct in the Nernst equation. A general rule can be formulated as

*The value of* n *in the Nernst equation equals the molar electronic charge transmitted between the two half-cells of an electrochemical cell.*

In the half-cell (of the copper concentration cell) where the concentration increases, copper metal is oxidizing to the ion. In the half-cell where the concentration decreases, copper ions are reduced and plate out as the metal. The half-reaction is

$$Cu^{2+} + 2\ e^- \rightleftharpoons Cu^0$$

For each copper ion that appears in one side, *two* electrons are transported to the other side where they reduce a copper ion. As a result, the value of $n$ is two. A more difficult situation is illustrated in the following example.

---

**EXAMPLE**

Calculate the potential expected from the following cell.

$$Tl|Tl^+(1\ M)\|Fe^{3+}(0.1\ M),\ Fe^{2+}(1\ M)|Pt$$

**SOLUTION**

This calculation involves the potential of an electrochemical cell not at the standard conditions. The reaction for the diagrammed cell is written

$$Fe^{3+}(0.1\ M) + Tl^0 \rightleftharpoons Fe^{2+}(1\ M) + Tl^+(1\ M)$$

The assignment of the product side of the stoichiometric equation can be made from the illustrated cell by recalling an **R** mnemonic: **R**ight (half-cell) = **R**eduction.

*First step.*   Calculate the standard potential for the system.
The cell at standard conditions is written

$$Tl|Tl^+(1\ M)\|Fe^{3+}(1\ M),\ Fe^{2+}(1\ M)|Pt$$

The potential for this cell (all the reagents at unit activity) was calculated earlier from the sum of two half-reactions. The reaction and potential are

$$Fe^{3+} + Tl^0 \rightleftharpoons Tl^+ + Fe^{2+}; \qquad E^\circ = +1.11\ V$$

*Second step.*   Decide on the value of $n$.
Since the ions in the dilution step are tripositive, should the value of $n$ in the Nernst equation be 3? The answer is no, $n$ is 1. To decide this, we write the half-reactions for both sides.

$$Fe^{3+} + e^- \rightleftharpoons Fe^{2+}$$
$$+ \qquad Tl^0 \rightleftharpoons Tl^+ + e^-$$
$$\overline{Fe^{3+} + Tl^0 \rightleftharpoons Tl^+ + Fe^{2+}}$$

For each ion of Fe(III) converted to Fe(II), one electron must be transferred from the other half-cell. The electron is generated from the oxidization of an atom of metallic thallium. The value of $n$ is 1 in the Nernst equation.

*Third step.*   Calculate the concentration contribution.
The Nernst equation for the cell is

$$E_{cell} = E^\circ_{cell} - \frac{0.059}{n} \log \frac{[Fe^{2+}][Tl^+]}{[Fe^{3+}]}$$

Substituting the appropriate values, we obtain

$$E_{cell} = 1.11\ V - \frac{0.059}{1} \log \frac{[1][1]}{[0.1]}$$

and

$$E_{cell} = 1.11 \text{ V} - 0.059 \, (1) = 1.05 \text{ V}$$

**EXAMPLE**

What is the value of $n$ to be used in the Nernst equation for the following reaction?

$$\text{Sn}^0 + 2 \text{ Fe}^{3+} \rightleftharpoons \text{Sn}^{2+} + 2 \text{ Fe}^{2+}$$

**SOLUTION**

The two half-reactions and their sum are

$$\text{Sn}^0 \rightleftharpoons \text{Sn}^{2+} + 2 \text{ e}^-$$
$$\underline{2 \times [\text{Fe}^{3+} + \text{e}^- \rightleftharpoons \text{Fe}^{2+}]}$$
$$\text{Sn}^0 + 2 \text{ Fe}^{3+} \rightleftharpoons \text{Sn}^{2+} + 2 \text{ Fe}^{2+}$$

The number of moles of electrons transferred between half-cells per mole of reaction is two, which is the value of $n$. The Nernst equation for this reaction is

$$E_{cell} = E_{cell}^\circ - 0.029 \, \log \frac{[\text{Fe}^{2+}]^2[\text{Sn}^{2+}]}{[\text{Fe}^{3+}]^2}$$

Note: If one chooses $n$ based on the reaction

$$\tfrac{1}{2} \text{ Sn}^0 + \text{Fe}^{3+} \rightleftharpoons \tfrac{1}{2} \text{ Sn}^{2+} + \text{Fe}^{2+}$$

the result of the calculation is the same.

## pH and the Nernst Equation

This single chapter has covered the methods of calculations associated with both acid–base and oxidation–reduction equilibria. As you have seen, there are, in fact, only a few rules needed to set up the calculations associated with any equilibrium. The rules concern the reversal and addition of stoichiometric equations and how the equilibrium constant and voltage of the sum equation are related to those of the reactions added.

So far, acid–base and redox equilibria have been treated separately even though they are closely related. The final juncture between acid–base and redox chemistry is presented here, where we discuss how changing the pH alone can change the EMF of a redox reaction. Nevertheless, the protons themselves are not oxidized or reduced.

In other words, changes in the concentrations of species other than the redox couple can alter the EMF associated with it. We can use the Nernst equation to describe this effect. Let us consider a redox half-reaction in which protons must be included in the equilibrium constant.

A well-studied example of such a redox reaction involves vanadium. The following equation has been found to describe the reduction of vanadate, $\text{VO}_2^+$, in acid solution:

$$\text{VO}_2^+ + 2 \text{ H}^+ + \text{e}^- \rightleftharpoons \text{VO}^{2+} + \text{H}_2\text{O} \tag{2-55}$$

In the stoichiometric equation, the formulae for the vanadium species do not include a number of water molecules of hydration that are present. For instance, a possible form of $\text{VO}_2^+$ binds four additional waters, forming $[\text{VO}_2(\text{H}_2\text{O})_4]^+$. The ambiguity in the formulae is similar to that for hydrated protons.

How can the equilibrium be shifted by altering the solution pH, and why? To describe the expected shift qualitatively only requires reference to Le Chatelier's principle, which states that an equilibrium shifts so as to remove the changes imposed on it. Thus, if the pH is lowered ([$H^+$] raised), the extra protons in the solution will force the reaction 2-55 to the right. The vanadium is forced toward its reduced form, $VO^{2+}$. In other words, in a more acidic solution, the reduced form of vanadium is more stable.

The explanation of the above chemistry is that protons compete with both oxidized and reduced forms of vanadium for the oxygens. However, the reduced form of vanadium, $VO^{2+}$, holds its oxygens less strongly than $VO_2^+$. Thus, when an oxygen is removed from a vanadium by combining with the protons, the system is more stable if the weaker-binding $VO^{2+}$ loses them.

The following calculation describes this phenomenon quantitatively. In the standard state

$$VO_2^+ + 2\,H^+ + e^- \rightleftharpoons VO^{2+} + H_2O; \quad K = \frac{[VO^{2+}]}{[H^+]^2[VO_2^+]}; \quad E° = 1.00\text{ V} \quad \textbf{(2-56)}$$

Under the standard conditions, the activity of hydrogen, $a_{H^+}$, equals unity; the pH = 0. The Nernst equation for reaction 2-56 is

$$E = E° - \frac{0.059}{1} \log([VO^{2+}]/[H^+]^2[VO_2^+]) \quad \text{at 25 °C}$$

With the use of the rules of logarithms (Supplement 1A), this can be written equivalently as

$$E = E° - \frac{0.059}{1} \{\log([VO^{2+}]/[VO_2^+]) + \log[H^+]^{-2}\} \quad \textbf{(2-57a)}$$

If we assume $a_{H^+} = [H^+]$, then from the definition of pH $= -\log[H^+]$, the last term in the curly brackets can be expressed as shown below.

$$E = E° - \frac{0.059}{1} \{\log([VO^{2+}]/[VO_2^+]) + 2\text{ pH}\} \quad \textbf{(2-57b)}$$

How much does the potential change with a shift in pH? Since $E°$ is a constant, the entire shift appears in the logarithmic term of Eq. 2-57. More specifically, the entire pH dependence is accounted for by the last term inside the curly brackets, 2 pH. Therefore the change in potential, $\Delta E$, that results from a change in pH *alone,* $\Delta$pH, should be

$$\Delta E = (-0.059/1) \cdot 2\,\Delta\text{pH} = -0.118\,\Delta\text{pH}$$

This calculation indicates that the potential of the vanadium couple is expected to *decrease* 118 mV per pH unit. In other words, for each 10-fold increase in [$H^+$] — equivalent to $\Delta$pH $= -1$ — the potential of the half-cell will increase by 118 mV. An increase in potential means that the equilibrium 2-55 lies further to the right; it is *easier* to reduce the vanadium(V). This is the same conclusion obtained from considering Le Chatelier's principle, but we have calculated a quantitative value for the change with pH.

The term **per decade** means "a factor of 10"; thus another way to state the change is to say that the potential changes 118 mV per decade change in $[H^+]$. Note that both directions of change are included: both raising and lowering $[H^+]$.

## 2.10 The Formal Potential

The Nernst equation correctly describes voltage changes of electrochemical cells or half-cells only if thermodynamic equilibrium constants $K_{act}$ are used. The EMF depends on the activities of the components of the redox couple.

Most often, however, an equilibrium potential is measured in complicated solutions in which numerous ionic species, known and/or unknown, are present and could influence the activities of the redox pair. A biological fluid presents such conditions. Biological solutions can be so complicated that it is not possible to understand fully all the effects of the thousands of species present. All these effects make it difficult or impossible to compare the measured potentials with the standard potentials stored in our lists. Even relatively simple solutions such as acid-digested samples must be considered complicated by this criterion.

To avoid these problems, an electrochemical equilibrium can be characterized by a **formal potential** $E^{\circ\prime}$ relative to the hydrogen standard. The prime sign (′) indicates that not all the concentrations are at unit activity. When the general reaction is

$$\text{oxidized form} + n\,e^- = \text{reduced form}$$

then the formal potential is defined as the potential measured when

$$\frac{[\text{reduced form}]}{[\text{oxidized form}]} = 1 \qquad (2\text{-}58)$$

This is a ratio of concentrations, not activities. Further, the concentrations are not necessarily 1 M.

If the EMF published for a redox couple (half-cell) is a formal potential, the detailed conditions under which it was measured *must* accompany the measured value. Otherwise the data could be worthless, the conditions being unknown to the readers. For a wide range of analytical methods, the formal potential is more useful than the standard potential because it is a quantity that can be measured directly. The variations in the formal potential for two representative half-reactions are shown in Table 2.3. The variations result from changes in the solution environment including effects of ionic strength, complex formation, and pH. The largest effects tend to be from pH changes. This point is illustrated in the following example.

---

**EXAMPLE**
What is the expected value of $E^{\circ\prime}$ for the vanadium(V)/vanadium(IV) couple at pH 3 in a solution in which the total vanadium concentration is 0.02 M? Ignore possible effects of ionic strength and complex formation.

**Table 2.3**  Formal Potentials of Iron and Dichromate Half-reactions in Various Solutions

| Acid and Concentration (M) | $E^{\circ\prime}_{Fe^{+++}/Fe^{++}}(V)^*$ | $E^{\circ\prime}_{Cr_2O_7^-/Cr^{+++}}(V)^*$ |
|---|---|---|
| HCl | | |
| 0.1 | 0.73 | 0.93 |
| 0.5 | 0.72 | 0.97 |
| 1 | 0.70 | 1.00 |
| 3 | 0.68 | 1.08 |
| H$_2$SO$_4$ | | |
| 0.1 | 0.68 | 0.92 |
| 0.5 | 0.68 | 1.08 |
| 4 | 0.68 | 1.15 |
| HClO$_4$ | | |
| 0.1 | 0.735 | 0.84 |
| 1 | 0.735 | 1.025 |
| Standard Potential | 0.771 | 1.33 |

\* These values include voltages that arise in parts of the measuring apparatus: the junction potentials, which are discussed in Chapter 12.

*Source:* G. F. Smith, *Anal. Chem.* **1951**, *23*, 925.

## SOLUTION

Given that the reaction is described by Eq. 2-56,

$$VO_2^+ + 2\,H^+ + e^- \rightleftharpoons VO^{2+} + H_2O; \qquad K = \frac{[VO^{2+}]}{[H^+]^2[VO_2^+]}; \qquad E^\circ = 1.00 \text{ V} \quad \textbf{(2-56)}$$

The formal potential is defined at the point where the ratio of the oxidized and reduced vanadium species is unity, Eq. 2-58. Since the total concentration is 0.02 M, then the formal potential will be measured when $[VO_2^+] = [VO^{2+}] = 0.01$ M. Thus, from Eq. 2-57,

$$E^{\circ\prime} = E^\circ - (0.059/1) \log([\,0.01]/[0.01][10^{-3}]^2)$$
$$= 1.00 - 0.059 \cdot (+6) = 0.65 \text{ V}$$

The formal potential of vanadium in an aqueous solution at pH 3 with 0.02 M total vanadium is expected to be about 0.65 V versus NHE. This compares with a standard potential of 1.00 V.

Note: We assumed that the total vanadium concentration has no effect on the value calculated, which will only be the case if the vanadium species involved have the same stoichiometric coefficients. (Their concentrations are raised to the same power.) In addition, effects of ionic strength and complex formation were ignored. If the solution were made with pure water, a vanadium salt, and an acid, both the ionic strength and formation of coordination complexes would depend on the identity of the acid—effects such as are seen in Table 2.3.

One final point of nomenclature: If the electrochemical potential is measured in an experiment in which the ratio

$$[reduced]/[oxidized] \neq 1$$

the measured potential is usually written as $E$. Only when the data can be connected[6] to a standard potential $E°$ or a formal potential $E°'$ should these symbols be used. The lack of a superscript standard – state symbol ($°$) or the presence of a prime sign ($'$) should alert you to look for a listing of the specific conditions of the measurement.

## Suggestions for Further Reading

A carefully written textbook mostly of methodical equilibrium calculations ignoring, in general, activity effects.
Ramette, R. W. "Chemical Equilibrium and Analysis"; Addison–Wesley: Reading, MA, 1981.

A short book that shows how ionization constants are determined and the necessary care required.
Albert, A.; Serjeant, E. P. "The Determination of Ionization Constants: A Laboratory Manual," 3rd. ed.; Chapmann and Hall: London and New York, 1984.

Extensive tables of $pK_a$ values and references printed under the auspices of the IUPAC (International Union of Pure and Applied Chemistry).
Serjeant, E. P.; Dempsey, B. "Ionisation Constants of Organic Acids in Aqueous Solution"; Pergamon Press: Oxford and New York, 1979.
Perrin, D. D. "Ionisation Constants of Inorganic Acids and Bases in Aqueous Solution," 2nd. ed.; Pergamon Press: Oxford and New York, 1982.
Hogfeldt, E. "Stability Constants of Metal-ion Complexes, part A: Inorganic Ligands," 1st. ed.; Pergamon Press: Oxford and New York, 1982
Perrin, D. D. "Stability Constants of Metal-ion Complexes, part B: Organic Ligands," 1st. ed.; Pergamon Press: Oxford and New York, 1979.

Extensive tables of stability constants printed under the auspices of the Chemical Society (London).
Sillen, L. G.; Martell, A. E. "Stability Constants of Metal-ion Complexes"; The Chemical Society: London, 1964.
Sillen, L. G.; Martell, A. E. "Stability Constants of Metal-ion Complexes, Supplement No. 1"; The Chemical Society: London, 1971.

A short, more advanced review article.
McBryde, W. A. E. "Spectrophotometric Determination of Equilibrium Constants in Solution," *Talanta* **1974**, *21*, 979–1004.

A review article illustrating that there is always more to learn about "simple" equilibria.
Covington, A. K. "Potentiometric Titrations of Aqueous Carbonate Solutions," *Chem. Soc. Revs.* **1985**, *14*, 265–281.

---

[6] Connected means by extrapolation, interpolation, direct measurement, and/or including corrections.

## Problems

### Acid–Base Equilibrium

2.1    Write the mass action expressions for the following reactions.
- a.   $CH_3OH(g) \rightleftharpoons CO(g) + 2\ H_2(g)$
- b.   $CO(g) + H_2O(g) \rightleftharpoons CO_2(g) + H_2(g)$
- c.   $[Co(NH_3)_6]^{++} + H_2O \rightleftharpoons [Co(NH_3)_5(OH)]^{++} + NH_4^+$
- d.   $N_2(g) + 4\ H_2O \rightleftharpoons 2\ NO_2(g) + 4\ H_2(g)$
- e.   $CaCO_3(s) \rightleftharpoons CaO(s) + CO_2(g)$
- f.   $NH_3(g) + H_2S(g) \rightleftharpoons NH_4HS(s)$

2.2    Write the mass action expressions for the following reactions.
- a.   glucose $+ 3\ OH^- + I_3^- \rightleftharpoons$ gluconate$^- + 3\ I^- + 2\ H_2O$
- b.   azobenzene $+ 4\ Cr^{2+} + 4\ H^+ \rightleftharpoons 2$ (analine) $+ 4\ Cr^{3+}$

2.3    A table is found in a handbook which lists "Dissociation constants for water-soluble organic bases." Since a base does not dissociate protons, but on the contrary takes them up, you become confused. The "base dissociation constant" for sodium barbiturate (NaBar) is given as $K = 1.08 \times 10^{-10}$. You look up the $pK_a$ of the conjugate acid, and it is 4.035. Since the reaction is not a dissociation of a base, write the reaction that goes along with $K$.

2.4    Determine the pH of a 0.15-M strychnine solution made by adding pure strychnine to water. Strychnine has a $pK_a$ of 6.00.

2.5    Some enzymatic assay procedures produce phenol as the product. Phenol is a monoprotic acid with $pK_a = 9.85$ at 25 °C. What is the pH of a 0.01-M phenol solution?

2.6    Solubilities of fatty acids, $CH_3(CH_2)_nCOOH$, in water increase when the acids are in their charged form, $CH_3(CH_2)_nCOO^-$. Would the solubilities increase, decrease, or remain the same as the pH of an aqueous solution is lowered?

2.7    Assume for the following question that the activity coefficients of the substance behaves similarly to those shown in Table 2.1; the higher the ionic strength, the lower the $\gamma_\pm$. Also, assume each ion is affected equally.
Nonesuch acid dissolves in water with the following reaction,

$$HNs \rightleftharpoons H^+ + Ns^-; \qquad K_a = 0.01$$

- a.   If a nonreactive salt such as $KNO_3$ is added to make a 0.01-M nonsuch acid solution 0.1 M in the salt, will the value of $K_a$ decrease, stay the same, or increase?
- b.   Under the conditions of (a), will the value of $[H^+]$ increase, stay the same, or decrease?

2.8    Two nonesuch acid solutions were made (HNs; $K_a = 0.01$, $\mu = 0$). Solution I has 0.05-M acid. Solution II has 0.25-M acid. Both were made from pure acid and pure distilled water.
- a.   Assuming $K_a$ for zero ionic strength applies, what value do you expect for the ionic strengths of solution I and solution II?
- b.   Will the value of the thermodynamic equilibrium constant $K_{a,thermo}(I)$ be greater than, equal to, or less than $K_{a,thermo}(II)$?

c.  Will the percent dissociation of acid in solution I be greater than, equal to, or less than the percent dissociation of acid in solution II?
d.  Will $[H^+]_I$ be greater than, equal to, or less than $[H^+]_{II}$?
e.  Will $[HNs^-]_I$ be greater than, equal to, or less than $[HNs^-]_{II}$?
f.  Assume that Eq. 2-16 for the single-ion activity coefficients applies to the solutions. What do you expect the value of $\gamma_\pm$ for each solution to be?
g.  Given the value of $K_a$, what do you expect the values of $K_{a,\text{thermo}}$ to be for the two solutions? Assume all types of activity coefficients are equal and that only a single iteration provides sufficient precision in the calculation.

2.9    Citric acid is a triprotic acid with $pK_a$ of 3.13, 4.76, and 6.40 for respective dissociations. Determine the following ratios of the citrate species

$$A = [H_3Cit]/[H_2Cit^-]$$
$$B = [H_2Cit^-]/[HCit^-]$$
$$C = [HCit^-]/[Cit^{3-}]$$

at
a.  pH 1.00
b.  pH 4.76
c.  pH 7.00

2.10    Assume you have 100 mL of each of the three solutions of problem 2.9 (pH 1.00, 4.76, and 7.00) and dilute each 200 mL with a salt solution so as to keep the ionic strength constant. For each solution,
a.  Would the ratio A increase, decrease, or remain the same?
b.  Would the ratio B increase, decrease, or remain the same?
c.  Would the ratio C increase, decrease, or remain the same?

2.11    $HNO_3$ is a strong acid.
a.  Calculate the pH you expect in a 0.1-M nitric acid solution.
b.  Calculate the pH you expect in a $1 \times 10^{-8}$-M nitric acid solution.

2.12    Calculate the concentrations of $H_2CO_3$, $HCO_3^-$, and $CO_3^-$ in a solution with 0.10 M total carbonate when the solution pH = 10.00.

2.13    Aspirin is acetylsalicylic acid (abbreviated Hasal; formula $C_8H_7O_2COOH$; F.W. 180.15). At 25 °C it has $K_a = 3.30 \times 10^{-4}$.
a.  If four aspirin tablets each weighing 32 mg were dissolved in 50.0 mL water, what pH do you expect the solution to have?
b.  What percentage difference is there between the exact solution and a solution calculated with the usual approximation method? Express the answer both as $\%[H^+]$ and %pH.

2.14    Ammonia reacts with water with the reaction

$$NH_3 + H_2O \rightleftharpoons NH_4^+ + OH^-: \qquad K_b = 1.8 \times 10^{-5}$$

a.  Calculate the pH you would expect from a 0.10-M solution of ammonia in water.
b.  What percentage difference is there between the exact algebraic solution and one calculated with the usual approximation methods?

    c.  What percentage of the ammonia is in the form $NH_3$, and what percentage is in the form $NH_4^+$? Is the solution ammonia water or ammonium hydroxide?

2.15   A phosphoric acid solution, $H_3PO_4$ in water, was neutralized with ammonia to a pH of 8.20. Using the $K_a$ values for phosphoric acid from Appendix II and the $K_b$ value shown in problem 2.14,

    a.  What is the value of the ratio $NH_4^+/NH_3$?

    b.  Which of the following species make up more than 1% of the total phosphate at pH 8.20: $H_3PO_4$, $H_2PO_4^-$, $HPO_4^-$, $PO_4^{3-}$?

2.16   Highly purified $(NH_4^+)_3PO_4$ was added to water until its concentration was 0.01 M. What do you calculate the pH of such a solution to be at 25 °C? Relevant $pK_a$ and $pK_b$ values are listed in Appendix II.

### Redox Equilibria

2.17   The following redox reactions occur in acid solution. Balance the equations.

    a.  $ClO_3^- + I^- \rightarrow Cl^- + I_2$

    b.  $Zn(s) + NO_3^- \rightarrow Zn^{++} + NH_4^+$

    c.  $ReO_2 + Cl_2 \rightarrow HReO_4 + Cl^-$

2.18   The following redox reactions occur in basic solution. Balance the equations.

    a.  $Al^0 + H_2O \rightarrow Al(OH)_4^- + H_2$

    b.  $I_2 + Cl_2 \rightarrow H_3IO_6^- + Cl^-$

    c.  $NiO_2 + Fe^0 \rightarrow Ni(OH)_2(s) + Fe(OH)_3(s)$

2.19   a.  What is the reduction potential for $Ni^{++} + 2e^- \rightleftharpoons Ni^0$ if the standard potential was set on a scale based on lead:

$$Pb^{++} + 2e^- \rightleftharpoons Pb^0; \qquad E° = 0.00 \text{ V exactly?}$$

    b.  On the "Standard Lead Electrode" scale, what would be the potential for the following two cells?

$$Ni|Ni^{++}(a=1)||H^+(aq,a=1)|H_2(a=1)|Pt$$
$$Zn|Zn^{++}(a=1)||Ni^{++} (a=1)|Ni$$

2.20   Consider the abstract electrochemical reaction

$$O + e^- \rightleftharpoons R \qquad \text{at 25 °C}$$

O stands for the oxidized form and R stands for the reduced form. Initially, a solution has equal concentrations of the oxidized and reduced species.

    a.  If the solution is changed from its initial conditions until

$$[O]/[R] = 0.10,$$

how much, in mV, will the potential change? Will the change be to more positive or negative voltage?

    b.  If the solution is changed from its initial conditions until

$$[O]/[R] = 100.0,$$

how much, in mV, will the potential change? Will the change be to more positive or negative voltage?

c.  If the solution is changed from its initial conditions until

$$[O]/[R] = 0.010$$

how much, in mV, will the potential change? Will the change be to more positive or negative voltage?

2.21  A voltaic cell is constructed with the following overall reaction. Species are all in their standard states.

$$Mg(s) + Ag^+(aq) \rightarrow Mg^{++}(aq) + Ag(s) \qquad \text{(unbalanced)}$$

a.  Balance the reaction equation.
b.  Write the equations for the individual electrode reactions.
c.  With the reaction occurring as written above, what is the cell voltage?
d.  Would the reaction proceed spontaneously as written?

\* 2.22  The following reactions were run as a voltaic cell.

$$Co^{++} + 2\,e^- \rightleftharpoons Co^0$$
$$Ni^{++} + 2\,e^- \rightleftharpoons Ni^0$$

Initially both half-cells were at the standard conditions. Then, the electrodes were connected and the reaction allowed to proceed until the cell's potential was zero. What are the ionic concentrations in each half-cell when $E_{cell} = 0.0$ V at 25 °C. Assume activity = concentration.

\* 2.23  The following cell is set up.

$$Pt|Fe^{+++}(1\ M),\ Fe^{++}(1\ M)||Fe^{+++}(1\ M),\ Fe^{++}(1\ M)|Pt$$

a.  What will the cell potential be?
b.  If, in the left half-cell, the $Fe^{+++}$ concentration is changed to 0.1 M, does the cell potential become more positive or more negative?
c.  What is the numerical value of the voltage change in part b?
d.  If, in the right half-cell, the $Fe^{+++}$ concentration is changed to 0.1 M, does the cell potential become more positive or more negative?
e.  If, in one left half-cell, the $Fe^{++}$ concentration is changed to 0.1 M, does the cell potential become more positive or more negative?
f.  What is the numerical value of the voltage change in part e?

2.24  The following reaction takes place in solution.

$$H_2O_2 + 2\,H^+ + 2\,e^- \rightleftharpoons 2\,H_2O; \qquad E^\circ = 1.776\ V$$

a.  What will the half-cell potential be at pH = 4.0?
b.  What will the half-cell potential be at pH = 6.0?
c.  Rewrite the reaction for basic solutions. Does $E^\circ$ change (yes or no)?

---

\* Denotes more involved problems.

2.25   The following information is known.

$$Ag^+ + e^- \rightleftharpoons Ag(s); \qquad E° = 0.799 \text{ V}$$
$$AgCl(s) + e^- \rightleftharpoons Ag(s) + Cl^-; \qquad E° = 0.222 \text{ V}$$

What is the voltage equivalent to the solubility product of $AgCl(s)$?

2.26  a.   For the reaction at 25 °C,

$$NO_3^- + 3\,H^+ + 2\,e^- \rightleftharpoons HNO_2(aq) + H_2O$$

at pH 5.0, the formal potential is 0.50 V with respect to SHE. What is the value of $E°$?

b.   Calculate the value of $E°$ for the reaction

$$SeO_4^- + 4\,H^+ + 2\,e^- \rightleftharpoons H_2SeO_3 + H_2O$$

if the value of $E°'$ is 0.56 V with respect to SHE at pH 5.0 at 25 °C?

2.27   Can a formal potential be defined for the following reaction?

$$MnO_2(s) + 4\,H^+ + 2\,e^- \rightleftharpoons Mn^{++} + 3\,H_2O$$

2.28   The potential is given for a silver chloride reaction

$$AgCl(s) + e^- \rightleftharpoons Ag(s) + Cl^-; \qquad E° = +0.222 \text{ V}$$

What is the solubility product of $AgCl$?

2.29   An electrochemical cell is composed of a saturated calomel electrode ($E = 0.2412$ vs. SHE) and a metallic cadmium electrode in contact with a solution that is 0.100 M in cadmium ion in the presence of 1 M cyanide. The complex $Cd(CN)_4^-$ forms. The *dissociation* constant for the tetracyano complex is $1.4 \times 10^{-19}$ under the conditions of the experiment.

a.   What is the value of $[Cd^{++}]$ (not $Cd(CN)_4^-$), assuming all activity coefficients are unity?

b.   What is the potential of the full cell written as a cadmium oxidation?

2.30   An electrochemical cell is set up between iron ($Fe^{+++}/Fe^{++}$) and titanium ($TiO^{++}/Ti^0$). The reactions that can occur in the half-cells are, respectively,

$$Fe^{+++} + e^- \rightleftharpoons Fe^{++} \qquad E° = 0.77 \text{ V}$$
$$TiO^{++} + 2\,H^+ + 4\,e^- \rightleftharpoons Ti^0 + H_2O \qquad E° = 0.1 \text{ V}$$

a.   Under standard conditions, what is the cell potential, written so it will proceed spontaneously?

b.   If the concentration of $Fe^{++}$ is reduced to 0.15 M, what do you calculate the cell potential will be?

c.   If the concentration of $Fe^{+++}$ is reduced to 0.15 M, what do you calculate the cell potential will be?

d.   If the original conditions exist and the pH of the titanium half-cell is raised from 0 to 2, what do you expect the cell potential to be then?

e.   If the conditions of part d exist, and then the pH of the iron half-cell is raised from 0 to 2 as well, what do you expect the cell potential to be then?

# Diprotic Acid Calculations without Approximations

To find the pH of a $10^{-7}$-M strong acid in water when the chemistry could not be simplified to find one principal equilibrium system, a quadratic equation was obtained and solved for $[H^+]$. For the carbonate system, an equation to the fourth power in $[H^+]$ must be solved to find a general algebraic solution describing the stoichiometry over a range of conditions. This equation is derived here mainly to illustrate how important it is to discern the principal equilibrium or, perhaps, the two joint principal equilibria that determine the concentration of the species of interest. Although the carbonate system is used, the equation holds for diprotic acids.

## Bicarbonate as a Function of pH

Before deriving the most general equation, let us first derive a simpler one expressing $[HCO_3^-]$ as a function of $[H^+]$ for a carbonate solution with a total carbonate concentration $C_{tot}$. Assume the solution is made by mixing portions of 0.10 M $H_2CO_3$, $NaHCO_3$, and/or $Na_2CO_3$ stock solutions so that $C_{tot}$ always equals 0.10 M. The concentration of ionic sodium is then fixed by the relative volumes of the three stock solutions used to make up the final solution.

There are seven species in the solution: $H_2O$, $H^+$, $OH^-$, $Na^+$, $CO_3^=$, $HCO_3^-$, and $H_2CO_3$. We assume that the water activity remains unity, leaving six species. Since $[Na^+]$ is determined by the initial recipe, that leaves five species, which need five equations connecting their concentrations to ensure a unique solution. The stoichiometric equations for these five species are

$$H_2CO_3 \rightleftharpoons H^+ + HCO_3^- \tag{2-42}$$

$$HCO_3^- \rightleftharpoons H^+ + CO_3^= \tag{2-43}$$

$$H_2O \rightleftharpoons H^+ + OH^- \tag{2-30}$$

There are three mass-action expressions:

$$K_1 = \frac{[H^+][HCO_3^-]}{[H_2CO_3]} \tag{2-45}$$

$$K_2 = \frac{[H^+][CO_3^=]}{[HCO_3^-]} \tag{2-46}$$

$$K_w = [H^+][OH^-] \tag{2-33}$$

Since Eqs. 2-42 and 2-43, the two stoichiometric equations for carbonate, contain two species in common, $[H^+]$ and $[HCO_3^-]$, the information in the two equations can be reduced to one equation, the conservation equation

$$C_{tot} = [H_2CO_3] + [HCO_3^-] + [CO_3^=] \tag{2-44}$$

Since the pH of the solution is given, not only are the magnitudes of $[H^+]$ and $[OH^-]$ fixed, but $[Na^+]$ is fixed as well. This is seen in the charge-balance equation,

$$[H^+] + [Na^+] = [OH^-] + 2[CO_3^=] + [HCO_3^-] \qquad \text{(2A-1)}$$

Here $[H^+]$ is given, which fixes $[OH^-]$ through $K_w$, Eq. 2-33. Then

$$[Na^+] = 2[CO_3^=] + [HCO_3^-]$$

which is determined by the composition of the final solution.

The remainder of the derivation is algebra. We want to substitute into Eq. 2-44 so that it contains only $[HCO_3^-]$, which is the dependent variable. This substitution is done by using rearranged Eqs. 2-45 and 2-46.

$$[H_2CO_3] = [H^+][HCO_3^-]/K_1 \qquad \text{(2-45)}$$

$$[CO_3^=] = K_2[HCO_3^-]/[H^+] \qquad \text{(2-46)}$$

With substitution for $[H_2CO_3]$ and $[CO_3^=]$, Eq. 2-44 becomes

$$C_{tot} = [H^+][HCO_3^-]/K_1 + [HCO_3^-] + K_2[HCO_3^-]/[H^+] \qquad \text{(2A-2a)}$$

This is the desired equation albeit in a different form. By dividing both sides of Eq. 2A-2a by $[HCO_3^-]$ and rearranging, we obtain

$$[HCO_3^-] = \frac{C_{tot}}{([H^+]/K_1) + 1 + (K_2/[H^+])} \qquad \text{(2A-2b)}$$

---

**EXAMPLE**
What is $[HCO_3^-]$ in a 0.10-M carbonate solution with a pH of 6.38?

**SOLUTION**
$K_1 = 4.2 \times 10^{-7}$, and $K_2 = 4.8 \times 10^{-11}$. At pH 6.38, $[H^+] = 4.2 \times 10^{-7}$ M. Substituting the appropriate values into Eq. 2A-2b, we find

$$[HCO_3^-] = \frac{0.10}{(4.2 \times 10^{-7}/4.2 \times 10^{-7}) + 1 + (4.8 \times 10^{-11}/ 4.2 \times 10^{-7})} = 0.050$$

which is the same result calculated with a single, simple equilibrium.

---

## A General Equation

To derive a general equation for this system, we begin in the same way as above but have an unknown pH and, hence, an unknown value of $[Na^+]$ and $[OH^-]$. The sodium concentration is an independent variable.

There are a number of different algebraic routes to the final, unique equation. However, the general direction toward its derivation can be ascertained since we want to find $[H^+]$ in terms of $K_1, K_2, K_w, C_{tot}$, and $[Na^+]$; thus we substitute only these five factors and $[H^+]$ into the charge-balance equation 2A-1. Part of the derivation of the general equation has already been done in expressing the conservation equation as a function for $[HCO_3^-]$. Since $[HCO_3^-]$ has been used there, we shall proceed by rewriting the charge-balance terms also as functions of bicarbonate.

$$[H^+] + [Na^+] = K_w/[H^+] + 2(K_2[HCO_3^-]/[H^+]) + [HCO_3^-] \quad \text{(2A-3a)}$$

To obtain Eq. 2A-3a, Eq. 2-33 was used to replace $[OH^-]$ by a term composed of $[H^+]$ and an equilibrium constant only.

Equation 2A-3a is easier to work with if both sides are multiplied by $[H^+]$, and the terms collected.

$$[Na^+][H^+] + [H^+]^2 = K_w + (2K_2 + [H^+])[HCO_3^-] \quad \text{(2A-3b)}$$

Into this equation is substituted a rearranged form of the conservation equation, Eq. 2A-2,

$$[HCO_3^-] = \frac{[H^+]K_1 C_{tot}}{[H^+]K_1 + K_2 K_1 + [H^+]^2} \quad \text{(2A-2)}$$

After the substitution, a moderately messy rearrangement produces the general equation.

$$[H^+]^4 + (K_1 + [Na^+])[H^+]^3 + (K_1 K_2 + [Na^+]K_1 - K_w - K_1 C_{tot})[H^+]^2 + \\ + ([Na^+]K_2 - K_w - 2K_2 C_{tot})K_1[H^+] - K_1 K_2 K_w = 0 \quad \text{(2-45)}$$

This equation holds for any dibasic acid in water made by mixing monocation salts of the anionic forms of the acid: in generalized terms, mixing $H_2L$, $MHL$, and $M_2L$. Different equations must be derived for any other systems such as one made by adding HCl to a solution of $NaCO_3$. It seems that such equations certainly can obfuscate the chemistry.

## SUPPLEMENT 2B

# Methods for Simulating Simultaneous Equilibria

## 2B.1 Background

When a molecule that is stable by itself in solution binds to a metal ion, the compound that is formed is called a **coordination complex**. The nomenclature of this area is reviewed in Figure 2B.1. The strength of the molecule–metal-ion binding depends on the conditions under which the equilibrium exists. In addition to the ionic strength affecting the equilibrium somewhat, pH changes produce significant effects. The reason that the pH effect is so large is that the protons compete with the metal for the ligand.

It is worthwhile to go through a calculation of the simultaneous equilibria involved to get an idea of the factors producing changes in the equilibrium and how truly complicated such multiple equilibria are. The principles of calculations that mimic

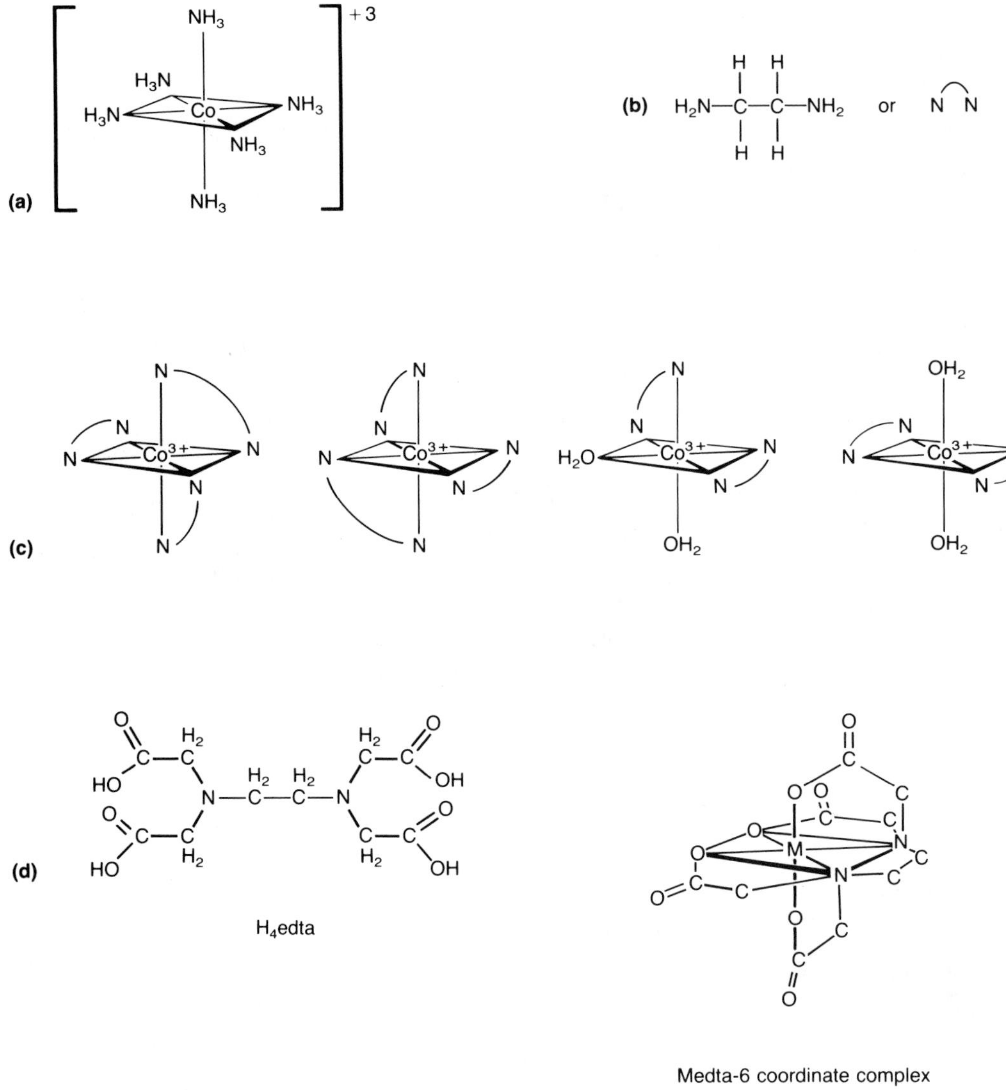

Medta-6 coordinate complex

**Figure 2B.1** A molecule that can bind a metal ion at one site is called a **coordination agent**, or **ligand.** Ammonia is a ligand. The compound formed is called a **coordination complex.** The structure of a cobalt–ammonia coordination complex is shown in (a). However, the compound ethylenediamine, in (b), is a **chelating agent**, or **chelon.** A chelating agent is a molecule that can bind to a metal ion at two or more sites simultaneously. The compound formed between an ion and a chelon is called a **chelate.** Ethylenediamine in water forms stable chelates with $Co^{3+}$ with the structures shown in (c). Each ethylenediamine binds at two sites on the cobalt ion. Figure (d) shows the chemical structure of $H_4$edta. A number of different metal ions bind with edta$^{4-}$ as shown. However, this is not always the structure. Some complexes are four coordinate with two acetates dangling free. Other metal ions can be bound at seven or eight coordination sites and, thus, bind all six edta sites and still bind other groups. The structures of a few other chelating agents are shown in Table 10.4.

**Table 2B.1** Stoichiometric Equations and Equilibrium Constants for edta

The equilibrium constants are dissociation constants for the protonations and formation constants for the metals with $edta^{4-}$.

$$H_4edta \rightleftharpoons H^+ + H_3edta^- \qquad \log K_1 = -2.0$$

$$H_3edta^- \rightleftharpoons H^+ + H_2edta^{2-} \qquad \log K_2 = -2.67$$

$$H_2edta^{2-} \rightleftharpoons H^+ + Hedta^{3-} \qquad \log K_3 = -6.16$$

$$Hedta^{3-} \rightleftharpoons H^+ + edta^{4-} \qquad \log K_4 = -10.26$$

$$VO^{2+} + edta^{4-} \rightleftharpoons VOedta^= \qquad \log K\,(V^{IV}) = 18.0$$

$$VO_2^+ + edta^{4-} \rightleftharpoons VO_2edta^{3-} \qquad \log K\,(V^V) = 15.55$$

*Source:* L. B. Sillen and A. E. Martell, "Stability Constants of Metal–Ion Complexes," Supplement 1, Spec. Publ. No. 25; The Chemical Society: London, 1971.

both simple and complicated simultaneous equilibria are the same. However, there are some tricks that help to get us through.

In this section, the binding of one of the most useful coordination agents, ethylene-diaminetetraacetate *(edta)*, is used to illustrate such calculations. The general structure of a metal–edta complex is shown in Figure 2B.1d. The metal to which we shall consider it binding is vanadium, which in the presence of air can be in one of two different oxidation states. The applicable equilibria and equilibrium constants are listed in Table 2B.1.

In a vanadium–edta solution there will be free and bound (by ligand) vanadium and specific concentrations of free and bound (to vanadium) ligand. We attempt to describe these concentrations correctly as they depend on pH by accounting for all the important equilibria that simultaneously occur. This calculation, in effect, is reversing the course through which the equilibrium constants were derived from experimental data.

Two of the reasons that edta and vanadium are used as an example is that the equilibria are relatively well understood, and the necessary equilibrium constants are available. These two criteria often do not apply to samples for analysis, and such calculations can only be used for guidance. As an example, the quantities and chemistry denoted in Table 2B.1 were found under conditions of constant ionic strength. Usually this is done by running experiments in solvents containing 0.1 to 1 M inert 1 : 1 electrolyte such as sodium perchlorate or potassium chloride. If an analyte or interferent is to be chelated and does not reside under the same conditions after sample pretreatments (which is likely!), the literature values do not apply exactly to the system. Even if there are not unaccounted equilibria, the error in the calculation can range up to a factor of five due to ionic strength effects alone.

As a final caution, note that coordinating agents such as edta are not highly specific. This can be seen in Table 2B.2 in which formation constants for some edta–metal complexes are listed. As a result, if edta is used for titration (Chapter 10), significant effort must often be expended to eliminate interferences in all but the simplest solutions such as those containing mostly $Ca^{++}$ or $Mg^{++}$ with less than, say, 1% (of

**Table 2B.2**  Formation Constants for edta$^{4-}$ with Various Metals

Temperature at 20 °C unless noted. The general reaction is
$M^{+n} + edta^{4-} \rightleftharpoons [Medta]^{n-4}$.

| Metal Ion | Log of Formation Constant | Conditions (M) |
|---|---|---|
| Ag$^+$ | 7.72 | 0.1 KNO$_3$ |
| Al$^{+++}$ | 16.7 | 0.1 KNO$_3$ |
| Bi$^{+++}$ | 28.8 | 0.1 NaClO$_4$ |
| Ca$^{++}$ | 10.7 | 0.1* |
| Cd$^{++}$ | 16.9 | 0.1 KClO$_4$ |
| Ce$^{+++}$ | 15.5 | 0.5 NH$_4^+$, 22° |
| Co$^{++}$ | 16.5 | 0.1 KClO$_4$ |
| Co$^{+++}$ | 40.6 | 0.2 KNO$_3$, 25° |
| Cu$^{++}$ | 18.9 | 0.1 KClO$_4$ |
| Fe$^{+++}$ | 25.1 | 0.1 NaClO$_4$ |
| Hg$^{++}$ | 21.8 | 0.1 KNO$_3$ |
| Mg$^{++}$ | 8.7 | 0.1 KNO$_3$ |
| Mn$^{++}$ | 14.0 | 0.1 KNO$_3$ |
| Ni$^{++}$ | 18.4 | 0.1 KClO$_4$ |
| Pb$^{++}$ | 18.3 | 0.1 KClO$_4$ |
| Pu$^{+++}$ | 25.8 | 0.1* |
| Sn$^{++}$ | 18.3 | 1 NaClO$_4$ |
| UO$_2^+$ | 10.4 | 0.1 NH$_4$Cl |
| Zn$^{++}$ | 15.9 | 0.2 KNO$_3$ |
| Zr$^{4+}$ | 27.7 | 1 NaClO$_4$ |

* Unnamed salt used.

*Source:* L. B. Sillen and A. E. Martell, "Stability Constants of Metal–Ion Complexes," Supplement 1, Spec. Publ. No. 25; The Chemical Society: London, 1971.

Ca$^{++}$ + Mg$^{++}$) other metal ions. It is no surprise that metals are determined now more by spectrometric methods (Chapters 17 and 18), which tend to be far more specific than titration.

## Overview

The main trick in simplifying the multiple simultaneous equations of ligands and metals is to calculate using only one equilibrium equation which has been "corrected" for all the other simultaneous equilibria by multiplying the equilibrium constant by a numerical factor. In the case of edta$^{4-}$ and a metal, only the single equation

$$M^{+n} + edta^{4-} \rightleftharpoons [Medta]^{n-4} \tag{2B-1}$$

is used directly. However, there are six other edta species present: $H_6edta^{++}$, $H_5edta^+$, $H_4edta$, $H_3edta^-$, $H_2edta^{2-}$, and $Hedta^{3-}$. There are six competing equilibria involved.

$$H^+ + [H_{n-1} edta]^{n-5} \rightleftharpoons [H_n edta]^{n-4}; \qquad n = 1 \text{ to } 6 \qquad \text{(2B-2)}$$

These can all be accounted for in the mass-action expression

$$K_4 = \frac{[H^+][edta^{4-}]}{[Hedta^{3-}]}$$

by finding the *fraction* of the total of all seven edta species that is in the form $edta^{4-}$. (For the calculations done below, the formation of $H_5edta^+$ and $H_6edta^{++}$ — $K_{dissoc}$ 0.11 and 0.55, respectively — will be ignored since at pH above 1, less than 1% of the total edta is in those forms.)

The same procedure can be done for the metal–ion equilibrium to compensate for any forms other than $M^{n+}$ that may be present. For instance, one possibility is the formation of hydrolysis products such as $M(OH)^{(n-1)+}$, where one proton has been removed from the waters of hydration. The effective concentration of $M^{n+}$ is lowered when hydroxide competes against the chelating agent for the metal. How to account for each of these effects — protonation of $edta^{4-}$ and hydrolysis of the metal — is shown in the sections following.

## 2B.2  Correcting for the Protonation of $edta^{4-}$

The fraction of the total chelate that is in the form $edta^{4-}$ is found as follows. The calculation requires, as usual, all the mass-action expressions and the conservation relationship. The following abbreviations are used for edta species.

| | |
|---|---|
| $H_4edta$ | $H_4L$ |
| $H_3edta^{3-}$ | $H_3L$ |
| $H_2edta^{2-}$ | $H_2L$ |
| $Hedta^{3-}$ | $HL$ |
| $edta^{4-}$ | $L$ |

The total amount of edta present *in all forms* is called $C_L$, and

$$C_L = [H_4edta] + [H_3edta^-] + [H_2edta^{2-}] + [Hedta^{3-}] + [edta^{4-}]$$

or

$$C_L = [H_4L] + [H_3L^-] + [H_2L^{2-}] + [HL^{3-}] + [L^{4-}] \qquad \text{(2B-3)}$$

You will recognize this as the conservation equation for edta.

Our initial aim is to obtain an expression for the pH dependence of $[edta^{4-}]$. The process is *exactly equivalent* to solving the five mass-action equations and conservation equation just as you have done before for simpler simultaneous equilibrium systems with fewer species involved.

From the equations in Table 2B.1 for acid dissociation of edta, the following is known.

$$K_1 = \frac{[H^+][H_3L^-]}{[H_4L]} = 10^{-2.0}$$

$$K_2 = \frac{[H^+][H_2L^{2-}]}{[H_3L^-]} = 10^{-2.67}$$

$$K_3 = \frac{[H^+][HL^{3-}]}{[H_2L^{2-}]} = 10^{-6.16}$$

(2B-4a)

$$K_4 = \frac{[H^+][L^{4-}]}{[HL^{3-}]} = 10^{-10.26}$$

The mass-action expressions can be rearranged to obtain expressions of concentrations of $[H_nL]$ in terms of $K_i$ and $[H^+]$.

$$[H_3L^-] = K_1[H_4L]/[H^+]$$
$$[H_2L^{2-}] = K_2[H_3L^-]/[H^+]$$
$$[HL^{3-}] = K_3[H_2L^{2-}]/[H^+]$$
$$[L^{4-}] = K_4[HL^{3-}]/[H^+]$$

(2B-4b)

Notice the repetition in Eqs. 2B-4b: The left-hand side of each equation has the same term as the right-hand side of the one below. Therefore, by simple algebraic substitution,

$$[H_3L^-] = K_1[H_4L]/[H^+]$$
$$[H_2L^{2-}] = K_1K_2[H_4L]/[H^+]^2$$
$$[HL^{3-}] = K_1K_2K_3[H_4L]/[H^+]^3$$
$$[L^{4-}] = K_1K_2K_3K_4[H_4L]/[H^+]^4$$

(2B-5)

Note that all four equations have the factor $[H_4L]$ on the right. Substituting Eqs. 2B-5 (derived from the mass-action expressions) into Eq. 2B-3 (the conservation relation) and then factoring out the common term leaves

$$C_L = [H_4L]\left(1 + \frac{K_1}{[H^+]} + \frac{K_1K_2}{[H^+]^2} + \frac{K_1K_2K_3}{[H^+]^3} + \frac{K_1K_2K_3K_4}{[H^+]^4}\right)$$

(2B-6)

A useful change in nomenclature is to express each species of edta as a fraction of the whole. This is done by dividing the concentration of each by $C_L$ and the fractions are abbreviated as $\alpha$-values.[7]

$$\alpha_4 = [H_4L]/C_L$$
$$\alpha_3 = [H_3L^-]/C_L$$
$$\alpha_2 = [H_2L^{2-}]/C_L$$
$$\alpha_1 = [HL^{3-}]/C_L$$
$$\alpha_0 = [L^{4-}]/C_L$$

(2B-7)

[7] Some workers use $\alpha_i$ to represent the inverse $1/\alpha_i$.

The numerators of the $\alpha_i$ are the concentrations of specific species; the denominator is the total concentration as given by Eq. 2B-3 or 2B-6.

To find the numerical value that allows [edta$^{4-}$](abbreviated[L$^{4-}$]) to account for the other edta species present, we derive the algebraic expression for the value of $\alpha_0$. The expression may appear to be formidable even though it is the same as the last of Eqs. 2B-7. To begin,

$$\alpha_0 = [L^{4-}]/C_L$$

and from Eq. 2B-5,

$$[L^{4-}] = K_1 K_2 K_3 K_4 [H_4L]/[H^+]^4$$

Therefore

$$\alpha_0 = \frac{K_1 K_2 K_3 K_4 [H_4L]/[H^+]^4}{[H_4L]\left(1 + \dfrac{K_1}{[H^+]} + \dfrac{K_1 K_2}{[H^+]^2} + \dfrac{K_1 K_2 K_3}{[H^+]^3} + \dfrac{K_1 K_2 K_3 K_4}{[H^+]^4}\right)} = \frac{[L^{4-}]}{C_L}$$

After carrying out some simplifying algebra, we obtain a somewhat more convenient equation for $\alpha_0$.

$$\alpha_0 = \frac{K_1 K_2 K_3 K_4}{[H^+]^4 + K_1[H^+]^3 + K_1 K_2[H^+]^2 + K_1 K_2 K_3[H^+] + K_1 K_2 K_3 K_4} \quad \text{(2B-8)}$$

By substituting the appropriate values of [H$^+$] into the equation, we can find the fraction of edta in the form edta$^{4-}$ at any pH.

---

**EXAMPLE**
What is the fraction of edta that is in the form edta$^{4-}$ at pH 7.4?

**SOLUTION**
The value of [H$^+$] at pH 7.4 is $3.98 \times 10^{-8}$. From the log $K_i$ values in Table 2B.1, the equilibrium constants that can be substituted into Eq. 2B-8 are

$$K_1 = 10^{-2}; \quad K_2 = 2.14 \times 10^{-3}; \quad K_3 = 6.9 \times 10^{-7}; \quad K_4 = 5.5 \times 10^{-11}$$

The equilibrium-constant products are

$$K_1 = 1 \times 10^{-2}$$
$$K_1 K_2 = 2.14 \times 10^{-5}$$
$$K_1 K_2 K_3 = 1.48 \times 10^{-11}$$
$$K_1 K_2 K_3 K_4 = 8.1 \times 10^{-22}$$

Therefore

$$\alpha_0 = \frac{8.12 \times 10^{-22}}{2.5 \times 10^{-30} + 6.3 \times 10^{-23} + 3.39 \times 10^{-20} + 5.89 \times 10^{-19} + 8.1 \times 10^{-22}}$$
$$\alpha_0 = 1.30 \times 10^{-3}$$

---

Incidentally, the sum of all the $\alpha$-values is unity.

$$1 = \alpha_4 + \alpha_3 + \alpha_2 + \alpha_1 + \alpha_0$$

This equation is derived by dividing both sides of Eq. 2B-3 by $C_L$ and comparing the terms on the right side with Eqs. 2B-7. By carrying out such calculations at a number of pH values, graphs of the fractional concentrations, $\alpha_i$, versus pH can be constructed. Two different ways these are plotted are illustrated in Figure 2B.2.

### Rules for Writing $\alpha$-Expressions

As formidable as the $\alpha$-expressions such as Eq. 2B-8 appear, they are quite regular, and it is not necessary to go through so much algebra to find the correct form. The rules are

1. *All $\alpha$ in the system have identical denominators.*
2. *The denominator is a decreasing power series in $[H^+]$ with the following properties:*
   a. *The power series begins with $[H^+]^N$, where N is the total number of protons that can dissociate.*
   b. *The first term has no factor, and the factor for each successive term contains an additional stepwise dissociation constant.*
   c. *There will be $(N + 1)$ terms, the last of which is $K_1 K_2 \ldots K_N$.*
3. *Each term in the denominator is proportional to the concentration of one of the species present. Thus the first term forms the numerator of $\alpha_N$, the second for $\alpha_{N-1}$, and so on, finishing with the numerator for $\alpha_0$ being $K_1 K_2 \ldots K_N$.*

---

**EXAMPLE**

Write the $\alpha$-term for $PO_4^{3-}$ and find its concentration in a 0.10-M solution of phosphate at pH 4.0. The $pK_a$ values for $H_3PO_4$ are 2.23, 7.21, and 12.32.

**SOLUTION**

Following the rules laid out above, we find that the denominator is

$$[H^+]^3 + K_1[H^+]^2 + K_1K_2[H^+] + K_1K_2K_3$$

The $\alpha_i$ ($i = 3$, 2, 1, and 0) values refer, respectively, to $H_3PO_4$, $H_2PO_4^-$, $HPO_4^-$, and $PO_4^{3-}$. Therefore

$$\alpha_0 = \frac{K_1K_2K_3}{[H^+]^3 + K_1[H^+]^2 + K_1K_2[H^+] + K_1K_2K_3}$$

Since pH 4.0 is about midway between the $pK_a$ values 2.23 and 7.21, we expect $H_2PO_4^-$ to be by far the major species present. Thus the only term in the denominator that needs to be calculated is $K_1[H^+]^2$. The result is

$$\alpha_0 = 10^{-11.53}, \qquad \text{pH 4.0}$$

Since the $\alpha$-values are proportional to the fraction of each species,

$$[PO_4^{3-}] = C_L \cdot \alpha_0 = 0.1 \times 2.95 \times 10^{-12} = 2.95 \times 10^{-13} \text{ M}$$

---

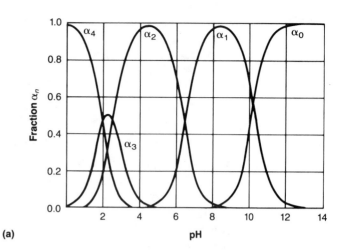

**(a)**

**Figure 2B.2**    (a) The values of $\alpha_n$ as a function of pH for $H_n\mathrm{edta}^{n-4}$. The curves were calculated as described in the text. The values of the equilibrium constants were determined in 0.1-M ionic strength solution. As a result, the calculated values are for the same conditions. (b) Plot of the logarithm of $\alpha_n$ as a function of pH. Each type of plot emphasizes a different portion of the curves. The linear plot is vertically spread out most for the major species, whereas the logarithmic plot is spread out most for the minor species.

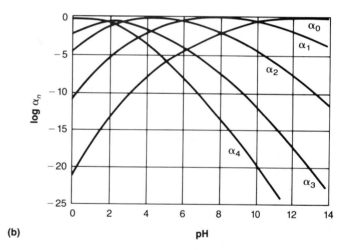

**(b)**

## 2B.3  Determining the Effective Formation Constant

The most common application of edta is to quantify a metal by titration (Chapters 9 and 10). However, only the total quantity of the chelating agent present is measured experimentally. This is signified below for edta by the symbol edta(all forms). For a general metal M, the equilibrium that can be applied is

$$\mathrm{edta(all\ forms)} + M \rightleftharpoons \mathrm{Medta} \quad ; \quad K_{\mathit{eff}} = \frac{[\mathrm{Medta}]}{[\mathrm{edta(all\ forms)}][M]}$$

In other words, we use an *effective* equilibrium constant $K_{\mathit{eff}}$ that applies to experimentally measured values. For vanadium(V), the stoichiometric equation that represents the system is

$$VO_2^+ + edta(all\ forms) \rightleftharpoons VO_2edta^{3-}; \qquad K_{eff}(V^V) = \frac{[VO_2edta^{3-}]}{[edta(all\ forms)][VO_2^+]}$$
(2B-9)

On the other hand, published equilibrium constants involve a specific species of the ligand. For the example of vanadium(V), Table 2B.1 lists

$$VO_2^+ + edta^{4-} \rightleftharpoons VO_2edta^{3-}; \qquad \log K(V^V) = 15.55 \qquad \textbf{(2B-10)}$$

The problem at hand is to relate these two equilibrium constants $K_{eff}(V^V)$ and $K(V^V)$.

If the pH of the solution is high enough, then the formation constant $K(V^V)$ applies since all the edta would be $edta^{4-}$. However, at lower pH, any vanadium present in the solution will not bind as strongly as it would at higher pH values. The protons compete against the metal for the edta ligand. As you saw from the calculation in the previous section, at pH 7.4 only about one part in 800 of the edta added is in the form $edta^{4-}$, the chemical form that appears in Eq. 2B-10. The great majority of edta is in other forms: 95% of the edta ligand will be in the form $Hedta^{3-}$ and 5% will be $H_2edta^{2-}$.

But note that the only difference in the mass-action expressions for Eqs. 2B-9 and 2B-10 is the quantity [edta(all forms)] and $[edta^{4-}]$. Since

$$[edta^{4-}] = [edta(all\ forms)] \cdot \alpha_0$$

then from substitution, the following important relation is found:

$$K_{eff}(V^V) = K(V^V) \cdot \alpha_0 \qquad \textbf{(2B-11)}$$

A more convenient form of this equation is

$$\log K_{eff}(V^V) = \log K(V^V) + \log \alpha_0 \qquad \textbf{(2B-12)}$$

---

**EXAMPLE**

Taking account of the competition of vanadium with $H^+$ for the ligand edta, what is the concentration of free $VO_2^+$ in a solution that has a total of 5 mM $VO_2^+$ and 10 mM total edta (a large excess) in it?

**SOLUTION**

Using Eq. 2B-12, the information in Table 2B.1, the value of $\alpha_0$ at pH 7.4 from the first example in the previous section, and the value of $\log K(V^V)$, we get

$$\log K_{eff}(V^V) = 15.55 + \log(1.30 \times 10^{-3})$$
$$= 15.55 + (-2.89) = 12.66$$
$$K_{eff}(V^V) = 4.57 \times 10^{12}$$

where

$$K_{eff} = \frac{[VO_2edta^{3-}]}{[edta(all\ forms)][VO_2^+]}$$

Given the very large formation constant—which means that essentially all the vanadium is coordinated as $VO_2edta^{3-}$—one of many ways to solve this equation is to let $[VO_2^+] = X$. Then the effective equilibrium constant expression can be written as if 5 mM vanadium complex dissociates in a solution containing 5 mM residual edta(all forms).

$$4.57 \times 10^{12} = K_{eff} = \frac{[\,0.005 - X\,]}{[0.005 + X][X]}$$

Since $X$ will probably be quite small relative to 0.005 M, this can be approximated quite well by

$$4.57 \times 10^{12} = [0.005]/([0.005][X]) = 1/[X]$$

and

$$[X] = [VO_2^+] = 2.2 \times 10^{-13} \text{ M}$$

With $[X]$ this small, the approximation is quite good.

---

## 2B.4  Correction for Hydrolysis

In a similar manner, if any of the vanadate is in a form other than $VO_2^+$, the effective binding constant can be modified to account for simultaneous equilibria involving protons by using $\alpha$-values. Another form of vanadate arises due to hydrolysis. The reaction is

$$H_2O + VO_2^+ \rightleftharpoons VO_3^- + 2\,H^+; \qquad \log K_h = -7.0 \qquad \textbf{(2B-13)}$$

which is the predominent reaction of vanadium hydrolysis at pH 7.4. Evaluation of the mass-action expression leads to the ratio

$$[VO_2^+]/[VO_3^-] = 10^{-7.8} = 1.58 \times 10^{-8}; \qquad \text{at pH 7.4}$$

This expression says that almost all the vanadium exists as $VO_3^-$ at pH 7.4. The equation expresses a competition for the vanadate, $VO_2^+$, between the solvent, water/ hydroxide, and the edta ligand. This competition will result in an increase in the total amount of vanadate in its various chemical forms that is *not* bound to edta.

To compensate $K_{eff}(V^V)$ and account for this competition, we proceed in a manner similar to accounting for protonation of the ligand. We use a term $M_h$ which is the fraction of vanadium that is $VO_2^+$.

$$M_h = \frac{[VO_2^+]}{[VO_2^+] + [VO_3^-]} \approx \frac{[VO_2^+]}{[VO_3^-]} \qquad \textbf{(2B-14)}$$

The approximation in Eq. 2B-14 is excellent in this pH range because of the relatively small amount of $VO_2^+$. Therefore,

$$[VO_2^+] \approx M_h[VO_3^-] \approx M_h[V^V]$$

where $[V^V]$ represents the concentration of V(V) in all forms.

By substitution into the equation for $K_{eff}$, we have

$$M_h \cdot K_{eff} = \frac{[VO_2\text{edta}^{3-}]}{[\text{edta(all forms)}][V^V]}$$

Equation 2B-12 is then modified to include the $M_h$ term, which accounts for the hydrolysis.

$$\log K_{eff}(V^V) = \log K(V^V) + \log \alpha_0 + \log M_h \qquad \textbf{(2B-15)}$$

At pH 7.4, the value of $\log M_h$ is $-7.8$. Therefore

$$\log K_{eff} = 15.55 - 2.89 - 7.8 = 4.9$$

---

**EXAMPLE**

What is the concentration of uncomplexed vanadium(V) in a pH 7.4 solution of edta containing a total of 5 mM vanadate and 10 mM total edta in it?

**SOLUTION**

With compensation for hydrolysis,

$$\log K_{eff} = 4.9 \qquad \text{therefore} \qquad K_{eff} = 8 \times 10^4$$

And, as in the previous example,

$$8 \times 10^4 = K_{eff} = \frac{[0.005 - X]}{[0.005 + X][X]}$$

However, in the equation,

$$[X] = V_{total}^V = [VO_2^+] + [VO_3^+]$$

Again, since $[X]$ is expected to be small, we try the approximation

$$8 \times 10^4 = 1/[X]$$
$$[X] = V_{tot}^V = 1.3 \times 10^{-5} \text{ M}$$

Solving with an approximation is still satisfactory.

Note: We calculate that the vanadium is almost eight orders of magnitude less complexed after we made the correction for hydrolysis. Errors due to differences in ionic strength become inconsequential by comparison. For most metals, this correction *must* be made to have any hope of approximating the chemistry.

---

The above example shows how the concept of effective equilibrium constants can be used to simplify complicated equilibrium calculations by separately treating the various equilibria that simultaneously affect the components of the solution. You have seen how the formation constant for $VO_2^+$ with edta$^{4-}$ contains only a small part of the information needed to reproduce the correct concentrations of species in this relatively simple chemical system composed only of vanadate, water, and edta as the reacting chemicals (ignoring spectator ions). Similar mathematical treatments can be carried out for reactions such as for zinc with ammonia, which is treated in Chapter 10.

However, such calculations as these do not reflect the real solution properties unless at least all the *major* reactions are included. The possible errors are huge. For the solution considered in the examples above—5 mM total vanadate and 10 mM total edta—with no corrections for "side reactions," the uncomplexed vanadium concentration is calculated to be $2.8 \times 10^{-16}$ M. Considering the acid–base reactions, this concentration rose to $2.2 \times 10^{-13}$ M. And finally, including the hydrolysis of the metal ions, the free vanadate rises to $1.3 \times 10^{-5}$ M.

## Postscript about Titrations

A requirement of titrations is that the reaction of the titrant with the analyte must go to completion. (This requirement is discussed in Chapter 9 in more detail.) The vanadium–edta reaction at pH 7.4 might seem, at a first glance at the formation constant for $VO_2edta^{3-}$, to be a fine one on which to base a titration of vanadium. But it will not be as good as it might initially appear; the equilibrium of the complexation reaction does not lie as far toward completion as might be desired. Even with a large excess of edta titrant, the uncomplexed vanadium(V) is relatively high.

A reasonable question is, Why not just change the pH to a point where the reaction is more complete? In the case of vanadium, pH 7.4 is near to the optimum pH: near the pH with the largest $K_{eff}$. At lower pH values, the ligand protonation becomes more important, and less vanadium binds. (In Eq. 2B-15, the value of $\alpha_0$ becomes smaller.) At higher pH values the hydrolysis competes even more effectively, and, again, less vanadium binds. (In Eq. 2B-15, $M_h$ becomes more negative.) An optimum pH exists with practically all the metals listed in Table 2B.2. The only ones that do not undergo significant hydrolysis (pH dependent!) are $Ca^{++}$ and $Mg^{++}$. At low pH, few metal ions can compete effectively with $H^+$ for the ligand.

## SUPPLEMENT 2C

# An Alternate Method to Balance Redox Equations

One method of balancing redox equations was described in Section 2.8. An alternative is presented here. This method requires that oxidation numbers be assigned to the atoms in the compound to be oxidized or reduced. The assignments are carried out according to the following rules.

A. *Elements are always zero.*

B. *Hydrogen is $+1$ unless it is in a metal hydride in which case it is $-1$.*

C. *Oxygen is $-2$ unless it is in a peroxide, $O_2^=$, or superoxide, $O_2^-$.*

D. *Alkali metals (Na, K, and so forth) in compounds are assigned $+1$; alkaline earths (Be, Ca, and so forth) are assigned $+2$.*

E. *Halogens (F, Cl, Br, and I) are $-1$ unless combined with oxygen or another halogen. When combined with oxygen, the oxidation number is assigned using oxygen as $-2$. When combined with another halogen, the element closest to the top of the periodic table is assigned an oxidation number of $-1$. (Example: BrCl has Br $+1$ and Cl $-1$.)*

> F.  *All other values are assigned to make the charge on the compound*
>     *add to zero or, in the case of a polyatomic ion, equal the total ionic*
>     *charge.*

There are six steps to follow when balancing a redox equation.

> 1.  *Assign oxidation states.*
> 2.  *Write half-reactions.*
> 3.  *Balance half-reactions with regard to $e^-$ and the atoms that change*
>     *oxidation state.*
> 4.  *Balance the charge on each side of the half-reactions by adding $H^+$ in*
>     *acid or $OH^-$ in basic solution.*
> 5.  *Balance the half-reactions with regard to hydrogen and water.*
> 6.  *Multiply the half-reaction equations by the appropriate factors to ob-*
>     *tain equal numbers of $e^-$ in each and add.*

---

**EXAMPLE**

Balance the following redox equation of the oxidation of oxalate by bromate in acid solution.

$$H_2C_2O_4(aq) + KBrO_3 \rightarrow CO_2 + KBr + H_2O \qquad \text{(unbalanced)}$$

**SOLUTION**

Step 1.  Assign oxidation states.

$$
\begin{array}{ccccccc}
 & & & \overset{\displaystyle -6}{\overbrace{\phantom{KBrO_3 \qquad CO_2}}} & & & \\
H_2C_2O_4 & + & KBrO_3 & \rightarrow & CO_2 & + \; KBr & + \; H_2O \\
\underset{\underbrace{\phantom{+1 +3 -2 \qquad +4 -2}}}{+1\ +3\ -2} & & +1\ +5\ -2 & & +4\ -2 & +1\ +1 & +1\ -2 \\
 & & & +1 & & &
\end{array}
$$

As shown, the reaction involves oxidation of carbon from $+3$ to $+4$ with concomitant reduction of bromine from $+5$ to $-1$.

Step 2.  Write half-reactions.

$$H_2C_2O_4 \rightarrow CO_2$$
$$KBrO_3 \rightarrow KBr$$

Step 3.  Balance with respect to $e^-$ and the atom changing oxidation state.

$$H_2C_2O_4 \rightarrow 2\, CO_2 + 2\, e^-$$
$$6\, e^- + KBrO_3 \rightarrow KBr$$

Step 4.  Balance charge with $H^+$ since this is in acid.

$$H_2C_2O_4 \rightarrow 2\, CO_2 + 2\, H^+ + 2\, e^-$$
$$6\, H^+ + 6\, e^- + KBrO_3 \rightarrow KBr$$

Step 5.  Balance hydrogen and oxygen with water.

$$H_2C_2O_4 \rightarrow 2\,CO_2 + 2\,H^+ + 2\,e^-$$
$$6\,H^+ + 6\,e^- + KBrO_3 \rightarrow KBr + 3\,H_2O$$

Step 6.  Multiply to obtain equal $e^-$ and add.

$$3 \times [H_2C_2O_4 \rightarrow 2\,CO_2 + 2\,H^+ + 2\,e^-]$$

$$+ \qquad 6\,H^+ + 6\,e^- + KBrO_3 \rightarrow KBr + 3\,H_2O$$

$$\overline{3\,H_2C_2O_4 + \cancel{6\,H^+} + \cancel{6\,e^-} + KBrO_3 \rightarrow 6\,CO_2 + \cancel{6\,H^+} + \cancel{6\,e^-} + KBr + 3\,H_2O}$$

After canceling the 6 $H^+$ and 6 $e^-$ occurring on both sides, the final equation becomes

$$3\,H_2C_2O_4 + KBrO_3 \rightarrow 6\,CO_2 + KBr + 3\,H_2O \text{ (balanced)}$$

---

## SUPPLEMENT 2D

# Energy – Voltage Relationships

Associated with a chemical reaction is its Gibbs free energy $\Delta G$. If the species involved are in their standard states, the associated free energy is denoted by $\Delta G°$. For an electrochemical reaction, which involves transfer of charge across a potential, the electrochemical potential is related to the molar free energy by

$$\Delta G = -n\mathscr{F}E \qquad\qquad \textbf{(2D-1)}$$

With $n$ the number of moles of electrons, $n\mathscr{F}$ denotes a quantity of electrical charge since $\mathscr{F}$ represents 96,486 coulomb $mol^{-1}$. If this negative charge is transferred across a voltage from 0 V to $+E$ V, the motion is spontaneous, and the magnitude of the work equals $|\Delta G|$. Spontaneous reactions have $\Delta G° < 0$. The minus sign in Eq. 2D-1 keeps the convention consistent.

If the species are in their standard states, Eq. 2D-1 becomes

$$\Delta G° = -n\mathscr{F}E° \qquad\qquad \textbf{(2D-2a)}$$

or

$$E° = -\Delta G°/n\mathscr{F} \qquad\qquad \textbf{(2D-2b)}$$

The manipulations of the voltages associated with reactions follows in a straightforward way from the relationship between $\Delta G$ and $E$. Recall the two rules:

a. If the direction in which a half-reaction is written is reversed, the sign of its half-cell potential is reversed.

b. If the half-reaction is multiplied by a positive number, its voltage remains *un*changed.

### Reversal of a Reaction

As an example, the reduction of Fe(III) to Fe(II) under standard conditions is used.

$$Fe^{+++} + e^- \rightleftharpoons Fe^{++} \qquad E° = -\Delta G°/\mathscr{F} \qquad \text{(2D-3)}$$

Since $n = 1$, the voltage is related to the free energy as shown. If this reaction is reversed,

$$Fe^{++} \rightleftharpoons Fe^{+++} + e^-$$

then the free energy changes sign and becomes $-\Delta G°$. From Eq. 2D-2, the sign of $E°$ changes as well.

$$Fe^{++} \rightleftharpoons Fe^{+++} + e^- \qquad -E° = -(-\Delta G°)/\mathscr{F}$$

Thus, when we reverse a reaction, the sign of the potential is reversed as well.

### Multiplication of a Reaction

If the Fe(III)/Fe(II) half-reaction is multiplied by 2, it becomes

$$2\,Fe^{+++} + 2\,e^- \rightleftharpoons 2\,Fe^{++} \qquad \text{(2D-4)}$$

When two times the amount of reactants and products react, the free energy doubles — algebraically, $\Delta G°_{2Fe} = 2\,\Delta G°_{Fe}$. However, as reaction 2D-4 shows, $n = 2$. Therefore

$$E° = -2\,\Delta G°/2\mathscr{F} \qquad \text{for Eq. 2D-4}$$

or

$$E° = -\Delta G°/\mathscr{F}$$

which is the same as for the original reaction, Eq. 2D-3. The potential $E°$ does not change simply by multiplying a reaction. It depends on the identities of the reactants and products only. Thus, when we multiply a reaction by a constant value, the potential remains the same.

Both rules hold even when the reactions are not done under standard conditions; the results here are general.

### Supplement Problems

S2.1   Given that

$$H_2O + VO_2^+ \rightleftharpoons VO_3^- + 2\,H^+; \qquad \log K_h = -7.0$$

Show that at pH 7.4

$$[VO_3^-]/[VO_2^+] = 10^{7.8} = 6.3 \times 10^7$$

S2.2   An attempt was made to used edta as a titrant for vanadate for a complexometric titration at pH 7.4. At the end point, the number of moles of edta added should equal the number of moles of vanadate since a 1:1 complex forms. If the analyte vanadium is 1 mM, calculate what fraction of the vanadium is left uncomplexed at the expected end point using each of the following assumptions.

 a. Assume that no protonation or hydrolysis occurs.
 b. Assume that protonation of the edta occurs, but no hydrolysis occurs.
 c. Assume that both protonation of the edta and hydrolysis of the vanadate occurs.

S2.3 At pH 6, what is the value of $\alpha_2$ for edta? Do the calculation to three decimal places. What are the units of $\alpha_2$?

S2.4 Since pH 7.4 does not seem to be too promising for a titration of vanadate with edta, you want to try higher and lower pH values. To see if it is worth trying under these conditions, answer the following questions for pH 2 and pH 12.

 a. What fraction of edta is in the form edta$^{4-}$ at the two pH values?
 b. What is the value of the ratio $[VO_3^-]/[VO_2^+]$ at the two pH values?
 c. What is the value of $K_{eff}$ at the two pH values, taking account of both protonation of the edta and hydrolysis of the vanadium?
 d. Will the conditions be better at either pH 2 or pH 12 than at pH 7.4?

# Part II
# FOUNDATIONS OF ANALYSIS

# 3

# Elementary DC Electronics

## 3.1 Introduction

Contemporary chemical analysis would not exist without the capabilities made possible by the intelligent use of electronic instruments. These instruments may be divided into three parts with different functions, although the divisions are not sharp. In the first part of the instrument are devices that can convert into electrical signals the changes in the intensity of light, temperature, chemical concentration, pressure, or volume. These are the **transducers.** They comprise all or part of instrumental **detectors.** In most cases, they are now far more sensitive and respond more uniformly to the changes than our own senses.

Another part of the instrument, the signal processor, converts the information — the signal — obtained from the detectors into a different form for output. This information is processed as changes in electrical current or voltage. The transformation may be to an output we find easy to use in our analysis, for example, a digital display. However, often the capabilities of signal processors include discriminating between information of interest and superfluous signals.

The third division includes the devices used mostly to manipulate, convert, and store the data. These are the computers. They are not essential like the transducers. Also, they would be useless unless the information from the detector were converted by the signal processor electronics into a form that they could manipulate. However, computers are indispensable for the high productivity of contemporary chemical analysis. Quite simply, without the computer's ability to manipulate data and automate the operation of the instruments, the price of a chemical analysis would be far higher than it is.

In this text, we shall not dwell on the capabilities of computers. The signal processors are described only in general ways and mostly in the Supplements of the chapters. However, the transducers and detectors are discussed in some detail. Their limitations and capabilities determine the requirements, advantages, and drawbacks of the analytical methods in which they are used. After all, if the chemistry of the sample causes the detector to respond to the wrong chemical species, even sophisticated instruments and the computers, which might be attached, will not make the data usable for accurate and precise chemical analyses. However, computers may make it *easier* to manipulate the data from large numbers of experiments. And that is a great help in making corrections and figuring out why an analysis may not be working. Computer-operated instruments may also be somewhat more precise in repetitive analyses. But the sampling, sample preparation, and other chemistry-related steps must be well done first; the motto of the computer also applies to computers in chemical analysis: "garbage in, garbage out."

To understand the operation of the detectors and the way they transform the chemical information, a certain amount of knowledge of electronics is desirable. A short introduction to this topic is presented in this chapter. The content is limited here to the behavior of circuits and devices in which the electrical currents always flow in the same direction. This type of current is called direct current or DC. Included in this classification are currents that vary in time but do not change in their direction of flow. By convention, current is defined to be flowing from positive to negative potentials.

## 3.2 DC Current, Voltage, and Resistance

No matter how complicated any electronic device appears, it is set up to manipulate the flows of electric currents. The magnitude of the current flow depends on the electrical potential (the voltage) and the resistance to its flow. The manner in which these three quantities are related is given, algebraically, by the equation for Ohm's law:

$$V = IR \quad \text{or} \quad I = \frac{V}{R} \tag{3-1}$$

where

$V$ is the potential in volts,
$R$ is the resistance in ohms,
$I$ is the current in amperes.

The equation on the left says, "With a current passing at the rate $I$ through a resistance $R$, a voltage drop of magnitude $IR$ appears across the resistor." (This

**Figure 3.1**   Schematic diagram of a circuit for verifying Ohm's law. The symbols of the schematic are defined in Supplement 3A. The current flows from − to + inside the battery and from + to − through the circuit and across the resistor. The wires are assumed to be perfectly conducting. In the circuit, $V_{\text{measured}} = V_{\text{voltage source}}$

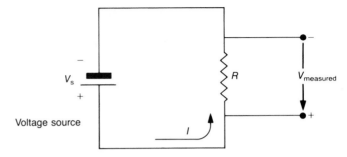

voltage is often called the *IR* drop.) Another way of reading the equation is, "The voltage drop developed across a resistance *R* exactly equals the voltage of the source." A circuit to verify this relationship is illustrated in Figure 3.1.

Figures such as these are called **schematics,** or **schematic diagrams,** of the circuit. If you are not familiar with the electrical circuit symbols in the figure, a list is provided in Supplement 3A.

In the following discussion, if you have trouble understanding the behavior of the various parts of the circuits, you may find it easier to think about the flow of a fluid instead of electrical current. The behaviors of the electric current and the fluid flow are analogous. The voltage is the same as a pressure. Resistance has a similar meaning in both cases; for example, there is more resistance to flow through a small pipe than through a large one. (Compare sucking a thick milkshake through a large and a thin straw.) The amount of electric charge is analogous to the volume of fluid. Thus the flow of electric current (charges flowing per unit time) is similar to a fluid flow rate (volume of fluid flow per unit time).

## 3.3 Power

You are familiar with the idea that if an electrical charge is moved across a voltage, then energy is involved. The energy units are electron volts ($1 \text{ eV} = 1.602 \times 10^{19} \text{ J}$).

An electric current flowing through a resistor loses its energy as heat. The usual way to express this energy loss is as a *rate of production* of energy to the environment: the **power.** This is written algebraically as

$$\text{power} = P = VI \tag{3-2}$$

For the resistor, when

*V* is the voltage across the resistor in volts (V), and
*I* is the current flowing through it in amperes (amps or A)
    [1 ampere = $6.241 \times 10^{18}$ electrons/second],

then *P* is in units of watts (W) [1 watt-second = 1 joule].

Equation 3-2 defines the power produced as we see it from the outside of the resistor: the rate of **power dissipation.** This term means that the energy is lost from the circuit to the surroundings (for instance, air or fingers). We could feel the resistor heat up.

Alternate equations to express the power dissipation can be written by using Eqs. 3-1 and 3-2 together.

$$P = I^2 R = V^2 / R \tag{3-3}$$

## 3.4 Kirchhoff's Laws

Two more rules are needed to understand the functions of simple DC circuits and how the circuits are constructed to manipulate the voltages and currents used in analytical instruments. Actually, the rules are corollaries of two fundamental physical laws: conservation of charge and conservation of energy. The rules for DC circuits are called Kirchhoff's laws. They are

1. The algebraic sum of the currents at any junction in a circuit is zero. In mathematical shorthand, the first law is the same as $\sum_n I_n = 0$ (conservation of charge). See Figure 3.2 for an illustration.
2. The algebraic sum of the potential differences around any complete loop of a circuit is zero. In mathematical shorthand, the second law is written $\sum_n V_n = 0$.

We have already seen an example of the second law in the circuit of Figure 3.1. If you trace the circuit from any point, going completely around the loop, and add the voltages that you pass, the sum must be zero. You might want to test this idea by tracing around some other circuit schematics in this chapter.

### Resistors in Series: Voltage Dividers

An example of the use of the first law follows. We consider two circuits consisting of a battery and two resistors. As shown in Figures 3.3 and 3.4 (p.114), there are two ways to connect the resistors together in the circuit. These are **in series** (end-to-end) and **in parallel** (front ends connected together). Examples and rules for adding the resistances of resistors in series and in parallel are discussed next.

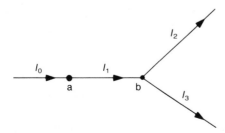

**Figure 3.2**    Illustration of Kirchhoff's first law: The algebraic sum of the currents at any junction in a circuit is zero. Illustrated are two junctions, a and b.

At junction a, $I_0 = -I_1$ or $I_0 + I_1 = 0$.
At junction b, $I_1 = -I_2 - I_3$ or $I_1 + I_2 + I_3 = 0$.

The arrowheads indicate the direction of positive current flow in the wires.

**Figure 3.3**   The application of Ohm's law to two resistors in series. The current is the same across both resistors. The total voltage across the two resistors equals the sum of the voltages across each. The wires are assumed to be without resistance. The current flows in the direction of the arrow labeled *I*. Equations describing the relationships between the voltages, resistances, and current appear in the text.

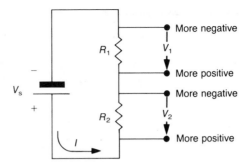

Placing resistors in series means that they are connected end-to-end, as shown in the schematic of Figure 3.3. This is a circuit called a **voltage divider**. The voltage divider does what its name suggests: It provides a potential that is some fraction (less than 1) of a reference voltage. Voltage dividers are found in almost every useful electronic circuit. We can discover the circuit's behavior by doing a straightforward calculation based on Ohm's law:

$$V_1 = IR_1$$
$$V_2 = IR_2$$
$$V = IR_{\text{total}}$$

The sum of the first two equalities is

$$V_1 + V_2 = I(R_1 + R_2) = V$$

The sum $V_1 + V_2$ must equal the total voltage $V$, because the wires are assumed to be without resistance. As a result,

$$R_{\text{total}}(\text{series}) = R_1 + R_2$$

And in general, for any number of resistors in series,

$$R_{\text{total}}(\text{series}) = R_1 + R_2 + \cdots + R_n$$

A further, useful equation is derived from the same set of equations. From simple proportions,

$$\frac{V_1}{V} = \frac{IR_1}{IR_{\text{tot}}} \qquad \text{or} \qquad \frac{V_1}{V} = \frac{R_1}{R_{\text{tot}}} \tag{3-4}$$

The same kind of relationship is true for $V_2$. We can substitute for $R_{\text{tot}}$ ($= R_1 + R_2$) in Eq. 3-4 and obtain the **voltage divider formula,**

$$\frac{V_1}{V} = \frac{R_1}{R_1 + R_2} \qquad \text{or} \qquad V_1 = V\left(\frac{R_1}{R_1 + R_2}\right) \tag{3-5}$$

This says that the fraction of the total voltage that is measured across a resistor in series is proportional to that resistor's fraction of the total resistance.

### Resistors in Parallel

Placing resistors in parallel means that their ends are connected together such that they all have the same voltage across them. This situation is illustrated in the schematic in Figure 3.4. We shall derive the general equation for the total resistance of this circuit. To do so, we need to analyze the currents flowing in the circuit. From Kirchhoff's first law, we have

$$I = I_1 + I_2$$

From Ohm's law $I = V/R_{total}$, $I_1 = V/R_1$, and $I_2 = V/R_2$. Thus

$$\frac{V}{R_{tot}} = \frac{V}{R_1} + \frac{V}{R_2}$$

or

$$\frac{1}{R_{total}} = \frac{1}{R_1} + \frac{1}{R_2}$$

If we generalize this equation, we derive the general equation for any number of resistors in parallel,

$$1/R_{total}(\text{parallel}) = 1/R_1 + 1/R_2 + \cdots + 1/R_n$$

The uses of these circuits in simple ammeters, ohmmeters, and voltmeters are outlined more fully in Supplement 3B.

### Wheatstone Bridge Circuit

Constructing electronic circuits that can measure changes in voltage of 1 part in 10,000 is not easy. And the difficulties increase rapidly when we want to measure even smaller relative changes. A way to overcome this problem is to measure *differences* in voltages. For instance, it is easier to determine precisely whether a voltage is 0.0000, 0.0001, or 0.0002 than to determine whether it is 2.0000, 2.0001, or 2.0002. In the second case, the measurement must be made with a stability of at least a few parts in $10^5$ so that 1 part in 20,000 can be measured. In the first case, the stability need only be in the range of a few parts in ten.

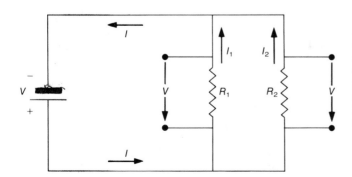

**Figure 3.4**    The application of Ohm's law to two resistors in parallel. The voltage across both of the resistors is the same. The current through each resistor differs and depends on the inverse of the individual resistances. The wires are assumed to be without resistance. The current flows in the direction of the arrow labeled *I*. Equations describing the relationships between the voltage, resistances, and currents appear in the text.

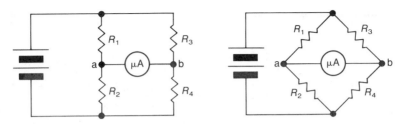

**Figure 3.5**    Illustration of a Wheatstone bridge circuit. Two different schematic forms of the same circuit are shown. Note that $R_1$ and $R_2$ are in series, as are $R_3$ and $R_4$. However, the *sets* are in parallel connected to the battery.

Let $I_{12}$ be the current flowing through resistors $R_1$ and $R_2$, and similarly for $I_{34}$ and $R_3$ and $R_4$. Then, from Ohm's law

$$V = I_{12}(R_1 + R_2) = I_{34}(R_3 + R_4)$$

There are a number of ways to express the condition for obtaining a null. Among these are

$$I_{a,b} = 0 \quad \text{or} \quad V_a = V_b \quad \text{or} \quad V_2 = V_4$$

As shown in the main text, the condition for balance is

$$\frac{R_3}{R_4} = \frac{R_1}{R_2}$$

A **Wheatstone bridge** circuit allows us to measure precisely such differences in either voltage or resistance. The schematic for the circuit is shown in Figure 3.5. The schematic diagram is drawn in two different ways. Both contain exactly the same components *connected* in exactly the same way. However, it is quite common to find the same circuit with differences in the *arrangement* of the elements of its schematic form. Whenever a **bridge circuit** is mentioned, it will be similar in organization to the schematic of Figure 3.5.

A bridge circuit will be the central part of the detectors for many gas chromatographs (Chapter 15). Instruments for thermal analysis also utilize a bridge circuit as do some spectrometers (Chapter 17) and electrical conductivity meters (Chapter 12). In addition to its usefulness in a number of detectors, the circuit illustrates the principles of combining resistors in series and parallel at the same time. The circuit can be thought of as two voltage dividers connected in parallel. That is the way the circuit will be analyzed.

One way to use a bridge circuit is to compare potentials by measuring a *null* precisely. This null point is reached when there is no current (and hence no potential) across a wire connected between points such as a and b in Figure 3.5. The problem that will be solved here is to find the values of $R_1$, $R_2$, $R_3$, and $R_4$ that will cause a null current. This lack of current is detected by a microammeter.

**EXAMPLE**
Find the resistance values for a null current in the bridge circuit.

**SOLUTION**
A null condition exists only if there is no potential difference between points a and b. In that case, the voltage drop across resistors $R_2$ and $R_4$ must be the same, since both resistors are at the

same potential at points a and b ($V_a = V_b$). The voltages across resistors 2 and 4 are also equal ($V_2 = V_4$). According to Eq. 3-5 for voltage dividers,

$$\left(\frac{R_2}{R_1 + R_2}\right) \cdot V = V_a = V_b = \left(\frac{R_4}{R_4 + R_3}\right) \cdot V$$

The value $V$ cancels from the right and left sides. So, inverting the values in brackets, we have

$$\frac{R_1 + R_2}{R_2} = \frac{R_3 + R_4}{R_4}$$

With straightforward algebraic manipulation, this becomes

$$\frac{R_1}{R_2} = \frac{R_3}{R_4} \tag{3-6}$$

This is the condition for a null.

---

How is this formula used in practice in the bridge circuit? In general, three of the resistors have known values. Of these three, one is adjustable. The fourth is variable and is part of the transducer of the instrument. The adjustable resistor is set so that the four resistors have relative values, which follow the relation shown in Eq. 3-6. A null is observed, and the bridge is said to "be in balance." No current flows between points a and b. The fourth resistor, then, controls the current measured with the microammeter.

For a number of contemporary instruments, the response to the experiment will be a voltage proportional to the current flowing between points a and b. A general name for such an output is the **error signal.** In general, the bridge is set to the null (or near null). Then the output error signal can be used directly in the analysis. The result is best when the error signal is linearly proportional to the chemical concentration of an analyte.

## 3.5 Amplifiers and Amplification

In addition to being a wave, light can be considered to be composed of particles, photons with energies $E = h\nu$. It is possible to measure experimentally the arrival of a single photon of visible light at a transducer. In the visible light region, a single photon has an energy in the range $4 \times 10^{-19}$ joules. In contrast, a pocket calculator with a liquid crystal display uses about 1 mW of power. The energy contained in the photon could power the calculator for $4 \times 10^{-16}$ s. A display for this short a time would hardly allow an electronic device to convey the occurrence to us. Thus, to register the photon's arrival so that we can know about the event, the signal must be amplified. For a photon counter, the small, short-term current from the transducer must be amplified in order to change a digital display, flash a light, or move a meter's needle. All these require far more power for a longer time than is obtainable from the single photon.

**Figure 3.6** Schematic figure for amplifiers. For both symbols, the line intersecting at an apex represents the output connection. A line connecting on the vertical side opposite represents an input. The + sign on the input indicates the noninverting input; a positive change in input potential will produce an amplified positive change at the output. The − sign indicates an inverting input; a positive change in input potential will produce an amplified negative change at the output. The number +6 indicates amplification by a factor of 6.

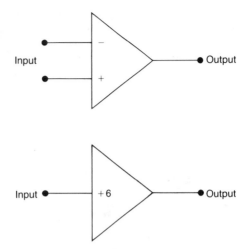

The amplification is done by an amplifier. When the signal is applied at its input, the modified signal appears at its output. One of the most important requirements of an amplifier is that the output precisely and accurately follows the input. Ideally, if the input signal changes from $I_0$ to $I_1$ and then to $2.00000\,I_1$, the output should change from $O_0$ to $O_1$ to $2.00000\,O_1$. If it does so, we say that the amplifier possesses high **linearity** in that range. The linearity of response of an instrument is directly dependent on the linearity of the amplifiers in it. If the response of the amplifier is not linear, it is said to be nonlinear in its response. Real amplifiers approach linearity only over a finite range of the signal. We seek to make this range as wide as possible and the response as linear as possible.

The amount of amplification obtained from an amplifier can be expressed quantitively by the ratio of the power produced at the output and the power applied at the input. This ratio is called the gain:

$$\text{gain} = \frac{P_{\text{output}}}{P_{\text{input}}} \tag{3-7}$$

When the details of an amplifier circuit are not shown in a schematic diagram, one of two simple symbols for amplifiers is used. These are illustrated in Figure 3.6. Both are triangles, but with either two or three lines connected. The line intersecting at an apex is the symbol for the output connection. On the vertical side opposite, there can be one or two lines representing inputs. If only one input is shown, then the transformation of the signal should be noted nearby. As shown in the figure, the +6 represents the operation of multiplication of the signal by six. Labeling the transformation is not always done, and when the information is not provided, inferring the operation of the circuit from the diagram can be extremely difficult.

When two lines connect to the vertical side, they are usually labeled + and −. This format is used when somewhat more information about the circuit is included in the schematic, such as the resistors and capacitors connected to the amplifier. The signs indicate the **polarity** of the output relative to the input. The input marked with a +

**Table 3.1**  Some Signal Transformations That Can Be Accomplished with Amplifiers

Amplification (multiplication by a constant $> 1$)

Attenuation (multiplication by a constant $< 1$)

Addition

Subtraction

Isolation of one part of a circuit from another (often multiplying the signal by $+1$ or $-1$)

Integration

Differentiation

**Table 3.2**  Definitions of Capacitance and Inductance ($e$ in volts, $t$ in seconds, and $i$ in amperes)

The voltage across an inductor is

$$e_L = L \frac{di_L}{dt}$$

The inductance $L$ is in henrys.

The current across a capacitor is

$$i_C = C \frac{de_C}{dt}$$

The capacitance $C$ is in farads.

indicates it to be a **noninverting input;** as the input voltage applied to this input becomes greater, so does the output voltage. The input marked with a $-$ is the **inverting input;** as the voltage applied to the inverting input becomes more positive, the output goes more negative.

Amplifiers are versatile elements of electronic signal processors and are not limited to linear amplification. By suitable circuit design, a number of other operations on the signal are possible. Some of these transformations are listed in Table 3.1.

Amplifiers can be built that can subtract one signal from another. An important application of this is in a special type of amplifier called a **lock-in amplifier.** The basis for its operation is described in Supplement 18C.

## 3.6 Time-Dependent Responses

If the current that flows through an electrical circuit is not constant, then the time-dependent response of the system must be considered. Two different elements of a circuit cause time-dependent behavior. One is a capacitor, which can store charge. The other is an inductor, which resists a change in voltage. The time-dependent response of a circuit is characterized by its **capacitance** and **inductance** as well as resistance.

The definitions of capacitance and inductance are listed in Table 3.2. When the voltage and current change with time, lower-case letter abbreviations are used: A time-dependent voltage is $e$, and a time-dependent current $i$. Do not confuse the voltage $e$ with the natural logarithm base $e$.

Following are two examples of the response to a time-dependent voltage change. These are illustrated in Figures 3.7 and 3.8. The voltage change used in the examples is an extremely rapid jump in potential: a voltage step.

In each case, the voltage change of the output is significantly different in time from the input voltage step. Specifically, instead of duplicating the input voltage step, the voltage at the output rises in an exponential manner. In each case, the rise is characterizable by a **time constant.** The time constant quantifies how quickly or sluggishly

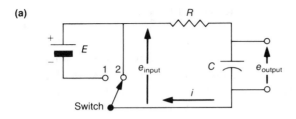

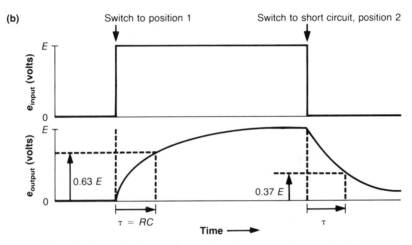

**Figure 3.7**   (a) Schematic diagram of a circuit containing a resistor and capacitor in series. (b) Illustration of the response of such a circuit to a step jump in the voltage applied to the circuit. The voltage across the capacitor is changed by applying a step change on the input. This is measured as $e_{input}$ and shown on the top axis. Later the potential is short circuited, resulting in a step drop to zero of $e_{input}$. As shown on the bottom axis, the resulting potential across the capacitor shows an exponential growth and decay with a characteristic time constant. The response of the circuit is described algebraically by

For the rise, $e_{output} = E(1 - e^{-t/RC})$; $t = 0$ at step up
For the fall, $e_{output} = E e^{-t/RC}$; $t = 0$ at step down

The time constant for a circuit of this type is

$$\tau = R\,C$$

$\tau$ has units of seconds if $C$ is in farads and $R$ is in ohms.
The time constant is defined as the time

For the rise, when $e_{output} = E(1 - e^{-1}) = E(1 - 0.37) = 0.63\,E$
For the fall, when $e_{output} = E e^{-1} = 0.37\,E$

the output responds to a change in input. The time constant is the time at which the output reaches $0.632 \,(= 1 - 1/e)$ of the total jump in the output. The time constant characterizes an exponential rise in a way similar to a half-time in kinetics. (See Supplement 1B.) The important similarities are outlined in Table 3.3.

Two other names for the time constant $\tau$, which expresses the rate of change of the output, are the instrument **response time** or the **rise time**. All three are equal but used in different contexts. Such time constants are important in characterizing the response of transducers alone as well as the electronics of instruments.

This response time depends on the magnitude of the capacitance, inductance, and resistance of each of the components of the circuit. For these simple circuits, the value can be calculated from the formulae shown in the figures. The calculation of the response time for more complex circuits as a function of the various components is beyond the level of this text. A beginning point for further study of this topic can be found in the texts in the Suggestions for Further Reading list for this chapter.

One use of this predictable time-dependent response is in a design for one type of **analog-to-digital** converter, abbreviated as A/D converter or ADC. This is one of the

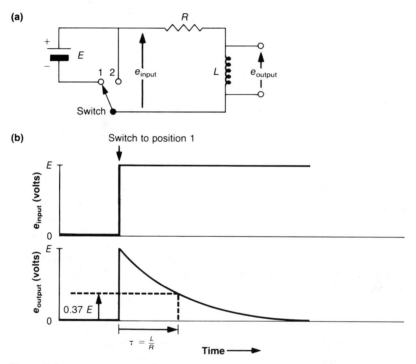

**Figure 3.8**    (a) Schematic diagram of a circuit containing a resistor and inductor in series. (b) Illustration of the response of such a circuit to a step jump in the voltage applied to the circuit. The voltage across the inductor is changed by applying a step change on the input. This is measured as $e_{input}$ and shown on the top axis. There is a resulting exponential decay with a characteristic time constant as shown on the bottom axis. The response of the circuit is described algebraically by

$$e_{output} = E\, e^{-(R/L)t}$$

The time constant is defined as the time when

$$e_{output} = 0.37\, E$$

The time constant is

$$\tau = L/R$$

$\tau$ has units of seconds when $L$ is in henrys and $R$ is in ohms.

**Table 3.3**   Analogies between Kinetics and Electronics

| Kinetics | Electronics |
|---|---|
| The half-time for $A = A_0 \exp(-kt)$ is when $A = (1/2)A_0$. | The time constant for $e_0 = E[1 - \exp(-t/RC)]$ is when $e_0 = (1 - 1/e)E$. |
| This occurs when $\exp(-kt) = 0.5$. $\exp(-kt_{1/2}) = 0.5$ $t_{1/2} = (\ln 2)/k = 0.69/k$ | This occurs when $\exp(-t/RC) = 1/e$. $\exp(-\tau/RC) = 0.368 = e^{-1}$ $t = (\ln e)RC = RC$ |

more important parts of contemporary computer-driven instruments. Its importance comes from the incompatible nature of a *digital* computer, which can deal only with on-or-off voltages, and the nature of most transducers. The voltage or current output of transducers is generally an *analog* quantity; it follows the concentration of some material, which can vary over a continuous range. The continuous voltage or current value must be interpreted for the computer. The operation of one type of ADC is described in Supplement 3C.

## Suggestions for Further Reading

A good place to go next for more analog (DC and AC) and digital electronics is to one of three good introductory textbooks of electronics.

Bassos, B. H.; Ewing, G. W. "Analog and Digital Electronics for Scientists," 2nd. ed.; Wiley: New York, 1980.

Diefenderfer, J. "Principles of Electronic Instrumentation," 2nd. ed.; Saunders: Philadelphia, 1979.

Malmstadt, H. V., *et al.* "Electronics and Instrumentation for Scientists"; Benjamin/Cummings: Menlo Park, CA, 1982.

A short review article showing how bridge circuits are now used in temperature sensor systems. It shows how slightly nonlinear resistances are treated. Straightforward but not for the faint-hearted.

Bentley, J. P. "Temperature sensor characteristics and measurement system design," *J. Phys. E: Sci. Instr.* **1984,** *17*, 430–439.

## Problems

3.1   A diode transducer that converts light to electrical current is called a photodiode. The photodiode output is linear up to a current of 3 mA. Assume this is to be used with an amplifier that responds to voltage and is linear for voltages up to 4.2 V. What size resistor is needed in the circuit in Figure 3.1.1 so that the full linear range of the photodiode and the amplifier are used?

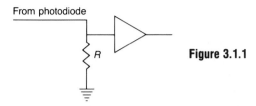

Figure 3.1.1

**3.2** The circuit shown in Figure 3.2.1 is used in conjunction with a transducer for low light intensity measurements. The voltage applied by the power supply is 900 V and $R = 100$ kΩ.
    a. What is the current flow through the circuit?
    b. What is the voltage between connections 1 and 2? Between 5 and 6? Between 0 and 4?
    c. Answer part b if $R = 10$ kΩ.

Cathode connection

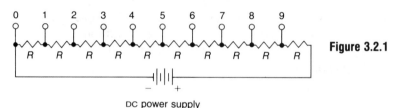

Figure 3.2.1

DC power supply

**3.3** One way to determine a potential from a transducer is to apply a precisely known voltage against it and determine when no current flows with a sensitive ammeter. Such a circuit is shown in Figure 3.3.1. The known voltage is called a bucking voltage. In the bucking circuit, the variable resistor with the wiper between $R_1$ and $R_2$ can be set with any ratio of $R_1/R_2$ from zero to infinity. Assume $V = 9$ V exactly. The grounds are connected.
    a. At what value of $R_1/R_2$ is the input to the ammeter equal to 0 V?
    b. At what value of $R_1/R_2$ is the input to the ammeter equal to 9 V?
    c. If the transducer voltage is 2.32 V, at what value of $R_1/R_2$ will the voltage be exactly compensated so that no current flows?

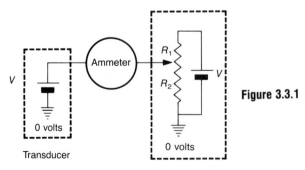

Figure 3.3.1

Bucking voltage circuit

**3.4** The circuit in Figure 3.4.1 is found as part of an amplifier. Specifically, it controls the level of amplification. What is the total resistance when the following are connected at the same time?

a. A alone?
b. A and B?
c. A and B and C?

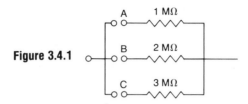

**Figure 3.4.1**

3.5  a. Find the time constant for the circuit in Figure 3.5.1 when $R = 300 \text{ k}\Omega$ and $C = 10 \text{ pF}$.
b. If $C$ is held constant, what value of $R$ would be needed for the time constant to increase to 0.1 s?
c. If $R$ is held constant, what value of $C$ would be needed for the time constant to increase to 0.1 s?

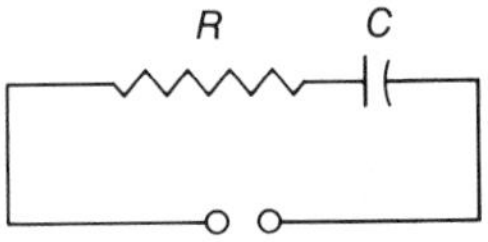

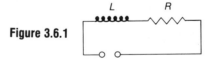

**Figure 3.5.1**

3.6  a. Find the time constant for the circuit in Figure 3.6.1 when $R = 300 \text{ k}\Omega$ and $L = 10 \text{ }\mu\text{H}$.
b. What value of $R$ would be needed for the time constant to increase to 0.1 s?

**Figure 3.6.1**

3.7  A real battery actually behaves as a voltage source and a series resistance, as shown in Figure 3.7.1. An old 1.5-V dry cell, when connected to a wire with insignificant resistance, provides only 5 mA of current. What is the internal resistance?

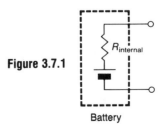

**Figure 3.7.1**

Battery

3.8 The circuit of Figure 3.8.1 is charged to 10 V across the capacitor ($C = 0.01 \ \mu F$; $R = 10 \ k\Omega$) and then is discharged by flipping the switch.

    a.  What time is needed to discharge the capacitor to 10% of its original voltage?

    b.  If the capacitor is charged to 20 V, what time is needed to discharge the capacitor to 10% of its original voltage?

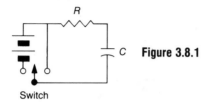

**Figure 3.8.1**

Switch

3.9 One way of creating a linear voltage change is to charge or discharge a capacitor for a time that is short compared with the time constant of the circuit. For the circuit in Figure 3.9.1, what is the value of $R$ needed to make the capacitor charge from 0 to 5% of $V_{app}$ at exactly 50 ms after the switch is closed if $C = 0.1 \ \mu F$?

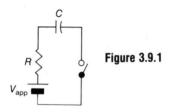

**Figure 3.9.1**

## SUPPLEMENT 3A

# Schematic Symbols in Electronics

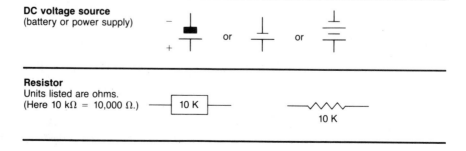

| | | | | |
|---|---|---|---|---|
| **DC voltage source** (battery or power supply) | − $\underset{+}{\|}$ | or | | or |

**Resistor**
Units listed are ohms.
(Here 10 k$\Omega$ = 10,000 $\Omega$.)    10 K            10 K

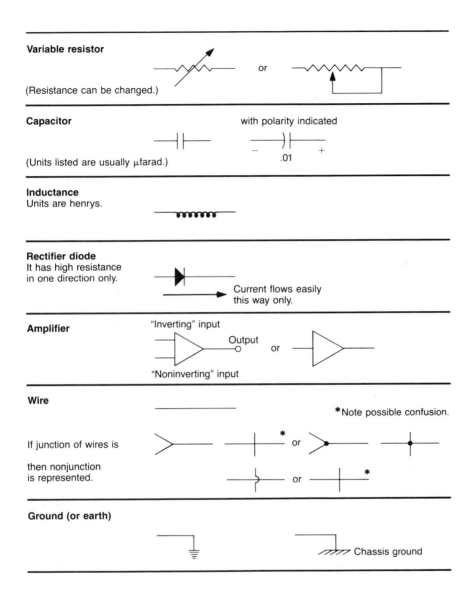

**Variable resistor**

(Resistance can be changed.)

or

**Capacitor**

with polarity indicated

(Units listed are usually μfarad.)

− .01 +

**Inductance**
Units are henrys.

**Rectifier diode**
It has high resistance
in one direction only.

Current flows easily
this way only.

**Amplifier**

"Inverting" input

Output

or

"Noninverting" input

**Wire**

＊Note possible confusion.

If junction of wires is

then nonjunction
is represented.

＊ or

or ＊

**Ground (or earth)**

Chassis ground

---

## SUPPLEMENT 3B

# Voltmeters, Ammeters, and Ohmmeters

Numerous assay methods require the precise measurement of electrical resistance, voltage, and/or current. In this supplement we show how a meter that is able to measure one of these can be used to measure all three. The electrical circuit behavior

**(a)**

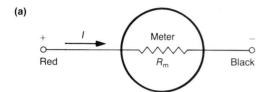

**(b)**

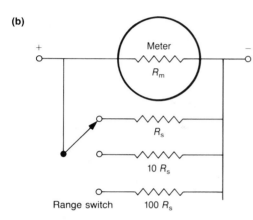

**Figure 3B.1**    Illustration of (a) the circuit of a meter that responds with a displacement proportional to current and (b) how it can be placed in a circuit to produce a meter that responds to different ranges of current. To obtain the most accurate measurement, $R_m$ should be as small as possible.

described by Ohm's law is the key in the transformation. Any contemporary instrument uses far more sophisticated circuits to do the job with greater precision ($\pm 0.1\%$ or better). The following examples are presented to illustrate the major principles involved.

Let us assume that the measuring device produces an output that is linearly related to current. A d'Arsonval meter is such a transducer; the meter's needle is rotated in proportion to the current flowing through the meter.[1] The general name for such a meter is a **galvanometer.** Such meters are commonly found on consumer electronics equipment, not just in specialist laboratories. However, they are rapidly being replaced with solid-state devices such as digital displays. The digital response is linear with applied voltage as is the meter's needle. (Such a digital device divides up the possible voltages into small steps with magnitudes of a unit change in the last digit. It is inherently no more precise than a continuous scale on a meter.)

As illustrated in the schematic diagram of Figure 3B.1, to measure a current, it is necessary only to connect the meter to the circuit in series with the current flow. The meter introduces some resistance $R_m$ into the circuit. The dial of the meter can be calibrated directly in amps (A) or milliamps (mA) depending on the sensitivity of the meter. (More sophisticated measuring devices can measure down to the nA and even pA range.) If a number of different current ranges are wanted, then a parallel resistor can be placed in the circuit to produce a smaller current through the meter, but one that is linearly proportional to the one being measured.

---

[1] This effect is due to the magnetic field generated by the current that flows through a small wire coil.

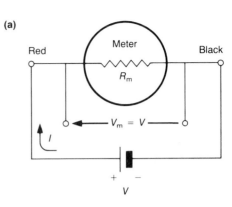

**Figure 3B.2**    Illustration of (a) the circuit of a meter that responds with a displacement proportional to current and (b) how it can be placed in a circuit to produce a meter that can be used to measure voltage. Also shown is an arrangement (a voltage-divider circuit) that allows a number of different ranges of voltage to be measured. The equation applies to the meter connected as shown. To obtain the most accurate measurement, the meter resistance should be as large as possible.

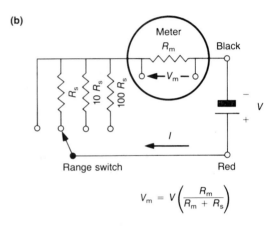

The meter can be converted to a voltmeter in the following way. The circuit is shown in Figure 3B.2. Since the circuit's behavior can be described by Ohm's law, whenever a current flows through the meter's coil, a voltage must appear across it. The voltage across the coil is

$$V_m = I_m R_m$$

Conversely,

$$I_m = \frac{V_m}{R_m}$$

and so the current following through the meter produces a measure of the voltage across it. This voltage can be calculated if the meter's resistance is known. Then, with this calibration, the ammeter is converted to a voltmeter by drawing a new scale on the face of the meter—one for voltage. If a number of different voltage ranges are wanted, a resistor can be placed in series with the meter. The circuit is then a voltage divider with the meter forming one of the elements.

A resistance-measuring device can also be made, but it requires an internal voltage supply for operation. A battery can serve quite well. The circuit for a resistance meter

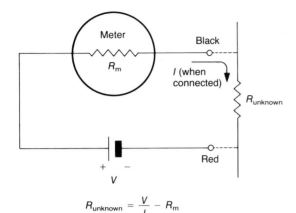

**Figure 3B.3**    Illustration of the circuit of a meter that responds with a displacement proportional to current and how it can be placed in a circuit to produce a meter that can be used to measure resistance. With a known value of $V$ and $R_m$, a nonlinear resistance scale can be constructed for the meter face.

is shown in Figure 3B.3. To determine a resistance, the wire leads are connected together, which produces the full battery potential across the meter. This point can be noted as zero ohms. The current that passes through the meter is then

$$I_0 = \frac{V}{R_m}$$

When the unknown resistance $R_{unk}$ is connected across the leads, it is placed in parallel with $R_m$. The meter then registers $I_{unk}$:

$$I_{unk} = \frac{V}{R_{unk} + R_m}$$

The ratio between $I_0$ and $I_{unk}$ is

$$\frac{I_{unk}}{I_0} = \frac{V/(R_{unk} + R_m)}{V/R_m}$$

This equation can be rearranged to

$$\frac{I_{unk}}{I_0} = \frac{R_m}{R_{unk} + R_m}$$

or

$$R_{unk} = R_m\left(\frac{I_0}{I_{unk}} - 1\right)$$

The current (and hence the needle position) is nonlinear with the change in $R_{unk}$, however.

Notice that the accuracy of all three types of measurements is limited by two factors. The first is the accuracy with which the resistance values are known. The second is the limit of accuracy of our ability to read the scale. Even if the instrument has a digital scale that can be read to 5 significant figures ($\pm 0.01\%$), the last few digits would be meaningless if the resistances were only known to 1% accuracy.

## SUPPLEMENT 3C

# Analog-to-Digital Converters

Transducers of analytical instruments used in the great majority of techniques produce a voltage or current proportional to a continuously variable (as opposed to stepwise variable) chemical concentration or physical property such as pressure. The output of the transducer follows the variation in the quantity to which it is sensitive. This is an **analog output.**

On the other hand, digital computers operate with discontinuous or discrete jumps in electrical potential. Information is transferred by a series of off–on switching. The output from the analog transducer is incompatible with the needs of the digital operating system. To get the chemical information into a digital form from the transducer, it is necessary to use an **analog-to-digital converter,** abbreviated ADC or A/D converter. It is an essential part in computerized instruments.

**Figure 3C.1**  Illustration of the operation of one type of analog-to-digital converter. Here the voltages of two different parts of the circuitry are shown as they vary with time. (Top) After a capacitor is charged to the input voltage, the decay of voltage in a circuit consisting of a capacitor and resistor is monitored. The voltage decreases with a characteristic time constant. The curvature here is greatly exaggerated. At short times, the voltage drops nearly linearly, and the counting time is directly proportional to the measured potential $V_{in}$. (Bottom) At the same time that the capacitor begins to discharge, a register records the number of pulses that are made before the preset voltage is reached. The number of counts depends on the input voltage as shown in the figure. A digital value — the number of pulses counted — can then be transmitted to the digital computer and stored. The actual pulse rate is relatively much faster than illustrated. The voltage $V_{in}$ is sampled approximately 50 times/second for this design.

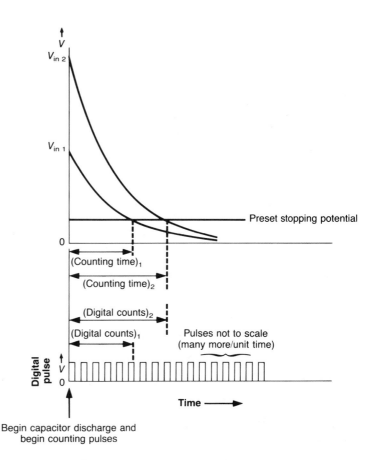

The manner of operation of one type of analog-to-digital converter is described below and in Figure 3C.1. The converter responds to the analog voltage at its input. Its output is a digital number that is proportional to the input voltage.

The ADC circuit requires a short period of time to carry out the conversion. It operates intermittently: The input potential is sampled, some time is spent converting the analog signal, the output is registered in the computer, then another voltage sampling follows, and so on. The analog voltage is sampled anywhere from about 50 times a second up to $10^8$ times a second, depending on the ADC system.

In the sampling period, the input is connected to a capacitor, and a current flows until the voltage across the capacitor equals the input voltage. The capacitor is then disconnected from the analog voltage source and connected to a second circuit where it begins to discharge with a known time constant. The circuit is similar to that shown in Figure 3.7. At the same time that the capacitor begins to discharge, a digital circuit begins to count pulses from a high-frequency pulse generator. When the voltage drops to a previously defined level, the digital-pulse counter is stopped. The number of counts up to the stopping time is proportional to the input analog voltage. There are more counts for larger voltages since it takes longer to reach the preset stopping voltage.

# 4

# Statistical Tests and Error Analysis

## 4.1 Introduction

"Analytical chemistry is not spectrometers, polarographs, electron microprobes, etc., it is experimentation, observation, developing facts, and drawing deductions." [Ref:· P. W. West, *Anal. Chem.* **1974,** *46(9),* 784A.]

When a conclusion is made, it is most useful to be able to say how certain it is. We all use phrases like "pretty sure," "very sure," "most likely," and "improbable" to express different degrees of certainty. However, in chemical analysis, more-quantitative statements of certainty are desirable. Replacements for the rather vague phrases can be made by using mathematical statistical tests.

In this chapter a number of these statistical methods for evaluation of experimental results are shown. The two main questions that we want to be able to answer are

How sure are you of the results you have obtained experimentally?

How sure are you that the value you have obtained is the same (or different) from the value obtained on the same sample at a different time or by another person?

The second question is quite important in the area of national or international regulation of the presence (or absence) of chemical substances. Information from numerous laboratories and analysts must be coordinated, and the analytical results must be reliable. If the results are unreliable, the regulation could be meaningless. What is desired is the interchangeability of results among laboratories. To this end, significant effort is expended to validate an analytical methodology. To validate means to determine that the methodology developed to determine a specific analyte (such as vanadium) in a specific type of sample (such as beef liver) produces comparable results for the majority of analysts.

The practical example in this chapter is a description of a trace analysis method and some of the problems encountered in an interlaboratory comparison of results from ten laboratories. In the quote, there are some terms that may be unfamiliar to you. One of these is **outliers.** To do an analysis, the measurement is normally made a number of times on portions of the same sample. The results for all the portions are then averaged. If the result for one portion has a value significantly far from the others, it is called an outlier and may be rejected as invalid. The validity of outliers can be tested by using statistical methods.

The specific assay mentioned in the quote is a test for traces of a highly carcinogenic (cancer-causing) material called aflatoxin. The samples are cacao beans, which are used to make chocolate. Each of the ten laboratories was sent a sample consisting of uncontaminated cacao beans to which was added a known, fixed amount of the aflatoxin. Each laboratory used the same analytical methodology and ran two separate analyses on its sample. The reported values are plotted in Figure 4.1.

An irreducible difference exists between supposedly identical measurements made in different laboratories. This point was recently demonstrated by a group of New Zealand government laboratories in attempting to minimize the discrepancies in values for blood alcohol between laboratories. The laboratories went to great pains to discover every source of error, even to the extreme of moving analysts from one laboratory to another. They found that an analyst increases his or her intra-analyst variability when moved to a different laboratory environment. They concluded that the only way to eliminate interlaboratory variability was to conduct all analyses in a single laboratory and, presumeably, by the same analyst. Under our legal system, this solution is impossible since defendants accused by laboratory evidence have a constitutional right to produce rebuttal evidence from any laboratory of their choice. Therefore, the important question to be answered in the evaluation of methods of analysis is how much allowance must be made for between-laboratory variability in interpreting the values produced by different laboratories. If the variability or error produced by the method is excessive—that is, it does not permit effective regulation as required by the statute—the method must be judged unacceptable for the intended purpose . . . .

Every analyst has implicit faith that if a method is followed exactly, the correct result will automatically be produced. However, our review of several hundred collaborative studies in which the samples were examined as true unknowns reveals that often 5 to 15% of the reported values are statistical outliers—values that are far outside the region where most of the other values reside. Outliers are produced by experienced chemists as well as by novices . . . .

Outliers are a fact of laboratory life, and allowance must be made for them. By definition, [their values] lie at the extreme points . . . of a series of analytical values. . . . they have a large influence on the . . . indices used to measure the [quality of the] performance of methods.

[Figure 4.1 shows the values of aflatoxin found by ten laboratories using the same assay method. One value is classified as an outlier.]

. . . There is one important statistical problem with outliers: What outlier test should be used? This is a complex statistical problem . . . . Chemists seem to have little difficulty in applying intuition and experience to this problem, but many statisticians are appalled at this approach. We hope to apply a number of outlier tests to a number of . . . collaborative studies to determine if any of the several dozen procedures described in the statistical literature is best suited for application to interlaboratory work.

[Ref: Horwitz, W. *Anal. Chem.* **1982,** *54,* 67A–76A.]

It should be clear from Figure 4.1 that the measured result of each analysis was not the quantity of aflatoxin that was added to the uncontaminated samples. The differences between the quantity of added aflatoxin and the quantities measured through experiment are due to errors in the analyses (assuming negligible error in preparing the samples).

**Figure 4.1**    Graph of the results of the interlaboratory study of the determination of aflatoxin in cacao beans. The vertical (dashed) line indicates the amount of aflatoxin added to uncontaminated beans by the supervisors of the study. Open circles indicate the values obtained for the two duplicate determinations made by each of the 10 laboratories taking part in the study. The circled value was deemed an outlier. Note that laboratory 9 produced two results that were quite close to each other but not to the true value; the results were precise but not accurate.

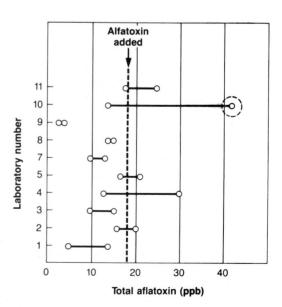

## 4.2 Finding Errors

One crucial part of improving chemical analyses is to be able to identify the places where errors can occur and be able to evaluate their magnitudes. To illustrate some factors of error detection and error analysis, let us consider a technique which may be familiar to you. The assay method is a titration. If you are not familiar with titrations from an earlier chemistry course, you may want to glance at the early sections of Chapter 9.

### A Titration Procedure

Some unknown powder was spilled in an accident and was found to burn the skin when touched. You are given a kilogram of the powder in a plastic bag. It is an acid, and you are asked to find out approximately how much sodium carbonate will be needed to neutralize the material that remains after scooping up the bulk. The carbonate will be mixed with water and used to wash away any residue.

Normally, to obtain an accurate analysis of the acidity of a powder sample, the sample should be dried in an oven at 110 °C. Most likely there will be some unknown amount of water associated with it which should be removed. The sample is "dried" when it no longer loses weight by heating at 110°. In this case, drying may not be needed for two good reasons. First, the weight of the residue of the spilled material may not be known that accurately anyway. And second, there may not be enough time to do the drying since the result is needed immediately.

Whether the sample is dried or not, a part of the sample (a few grams) is weighed accurately, and the weighed portion is added to water. Let us assume that the powder easily dissolves in water.

Next, a clean burette is filled with a solution of sodium hydroxide in water. The initial volume of liquid in the burette is measured by seeing where the top of the liquid lies on the volume scale on the glass.

Then some acid–base indicator solution (Section 9.8) is added to the solution of unknown acid in water. The hydroxide solution in the burette is added to the sample solution until the indicator just changes color. The volume of solution remaining in the burette is then read, and the amount of acid in the weighed sample is calculated.

In this procedure, as in *all* analyses, the possibility exists for making mistakes in *every* step of the analysis. We call these possible mistakes **sources of error.** For instance, a few possible sources of error in this analysis are misreading a volume measurement, incomplete drying of the sample (if dried), spilling or splashing some of the sample solution out of its container, and making an error in weighing the powder.

### Pulling Yourself Up by Your Bootstraps

How can you find out if you make a mistake? After all, no one knows the amount of acid in your sample. One method to use is to do the whole procedure more than one time. The jargon is you will **run** several **replicate samples** or **replicates.** All the replicates are measured out from the same original sample. If the answers are similar for two replicate runs, it would appear that no major errors were made in either run. If you did the determination three times from beginning to end on separate parts of the sample, you would be even more certain that no major mistakes were made. And running a fourth replicate would make you even more certain. This sort of intuitive certainty is expressed in statistics as a **confidence limit.** For the acidic powder, the confidence limit is found using all the values of acid content found from all the replicate runs. The confidence limit has the property that as more experiments are run, one becomes more certain that the next result is *going to be in the same range* as the previous ones. Thus, the statistical description of the confidence limit behaves in the same manner as your intuitive confidence in your average result.

## 4.3 Measuring Errors

To calculate a numerical value for the confidence limit—a probability that more results will have about the same value—a quantity called the **standard deviation** first must be calculated. The standard deviation is one measure of how closely the individual results agree with each other. The standard deviation is a statistically meaningful description of the scatter of the values determined in a series of runs.

It is important to recognize that a statistical measure of the quality of an assay can*not* be obtained unless a *series* of tests is carried out. If only one sample is analyzed, we must take the measured value as the result. However, if a number of replicate runs are done, no result will be exactly the same as any other. You saw this in the case reading on aflatoxin testing. If each run provides a different value, what value is to be reported as the result for the assay? The meaningful numerical result of a series of tests is the **arithmetic mean** of the individual results. The arithmetic mean of a series of values is found by taking the sum of the results and dividing it by the number of determinations in the series. This is expressed mathematically by

$$\overline{X} = \frac{\Sigma_i X_i}{N} \tag{4-1}$$

where

$\overline{X}$ represents the arithmetic mean,
$X_i$ represents the numerical result of the $i$th run, and
$N$ is the total number of runs.

Other names for the arithmetic mean are the **mean,** the **mean value, average,** and **average value.** Calculation of the mean for a specific example is illustrated in Table 4.1.

The example of Table 4.1 shows a statistical test of the quality of a technique for weighing a sample. To do the experiment, the sample was loaded onto the balance, weighed, and removed. This procedure was then repeated three more times. The results of the four weighings are shown in the first column of Table 4.1. The sum of

**Table 4.1**  Example of Statistics of Weight Determination

| Weighing Number ($n$) | Observed Weight ($X_i$) | Deviation from the Mean | |
|---|---|---|---|
| | | $x_i$ | $x_i^2$ |
| 1 | 3.1023 g | 0.0006 | $36 \times 10^{-8}$ |
| 2 | 3.1032 | −0.0003 | $9 \times 10^{-8}$ |
| 3 | 3.1027 | 0.0002 | $4 \times 10^{-8}$ |
| 4 | 3.1033 | −0.0004 | $16 \times 10^{-8}$ |
| $N = 4$ | Sum   12.4115 | | $65 \times 10^{-8}$ |
| | Mean, $\overline{X}$  3.1029 | Av. dev., $\bar{x}$  0.00038 | Variance $22 \times 10^{-8}$ |
| | | | $s = 0.0004_6^*$ |

* The integer 6 written as a subscript means that the error in this digit is greater than $\pm 1$.

the four weights is 12.4115 g. The mean weight $\overline{X}$ is then $12.4115/4 = 3.1029$ g, calculated according to Eq. 4-1.

Some further definitions follow. The **deviation from the mean** $x_i$ is defined for each individual measurement. It is the difference between each measured value $X_i$ and the mean value for all the measurements, $\overline{X}$. Algebraically,

$$x_i = X_i - \overline{X} \tag{4-2}$$

(As written here, it is common to use upper-case letters to refer to quantities that are measured and lower-case letters to refer to errors or differences.) As an example, representative calculations are made for the data in Table 4.1. In the third column, the calculated values of the deviations from the mean of the individual weighing measurements are shown.

From the deviations from the mean of the individual measurements, some indicators of the scatter of the data can be calculated. One measure of the reproducibility of the measurement is the **average deviation.** The average deviation is the average value of the individual deviations without regard for the sign of the term (+ or −). The average deviation is defined as

$$\bar{x} = \frac{|x_1| + |x_2| + |x_3| + \cdots + |x_N|}{N} \tag{4-3}$$

This measure is now seldom used in practice, but it does appear in the literature of analytical chemistry. The average deviation has no statistical meaning.

The **range** of the measurements is the difference in magnitude between the highest and lowest test results in a series.

$$\text{range} = w = X_{\text{highest}} - X_{\text{lowest}} \tag{4-4}$$

The range for the example shown in Table 4.1 is 0.0010.

The most commonly used (and statistically meaningful) measure of the reproducibility of a set of measurements is the standard deviation. The standard deviation is the square root of the **variance.** The variance is defined by Eq. 4-5.

$$\text{variance} = \frac{x_1^2 + x_2^2 + \cdots + x_N^2}{N - 1} \tag{4-5}$$

$$\text{standard deviation} = s = \sqrt{\frac{x_1^2 + x_2^2 + \cdots + x_N^2}{N - 1}} \tag{4-6}$$

Another statistical measure of the error is the **relative standard deviation.** This is the standard deviation as a fraction or percentage of the mean value. This is written algebraically as

$$\text{relative standard deviation} = \frac{s}{\overline{X}} \tag{4-7a}$$

or

$$\text{percent relative standard deviation} = \frac{s}{\overline{X}} \times 100 \tag{4-7b}$$

Occasionally the relative standard deviation is referred to as the **coefficient of variation.** This occurs mainly in the older literature, and the name *coefficient of variation* is no longer recommended for use. [*Anal. Chem.* **1968,** *40,* 2271.]

## 4.4  Absolute and Relative Measures

In the previous section, we defined the standard deviation and the relative standard deviation. The standard deviation is a number: an absolute measure. The relative deviation is a ratio of the standard deviation to the mean value.

A large number of quantities in analytical chemistry are such ratios. They are all called **relative** measures. Relative measures are unitless numbers. When you see the label *relative,* it will foretell a ratio of two numbers that have the same units.

The other quantities are called **absolute** measures. They will have units, such as g/mL.

## 4.5  Precision and Accuracy

So far in this chapter, only the spread of the values of analytical results has been considered. When the scatter of the experimentally determined values is small, we say the **precision** is high. A quantitative measure of scatter is the standard deviation of a set of measurements: The standard deviation is small when the precision of the experiment is high.

However, we have not yet considered whether the mean value, which is calculated from a series of experimental runs, is close to the amount of the species of interest that

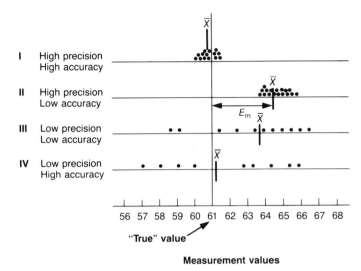

**I**   High precision
    High accuracy

**II**  High precision
    Low accuracy

**III** Low precision
    Low accuracy

**IV**  Low precision
    High accuracy

56 57 58 59 60 61 62 63 64 65 66 67 68

"True" value

**Measurement values**

**Figure 4.2**   Accuracy and precision illustrated. Four plots of the values of the results of four series of determinations. Dots indicate the values of results from individual experiments. Assume that a "true value" is known; this equals 61.

$\overline{X}$ is the mean value of each series of experiments I–IV.
$E_m$ is called the mean error.

The relative mean error is $E_m/\overline{X}$ or, as a percentage, $(E_m/\overline{X}) \times 100$.

is *actually present* in the sample. In other words, the precision of our data bears no necessary relation to the question, "How **accurate** is the analysis?"

The difference between precision and accuracy can be shown visually as in Figure 4.2. There, four different combinations of precision and accuracy are illustrated. The values from individual experiments are shown as dots along the *x*-axis. The first two plots show the results of determinations that are precise. However, the mean value of the second set of data is far from the "true value"; the mean value has a low accuracy, but the results have high precision. How can that be?

One possibility might be that in each sample there is some other component that reacts the same way as the assayed component: an interferent. This happened in the applications example in Chapter 1 where, in the method that was used to determine chlorine, the thiol in the samples reacted in the same way as the chlorine itself. Since each sample contained the same amount of thiol, the analysis produced a result that always was too high by a constant amount. No matter how precise the analytical methodology, the result was always inaccurate.

The difference between the "true value" of the analyte and the mean value of a series of analytical results is called the **mean error,** here abbreviated $E_m$. This kind of error is in the group called **determinate errors.** Note that the word is not determin*ant* but determin*ate*. Determinate means to originate from a fixed cause. On the other hand, the errors for which we use the standard deviation as a measure are assumed to be **random errors,** which means they originate in arbitrary or indeterminate processes. Random errors and randomness are discussed further in Section 4.6.

The mean error has units that are the same as the measurement, for instance, g/L. When the mean error is expressed as a percentage of the true result, it is called the **relative error.** As *relative* suggests, the relative error is a unitless number. Do not confuse this with the relative standard deviation of the random error.

An important point to understand is that from a purely statistical point of view, there is no such idea as a mean error because the "true result" or "true value" is never known. [For example, see E. F. McFarren et al., *Anal. Chem.* **1970,** *42*, 358–365.] However, we shall assume that we can approach the "truth" by suitable clever experimentation.

An example of such an approach would be to mix completely, and in the same ratios, highly purified materials to produce a material that is the same as the samples being analyzed. Various known "artificial" samples can be tested in exactly the same way as the unknown until a mixture is found that gives results the same as the unknown sample. As you might suspect, such an approach is tedious.

Easier experimental approaches to determine precision and accuracy are possible. Before describing these, some ideas of randomness and probability are needed and are described in the next section.

## 4.6 The Normal (Gaussian) Distribution

When a number of events or errors occur at random, it means that they are independent of each other. An example of random events is the appearance of heads or tails after flipping a (fair) coin. Half of the time the coin will land heads and half of the time tails. The result of one flip will have no effect on the next: They are independent (unless one flip bends the coin).

When the errors in a measurement are random, the values tend to be distributed in a characteristic manner on either side of the mean value. The following shows one way that results can be treated to illustrate the statistical nature of this distribution.

First, a series of runs are made using replicate samples of the same material. Such a procedure is similar to weighing some mass repeatedly, as described earlier in Section 4.3. From the results obtained, the mean value can be calculated, and following that, the deviations from the mean for each measurement can be determined. Let us assume for our illustration that the measurement was done 40 separate times. This number of runs might be done when validating a new method of analysis. For routine procedures, seldom are more than four replicate samples used.

As illustrated in Figure 4.3 the data were collected and evaluated after the first 5, the first 10, and finally the full 40 replicate determinations. Next, each of the experimental values is classified into a group depending on how far it is from the mean value. Each group is equal in its range of error—the numerical difference between the highest and lowest values in the group. The data are then plotted on a bar graph, such as that shown in Figure 4.3a. The graph shows the number of experiments in which the results fall into each of the ranges versus the value of the experimental result. Such a plot is called a **histogram.** The histograms for the results after 5, 10, and 40 samples are illustrated. The mean value is 100 ppm, and each bar is 2 ppm wide: equal to the range of values it contains.

In the histogram, the area of each bar is proportional to the number of experiments giving results in that bar's range. In addition, since the bars are all equal in width, the height of each bar is also proportional to the occurrence within its range. As a result, the scale on the left can be labeled *number of experiments.* In addition, the total area of all the bars is proportional to the total number of experiments: here 5, 10, and 40.

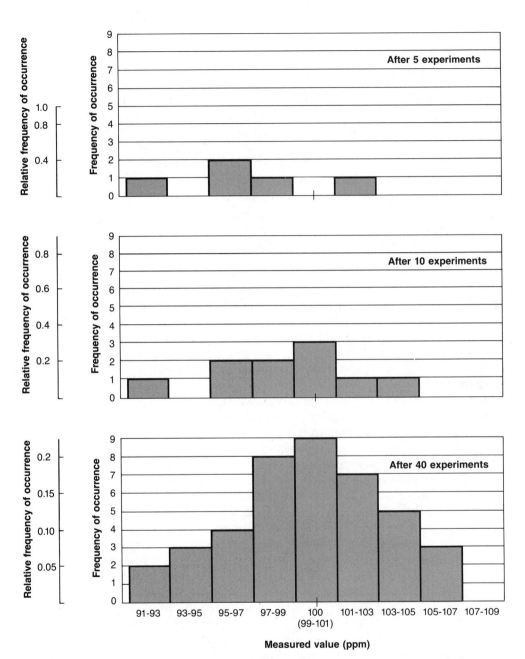

**Figure 4.3a**   Graphs of the results of a series of 40 experiments classified into sets of adjacent values. Each bar represents a 2-ppm range. The height of each represents the number of results in each range. Two different ordinates are shown: the frequency of occurrence and the relative frequency of occurrence. The mean value is 100 ppm. As labeled on the histograms, the results are shown after 5, 10, and 40 experiments are run.

**Figure 4.3b** For a "very large" number of results, the distribution of the results approaches a Gaussian curve characterized by the standard deviation $\sigma_m$ and a peak at the mean value. When we scale the curve so that the area under it is 1 (a **normalized** Gaussian curve), then the peak height becomes 0.399. For the normalized curve, the value of the vertical axis at $\sigma_m$ is 0.242, at $2\sigma_m$ 0.054, and at $3\sigma_m$ 0.0044. After the curve is established from the results of a number of replicate analyses, we expect that for *future* experiments on the same material:

68.3% of the results will lie within $\sigma_m$, and
95.4% of the results will lie within $2\sigma_m$.

These percentages are equal to the areas under the normalized Gaussian curve between $\pm\sigma_m$ and $\pm2\sigma_m$, respectively.

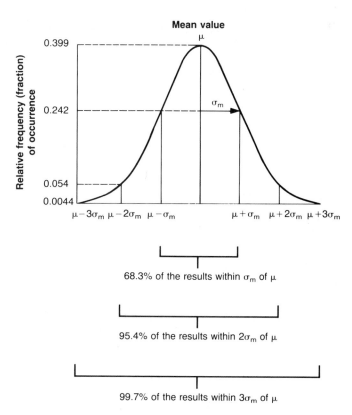

Another way to treat the data is to set the sum of the areas of all the bars equal to unity. Then the left scale becomes the *proportion of experiments run* or *relative frequency of occurrence.* The area of each bar will then be proportional to the *fraction* of the experiments that have the error in the given range. (Note in Figure 4.3a that the relative frequency of occurrence scale expands as the number of experiments increases.)

Another characteristic of the histograms is the *tendency* of the bar heights to decrease away from the mean. The number of errors in each group *tends* to decrease as the group's deviation from the mean increases (as the group lies further from $\overline{X}$). This is seen most clearly in the histogram for 40 replicate runs. However, as seen in the histograms for 5 and 10 replicates, the expected regular distribution is not necessarily observed after only a few results are measured.

The statistical treatment of random errors is only slightly different from plotting a histogram and calculating the relative frequencies of occurrence as they depend on the deviation. The main difference is that the statistical error is calculated *as if* there were an infinite number of experiments: Each range of error is infinitely narrow. In that case, the distribution of random errors about the mean value can be described quite well by a **Gaussian error** curve, as illustrated in Figure 4.3b. Other names for the curve include the **normal error** curve or **normal distribution** curve.

The *x*-axis of the plot is a scale of experimental values. The mean value is at the peak of the curve. The height of the curve is a plot of the probability that an experimental result will have the value noted on the *x*-axis. Note that the errors can be both positive and negative from the mean value. The mean value for the Gaussian curve is labeled $\mu$. It is the value of $\overline{X}$ in the limit of a very large number of trials.

## Properties of Gaussian Distributions

The Gaussian curve is characterizable by a parameter $\sigma_m$, which, as you will see, can be related to *s*, the standard deviation for a small number of runs (defined in Eq. 4-6). The parameter $\sigma_m$ is also called a standard deviation but for a very large number of runs. The equation describing the Gaussian curve is

$$f(x) = \text{Gaussian} = \frac{1}{\sigma_m \sqrt{2\pi}}\, e^{-[(x-\mu)^2/2\sigma_m^2]} \qquad (4\text{-}8)$$

where

    *x* is the experimental result value,
    $\mu$ is the mean value and is at the peak,
    $\sigma_m$ is a parameter describing the relative width of the curve, and
    $(x - \mu)$ is the deviation from the mean for a result.

The detailed form of the equation is not important for now. If you are familiar with Gaussian curves, you know that $\sigma_m$ is usually written as simply $\sigma$. The appropriate subscript is kept here to help lessen possible confusion later.

The power of the statistical viewpoint of error is its ability to relate the error of a *series* of runs to the error of *individual* experiments in the series. Such a relationship is an important key in understanding chemical analysis results.

In the statistical treatment of error, the Gaussian curve itself is not as important as the *area* under the curve. In the same way as for the histograms in Figure 4.3a, the area under the Gaussian curve can be associated with the total number of experiments that are run, from "very large" to infinite. Therefore any *fraction* of the area under the curve will refer to a *fraction* of the total number of experiments.

Now we can connect the standard deviation (Eq. 4-6) to the statistical measurement of certainty. For a "very large" number of identical experiments, the distribution of random errors will fall approximately on a Gaussian curve centered at the mean value. We then *expect* that for *future* experiments:

    68.3% of the results will lie within 1 standard deviation of the mean;
    95.4% of the results will lie within 2 standard deviations of the mean.

The fractional areas corresponding to these ranges are indicated in Figure 4.3b. Bear in mind that the relationships apply for a number of experiments far larger than the number we usually do in a series of replicate samples.

## 4.7 The Confidence Limit

Earlier, an argument was presented appealing to your intuitive ideas of confidence: When more runs are made in a series of determinations, you can become more certain of the calculated mean value. However, this increase in assurance slows down beyond a certain number of replicates. Further repetition will merely reinforce the feeling that the analysis is getting incredibly boring.

In other words, one determination alone will be without much certainty. Two determinations help. But only when three or more are made can you begin to be more sure of the result. The statistical calculation of the *confidence limit* quantitates these intuitive ideas. In addition, the statistical calculations allow us to determine how many replicates are needed to make us confident that the mean value will *remain* within a specified *range* of values.

The essence of the calculation of the confidence limit lies in relating the standard deviation of a single result, $\sigma$, to the standard deviation of the mean value, $\sigma_m$. The relationship is illustrated in Figure 4.4. For an analysis, it is $\sigma_m$ that we wish to know. Another name for the standard deviation $\sigma_m$ is the **error of the mean.** Do not confuse this name with the *mean error* or *deviation from the mean,* which were defined earlier.

### Confidence Limits When $\sigma$ Is Known

In a simple case, let us *assume* that we know a value $\sigma$, the average random error associated with *each replicate.* The relationship between the error of the mean and the average random error associated with each measurement is

$$\sigma_m = \frac{\sigma}{\sqrt{N}} \tag{4-9}$$

**Figure 4.4**  These curves illustrate how the standard deviation of the mean, $\sigma_m$, varies with the number of experiments run, $N$, where each experiment has a random error of magnitude $\sigma$. The value of $\sigma_m$ decreases as the number of trials increases but only as $N^{1/2}$. The reason that $\sigma_m$ can be less than $\sigma$ is that the *average* of the individual errors tends toward zero as more experiments are run.

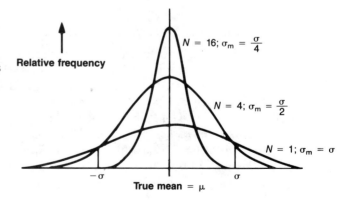

where $N$ is the number of measurements made, each with a random error $\sigma$. Simply stated, the intuitive ideas of increasing certainty with increasing replicate runs show up in the statistics as a square root of the number of experiments. So, according to statistics laws, you can be only twice as certain of the mean value if you run 16 replicates as if you run four replicates. In addition, this square root dependence on $N$ also agrees with our intuition that eventually there are diminishing returns in extra precision as the number of runs increases—double the precision for four times the number of runs.

After studying Figure 4.4, you might reasonably inquire how the error of the mean $\sigma_m$ can ever be less than $\sigma$, the average random error of each determination. In answer, recall that the result of each separate determination can be randomly high or low compared with the "true value." Therefore, as more replicates are run, the random errors tend to cancel; the average of the individual errors approaches zero. Equation 4-9 states that the errors tend to average to zero only as $N^{1/2}$, however.

From the simple idea expressed in Eq. 4-9, we can make a quantitative declaration of confidence. This derives from the relationship of $\sigma_m$ and the area under the Gaussian curve, as illustrated in Figure 4.3b. The statistical theory tells us that we can be 68.3% confident that another measurement *will* be within $\sigma_m$. And we can be 95.4% sure that *future* measurements will lie within $2\sigma_m$ of the mean value that has been determined from the measurements *already made.*

The most common confidence limit or **confidence interval** stated is at the 95% level. The interval containing 95% of the area under the curve must lie within something less than $2\sigma_m$, since $2\sigma_m$ includes 95.4% of the results. In fact, the interval is between plus and minus 1.96 $\sigma_m$. Therefore,

$$\text{mean} \pm 95\% \text{ C.L. for } \mu = \overline{X} \pm 1.96 \left(\frac{\sigma_m}{\sqrt{N}}\right) \qquad \text{when } \sigma \text{ is known} \quad \textbf{(4-10)}$$

This is written as 95% C.L. and spoken of as "the 95% confidence limit."

### Confidence Limits When $\sigma$ Is Unknown

Equation 4-10 does *not* apply to the usual chemical analysis, because the value of the random error for each experiment, $\sigma$, is seldom known. But Eq. 4-10 can be modified so it applies to the situation where $\sigma_m$ and $\sigma$ are *both* unknown. As you might suspect, we must be less certain in this case. To reflect this decreased certainty, a factor $t$ is substituted for 1.96 in Eq. 4-10. The value of $t$ is always greater than 1.96. In addition, as noted earlier, seldom is $N$ "very large" for replicate analyses. Therefore, we must substitute $s$ (Eq. 4-6) for $\sigma_m$ in Eq. 4-10. So when $\sigma$ is unknown, the following equation is used:

$$\text{mean} \pm 95\% \text{ C.L. for } \mu = \overline{X} \pm t \left(\frac{s}{\sqrt{N}}\right) = \overline{X} \pm Q \cdot s \qquad \text{when } \sigma \text{ is unknown}$$

$$\textbf{(4-11)}$$

**Table 4.2**  Values of $t^*$ and $Q$ to Use with Equation 4-11

| Number of Measurements Used to Determine $\overline{X}$ $N$ | $t$ for 95% C.L. | $Q = t/\sqrt{N}$ |
|---|---|---|
| 2 | 12.71 | 8.99 |
| 3 | 4.30 | 2.48 |
| 4 | 3.18 | 1.59 |
| 5 | 2.78 | 1.24 |
| 6 | 2.57 | 1.05 |
| 7 | 2.45 | 0.936 |
| 8 | 2.36 | 0.831 |
| 9 | 2.31 | 0.77 |
| 10 | 2.26 | 0.71 |
| 20 | 2.09 | 0.467 |
| 30 | 2.04 | 0.372 |
| 120 | 1.98 | 0.180 |
| $\infty$ | 1.96 | — |

* For tables of $t$ for other probability values of the C.L.—80, 90, and 99%—see Appendix VI.

Values for $t$ and $Q$ are shown in Table 4.2. The use of $Q$ only simplifies the calculations: It provides no new information. Notice that $t$ is quite large for small $N$ values but approaches 1.96, as in Eq. 4-10, as the number of trials $N$ becomes very large. The trend in the values of $t$ means that as $N$ becomes larger, it does not matter that the magnitudes of $\sigma$ are unknown, since the errors tend to cancel and tend to the same limit as with known $\sigma$-values. In this way, the standard deviation $s$—as defined in Eq. 4-6—smoothly transforms to $\sigma_m$.

**EXAMPLE**
Calculate the 95% C.L. for the example of the four weighings listed in Table 4.1.

**SOLUTION**
The value of $s$ as shown in Table 4.1 is $0.0004_6$. Since the value of $\sigma$, the error, is unknown, we use Eq. 4-11. Therefore,

$$95\% \text{ C.L.} = 0.0004_6 \cdot 1.59 = 0.0007$$

and the result would be reported as

$$\text{weight} = 3.1029 \pm 0.0007 \text{ g} \quad (\text{mean} \pm 95\% \text{ C.L.})$$

One final point for this section: Remember, both the standard deviation of the mean and the confidence limit are measures of precision only and *not* accuracy.

## 4.8 Controls, Blanks, and Accuracy

So far, only the precision of a series of replicate measurements has been considered in detail. The level of precision of the series is determined by the magnitude associated with the measurements and the number of measurements made. In this section we consider the cause and prevention of determinate errors. Determinate errors are errors in the analytical method; their presence or absence and magnitude determine the accuracy of analytical measurements.

High accuracy means that the "true value" and the mean value $\overline{X}$, which is experimentally determined, are close in value. Poor accuracy or low accuracy means that the value of the mean differs significantly from the "true value." As noted earlier, this difference is called the mean error $E_m$ and is illustrated in Figure 4.2. Do not confuse it with the deviation *from* the mean, defined in Eq. 4-2.

Errors that cause inaccuracy must be either eliminated in the experimental technique or at least accounted for by a correction in an analysis. Otherwise, the assay results will not correspond to the actual amount of the component being determined. This idea may seem elementary, and perhaps it is. However, in practice it can be difficult to account for all the major causes of inaccuracies. The practical example at the beginning of the chapter illustrates some of the obstacles.

Determining the accuracy of a methodology is part of the verification process. To develop any new analytical methodology, *known* samples of different sizes are assayed. Such samples, which contain known amounts of the species to be assayed, are the **controls** or **standards.** If the analytical results from the controls agree with the known content, the accuracy of the methodology is confirmed. This procedure is called **calibrating** the method. Of course the assay must be shown to work accurately with the control samples before trusting the results with unknowns.

One other type of sample is essential to calibrate any method: the **blank.** A blank sample contains all the components in the sample *except* the assayed species. The blank sample is passed through all the steps of the procedure just as a regular sample. The results from a blank sample allows us to investigate interferences which affect the accuracy.

Through the analysis of control and blank samples, any causes of poor accuracy can be found. Then, with wise use of descriptive chemistry, changes in the technique can be found to eliminate or reduce the errors. If necessary, appropriate corrections can be determined. A procedure to eliminate determinate error will illustrate this. The problem to which this procedure is applied involves finding an accurate analytical methodology to determine aluminum.

### The Assay for Aluminum

The following illustration was adapted with permission from A. A. Benedetti–Pichler, *Ind. Eng. Chem., Anal. Ed.* (precurser to *Analytical Chemistry*) **1936,** *8*, 373.

The problem, which we shall study in detail, involves the development of a **gravimetric** assay for aluminum. *Gravi/metric* (weight/measure) methods (Chapter 8) are those in which the analyte is transformed into a solid that is weighed. The weight that

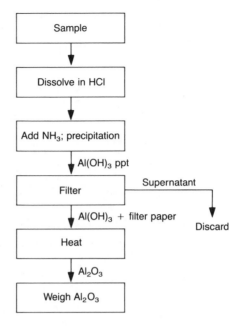

**Figure 4.5**  Diagram of the steps in determining aluminum by a gravimetric method.

is measured should be proportional to the amount of the analyte present. Currently, it is much more likely that an assay for aluminum would be done using one of a number of possible spectrometric methods.

The flow chart in Figure 4.5 illustrates the general steps in the procedure. The results, which we shall examine in detail, are listed in Table 4.3.

Two further definitions are needed before proceeding. Determinate errors, which cause inaccuracy, may be classified into two groups: **constant errors** and **proportional errors.** Constant errors are those that are the same magnitude even though the sample size changes. Proportional errors are those which have a magnitude that is directly proportional to the size of the sample; as the sample gets larger, so does the size of the error. As illustrated below, by using different-sized control samples, we can evaluate the proportional and constant errors.

The assay is to be used to determine the amount of aluminum in a sample of $AlK(SO_4)_2 \cdot 12\ H_2O$. This is alum. The samples were dissolved in a fixed volume of HCl solution. The aluminum was precipitated as a hydroxide, $Al(OH)_3$, by using an ammonia solution. Following that, the aluminum hydroxide was filtered from the solution with filter paper. This paper was placed in a crucible. The crucible had previously been dried at high temperature, cooled, and weighed. The crucible with the filter paper and precipitate was heated until the aluminum hydroxide was converted to alumina, $Al_2O_3$. The filter paper produces an ash under these conditions, and the weight of this ash was reproducible. The weight of the ash together with the weight of the crucible were subtracted from the total weight to give the weight of the alumina that remained in the crucible. It probably seems completely obvious that the weights of the crucible and filter-paper ash must be subtracted from the total

**Table 4.3** Results and Errors in the Determination of Aluminum as $Al_2O_3$*

| Col 1 | Col 2 | Col 3 | Col 4 | Col 5 | Col 6 | Col 7 | Col 8 | Col 9 | Col 10 |
|---|---|---|---|---|---|---|---|---|---|
| Size of Sample of AlK(SO₄)₂·12 H₂O (g) | Expected $Al_2O_3$ (g) | $Al_2O_3$ Found Using Reagent Grade $NH_3$ | | | $Al_2O_3$ Found Using Freshly Distilled $NH_3$ | | | $Al_2O_3 \times 0.9903$ | |
| | | Weight (g) | Abs. Error (g) | Rel. Error (%) | Weight (g) | Abs. Error (g) | Rel. Error (%) | Weight (g) | Abs. Error (g) |
| 1.0000 | 0.1077 | 0.1288 | 0.0211 | 16.4 | 0.1087 | +0.0010 | 0.9 | 0.1076 | −0.0001 |
| 2.0000 | 0.2154 | 0.2384 | 0.0230 | 9.7 | 0.2178 | +0.0024 | 1.0 | 0.2154 | +0.0000 |
| 3.0000 | 0.3231 | 0.3489 | 0.0258 | 7.4 | 0.3258 | +0.0027 | 0.8 | 0.3226 | −0.0005 |
| 4.0000 | 0.4308 | 0.4588 | 0.0280 | 6.1 | 0.4352 | +0.0044 | 1.0 | 0.4310 | +0.0002 |

$s = 0.0003$

* Adapted from A. A. Benedetti–Pichler, *Ind. & Eng. Chem.*, Anal. Ed. (Precursor to *Anal. Chem.*) **1936**, 8, 373.

measured weight to find the weight of the alumina. However, we could classify this operation as a correction, albeit an obvious one.

The first column in Table 4.3 shows the sizes of the control samples that were taken. The amount of aluminum in each is calculated as if it all were in the form of the final product $Al_2O_3$. This is 1.8895 times the weight of the aluminum itself. The weight of $Al_2O_3$ is shown in the second column. If the determination were exact, these numbers would correspond to the amount of $Al_2O_3$ that would be weighed at the end of the procedure.

The first set of experiments produced the weights shown in column 3. In order to consider the problems here, both the relative and the absolute errors are calculated. These are in columns 4 and 5, respectively. From perusal of these numbers, it appears clear that there is a relatively *constant* error associated with the assay. The error is not a proportional error; if it were, the relative error would be about the same for all the samples. Be sure you understand this before you continue.

The cause of the error was investigated. The fact that the error is constant implies that some reagent that is added in a constant amount is at fault. The ammonia added was about the same in each sample; the quantity of HCl that was neutralized by the whole solution was nearly the same in each. Relatively little of the basic ammonia solution was needed to neutralize and precipitate the much smaller amount of aluminum in each sample. Therefore, there is a good possibility the problem originates in the ammonia solution.

But what can be the problem with the ammonia solution? The fact that the blank had no excess weight (beyond the filter-paper ash) was a key point in solving the problem. The basic ammonia solution had dissolved some of the silica of the bottle in which it had been stored. In a more neutral solution, the silicon precipitated as silicic acid (formula $SiO_2 \cdot (H_2O)_x$) along with the aluminum hydroxide. The descriptive chemistry of silica includes the fact that it will not precipitate out of solution as easily alone as with aluminum hydroxide present. Thus the silica acted as an interferent which had to be removed. This discovery depended on a good knowledge of the chemistry of the reagents used.

The problem was overcome only by changing the analytical procedure. The ammonia was distilled immediately before using it. Since the silica is not volatile under the distilling conditions, the interference was removed. Incidently, these days, we would store the ammonia in containers made of more inert material such as polyethylene.

With the freshly distilled ammonia, the same series of experiments were rerun with the results shown in column 6. Again, the absolute and relative errors are calculated and are shown in columns 7 and 8. The absolute error values are plotted in Figure 4.6 as well. We can see that with the fresh $NH_3$ the absolute error seems to be about proportional to the sample size. This may show up more clearly in Figure 4.6a. In agreement with a proportional error, the relative error is approximately constant.

With this error, can the assay now be used? There are a number of considerations in deciding the answer. It is possible that the assay as it stands could be used if the accuracy wanted is poorer than 1%. But a greater accuracy can be obtained without further experimental work. A simple correction can be applied to the results.

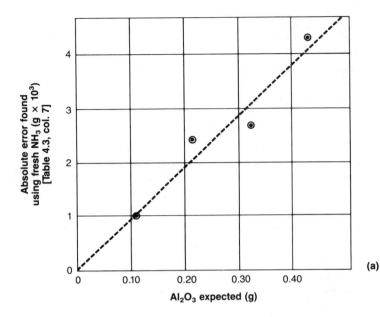

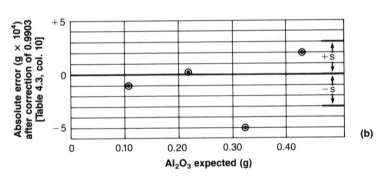

**Figure 4.6**  Illustration of the trends in errors found when developing a gravimetric assay for aluminum.
(a) Plot of the absolute error versus the expected weight of $Al_2O_3$ for the four samples tested. The trend of the points illustrates the effect of a proportional error associated with the determination. Do not confuse the absolute error with an absolute value of a number.
(b) Plot of the absolute error after a correction of $-1.0\%$ is applied to the results. (Each value was multiplied by 0.9903.) The error values are approximately randomly distributed about the zero–error (perfect accuracy) line. The magnitude of the standard deviation $s$, calculated from the four results, is indicated by the error bar at the right. Note that the final precision, after including the correction, is in the range of a few parts per thousand.

Notice that the relative errors reported in column 8 are all positive. If the errors were random, we would expect that there would be as many positive as negative values. The slope of the line in Figure 4.6a indicates that there is still about 1% proportional error. A correction can be made by multiplying each of the results by 0.9903. The resulting errors are listed in column 10. This appears to be random error, with as many positive as negative values with the same magnitudes. The precision of the method is demonstrated with a calculation of the standard deviation.

Another way to visualize the randomness is to see the random distribution of points about the zero line in Figure 4.6b. The 1% correction that was made eliminates the **bias of the method.** The bias in the assay method described above is that the results are 1% too high.

Statistical methods exist which can be used to test the possible presence of determinate error. However, they are beyond the level of this text.

In conclusion, this example shows the differences in the types of errors of accuracy. If the absolute errors of the controls have constant values, a constant error is present. If the relative error values are the same, a proportional error is present. It is only in ways such as these, with careful testing of procedures with blanks and calibrating controls and, perhaps, with a correction for inherent bias, that an effective and *accurate* assay can be created.

## 4.9 Unacceptable Methodologies

Let us now consider briefly the problems when an analysis of a replicate series has poor precision. There are two cases, as shown in lines III and IV in Figure 4.2. From the results plotted, you now recognize that both determinations have wide confidence limits. No knowledge of the details or even the general test procedures is needed to make this evaluation.

It is also possible to evaluate the accuracy in the two cases. In Case III, the accuracy is low. In Case IV, the accuracy is high. However, such judgments could be made only if we knew the "true value" of the result beforehand. Any methodology that produces such a spread in results as Cases III and IV show could not be used on unknown samples. And we could not even put much reliance on attempts to separate out constant, proportional, or random errors. Any assay that has initial test results like those illustrated in Figure 4.2, Cases III and IV, would be worthless. Another methodology would need to be found.

## 4.10 Total Error

Much of this discussion of total error follows from McFarren, E. F. et al., *Anal. Chem.* **1970,** 358.

In a great number of cases, an analytical method is developed to be used by a wide variety of people under some regulatory rules or as an often-used, standardized test for some component of a material. Additionally, to enhance meaningful communication and comparisons, standard testing methods are often developed voluntarily by various groups. Nevertheless, each analyst has certain habits and biases that have developed over years of practice. These are not "bad" or "good" habits, but usually convenient ones for the individual involved. This individuality inevitably shows up in the results using standardized methods of analysis. These disparities can result from differences in reagents, from instrument calibration with control solutions, and in fact, from all the steps involved in an analysis.

As mentioned in the introductory example, the differences in the results between laboratories and individuals are inevitably larger than the level of precision achieved by an individual alone. Let us briefly consider the problem of defining the errors within collaborative studies.

**Table 4.4**    Results Used to Calculate the Total Error. True Value = 1.00 ppm.

| Analytical Method | No. Results | Mean Value | Mean Error | Std. Dev. | Rel. Err. |
|---|---|---|---|---|---|
| A | 25 | 1.10 ppm | +0.10 ppm | 0.05 | 10.0% |
| B | 25 | 0.90 ppm | −0.10 ppm | 0.05 | 10.0% |

For interlaboratory comparisons of this type, it is common to send out identical samples made of known, pure components. Thus a "true value" *is* known. This procedure was illustrated in the example at the beginning of the chapter. Knowledge of the true value of the analyte is necessary in such interlaboratory comparisons.

In doing the comparison, a useful measure of the error is the **total error.** This is a *relative* error and is stated as a percentage. The total error is defined by

$$\text{total error (\%)} = \frac{\text{absolute value of the mean error} + 2(\text{s. d.})}{\text{true value}} \times 100$$

$$= \frac{|E_m| + 2\sigma_m}{X_{\text{true}}} \times 100 \qquad (4\text{-}12)$$

Recall that the mean error was defined in Figure 4.2 and Section 4.7. It is the absolute difference between the mean experimental value and the true value, $|\overline{X} - X_{\text{true}}|$. Notice that the total error value includes both a measure of the precision (the standard deviation) as well as of the accuracy (the mean error) of the results.

---

**EXAMPLE**

Two different analytical methodologies were tried by 25 different laboratories with samples having 1.00 ppm of some material to be assayed. The mean value of the reported results were used to compare with the true value. The values shown in Table 4.4 were calculated from the data. What is the total error?

**SOLUTION**

The total error is found using Eq. 4-12. The calculation for both analytical methods, A and B, is the same.

$$\frac{0.1 + 2(0.05)}{1.00} \times 100 = 20\%$$

---

**Origins of Total Error**

Figure 4.7 shows three different ways for a series of results to possess the same 25% total relative error. As you will see, there are significant similarities to Figure 4.2. The treatment of interlaboratory error is not much different from the treatment of error in replicate samples done in the same laboratory by the same analyst. However, the total error incorporates both the measures of precision and accuracy.

In Case A of the figure, the graph indicates a large scatter of values. In Case B, there is a large relative error in accuracy, but the precision is good. And Case C illustrates a

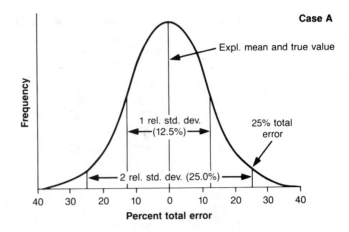

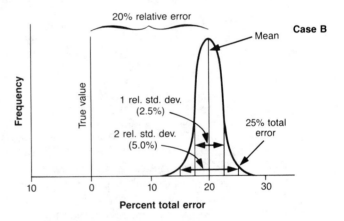

**Figure 4.7** Illustration of the concepts of total error. The total error is a combination measurement of both the precision and the accuracy of a series of determinations. Notice the similarities between these three situations and the cases illustrated in Figure 4.2. In this case, however, the abscissa is a relative error measurement rather than an absolute error measurement.

*Case A.* No relative error of the mean but a large relative standard deviation.

*Case B.* Large relative error of the mean but minimal relative standard deviation.

*Case C.* Relative error of the mean and relative standard deviation approximately equal.

[Reprinted with permission from McFarren, E. F.; Lishka, R. J.; Parker, J. H. *Anal. Chem.* **1970,** 358–365. Copyright 1970, American Chemical Society.]

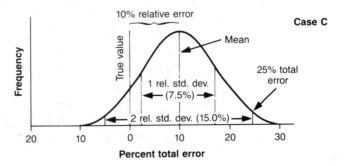

replicate series that has an approximately equal contribution from both the precision and the accuracy of the measurements.

The total error illustrated for Case A will be irreducible, since the error appears to be random. However, the situation in Case B clearly allows a correction for the bias of the method. After this correction is applied, the method will be significantly better. This is the same sort of correction as was done in the example of gravimetric determination of aluminum described earlier. The possibility of correcting for the bias in the results, such as illustrated for Case C, is questionable since the bias may not be separable. It is "hidden" under the scatter of the data. If a correction can be made, the total error will almost halve.

---

**EXAMPLE**

The following samples were tested by atomic absorption spectrometry (Chapter 18). Are the results acceptable by a criterion of total error of less than 25%?

The sample was 0.5 ppm $Mn^{++}$ which was sent to 14 laboratories to test. The mean value of the tests was 0.507 ppm. The standard deviation of the laboratories' results was 0.043 ppm.

**SOLUTION**

The mean error is $0.507 - 0.500 = 0.007$. Therefore, the total error is

$$\frac{0.007 + 2(0.043)}{0.500} \times 100 = 18.6\%$$

and the method is acceptable.

---

Such tests for determining useful and effective methodologies are more and more essential as communication with true understanding becomes more necessary and decisions require more reliable measures of the validity of the data. One of the corollaries of such studies is that simple but inherently less accurate methods may win out over more complex and accurate ones when interlaboratory collaboration is necessary. The more complex the method, the more probable it is that there will be greater interlaboratory variability.

## 4.11 Propagation of Uncertainty

So far in this chapter, you have read about the errors associated with the final results of a complete analysis, from sampling to numerical result. However, each analytical methodology consists of numerous steps. For example, consider the titration analysis mentioned in Section 4.2. Each step of the titration has some random error associated with it. In addition, each of the reagents has some error in its concentration. The steps in the titration were

1. First the sample was weighed and dissolved in water.
2. Then the proper amount of hydroxide solution was added; this required at least two readings of the burette contents.

3. The base addition was stopped when a color change was observed. This is yet another place for error.

4. Also, there were other, similar kinds of measurements made in preparing the hydroxide reagent solution, leading to an uncertainty in its concentration.

How do these random errors in each step contribute to the error in the final result? Can we calculate these contributions, and, if so, is this ability important? Following are answers to these questions.

Let me first relate a general picture of how a number of separate errors can affect a result in a particularly straightforward case. Consider that we wish to measure a 10-meter distance with a meter stick. However, the ruler is only 99 cm long, although the scale appears to be normal. By the time the distance is measured, there will be an error of a bit more than 10 cm in our result. The true 10 meters appears to be 10.10 "meters." Each error of the 11 measurements contributed to the larger total error.

Now, assume that there are three meter sticks, one a centimeter too long, one a centimeter too short, and one the correct length. A "friend" hands us one of these at random for each measurement. After the first measurement, we are uncertain of the length by 1 cm. After two measurements, we are less certain of the distance. It may be 2 cm off, 1 cm off, or correct. As you can see, the greater the number of measurement steps we take, the less certain we become of the true measurement. Each of the errors contributes its part to a final, larger possible uncertainty. Since each of the errors is carried along to the final measurement, the process is called **propagation of error.** I believe that the mathematical treatment of this propagation might be more clearly named *propagation of uncertainty.*

The mathematical treatment of propagated error is exceedingly important for determining the various methods that will be used in designing the methodology of an analysis. As should be observed in the following sections, the final error can depend on certain steps far more than others in the analysis. As a result, making a single step more precise can be highly effective in creating a more precise analytical methodology. Such thinking has a large part to play in planning the time and predicting the cost of an analysis as well as contributing to planning experiments in all quantitative sciences.

### Propagation of Absolute and Relative Random Error

Early in this chapter, two different measures of the error were defined: the absolute error and the relative error. The two are related.

$$\text{relative error} = \frac{\text{absolute error}}{\text{mean value}} \qquad \textbf{(4-13)}$$

The absolute error could be converted to a relative error by dividing by the mean (or the true value if it is known), as in Eq. 4-13. Conversely, the relative error can be converted to an absolute error measure by multiplying by either the mean or the true value.

**Table 4.5**   Examples of Propagation-of-Uncertainty Calculations

$R$ is the result to be calculated. Its absolute error is $r$. Each contributing factor $A$, $B$, and $C$ has an absolute error $a$, $b$, and $c$, respectively. The $s_A$, $s_B$, and $s_C$ are the standard deviations calculated from a finite number of trials, as in the example of Table 4.1.

| Sum and Differences (absolute error applies) | Products and Quotients (relative error applies) |
|---|---|
| For the equation: | |
| $R = A + B - C$ | $R = \dfrac{AB}{C}$ |
| Random errors: | |
| $s_R = \sqrt{s_A^2 + s_B^2 + s_C^2}$ | $s_R/R = \sqrt{(s_A/A)^2 + (s_B/B)^2 + (s_c/C)^2}$ |
| Determinate errors: | |
| Absolute     $r = a + b - c$ | Relative     $r/R = a/A + b/B - c/C$ |

Propagation-of-uncertainty calculations are straightforward even though the algebra may sometimes become messy. There are three simple rules to follow:

1.  *If the mathematical operation is addition or subtraction, the absolute error is used in calculations.*
2.  *If the mathematical operation is multiplication or division, the relative error is used to make the calculations. The absolute error is then calculated by multiplication of the relative error and the value of the result.*
3.  *All the terms of a sum must be of only one type: either absolute or relative error values. Absolute error values cannot be added to relative error values and vice versa.*

The equations for determining the random error of a sum or difference of values are shown in Table 4.5.

Note that if the error measure is a standard deviation, the result of the calculations will be a standard deviation.

---

**FIRST EXAMPLE**

As illustrated in Figure 4.8, a closed bottle containing a sample to be analyzed was weighed. The weight was 15.6784 g. A sample was poured from the bottle into a predried, weighed crucible, and the bottle with the remainder was reweighed. The second time, the weight was 15.5237 g. This technique is called **weighing by difference.**

Assume the weighings were done on the same balance, and the same person did both weighings. Further, assume the technique was the same as was done to obtain the data shown in Table 4.1. What is the error limit (standard deviation) that should be quoted along with the weight of the sample?

**SOLUTION**

The weight of the sample is (bottle with material) − (bottle less sample), or

$$\text{sample weight} = 15.6784 - 15.5237 = 0.1547 \text{ g}$$

The random error associated with each weight (0.0004 g) is found in Table 4.1, since the conditions were the same. We could write

$$(15.6784 \pm 0.0004) - (15.5237 \pm 0.0004) = 0.1547 \pm \text{s.d.}$$

Since the result is a difference of two uncertain numbers, we add the absolute errors of both, using the appropriate statistical formula from Table 4.5.

$$\text{s.d.} = (0.0004^2 + 0.0004^2)^{1/2} = 0.0005_6$$

The weight should be reported as

$$(0.1547 \pm 0.0006) \text{ g} = \text{weight} \pm \text{s.d.}$$

Another, shorter way to write this is

$$0.1547(6)$$

## Discussion of the First Example

Let us compare the statistically derived standard deviation, 0.0006 g, with the possible extreme values of the error. The extreme values occur when both errors are positive or both negative. For instance, if we simply added the two standard deviations—each being 0.0004 g or $-0.0004$ g—the weight and error would be

$$(0.1547 \pm 0.0008) \text{ g}$$

On the other hand, the other extreme would be when the standard deviations were opposite in sign, one being positive and the other negative. The sum would be zero. In that instance, the answer might be written

$$(0.1547 \pm 0.0000) \text{ g}$$

However, the statistical calculation says that it is *un*likely that both weighings will be in error by the extremes (here measured by the standard deviations). The weight of the sample weighed by difference is unlikely to have an error of zero, and also it is unlikely to have an error of 0.0008 g. Statistically, the standard deviation is between the two possible extremes, 0.0006 g.

This is a general result: The propagated error is always less than the error found by assuming the worst case for each factor. It is not likely that all the random errors will have the extreme value of the random error and, in addition, have them all in the

**Figure 4.8**    Steps in weighing a sample by difference.

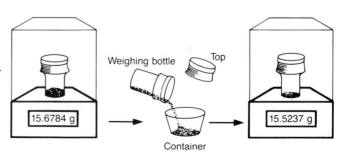

same sense. Both positive and negative deviations with different magnitudes are far more likely.

---

### SECOND EXAMPLE

The 0.1547-g sample discussed in the above example is heated to drive off some water of crystallization. The sample is heated in a crucible, and after heating, it is cooled and weighed. This procedure is repeated until, finally, two successive weighings are the same. This **constant weight** sample weighed 0.1234 g. What is the fraction of water in the sample and the error (standard deviation) in the measurement?

### SOLUTION

The fraction of water in the sample is simply

$$\frac{\text{weight of water}}{\text{weight of sample}} = \frac{0.1547 - 0.1234}{0.1547} = 0.202 \pm \text{s.d.} \qquad \textbf{(4-14)}$$

In the algebra of this calculation is the combination of a difference and a quotient. The three rules dictate that absolute errors are used for differences, relative errors are used for quotients, but the two types of error cannot be "mixed." So somewhere in the calculation, either the relative or absolute error will have to be transformed to the other type.

How can you approach this calculation? You could separate the quotient into two quotients (0.1547/0.1547 and −0.1234/0.1547), calculate the relative errors for each, and then calculate the error of the difference. This will yield the correct answer. However, this path is slightly more complicated than if you calculate the error of the numerator first and then of the quotient. Such a calculation is done next.

The error of the numerator is calculated in the same manner as done in the first example. However, as yet no error has been associated with 0.1547 and 0.1234. We must determine the values by taking account of the experimental details. Finding the weight (0.1547 g) of the original sample by difference required two weighings: the original sample in its container and then the weight of the remaining material.

In addition, it was implied in the description of the experiment that to find the final value (0.1234 g) the crucible was weighed alone and then with the dried sample. (Weighing of the crucible with the undried sample has no effect on the results here since these intermediate results are used only to ascertain the progress of drying.) Therefore, again *two* weighings were required to determine the value 0.1234 g. As a result, the associated error limits are the same for both weights, $\pm 0.0005_6$ g.

The numerator in Eq. 4-14 is then

$$0.0313 \pm \sqrt{2 \times (0.0005_6)^2} = 0.0313 \pm 0.0008$$

Equation 4-14 with errors then can be written

$$\text{fraction water} = \frac{0.0313 \pm 0.0008}{0.1547 \pm 0.0006}$$

For this simple quotient, manipulation of relative errors provides the overall error,

$$\left[\left(\frac{0.0008}{0.0313}\right)^2 + \left(\frac{0.0006}{0.1547}\right)^2\right]^{1/2} = 0.026$$

The relative error of the quotient (0.0313/0.1547) is a 2.6% relative error. To obtain the absolute error, we multiply the relative error by the value of the quotient.

$$\text{s.d.} = 0.026 \times 0.2023 = 0.005$$

The result (value $\pm$ s.d.) is then

$$\text{fraction that is water} = 0.202 \pm 0.005$$

## Propagation of Determinate Error

Determinate errors also propagate through the steps of an analysis. The measurement with a meter stick *known* to be 1 cm short is an example. Accounting for the effects of determinate errors is relatively straightforward mathematically. Expressions for calculating determinate errors in two representative equations are listed in Table 4.5. In Section 4.8, you have already dealt with the essentials of problems with determinate errors through the discussion of a gravimetric assay for aluminum. The main problem in analysis is not to make estimates of the propagation of determinate errors but to eliminate the errors altogether.

## Logarithms and Exponentials

There are no simple rules for the propagation of uncertainty in logarithmic and exponential forms which appear in formulas of chemical analysis. To treat these, it is necessary to use the calculus form of the error propagation. This subject is covered briefly in Supplement 4A.

## 4.12 Significant Figures (or Significant Digits)

Note in the second example in Section 4.11 that although there are four digits in each of the numbers in the equation, only three digits were written in the final answer, 0.202. This follows the correct practice of reporting only the digits that have meaning. In other words, the numerical result that is reported should communicate information about its error. The generally accepted convention is that the error of a measured value is indicated by error in the last digit *only*.

For instance, assume the result that is reported has a relative precision of 0.1% (1 ppt). Then we *could* write a report such as, "the material contains 23.4562687% of ingredient Z." However, given the relative error of the value, 0.1%, only some of the digits of 23.4562687 are significant — the rest are insignificant. Also, to a certain extent the retention of all the digits is misleading. Since the relative error is 0.1%, the value should be meaningful to $0.001 \times 23.4 = \pm 0.02$, and the result should be written 23.46% with or without a standard deviation listed. Note that the last digit was rounded up to a six. This follows the rules of **rounding off** when limiting the number of digits being reported for a number. Rounding off is described in the next subsection.

There are other, simple conventions that are often used when there is an error that is greater than $\pm 1$ in the last digit. The first convention is to signify that the digit is less certain than $\pm 1$ by writing it as a subscript. An example is $1.98_7$. This notation relates that the error is smaller than $\pm 0.01$ but greater than $\pm 0.001$.

Recall also a second convention shown earlier: Often, instead of writing 5.785 $\pm$ 0.003 (value $\pm$ s.d.), one writes 5.785(3). The digit enclosed in parentheses is the error in the last digit of the value.

### Rules of Rounding Off

The rules of rounding, which you may have learned as rules to be memorized, *follow* from the rules of error analysis and calculations of error propagation. A slavish following of the rounding-off rules should be unnecessary once you are familiar with error calculations.

The general rules for rounding are

1. *Retain no digits beyond the first uncertain one.*
2. *If the number in the place beyond the last significant one is less than 5, leave the figure as it is.*
3. *If the next digit is 6 or greater, add one to the last retained digit.*
4. *If the next digit is a 5, then round to the nearest even digit: 2, 4, 6, 8, or 0. This will cause a statistical 50:50 chance that you will round up or down. Thus no bias will be introduced by the rounding.*

**Significant Digits and Arithmetic Operations**    Rule 1 above states that no digits beyond the first uncertain digit should be retained. A discussion of the rules for determining the cutoff points follows.

As for propagation of error, there are two different cases in determining the significant figures in arithmetic operations: those involving addition and/or subtraction and those involving multiplication and/or division.

### Addition and Subtraction

*Rule: In addition and subtraction, retain only as many decimal places as there are in the number with the fewest digits to the right of the decimal.*

---

**EXAMPLE**

$$
\begin{array}{r}
21.2 \\
3.035 \\
\underline{0.12} \\
24.355 \quad \text{becomes} \quad 24.3
\end{array}
$$

---

The last two digits have no meaning, since at least one of the numbers that were summed has no significant figures in those places. Compare this with the rules for the calculation of the absolute error for sums and differences.

### Multiplication and Division

*Rule: The factor containing the least number of significant figures determines the number of significant figures in the product or quotient.*

This means that the *relative* error of the result cannot be less than the largest relative error of any factor.

It is common that the rules of significant figures for multiplication and division cause mistakes in the number of significant digits retained. For instance, the next example shows a situation where the rules of significant digits suggests writing an answer in a way that disagrees with propagation-of-error calculations.

---

**EXAMPLE**

How should the result of the multiplication $2.0 \times 43$ be written so as not to be misleading?

**SOLUTION**

By the rules of significant digits,

$$2.0 \times 43 = 86$$

The implication of this result is that the answer is $86 \pm 1$. However, we cannot be certain of the result better than to a relative error of 5% — $0.1/2.0 = 0.05$.

But $0.05 \times 86 = 4._3$, and the answer should be written

$$2.0 \times 43 = 86 \pm 4$$

Thus the rules for significant digits provide an answer that is quite misleading. It is more correct to write

$$86 \pm 4 \quad \text{or even} \quad 9 \times 10^1$$

than it is to write 86.

---

If in doubt, use the propagation-of-error calculations described in Section 4.11. The rules for significant digits were derived from them. Following are a few more examples of truncation and rounding of numerical results. Note that approximations to the uncertainties are used. In both examples, the relative error of the result is approximately the relative error of the *least* certain term.

---

**EXAMPLE**

$$31.1 \quad \times \quad 0.063 \times \quad 98.9 \quad = 193.77477 \text{ rounded to } 194$$
*Approximate*  
*uncertainty:* 1 in 300    1 in 60    1 in 100    round to 1 in 50 (or $\pm 2\%$)

**EXAMPLE**

1 in 1000    2 in 1000

$$\frac{961 \times 547}{0.053} = 9,918,245.2 \text{ rounds to } 9.9 \times 10^6$$

1 part in 50 (2%)      1 in 50 (or $\pm 2\%$)

---

Choosing the last significant digit after multiplication and division becomes easy to do in your head with practice. The approximations (such as 1 in 50 ≈ 1 in 60) work because large changes in uncertainty must occur before an additional digit is dropped. A full **order of magnitude** (meaning a factor of 10) in the uncertainty must occur (for instance, 0.2% versus 2%) before you decide whether the next digit in the answer should be dropped or not.

---

**EXAMPLE**

$$15 \times 346 \times 165 = 856,350.0$$

The largest error is 1 part in 15 (or ±7%), thus the error in the answer must be ±7%. The answer becomes

$$8.6(6) \times 10^5 \text{ or } 8._6 \times 10^5$$

The error would have to be around 12% (1 in 8—8 being the first digit of the product) before another digit would be dropped. On the other hand, in order to retain another digit, the relative error would have to be about 1 part in 856 or 0.1%.

---

**Logarithms**     Logarithms appear in numerous formulas of chemical analysis. In these cases, the rule for the number of significant digits is simple. When converting numbers to logarithms, use as many decimal places in the mantissa—the part of the logarithm following the decimal point—as there are significant digits in the number.

---

**EXAMPLE**

$$\log 12.35 = 1.0917$$

There are four significant digits in the number and four decimal places in the mantissa.

---

## 4.13 Discordant Data

In Figure 4.1, you saw a plot of data from analyses done at ten different laboratories. An analyst in each laboratory determined the content of aflatoxin in identical sample material. One of the resulting values was an outlier. The directors of the interlaboratory study rejected that single value from their sample.

The same kind of problem also arises in replicate determinations on a single sample done in a single location. Of a number of replicates, one result may seem to be quite far from the mean value and also far from all the other results. This is called a **discordant data** point or an outlier. Should it be kept as a part of the analysis or should it be ignored? There are two different approaches to answering this question: statistical and judgmental. Usually both skills are required.

The data in Table 4.6 illustrate this dual approach. The values in the table were determined from two sets of triplicate determinations. They were determined in the order shown on the first line. The reason that the triplicate determination was rerun was that the third value, 7.19, *seemed* discordant. This was a judgment of the person

**Table 4.6**   Results of Two Sets of
Triplicate Determinations

Results in order of determination:
    7.06, 7.04, 7.19, 7.10, 7.02, 7.09

Results in order of value:
    7.02, 7.04, 7.06, 7.09, 7.10, 7.19

doing the analyses. It may be that this analysis had been done many times on a wide variety of samples. With past experience and the results of all six determinations, the analyst might decide to ignore the 7.19-value. It may be outside of the range in which the random error usually lies. Or perhaps he remembered that some blue specks were in that sample, and they may have caused the problem.

However, we ourselves do not have as much experience with the analysis. (In fact, I have not even explained anything about it at all, except some numerical results.) So let us look at a statistical test to help us determine whether to reject the outlier. It is called **Dixon's test.** The following approach can be used to determine only if the extreme value is *statistically* subject to rejection. This is *not* a method to use mechanically but an aid in deciding whether to ignore the result of the specific run.

To see more clearly the distribution of the values obtained from the six replicates, they are rewritten in order of their values. Let $d$ equal the difference between the possible outlier value and the result closest to it in value. Recall the definition of the range $w$ from Eq. 4-4, the difference between the highest and lowest measured values.

For three to seven replicates, Dixon's test involves comparing the value of $r$,

$$r = d/w \qquad \text{(4-15)}$$

with an appropriate "critical value" of $r$, which is listed in Table 4.7. The value in the table depends on the total number of measurements made, $N$. Then the conclusion can be made as follows. If the calculated value of $r$ exceeds the applicable critical value, then there is a 10% chance that rejection is a wrong decision. This clumsy wording states the statistical conclusion. It may be easier to consider that if $r$ exceeds the applicable critical value, you are 90% confident that it should be rejected.

---

**EXAMPLE**

Using the Dixon text on the results tabulated in Table 4.6, determine whether the value 7.19 should be rejected as an outlier.

**SOLUTION**

For the results shown in Table 4.6,

$$d = 7.19 - 7.10 = 0.09$$

and

$$w = 7.19 - 7.02 = 0.17$$

so

$$r = 0.09/0.17 = 0.53$$

**Table 4.7**   Critical Value for $r$ for Rejection of Outliers with 1-in-10 Probability of a Wrong Decision*

$r = (X_{outlier} - X_{adjacent\ value})/w$

$w = \text{range} = X_{highest} - X_{lowest}$

| Number of Experiments ($N$) | $r$ critical Value |
| --- | --- |
| 3 | 0.94 |
| 4 | 0.76 |
| 5 | 0.64 |
| 6 | 0.56 |
| 7 | 0.51 |

* For a more complete table see the references cited here or the Youden and Steiner reference (their Table C.1.) cited at the end of the chapter. With $N$ larger, the criteria for $r$ and $w$ change.

*Sources:* W. J. Dixon. Biometrics **1953**, *9*, 74. R. B. Dean; W. J. Dixon. *Anal. Chem.* **1951**, *23*, 636.

According to Table 4.7, the critical value for six samples is 0.56. Thus, statistically, it is *not* suggested to reject the sample.

The analyst familiar with the assay may still wish to reject the 7.19-value anyway. If the value is thrown out, then a new mean and standard deviation must be calculated for the five remaining results. The result will then be

$$\text{5 points:} \qquad \overline{X} = 7.06_2 \qquad s = 0.03_3$$

This compares with the results from all the values

$$\text{6 points:} \qquad \overline{X} = 7.08_3 \qquad s = 0.06_0$$

Notice that the relative change in the mean, about 0.3%, is much less than the relative change in the standard deviation, which is almost a factor of two. This is typical for the effects of outliers (or almost-outliers) on the results of replicate determinations.

## 4.14 The Median

The **median** value of a set of numbers is the middle value. There are as many values above it as below. If the number of values is even, then the median value is the mean of the two middle values.

**EXAMPLE**

What is the median value of those in Table 4.6?

**SOLUTION**

$$\frac{7.06 + 7.09}{2} = 7.07_5$$

The median is a less reliable estimate of the correct measured value than is the mean. However, it is less sensitive to large errors in experimental values. Thus, when there is wide scatter in the data, the median may be better than the mean to estimate the experimental value. However, it is always preferable to run more samples and, even better, to isolate the cause of the scatter and correct the methodology.

## Suggestions for Further Reading

Rigorous mathematical treatment of estimation, least squares, and testing of hypotheses.
Hamilton, W. C. "Statistics in Physical Science"; Ronald Press: New York, 1964.

A readable monograph with both mathematical derivations and examples and suggestions for experimental design.
Youden, W. J. "Statistical Methods for Chemists"; Wiley: New York, 1951.

Completely mathematical treatment of title subject.
Green, J. R.; Margerison, D. "Statistical Treatment of Experimental Data"; Elsevier: Amsterdam, 1968.

A short, practical review of evaluating accuracy and precision in experimental results.
Mandel, J. In "Treatise on Analytical Chemistry," Pt. I, Vol. 1, 2nd ed.; I. M. Kolthoff and P. J. Elving, Eds.; Wiley Interscience: New York, 1978; Chap. 5.

A chapter relating the testing of error to practical problems.
Currie, L. A. In "Treatise on Analytical Chemistry," Pt. I, Vol. 1, 2nd ed.; I. M. Kolthoff and P. J. Elving, Eds.; Wiley Interscience: New York, 1978; Chap. 4.

An exhaustive analysis of the title subject.
Liteanu, C.; Rica, I. "Statistical Theory and Methodology of Trace Analysis"; Ellis Horwood: Chichester, 1980.

Derivation of the statistical treatments used in analysis along with listings of computer programs used to calculate some of them. Examples with references for numerous different analytical methods are included.
Moritz, P. In "Comprehensive Analytical Chemistry," Vol. 9; G. Svehla, Ed.; Elsevier: Amsterdam, 1981; Chap. 1.

An introductory level monograph of probability and statistics.
Young, H. D. "Statistical Treatment of Experimental Data"; McGraw–Hill: New York, 1962.

A compilation of statistical methods including interlaboratory comparison methods; complete and detailed.
Youden, W. J.; Steiner, E. H. "Statistical Manual of the AOAC"; Association of Official Analytical Chemists: Washington, D.C., 1975.

An easy to read, profusely illustrated book covering most aspects well. It includes worked solutions to chapter problems. British in tone.

Caulcutt, R.; Boddy, R. "Statistics for Analytical Chemists"; Chapman and Hall: London, 1983.

## Problems

**4.1**   For a single observation, $\sigma = 0.01$ cm. How many observations would be required to report a mean with $\sigma_{mean} = 0.001$ cm?

**4.2**   The results for seven determinations of the percentage of Cu in a sample were 39.3, 41.2, 40.4, 40.0, 41.1, 39.9, and 40.9 ppm Cu.
    a.   Compute the mean and its standard deviation.
    b.   What is the 95% confidence limit for the value of the mean being the true value?

**4.3**   If the determinate error of a sample is 0.2% and the determinate error for the constituent in the sample is $-0.2\%$, what is the relative determinate error?

**4.4**   The results of a set of determinations (in weight %) were

$$21.25, 21.27, 21.30, 21.23, 21.21$$

Find the mean, standard deviation ($s$), and 95% C.L. of the mean.

**4.5**   [Reprinted with permission from C. Palme; E. Jagoutz. *Anal. Chem.* **1977**, *49*, 717. Copyright 1977, American Chemical Society.] The following table contains data from five different research groups, each determining the content of a rock sample from the moon. The sample was collected by the Apollo 17 crew and was labeled as sample Mare Basalt 70215.

Major Element Data (wt%)

| Group | Mg | Al | Si | Ca | Ti | Fe |
|-------|------|------|-------|------|------|-------|
| 1 | 4.82 | 4.69 | 17.72 | 7.70 | 7.84 | 15.51 |
| 2 | 5.14 | 4.64 | 17.39 | 7.45 | 7.88 | 15.25 |
| 3 | 4.77 | 4.77 | 17.98 | 7.82 | 7.48 | 15.08 |
| 4 | 5.63 | 4.65 | 17.58 | 7.73 | 7.91 | 14.94 |
| 5 | 4.98 | 4.57 | 17.90 | — | 7.70 | 15.70 |
| 6 | 5.03 | 4.64 | 17.82 | 7.62 | 7.66 | 15.83 |

Minor Element Data (ppm)

| Group | P | S | K | Cr | Mn |
|-------|-----|------|-----|------|------|
| 1 | 500 | 1880 | 340 | 2950 | 2044 |
| 2 | 390 | 1800 | 330 | 2870 | 2170 |
| 3 | 430 | 1700 | 415 | 2670 | 2250 |
| 4 | 300 | — | 660 | 2800 | 2090 |
| 5 | — | — | 340 | 2680 | 1940 |
| 6 | 440 | 1620 | 407 | 2820 | 2020 |

    a.   Calculate the mean and standard deviation of the content for each of the elements.
    b.   Calculate the relative standard deviation (in %) for each of the elements.

c. Is the result for K from group 4 an outlier?

d. Which of the sets of minor element data has the largest range? Which has the largest relative range?

**4.6** A titration is done on a regular basis. The samples, all about 0.1 g, are weighed to the nearest 0.1 mg. Approximately 40 mL of titrant (liquid added from a burette) is needed to titrate each sample to the end point. The molarity of the titrant is known to $\pm 0.1\%$. The titrant volume is measured by reading a burette before adding any liquid and again after all the liquid has been added. The error in reading the burette is estimated to be a random error of 0.02 mL. The titration is finished when an indicator color change is seen. This color change occurs over a 0.02-mL volume.

a. Estimate the relative error of the result.

b. If a 0.1000-g sample contained 22.2% of element Z, what is the absolute error in the content of Z?

**4.7** The inscription on a 5-mL pipette says that it will deliver (to deliver is noted as TD on the pipette) 5.000 mL of water at 20 °C. A set of experiments is made to determine the errors of the operation. The water and pipette are held at 20°, and the following values of the volume were found by weighing the water delivered. The weights could be measured to 0.0002 g, so any error due to weighing can be ignored. Twenty runs were made.

Replicate measurements for a 5.000-mL TD pipette (in mL) gave the following results.

4.985, 4.981, 4.989, 4.970, 4.974, 4.981, 4.976, 4.988, 4.993, 4.973,
4.970, 4.985, 4.988, 4.982, 4.977, 4.982, 4.974, 4.988, 4.979, 4.985

a. Calculate the mean volume.

b. Calculate the range of the measurements.

c. Calculate the standard deviation.

d. Calculate the 95% confidence limit.

**4.8** The Nernst equation describes a relationship between a voltage and a chemical concentration expressed as an activity, $a_i$.

$$E = E^\circ + \left(\frac{RT}{nF}\right) \ln a_i$$

For $n = 1$, what is the relative error in $a_i$ for a 1-mV change in $E$ at 25 °C?

**4.9** A standard alloy sample from the National Bureau of Standards contains 57.85 wt% of chromium. Some 0.1-g samples were prepared by acid dissolution and dilution to 500.0 mL. Each sample was assayed eleven times. The following results were obtained.

Sample No. 1: 57.64, 58.07, 57.88, 57.79, 57.67, 57.79, 57.67, 57.73, 57.59, 57.81, 57.76

Sample No. 2: 57.88, 58.00, 57.61, 57.80, 57.56, 57.61, 57.51, 57.77, 57.55, 57.40, 57.67

a. Calculate the mean and relative s.d. for each sample.

b. How many assays would it be necessary to run on each sample to have the 95% C.L. $\pm 0.5\%$? Assume that $s = \sigma$, the value found in part a.

**4.10** A method for the determination of chloride tested on pure NaCl gave the following results.

| NaCl in the Sample (g) | | Error in Analysis | |
|---|---|---|---|
| *Used* | *Found* | *(in gm)* | *(in %)* |
| none | 0.0022 | | |
| 0.1000 | 0.1023 | | |
| 0.3927 | 0.3960 | | |
| 0.8295 | 0.8311 | | |
| 1.2976 | 1.3005 | | |

a.  Complete the table.
b.  What type of correction should be applied, and what is the value of it?
c.  After the correction is made, what is the nature of the residual error, and what is its *average* value?
d.  How large should the samples be to keep this residual error less than 0.1% of the true value after making the correction b?

4.11   [Ref: J. Isreeli; M. Pelavin; G. Kessler. *Ann. New York Acad. Sci.* **1960,** *87,* 636.] Figure 4.11.1 shows the result for an automated analysis of potassium in blood serum. Each peak represents the instrument response. The zero-response is at the initial horizontal line at the left. On the scale, 72.5 spaces represent 6 mM $K^+$.

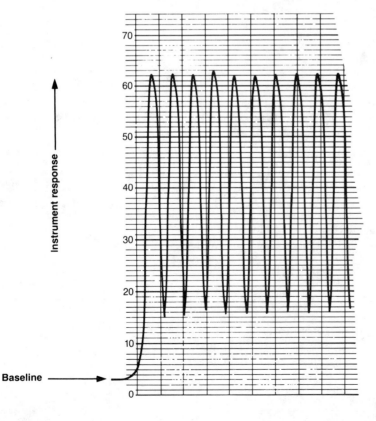

**Figure 4.11.1**   Data from an automated analysis of potassium in blood serum.

a. What is the relative standard deviation of the method for the results from the ten samples? Assume no baseline fluctuation occurs.
b. Estimate the relative standard deviation due to errors in reading the chart. What, then, is the relative standard deviation due to the instrumental method, excluding the chart reading?

4.12   [Ref: Application note 110. Yellow Springs Instruments Co. Inc., Yellow Springs, OH.] Several beers and wines were tested for ethanol content by two methods. One used an instrument that utilizes an enzyme–electrochemical method which involves only injecting a 25-$\mu$L sample into the sample chamber and reading the (precalibrated) ethanol content on a digital display within a minute or so. The other involves distilling a large volume of sample and measuring the density of the resulting distillate. The following results %(w/w) were found.

| Sample | By Electrochemical Analyzer | By Distillation and Density |
|--------|------------------------------|------------------------------|
| Beer A | 3.80 | 3.82 |
| Beer B | 4.32 | 4.35 |
| Beer C | 3.48 | 3.47 |
| Wine A | 10.60 | 10.72 |
| Wine B | 5.90 | 5.83 |
| Wine C | 8.49 | 8.58 |

a. For each of the six measurements, find the relative deviation between the two methods. (Does it matter to the results which set is considered to be "true?")
b. Calculate the average relative deviation between the two methods. Is there a significant (> |0.005|) *relative* bias in the method?

4.13   [Ref: Application note 100. Yellow Springs Instruments Co. Inc., Yellow Springs, OH.] Ten molasses samples were tested for "total sugar as invert" by two methods. One is a relatively fast instrumental method, the other a more tedious wet-chemical one. The following results were obtained.

| Sample | By Instrumental Analyzer | By Wet-Chemical Method |
|--------|---------------------------|-------------------------|
| A | 43.1 | 49.8 |
| B | 43.6 | 49.9 |
| C | 45.1 | 50.3 |
| D | 50.6 | 56.3 |
| E | 47.5 | 54.4 |
| F | 47.1 | 54.0 |
| G | 51.6 | 57.3 |
| H | 50.8 | 57.5 |
| I | 50.9 | 57.0 |
| J | 44.1 | 49.8 |

a. Assume the "true" value is that found by the wet-chemical method. For each of the samples A–J, calculate the relative deviation between the two methods.
b. Calculate the average relative deviation for all samples together for both methods. Is there a significant bias (> 0.5%) to the instrumental method?
c. Is enough information available to decide whether a relative or absolute adjustment should be made for any necessary correction?

d.  If there is a significant bias, what is the value of the correction you can make to the instrumental results? Express the answer as

instrumental result (+ or ×) value = true (wet-chemical) result

4.14   [Ref: A. Benedetti–Pichler. *Anal. Chem.* **1936**, *8*, 373.]
You want to analyze a sample for aluminum. The relative error inherent in the analyses for all components is 0.1%. The impurities that are found are Be, oxygen, and Sb. The three impurities are each 0.03% of the total sample. The mean aluminum content that is calculated is 99.09% from an aluminum assay. Is the amount of aluminum found more precisely from the aluminum determination or from the impurity determinations?

4.15   In the titration done as described in Section 4.2, a solution of base was used which was assumed to be 0.1019 *N* exactly. The result from the assay is found using the product

normality × other terms = result = $21.25 \pm 0.04$

In fact, the correct description was $(0.1019 \pm 0.0003)$ *N*. This is the mean ± s. d. Thus the standard deviation of the result must be less certain than was stated. The result, $21.25 \pm 0.04$, must be modified to account for the extra uncertainty.
a.  What is the new standard deviation of the result?
b.  What would be the new standard deviation if the normality were $0.1019 \pm 0.00005$?

4.16   Calculate the propagated uncertainty in the calculation

$$\frac{81.32(\pm 0.09) \cdot 0.1399(\pm 0.0002)}{-3.21(\pm 0.01)} - 22.3323(\pm 0.0001) = -25.8764 \pm (\quad)$$

Write the correct number of digits for the mean and the s.d.

4.17   What is the propagated error in the following calculation? The estimated error for each value is in parentheses. Give a numerical value where possible — otherwise leave in the *simplest* algebraic form.

$$Q = \frac{A(\pm 0.02)}{B(\pm 0.01) + C(\pm 0.05)} + D(\pm 16) - E(\pm 3) = \frac{A + (D - E)(B + C)}{B + C} \pm (\quad)$$

4.18   What is the propagated error in the following calculation? The estimated error for each value is in parentheses. Give a numerical value where possible — otherwise leave in the *simplest* algebraic form.

$$\frac{A(\pm 0.02) + B(\pm 0.1) - C(\pm 0.05)}{D(\pm 16)} = \frac{A + B - C}{D} \pm (\quad)$$

4.19   Given the general rule for determining significant figures, that the relative error of the calculated result be the same as that of the least accurate number, how should the value of

$$\log_{10}(1.125 \times 10^{13})$$

be reported? Certainly 13.05115252 is not correct.

4.20   *On total error.* A sample containing a known amount of DDT was sent out for analysis to 17 laboratories to be analyzed by gas chromatography. The samples contained 0.30 ppm DDT. The relative error of the reporting laboratories was 11.8%, and the relative standard deviation was 23.9%. What is the total error, and is the method acceptable within the criterion of less than 25% total error?

**4.21** *On total error.* [Reprinted from H. Adelman; S. W. Jenniss; S. A. Katz. *American Laboratory,* December 1981, 31. Copyright 1981 by International Scientific Communications, Inc.]

An interlaboratory cooperative analysis was carried out on NBS standard sewage sludge. Sixteen laboratories cooperated in the work. Each result is the mean value of 5 analyses from the same sample. Not all the laboratories assayed all the elements. Some of the results are listed below. The zeros are not significant in the values. They are used for ease of tabulation.

Individual Mean Results of Laboratories (Concentrations in ppm)

Aluminum 4200, 7620, 4660, 3800, 4580, 7600, 5790, 3150, 3690, 4270
Calcium 27000, 15400, 15800, 5900, 15900, 18500, 614, 8730, 12800, 17200
Copper 1050, 907, 963, 945, 933, 913, 936, 1030, 1080, 983, 893, 954, 830, 929
Iron 17000, 636, 691, 11400, 15700, 11700, 1130, 17900, 16400, 15600, 1410, 13400, 13200, 12600
Lead 441, 565, 563, 524, 546, 387, 496, 532, 451, 565, 484, 534, 418, 714
Nickel 200, 154, 178, 157, 172, 154, 92.8, 143, 172, 209, 152, 164, 184, 166, 180
Silver 65.0, 132, 84.4, 41.0, 50.3, 82.7, 52.3, 93.4, 50.7, 72.9, 51.9, 62.1
Zinc 1440, 1410, 50.2, 1060, 1150, 1060, 1150, 1320, 1350, 1230, 1300, 1230, 1030, 814

The reference values for the elements in the standard are listed below as mean (ppm) and (2 $\sigma$-range) (ppm).

Aluminum    4557    (2010-7110)
Calcium              not listed
Copper      1095      (831-1360)
Iron       15155   (3810-28500)
Lead         519      (305-733)
Nickel       198      (164-233)
Silver        80        (0-203)
Zinc        1323    (1190-1450)

a.  Calculate the mean and s.d. for each element.
b.  Calculate the total error for each except calcium.
c.  Are there any biases in the assay method for Ni or Zn? If so, in what direction?

### The following problems require calculus.

**4.22**  For the function $R = A - B + C$ and the formula for the total differential of $R$, write the algebraic expressions for the propagated determinate error (*not* random) with errors $a$, $b$, and $c$.

**4.23**  For the function $R = AB/C$, calculate the absolute and relative determinate errors if $a$, $b$, and $c$ are the errors in each factor.

**4.24**  Derive the expression for the relative error and absolute error for $R$ in terms of $A$.

  a.  $R = \ln A$;    b.  $R = e^A$;    c.  $R = 0.879 \ln A$;    d.  $R = 2.193/e^A$.

## SUPPLEMENT 4A

# Total Differential Error Analysis

The equations of Table 4.5 can be derived from the equation expressing the total differential of any function. Call the function $f(a_1, a_2, \ldots, a_n)$. We shall also use the abbreviation $F$ for the function. In the language of calculus, the total differential of a general function $f(a_1, a_2, \ldots, a_n)$ is given by

$$d[f(a_1, a_2, \ldots, a_n)] = \left(\frac{\partial F}{\partial a_1}\right) da_1 + \left(\frac{\partial F}{\partial a_2}\right) da_2 + \cdots + \left(\frac{\partial F}{\partial a_3}\right) da_n \quad \textbf{(4A-1)}$$

The quantities in parentheses are partial differentials. This means that the derivative of the function is taken with respect to one parameter while all the rest are held constant. Take as an example the function $f(A, B, C) = A^2BC$.

$$d(A^2BC) = (2ABC)\, dA + (A^2C)\, dB + (A^2B)\, dC \quad \textbf{(4A-2)}$$

In Eq. 4A-2, $dA$, $dB$, and $dC$ would be the estimated errors of the measurements used to determine the quantities $A$, $B$, and $C$. In this notation, the relative errors would be written

$$dA/A, \qquad dB/B, \qquad \text{and} \qquad dC/C$$

respectively, and $f(A, B, C)$ represents the result calculated from experiments.

Equation 4A-1 holds for the situation where the changes in the parameters are infinitesimal. In order to use a total differential error calculation that is applicable to chemical analysis, the errors in the factors $a_1$, $a_2$, and so forth must be able to be finite. For a finite change, the equation to be used is

$$\Delta[f(a_1, a_2, \ldots, a_n)] = \left(\frac{\partial F}{\partial a_1}\right) \Delta a_1 + \left(\frac{\partial F}{\partial a_2}\right) \Delta a_2 + \cdots + \left(\frac{\partial F}{\partial a_3}\right) \Delta a_n \quad \textbf{(4A-3)}$$

The value $\Delta f$ is the now the error of a result calculated from an equation that has factors $a_1, a_2, \ldots, a_n$.

## Supplement Problems

**S4.1**  For *random* errors $R$,

$$r^2 = (\partial R/\partial X)^2\, x^2 + (\partial R/\partial Y)^2 y^2 + \cdots$$

where $x$ and $y$ are the absolute standard deviations of $X$ and $Y$, and so on. Show that the relative error of a computed result is

$$(r/R)^2 = (x/X)^2 + (y/Y)^2 + (z/Z)^2$$

if

$$R = XY/Z$$

Note: The equation for the relative error is a general result.

S4.2   What is the random absolute error of a computed result if $R = X + Y - Z$? Use the nomenclature of problem S4.1.

# 5

# Sampling

## 5.1 Factors Involved in Effective Sampling

Suppose that a shipload of ore is delivered to a plant, and the average metal content of the load is needed to plan the refining process. When you look at the delivered material, you see different-colored pieces in a large range of sizes. The larger, individual pieces are striated, indicating that they are **heterogeneous;** they differ from one region to another. In addition, the larger rocks have rolled down to the bottom of the pile—at least on the outside where they can be seen. How can you sample the shipment so that the average metal content of the whole can be ascertained?

Or consider the mapping of a pollutant such as mercury among a system of streams in an area. You probably would want to know more than just how much mercury is in a sample scooped from a stream. You might want to know how the mercury is moved around. Is it dissolved in solution in the water? Is it on small particles that can settle out of the liquid? Is it contained in or on the surfaces of colloidal particles (particles so small that they remain suspended, essentially indefinitely)? Is the mercury in an organomercury compound that is volatile? (Such compounds form when the mercury interacts with bacteria.)

Another example might be a few microscopic, smooth spheres of a material that were found embedded in the threads of a sweater belonging to a person suspected of firing a gun in a murder. The spheres could be bullet propellant, but the suspect asserts that he is a welder, so they could have come from the sputtering of a welding rod. Is the material from the bullet propellant or from a welding rod? How do you find out?

These are just three typical problems that require the ideas and methods of sampling of materials for analysis. The sampling method is a significant and integral part of a chemical analysis. If the sampling method produces laboratory samples that are erratic or do not represent the true content of the material, its error will overwhelm errors from other parts of the analysis, and the determination could be meaningless.

The purpose of this chapter is to present some of the fundamentals of sampling methods. The details are too diverse and involved to present more than a few of them. As I hope you come to appreciate, the preparation of a sample often depends on a broad understanding of the descriptive chemistry of the substances being determined and the other elements and compounds present in the sample.

Deciding how to approach creating a sample for analysis depends on a number of factors:

1. The *size* of the bulk to be sampled, which can vary from shiploads to subcellular organelles a fraction of a micron in diameter.
2. The *physical state* of the fraction to be analyzed, for instance, crystalline solid, liquid, gas, or glassy solid.
3. The *chemistry of the material* to be assayed. If a specific molecule is sought, nothing can be done that would destroy or even inadvertently alter the substance before the assay itself is run.

Thus the method of sampling a substance and preparing it for testing is, inevitably, linked to the assay method(s) that are planned to be used. *The sampling method is an integral part of an analytical methodology.*

Sampling of the progress of a chemical reaction, whether it occurs in a flask in a laboratory, in a reactor in a process plant, or in a biological organism, can be done either intermittently or continuously. If the sampling is continuous, the "laboratory" must somehow be connected directly to the reaction. While this idea may seem rather trivial at first, it is not so simple if the sample is of a stack gas at 600 °C or a metal smelter around 2000 °C. In biology, there is a finite amount of sample that can be taken from an organism without harming it irreparably. And for an in situ test, the size of the **sensor** that can be placed at the location of interest may also be limited (for instance, in a needle in an arm vein). These are some of the challenges of sampling in chemical analysis.

## 5.2 Good Samples: Representative and Homogeneous

The two aims of the sampling methods are to ensure that the sample is **representative** of the material being analyzed and that the sample that is analyzed in the laboratory is **homogeneous.** The term *representative sample* has the same meaning in this context as in common usage: The content of the sample is the same overall as the material from which it is taken. *Homogeneous* means that the sample is the same throughout.

The more representative and homogeneous the samples are, the smaller will be the part of the analysis error that is due to the sampling step. As can be appreciated from the discussion of propagation of error in Chapter 4, an analysis cannot be more precise than the least precise operation. Quite often, the sampling is the limiting factor of both the accuracy and the precision of analyses.

### Making the Sample Representative

If a sample is not representative of the material to be analyzed, the results of the analysis cannot accurately reflect the content of the material. Almost all materials to be sampled, with the exception of smaller volumes of liquids or gases, are heterogeneous in composition. For example, a heterogeneous solid might be composed of a mixture containing particles of various sizes and varying compositions in different regions within individual particles. Stone is one such heterogeneous solid; it is composed of agglomerations of small crystals of varying compositions. Another heterogeneous material to be sampled might be a field of grass that is to be tested for natural nutrients such as phosphorous or for pollutants such as polychlorinated biphenyls (PCBs).

In all materials, one should take **random samples** and reduce them to a homogeneous fraction that represents the whole. The chemical mix in the sample should be a miniature replica of the contents of the whole from which it came.

Consider again the shipload of ore mentioned at the beginning of this chapter. How would an analytical sample be taken when the huge load is composed of large and small heterogeneous particles and chunks? One way is to remove portions of the ore as it travels up a conveyer to the plant. Consider some of the properties of the ore that become evident as it travels along the conveyer belt. The ore is shaken by vibration of the belt. When a solid with chunks of disparate sizes is shaken, the particles, from the size of dust to pieces weighing a few kilograms, usually separate into different locations according to size. It is likely that the different-sized particles have different compositions; in fact, differences in composition may be the reason the material fell apart to form particles of varying sizes as it was crushed and transported. So, to take a representative sample from the conveyer belt, we must sample all the way through a cross section of the material.

The next question is, how often are such samples taken? Are they obtained at regular or irregular times? Figure 5.1 demonstrates sampling from a conveyer belt over the 8-h period required to transport the entire shipload of ore. First we assume that a large number of samples were taken and analyzed to obtain a continuous measure of the metal content over time; the results of this continuous analysis are shown as the continuous line in Figure 5.1. It is evident that the contents varied over time.

What might we find if the sampling were done randomly and far less often: specifically, at the points shown by the arrows? In all, 16 collections are made and analyzed. Just by chance, the samples collected at short intervals coincide with the points where the ore is relatively low in the metal: from hours 1.5–2.5 and 5.5–6.5. The overall value — the average of the 16 measurements taken — will be significantly

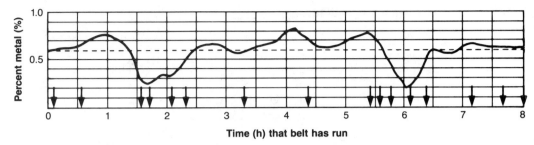

**Figure 5.1**    Plot of the true metal content in a cross section of ore versus the time at which the belt is sampled. The content fluctuates about a mean value of 6%. The arrows indicate 16 random times at which the ore could be sampled; the ore thus sampled will yield an assay value significantly below the true average value. Sixteen regularly spaced sampling times half an hour apart will produce a more accurate indication of the true average content.

low. If, however, the samples were taken at regular intervals of half an hour, the average would approach that of the true average of the pile.

Notice that to obtain a homogeneous sample in time, it seems reasonable (and can be justified by mathematical statistics) to use regular sampling times. But even with regular sampling intervals, there will still be some possibility of obtaining an imperfect representation of the whole. This problem will be more likely when fewer individual samples are taken. However, nearly perfect information will require nearly infinite effort.

The general rule for sampling of a large quantity of material on a conveyor belt is: *To obtain a random sample, the sampling should be regular in time and fixed in its form or method.* Recall that the whole cross section should be taken for each sample.

A similar general rule holds for a pile of material: To obtain a random sample, the sampling should be fixed in its method. An application of this rule is the method of *coning and quartering* in which a pile of material is divided into quarters (if not in reality then in imagination, since it is difficult to cut a 10,000-kg pile apart). If the conical pile is large, samples are taken from each quarter of the pile (north, south, east, and west). These samples are crushed and formed into a smaller, conical pile. This smaller pile is flattened and cut into equal quarters, and two opposite quarters are taken. The choice of the set of two quarters should be made in some random manner, such as by flipping a coin. The quarters are crushed further, mixed thoroughly, and repiled. This fixed procedure (cutting, taking opposite quarters randomly chosen, mixing and grinding, and repiling) is continued until a sample of the size needed for replicate samples in the laboratory is obtained. This technique minimizes the bias of the sampling.

If the original pile is so large that the sampling cannot be done on the total, since the core of the pile is not really available, you might expect that the regular removal of samples of the ore from the conveyor belt, where the whole amount is subject to sampling, produces a more representative sample than the preceding method. This has been shown to be true.

### Making the Sample Homogeneous

One of the treatments for a solid sample is to crush it, pulverize it, grind it, or otherwise render it into a thoroughly mixed powder. Particular methods used to reduce particle sizes are too diverse to consider here. This may be understood simply by considering a list of some of the solid materials that are analyzed, including muscle, rock salt, leather, wood, feathers, fish scales, jet turbine blades, limestone, seaweed, granite, tomatoes, hair, vitamin pills, bean pods, iron ore, rubber, cardboard, butter, and sheet steel.

Subdividing and mixing a material is done to increase the homogeneity of the sample. In fact, the smaller the particle size, the lower the error in analyzing a given weight of material. The reason for this result is shown in a simplified example following.

Consider a sample consisting of a number of small, spherical particles of uniform size and density. Some of the particles contain pure *Examplium* (an uncommon material discussed very seldom other than in this text); the rest of the spheres have none. The particles are well mixed and thus random in distribution, as illustrated in Figure 5.2. Let us assume that the particles of Examplium are 20% of the total number.

Also, assume we have an assay method for Examplium that is so precise we can ignore any random error in it. Then we can consider two extreme methods of sampling. First, we can take the whole sample and use it for the assay. Of course we will find 20% Examplium in it. However, because there can be only one sample, a statistical analysis of the data cannot be obtained. This is not a problem for a "perfectly" precise analytical procedure, but it is a significant drawback in the real world.

At the other extreme, only one particle is taken at a time, and an assay is run on it. In the majority of the assays, we shall find no Examplium at all. In others, we shall find 100% Examplium. These analyses would give a very large random error, which could be overcome only by running many more samples and finding the statistical mean value. After obtaining a large number of results, we would find the mean value of Examplium to be 20%. However, I suspect that after that many runs, most analysts would be too exhausted to care.

From the difficulties associated with these two extreme methods, you might believe that there must be a better "middle way." We should take a moderate number of individual samples, each made up of a number of randomly mixed particles. However, if the particles were large, a sample with enough particles for the assay might weigh many hundreds of kilograms. Small particles would enable us to maximize the number of particles in each reasonably sized sample, thus ensuring a minimum of

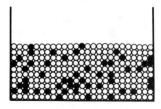

**Figure 5.2**   A material made up of particles of either pure Examplium (black) or filler (white). Of the total mass, 20% is Examplium. In the accompanying text, two extreme cases of sampling the material are considered: a single assay of the entire sample and assaying each particle one at a time.

variation in content between samples. In many ways, small particles produce better samples.

You might then ask, Why not always grind the particles as finely as possible and mix them well? The answer is that there is a trade-off for such grinding. The very small powders are more likely to be contaminated from the large amount of handling. They also can be hard to handle and transfer since they act like dust.

The samples should contain large numbers of particles for two reasons:

1. Each sample will be more representative of the material.
2. This will minimize the variation in content between individual samples.

You can attain a qualitative understanding of both conclusions by again considering the two extreme sampling methods described above.

Note that when only one particle was analyzed at a time, the measured content of Examplium was highly erratic — randomly either 0 or 100%. Thus the use of individual particles maximized the variation between individual samples, and no sample represented the whole material. This suggests that more particles in samples will lessen the variations between replicates (reason 2).

On the other hand, when all the material was analyzed at one time, the sample certainly was fully representative. Thus more particles in each replicate will make it more representative (reason 1).

But a reduction in particle size will allow more particles for a fixed weight of sample. Thus, having smaller particle sizes in a randomly mixed sample will create a better sample — one that has less random error of content and is more representative of the bulk. In the next section, these conclusions are justified in more detail mathematically.

## 5.3 The Binomial Distribution

Let us consider a model for a ground up, well-mixed sample. This model is still oversimplified compared with a real sample, but it will serve to illustrate the trends of the results of further subdividing a sample (as by grinding). You can believe the results of the calculation because it has been shown to reflect the true results of experiments for more than two centuries, ever since first done by Bernoulli. (After all, not all calculations can be believed. Some equations can agree internally but do not reflect the real world.)

We assume three simple, imaginary samples similar to the one illustrated in Figure 5.2. As before, each bottle contains balls of the same density. Here 20% of the balls are pure Examplium and 80% are pure filler. However, as illustrated in Figure 5.3, the balls of bottle A are twice the size of those in bottle B and 8 times larger than those in bottle C. Samples of equal mass will be taken, one sample from each bottle. However, because of the different masses of the balls, the number of particles in each sample varies.

The average concentration of Examplium in all three samples is 20%. However, not every sample will contain exactly 20% Examplium; some will contain more,

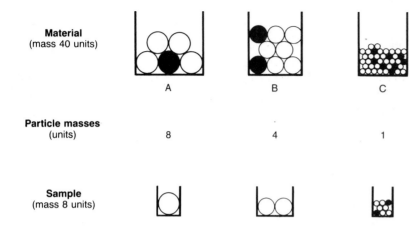

**Material**
(mass 40 units)

A    B    C

**Particle masses**
(units)    8    4    1

**Figure 5.3** An illustration showing the three imaginary experiments to demonstrate the effect of particle size on samples. The three materials A, B, and C contain particles with masses 8, 4, and 1, respectively. Black particles are Examplium; white particles are inert filler. All three materials contain 20% Examplium.

**Sample**
(mass 8 units)

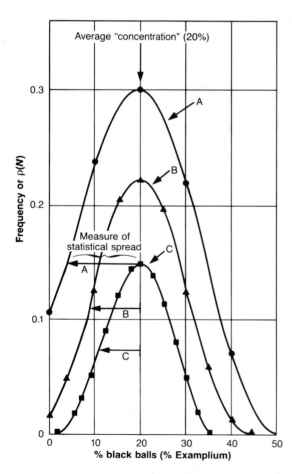

**Figure 5.4** Probability that a sample will have a given percentage of black balls in it. Of the total material, 20% is black. The material mass is constant at 40 units, and the masses of the samples are constant at 8 units. Three cases are shown: A, B, and C.

|  | A | B | C |
|---|---|---|---|
| Total number of particles in material | 5 | 10 | 40 |
| Particle mass | 8 | 4 | 1 |
| Number of particles in each sample | 1 | 2 | 8 |

Notice that the standard deviation (arrows) decreases with smaller particles.

These curves were calculated using the binomial distribution. The equation for this distribution is usually written

$$(p + q)^m = \binom{m}{0} q^m p^0 + \binom{m}{1} q^{m-1} p^1 +$$
$$+ \binom{m}{2} q^{m-2} p^2 + \cdots + \binom{m}{m} q^0 p^m$$

The shorthand notation $\binom{m}{r}$ stands for the quotient

$$_mC_r = \frac{m!}{r!(m-r)!}$$

Also, $p$ is the probability of obtaining an Examplium particle ($= 0.2$), and $q$ is the probability of choosing a filler particle ($= 0.8$) in a random draw of one particle from the whole batch. The first term of the series is the probability that in $m$ drawings no Examplium particles are drawn. The second term is the probability that one of the set of $m$ particles is composed of Examplium. And so on.

some less. A plot of the probability that a sample will have a given concentration of Examplium appears in Figure 5.4. The calculation of the curves is described in the figure caption.

The statistical spread of the values for each distribution is measured by the horizontal arrows for each plot. It should be clear from the graphs that the relative error is decreasing. In other words, the certainty of the mean increases when the sample with a fixed mass consists of a larger number of smaller particles.

Having smaller particles in a sample, then, has the same effect as doing a greater number of individual experiments. Both tend to make the measured mean value more precise. The main difference is that, usually, it is more trouble to analyze more replicates than to grind the particles smaller.

The quantitative improvement in precision that occurs upon reducing the particle size parallels that obtained by running more replicates. As you found in the last chapter, when more replicates are analyzed, the error of the mean will be reduced as the square root of the number of samples run: $\sigma_m \propto 1/\sqrt{N}$. It is the same for a reduction of particle size. The error expected from sampling would decrease by a factor of 10 if we reduced the average particle mass by a factor of 100 while keeping the masses of the samples constant.

## 5.4 Samples of Mixtures

It may seem reasonable at first that obtaining homogeneity in samples of liquids and gases should present no problem. However, experience proves otherwise for most real samples that are larger than those that can be put in a small container in a laboratory. Consider again the sampling of a flowing stream mentioned in the introduction to this chapter. Say that a bottle of the stream water is taken and stored for a short time. Once in the bottle, the water is no longer mixed by the stream's motion; so the particulate matter settles to the bottom, with the heavier and larger particles doing so more quickly. If the solids are to be included in the analysis, the bottle contents must be thoroughly mixed before a sample is taken out to be run. (Incidentally, a known fraction of a liquid sample, such as taken out with a pipette, is called an **aliquot,** from the Latin *some* or *several.*) If the particles are small enough, samples taken from a well-stirred solution will be homogeneous. If the particles are few and larger, sample homogeneity problems can arise, since one aliquot may have more or fewer of the particulates than another. Such variations in particulate content will lead to less-precise analyses due to the sampling technique. Finally, all the effects described for solid particles in liquids also occur in mixtures of liquids and solids suspended in gases.

## 5.5 Physical Separations in Sample Preparation

When sampling a heterogeneous material (solid, liquid, or gas) for analysis, the aim is to make it homogeneous. On the other hand, when one wants to analyze only one

**Table 5.1**  Examples of Quantitative Physical Separations Used in Analyses

Separate Material That at Ambient Temperature Is a . . .

| | Solid | Liquid | Gas |
|---|---|---|---|
| From a Solid | By density-floatation<br><br>By solubility differences | By heat and trap (e.g., water from $CaCO_3$)<br><br>By filtration | By melting solid and purging gas (e.g. radon from rock) |
| From a Liquid | By filtration (e.g., precipitate removal) | By distillation<br><br>By decantation of immiscible liquids | By purge and trap (bubble inert gas through liquid and trap gases at low temperature)<br><br>By gas permeable membranes |
| From a Gas | By filtration (suspended particle removal) | By filtration (liquid aerosol removal: often includes solid nucleus) | By differential diffusion (used in GC/MS couplers)<br><br>By selective freezing |

part of a complex physical mixture—to separate the heterogeneous parts—the separation may introduce error. For example, what if you want to analyze the dust in the air near a copper smelter? The dust can be separated from air with a paper filter, but depending on the qualities of the paper, you might report that only 1% of all particles larger than 1 $\mu$m escape through the filter. In this case, some smaller particles pass through the filter, and the sampling is deficient if they form a significant fraction of the solid.

The separation of dust from the gases in which it is suspended provides an example of a **physical separation.** Where the separation of sample components is clearly done through their physical differences, a variety of methods are available. Some representative examples are shown in Table 5.1. However, defining a separation as a physical separation or a chemical separation (chemical separations are discussed in Chapters 13–15) is not necessarily simple. Certainly, collecting a solid precipitate from a gas or liquid with filter paper is a physical separation. However, there are now available filters with pores so small that they separate molecules by their molecular size. At this level, the line between physical and chemical separations disappears.

As an example, consider some production method that uses bacteria or immobilized enzymes (enzymes chemically anchored to a solid support) to produce the desired products. Assume the molecule that is to be assayed has a molecular mass of 300. (Sugars are good examples.) Also assume that there is a large amount of protein in the solution. (A good example of a solution of sugars and proteins is beer.) The protein can be removed and prevented from interfering in the analysis for sugars by using a membrane filter that will not pass molecules above molecular mass 2000. Further details of a method that uses such a separation is presented in Supplement 12B.

Similarly, it is difficult to decide whether the separation of gaseous oxygen from water is a physical or a chemical separation. In this case, the gas can diffuse through a polymer membrane that will not allow the other components of a test solution to pass. (This kind of separation is used in gas selective sensors described in Section 12.4.)

As can be inferred from the preceding information, a physical/chemical separation is often made immediately before the assay is run and may even be done by part of the assay apparatus. This close relationship makes many instruments far more selective than are the assay methods on which they are based. In other words, the apparatus performs a combination of physical separation and assay functions. An example of such a selective method is shown in Supplement 5A.

## 5.6 Preconcentration and Predilution

Every instrumental and wet-chemical method of analysis possesses an **optimum range of testing.** This is the range in which the inherent errors of a method will be minimized. Some samples will be too concentrated or too dilute to be in the optimum range of the assay method of choice, so the assayed material must be diluted or concentrated to achieve a smaller error. It is most important for you to realize that *every* assay method has an optimum range. This limitation must be considered both in the sampling methods and in any further preparation of the samples before assaying.

One simple example of this limited range can be understood by recognizing the limitations of our eyes to discern changes in the intensity of the color of, say, a solution. Assume you have a glass of solution containing black dye. There is a minimum concentration below which you cannot see any difference between a sample of pure solvent and a sample of an extremely dilute solution of the dye. Also, there is a region of high concentration in which the sample will look identically black, even though the dye concentration might be changed by a factor of two or more.

The highly concentrated solution could easily be diluted to a point at which the change in color intensity is easy to see. The very dilute solution could be evaporated to a smaller volume until the color change can be seen (assuming the dye does not evaporate too).

We can dilute a sample immediately prior to placing it in the assaying instrument for analysis. This dilution might be done simply by adding more of the same solvent to the solution being used. On the other hand, the original sample collection might include dilution, as when a solution containing an anticlotting factor is added to a blood sample.

Preconcentration often is part of the sample-collection stage; the materials of interest can be removed from large amounts of the **sample matrix** during collection. Sample matrix is another name for the part of the material that is not of interest in an analysis. For example, the air matrix is eliminated when suspended particulate matter (more commonly called smoke) is caught on a filter; simultaneously, the process concentrates the particulate matter.

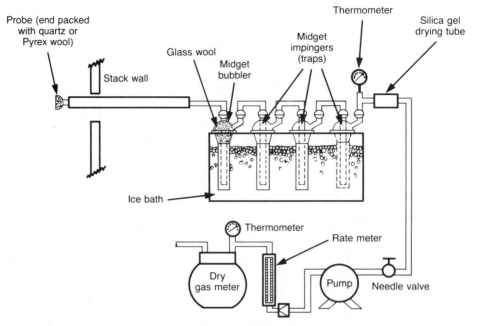

**Figure 5.5** An illustration of the gas-collecting apparatus approved by the EPA for quantitating sulfur dioxide passing up smokestacks. The gas is pulled through the apparatus at about 1 L/min by the pump. The glass wool keeps particulate matter (ash) out of the apparatus. The $SO_2$ is removed from the rest of the gases in the second and third traps (impingers), in each of which is 100 mL of 3% hydrogen peroxide, $H_2O_2$, in water. The peroxide oxidizes the $SO_2$ to sulfate. The first trap contains 80%(v/v) propanol : water which removes any acids and any $NH_3$ that might be in the gas sample ($NH_3$ would interfere with the assay). The fourth impinger contains silica gel, which is a drying agent — it removes water from the gas. A second drying tube (top right) ensures the dryness of the gas stream. The volume of the dried gas sample is measured by the "dry gas" meter. The thermometers are needed so that the volume of the gas can be converted to standard temperature. The total volume of a collected sample is about 20 L. The minimum detectable amount of $SO_2$ is about 3.4 mg $SO_2/m^3$. The method is effective up to at least 80,000 mg $SO_2/m^3$.

Preconcentration is also used when collecting gases for analysis. Figure 5.5 shows an apparatus used for the simultaneous collection, separation, and concentration of gaseous $SO_2$, sulfur dioxide. This apparatus satisfies the requirements of the official U.S. Environmental Protection Agency (EPA) method for sampling $SO_2$ in smokestack emissions. The concentration of $SO_2$ is done through its conversion into water-soluble sulfate ions, which is accomplished by oxidizing the $SO_2$ with hydrogen peroxide. With this process, the $SO_2$ in 20 L of gas is collected into about 100 mL of solution in less than an hour. Since other gases, such as $O_2$ and $N_2$, do not undergo the same reaction, the $SO_2$ is separated from them at the same time. More details are described in the figure caption.

Preconcentration in the sampling steps is also commonly associated with trace analysis, which is discussed in Chapter 7.

Incidentally, the analysis of the particles from the sweater mentioned on the first page of this chapter showed that they were, in fact, from a welding rod. The man was innocent. The sampling was not much problem, however, since individual particles could be analyzed by microanalytical techniques such as those mentioned in Chapter 7. The particles were collected by lifting them from the cloth on lengths of sticky tape.

## Suggestions for Further Reading

A practical guide to the sampling of bulk liquids and solids with further references.
Bicking, C. A. In "Treatise on Analytical Chemistry," Pt. I, Vol. 1, 2nd ed.; I. M. Kolthoff and P. J. Elving, Eds.; Wiley Interscience: New York, 1978, Chap. 6.

The best single, readily available book on the subject of sampling. It is equally divided between theory and practice. The theory is developed in mathematical language.
Smith, R.; James, G. V. "The Sampling of Bulk Materials"; Royal Society of Chemistry: London, 1981.

## Problems

5.1   Use the data of Figure 5.1 for this exercise.
   a.   Calculate, to the nearest 0.1%, the average concentration of the 16 samples taken at the random times indicated by the arrows.
   b.   Do the same as for part a, twice, for 16 samples taken at regular half-hour intervals. Begin at two different random times within the first half hour.
   c.   Compare the difference between the means of the regular-time samples and the difference between either of these with the mean of the random-time samples.

5.2   [Reprinted with permission from P. J. Murphy, *Anal. Chem.* **1979,** *51,* 1599. Copyright 1979, American Chemical Society.]
A new instrumental method to determine mercury in water is being validated against an older one. The new method consists of reducing ionic mercury to its atomic form and then removing it as mercury vapor from solution. The mercury vapor is purged from the water with air and passed into an instrument especially constructed to measure Hg in air. The results were as follows (triplicate determinations); 50 ng of Hg metal was added for each run.

| Aeration and Collection Time (s) | Hg Measured (ng) |
|---|---|
| 5 | 6 |
| 15 | 13 |
| 30 | 22 |
| 60 | 35 |
| 120 | 50 |
| 300 | 52 |
| 600 | 51 |

   a.   What is the minimum time that aeration is needed to obtain the most precise results?
   b.   Does additional time harm the results?

c. Assume that the collection times were 60 s exactly and that the fraction of Hg collected is completely reproducible. By what factor would you have to multiply the measured results to obtain the correct ones?

5.3   An analysis had to be done within an hour on a sample that clearly had a significant amount of water associated with it. Therefore, part of the sample was dried for an hour, and part of the sample, still undried, was assayed during the hour. The results were

Sample 1:   Weight of wet sample, 0.1362 g. Weight of dried sample, 0.1128 g.
Sample 2:   (Wet sample) percentage of sample that is the assayed material: 32.4%

What percentage of the sample is the assayed material, reported on a *dry basis?* (This means *as if the sample were dry.*)

5.4   A 10.45-m³ sample of gas (310 K, 745 mm Hg pressure) was passed through a series of traps that removed the water from the gas. The gas that came out after passing through the dehydration train was measured to be 273 K, 780 mm Hg pressure. Its volume was 8.73 m³. What was the percentage of water vapor (v/v) in the original sample of gas? Assume ideal behavior for all gaseous components.

5.5   A sample of gas was passed through a sampling train such as shown in Figure 5.5. A volume of 20.00 L was collected. The sample gas was at a temperature of 544 K and a pressure of 751 mm Hg.
a. What is the volume of the sample at STP (273 K, 760 mm Hg) assuming ideal gas behavior?
b. If the sample was found to contain 32.02 mg of $SO_2$, what was the $SO_2$ concentration (in g/m³) in the original sample and at STP.
c. The original sample gas was subsequently found to contain 20 vol % water. Correct the content of $SO_2$ to mg of $SO_2$ per m³ of *dry* gas at STP.

5.6   An assay method involves injecting a liquid sample into an injection port of the instrument, either manually, using a small syringe, or with an automatic injector. The following results were obtained. Assume that the precision of the method depends entirely on the injection process.

| Manual Injection | | Automatic Injection | |
|---|---|---|---|
| Injection No. | Instr. Response | Injection No. | Instr. Response |
| 1 | 96 | 1 | 1696 |
| 2 | 101.5 | 2 | 1685 |
| 3 | 100 | 3 | 1665 |
| 4 | 100.5 | 4 | 1648 |
| 5 | 98.5 | 5 | 1658 |
| 6 | 101 | 6 | 1678 |
| 7 | 100 | 7 | 1672 |
| 8 | 97.5 | 8 | 1663 |
| 9 | 98.5 | 9 | 1673 |
| 10 | 101.5 | 10 | 1677 |
| 11 | 100 | | |
| 12 | 101 | | |

   a. Calculate the relative standard deviations for experiments run under both conditions.

   b. Is the method more precise with the automatic injector or manual injection?

   c. Assume that the relative standard deviation of a final result using this method is 10.8% regardless of the injection method. Further assume that the total error is due entirely to sampling error and injection error. What is the relative error due to sampling alone in these two cases?

   d. Assume that you obtain the error found in part c (10.8%) using manual injection. If more precision were wanted, would you invest first in buying an automatic injector or in developing a better sampling method?

5.7   A drum of chemical waste was being analyzed for arsenic among other elements. The contents consisted of a top layer of water, a bottom layer of a water-immiscible liquid, and some metal-containing solid lumps at the bottom. The total volume of the contents was 55 U.S. gallons (1 U.S. gal = 3.785 L). The volume of the solid was 1.44 gal as measured by displacement. A 5.14-mL (wet) portion of the solid weighed 9.364 g when dried. The volume of the water was 62% of the liquid volume. The arsenic content of each part was determined in the same way with a method requiring a solid sample. A 10-mL aliquot of the water was placed in a foil cup, and the water was evaporated at low temperature. Subsequently, the remaining solids were dried at 110°; these weighed 83.2 mg. Similarly, 100 mL of the immiscible layer was evaporated, and the solids were redissolved in a small volume of acid and dried. Their weight was 55 mg. The bulk solid was ground to 5 mm and smaller particles. Part of this was dried and reground to a fine powder, which was used directly. The following results were obtained. The solids from the water sample were found to contain 220 ppm As. Those from the other liquid layer had 43 ppm As. The solid consisted of 157 ppm As.

   a. What is the total mass of arsenic contained in the barrel?

   b. What is the total mass of nonvolatile solids in the barrel?

   c. An alternate method to analyze the aqueous liquid for arsenic was also tested. The result was that the water contained 2.2 ppm (w/v) As. Were there any volatile (under the drying conditions) arsenic compounds present in the water, and, if so, what percentage of the total?

5.8   [Ref: EPA-600/7-/79-191.]
The Environmental Protection Agency lists a number of cancer-causing agents for which reliable analysis is important. These are called *polycyclic organic matter* or *POM*. Among these are benz(a)anthracene (crysene), benzo(b or k)fluoranthene, benzo(a)pyrene, and dibenz(a,h)anthracene. The following data were found when these compounds were added to wastewater samples and extracted into methylene chloride. The volumes refer to the water phase.

| Compound Added | Amount Added ($\mu$g/m$^3$) | Amounts Measured in Trials ($\mu$g/m$^3$) | | | | | | |
|---|---|---|---|---|---|---|---|---|
| Chyrsene | 1.0 | 0.64 0.70 0.45 0.72 0.84 0.75 0.61 | | | | | | |
|  |  | 0.92 0.80 0.59 0.63 0.86 0.58 0.77 | | | | | | |
| Benzofluoranthenes | 1.0 | 0.48 0.96 0.60 0.69 0.08 0.20 0.30 | | | | | | |
|  |  | 0.85 0.90 0.69 0.88 0.76 0.33 0.39 | | | | | | |
| Benzo(a) pyrene | 10 | 6.9 5.7 7.8 5.2 8.0 5.4 4.7 | | | | | | |
|  |  | 5.0 6.5 4.0 7.3 7.6 5.2 8.5 | | | | | | |
| Dibenz(a,h)anthracene | 50 | 32 22 37 39 40 30 17.5 | | | | | | |
|  |  | 44 22 24 27 34 37 46 | | | | | | |

    a.    Calculate the mean percent recovered of the amount added for each of the four compounds.

    b.    Calculate the relative standard deviations of the percentage recovered.

    c.    If a minimum relative error in the determination of each compound must be below 50% to be useful, which, if any, compound(s) cannot be determined with this sampling and extraction method as a part of the analysis?

**\*5.9**     [Ref: K. Ishii; K. Aoki. *Anal. Chem.* **1983**, *55*, 604.]

Atmospheric nitrogen monoxide (NO) and nitrogen dioxide (NO$_2$) were collected by pumping air through an absorbing solution. The solution contained a cobalt coordination compound — written Co(ligand) — which reacts with the NO in an equilibrium reaction. The reaction that occurs is

$$Co(ligand) + NO \rightleftharpoons Co(ligand)NO \qquad (5.9\text{-}1)$$

The equilibrium constant for this reaction is a mixed equilibrium constant.

$$K_P = \frac{[Co(ligand)NO]}{[Co(ligand)] \cdot P_{NO}} \qquad (5.9\text{-}2)$$

$$= 4.5 \times 10^7 \text{ atm}^{-1} \text{ at 35 °C in } o\text{-dichlorobenzene solvent}$$

Note from the units of $K_P$ that $P_{NO}$ must be in atmospheres. Assuming that an equilibrium occurs between the gas and the coordination compound, from Eqs. 5.9-1 and 5.9-2,

$$P_{NO} = \frac{[Co(ligand)NO]}{[Co(ligand)]_{initial} - [Co(ligand)NO]} \cdot \left(\frac{1}{K_P}\right) \qquad (5.9\text{-}3)$$

Assume that the outlet pressure of NO after the gas passes through the solution is given by Eq. 5.9-3. The pressure (in atmospheres) in the outlet stream is describable by the product of the atmospheric pressure, $P_{atmospheric}$, and the relative concentration of NO in the effluent stream, $C_{outlet}$, in ppm (v/v). Algebraically,

$$P_{NO} = C_{outlet} P_{atmospheric}$$

Assume that $P_{atmospheric}$ is 1.00 atm. The efficiency of the collection is defined by

$$\frac{\% \text{ collection efficiency}}{100} = 1 - \left(\frac{C_{outlet}}{C_{inlet}}\right)$$

Assume that $[Co(ligand)]_{initial} = 1$ mM in 5 mL of solution, and that the gas entering the trap is 9.8 ppm (v/v) NO. The rate of flow was 130 mL/min with the trap temperature held 35 °C.

    a.    If 1% changes in the volume of the gas can be ignored, what is the efficiency of gas collection as the gas flows in initially? ([Co(ligand)NO] vanishingly small.)

    b.    What number of moles of NO gas can be absorbed in the collector and still have the efficiency remain above 98%?

    c.    What volume of NO gas at STP can pass through the collector and still have the efficiency remain above 98%?

    d.    What volume of analyte gas at STP can pass through the collector and still have the efficiency remain above 98%?

    e.    Under the conditions of part d, how long will the 5-mL volume of absorber last?

**5.10**    The relative standard deviation (in percent) of the binomial distribution can be described mathematically as

$$\sigma_{rel} = 100\sqrt{\frac{(1-p)}{np}}$$

---

\* Asterisk indicates more involved problems.

To apply this equation to a mixed sample such as illustrated in Figure 5.2, $p$ is the fraction of particles composed of pure assayed component, and $n$ is the total number of particles in the sample. For an analysis, the relative deviation should be less than 0.1%. Assume that the sample is composed of perfect spheres of density 3 g/cm³, 10% of which are pure Examplium and 90% are inert filler.

    a.   What is the minimum number of particles needed in each sample to achieve a relative standard deviation of less than 0.1%?

    b.   What minimum weight (of sample in grams) will be needed to ensure that $\sigma_{rel} = 0.1\%$ if the diameters of the particles are 1 $\mu$m, 10$\mu$m, and 100 $\mu$m?

    c.   If the balance used for weighing the sample measures up to 300 g with an accuracy of 0.1 mg, which sample particle sizes would be usable?

5.11    Will the answers of problem 5.10 differ if the particles are cubes with edges equal to the sphere diameters? (This is equivalent to increasing the mass of each particle.)

5.12    Refer to Figure 5.12.1. A hollow pipe was inserted into the center of a full, perfectly cylindrical tank car to sample the liquid that was in it. In the laboratory, the tube was found to contain three immiscible fractions, each with exactly the same volume. For an analysis of the contents, the analyst wants to calculate the total amount of the assayed material in the tank. Therefore the analysis was done on each of the three immiscible fractions. If the top, middle, and bottom fractions contain 1.24%, 2.48%, and 1.83%, respectively, what is the percentage for the whole tank? [Mathematical hint: The area of a sector of a circle subtending the angle $\theta$ is $(\pi r^2 \theta / 360)$.]

**Figure 5.12.1**

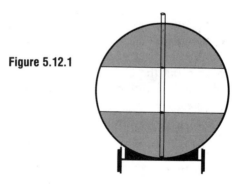

## SUPPLEMENT 5A

# Electrocatalytic Oxygen Analyzer

[Ref: U.S. EPA Continuous Air Pollution Source Monitoring Systems, EPA 625/6-79-005, 1979, pp. 5–27, 5–28.]

    This instrument is presented as an example in which a separation is an integral part of the assay instrument. To understand its operation, you need only know the

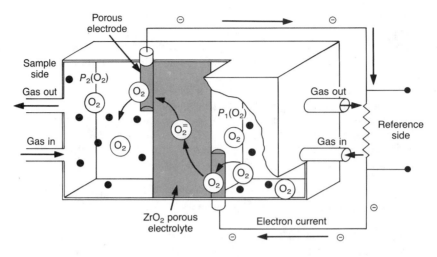

**Figure 5A.1**   Drawing of the structure and operation of an on-line oxygen analyzer. The $ZrO_2$ porous electrolyte selectively allows only $O_2^-$ ions to pass through to complete the circuit. As a result, only the electrical potential due to oxidation of $O_2^-$ and reduction of $O_2$ is measured. The potential is proportional to the ratio of the partial pressures of $O_2$ on the reference side (right) and sample side (left).

electrochemical behavior of a concentration cell as described by the Nernst equation (Section 2.9).

In this cleverly constructed apparatus, the concentration of oxygen is measured by determining the electrochemical potential of a concentration cell. However, the concentration difference is that of a gas and is expressed as a difference in partial pressures.

The selectivity of the apparatus depends on the selective solid membrane and the gas-porous (or gas-permeable) electrodes. Because of the high selectivity, the method can be used for **on-line** monitoring. *On-line* means that the sample is not collected and taken to a separate laboratory. Instead, the instrument is near the source of the sample, and the gas stream is sampled continuously. The sampling is done by passing part of the combustion gases through the chamber on the left side of the figure. The following quotation from the original reference describes the operation of the instrument. Figure 5A.1 illustrates the detector portion of the instrument.

> **Operation**   When sampling combustion gases, the partial pressure of the oxygen in the sample side will be lower than the partial pressure of oxygen in the reference side, which is usually that of air [used as the standard]. When such a cell is kept at a temperature of about 850 °C [a dull red heat], oxygen molecules on the reference side will pick up electrons at the electrode–electrolyte interface. The porous ceramic material of [zirconium oxide], $ZrO_2$, has the special property of high [diffusion rates] for oxygen ions . . . . [The high temperature increases this diffusion rate to a useful value.] The oxygen ions [formed electrochemically]

migrate to the electrode on the sample side of the cell [are oxidized], and emerge as oxygen molecules. The EMF [arising] from this process, expressed in terms of the oxygen partial pressures, is given as

$$E = \frac{RT}{4F} \ln \left( \frac{P\,(O_2)_{ref}}{P(O_2)_{sample}} \right)$$

The partial pressure of oxygen in the combustion gases is then measured by the output voltage of the instrument. This can easily be manipulated electronically to put out the partial pressure directly as on a digital meter. Note that the method only measures the relative oxygen difference. The air-standard's oxygen concentration must be known to obtain a numerical result.

# 6

# Sample Treatments, Interferences, and Standards

## 6.1 Sample Preparation

After a representative, homogeneous sample of an appropriate size for analysis is obtained, it must be brought into the correct form to be used with the chosen assay method. For instance, the sample may be an apparently solid substance, such as a piece of tuna. Perhaps an assay for mercury is desired, which may require that the tuna meat be dissolved in solution. There may be some difficulties associated with the dissolution. For instance, metallic mercury and many mercury compounds are volatile and may be lost if the sample is heated. Also, the sample of fish consists of both water and solid material. Should the amount of solids in the sample be determined? These are typical examples of the problems involved in devising a **sample preparation** method. Other names for the process are **sample treatment** or **sample pretreatment.**

Sample treatment must be planned so as to best achieve four goals: to maximize the **recovery,** minimize the interferences, optimize the concentrations, and minimize the cost.

1. *Maximum recovery.* Material should not be lost in the pretreatment, but if it is, some manner must be found to discover how much is being lost.

The quantitative measure for the loss is called the recovery of the procedure.

2. *Minimum interferences. All* assay methods respond, to a lesser or greater degree, to species of molecules, atoms, or ions other than the one (or ones) being assayed. These other materials are said to interfere in the assay. The processing of the samples must be done so as to eliminate as many interferents as possible. An assay method that is less prone to interference is said to possess greater **specificity.** When the specificity of an assay is improved, the sample preparation steps can often be significantly simplified. Thus the properties of the assay method and the appropriate sample preparation method are closely tied together. (Recall Figure 1.1.)

3. *Optimum concentrations.* The sample treatment method must be designed to bring the analyte into the working concentration range of the analytical method(s) being used. This was discussed briefly at the end of Chapter 5.

4. *Minimum cost.* A number of different methods of sample treatment are possible or even necessary. Proper selection usually balances the required precision and the total cost in time and materials needed for sample preparation and the assay together.

Sample pretreatment can be by far the most difficult and time-consuming portion of an analysis. As a result, a great deal of development work is done to find easier methods of sample treatment or more specific assays requiring less pretreatment.

## 6.2 Methods of Sample Pretreatment

In the previous chapter it was argued that sample homogeneity could be improved by reducing solids into smaller particles or powders. In certain assays, this is sufficient sample treatment since the assay method can use such a powder directly. Examples of such methods are thermal analysis, X-ray fluorescence, neutron activation, and a number of surface assay methods. (These topics are not covered in this text.) In fact, one of the great strengths of these assay methods is that they *can* be used for solid samples.

When a sample is solid but cannot be used in that form, a sample pretreatment is needed. In general, sample pretreatment is a series of physical and chemical changes designed to bring solid, liquid, or gaseous samples into the form needed for the assay. Sometimes this is quite easily done simply by changing the temperature: A heterogeneous solid might be melted into a homogeneous liquid or a liquid vaporized to a gas. Some general methods of sample pretreatment are covered below.

### Dissolution

If a liquid is required for an assay and a melted solid is not adequate or obtainable at a sufficiently low temperature, then a homogeneous solution must be made. The

**Table 6.1**  Solutions Used for Digestion

| Solution | Properties | Use |
|---|---|---|
| *Acid* | | |
| Hydrochloric | Mildly reducing | Metals more easily oxidized than hydrogen |
| Nitric | Oxidizing | Metals not reacting with HCl; oxidizes organic matter |
| Sulfuric | High boiling ($\sim 340$ °C) | Metals; destroys most organics |
| Perchloric | Strongly oxidizing | Metals; NOT TO BE USED WITH ANY ORGANICS OR INORGANIC REDUCING AGENTS |
| Hydrofluoric | Forms stable fluorides | Silica and silicates |
| *Base* | | |
| Sodium hydroxide | Strong base; oxidizing when concentrated | Aluminum and amphoteric oxides of Sn, Pb, Zn, Cr |

simplest way to form a homogeneous solution is, of course, to dissolve the solid in an appropriate solvent. Examples are the dissolution of sodium bicarbonate or sodium chloride in water, or the dissolution of organic materials such as anthracene (mothballs) or polymers such as polystyrene in benzene or other organic solvents. The process of dissolution may usually be hastened by some shaking or gentle heating to warm the solution.

### Digestion

If, as is common, most or all of a sample will not dissolve in a solution, then more vigorous conditions will be required. The process is called **digestion** and involves bringing the solid into solution with the aid of an acid, base, or oxidizing agent. Some digestion mixtures are listed in Table 6.1. The case study in Section 6.4 describes an elaborate, but necessary, digestion method for automobile exhaust catalysts; its digestion required a combination of sulfuric and phosphoric acids.

### Fusion

If solution digestion cannot break the solid down, a still more vigorous method that might work is **fusion.** The process involves heating the solid sample with about 10 to 20 times its weight of a **flux.** Fluxes are solids at room temperature. The solid flux is a strong oxidizing agent, or acid or base, which will form a homogeneous solution (a **melt**) with the sample material when they are heated together. The melt, upon cooling, hardens to a homogeneous solid, usually a glass. A number of different fluxes are listed in Table 6.2. Often, after fusion, the cooled melt can be dissolved in appropriate solvents.

Sometimes both acid digestion and fusion are needed in sample pretreatment. Sometimes only one of the two is necessary. Choosing the method requires knowl-

**Table 6.2**  Reagents for Flux Digestion

| Flux for Fusion | Properties | Use |
|---|---|---|
| Sodium carbonate | Basic | Silicates, insoluble sulfates |
| Sodium hydroxide | Basic | Oxides and sulfides<br>Ores Sn, Zn, Cr, Zr |
| Sodium peroxide | Alkaline and oxidizing | Ores of, e.g., Sb, Sn, Mo, V, Cr<br>Chrome steel |
| Potassium pyrosulfate | Acidic | Oxides of Fe, Co, Ni, Cr, Ti, W, Al<br>Steels |
| Potassium nitrate<br>Potassium chlorate | Used with $Na_2CO_3$<br>for alkaline and<br>oxidizing fluxes | Acid-insoluble metal sulfides |

edge of the descriptive chemistry of the materials being used, but making good first choices for a specific material comes only with experience.

In the case cited in Section 6.4, both digestion with acid and fusion with a flux are needed; the acid alone cannot digest the entire sample. The remaining solid must be heated in a crucible with potassium pyrosulfate flux, $K_2S_2O_7$, to digest it.

There are a number of reasons why a fusion may work when acid or oxidative digestion cannot:

1. The temperature of fusion is usually higher (300–1000 °C) than in concentrated aqueous acid solutions (about 100–200 °C).
2. The concentrations of the reagent flux (on the order of 10–20 M) are usually higher than in an acid solution (0.1–10 M).
3. In an aqueous solution, the water limits the maximum oxidizing power and the maximum acidity and basicity that can be obtained. In a flux, this restriction is not present.

There are limitations, or course, with any digestion method. The principal one is that as the digestion conditions become stronger and harsher, real ingenuity is required to find a container for the mixture that will not be dissolved itself. (This brings up the old question, If you had a universal solvent, what would you keep it in?) One way to bypass this problem is to use containers composed of elements that are not to be determined and also do not act as interferents. One commonly used material for such crucibles is purified zirconium.

There is another complication that must be considered in digesting a sample for analysis: One or more of the components that are to be assayed could be transformed into a volatile product by the digestion. If that happens, the component(s) could be lost as a gas, and any assay for it will produce a low result. If volatility is a potential problem, the material should be digested in a closed container to trap the gas. Such containers are usually constructed of thick-walled (2–5 cm) stainless steel and are called *digestion bombs.* The name is somewhat misleading, since they are constructed so strongly that they will not explode even though the pressure might rise to the range of that inside a high-powered rifle when it is fired.

Another digestion method involves elimination of organic components of the matrix while leaving inorganic ones. This is usually done by destruction of the organic compounds by oxidation, leaving the inorganic components in the resulting ash. This ash can then be digested further by one of the methods described above.

### Vaporization

A number of instruments used for assays require that the sample be introduced as a vapor. If the sample is a solid or liquid, the material must be vaporized into a gas—either atomic, molecular, or ionized (a plasma).

Some instruments operate under high vacuum. Thus, if the sample is volatile enough, the vaporization might be done simply by introducing the sample into the vacuum. Or, if necessary, the material can be heated. Heating to vaporize the sample will be necessary if the instrument operates at atmospheric pressures or above. A more esoteric method of vaporizing materials is to add the necessary energy by bombarding them with high-energy (in effect, high-temperature) particles such as argon atoms.

Vaporization is used to prepare samples for gas chromatography, mass spectrometry, and the atomic spectrometries (Chapters 15, 16, and 18, respectively). Details about vaporization methods are presented there, including the use of electric sparks and high-intensity laser light.

## 6.3 Quantitative Material Transfer and Correction for Losses of Samples

If some of the analyte is lost during the sample pretreatment, we say there is a reduced **recovery.** This is usually expressed as a *percent recovery.* The value is simply

$$\% \text{ recovery} = 100 \times \frac{\text{concentration of analyte from assay}}{\text{concentration of analyte in sample}} \tag{6-1a}$$

or

$$\% \text{ recovery} = 100 \times \frac{\text{weight of analyte from assay}}{\text{weight of analyte in sample}} \tag{6-1b}$$

Some causes for the losses are

    a. Adsorption of analyte onto container surfaces
    b. Evaporation of volatile samples
    c. Losses in unanticipated chemical side reactions
    d. Leaks in transfer systems for gases or liquids
    e. Accidents and carelessness

Throughout the sample treatment, careful attention must be given to ensure that only a minimum of the material is lost: The larger the loss, the larger the *probable* error in the result. However, a relatively low recovery need not condemn a method outright. If the recovery is constant, we can compensate for it just as we would for any constant determinate error.

For example, assume that a sample preparation method yielded an 87% recovery of the analyte every time. Then it would only be necessary to multiply the result by 1.149 to correct for the effect ($1.149 \times 0.87 = 1.00$).

Nevertheless, methods that have inherently low recoveries will *probably* be less precise. A simple comparison can illustrate the point. Assume that a sample preparation procedure generally produces 99% recoveries. From the opposite viewpoint, there is a 1% loss. If the loss doubles or halves, the recovery varies from 98% to 99.5%, a relative range of about 1.5%.

In comparison, consider a sample preparation method with poor recoveries — say, 60%. With this method, 40% of the analyte is lost. Again, if the loss doubles or halves randomly, the recoveries will vary from 40% to 80%, a relative range of about 66%. No determination with that kind of error can be useful. Thus there is an inherently larger possible imprecision associated with low recoveries. It is generally best to aim for the best analyte recovery possible.

## 6.4  Case Study of Sample Pretreatment

The case study in this section illustrates a sample preparation methodology illustrating a number of the points presented above. The problem to be solved is to find a methodology to determine the concentrations of platinum and palladium contained in used automobile exhaust catalysts. As you will see, there are a number of difficulties and a wide range of choices to be made. The procedure is outlined in Figure 6.1.

Note that there are two operations in which reagents are added to aid in ensuring quantitative transfer of the material being assayed. The first step involves addition of tellurium (Te) and the second, immediately preceeding the assay measurement, addition of lanthanum chloride. The purpose of each is to enhance the recovery of the analytes.

Tellurium is added as a **carrier.** A carrier is a substance that has a chemistry sufficiently similar to that of the analyte so that the analyte and carrier coprecipitate or the analyte ions *ad*sorb on the carrier precipitate surface. A greater bulk of precipitate is produced with the analyte as only a small part of the total. This larger mass of precipitate can be transferred more easily than the very small amount that would be produced by the quantity of analyte alone. Carrier precipitation is especially useful in cases similar to that in the example, in which the analytes are at low levels. For example, a relatively large 10-g replicate sample with 0.05% (w/w) platinum contains only 5 mg of platinum. If the 5 mg of Pt metal could be collected and formed into a cube, the cube would be only half a millimeter on a side — the size of a grain of table salt. (You might be interested in calculating what relative error in the assay would result if a single dust-sized cube 100 $\mu$m on a side were lost during a transfer.)

Lanthanum is added to the solution to ensure that all the ionic platinum and palladium are free in solution. For instance, among other unwanted reactions, the surfaces of glass containers may bind the metal ions and cause the measurement of

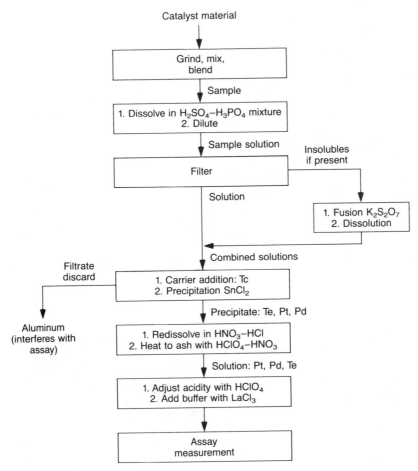

**Figure 6.1**    Flow chart of the procedure used to determine platinum and palladium. The terms *fusion, dissolution,* and *carrier* are defined and explained in Section 6.2. The addition of Te as carrier increases the bulk of precipitate without being itself an interferent. The addition of lanthanum just before the assay was found empirically to suppress interferences from the heavy metals such as the transition elements and other lanthanides. As is common, the assay itself is a relatively small part of the required methodology.

their concentrations to be low. The added lanthanum ions tend to bind to the surface more strongly than the platinum and palladium ions. The $La^{+++}$ replaces $Pt^{++}$ and $Pd^{++}$, causing them to be freed into solution. This is another example in which the knowledge of the descriptive chemistry of such solutions is needed to develop the method used to eliminate interferences and losses of materials.

For now, concentrate on the steps of the sample preparation in this case study. Note the care required to obtain good accuracy and precision, the necessity to separate various components, and the difficulty in dissolution of an extremely stable support material, alumina ($Al_2O_3$).

If you find difficulty with the names of specific assay methods, skip the specific properties of the techniques for now. Some of them are described in the chapters noted.

Note that the assay method itself, atomic absorption (Chapter 18), is relatively easy compared with the sample preparation. To do the assay, all that is needed is to feed the final, homogeneous liquid samples into the appropriate instrument and register the instrument output.

> In automotive catalysts, one or more of the platinum group metals (platinum, palladium, or rhodium) are deposited on alumina pellets or monolithic supports. . . . With the continuing development of catalytic converters to control automotive emissions, analytical methods are required to measure the noble metal concentrations and distributions in automotive catalyst materials.
>
> Because of the low levels of platinum (~0.05%), palladium (~0.02%), and rhodium (~0.005%), sensitive techniques are required. For most research catalysts, accuracy requirements of ±5% are sufficient. However, requirements for some research catalysts and economic considerations associated with production catalysts necessitate methods capable of higher accuracy (±2%).
>
> Many methods for the determination of platinum, palladium, and rhodium have been developed, but most have been concerned with analyzing geological materials containing trace amounts of these elements or evaluating reagents or techniques using synthetic solutions. Instrumental techniques such as X-ray fluorescence, X-ray diffraction, neutron activation, spectrography, and conventional atomic absorption spectrometry (AAS) [see, for the last two, Chapters 17 and 18] have been applied in the determination of platinum, palladium, and rhodium in a variety of materials, including alumina-base reforming [petroleum production] catalysts.
>
> Methods used to analyze reforming catalysts are not applicable to automotive catalysts; for the latter, noble metal concentrations are lower, catalyst substrates are often different, and potential interferences are expected for used automotive catalysts. Therefore, methods applicable to automotive catalysts were needed for sample dissolution and pretreatment prior to determination. . . .
>
> Selection of a representative sample of this type of catalyst material for chemical analysis is the first critical step in any method used to determine noble metal concentrations. . . .
>
> To determine the degree of homogeneity of the catalyst material, individual production catalyst pellets were analyzed by neutron activation analysis for platinum and palladium. . . .
>
> A statistical treatment of the data showed that selection of a 300-g catalyst sample would result in a relative standard deviation of approximately 0.25% of the amount present of both platinum and palladium. However, for some research catalysts such large samples are not always available for chemical analysis, and smaller samples must be taken.
>
> After the samples were selected, pellets were ground to pass through a 150-$\mu$m sieve and then blended. Water concentrations were determined by drying the ground samples in an oven at 125 °C for 3 h. Although water is not completely expelled from the alumina, a constant amount of water remains after the low temperature bake so that comparative data from several laboratories can be obtained.

Selection of the methods for analysis of the catalysts depended on accuracy and time requirements. When accurate ($\pm 2\%$) and precise determinations of platinum and palladium were required, a method employing large samples and chemical separations was used. For samples requiring less stringent accuracy requirements ($\pm 5\%$), rapid methods for the determination of platinum and rhodium were preferred. . . .

Various atomic absorption techniques applied to the determination of platinum and palladium appear in the literature. Each element can be measured in the presence of the other, and conditions for accurate measurement can be controlled easily once platinum and palladium are separated from the matrix material. . . . Figure [6.1] shows a schematic representation of the procedure. . . .

Radioactive tracers were used to evaluate the completeness of the tellurium – stannous chloride separation of the platinum and palladium from the solution of dissolved catalyst material. No significant amounts of platinum and palladium were lost. . . .

Because no acceptable catalyst standards exist, platinum and palladium were determined in three catalyst samples in an interlaboratory cooperative study. . . .

[Refs: N. M. Potter; W. H. Lange, *Am. Lab.* **1981**, *13*, 81–91. N. M. Potter, *Anal. Chem.* **1976**, *48*, 531–534.]

## 6.5 Standard Calibration

When low recoveries do not arise from clumsy mistakes, such as we all make from time to time, but result from errors in the methodology, improvements of the sample treatment method may be highly desirable. Improving the methodology requires the use of standards. In Chapter 4, you read how standard samples were used to test a method for the determination of aluminum. That set of experiments tested the reagents and chemistry of a relatively simple sample, such as now might be used in an introductory lab course.

Most realistic samples are far more complicated, as indicated by the case study just described. As a result, more sophisticated standardization techniques must be used. Before discussing some of the different standardization procedures, let us digress on the nature of analytical measurement and types of standards.

Chemical analysis is concerned with measurement, which is usually indirect. Direct measurements are those such as mass, time, volume, and voltage. Values found from indirect measurements are calculated from directly measured values such as the four just listed.

In an analysis, a direct measurement of the mass or concentration of the assayed component is seldom carried out. It is easier to correlate analyte concentration with some other measurement such as color intensity. However, it is necessary to **calibrate** such indirect measures. To calibrate literally means to determine the caliber or quality of some methodology. More specifically, in the context of a chemical analysis, calibrate means "to ascertain the relationship between the content of the sample and the response of the assay method." Do not confuse calibration with validation. The latter term denotes ascertaining the usability of a specified analytical methodology among a number of laboratories and analysts.

Chemical standards are, then, standards of chemical *content.* This is in contrast to physical standards, which are used as standards of the characteristics of materials: for example, elasticity, color, and viscosity, as well as the more basic mass, time, length, and voltage. Chemical standards are pure substances, mixtures, solutions, gases, or materials such as alloys or biological substances that are used to calibrate and validate all or part of the methodology of a chemical analysis. In Supplement 6A are described some of the types of standards available.

## 6.6  Standards: External and Internal

Standards of chemical content are used to determine pretreatment recoveries and to calibrate assays. The methods of standardization can be separated into two types: **external standards** and **internal standards.**

An external standard is one that is analyzed separately from the replicate unknowns being tested. A series of such standards contain varying amounts of the analyte, together with a matrix that is similar or identical to that of the samples.

An internal standard is added to each replicate sample and blank. This addition is also known colloquially as *spiking* a sample. These standardization methods are described below.

### External Standards

External standards can be used to calibrate an assay method when the components of the matrix do not interfere. Note that after any pretreatment, the matrix will be composed of components of the sample as well as of any reagents required for pretreatment. External standards can also be used if the analyst has enough control of the conditions so that the interferents' contribution to the measurement can be kept constant.

Let us look back at the practical example described in Section 6.4 in which the concentration of platinum was determined. This illustrates the use of an external standard—or, more correctly, an external standard *series.* The calibration of the machine response was made by using standards containing different concentrations of platinum in solutions which contained the same matrix components as were in the unknowns: for example, the same concentrations of lanthanum and the acid used for digestion. To emphasize the point: It is important to keep constant the concentrations of *all* the other possible species that could influence the assay measurement. Only the concentration(s) of the assayed species should change in a series of standard solutions.

The manner in which a calibration can be done is shown in the following example. Here, *calibration* refers to relating the instrument output to the analyte concentration. At this time you need not know the principle of operation of the equipment. For

**Table 6.3**   Result of Assay on a Series of Exernal Standards

| Concentration of Standard (ppm) | Instrument Response (arbitrary units) |
| --- | --- |
| 0 | 0.0 |
| 190 | 31.7 |
| 410 | 68.3 |
| 580 | 96.6 |

platinum in the practical example of Section 6.4, the instrument was an atomic absorption spectrometer. This instrument can be treated as a "black box," which means that a sample is introduced into it, and its response is measured. (For those of you who dislike such mysterious operations, some details of this black box can be found in Chapter 18.)

---

**EXAMPLE**

An atomic absorption assay was done on catalyst samples prepared as described in Section 6.4. A set of standards was made for platinum. The concentrations and the instrument response to each standard are shown in Table 6.3. The instrument response—such as could be read from a gauge or digital output—is given in *arbitrary units*. We do not care at this time what the units of the numbers are.

To see the trend in the output more clearly, a graph can be made of the instrument response versus the concentration of the standards. This graph is shown in Figure 6.2.

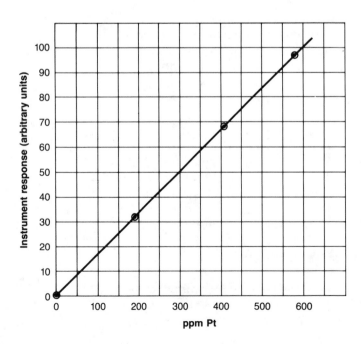

**Figure 6.2**   Plot of a calibration curve from a blank and three standards. The instrument was adjusted so that the blank response was zero on the arbitrary linear scale. The instrument response is linear to the analyte, although it need not necessarily be so.

**Table 6.4**  Results of Assay on Replicate Samples: Instrument Response

| Sample Number | Instrument Response |
|---------------|---------------------|
| 1 | 64.3 |
| 2 | 63.5 |
| 3 | 64.5 |
| 4 | 63.3 |

Immediately after the instrument was calibrated using the standards, and *without changing the procedure used for the assay,* the four samples were run and the output noted. The assay results for a series of four samples from the same catalyst are seen in Table 6.4.

We are now able to relate the instrument response to the content of platinum in the sample. From the results of the calibration of the instrument with platinum standards, we see that the response of the atomic absorption spectrometer is *linear.* This means that the numbers presented by the instrument are directly proportional to the concentrations of the platinum in the standards. The meaning of the linear response can be expressed as an equation:

$$\text{instrument output} = \text{constant} \times \text{concentration} \qquad \textbf{(6-2a)}$$

The constant factor will be specific for the following factors:

1. The instrument
2. The conditions of the instrument at the time of the determinations
3. The other materials present in the sample matrix after the sample treatment

This constant is, you see, *very* specific. It depends on all the conditions of the assay, on all compounds present in the sample, and on chemical species introduced through the pretreatment prior to the actual running of the instrumental assay. If any of the conditions or preparation steps were changed, the instrument would have to be recalibrated and, perhaps, new standards prepared.

Using the calibration graph shown in Figure 6.2, we can convert these readings to concentrations. First, the value of the calibration constant is determined. This can be done with a least-squares calculation, which is a built-in function of many hand calculators. If you want to check the result in this example, do not neglect the point at ppm 0 with instrument response 0.

Thus, from the calibration data of Table 6.3, Eq. 6-2a becomes

$$\text{response (arbitrary units)} = 0.167 \times \text{concentration of Pt (ppm)} \qquad \textbf{(6-2b)}$$

From Eq. 6-2b and the results listed in Table 6.4, the solution concentrations can be found. These are listed in Table 6.5 (p. 204).

Carefully formulated standards can help eliminate a problem caused by possible difficulties in removing interferents during the sample pretreatment. Commonly, all the possible interferents cannot be eliminated economically during the sample preparation. But it is possible to calibrate the effects of the interferents and correct for them. This point will be pursued further in Chapter 18.

We have looked in detail at only two examples of how standards are used to calibrate an assay: those describing the gravimetric assay for aluminum (Section 4.8) and platinum in automobile catalysts (Section 6.4). However, the strategy is the same

**Table 6.5**  Results of Assay on Replicate Samples:
Concentration

| Sample Number | Solution Concentration (ppm) |
|---|---|
| 1 | 386 |
| 2 | 381 |
| 3 | 387 |
| 4 | 380 |

for all assay methods in daily use. To see the similarities in strategy, it may be worthwhile to compare the use of standards in developing the assays for platinum and for aluminum. While there are significant differences between the instrumental assay for platinum and the classical gravimetric assay, the standards occupy the same place in the analyses: ascertaining and improving the accuracies of the methodologies.

### Internal Standards: Standard Additions or Spiking

There are three situations in which standardization is done by adding the standard material to the samples themselves:

A.  When the solid or liquid matrix of a sample is either unknown or so complex that an external standard cannot be used with confidence
B.  When the chemistry of the pretreatment or of the assay method is complex or highly variable
C.  When the assay depends on highly precise instrumental conditions that are difficult to control

As an example, situations B and C both occur when an assay involves injecting the sample into a flame. The chemistry is quite complicated, and the flame conditions are sometimes difficult to regulate with the desired precision.

The standard addition method is done in the following manner.

1.  An assay measurement is made on the sample being tested.
2.  A known amount of the *analyte* is added (spiking of the sample).
3.  The sample, modified by spiking, is then assayed again under the same conditions.
4.  Further spiking and remeasuring—steps 2 and 3—are often repeated a few times.
5.  The amount of the original constituent is calculated from the data points by extrapolation.

A description of the procedure for a copper determination follows.

---

**EXAMPLE**

Three aliquots of 5 mL each are taken from the same original sample solution. Nothing is added to the first aliquot, and 5 $\mu$L of a 100 ppm solution of copper chloride standard is added

**Table 6.6** Outline of a Copper Assay Using Standard Addition

| 5-mL Aliquot Number | Standard 100-pm $Cu^{++}$ Added ($\mu$L) | Final Concentration of $Cu^{++}$ Spiked (ppm) |
|:---:|:---:|:---:|
| 1 | 0 | — |
| 2 | 5 | 0.10 |
| 3 | 10 | 0.20 |

to the second. To the third aliquot is added 10 $\mu$L of the 100-ppm copper chloride solution—twice the amount added to the second aliquot. This procedure is outlined in Table 6.6, and the experimental results are illustrated in Figure 6.3. Ignoring the small change in volume for now, you can calculate that the concentrations of the standard-solution spikes are diluted 1000-fold and 500-fold for aliquots 2 and 3, respectively. The spiked copper amounts to 0.10 ppm for the second sample and 0.20 ppm for the third. In Figure 6.3, the instrument response (in arbitrary

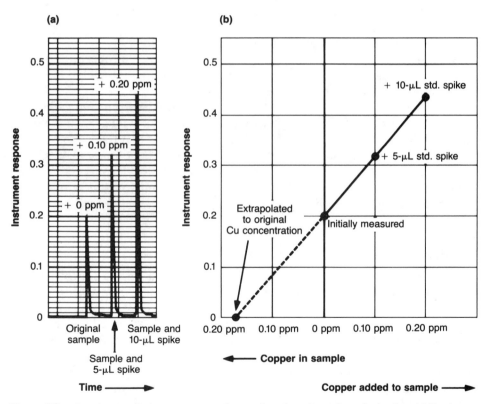

**Figure 6.3** Data from a set of measurements of a sample and two internal standard spikes. (a) The data as they appear on a chart recorder. The peak heights are used as the measure of the instrument response. (b) Plot of the instrument response versus concentration. The sample concentration is unknown and is plotted at "0." The two spikes add 0.1 ppm and 0.2 ppm of Cu, respectively. A linear extrapolation back to zero-response yields the content of the sample. Among the assumptions of the method are that the instrument response is linear over the entire range tested and that the instrument response reads zero when no Cu is present in the sample matrix.

units) is plotted versus the amount of copper added. Each data point corresponds to a measurement. The "zero" (on the horizontal axis) is the instrument response when the original solution was measured. The other points are the measurements on the spiked aliquots. The higher responses are due to the sum of the copper in the original solution and the copper of the spike.

In this case, a linear extrapolation can be made, and the intercept on the horizontal axis is the calculated copper content of the original sample. Here, the concentration of copper is 0.17 ppm. More details are presented in the figure caption.

---

One drawback of the standard addition method is that spiking must be done for each sample that is used. Another is that a blank sample is not available. Thus the method is best used with a highly selective assay.

The benefit is that the procedure allows compensation for quite complex effects. However, the concentration range over which instrument response is linear tends to be quite narrow — less than an order of magnitude (factor of 10). Thus the pretreatment must include bringing all samples into this narrow working range. At least part of the reason for the narrow linear range is the presence of the complicated chemical effects that caused the spiking to be needed in the first place. Sometimes it seems that easy winning is continually just out of reach!

When the instrument response is nonlinear, calculating the concentration in the original sample becomes more complicated than this simple extrapolation. In addition, the result is significantly less certain.[1]

The standard addition method is more commonly used with emission spectroscopies such as atomic emission (Chapter 18) and with electrochemical methods (Chapter 12). Another important use for spiking is in methods that use isotopes as standards. These may be radioactive isotopes or stable isotopes with, for instance, mass spectrometry (Chapter 16). A similar method is used for determining percent recoveries in sample preparation methods. A practical example of this is quoted and discussed in Supplement 6B.

## Internal Standards That Differ from Analytes

It is sometimes useful and possible to add to a sample a standard material that is chemically different from any analyte. This kind of standard is also called an internal standard. In order to use such a method, it is necessary to have an assay method that produces an output response for both the analyte and the standard under the same conditions. The measure of content is then calculated from the *relative* response to the analyte and to the separate standard. In this way it is possible to use the output response to the standard material to compensate for any variations in the steps of the determination. For instance, if the standard is added to the original sample, and only half of the standard is found at the end of the analysis, then we can assume that half of the analyte is lost as well. In essence, the percent recovery is determined simultaneously with the analyte.

---

[1] See Beukelman, T. E.; Lord, S. S., Jr. *Appl. Spec.* **1960**, *14*, 12-17.

This type of internal standard is most used in two situations:

1. When there is significant, unavoidable variability in some part of the analytical methodology
2. When high precision is desired

It is used with almost all the methods presented in this text. Exceptions are a number of titrimetric methods (Chapters 9 and 10) and kinetic methods (Chapter 11). A practical example of the use of such internal standards is described in Supplement 6B.


## Suggestions for Further Reading

A guide to standards for all aspects of analysis: with references.
Chalmers, R. A. In "Comprehensive Analytical Chemistry," Vol. 3; G. Svehla, Ed.; Elsevier: Amsterdam, 1975, Chap. 2.

A chapter on the methods of sample preparation for inorganic compounds.
Bogen, D. C. In "Treatise on Analytical Chemistry," Pt. I, Vol. 5, 2nd ed.; I. M. Kolthoff and P. J. Elving, Eds.; Wiley-Interscience: New York, 1982, Sect. F, Chap. 1.

A chapter on both dry and wet methods of preparing samples composed of organic compounds.
Dunlop, E. C.; Ginnard, C. R. In "Treatise on Analytical Chemistry," Pt. I, Vol. 5, 2nd ed.; I. M. Kolthoff and P. J. Elving, Eds.; Wiley-Interscience: New York, 1982, Sect. F, Chap. 2.

Two review articles in which expert authors critically survey the methods noted in the titles.
Sulcek, Z.; Dolezal, P. P. J. "Decomposition Procedures in Inorganic Analysis." *Critical Reviews in Analytical Chemistry* **1977**, *6*, 255-323.
Uriano, G. A.; Gravatt, C. C. "The Role of Reference Materials and Reference Methods in Chemical Analysis." *Critical Reviews in Analytical Chemistry* **1977**, *6*, 361-411.

A short article describing the U. S. National Bureau of Standards reference materials program.
Alvarez, R.; Rasberry, S. D.; Uriano, G. A. "NBS Standard Reference Materials: Update 1982." *Anal. Chem.* **1982**, *54*, 1225A-1244A.

## Problems

The first three problems refer to the data in Tables 6.3, 6.4, and 6.5 and Figure 6.2.

6.1    Using the values found in Table 6.4, calculate the mean, standard deviation, and 95% confidence limit of the concentration of platinum in the sample.

6.2    What would the concentration of platinum be if the instrument output were 54.7 on the same scale that was determined from the standards?

6.3    Upon review of the laboratory notebook recording the experiment, it was found that the standards were made up incorrectly. The actual values of the concentrations of the standards are 32.43% lower than reported. Calculate the correct concentration of the platinum found in the sample.

6.4    The following values were found for the palladium standards when an instrumental assay similar to that used for platinum was used.

| Concentration of Pd in Standard (ppm) | Instrument Reading (arbitrary units) |
| --- | --- |
| 00 | 14.1 |
| 80 | 27.7 |
| 205 | 48.4 |
| 325 | 68.6 |

The instrument reading for a sample with unknown Pd was 56.0 units. Plot the calibration curve and find the concentration of the sample.

6.5    This exercise involves deriving a mathematical expression to calculate the concentration of analyte using the method of standard additions. The machine response is found to be linear with concentration.

If the machine response is $A_i$, for the concentrations $C_i$, then

$$A_{\text{sample}} = k \, C_{\text{sample}}$$

where $k$ is the factor that relates the concentration to the instrument response. Similarly, after a spike,

$$A_{\text{sample + spike}} = k \, (C_{\text{sample}} + C_{\text{spike}})$$

Solve these two equations for $C$(sample) in terms of $C_{\text{spike}}$, $A_{\text{sample}}$, and $A_{\text{sample + spike}}$.

6.6    This problem refers to the copper determination described in Section 6.6. The volume of a 5-$\mu$L spike added to the 5.000-mL sample was not taken into account in calculating the copper concentration. Not accounting for this volume change introduces a determinate proportional error.
   a.    The spike was not exactly 0.10 ppm; calculate the true concentration of the spike accounting for the volume change.
   b.    The random error of the instrument is found to be 0.5% relative error. What is the ratio of the random relative error and the relative determinate error arising from ignoring the volume change?

6.7    [Data courtesy of Yellow Springs Instrument Co., Yellow Springs, OH.]
Figure 6.7.1 illustrates the data for a linearity check of a membrane used in an electrochemical method to determine alcohols directly. The instrument is described in Supplement 12B. Product lot 9208 is being tested. The output from the instrument is plotted on the vertical axis with time on the horizontal axis. The chart moves at 1 in. min$^{-1}$. The output of the chart recorder is set so that each of the 20 divisions equals 25 mV. The instrument output is set to have 1 mV = 1 mg/dL$^{-1}$ for the 5-$\mu$L injection of sample. The run starts from the left. The zero of the instrument is set on the bottom line. A 200-mg/dL$^{-1}$ sample is injected, and the instrument is allowed to respond fully. The plateau is taken as the instrument reading. The sample is washed out and another injected until the full series is run. The volumes and concentrations of each sample are noted on the chart. Answer the following questions.

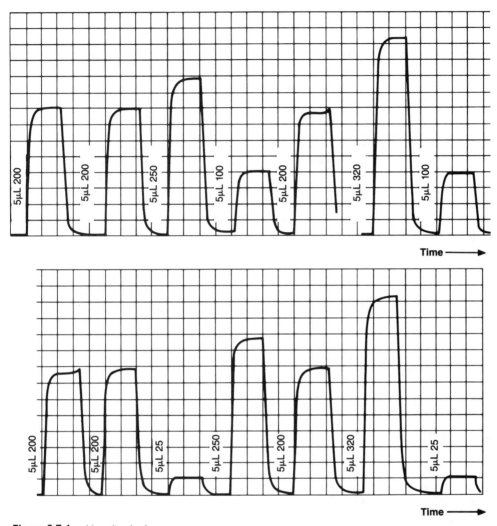

**Figure 6.7.1**    Linearity check.

    a.    For the third and fourth 200-mg/dL$^{-1}$ samples, there is a slight hop in the plateau voltage. How will you treat these values?

    b.    For the six 200-mg/dL$^{-1}$ samples, what is the mean (in mV) and standard deviation of the measurements?

    c.    Is the response linear over the entire range tested (25–320 mg/dL$^{-1}$)?

    d.    If the range is not linear, at what concentrations does it deviate by more than one relative standard deviation (as determined from part b) from the values expected by extrapolating the linear portion of the plot?

6.8    [Ref: Fu-sheng W.; Fang, Y. *Talanta* **1983,** *30,* 190.]

A determination of silver in waste water was done with the method developed by spiking silver into the sample. A sample originally containing no silver was run six times with the following protocol. Each sample of 100.0 mL had added to it 200 $\mu$g of silver and was well mixed. For each run, a 5.00-mL aliquot was used. This was digested, and the following results were obtained.

$$\mu g\ Ag\ found:\ 8.8,\ 9.2,\ 9.2,\ 9.1,\ 9.1,\ 8.9$$

a. What is the mean recovery of the method?
b. After correction for the recovery, what is the relative standard deviation of the analysis?
c. Without correction for the recovery, what is the relative standard deviation of the analysis?
d. If the assay for silver has a relative standard deviation of 0.1%, what is the relative standard deviation due to the recovery?
e. If the assay for silver has a relative standard deviation of 5%, what is the relative standard deviation of the sample preparation procedure?

6.9  [Ref: Liddell, P. R. *Am. Lab.,* March 1983, *15*, 111. Copyright 1983 by International Scientific Communications, Inc.]
The following data were found in a determination of chromium in water. The protocol was to run a blank, the sample, and then the sample with two internal addition spikes. Three replicate runs were made for each sample. Incidentally, this was all done automatically using a computer-controlled instrument. The significantly different responses to the spikes were caused by the other matrix components.

|  |  | Instrument Readings | | |
| --- | --- | --- | --- | --- |
| Solution | Conc. Cr (ng/mL) | #1 | #2 | #3 |
| Blank | 0.000 | 0.005 | 0.009 | 0.005 |
| Sample 1 | — | 0.054 | 0.058 | 0.058 |
| Addition 1 | +2.000 | 0.199 | 0.191 | 0.188 |
| Addition 2 | +4.000 | 0.305 | 0.310 | 0.318 |
| Blank | 0.000 | 0.000 | 0.000 | 0.000 |
| Sample 2 | — | 0.537 | 0.548 | 0.531 |
| Addition 1 | +2.000 | 0.666 | 0.659 | 0.667 |
| Addition 2 | +4.000 | 0.790 | 0.784 | 0.805 |
| Blank | 0.000 | 0.001 | 0.003 | 0.000 |
| Sample 3 | — | 0.086 | 0.083 | 0.078 |
| Addition 1 | +2.000 | 0.214 | 0.213 | 0.212 |
| Addition 2 | +4.000 | 0.341 | 0.350 | 0.337 |

Calculate the concentrations of Cr in the three samples, assuming that the means of the measured values are exact. Make the blank correction if necessary.

6.10  [Ref: Application note 101. Yellow Springs Instruments Co., Yellow Springs, OH.]
A 0.750-g sample of corn syrup was assayed for dextrose with an instrument that specifically measures dextrose using an enzyme assay. (The instrument is described in Supplement 12B.) The sample was diluted in a volumetric flask to 100.0 mL. A 25-$\mu$L sample was injected into the instrument which read 373 mg/dL for the sample. The instrument had previously been calibrated with a 200-mg/dL standard using the same 25-$\mu l$ sample injector. What is the content as %(w/w) dextrose in the corn syrup?

6.11  [Ref: Application note 105. Yellow Springs Instruments Co., Yellow Springs, OH.]
A 1.596-g sample of a flavoring agent was assayed for dextrose with an instrument that specifically measures dextrose and records the result on a digital readout panel. The instrument was calibrated with a 200-mg/dL standard using a 25-$\mu$L sample injector. The sample was diluted in a volumetric flask to 100.0 mL. With the same sample injector as for the standard solution, a 25-$\mu$L sample was introduced into the instrument. The output read 23 mg/dL for the sample. What is the content as %(w/w) dextrose in the flavoring agent?

6.12  [Ref: Data reprinted from Gilbert, W.; Behymer, T. D.; Casteneda, H. B. *Am. Lab.,* March 1982. Copyright 1982 by International Scientific Communications, Inc.]
A method was developed to determine oxygen dissolved in water. The method was tested to quantify any interferences that might be present. The following data were obtained from solutions which all contained 3.5 mg/L dissolved oxygen. Each result is the average value of between two and four replicates.

| Potential Interferent | Interferent Conc. (mg/L) | Av. Oxygen Found (mg/L) |
|---|---|---|
| Chromate ($CrO_4^-$) | 0.1 | 3.4 |
|  | 1.0 | 3.9 |
|  | 10 | 6.4 |
| Hypochlorite ($OCl^-$) | 1.0 | 3.5 |
|  | 10 | 4.9 |
|  | 100 | 7.8 |
| Thiosulfate ($S_2O_3^-$) | 1.0 | 3.4 |
|  | 10 | 3.5 |
|  | 100 | 3.4 |

a.  Which of the three compounds show an interference? Is it a positive or negative interference (raising or lowering the result)?
b.  Plot the result as it changes with a possible interferent. That is, plot $O_2$ concentration versus the interferent concentration. Calculate the best value for a correction factor that should be applied as $F_c$ in the equation

$$[O_2]_{true} = [O_2]_{apparent} - F_c[\text{interferent}]$$

Do this for each interferent using its concentration in mg/L.
c.  What is the value of $F_c$ for a species that does not interfere?

6.13  [Ref: Application note 102. Yellow Springs Instruments Co., Yellow Springs, OH.]
Table sugar (sucrose) molecules can be enzymatically broken down to produce fructose and dextrose (also called glucose). The reaction can be written

$$\text{sucrose} \xrightarrow{\text{invertase enzyme}} \text{dextrose} + \text{fructose}$$

The following procedure was followed to determine the sucrose content of sweet potatoes. A 78.7-g sample of sweet potato was reduced to juice, and the juice was collected in a beaker. The juicing machine was washed three times with 100-mL portions of sodium phosphate buffer with 2 to 3 min between rinsings. The juice and washings were poured into a 500-mL volumetric flask, and the containers were rinsed with several 10-mL portions of buffer, which were added to the flask. Buffer was then added "to the mark," which indicates the point where the volume is 500.0 mL. The sample was then refrigerated for 1 hour. After an hour, a 3-mL aliquot of the sample solution was taken, and the invertase enzyme was added to it.

The sample that was not enzymatically decomposed was assayed for free dextrose. The content was found to be 208 mg/dL. After 20 min, the sample with invertase was assayed and found to contain 458 mg/dL dextrose. In each case, a 25-$\mu$L sample is used.

a. How many milligrams of dextrose are due to the decomposition of sucrose in the sweet potato?

b. Given the molecular masses of dextrose (180.16) and sucrose (342.3), what is the percent sucrose (w/w) in the potatoes?

6.14   With the procedures described in problem 6.13 and the sample of 1.596 g of flavoring agent as described in problem 6.11, the following results were obtained.

Dextrose in untreated sample, 23 mg/dL

Dextrose from sucrose in sample treated with invertase enzyme, 287 mg/dL

a. What is the content of sucrose in the sample expressed as %(w/w)?

b. What is the total mass of sucrose in 1000 kg of the agent?

6.15   [Data reprinted with permission from Murphy, P. J. *Anal. Chem.* **1979,** *51*, 1599. Copyright 1979 American Chemical Society.]

A new instrumental method to determine mercury in water is being validated against an older one. The new method consists of first reducing ionic mercury to its atomic form and then removing it as mercury vapor from the aqueous solution. The mercury vapor is purged from the water with air and passed into an instrument especially constructed to measure Hg in air. Possible interferents were investigated. The following results were obtained on water solutions of Hg containing the interferents. The figures in parentheses are the percentages by weight of the reagents.

| Interferent | Hg Added (ng) | Hg Measured (ng) | |
|---|---|---|---|
| | | *Old Method* | *New Method* |
| Acetone (20%) | 100 | >1000 | 95 |
| Sodium sulfite (10%) | 100 | 220 | 100 |
| Sodium thiosulfate (10%) | 100 | 750 | 102 |
| Pyridine (5%) | 100 | >1000 | 91 |
| Ammonium hydroxide (20%) | 100 | >1000 | 96 |

a. Assume that the inherent precision of the assay methods is ±5% at this concentration level. For each of the potential interferents, and for the old method, what is the nature of the interference—positive or negative (causing the result to appear too high or too low, respectively)?

b. For the new method, is there any interference from any of these compounds?

c. With the information given can you tell whether the interference in the new method arises from the sample preparation step for any of the interferents?

6.16   [Ref: Beukelman, T. E.; Lord, S. S., Jr. *Appl. Spec.* **1960,** *14*, 12.]

With the method of standard additions, sometimes it is useful to use a logarithmic extrapolation. For instance, sometimes the instrument response is

$$I = K c^m$$

where $I$ is the response, $c$ the concentration, and $K$ and $m$ constants. The form of the equation that is more useful is

$$\log I = m \log c + \log K$$

If log $I$ is plotted versus log $c$, a straight line results with slope $m$ and ordinate intercept log $K$. The values of $m$ and $K$ will depend on the experimental conditions.

Assume that two solutions with spiked concentrations $(c + \Delta_1)$ and $(c + \Delta_2)$ are run. The instrument responses are $I_1$ and $I_2$, respectively. Now define

$$A = \log(I_2/I)/\log(I_1/I)$$

a.  Show that

$$A = \log[1 + (\Delta_2/c)]/\log[1 + (\Delta_1/c)]$$

This equation can be used to find $c$ from the three measured values $I$, $I_1$, and $I_2$ alone.

b.  An assay for calcium was run with the following results:

| Sample | Response (corrected for background) |
|---|---|
| $c$ | 22.5 chart divisions |
| $c + 20$ ppm $Ca^{++}$ | 44.1 chart divisions |
| $c + 80$ ppm $Ca^{++}$ | 106.8 chart divisions |

Calculate $c$ with a linear extrapolation and with the logarithmic equation and compare the results.

## SUPPLEMENT 6A

# Standard Reference Materials

[Ref: NBS Special Publication 260. NBS Standard Reference Materials Catalog, U. S. Dept. of Commerce, 1981.]

The following is a quotation from the introduction to the U. S. National Bureau of Standards Catalog.

The National Bureau of Standards issues over 1000 different materials through its Standard Reference Materials Program. These materials are primarily Standard Reference Materials (SRMs) certified for their chemical composition, chemical property, or physical property, but also include Research Materials (RMs) and Special Reference Materials (GMs). All SRMs, RMs, and GMs bear distinguishing names and numbers by which they are permanently identified. Thus each SRM, RM, or GM bearing a given description is identical (within the required or intended limits) to every other sample bearing the same designation — with the exception of individually certified items, which are further identified by serial number.

The first materials issued by NBS were called Standard Samples and consisted of a group of ores, irons, and steels certified for their chemical composition. Since the mid-1960s these materials have been issued as Standard Reference Materials, and cover a wide range of chemical and physical properties and an equally wide range of measurement interests.

**Definitions**   The different terms, SRM, RM, or GM, are used to indicate differences in the types of information supplied and in the purposes for which the material is intended.

*Standard Reference Materials* have been characterized by the National Bureau of Standards for some chemical or physical property and are issued with a Certificate that gives the results of the characterization. These results are obtained by one of the three established routes of certification, i.e., measurement of the property using: (1) a previously validated reference method, (2) two or more independent, reliable measurement methods, or (3) an *ad hoc* network of cooperating laboratories, technically competent, and thoroughly knowledgeable with the material being tested. These routes are described in detail in "The Role of Standard Reference Materials in Measurement Systems," NBS Monograph 148, 54 pages (Jan. 1975). SRMs are defined as being well-characterized and certified materials produced in quantity to improve measurement science. They are prepared and used for three main purposes: to help develop accurate methods of analysis (reference methods); to calibrate measurement systems used to: (a) facilitate the exchange of goods, (b) institute quality control, (c) determine performance characteristics, or (d) measure some property at the limit of the state-of-the-art; to assure the long-term adequacy and integrity of quality-control processes. In these ways, SRMs help ensure the compatibility and accuracy of measurements in many facets of national life—from science and technology to trade and commerce.

*Research Materials,* unlike SRMs, are not certified. Instead of a Certificate, RMs are issued with a "Report of Investigation," the sole authority of which is the NBS staff member who authored the report. An RM is intended primarily to further scientific or technical research on that particular material. The principal consideration in issuing an RM is to provide a homogeneous material so that investigators in different laboratories are assured that they are investigating the same material.

*Special Reference Materials* differ from both SRMs and RMs in that NBS does not participate in the characterization of these materials. GMs are reference materials produced and certified or guaranteed by other government agencies, standards bodies, or other non-profit organizations. When deemed to be in the public interest and when alternate methods of national distribution do not exist, NBS acts as the distributor for such materials. This service is available to all organizations that qualify and have reference materials that would help solve a national measurement problem.

## SUPPLEMENT 6B

# Determining Percent Recovery with Internal Standards: A Case Study

Internal standards can be used to determine percent recoveries. In this case study, you will read about the changing matrix of the sample causing changes in recoveries.

The example quoted below describes the development of a method for removing and concentrating small amounts of organic compounds from water samples. The basis of this collection technique involves passing a gas, such as nitrogen, through a solution that contains volatile organic compounds in low concentrations. As the nitrogen bubbles through the solution, it picks up the organic compounds. The gas, which is called the carrier gas, is said to **purge** the solution. After its passage through the solution, the gas, now containing the purged organics, is then directed into a cold tube called a *cold trap.* This trap allows the nitrogen to pass through but retains the more easily frozen organic compounds. After a time (determined by previous testing and on the order of minutes), the gas flow is stopped. Some fraction of the total of each volatile organic has now been carried to and concentrated in the trap. Next, the trap is heated as rapidly as is convenient, and the volatile organics are driven out into the instrument being used for the assay. This is usually a gas chromatograph (Chapter 15), which for now is another "black box."

Not all the organics are totally recovered, and the recovery of each may vary. Some factors affecting the recoveries are the analyte concentrations and the identities and amounts of other materials in the matrix.

The following quote comes from the authors' description of their standardization procedure and their determination of the percent recoveries for various samples in various waters. The waters that are mentioned are described in the caption to Figure 6B.1.

Notice the care required to choose a solvent for the standards — the analysts choose ethylene glycol ($HOCH_2CH_2OH$). Further care is taken to prevent any changes in the concentrations of the standards resulting from evaporation; they are stored refrigerated.

## Recoveries of Volatile Pollutants from Water

Since each of the four types of water [tested] contained at least some measurable purgeables, it was necessary to prepurge each of the waters to remove these volatiles prior to fortification with those compounds [to be tested.] [See the partial list in Table 6B.1.]

[Ethylene glycol was used as the solvent to store and inject the standards.] . . . Ethylene glycol has the following [beneficial] features: (1) it is a satisfactory solvent for volatile organics . . . ; (2) it is infinitely miscible with water; (3) it does not [itself] purge from water at room temperature and hence [does not interfere with the analysis]; (4) standards of volatiles in the relatively

**Table 6B.1**  Volatile Water-Soluble Organics
That Were Tested

| | |
|---|---|
| Methylene chloride | Dibromochloromethane |
| Chloroform | Benzene |
| Hexachloroethane | Ethylbenzene |
| Trichloroethylene | Acrylonitrile |

viscous ethylene glycol are stable for extended periods of time. . . . Standards were stored in 1-dram vials filled to the top and fitted with screw caps and septa. When not in use, they were stored at 4 °C in a refrigerator.

*Recovery.* Either unsatisfactory or erratic results were obtained using more or less conventional approaches to determine recoveries. The procedure adopted involved the application of intense heat . . . to the purge vessel. Conditions of heat and time were selected which, when exceeded, resulted in essentially no increase in recovery. . . . Two internal standards—bromo-chloromethane and 1,4-dichlorobutane—were also included in the validation study.

A comparison of recoveries reported in the literature for the same compounds by different workers suggests that a significant error . . . may arise from inaccurate recovery determinations, or, at best, demonstrates the dependence of recovery on the equipment used. . . .

. . . ordinarily, standard compounds are purged only from a single reference water [solution]. In any other aqueous matrix, the recoveries of all compounds are

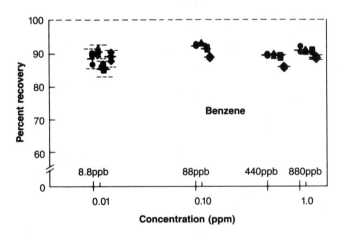

**Figure 6B.1**    Recoveries of benzene and tetrachloroethylene versus concentration. In the two graphs are shown the experimental results from four different waters: ● high-purity water; ▲ water from Lake Huron; ■ effluent waste water from a chemical plant after treatment; ◆ effluent waste water from a chemical plant.

The short horizontal lines show the mean recoveries. Where three determinations were made, two standard deviations on either side of the mean are marked by short dashed lines. For the two compounds here, there is a slight drop of a few percent in recovery from the samples of effluent waste water in comparison with the purer waters. This shows the adverse effect on their purgeability by other matrix components.
[Ref: Reprinted from Ramstad, R.; Nestrick, T. J.; Peters, T. L. *Am. Lab.* **1981,** *13*, 65-73. Copyright 1981 by International Scientific Communications, Inc.]

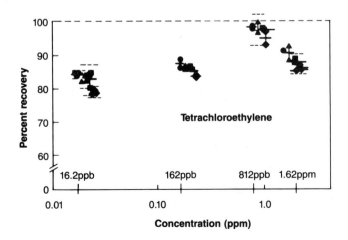

assumed to vary in direct proportion to the recovery of the internal standard(s). . . . As will be shown, the recoveries are only slightly matrix-dependent. Regardless of whether quantitation is by internal or external standard, the inclusion of a purgeable internal standard is a practical means to evaluate system performance. . . .

[Some results are shown in Figure 6B.1 and are described in the figure caption.]
[Ref: Ramstad, T.; Nestrick, T. J.; Peters, T. L. *Am. Lab.* **1981,** *13*, 65 – 73.]

# 7

# Sample Size and Major, Minor, Trace, and Ultratrace Components

7.1 Sample Classification by Size and Analyte Level
7.2 Limits of Trace Analysis
7.3 Trace, Ultratrace, and Microanalysis Compared
7.4 Four Cases

## 7.1 Sample Classification by Size and Analyte Level

Objects considered for analysis come in all sizes. On the large end of the range is the moon, samples of which have been subjected to extensive elemental analysis.[1] From the moon, relatively large samples ($\sim$ kg) could be obtained. A major problem was choosing the rock samples to bring back to earth to be analyzed. From these collected rocks, laboratory samples on the order of $0.1 - 1$ g-**macro** or **normal** samples— could be taken for analysis.

It is not hard to find an object to be sampled that is significantly smaller—for instance, a single small grain of powder found in the pocket of a suspected bank robber. The grain might be from some powder that was especially formulated for use as a marker for bank notes. If so, the powder would have been sprinkled on the money before it was handed over by a teller. The size and mass of the single powder particle would place it in the class of **micro** samples, which are analyzed with the techniques of **microanalysis.**

The size of a laboratory sample should not be confused with the concentration level of analyte in a sample. A descriptive classification scheme is presented in Figure 7.1.

---

[1] A brief review of the moon-rock analysis: Morrison, G. H. *Anal. Chem.* **1971,** *43*(7), 23A–31A.

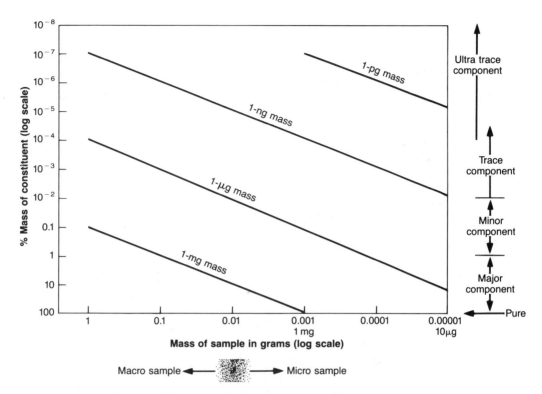

**Figure 7.1**   Illustration showing the names given to the ranges of sample size and ranges of analyte content. The horizontal axis has a logarithmic scale of sample size. The samples are divided into macro- and micro-analytical regions. On the vertical axis is a logarithmic plot of the percentage (w/w) of a constituent in a sample. A sample is classified by the *relative* amount of an element or compound in it. The component can range from **major** (1–100%), through **minor** (0.01–1%), and **trace** (less than 0.01% = 100 ppm), to **ultratrace** (in the range of parts per billion or less).

The meanings of the words *trace* and *ultratrace* are still changing because of technical advances. Determinations of certain compounds and elements can be done in the part-per-trillion range ($10^{-12}$ of the sample, or $10^{-10}$%). The total or absolute amount of a component present depends on the size of the sample and the component's fraction of the whole. As a result, lines of constant analyte mass run diagonally across the graph. [Modified from Sandell, E. B. "Colorimetric Determination of Trace Metals," 3rd. ed.; Interscience: New York, 1959.]

The sample masses are divided into macro and micro with a vague boundary. The concentrations range from major through minor to trace and ultratrace. The figure caption contains further details.

However, chemical analysis is a dynamic area, and the term *trace analysis* is changing its meaning even now. Such changes are reflected in the following quote of comments by A. H. Hayes who was, at the time, the commissioner of the U. S. Food and Drug Administration.

> The food safety law is really not that at all. It really is a series of laws, some of them old, some of them new. . . . When some of these laws were written, . . . scientist[s] would tell you, for a particular chemical, that they could

find one part per thousand. . . . [This] means that zero, when you're talking about zero risk  . . . [was] one part per thousand and one. Now, for some of these chemicals, 20 years after . . . we can find one part per billion or one part per trillion. . . . If you can find one part per trillion, what is the new definition of zero risk, one part per trillion and one? . . . Do we really believe that the risk of zero being one part per thousand is the same as one part per trillion? [Quoted in *Chem. & Eng. News,* Jan. 18, 1982.]

How much is a part per billion (ppb)? How much is a part per trillion (pptr)? These concepts may be easier to understand by considering some concrete examples.

Light travels approximately 0.3 m (about 1 ft) in one-billionth of a second.
The mean distance from the earth to the sun is about $1.5 \times 10^8$ km. One part per billion of that distance is 150 m, or 460 ft.
A gallon (U.S.) is equal to 3.8 L; so a 55-gallon oil drum contains about 200 L. The mass of water in a full drum is thus about 200 kg. If one grain of table salt (mass about 0.2 mg) were dissolved in the barrel of water, the salt concentration would be about 1 ppb.
A part per trillion of the surface of the earth is a square patch 12.3 m (40 ft) on a side.
A part per trillion of a 70-kg (154-lb) person weighs 70 ng. This is the weight of a piece of hair about 0.01 mm (10 $\mu$m) long.
To dilute 1 g of salt (or any other substance) to a part per trillion (w/v) water solution would require about $10^9$ L of water. This quantity of water would require a cube 100 m on a side to contain it.

I think that you will have to agree that the ability to quantitate a few parts per billion or parts per trillion of some substance in any material must rank among the most impressive achievements of chemical technology.

## 7.2  Limits of Trace Analysis

As the content of a component decreases into the ppm and ppb range, eliminating interferences from the sample matrix becomes more and more difficult. Recall the difficulties of determining trace levels of aflatoxins in peanuts, which were mentioned in Section 4.1 Such difficulties become even more acute and important when governments want to regulate concentrations of substances even into the ppb range. Disagreements among laboratories may increase until meaningful exchanges of information are not possible. An illustration of the trend is shown in Figure 7.2. Nevertheless, methods for the elimination of these problems are constantly being developed by chemists who approach them with great care and a firm knowledge of the chemistry of the trace components.

Another problem when carrying out trace analyses is that during the sample preparation, contamination becomes a problem. Every reagent, such as an acid used for digestion, must contribute as little interference as possible. Even the containers

**Figure 7.2** A plot illustrating the finding that in general an increase in interlaboratory errors occurs with decreasing concentrations of analyte. The horizontal axis is a logarithmic plot of concentration; the vertical axis is the percent relative standard deviation of the results. This plot should be interpreted as showing a trend and not hard, fast conclusions. Note that in the ppb range, the 95% confidence limit approaches the value of the mean. [From Horowitz, W. *Anal. Chem.* **1982**, *54*, 67A–76A.]

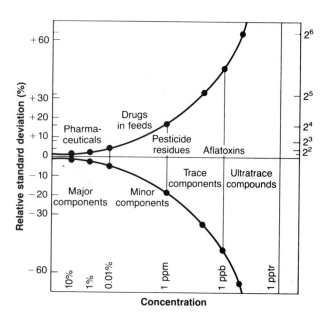

made of glass or plastic can contaminate solution samples. The introduction of interfering impurities will set a lower limit on our ability to ascertain whether or not some substance is present in the sample.

Tables 7.1a and 7.1b list the levels of some trace impurities in acids used for sample preparation. The amounts are stated in ng/mL, which is equivalent to ppb (w/v). An impurity concentration of 100 ppb may not seem very high; however, if the quantity of acid needed for digestion is 100 times the sample weight, 100 ppb of a species in the acid would be equivalent to 10 ppm in the sample.

Analyses at the lowest concentration levels require that the spaces in the laboratory where samples are handled be clean and filled with filtered air. In other words, the instrument used for the low-level assay is often *not* the limiting factor, as the example quoted below will demonstrate.

Two terms that are closely associated with measurements in trace and micro analyses are the **sensitivity** and the **detection limit.** Their definitions in a chemical context are similar to those in our everyday language. If an instrument (or assay method) is more sensitive, it is able to measure smaller changes in content or concentration. One measure of sensitivity is the slope of the working curve (of response versus concentration) at the concentration of interest. It is now the recommended definition (by the IUPAC) for all analytical methods. With this definition, a *larger* number (in the same units) means a method is more sensitive. Unfortunately, in the past, some measures of sensitivity were constructed to indicate the amount of analyte that was needed to obtain a fixed response. In this latter context, a *smaller* number means the method is more sensitive. One must approach the word in context with caution.

**Table 7.1a**  Concentrations of Impurities Found in Acids

| Element | Sample | Concentration [ng/mL = ppb(w/v)] | | | |
| | | *Supplier* *1* | *2* | *3* | *4* |
|---|---|---|---|---|---|
| *In nitric acids* | | | | | |
| Mn | 1 | 10.2 | 0.9 | 29.7 | |
| | 2 | 6.0 | 7.3 | 49.0 | |
| | 3 | 8.1 | 2.2 | 16.5 | |
| | 4* | 0.6 | — | — | |
| Cl | 1 | 65 | 53 | 259 | |
| | 2 | 79 | 66 | 330 | |
| | 3 | 55 | 59 | — | |
| | 4* | 50 | 500 | 80 | |
| Na | 1 | 66 | 109 | 1465 | |
| | 2 | 50 | 103 | 1400 | |
| | 3 | 98 | 124 | — | |
| | 4* | 30 | 500 | 100 | |
| Cu | 1 | 19 | 13 | 355 | |
| | 2 | 12 | — | — | |
| | 3 | 13 | — | — | |
| | 4* | 13 | 5 | 0.7 | |
| *In hydrofluoric acids* | | | | | |
| Mn | 1 | 0.3 | 0.2 | 0.3 | 1.0 |
| | 2 | 0.3 | 0.5 | 0.3 | 1.1 |
| | 3 | 0.6 | — | 0.2 | 0.9 |
| | 4* | 0.5 | — | — | — |
| Na | 1 | 12 | 11 | 12 | 129 |
| | 2 | 23 | 15 | 27 | 140 |
| | 3 | 16 | 14 | 17 | — |
| | 4* | 20 | — | — | 100 |
| Cl | 1 | 175 | 270 | 800 | 281 |
| | 2 | 180 | 472 | 805 | 287 |
| | 3 | 182 | — | 795 | 242 |
| | 4* | 1000 | 100 | — | 80 |
| *In acetic acids* | | | | | |
| Mn | 1 | 7.0 | 0.5 | 2.4 | 2.4 |
| | 2 | 13.3 | 0.5 | 2.7 | 1.5 |
| | 3 | — | — | 3.0 | 2.0 |
| | 4* | 1 | 5.0 | — | — |

*(continued)*

**Table 7.1a**  *(continued)*

| Element | Sample | Supplier | 1 | 2 | 3 | 4 |
|---------|--------|----------|-----|-----|-----|-----|
| | | | Concentration [ng/mL = ppb(w/v)] | | | |
| Na | 1 | | 903 | 90 | 359 | 35 |
| | 2 | | 916 | 24 | 359 | 39 |
| | 3* | | 400 | 50 | — | — |
| Cu | 1 | | 66 | 4.8 | 3 | — |
| | 2* | | 30 | 9 | 100 | 5 |
| Cl | 1 | | 655 | 101 | 187 | 133 |
| | 2 | | 799 | 377 | 159 | 66 |
| | 3 | | — | — | — | 58 |
| | 4* | | 300 | 300 | — | 500 |

* Concentration reported by supplier.

*Source:* Reprinted with permission from Mitchell, J. W.; Luke, C. L.; Northover, W. R. *Anal. Chem.* **1973,** *45,* 1503–1506. Copyright 1973 American Chemical Society.

**Table 7.1b**  Representative Inorganic Impurity Levels in Common Concentrated Reagent Acids. (Manufacturers' analyses in ng/g. Zeros are place holders only.)

| Impurity | Impurity Level (ng/g) | | |
|----------|-----------------|--------|------------------------|
| | Reagent Grade | Better Grade | Ultrahigh-purity Grade |
| *In hydrochloric acid* | | | |
| Ag | — | 50 | 0.1 |
| Al | — | 50 | 10 |
| As | 20 | 5 | 1 |
| B | — | 100 | 30 |
| Ba | — | 50 | <10 |
| Br | 50,000 | 500 | 400 |
| Ca | — | 50 | 1 |
| Cd | 50 | 5 | <1 |
| Co | 50 | 5 | <0.1 |
| Cr | 50 | 50 | 0.2 |
| Cu | 200 | 5 | 0.2 |
| Fe | 200 | 50 | 2 |
| K | — | 50 | <5 |
| Pb | — | 5 | 0.5 |
| Mg | — | 10 | 1 |
| Mn | 50 | 5 | 0.1 |

*(continued)*

**Table 7.1b**   *(continued)*

| Impurity | Impurity Level (ng/g) | | |
|---|---|---|---|
| | Reagent Grade | Better Grade | Ultrahigh-purity Grade |
| Hg | — | 5 | 0.2 |
| Na | — | 500 | 20 |
| Ni | 200 | 10 | 0.2 |
| Si | — | 100 | 200 |
| Sn | — | 50 | <0.5 |
| Sr | — | 50 | <1 |
| Zn | 500 | 5 | <1 |
| Ammonia | 3000 | — | 1000 |
| Phosphate | — | 50 | 20 |
| Sulfate | 800 | 500 | 300 |
| Sulfite | 1000 | 800 | 200 |
| *In nitric acid* | | | |
| Ag | — | 10 | <0.05 |
| Al | — | 50 | 2 |
| As | 5 | 5 | <1 |
| B | — | 50 | 2 |
| Ba | — | 50 | <10 |
| Ca | 5000 | 50 | 2 |
| Cd | 10 | 50 | <1 |
| Cl | 200 | 100 | 100 |
| Co | — | 5 | <0.1 |
| Cr | 100 | 50 | 0.3 |
| Cu | 50 | 5 | 0.1 |
| Fe | 200 | 20 | 0.5 |
| K | — | 50 | <5 |
| Pb | 50 | 5 | 0.5 |
| Mg | — | 50 | 0.3 |
| Mn | — | 5 | 0.05 |
| Hg | — | 5 | <0.1 |
| Na | — | 500 | 30 |
| Ni | 50 | 10 | 0.2 |
| Si | — | 1000 | 10 |
| Sn | — | 50 | <0.5 |
| Sr | — | 50 | <1 |

*(continued)*

**Table 7.1b** *(continued)*

| Impurity | Impurity Level (ng/g) | | |
|---|---|---|---|
| | Reagent Grade | Better Grade | Ultrahigh-purity Grade |
| Zn | 100 | 10 | <1 |
| Phosphate | 200 | 100 | <50 |
| Sulfate | 500 | 500 | <100 |
| *In sulfuric acid* | | | |
| Al | — | 30 | 3 |
| As | 10 | 5 | <1 |
| B | — | 50 | 3 |
| Ba | — | 50 | <10 |
| Ca | — | 50 | 10 |
| Cd | 100 | 5 | <1 |
| Cl | 200 | 100 | 40 |
| Co | 100 | 5 | 0.2 |
| Cr | 100 | 10 | 0.5 |
| Cu | 500 | 5 | 1 |
| Fe | 200 | 50 | 10 |
| K | — | 50 | <5 |
| Pb | 100 | 10 | 0.5 |
| Mg | — | 10 | 3 |
| Mn | 100 | 5 | 0.3 |
| Hg | 5 | 5 | 1 |
| Na | — | 500 | 200 |
| Ni | 100 | 5 | 0.5 |
| Si | — | 500 | 20 |
| Sn | — | 10 | 0.5 |
| Sr | — | 10 | <1 |
| Zn | 100 | 5 | 2 |
| Ammonia | 2000 | — | 200 |
| Phosphate | — | 500 | 10 |
| Nitrate | 1000 | 200 | 40 |

The detection limit is a limit below which a particular instrument or analytical method cannot be used to observe the presence or absence of some analyte in a sample. The magnitude of the detection limit depends on the following factors:

1. The identity of the element or molecular species to be determined
2. The identity of the other components in the sample

3.   The quality of the reagents required for the sample preparation
4.   The instrument or assay method used

The dependence of detection limits on the specific assay method and the element assayed is illustrated in Table 7.2; the numbers are associated with the lowest levels of each element that can be detected with each method. The **characteristic concentration** is a closely related number. The detection limit will be defined more precisely in Section 17.12 and Supplement 18B, but for now, you can think of them as approximately comparable quantities. From the information in the table, notice that for different elements the concentration range of an individual technique can vary by well over a factor of a thousand; and, for a given element, the concentration limit can vary among techniques by over a factor of a million. However, as you may already appreciate, the selection of an assay method to use depends on many criteria in addition to the information in Table 7.2.

Perhaps the following example will provide you with some idea of the capability of contemporary analytical technology and the difficulty in defining the detection limit in practical situations. One term needs to be defined: *absorbance units.* For the time being, it is only necessary to know that this is a unit of the output scale of the instrument. Most instruments that measure absorbance produce an output in the range from 0 to 2 absorbance units. If your eyes could register in absorbance units, a gray–colored solution registering 3 absorbance units would look black; a solution registering 0.01 units would appear colorless. A more detailed definition of absorbance is presented in Chapter 17.

At this time, all you need to know about the assay method, graphite-furnace atomic absorption spectrometry, is that the sample is placed inside a hole a few millimeters in diameter in a tube composed of graphite. The method is described more fully in Chapter 18.

It's a beautiful late summer day, warm and with a gentle breeze. Any lab window or door that can be propped open is. Mother Nature has begun her work to ensure an abundance of next year's plant life. That means pollen—a completely innocuous substance to a human without hay fever. A typical grain of pollen may be about 10 $\mu$m in diameter, and have a density of 0.8 [g/cm$^3$], and contain about 5% magnesium. A rough calculation gives a total magnesium content of $7 \times 10^{-11}$ g. The sensitivity[2] [of graphite-furnace atomic absorption spectrometry] for magnesium is about $10^{-14}$ g. This means that $10^{-14}$ g of magnesium produces 0.0044 absorbance units. [This is a small but measurable value.] Thus, if a single pollen grain were to get into [the sample chamber of the instrument], it alone would produce a magnesium signal of 31 absorbance units. That translates into a very off-scale reading on any instrument ever conceived.

Obviously it makes no sense to do a magnesium analysis during hay fever season. Nor is it practical to do it any other time either since similar effects, though not quite so gross, are produced by fungal spores or bacteria, and they are present all year.

[Reprinted from L. Morgenthaler, in *Spectroscopy,* Series II, Vol. II, 105–120; International Scientific Communications: Fairfield, CT; Copyright 1977.]

---

[2] This number is now called the *characteristic mass.*

**Table 7.2** Characteristic Concentrations* and Detection Limits* of Analytical Techniques for Elements in Parts per Billion (Parts per 10⁹) (Abbreviations and conditions are explained in the key.)

| Element | Char. Conc. MAS | Det. Lim. DPP | Det. Lim. ASV | Char. Conc. AAS | Char. Conc. ETA–AAS | Det. Lim. ICP–ES | Det. Lim. ICP/MS | Det. Lim. SSMS | Det. Lim. NAA | Det. Lim. XRF |
|---|---|---|---|---|---|---|---|---|---|---|
| Ag | 15 | 60 | 0.005 | 20 | 0.2 | 7 | 0.04 | — | 0.1 | 1300 |
| Al | 3 | 0.03 | 0.5 | 30 | 2 | 3 | 0.1 | — | 1 | 1600 |
| As | 15 | 1 | 0.5 | 100 | 2 | 60 | 0.4 | 0.6 | 1 | 300 |
| Au | 10 | 2 | 0.5 | 100 | 0.5 | 30 | 0.08 | — | 2 | 600 |
| B | 1 | — | — | 15000 | 100 | 3 | 0.08 | — | — | — |
| Ba | — | — | — | 200 | 10 | 0.2 | 0.02 | 2 | 3 | 100 |
| Be | 3 | — | — | 10 | 0.2 | 0.2 | 0.1 | 0.08 | — | 300 |
| Bi | 20 | 2 | 0.005 | 300 | 2 | 100 | 0.06 | 2 | — | 600 |
| Br | 20 | 0.03 | — | — | — | — | — | 2 | 0.2 | 600 |
| Ca | 20 | 15 | — | 20 | 1 | 0.02 | 5 | 0.3 | 200 | 100 |
| Cd | 15 | 0.5 | 0.005 | 15 | 0.05 | 2 | 0.07 | 7 | 10 | 200 |
| Ce | 60 | — | — | — | — | 10 | 0.01 | — | — | 200 |
| Cl | — | 0.3 | — | — | — | — | — | 0.4 | 2 | 200 |
| Co | 15 | 1 | 0.05 | 70 | 1 | 3 | 0.01 | 0.5 | 10 | 30 |
| Cr | 15 | 7 | — | 60 | 0.5 | 2 | 0.02 | 0.5 | — | 200 |
| Cs | — | — | — | 200 | — | — | 0.02 | — | 200 | 200 |
| Cu | 10 | 0.5 | 0.01 | 60 | 1 | 2 | 0.03 | 2 | 2 | 300 |
| Dy | 60 | — | — | 250 | — | 15 | 0.04 | — | 0.05 | — |
| Er | 60 | — | — | 300 | — | 3 | 0.02 | — | 2 | — |
| Eu | 60 | — | — | 200 | — | 0.5 | 0.02 | — | 0.005 | — |
| F | — | — | — | — | — | — | — | 0.2 | 20 | 1000 |
| Fe | 20 | 1 | 0.1 | 60 | 1 | 7 | 0.2 | 0.5 | — | 300 |
| Ga | 15 | — | — | 1000 | 5 | 10 | 0.08 | — | 2 | — |
| Gd | 60 | — | — | — | — | 7 | 0.04 | — | 3 | — |
| Ge | 3 | — | — | 1000 | — | 15 | 0.08 | — | 3 | — |
| Hg | 25 | — | 0.2 | 2200 | 100 | 60 | 0.08 | 7 | 3 | 300 |
| Ho | 60 | — | — | 300 | — | 15 | 0.01 | — | 0.1 | — |
| I | — | 0.03 | — | — | — | — | 0.01 | 1 | 0.2 | 100 |
| In | 15 | 0.5 | 0.01 | 300 | 10 | 60 | 0.01 | — | 0.1 | — |
| Ir | 30 | — | — | — | 60 | 200 | 0.06 | — | 0.1 | — |
| K | — | — | — | 20 | 0.5 | 60 | — | 0.3 | 10 | 30 |
| La | 60 | — | — | — | — | 3 | 0.01 | — | 1 | 200 |
| Li | — | — | — | 20 | 1 | 1 | 0.06 | 0.06 | 2 | — |
| Lu | 60 | — | — | — | — | 15 | 0.01 | — | 0.05 | — |
| Mg | 15 | 15 | — | 3 | 0.03 | 0.2 | 0.1 | 0.2 | 10 | 600 |
| Mn | 1 | 3 | — | 30 | 0.2 | 1 | 0.04 | 0.4 | 0.05 | 100 |
| Mo | 30 | 6 | — | 300 | 5 | 3 | 0.08 | 3 | 60 | 600 |
| Na | — | — | — | 6 | 2 | 0.5 | 0.06 | 0.2 | 1 | — |
| Nb | 15 | — | — | — | — | 3 | 0.02 | — | 3 | — |
| Nd | 60 | — | — | — | — | 7 | 0.01 | — | 20 | — |
| Ni | 5 | 1 | 7 | 70 | 0.1 | 7 | 0.03 | 2 | 30 | 100 |
| Os | 60 | — | — | 300 | — | 20 | 0.01 | — | 15 | — |
| P | 1 | — | — | — | — | 60 | — | 0.3 | 60 | 600 |
| Pb | 25 | 1 | 0.01 | 200 | 0.5 | 7 | 0.02 | 3 | — | 600 |
| Pd | 6 | — | — | 100 | 10 | 15 | 0.06 | — | 1 | — |
| Pr | 60 | — | — | — | — | 30 | 0.01 | — | 0.5 | — |
| Pt | 200 | — | — | 600 | 20 | 30 | 0.08 | — | 10 | 600 |
| Rb | — | — | — | 100 | — | — | 0.02 | — | 100 | — |
| Re | 30 | — | — | — | — | 60 | 0.06 | — | 0.1 | — |
| Rh | 200 | — | — | 200 | 100 | 7 | 0.02 | — | 0.1 | — |
| Ru | 30 | — | — | 300 | — | 200 | 0.05 | — | 60 | — |
| S | 3 | — | — | — | — | — | — | — | — | 300 |

*(continued)*

**Table 7.2**    *(continued)*

| Element | Char. Conc. MAS | Det. Lim. DPP | Det. Lim. ASV | Char. Conc. AAS | Char. Conc. ETA–AAS | Det. Lim. ICP–ES | Det. Lim. ICP/MS | Det. Lim. SSMS | Det. Lim. NAA | Det. Lim. XRF |
|---|---|---|---|---|---|---|---|---|---|---|
| Sb | 25 | 1 | 0.02 | 300 | 1 | 200 | 0.02 | 2 | 3 | 30 |
| Sc | 7 | — | — | 200 | — | 3 | 0.08 | — | 10 | — |
| Se | 25 | 0.01 | — | 100 | 2 | 60 | 1 | 2 | 3 | 300 |
| Si | 10 | — | — | 2000 | 7 | 10 | 10 | 0.2 | 60 | 100 |
| Sn | 10 | 1 | 0.02 | 1000 | 0.4 | 30 | 0.03 | 4 | 200 | 20 |
| Sr | — | — | — | 100 | 1 | 0.1 | 0.02 | — | 60 | 300 |
| Ta | 15 | — | — | — | — | 100 | 0.02 | — | 60 | — |
| Tc | 15 | — | — | — | — | — | — | — | — | — |
| Te | 30 | — | — | 200 | 10 | 100 | 0.04 | — | 15 | — |
| Ti | 10 | 3 | — | 100 | — | 2 | 0.06 | — | 60 | 100 |
| Tl | 20 | 3 | 0.01 | 200 | 3 | 200 | 0.05 | 4 | — | — |
| U | 30 | 3 | — | — | 1 | 30 | 0.02 | — | 1 | 100 |
| V | 15 | — | 1 | 700 | 1 | 7 | 0.03 | 0.4 | 0.1 | 100 |
| W | 60 | — | — | — | — | 7 | 0.06 | — | 1 | — |
| Yb | 60 | — | — | 100 | 3 | 1 | 0.03 | — | 2 | — |
| Zn | 7 | 1 | 0.01 | 7 | 0.02 | 2 | 0.08 | 2 | 20 | 300 |
| Zr | 200 | — | — | — | — | 3 | 0.03 | — | 300 | 200 |

* As a result of historic practices, the detection limits and characteristic concentrations are calculated in a variety of ways depending on the assay method involved. The detection limits and characteristic concentrations apply to ideal samples. This means that the samples have no untoward matrix effects and no interferences. The values are for samples directly assayed by the various methods without any preconcentration. If preconcentration treatments could be done without adding interferents, then, in theory, even lower concentrations of analytes in the samples could be determined. For realistic samples, detection limits are about 100 times higher than those noted here.

*Key*

—: means either not a method for trace (less than ppm) analysis of the element or not reported.

MAS: Molecular Absorption Spectrometry in conjunction with colorimetric reagents. The values correspond to an absorbance of 0.025 for a 1-cm pathlength with a sample consumption of 1 mL.

DPP: Differential Pulse Polarography, assuming a sample consumption of 10 mL.

ASV: Anodic Stripping Voltammetry, with a sample size of 10 mL and a plating time of 10 min. Various media are used for electrodes. Ni and Co are determined by cathodic stripping of their oxides on Pt.

AAS: Atomic Absorption Spectrophotometry, with flame atomization and pneumatic nebulizer. The values correspond to an absorbance of 0.0025 with a 10-cm path and 1 mL of sample.

ETA–AAS: Electrothermal Atomization–Atomic Absorption Spectrometry. The values correspond to an absorbance of 0.025 with a 50-$\mu$L sample.

ICP–ES: Inductively Coupled Plasma–Emission Spectrophotometry. Values are for a pneumatic nebulizer.

ICP/MS: Inductively Coupled Plasma/Mass Spectrometry. This is with a 10-sec integration. All ions determined in the positive ion mode.

SSMS: Spark-Source Mass Spectrometry, with photographic detection assuming a 100-$\mu$L sample. Solid-state mass spectrometry is less sensitive.

NAA: Neutron Activation Analysis, with a sample volume of 100 $\mu$L, thermal neutron flux of $10^{12}$n cm$^{-2}$s$^{-1}$, and irradiation time of 10 h followed by immediate counting.

XRF: X-ray Fluorescence Spectrometry, with a dispersive instrument for a sample of 1 mL over a surface area of 7–9 cm$^2$.

*Sources:* Commission on Microchemical Techniques and Trace Analysis. *Pure & Appl. Chem.* **1982,** *54,* 1565–1577. Morrison, G. H. *CRC Crit. Rev. Anal. Chem.* **1979,** *33,* 287–320. Manufacturer's literature for ICP/MS.

## 7.3 Trace, Ultratrace, and Microanalysis Compared

Earlier in this chapter, you read that if a single grain of sodium chloride is dissolved in a barrel of (ultrapure) water, the solution would become about 1 ppb in sodium chloride. It should be clear that both the salt grain and the barrel of salt solution contain exactly the same amount of sodium and chlorine. Chemical analysis should show the same total sodium and chlorine content in either case. However, the approach to the analysis of these two samples will be different, and each presents its own problems. Analyzing the grain requires a methodology that can be used for small, concentrated samples—the techniques of microanalysis. Determining the sodium and chlorine in the barrel of water requires a different methodology, that of trace analysis.

Within the classification scheme of Figure 7.1, trace analysis can be defined as analysis for components present at levels less than 1 ppm. Ultratrace analysis may be somewhat arbitrarily defined as being for total component concentrations below approximately a part per billion (per $10^{-9}$). Neither of these definitions is the last word because of continuing development in analytical technology. As noted in the quote in Section 7.1, not too long ago trace analysis was considered to be analyzing at the parts-per-thousand level.

In trace analysis and microanalysis, the assay methodology must be able to quantify small *amounts* of material. This may seem obvious only for microanalysis; but the same criterion is necessary for trace analysis so that the samples do not become unmanageably large. Indicative of this ability, a number of commercially available instruments allow determination of components in the picogram range ($10^{-12}$ g) for samples with low levels of interfering species.

Among the techniques used for element determination are anodic stripping voltammetry (Chapter 12), atomic absorption spectrometry (Chapter 18), neutron activation analysis, X-ray fluorescence, and a number of different surface-sensitive spectrometries. Also included is mass spectrometry (Chapter 16) combined with an internal standard of a stable isotope or mass spectrometry with a high-temperature (white heat) source.

To determine concentrations of organic compounds present at trace levels, gas chromatography (Chapter 15) in conjunction with mass spectrometry (Chapter 16) is the advanced method of choice. This powerful analytical combination is almost always abbreviated GC/MS.

With these contemporary instruments, it is relatively easy to obtain an analysis of 1 $\mu$g of pure sodium chloride taken from a 200-$\mu$g crystal of salt (the approximate crystal size of table salt). Being a pure material, there is little problem of interferences from the sample itself. However, the task is far more difficult if that 1-$\mu$g sample is dissolved in a liter of solution! To the author's knowledge, a liter sample will not fit all at one time into any commercial analytical instrument. And even if it would, the matrix components, present in amounts that are $10^6$, $10^9$, or even $10^{12}$ times the amount of analyte, would surely cause significant interference even if no additional interferents are introduced in the sample collection and preparation steps of the analysis.

## Preconcentration of Trace Analytes

Because of these difficulties, for trace analysis it is necessary to concentrate and at least partially separate the component that is to be assayed from the interferents in the bulk matrix. This is **preconcentration.** A large number of different preconcentration methods are used and are carried out as part of the sample preparation steps.

One representative technique for preconcentration in trace analysis is called **electrodeposition.** The operation is illustrated in Figure 7.3. Many elements in their metallic form dissolve in liquid mercury metal, forming solutions called amalgams. Among these metals are Ag, Au, Cd, Co, Pb, Sn, and Zn. If the elements are in solution as their ions ($Ag^+$, $Au^+$, and so on), they can be converted to their metallic forms by chemical reduction. This is easily done in an electrochemical cell using mercury metal as the cathode at which the ions are reduced. Once the ions are converted to their metallic forms, they can then easily dissolve in the adjacent mercury. However, when the liquid mercury metal is in contact with water, a number of ionic species are not reduced; they therefore remain dissolved in the water. Among

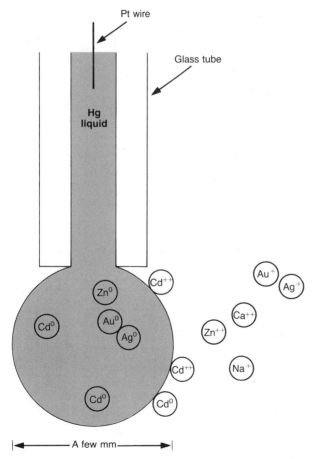

**Figure 7.3**   Diagram of the process involved in concentration by electrodeposition. Metal ions are reduced at the surface of the mercury-drop electrode and then dissolve in the mercury to form an amalgam. The maximum possible preconcentration depends on the volume of the mercury drop relative to the volume of the ionic solution. Ions that cannot be reduced to the metallic state will not be taken up.

these are $Ca^{++}$, $Cs^+$, $Na^+$, $Rb^+$, and $Sc^{++}$. The different chemistries of the above two sets of metal ions allow their separation electrochemically using mercury as a cathode.

A clever trick is to let the volume of the mercury cathode be much smaller than the volume of the water containing the dissolved sample. Then, if the electrochemical reduction proceeds for a long enough time, most of the mass of the metals that amalgamate (form amalgams) will dissolve in the drop of mercury and become highly concentrated there.

Thus, in one operation—electrochemical reduction at a mercury cathode—components can be both separated and concentrated. Such a method is used quite often when $Ca^{++}$, $Na^+$, $K^+$, and $Mg^{++}$ are present in large excess and elements such as Cd and Zn are being determined. Two substances for which this preconcentration method is used are seawater and blood. Of course, this method can be used in reverse, that is, to remove a matrix of metals such as Zn and Cd from the trace metals such as Ca and Mg that might be present.

There are a number of other techniques that can be used to separate and concentrate trace elements from the matrix materials. Since these utilize the methods covered in the sections devoted to separations (Chapter 13), their principles will be covered there. A clear, short review exists: Leyden, D. E.; Wegscheider, W. L. *Anal. Chem.* **1981,** *53*, 1059A–1065A.

## 7.4 Four Cases

The remainder of this chapter is devoted to four illustrative cases of trace and microanalysis. These examples are near current limits of analytical technology:

1. A sample of a few grams of a high-purity material for which the purity is to be assayed
2. A small (mg) solid sample to be analyzed for its elemental composition using the techniques of microanalysis
3. A micro sample to be analyzed for organic components
4. A solid for which the distribution of elements in its structure is desired

### Case 1: *A Sample of High Purity*

To carry out a determination with a *relative* error of less than 0.1% (one part per thousand) is an exceedingly difficult task. How, then, can you answer the following question: Is this silicon for use in making microcircuit chips 99.99997% or 99.99999% pure? If we were to test for silicon at 100% using a precise, carefully done assay, with enough different replicates to obtain good statistics, the result might be stated: This silicon is $(100.0 \pm 0.1)\%$ pure. But that result, as good as it is, does not answer the original question. In fact, it is about 10,000 times too crude to be useful.

To determine the purity of a highly pure material requires a different approach. What is done is to find the concentrations of all the trace impurities present: that is, all of the sample that is not silicon. It is certain that they are present only in trace amounts because the total concentration of all of them is about a few hundred ppb. Then, the total concentration of the impurities is *subtracted* from 100% to obtain the level of purity of the main component, in this case, silicon. Interestingly, the determination of high purity requires the use of trace element methodology.

To do this kind of analysis economically, it is necessary to determine simultaneously the levels of a number of trace elements. Some methods that are used for such analyses are mass spectrometry (Chapter 16) using the solid directly as the sample, emission spectrometry (Chapter 18), instrumental neutron activation analysis, and X-ray fluorescence spectrometry.

### Case 2: *A Milligram Sample To Be Analyzed for Its Elemental Composition*

Another point to consider in any kind of analysis is whether or not you want the original sample back. (Would you want to digest a part of a priceless Egyptian statuette?) If you want the sample intact, then it is necessary to use **nondestructive** analytical methods which cause no visible change in the sample. It *may* be modified on the atomic or molecular level, but these changes are not easily detectable. To cause as small a change as possible in the original sample, nondestructive tests are carried out with assay methods that are also used for micro or trace analysis. One method that is useful for both micro and trace analyses is neutron activation analysis. Two other microanalytical techniques are the electron microprobe and X-ray photoelectron spectroscopy. Further information on these can be found in the references at the end of this chapter.

A larger number of assay techniques require destruction of the sample. Any methodology that requires the digestion or fusion of samples must be classified as a **destructive** analytical methodology. For example, earlier in this chapter you read about the measurement of magnesium in a single pollen grain. The assay method required that the sample be vaporized by rapid heating. The sample was destroyed.

When the sample is in the micro range, you might guess that an analysis would be easier if the sample is not digested or diluted in any way. After all, why add interferents from handling and pretreatments with an impure acid? With the abilities of contemporary instrumentation, micro determinations (while often expensive) are generally better than trace determinations.

Samples for microchemical analysis arise in, among other areas, forensic analysis. (*Forensic* means suitable for a use in a law court.) Forensic analysts not only must run careful analyses, but also must take into account the rules of legal evidence by keeping track of the material that is analyzed. It is convenient to use nondestructive analytical techniques so the evidence can be kept if any legal questions arise.

The use of a nondestructive microanalytical technique for elemental analysis in a forensic laboratory is shown in the following quotation. The method is based on the recording of X-rays emitted by a sample that is bombarded with electrons.

> . . . $119,000 worth of selenium was stolen in an armed robbery. Clothing from a person believed to have been involved in the attack was examined in the Laboratory and a small fragment about 0.2 × 0.2 mm found on one of the shoes.

[Micro]analysis showed it to be pure selenium.
[Ref: Williams, R. L. *Anal. Chem.* **1973,** *45,* 1076A–1089A.]

As usual, to achieve a precise and accurate quantitation, the analytical methodology used here requires that proper standards be used. Although it may seem odd, when the sampled mass is quite small relative to the total sample, some care must be taken to ensure that the sampling of even a small fragment be representative. The same statistical problems of ensuring a representative sample arise just as they do in sampling a shipload of metal ore. However, in the case of the selenium fragment, it may be more difficult to *find* the original material to be analyzed. Notice, though, that no matter how large or small the sample, the methodology as shown in Figure 1.1 is followed to define and solve the problem.

Some of the more powerful analytical techniques, such as that used in the example above, can be used to quantitate composition with good precision for samples of less than $10^{-10}$ g. This number means that a one-milligram sample contains more than ten million times the amount of material needed for analysis. To get an idea of this scale, consider that a one-milligram sample with a specific gravity of four would fill a cube about 0.6 mm on a side.

### Case 3: *Micro-organic Analysis*

For micro, ultramicro, and trace chemical analysis of organic compounds, both man-made and biological, the methods of choice are mass spectrometry (Chapter 16) and gas chromatography (Chapter 15), or the combination of the two methods—gas chromatography/mass spectrometry (Chapter 16).

Our third case is, again, from forensic science and involves the analysis of the organic components of a sample of paint. The assay method used is gas chromatography alone. Here, the sample preparation method involves vaporizing $10–20$ $\mu$g of paint rapidly and reproducibly. At high temperatures, the organic-polymer components of the paint break down into smaller molecular fragments: This breakdown is called **pyrolysis** (literally, cutting by fire or heat.) The results from two samples are shown in Figure 7.4. Some experimental details are described in the figure caption. The example of the use of these **pyrograms** involves a murder.

   . . . the pyrogram was obtained of a black speck of material found embedded in the skull of a 38 year old woman who had died from head injuries. It turned out to be that of a styrenated alkyd, a coating used for tools; it matched similar pyrograms from paint on the tool kit of an abandoned car from which the [jack] was missing. There is no need to point out the significance!
   [Ref: Williams, R. L. *Anal. Chem.* **1973,** *45,* 1076A–1089A.]

### Case 4: *A Solid and the Distribution of Elements in Its Structure*

In Case 3, you read about analyzing exceedingly small samples (less than nanogram amounts of microgram samples). It may now seem reasonable to ask: If such small samples can be analyzed, is it possible to do the same type of microanalyses at

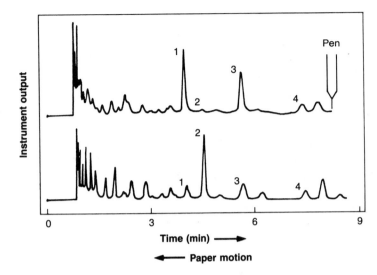

**Instrument output**

**Time (min)** ⟶

◀— **Paper motion**

**Figure 7.4**  *Pyrograms* from heating solid organic polymers. These are plots of the instrument output (vertical axis) versus time (horizontal axis). The instrument output is proportional to the amount of each component that passes through a detector. Each organic species arrives at the detector at a time that is characteristic of the species. The numbers identify the peaks that result from the same chemical species. The two samples are polymers of paint that are composed of the same two types of monomers, the only difference being in the ratio of the monomers in the original mix. The difference in composition results in the different heights of the corresponding peaks. In this way, unknown polymer resins can be characterized even down to their specific compositions if the experiments are carefully done with good standards. [Reprinted with permission from Williams, R. L., *Anal. Chem.* **1973**, *45*, 1076A–1089A. Copyright 1973 American Chemical Society.]

different places on a larger sample and determine the *locations* of various elements? The answer to the question is yes, and it is done in a manner analogous to the production of a picture on a television screen. What is done is to take samples regularly over the surface of a material and plot the results as a function of the position on the sample as illustrated in Figure 7.5. This result — the lead distribution in a bone — was neither simple nor inexpensive to obtain experimentally. The case history follows.

> Approximately 2 mm thick transverse sections cut from the femur of two autopsy cases were used for the study. Both persons had been working in different occupations in a metal smeltery in Sweden. The reference worker had been employed for 27 years during which time he worked with roasting and refining of arsenic and copper. He had not been exposed to excessive amounts of lead. He had retired from his work 8 years before his death. The lead exposed case had been working in the lead refinery of the same metal smeltery . . . [for] at least 21 years. The patient had experienced clinical symptoms of lead intoxication in periods, the first time about 20 years before his death. He had been away from lead exposure for more than 2 months before his death. The sections of femur were [frozen and then while still frozen] dried in a vacuum. . . .
> [Ref: Lindh, U.; Brune, D.; Nordberg, G. *Sci. Tot. Environ* **1978**, *10*, 31–57.]

The rest of the analysis was done as follows.

1.  The chemically localized sampling of the bone was done by focusing a beam of protons from a particle accelerator onto spots about 1/20th of a mm apart.
2.  The X-rays that characterize lead were separated from the X-rays from

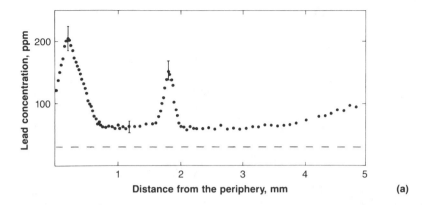

(a)

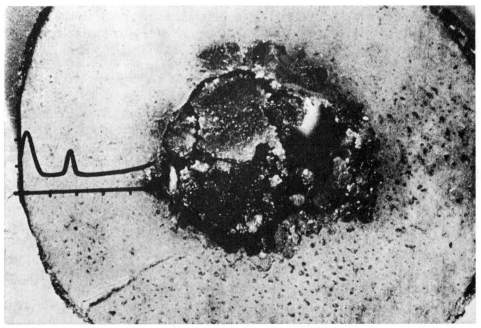

(b)

**Figure 7.5** (a) Lead distribution in a human femur. The concentration of lead is plotted as a function of the distance from the periphery of the bone. The dashed curve is the mean of the concentration in the reference bone. The dotted curve displays the distribution through the bone of the poisoned individual. (b) The same plot as in Figure 7.5a but superimposed on the bone to show the distribution relative to the cross-section structure. The central dark area is the bone marrow cavity.

most of the other elements present. This is analogous to the separation of colors in the visible spectrum.

3. Corrections were made for interference by the X-rays from other elements in the matrix that could not be separated in step 2. This is done in order to measure the intensity (of the "color") due to lead alone.

4. The intensity of these X-rays, corrected for the interference, were shown to depend on the amount of lead present. Standards were used to determine the correlation of intensity of the X-rays with the concentration of lead.

5. A blank specimen of a similar piece of bone was run.

6. The results of the determination were plotted as shown.

7. The conclusions were made that the method is usable to determine spatial distribution of trace lead concentrations in bone and has the potential for measuring the history of lead exposure.

This research was done to test a new method of lead detection. If the usual methodology of an analysis (Figure 1.1) were not followed, the data could establish the conclusion without holes in the argument.


## Suggestions for Further Reading

A survey of the different techniques usable for trace organic analysis. Some experimental methods of sample handling are included.
Beyermann, K. "Organic Trace Analysis"; E. Horwood: Chichester, and Halsted Press: New York, 1984.

A guide to the methods used in trace analysis.
Skogerboe, R. K.; Morrison, G. H. In "Treatise on Analytical Chemistry," Pt. I, Vol. 9; I. M. Kolthoff and P. J. Elving, Eds.; Wiley–Interscience: New York, 1971, Chap. 100.

An annual series of review articles on organic and inorganic trace analysis and techniques.
Lawrence, J. F., Ed. "Trace Analysis," Vol. 1; Academic Press: New York, 1980.

An expert in the field of trace analysis reviews the literature.
Morrison, G. H. "Elemental Trace Analysis of Biological Materials." *Critical Reviews in Biochemical Analysis* **1979**, *8*, 287–320.

Mostly a qualitative description with references to the literature. Now somewhat out of date.
Tolg, G. "Extreme Trace Analysis of the Elements." *Talanta* **1972**, *19*, 1489–1521.

Despite the title, this book is mostly an exhaustive coverage of liquid–liquid extraction and ion-exchange methods for inorganic preconcentration.
Minczewski, J.; Chwastowska, J.; Dybczynski, R. "Separation and Preconcentration Methods in Inorganic Trace Analysis"; E. Horwood: Chichester, and Halsted Press: New York, 1982.

A fascinating, short review of how to handle liquid samples of nanoliter size.
Bowman, R. L.; Vurek, G. G. "Analysis of Nanoliter Biological Samples." *Anal. Chem.* **1984**, *56*, 391A–305A.

## Problems

7.1   Assume you are analyzing a 2-g biological tissue sample for the presence or absence of Ni in the 0.1 ppb (w/w) range. The tissue was obtained with a metal scalpel which is 10% Ni.
   a.   What mass of the scalpel rubbing off on the sample would contribute a quantity of nickel to give a positive (0.1 ppb) result?
   b.   Assume that the density of the steel is 7.0 g cm$^{-3}$ and that the scalpel lost the material in a single, perfectly cubic chip. What is the length of the cube edge?

7.2   Dust picked up on a filter in your laboratory is analyzed. The following results are obtained for the metal content.

| Element | Content (wt%) |
|---------|---------------|
| Ca      | 10            |
| Si      | 5             |
| Fe      | 3             |
| Al      | 1.5           |
| Ni      | 1.5           |
| K       | 1             |
| Mg      | 1             |
| Cu      | 0.5           |
| Mn      | 0.5           |
| Others  | Traces        |

Assume that a dust particle that is a perfect sphere with a diameter of 100 $\mu$m and a density of 2 g cm$^{-3}$ lands in your 10-mL acidified sample and dissolves.
   a.   If you were to determine the sample's iron content in the range of 10 ppm, would you see the effects of the presence of this single dust particle?
   b.   If you were to assay a sample for trace Si at the 1-ppm level, what size spherical dust particle would contain an amount of Si equal to that contained in your sample?
   c.   Could you effectively analyze a 1-g sample for Mn in the ppb range if the sample were collected with one spherical 100-$\mu$m dust particle (with the composition listed in the table) on it?

7.3   Activated charcoal can be used to concentrate trace metals. The metals are adsorbed in basic solution and desorbed with acid. The solution that is used to desorb the ions then is analyzed. Under the experimental conditions, the charcoal was found to have the recoveries shown in the table. In addition, some impurities wash out from the charcoal, the amounts of which depend on the quantity of charcoal used. The impurity levels in the charcoal are listed below.

| Element Adsorbed | % recovery | Impurities in charcoal ($\mu$g/g) |
|------------------|-----------|-----------------------------------|
| Zn               | 85        | <1                                |
| Cu               | 96        | 18                                |
| Ni               | 98        | 19                                |
| Mn               | 98        | 150                               |
| Ag               | 92        | <0.2                              |
| Cd               | 96        | <0.1                              |
| Pb               | 92        | 2.3                               |

0.50 g of charcoal is used per sample. The charcoal was acid-washed so that 10-mL blanks showed a background that was no greater than 1% of the *total* of each impurity present in the charcoal. (This means that no more than 1% of each impurity will be leached from the charcoal into the acid solution during the analysis.) Final solution volumes are made to 10.0 mL. The original solutions which were passed through a bed of the charcoal were 500 mL. The metals are adsorbed from this.

a. If the recoveries are 100%, how much more concentrated is the eluted solution than the sample?
b. What are the apparent background concentrations (in ppb w/v) in the 500-mL blank if 1% of the charcoal impurity elements wash out with the 10-mL desorbing solution?
c. Samples with 10 ppb of each of the elements were used as standards. The analytical method is considered to be usable if the samples have five times the concentrations of the background from the charcoal. Can all the metals be analyzed at the 10-ppb range?

7.4   Assume that a trace analytical method has an inherent relative standard deviation of 2% for the random errors. You have a sample that has a low content of the component to be assayed. You measure a number of blanks and a number of samples and find that the sample measurement is only 25% above the blank level. In other words, when a sample content is assayed, the measurement (on average) consists of a background value that is 80% of the measure, whereas the sample content is 20% of the whole.

a. What is the relative standard deviation of the sample content which is determined? Note that the 2% RSD applies to the background measurements alone as well as to those with the sample.
b. What is the RSD of the sample content alone if the sample measurement if 100% above the blank?

7.5   [Table reprinted with permission from Jacobs, F. S.; Ekambaram, V.; Filby, R. H. *Anal. Chem.* **1982**, *54*, 1240. Copyright 1982 American Chemical Society.]

The accompanying table shows the results of determinations of the inorganic content of commonly used organic solvents.

Upper Limits Found of Trace-Element Concentrations in Representative Organic Solvents (in ng/L). (Zeros are place holders only.)

| Element | Hexane | Toluene | Chloroform | Methanol | Ether |
|---------|--------|---------|------------|----------|-------|
| As | 39 | 200 | 16 | 4 200 | 9.1 |
| Br | 10 | 230 | 13 | 140 | 50 |
| Ce | 200 | 120 | 200 | 200 | 200 |
| Cl | 20 000 | 10 000 | 40 000 | 1 800 | 30 000 |
| Co | 90 | 72 | 90 | 90 | 90 |
| Cr | 170 | 6 700 | 70 | 490 | 140 |
| Cs | 30 | 30 | 30 | 30 | 30 |
| Eu | 20 | 20 | 20 | 20 | 20 |
| Fe | 7 000 | 71 000 | 10 000 | 7 000 | 5 000 |

*(continued)*

Upper Limits Found of Trace-Element Concentrations in Representative Organic Solvents (in ng/L). (Zeros are place holders only.) *(continued)*

| Element | Hexane | Toluene | Chloroform | Methanol | Ether |
|---------|--------|---------|------------|----------|-------|
| Ga | 30 | 300 | 40 | 200 | 50 |
| Hf | 20 | 20 | 20 | 20 | 20 |
| Hg | 0.7 | 59 | 25 | 0.7 | 0.3 |
| K | 3 000 | 250 000 | 10 000 | 9 000 | 1 000 |
| La | 0.6 | 100 | 40 | 50 | 4 |
| Mn | 500 | 4 200 | 700 | 700 | 40 |
| Mo | 400 | 2 000 | 400 | 1 000 | 400 |
| Na | 4 800 | 390 000 | 2 500 | 99 000 | 580 |
| Sb | 4 | 4 | 8 | 40 | 40 |
| Sc | 2.6 | 22 | 2.2 | 1.1 | 0.5 |
| Se | 10 | 10 | 60 | 7 | 80 |
| Sm | 2 | 7.6 | 1.9 | 5 | 3 |
| Tb | 5 | 6 | 6 | 5 | 5 |
| Th | 20 | 35 | 30 | 20 | 20 |
| V | 9 000 | 8 000 | 20 000 | 10 000 | 20 000 |
| Zn | 40 | 17 000 | 200 | 2 700 | 580 |
| Zr | 200 | 600 | 1 000 | 700 | 2 000 |
| Total inorganics ($\mu$g/L) | 45.3 | 759 | 85.0 | 137 | 37.1 |

Inorganic trace elements in oil from oil sand are to be determined. The method involves extracting with an organic solvent followed by a multielement quantification. The elements of most importance are Co, Fe, K, Na, V, and Zn.

a. Assume that the extraction of the oil is done by heating and stirring the oil sand with a fixed volume of solvent and that the extraction is complete. Assume that the instrumental assay is done directly on the solution. If there is no interference in the assay for Zn, which would be the best solvent to use for determining only Zn?

b. If there is a large interference from Na in the assay method, which solvent(s) would you choose as being the best?

c. If the interference from Na is fully correctable, which solvent would be the best to use?

d. Assume that the solubilities of the organic components of the sample are about ten times greater in methanol than in any other solvent. As a result only one-tenth as large a solvent volume is needed for the extraction. How would you answer questions b and c in this case?

e. Assume you are doing analyses for all six important elements, and all six occur in an approximately the same range of concentration, and the same volume of each solvent would be used for extraction. Are there any of the solvents you would not use because of significant background contributions?

f.  Can you decide on the best solvent to use without determining the general range of each of the six most important elements?

7.6   The bitumen (organic phase) was extracted from ten grams of Athabasca oil sand with 1000 mL of toluene using a Soxhlet extractor.[3] The solvent was removed with a rotary evaporator, and the bitumen was analyzed for the trace elements with neutron activation analysis. The values measured for nine elements are shown in the table along with the standard deviations of the measurements.

| Element | Observed Concentration (ng/g) |
|---------|-------------------------------|
| Br      | $276 \pm 20$                  |
| Co      | $239 \pm 28$                  |
| Fe      | $242{,}000 \pm 17{,}000$      |
| Mn      | $4760 \pm 210$                |
| Hg      | $31 \pm 5$                    |
| Cr      | $854 \pm 70$                  |
| K       | $44{,}300 \pm 2900$           |
| Na      | $21{,}200 \pm 1600$           |
| Zn      | $2730 \pm 140$                |

a.  Using the data in the table in problem 7.5 (and assuming that the upper-limit numbers in the table are true and exact), correct the data for background.
b.  Are any of the corrected values outside the $\pm 2\sigma$ range for the observed values?
c.  Does the value for the Na content of the sample have any meaning under the conditions of the analysis?

---

[3] A Soxhlet extractor allows multiple extractions to be done using the same (smaller) volume of solvent repeatedly. The solvent is redistilled numerous times into a fiber filter cup that contains the sample.

# Part III

# WET-CHEMISTRY ASSAY METHODS

# 8

# Gravimetric Analysis

## 8.1 Overview of Gravimetric Methods

Gravimetric analysis, as its name suggests, depends on the measurement of weight to determine the quantity of analyte in a sample. Because of the capabilities of weighing technology, the potential precision of gravimetric analyses might be in the range of 1 part in 10,000 or even 1 part in 100,000. For instance, it is possible to weigh a one-milligram sample to the nearest microgram and weigh a one-gram sample to the nearest ten-microgram amount. However, it is necessary to have a highly pure and physically homogeneous material for this level of precision to be reached in practice. The chemical purity and homogeneity of the weighed material limit the precision of gravimetric analyses.

There are a number of different assay methods that fall under the classification of gravimetric methods. One of these is **electrodeposition.** This technique involves the electrochemical conversion of metal ions in solution into the metallic form which plates out onto the cathode. The amount of metal present is then found by the difference in weight of the cathode before and after the plating operation. Electrodeposition is generally used to determine the quantity of a specific metal — for instance, cadmium — in a macro sample. This technique is described in Section 12.6.

A second gravimetric method involves measuring a sample's changes in mass upon heating at a controlled rate in a controlled atmosphere. This is called **thermogravimetry.** Uses for thermogravimetry include the determination of water content of numerous types of samples and the carbon and ash content of coal.

A third gravimetric method involves determining the content of analyte in a sample by measuring the mass of a precipitate that forms from ions in aqueous solution. An example is the determination of sulfate by precipitation of barium sulfate:

$$Ba^{++} + SO_4^= \rightleftharpoons BaSO_4(s) \tag{8-1}$$

The precipitate is formed by addition of excess barium to the sample solution. The precipitated barium sulfate is separated from the **mother liquor** (the remaining solution), washed, dried, and weighed. The weight of barium sulfate is then related to the original sulfate concentration of the sample. Such gravimetric precipitation assays are the topic of this chapter.

To a significant degree, gravimetric precipitation analyses have been replaced by instrumental methods of analysis. With appropriate internal standards and automated sample manipulation, a number of different instrumental methods can achieve 0.1–0.3% precisions, which equal the common practical limit to these gravimetric methods. In addition, titration methods, especially with automatic titrators, can achieve the same level of precision with less manual work.

## Weighing Technique

The major difficulty with gravimetric precipitation assays arises in the production of a chemically pure, homogeneous precipitate with 100% recovery. In addition, the assay method requires reproducible weighings. Precise weighings require, in turn, accounting for the following possible problems.

A.  The weighed precipitation products must contain reproducible amounts of water. This is often achieved by heating the wet, washed precipitate at an appropriate temperature in an open container in an oven. To ensure that the moisture content remains constant, the sample is covered before cooling. The reason for covering the dried precipitate is that the material may be **hygroscopic** (note: *not hydro*scopic), which means attracting or absorbing moisture from the air. If such a material remains in contact with a humid atmosphere, it continually gains weight as water is taken up. Some compounds take up so much water from the air that the solid dissolves and becomes a concentrated aqueous solution; such materials are called **deliquescent** compounds. Precipitates for gravimetric analyses that are deliquescent or hygroscopic should be avoided if at all possible.

B.  If a sample or precipitate is weighed in a closed container, the pressure inside the container may differ from that in the surrounding room. For instance, if the container was closed when warmer than ambient temperature, the external pressure will be lower, and the container will be buoyant in the air like a ship is buoyant in water. The measured mass will be less than it should be. The internal and external pressures must be equalized to obtain a precise weighing.

C.  In order to obtain precise weighings, containers cannot be touched by bare fingers since oil and dirt (fingerprints) will be deposited. These deposits will increase the measured weight. Similar additions to the mass may also occur if the container is placed on an unclean surface.

D.  A fingerprint or other material inadvertently deposited on a weighing container must be cleaned off. However, rubbing the container with a lintless cloth or paper may also cause the container to become electrostatically charged. Static charges can cause the container to be attracted to or repelled from the surrounding parts of a balance and cause a mistake in weighing.

In the remainder of this chapter, it is assumed these difficulties with obtaining precise weights have been eliminated.

### Desirable Precipitate Properties

For analyses using quantitative precipitation, the following properties of the precipitate are desired:

1.  The solubility should be low enough so that losses are negligible in the concentration range of the analyte.

2.  The physical properties should allow it to be easily filtered from the mother liquor, washed, and transferred quantitatively from container to container. For instance, a gelatinous substance which sticks to glass would be undesirable. Another unwanted precipitate is one in which the particles are too small to be collected quantitatively from the solution either by filtration or centrifugation.

3.  The material should be stable and of definite composition in its weighable form. For instance, the unstable $Al(OH)_3$ precipitate can be converted by heating to stable $Al_2O_3$ to be weighed as was described in Section 4.8.

4.  The precipitant should have a chemistry that is as selective as possible to minimize interference. The precipitated species may be the analyte itself or a chemical species transformed from the analyte, for instance, $SO_4^=$ from $S^=$. A large body of descriptive chemistry is available to aid in making this choice.

5.  A small mass of the analyte should yield a large mass of precipitate, thus maximizing the assay sensitivity. A good example is when cobalt is reacted with the organic reagent 1-nitroso-2-naphthol to form the precipitate $Co(C_{10}H_6O_2)_3 \cdot 2 H_2O$, which weighs 10.3745 times the original cobalt.

Of these properties, numbers four and five are less important than the first three. To obtain the optimum properties, the conditions of precipitation must be chosen carefully.

## 8.2 The Mechanism of Precipitation of Ionic Compounds

Since the chemistry of precipitation affects the physical and chemical properties of precipitates, let us look at a representative mechanism of precipitation of ionic compounds from homogeneous solutions. Precipitate formation is a dynamic process—a reaction that is proceeding from an unstable solution toward an equilibrium between solid and solution. The precipitation can be separated into a set of ordered stages as illustrated in Figure 8.1. The stages are as follows.

A. **Supersaturation** is achieved by adding precipitant to the solution. Supersaturation occurs when the solution phase contains *more* of the dissolved precipitate than *can* be present at equilibrium. The greater this disparity, the greater the relative supersaturation.

B. **Nucleation** occurs next. The process of nucleation involves formation of the smallest particles of precipitate (the nuclei) that will grow spontaneously from the supersaturated solution.

C. **Crystal growth** follows. This process consists of the deposition of ions from the solution onto the *surfaces* of the solid particles present. The precipitation nuclei grow to form **colloids** or **colloidial particles** which, by definition, are 1–100 nm in diameter.

D. **Aggregation** of the solid particles *may* begin at this point. Particles aggregate by colliding and sticking together. It is a process that competes with crystal growth to form larger particles from the solution. The structure of an illustrative aggregated particle is shown in Figure 8.2e on p. 249. Such aggregates will trap mother liquor in voids between the composite particles. This trapping is called **occlusion.**

E. Aggregation and crystal growth compete. Depending on the conditions, the **final precipitate** may be composed of coarse crystals, aggregates of fine crystals, and/or aggregates of colloids. The paths to each are illustrated in Figure 8.1. The lines representing the competitive routes between crystal growth (vertically downward) and aggregation (horizontally to the right) are wavy to show randomness. This reflects the complexity of the precipitation process. The physical and chemical properties of the solution and precipitate are both changing with time and are heterogeneous through the solution. As a result the kinetic processes, which depend on solution composition, are heterogeneous as well. A number of different stages of precipitation will most likely be occurring simultaneously in a single solution.

    The following example illustrates this point. Consider, for instance, adding a drop (about 0.05 mL) of a 0.1-M barium chloride solution to a beaker containing 50 mL of solution 1.0 mM in sulfate. (The solubility product $K_{sp}(BaSO_4) = 1.3 \times 10^{-10}$. A saturated solution of 1 mM sulfate will contain approximately $1.3 \times 10^{-7}$ M $Ba^{++}$.) In the region at the edge of the added drop, the solution will be highly supersaturated in $BaSO_4$. When the solution is stirred, the drop disperses, and the level of

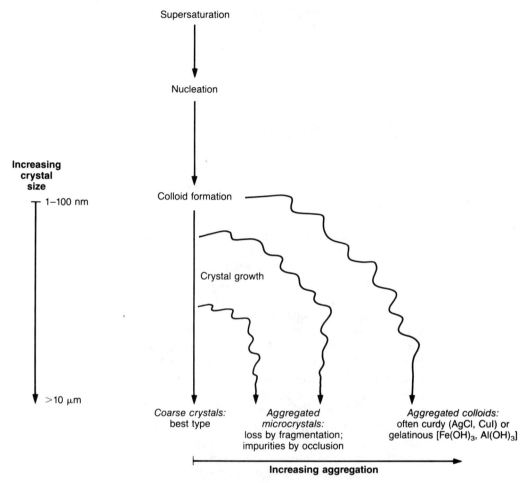

**Figure 8.1**   Diagram of the stages of crystallization from solution. After colloid formation, two competing processes occur: crystal growth by deposition of mass at crystal surfaces and aggregation of already-formed crystals of various sizes. A wavy line indicates that the path of formation of the precipitate particles is a complicated one that depends on conditions near the surfaces of the growing crystals.

supersaturation decreases as the local concentrations of $Ba^{++}$ decrease. Eventually the supersaturation *would* become homogeneous throughout the solution with $1 \times 10^{-4}$ M $Ba^{++}$ *except* that both barium and sulfate are continually being removed from the solution to form the solid precipitate. Finally, the solution will achieve equilibrium, but only the last-formed precipitate will arise under near-equilibrium conditions. The majority of the precipitation occurs under highly variable circumstances.

By controlling the conditions under which the nucleation and subsequent crystal growth and aggregation occur—mainly through adjusting the chemical concentrations and temperature—the desired precipitate properties are optimized. The best form for precipitates for gravimetric assays are coarse crystals because they tend to have a minimum of contamination and are easiest to filter and transfer quantitatively—actions that are usually necessary at least once during an assay procedure.

## 8.3 Coprecipitation of Interferences

Crystal growth (step C in the list above) is a two-step reaction. First, the ions must migrate from the solution to the surface of the growing crystal. The migration rate depends on the identities of the ions, the temperature, and the rate of stirring of the solution (whether by manual stirring or simply convection). Following their arrival at the crystal surface, the ions must fit into the lattice. The rate of fitting into the lattice depends on the concentrations and identities of the ions and the chemical characteristics of the surface. These characteristics include such properties as net charge on the surface and whether there are surface imperfections on the atomic level such as ridges or steps. It is a highly complicated process which is still not clear in detail.

Crystal growth is also fast. For instance, consider a crystal of sodium chloride which grows at a visible rate—say 1 mm an hour. This means that over a 1 mm$^2$ surface 1 mm$^3$ of sodium chloride is deposited each hour. From the density of NaCl ($2.165 \, g/cm^3$), its formula weight (58.45), and Avogadro's number ($6.022 \times 10^{23}$), we can calculate that in one hour, $2.23 \times 10^{19}$ ions of sodium and an equal number of chloride ions deposit on that mm-square area. This means that about $6.2 \times 10^{15}$ — six million billion—ions of each deposit there each second! You might expect that if some other ions, similar in size and charge, were present, they could easily become incorporated into the NaCl lattice. This is the case, and incorporation of such impurities is called **coprecipitation**. The coprecipitants may be distributed in different ways depending on the crystallization conditions. These distributions are illustrated in Figure 8.2 and are described next.

The closer the crystal and ionic structures of the possible impurities match those of the precipitate (the **host**), the easier it is for the impurities to coprecipitate. (The similarity in crystal structures is called **isomorphism:** Greek for *same form*.) When the isomorphism is close, the contaminating species will be relatively evenly distributed in the host crystal as shown in Figure 8.2b. Isomorphic impurities cannot, in general, be separated from the host by fractional precipitation. Examples of isomorphic precipitates are NaCl, KCl, KBr, and KI. All four form the same type of crystalline lattice (cubic) with atomic sites at the corners of cubes 563, 626, 659, and 710 pm on an edge, respectively. KCl forms good mixed crystals with KBr, its closest isomorph, but less mixed ones with NaCl and KI in which the atomic spacing is less similar. Another example is Pb$^{++}$ in the presence of BaSO$_4$. The lead will be found substituted for barium ions in the lattice. This is interpreted as lead sulfate dissolved in or coprecipitated with the barium sulfate.

Nevertheless, if the ionic structures differ, crystal isomorphism alone is not sufficient to cause coprecipitation. For example, even though KCl and PbS both are

**Figure 8.2** Various types of impurity distributions shown as two-dimensional representations. The impurity locations are indicated by the dots.
a. An ideal pure crystal has no impurities.
b. Isomorphic impurities are distributed throughout the crystal mass. Since the chemistry of the impurities is closely similar to that of the host, it is extremely hard to remove them.
c. Surface adsorbed impurities are excluded from the bulk crystal but bind to surface sites. The binding can range from weak (washed off easily) to extremely strong.
d. Impurities can be trapped inside faults of a crystal. These form when surface-adsorbed impurities are subsequently enclosed by *further* crystal growth. The impurity can be removed only by reforming the crystal.
e. Impurities and the mother liquor can be trapped in interstices between crystallites as they aggregate. If an interstitial volume is open to the solution, the impurities in it can be washed out. Otherwise, the impurity can be removed only by reforming the structure.

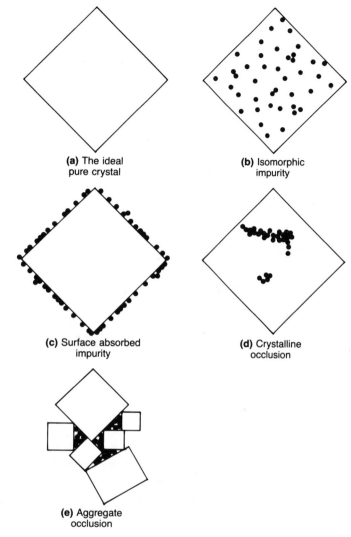

**(a)** The ideal pure crystal

**(b)** Isomorphic impurity

**(c)** Surface absorbed impurity

**(d)** Crystalline occlusion

**(e)** Aggregate occlusion

isomorphic, with atomic spacings 626 and 597 pm, respectively, they do not form mixed crystals due to the differences in ionic charges—a uni-univalent salt versus a bi-bivalent salt.

A second type of impurity structure is **adsorption** on the surface of the host precipitate. This type of structure is shown figuratively in Figure 8.2c. In adsorption (note: *ad-* and not *ab-*), the impurity, while not fitting into the lattice, can bind relatively strongly to the surface of the precipitate particles. Surface adsorption is significant only when a host precipitate has a large surface area such as is found in the gelatinous hydrated oxides, for instance, $Al(OH)_3$ and $Fe(OH)_3$. Adsorption seldom needs to be considered for precipitates that are crystalline or are aggregates of crystals.

A third type of impurity distribution, called **occlusion,** is illustrated in Figures 8.2d and e. The impurities are trapped inside of cavities within the precipitate. If the precipitate is *crystalline,* the occluded volumes tend to be along flaws and imperfections of the crystal. Crystalline occlusions occur when the impurities are entrapped during a rapid crystallization. During a slow crystallization, the adsorbed, nonisomorphic impurities tend to be excluded from the crystal matrix, leaving the bulk crystal uncontaminated. However, during a rapid crystallization, the impurities may not **desorb** (leave the surface—the opposite of *adsorb*) rapidly enough and so are covered over and entrapped in localized regions. (See Figure 8.2d.) On the other hand, if the precipitate is an *aggregate,* the impurities are entrapped in the volumes between the particles. Aggregate occlusions not only can trap surface contaminants, but the occlusion volumes can be so large that the mother liquor is trapped as well, as illustrated in Figure 8.2e. Even nonadsorbable contaminants dissolved in the mother liquor can be occluded.

## 8.4 Minimizing Contamination of Precipitates

Obtaining a precipitate in the form of large, perfect crystals will minimize the problems with impurities. The experimental precautions that tend to enhance such crystal formation are the following.

a. Have a dilute solution. This slows the crystal-growth process, allowing time for the exclusion of impurities from the solid matrix and minimizing chances for occlusion. The concentration level at which the crystallization can occur depends on the solubility of the precipitate under the conditions of the reaction. More soluble precipitates require more concentrated solutions.

b. Add the precipitant slowly. This procedure keeps supersaturation at a minimum and slows crystal growth.

c. Stir the solution well. This keeps the supersaturation constant throughout the solution and keeps the crystal growth as even as possible.

d. Keep the temperature high. A higher temperature speeds the equilibration processes. If the surface equilibrium excludes contaminants, high temperatures will speed the rates and reduce both occlusion and surface adsorption.

e. After the precipitate forms, allow the solution to stand for a long time to allow the solid to reach equilibrium with the mother liquor. This period is called **digestion.** A number of different processes occur during the digestion period. The mass of small crystals decreases and larger ones form; the surface area decreases and occlusions are eliminated as the small crystals dissolve. Also, imperfect regions of crystals become more perfect; the imperfect regions are less stable, so this is a spontaneous process. Concomitantly, impurities are released.[1]

Thus impurities can be removed from precipitates by reforming them. As was described, reformation does *not* necessarily mean that the precipitate is completely redissolved and then reprecipitated, although this may sometimes be done.

After obtaining the best crystalline material possible, separating the precipitate from the mother liquor generally involves filtering. The precipitate is held up by the filtering medium, such as filter paper or a glass frit.[2] The residual liquid is then removed by washing. Washing the precipitate is not as simple an operation as you might expect.

Special washing conditions are generally included with the precipitation instructions for specific gravimetric precipitation protocols. The reason that special conditions are needed are threefold:

To minimize redissolving any part of the precipitate

To minimize breaking up any coagulated precipitates into their composite microcrystals or colloids, which would then pass through the filter[3]

To minimize any chemical changes such as those causing further precipitation from the residual mother liquor

For example, special washing conditions are needed for $MgNH_4PO_4 \cdot 6 H_2O$, which can be used to determine either magnesium or phosphate. If washed with water alone, the material tends to hydrolyze to $MgHPO_4 \cdot n H_2O$. To avoid this problem, the precipitate must be washed with dilute $NH_3$ solution.

A second example involves washing the solid after precipitating chloride as AgCl with $AgNO_3$. For three reasons, it is necessary to wash with a dilute, cold $HNO_3$ solution. First, it avoids breaking up the aggregated colloid which forms the precipitate. Water alone would wash off the electrolyte that is essential for coagulation. Second, a cold wash solution is needed because the solubility of AgCl rises rapidly with temperature. And third, any adsorbed Ag, K, and Na nitrates are replaced by the nitric acid, which volatilizes upon drying the precipitate. The adsorbed metal nitrates would not volatilize, and due to the large surface area, they would lead to a high weight.

One other significant source of error has not yet been discussed: the filter paper or fritted glassware that is used to separate the mother liquor from the precipitate. Which of these is selected depends on the precipitation assay method chosen.

---

[1] There is a specific term used when an impurity ion is released from a precipitate and the correct ion is incorporated: this is **metathesis.** An example is substitution of $Ag^+$ for impurity $Na^+$ in solid AgCl after precipitating $Cl^-$ with $AgNO_3$.

[2] A glass frit (or fritted glass) is a porous glass plate made by fusing small glass particles together at temperatures below the melting point of the glass. Larger particles produce coarser frits.

[3] This is called **peptization.** The breakup can be done mechanically or through chemical action such as changes in ionic strength between the mother liquor and the wash.

If fritted glassware is used to both retain and weigh the precipitate, the glassware's weight must be the same before being used and after the precipitate is washed and dried. Therefore, the glassware must be cleaned, dried, and accurately weighed before using it in the assay. The drying is done at the same temperature at which the precipitate will be dried and is repeated until **constant weight** is achieved — that is, until the glassware weighs the same after each of two sequential drying periods.

On the other hand, filter paper cannot be dried to constant weight since it begins to decompose at temperatures as low as 100 °C. Therefore, it must be removed from the precipitate by oxidizing it at high temperatures — as high as 1000 °C. The process of oxidizing the paper is called **ashing,** and special low-ash papers are used to reduce the interference from the ash to minimum values. Ashing the filter paper cannot be done if the final form of the precipitate is not stable at such high temperatures. Stability in this case means both unchanging in stoichiometry and being nonvolatile. Most such materials are oxides. Examples when filter paper ashing may be used are with precipitated $Al(OH)_3$ which is converted to $Al_2O_3$ and with precipitated $Fe(OH)_3$ which is converted to $Fe_2O_3$ in the ashing step.

## 8.5 Stoichiometry and Gravimetric Factors

Consider that the material to be analyzed has been sampled, and the sample has been prepared for the precipitation assay and the analyte precipitated. Perhaps the precipitate has been digested. This was followed by separating, drying, and weighing the precipitate. The data that have been obtained consist of two weights: the weight of each sample and, for each sample, the weight of either the precipitate or a modified form of it. It is now necessary to find how the weight of the product is used to calculate the percentage of the sample that is the analyte. This calculation requires knowledge of the stoichiometry of the reaction(s) involved and straightforward stoichiometric calculations.

A number of examples follow. All the calculations involve finding:

1. The number of grams of analyte, given the number of grams of precipitate
2. The percentage of the sample that is the given analyte

These two steps should be recognizable within the calculations.

---

**EXAMPLE**

Calculate the number of grams of elemental sulfur (a.w. 32.064) that are expected from reacting 1.3423 g $KMnO_4$ (f.w. 158.04) with an excess of $H_2S$ in a sulfuric acid solution. The reaction is

$$5\ H_2S + 2\ KMnO_4 + 3\ H_2SO_4 \rightarrow K_2SO_4 + 5\ S(s) + 2\ MnSO_4 + 8\ H_2O$$

**SOLUTION**

From the stoichiometry,

2 moles of $KMnO_4$ produces 5 moles of precipitated surfur, S(s)

In the general case, where $y$ moles of $KMnO_4$ are used,

$$y \text{ moles of } KMnO_4 \text{ produces } \frac{5}{2} y \text{ moles of precipitated sulfur, S(s)}$$

Numerically, from the data given,

$$\left(\frac{1.3423}{158.04}\right) \text{ moles of } KMnO_4 \text{ produces } \frac{5}{2} \cdot \left(\frac{1.3423}{158.04}\right) \text{ moles of sulfur}$$

Note that the quantities in parentheses are the same.
Since one mole of sulfur weighs 32.064 g, we expect

$$\frac{5}{2} \cdot \left(\frac{1.3423}{158.04}\right) \cdot 32.064 \text{ g} = 0.6808 \text{ g of sulfur}$$

## EXAMPLE

A metal ore sample is digested with concentrated $HNO_3$ together with $KClO_3$. This treatment dissolves the ore and simultaneously converts the elemental sulfur present to sulfate. The nitrate and chlorate are removed by repeatedly adding concentrated HCl and heating to dryness. The sulfate is then precipitated as $BaSO_4$, which is separated, dried, and weighed. If the ore sample weighed 1.1809 g, and the dried $BaSO_4$ weighed 0.6068 g, what is the percentage sulfur in the original sample of ore?

## SOLUTION

From the stoichiometry,

$$1 \text{ S produces } 1 \text{ SO}_4^- \text{ which is precipitated as } 1 \text{ BaSO}_4$$

In other words

$$1 \text{ mole of S produces } 1 \text{ mole of precipitated BaSO}_4$$

Other information we need is

$$\text{a.w. sulfur} = 32.064$$
$$\text{f.w. } BaSO_4 = 233.40$$

Using reasoning similar to that in the example above, we can write

$$\left(\frac{0.6068}{233.40}\right) \text{ moles of } BaSO_4 \text{ comes from } 1 \cdot \left(\frac{0.6068}{233.40}\right) \text{ moles of sulfur}$$

The value 1 is the **stoichiometric factor.**
Using the weight of a mole of sulfur, we find that

$$1 \cdot \left(\frac{0.6068}{233.40}\right) \cdot 32.064 \text{ g} = 0.0834 \text{ g of sulfur}$$

The percentage of sulfur in the sample is then

$$\% \text{ S} = \frac{\text{weight of sulfur}}{\text{weight of sample}} \cdot 100 = \frac{0.0834 \text{ g}}{1.1809 \text{ g}} \cdot 100 = 7.06\% \text{ S}$$

In the second example, we could have written a single arithmetic expression that takes all the numerical values into account. It is

$$\% \text{ S} = \frac{0.6068 \cdot 32.064 \cdot 100}{233.40 \cdot 1.1809} = 7.06\% \text{ S} \tag{8-2}$$

Another way to write Eq. 8-2 would be to separate it into terms, each of which represents one step of the calculation.

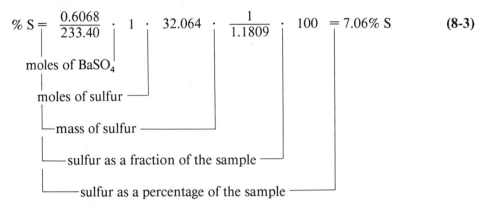

$$\% \text{ S} = \frac{0.6068}{233.40} \cdot 1 \cdot 32.064 \cdot \frac{1}{1.1809} \cdot 100 = 7.06\% \text{ S} \qquad (8\text{-}3)$$

moles of $BaSO_4$

moles of sulfur

mass of sulfur

sulfur as a fraction of the sample

sulfur as a percentage of the sample

Of course, there is no new information in this rearrangement of the equation.

However, neither Eq. 8-2 nor Eq. 8-3 is the most convenient form for the equation when the same gravimetric precipitation assay is done repeatedly on different samples. In such a case, the desired information — a percentage of the same component in each sample — is derived from two measured masses: the mass of each sample and the mass of the resulting precipitate. The formula weights and stoichiometry remain fixed. To simplify Eq. 8-3, we can rearrange it to include a **gravimetric factor** which contains the chemical information. Thus,

$$\% \text{ S} = 0.6068 \cdot 1 \cdot \frac{32.064}{233.40} \cdot \frac{1}{1.1809} \cdot 100 = 7.06\% \text{ S} \qquad (8\text{-}4)$$

weight of precipitate obtained | gravimetric factor | $\left(\begin{array}{c}\text{weight of}\\ \text{sample}\end{array}\right)^{-1}$ | % multiple

The gravimetric factor (G.F.) is specific for a procedure, and tables of gravimetric factors can be found in a number of chemistry handbooks. For sulfur determined by precipitation as $BaSO_4$, the gravimetric factor is 0.1374. That is

$$1 \cdot \frac{32.064}{233.40} = 0.1374 \qquad (8\text{-}5)$$

and Eq. 8-4 would be written simply as

$$\% \text{ S} = \frac{\text{weight of } BaSO_4}{\text{weight of sample}} \cdot 0.1374 \cdot 100 \qquad (8\text{-}6)$$

with only the masses found from the two weighings substituted for each sample.

A general equation can be written for the percentage of analyte:

$$\% \text{ analyte} = \frac{\text{mass of weighed material}}{\text{mass of sample}} \cdot \text{gravimetric factor} \cdot 100 \qquad (8\text{-}7)$$

**EXAMPLE**

Calculate the gravimetric factor for an analysis for chromium as $Cr_2O_3$ if the chromium is converted to $BaCrO_4$ for analysis.

**SOLUTION**

1 mole of $BaCrO_4$ is formed from $\frac{1}{2}$ mole of $Cr_2O_3$
f.w. $BaCrO_4 = 253.33$
f.w. $Cr_2O_3 = 151.99$
Let $y$ be the weight of $BaCrO_4$ found

The gravimetric factor would appear in the expression

$$\% \ Cr_2O_3 = \frac{\text{weight of BaCrO}_4}{\text{weight of sample}} \cdot \text{G.F.} \cdot 100$$

Reproducing the whole argument,

$$\frac{y}{253.33} \text{ moles of BaCrO}_4 \text{ comes from } \frac{1}{2} \cdot \frac{y}{253.33} \text{ moles of Cr}_2O_3$$

The value $\frac{1}{2}$ is the stoichiometric factor.

$$\% \ Cr_2O_3 = \frac{\text{Wt. Cr}_2O_3}{\text{Wt. sample}} \cdot 100$$

and

$$\% \ Cr_2O_3 = \frac{y}{\text{Wt. sample}} \cdot \frac{151.99}{2 \cdot 253.33} \cdot 100$$

Thus, by comparing this equation with Eq. 8-7, we find that the gravimetric factor for this procedure will equal

$$\frac{1}{2} \cdot \frac{151.99}{253.33} = 0.3000$$

From the above example, we can observe that a more straightforward way to calculate a gravimetric factor would be to substitute into the general equation

$$\text{gravimetric factor} = \text{stoichiometric factor} \cdot \frac{\text{f.w. analyte}}{\text{f.w. weighed material}} \quad \textbf{(8-8)}$$

**EXAMPLE**

Oxalate can be determined by precipitation with excess $Ca^{++}$ to form $CaC_2O_4$. This is subsequently heated and decomposed to $CaCO_3$ with evolution of gaseous CO. What is the gravimetric factor for this assay method?

**SOLUTION**

f.w. $C_2O_4^-$ (analyte) $= 88.02$
f.w. $CaCO_3$ (weighed) $= 100.09$

From the information above, the stoichiometry is such that

$$z \text{ moles of CaCO}_3 \text{ comes from } 1 \cdot z \text{ moles of C}_2O_4^-$$

After substitution into Eq. 8-8, the gravimetric factor for determination of oxalate by precipitation as $CaC_2O_4$ and ignition to $CaCO_3$ is found to be

$$\text{gravimetric factor} = 1 \cdot \frac{88.02}{100.09} = 0.8794$$

**EXAMPLE**

A 0.8680-g sample of a powdery white material containing oxalate was dissolved, and the oxalate was precipitated as $CaC_2O_4$, which was subsequently filtered and ignited to $CaCO_3$. The $CaCO_3$ was weighed after cooling to ambient temperature in a desiccator. Its weight was 0.1794 g. What is the percentage oxalate in the original sample?

**SOLUTION**

From Eq. 8-7 and the information in the previous example,

$$\% \text{ oxalate} = \frac{0.1794 \text{ g}}{0.8680 \text{ g}} \cdot 0.8794 \cdot 100 = 18.17\%$$

The calculation becomes quite simple indeed.

## 8.6 Solubility Calculations and Competing Reactions

### Solubility and Ionic Strength

As you have read in Chapter 2, solubility equilibria are treated mathematically the same as any other equilibrium. There is a simplifying factor in that the activity of the pure solid is always unity and, thus, does not appear in the activity equilibrium constant. The activity solubility product for a general solid $A_mB_n$ is

$$K_{sp} = (a_A)^m (a_B)^n \qquad (8\text{-}9)$$

In the calculations made here, the molar concentrations will be used instead of activities, for example,

$$K_{sp} = [A]^m [B]^n \qquad (8\text{-}10)$$

As a result, the numerical values which are calculated will *not* conform precisely to experimental values except for dissolution in pure solvent. With the addition of salts that will precipitate with neither $Ba^{++}$ nor $SO_4^-$, the solubility of $BaSO_4$ *increases*, as can be seen in Figure 8.3. In this range of inert electrolyte concentration — a few millimolar — the solubility of $BaSO_4$ changes by factors up to two or three.

These changes in solubility are due to activity effects. Mathematically, they can be accounted for by changes in activity coefficients of the precipitated species — here $Ba^{++}$ and $SO_4^-$ — with ionic strength. The mathematics will not be pursued further here. However, it should be noted that some margin of error must be allowed if calculations are made from data taken at ionic strengths different from that of the assay.

**Figure 8.3** The variation in equilibrium solubility of $BaSO_4$ at 25 °C in solutions of inert electrolytes present at different concentrations. The nonspecific ion effects, as seen for these salts, will be similar for other salts of the same type; KCl for uni-univalent salts, $MgCl_2$ for uni-bivalent salts, and $LaCl_3$ for uni-trivalent salts. These are ionic strength effects. [Data from Linke, W. F. "Solubilities," 4th ed.; American Chemical Society: Washington, D.C., 1958; Vol. I, p. 392.]

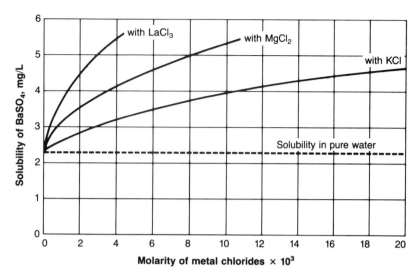

## Solubility and Solubility Products

Equilibrium calculations of solubilities involve the solubility product and any simultaneous equilibria that might change the concentrations of the ionic species in the solubility product expression. Such calculations merely reverse the process of deriving equilibrium constants from the solubility data. Thus a great deal of information can be stored in a table of solubility products. Appendix I contains a short list of $K_{sp}$ for salts dissolved in pure water. The concentration units for the ions are mol $L^{-1}$.

---

**EXAMPLE**

Calculate the solubility product for barium sulfate, given that the solubility of $BaSO_4$ at 25 °C in pure water is 2.23 mg $L^{-1}$.

**SOLUTION**

A saturated solution of $BaSO_4$ contains

$$(2.23 \times 10^{-3} \text{ g L}^{-1})/(233.40 \text{ g mol}^{-1}) = 9.554 \times 10^{-6} \text{ mol L}^{-1}$$

The molar solubility product at 25 °C is, then,

$$K_{sp} = [Ba^{++}][SO_4^-] = [9.554 \times 10^{-6}]^2 = 9.13 \times 10^{-11}$$

for $BaSO_4$ in pure water.

---

**EXAMPLE**

The log of the solubility product at 25 °C and zero ionic strength of lanthanum iodate, $La(IO_3)_3$, is $-10.92$. [Ref: Firsching, F. H.; Paul, T. R. *J. Inorg. Nuclear Chem.* **1966**, *28*, 2414.] Assume that the salt completely dissociates when dissolved. What is the solubility of lanthanum iodate under these conditions?

**SOLUTION**

The formula weight of $La(IO_3)_3$ is 663.68. From the logarithm of $K_{sp}$ we calculate

$$K_{sp} = 10^{-10.92} = 1.20 \times 10^{-11}$$

And

$$K = [La^{+++}][IO_3^-]^3$$

Upon dissolution, there are three times as many iodate ions as lanthanum ions. Thus,

$$1.20 \times 10^{-11} = (X)(3X)^3 = 27\,X^4$$

Yielding

$$X = 8.17 \times 10^{-4}\ \text{mol L}^{-1}$$

This is the molar concentration of lanthanum iodate in solution. An $8.17 \times 10^{-4}$ molar solution contains $0.542$ g L$^{-1}$ lanthanum iodate since

$$8.17 \times 10^{-4}\ \text{mol L}^{-1} \times 663.63\ \text{g mol}^{-1} = 0.542\ \text{g L}^{-1}\ La(IO_3)_3$$

Note that it has been assumed that $[La^{+++}]$ stands for the concentration of all the species of lanthanum in solution both in the calculation of $K_{sp}$ from the original solubility data and in this calculation. In reality, highly charged ions such as $La^{+++}$ generally exist in a number of different hydrolyzed forms such as $La(OH)^{++}$ and $La(OH)_2^+$. In other words, it has been implicitly assumed that

$$[La^{+++}] = [La^{+++}] + [La(OH)^{++}] + [La(OH)_2^+] + \cdots$$

## Changing Solubility by Adding a Common Ion

If a solution is saturated with barium sulfate which dissolves with the reaction

$$BaSO_4(s) \rightleftharpoons Ba^{++} + SO_4^-;\qquad K_{sp} = 9.12 \times 10^{-11}\ (25\ °C)$$

and sodium sulfate is added to the solution with a minimum volume change, the reaction is driven to the left. (Sodium sulfate completely dissociates under these conditions.) Similarly, if excess $BaCl_2$ were added, the reaction would be forced to the left as well. This is the well-known effect of adding a common ion to systems at equilibrium. Note that in both cases, some ionic strength effects will occur. However, the common ion effect will be much larger in this case—where no important reactions occur except the precipitation of $BaSO_4$.

---

**EXAMPLE**

What is the calculated solubility (in mg L$^{-1}$) of $BaSO_4$ in 1.0 mM $Na_2SO_4$ at 25 °C? At the same temperature, 2.23 mg $BaSO_4$ dissolves in a liter of water. Ignore the effects of changes in ionic strength.

**SOLUTION**

As calculated earlier, $K_{sp} = 9.13 \times 10^{-11}$. In the original solution, $[SO_4^-] = 9.55 \times 10^{-6}$ M. The sulfate originating from the precipitate can be ignored relative to that from the sodium sulfate. The calculation is done as follows.

$$K_{sp} = [Ba^{++}][SO_4^-] = 9.13 \times 10^{-11}$$
$$[Ba^{++}][0.001] = 9.13 \times 10^{-11}$$
$$[Ba^{++}] = 9.13 \times 10^{-8}\ \text{M}$$
$$9.13 \times 10^{-8}\ \text{mol L}^{-1} \cdot 233.40\ \text{g mol}^{-1} = 0.021\ \text{mg L}^{-1}$$

The solubility of $BaSO_4$ has been reduced by over a factor of 100 by the addition of the sodium sulfate.

Note: The ionic strength effect of $Na_2SO_4$ will be approximately the same as that due to $MgCl_2$ as seen in Figure 8.3. Thus the calculation here will be in error by about 30–40% from what should be observed.

## EXAMPLE

A liter of a solution contains 10.0 ppm $SO_4^=$. What fraction of the original sulfate ion remains in the solution if 10.0 mL of 1.00 M $BaCl_2$ is added to the solution? The temperature of the solution is 25 °C.

## SOLUTION

In this case, we ignore the volume change (1.0%) due to the addition of the barium solution and assume the solution specific gravity is unity.

10 ppm of $SO_4^=$ corresponds to

$$\frac{10 \times 10^{-6} \cdot 1000 \text{ g L}^{-1}}{96.06 \text{ g mol}^{-1}} = 1.04 \times 10^{-4} \text{ M } SO_4^=$$

If the $BaCl_2$ remained unreacted, its diluted concentration would be

$$[Ba^{++}] = \frac{10.0}{1000} \cdot 1.00 \text{ M} = 0.010 \text{ M}$$

At most, only about 1% of this could react with the sulfate present, so this may be considered the final concentration of $Ba^{++}$. Thus,

$$K_{sp} = 9.12 \times 10^{-11} = [Ba^{++}][SO_4^=]$$
$$9.12 \times 10^{-11} = [0.010][SO_4^=]$$
$$[SO_4^=] = 9.1 \times 10^{-9} \text{ M}$$

Since the original sulfate contained in the solution was $1.04 \times 10^{-4}$ M, only the fraction

$$\frac{9.1 \times 10^{-9}}{1.04 \times 10^{-4}} = 8.8 \times 10^{-5}$$

remains in solution. Quantitative precipitation—arbitrarily defined as greater than 99.99% precipitated—is expected.

Note: At 25 °C the density of pure water is 0.99707 g cm³, which means that 1000.0 g of water will occupy 1.0029 L. The density of a 10-ppm ionic solution will differ negligibly from this value. However, the assumption of unity for the specific gravity at 25 °C introduces an error of approximately 0.3%. This is small relative to the error introduced into the calculation by ignoring ionic strength effects. The ionic strength effect of $BaCl_2$ will be nearly the same as $MgCl_2$ shown in Figure 8.3. Thus the calculation is expected to be in error by a factor in the range 2 to 3.

## Changing Solubilities by Competitive Equilibria

The solubility product describes the relationship between a precipitate in contact with a solution containing its constituent ions. But consider the precipitate $Mg(OH)_2$. The concentration of hydroxide $[OH^-]$ depends on the pH of the solution, and so the precipitation of $Mg(OH)_2$ depends on the pH as well. Any calculation concerning the solubility of $Mg(OH)_2$ must involve competitive equilibria of the type you have seen in Chapter 2.

---

**EXAMPLE**

$Mg(OH)_2$ has a $K_{sp} = 7.1 \times 10^{-12}$ at zero ionic strength and 25 °C. Assuming equilibrium is reached, will a precipitate of $Mg(OH)_2$ form in a solution of pH 9.0 (buffered) containing 1.0-mM $Mg(NO_3)_2$? Assume that ionic strength effects can be neglected. Under these conditions, magnesium nitrate is completely dissociated in water.

**SOLUTION**

At a pH of 9.0, the pOH is 5.0 and $[OH^-] = 1.0 \times 10^{-5}$ M.

Then, under these conditions,

$$[Mg^{++}][OH^-]^2 = [1.0 \times 10^{-3}][1.0 \times 10^{-5}]^2 = 1.0 \times 10^{-13}$$

The ion product is smaller than $K_{sp}$, and the precipitate will *not* form even under equilibrium conditions. The factor of 70 difference between the $K_{sp}$ and the ion product ensures that this result will be correct even accounting for errors due to ionic strength effects.

Note: The calculation is made with the implicit assumption that the species used to buffer the solution does not bind to the $Mg^{++}$ ions.

---

In a similar way, the formation of coordination complexes with the ions in solution can shift the precipitate equilibrium toward dissolution. For instance, if ammonia is added to a solution containing silver chloride, the silver ions will react to form the diamino silver complex.

$$Ag(NH_3)_2^+ \rightleftharpoons Ag^+ + 2\,NH_3; \qquad K_{instability} = 6.0 \times 10^{-8}$$

Thus the ammonia competes for $Ag^+$ with the AgCl precipitate.

---

**EXAMPLE**

A precipitate AgCl(s) is placed in contact with a solution containing an *initial* concentration of 0.10-M $NH_3$. What is the final concentration of $NH_3$ in the solution? What is the solubility of AgCl (in mol $L^{-1}$) in the solution?

$$K_{sp}(AgCl) = 1.7 \times 10^{-10}$$
$$K_{inst}(Ag(NH_3)_2^+) = 6.0 \times 10^{-8}$$
and for $NH_3 + H_2O \rightleftharpoons NH_4^+ + OH^-$, $K_b = 1.8 \times 10^{-5}$

**SOLUTION**

As for other simultaneous equilibria, it is necessary to write the total reaction for dissolution and formation of the silver–ammonia complex as the sum of the two competing reactions.

| | |
|---|---|
| $AgCl(s) \rightleftharpoons Ag^+ + Cl^-$ | $K_{sp} = 1.7 \times 10^{-10}$ |
| $Ag^+ + 2\,NH_3 \rightleftharpoons Ag(NH_3)_2^+$ | $K = 1/K_{inst} = 1.67 \times 10^7$ |
| $AgCl(s) + 2\,NH_3 \rightleftharpoons Ag(NH_3)_2^+ + Cl^-$ | $K = K_{sp}(1/K_{inst})$ |

Then, the mass-action expression for the sum equation is

$$\frac{[Ag(NH_3)_2^+][Cl^-]}{[NH_3]^2} = 2.8 \times 10^{-3}$$

Let $X$ equal the molar concentration of $Ag(NH_3)_2^+$ (and $Cl^-$). So,

$$\frac{X^2}{(0.10 - 2X)^2} = 2.8 \times 10^{-3}$$

The 2 in the denominator arises because two ammonia molecules are needed for each complex.

Taking the square root of both sides and solving for $X$ gives

$$X = 4.8 \times 10^{-3} \text{ M}$$

which is the concentration of the diamino silver complex and $Cl^-$. This compares with a solubility in water alone for $AgCl(s)$ of $1.3 \times 10^{-5}$ M. The ammonia increases the solubility by a factor of nearly 400.

Note: Three assumptions were made in the above calculation. First, $[Ag^+]$ was negligible compared with $[Ag(NH_3)_2^+]$. Second, the monoamino complex $[Ag(NH_3)^+]$ was negligible. And third, $NH_3$ was the form of ammonia in the solution and, thus, $[NH_4^+]$ could be neglected. These assumptions will be shown to be reasonable in the next three sample calculations.

## SAMPLE 1

Show that $[Ag^+]$ is small (less than 0.1%) relative to $[Ag(NH_3)_2^+]$ under the conditions.

## CALCULATION

Using the data

$$Ag(NH_3)_2^+ \rightleftharpoons Ag^+ + 2\,NH_3; \qquad K_{instability} = 6.0 \times 10^{-8}$$

then

$$\frac{[NH_3]^2[Ag^+]}{[Ag(NH_3)_2^+]} = 6.0 \times 10^{-8}$$

We can calculate that for a solution 0.10 M in $NH_3$,

$$\frac{[Ag^+]}{[Ag(NH_3)_2^+]} = 6.0 \times 10^{-6}$$

Essentially all the silver is tied up as the diamino silver complex, even allowing for moderate changes in the free-ammonia concentration.

## SAMPLE 2

Show that the monoamino complex $Ag(NH_3)^+$ can be neglected in this analog calculation.

## CALCULATION

We need the data for the stepwise equilibrium constant for the silver–ammonia complex. This is

$$Ag(NH_3)_2^+ \rightleftharpoons Ag(NH_3)^+ + NH_3; \qquad K_1 = 1.4 \times 10^{-4}$$

Given an ammonia concentration of 0.10 M,

$$\frac{[Ag(NH_3)^+]}{[Ag(NH_3)_2^+]} = 1.4 \times 10^{-3}$$

From this ratio, it can be seen that $[Ag(NH_3)^+]$ is only about one part per thousand of the total complexed silver. It is negligible relative to the concentration of the disubstituted silver complex $[Ag(NH_3)_2^+]$.

## SAMPLE 3

Show that $NH_3$ is the form of ammonia in the solution and, thus, $[NH_4^+]$ can be neglected under the conditions of the solution.

## CALCULATION

The data needed are

$$NH_3 + H_2O \rightleftharpoons NH_4^+ + OH^-; \qquad K_b = 1.8 \times 10^{-5}$$

For a total ammonia concentration of 0.10 M,

$$0.10 = [NH_3] + [NH_4^+]$$

and

$$\frac{[NH_4^+][OH^-]}{[NH_3]} = 1.8 \times 10^{-5}$$

Let $X$ be $[NH_4^+] = [OH^-]$, then

$$\frac{X^2}{0.10 - X} = 1.8 \times 10^{-5}$$

and $X = 0.013 = [NH_4^+]$

$$\frac{[NH_4^+]}{[NH_3]} = 0.15$$

This ratio suggests that there is some competition for the ammonia from the protons in solution under the conditions.

---

The presence of moderate fraction of ammonia as $NH_4^+$ will affect the value of the solubility of AgCl which was calculated. The competition of $H^+$ with $Ag^+$ for the ammonia will tend to decrease the calculated AgCl solubility somewhat. However, within the limits of neglected ionic strength effects, making the correction will probably not be worthwhile.

However, another competitive reaction is suggested by sample calculation 3. The concentration of hydroxide is equal to $[NH_4^+]$. The pH of the solution is expected to be about 12. Will any silver hydroxide form? A calculation suggesting the answer to this question is made next.

---

## SAMPLE 4

Will any silver hydroxide form in the solution 0.10 M in ammonia with $[OH^-] = 0.013$ M?

## CALCULATION

The data needed are

$$K_{sp}(AgOH) = 6.8 \times 10^{-9} \text{ in 1-M NaClO}_4 \text{ solution at 25 °C}^4$$

---

[4] [Ref: Sillen, L. G.; Martell, A. E. "Stability Constants of Metal-ion Complexes," Supplement No. 1, Special Publication No. 25; The Chemical Society: London, 1971.] This is probably the best available value for $K_{sp}$ of AgOH. This $K_{sp}$ was measured in a solution 1 M in NaClO$_4$, whereas the other $K_{sp}$ values, which have been quoted from the same source, are extrapolated to zero ionic strength. $K_{sp}$ data on similar compounds run at a range of ionic strengths suggest an error of about a factor of two will result. However, this is a smaller difference than the usual disagreement among various workers.

From the solubility calculation, the total silver concentration in solution was found to be $4.8 \times 10^{-3}$ M.

We have found that almost all this silver is complexed with ammonia. Then, $[Ag(NH_3)_2^+] = 4.8 \times 10^{-3}$ M. But from sample calculation 1 we found

$$\frac{[Ag^+]}{[Ag(NH_3)_2^+]} = 6.0 \times 10^{-6}$$

Thus, substituting for the concentration of the diamino complex,

$$\frac{[Ag^+]}{[4.8 \times 10^{-3}]} = 6.0 \times 10^{-6}$$

and

$$[Ag^+] = 2.9 \times 10^{-8} \text{ M}$$

Under the conditions of the experiment,

$$[Ag^+][OH^-] = [2.9 \times 10^{-8}][0.013] = 3.8 \times 10^{-10}$$

which is an order of magnitude less than the solubility product. The ammonia in the solution prevents the silver hydroxide precipitate from forming.

## Suggestions for Further Reading

A three-volume treatise on the gravimetric analysis, including illustrations of the experimental equipment. The parts on precipitation analysis encompass the field. Methods for most elements are also included.

Erdey, L. "Gravimetric Analysis"; Pergamon: Oxford, 1963, Vol. I; Macmillan: New York, 1963, Vol. II; Macmillan: New York, 1965, Vol. III.

## Problems

8.1    Sulfate can be determined by gravimetric precipitation assay weighing it as $BaSO_4$. The precipitate is filtered through filter paper. The filter paper is removed by ashing in a covered crucible. However, under these conditions, carbon from the paper may reduce the $BaSO_4$ through the reaction

$$BaSO_4 + 4\, C \xrightarrow{\text{heat}} BaS + 4\, CO(g)$$

a.  Will this reaction tend to cause the final result of the analysis to be too high, too low, or will it remain correct?

b.  If the above reaction is not too extensive, continuing the ignition in air causes the reaction

$$BaS + 2\, O_2(g) \xrightarrow{\text{heat}} BaSO_4$$

If this reaction occurs for all the BaS, will the final result of the analysis be too high, too low, or will it remain correct?

8.2    In the determination of Cl as AgCl, the precipitated AgCl can be decomposed partially by light. The reaction is

$$2\, AgCl \xrightarrow{\text{light}} 2\, Ag^0 + Cl_2(g)$$

If this reaction occurs while the precipitate is still in contact with the mother liquor, the chlorine gas reacts with water to form $Cl^-$ again. If the photoreaction occurs after separating the AgCl from the mother liquor, the chlorine gas can escape into the air.

    a.   If the photoreaction occurs while the AgCl is in contact with the mother liquor, will it tend to make the final result too high, too low, or leave it the same?

    b.   If the photoreaction occurs while the AgCl is dry, will it tend to make the final result too high, too low, or leave it the same?

8.3    Calculate the gravimetric factors for the following analyses. They should be calculated to four significant figures.

    a.   Sr precipitated with $H_2SO_4$ and weighed as $SrSO_4$

    b.   SrO dissolved in acid, precipitated with $H_2SO_4$, and weighed as $SrSO_4$

    c.   Cu precipitated with salicylaldoxime and weighed as $Cu(C_7H_6NO_2)_2$

    d.   Mn precipitated with KOH and weighed as $Mn_3O_4$

    e.   Lu precipitated with $H_2C_2O_4$ and weighed as $Lu_2O_3$

8.4    $Eu(IO_3)_3$ has a $K_{sp}$ of $4.8 \times 10^{-12}$ at 25 °C and zero ionic strength. How many milligrams of europium iodate will dissolve in 100 mL of water at 25 °C?

8.5    At 18 °C, $1.8 \times 10^{-4}$ g of $Bi_2S_3$ dissolves in a liter of water.

    a.   What is the molarity of the saturated solution in contact with the solid?

    b.   What is the $K_{sp}$ of the compound?

8.6    $Eu(IO_3)_3$ has a $K_{sp}$ of $4.8 \times 10^{-12}$ at 25 °C and zero ionic strength. Under these conditions, the $HIO_3$ is completely dissociated. Assume no hydrolysis of Eu occurs, that is, no $Eu(OH)^{++}$ or other hydroxides form.

    a.   A quantitative precipitation is defined as one that is 99.99% complete. What concentration of $HIO_3$ is needed to precipitate quantitatively a 10.0-mM europium solution?

    b.   Could you answer the question in part a without the europium concentration being given?

    c.   If you wanted to carry out a europium gravimetric assay on an unknown, how would you determine whether enough acid was added to provide quantitative precipitation?

8.7    Four hydroxide precipitates have the $K_{sp}$ values tabulated below.

| ppt | $K_{sp}$ |
| --- | --- |
| $Ba(OH)_2$ | $5.0 \times 10^{-3}$ |
| $Ca(OH)_2$ | $1.3 \times 10^{-6}$ |
| $Cu(OH)_2$ | $1.6 \times 10^{-19}$ |
| $Mn(OH)_2$ | $2 \times 10^{-13}$ |

Assume no metal hydrolysis occurs at any pH (not true in fact). Assume ionic strength effects are equal for all the ions.

    a.   Assume all four metal ions, $Ba^{++}$, $Ca^{++}$, $Cu^{++}$, and $Mn^{++}$, are together in an acid solution, and each is 1 mM. KOH is then slowly added with vigorous stirring. Which metal hydroxide will precipitate first? What is the order of precipitation after that?

b.  At what pH will the first hydroxide precipitate form, assuming equilibrium conditions? Does the solution need to be basic to precipitate hydroxides?

8.8  The listed precipitates have the following $K_{sp}$ values.

| ppt | $K_{sp}$ |
| --- | --- |
| Im(OH) | $6.0 \times 10^{-12}$ |
| Un(OH)$_2$ | $6.0 \times 10^{-12}$ |
| Ev(OH)$_3$ | $6.0 \times 10^{-12}$ |
| My(OH)$_4$ | $6.0 \times 10^{-12}$ |

These are, respectively, the hydroxides of improbablium (Im), unlikelium (Un), evanescentium (Ev), and mythium (My).

a.  Assume all four metal ions, $Im^+$, $Un^{2+}$, $Ev^{3+}$, and $My^{4+}$ are together in an acid solution, and each is 1 mM. KOH is then slowly added with vigorous stirring. What is the order of precipitation of these four metal ions?

b.  At what pH will each of the hydroxides just begin to precipitate — assuming equilibrium conditions, no metal hydrolysis, and negligible ionic strength effects?

8.9  These calculations involve the same $K_{sp}$ values given in problem 8.8. Again, the metal ions are each present at 1 mM.

a.  The pH is adjusted to the point where 99.99% of the $Im^+$ is in the form Im(OH) under equilibrium conditions. At that pH, what percentage of the $Un^{++}$ present will be precipitated?

b.  The pH is adjusted to the point where 99.99% of the $Ev^{3+}$ is in the form Ev(OH)$_3$ under equilibrium conditions. At that pH, what percentage of the $My^{4+}$ present will be precipitated?

*8.10  A 50.0-mL portion of 0.100-M AgNO$_3$ is mixed with a 25.0-mL portion of 0.156-M K$_2$CrO$_4$ at 25 °C. $K_{sp}(Ag_2CrO_4) = 1.29 \times 10^{-12}$.

a.  Calculate the concentrations to be expected in the equilibrium solution of $K^+$, $NO_3^-$, $CrO_4^-$, and $Ag^+$ ignoring ionic strength effects.

b.  The Ag$_2$CrO$_4$ is slightly soluble. Given the concentration expected for $CrO_4^-$ calculated above, what fraction is due to $CrO_4^-$ from soluble Ag$_2$CrO$_4$? Can this quantity be ignored (less than one part per thousand of the $CrO_4^-$)?

c.  Under the final conditions, calculate what fraction of the total silver is expected to be left in solution.

*8.11  100.0 mL of a solution is 0.0100 M in Ba(NO$_3$)$_2$ and 0.0100 M in Pb(NO$_3$)$_2$. To this solution was added a slight excess of $SO_4^-$ over the $Ba^{++}$ consisting of 101.0 mL of 0.0100-M H$_2$SO$_4$. Under the conditions present,

$$K_{sp}(BaSO_4) = 1.00 \times 10^{-10}$$
$$K_{sp}(PbSO_4) = 1.70 \times 10^{-8}$$

Assume that $H^+$ does not bind to $SO_4^-$, that ionic strength effects can be ignored, and that the system is at equilibrium.

a.  Calculate the concentrations you expect to find for $Pb^{++}$, $Ba^{++}$, and $SO_4^-$ in the final solution.

b.  Calculate the expected composition of the precipitate and report it as mole fraction BaSO$_4$ and mole fraction PbSO$_4$ (if any is present).

---

\* Asterisks indicate more involved problems.

# 9

# General Introduction to Titrations: Neutralization Titrations

## 9.1 Requirements for Titrations

**Titration** is a method in which a volume of a standard(ized) solution is added to an unknown solution to ascertain the titer of some component of the unknown. *Titer,* a word now more or less obsolete, means the weight of a substance equivalent to a unit volume of solution. As an equation,

$$\text{titer} = \frac{\text{grams of substance}}{\text{liters of solution}}$$

Since an unknown amount of some substance is calculated from a known *volume* of added solution,[1] the method of titration falls under the classification of **volumetric** methods. Precision in the range of 0.1% is possible with moderate care.

---

[1] The titrant can also be added through an electrochemical reaction in the analyte solution. These coulometric titrations are described in Section 12.8.

The requirements for volumetric titrimetric assays include the following:

A. A **titrant** solution with a known concentration of reagent (a standard-ized reagent). It must react to completion with the analyte with a *reproducible stoichiometry* and an adequate reaction rate.

B. A technique to measure the volume of the titrant solution to the desired accuracy.

C. A technique to measure the sample weight or volume to the desired accuracy.

D. If the reagent–analyte reaction is not sufficiently specific to eliminate effects of the matrix components, a pretreatment to remove interferents.

E. A technique to measure when the titrant–analyte reaction has just used up all the analyte; the moles of titrant added are equivalent to the moles of analyte. This is the **equivalence point.** The volume of titrant added up to this point is the best quantity to use in calculating the analyte content. However, to discern the equivalence point, some phys-ical change must be observed. The center of that change is the **end point** of the titration. Ideally, the end point and the equivalence point coincide. If they do not, the difference is called the **titration error.**

## 9.2 A Case Study

The practical example of a titrimetric method involves determining three different sulfur-containing species, which can appear as a result of reactions of sulfur dioxide. The three species are sulfate ($SO_4^-$), sulfite ($SO_3^-$), and dithionate ($S_2O_6^-$).

The strategy for doing a titration for three different sulfur-containing species is as follows.

a. Determine the sum of sulfate and sulfite.
b. Determine sulfate alone.
c. Determine the sum of sulfate, sulfite, and dithionate.

From these three sets of determinations, the content of each of the three different species can be calculated. Note that the content of sulfite and dithionate are deter-mined indirectly by difference. As you will see, the ability to do these three different determinations requires different sample pretreatment methods.

By way of introduction to the quoted example, let us see how each of the five requirements, A through E, of a titrimetric analysis are satisfied.

The titrant for all titrations was a standardized $Pb(ClO)_4$ solution. The perchlorate is a spectator ion. The reactions that occur and go to completion are precipitation reactions. These are

$$Pb^{++} + SO_4^= \rightleftharpoons PbSO_4(s)$$
$$Pb^{++} + SO_3^= \rightleftharpoons PbSO_3(s)$$

The stoichiometries of these reactions are reproducible, as well as being simple and well understood. Also, the precipitates form essentially immediately as the titrant is added. A titration in which precipitates are formed between the analyte(s) and the titrant reagent are called, reasonably, **precipitation titrations.**

The titrant volume is measured using a burette.

The standard samples were made by weighing solid salts and dissolving them in solution. The sample of a hydrocarbon fuel was prepared by weighing the liquid sample and then burning it enclosed in a container with thick (cm or more) steel walls: a combustion bomb. The products of the combustion were then analyzed. (The sample pretreatment is integrally related to the strategy of the titration as described above.)

The end point was determined by using a selective sensor. The sensor—an electrode (discussed in more detail in Chapter 12)—responds only to the amount of $Pb^{++}$ in solution. From the sensor's response can be found the end point for the sulfate and/or sulfite reaction with lead ion.

One more factor worth noting is that the titrations described below are carried out in 1:1 methanol:water solution. Titrations are not limited to water, and some of the most useful have no water present at all. (These are the **nonaqueous** titrations.) Titrations such as these, done in 50:50 methanol:water, are titrations in **mixed solvents.**

Sulphur dioxide is one of the most ubiquitous pollutants in ambient air. Its oxidation in clean, dry air is slow, but certain transition metals such as iron, copper, and manganese, which are quite common constituents of urban atmospheres, may enhance it. The oxidation of sulphur dioxide may lead to dithionate ($S_2O_6^=$) or sulphate . . . .

There are few satisfactory analytical techniques for determination of dithionates or mixtures of dithionates with other sulphur-containing anions . . . .

The present paper describes the analysis of mixtures of sulphate, sulphite, and dithionate, by potentiometric titration with lead perchlorate.

An aliquot of sample is titrated with lead perchlorate to obtain the sum of sulphate and sulphite. Another aliquot is acidified with perchloric acid, and the sulphur dioxide produced is removed from the solution by passage of nitrogen. The sulphate is then titrated with lead perchlorate. A third aliquot is oxidized with hydrogen peroxide to convert sulphite into sulphate and then with concentrated nitric acid to oxidize dithionate to sulphate, and the total sulphate is titrated. All of the titrations are done at pH 4 in 50% aqueous methanol. The error and precision of each titration are both 2–5% for the sulphate concentration range $3 \times 10^{-3}$–$6 \times 10^{-4}$ M . . . .

[The technique has also] been applied to the determination of sulphur in some fuels derived from refuse.

**Procedures** (a) Determination of sulphate and sulphite. Pipette a 20-mL sample into a 100-mL beaker. Adjust the pH to 4 with sodium hydroxide solution or perchloric acid. Add 25 mL of methanol, immerse the electrodes in the solution, start stirring at the maximum speed at which air bubbles are not formed, and titrate with 0.01 M lead perchlorate.

(b) Determination of sulphate. Pipette a 20-mL sample into a 100-mL beaker. Add 1.0 mL of 3 M perchloric acid and bubble nitrogen vigorously through the solution for 15–20 min. Adjust the pH to 4 with 5-M sodium hydroxide, add 25 mL of methanol, and continue as in (a).

(c) Determination of sulphate, sulphite and dithionate. Pipette a 20-mL sample into an Erlenmeyer flask. Add 2 or 3 drops of 0.025 M sodium hydroxide and 0.5 mL of 30% hydrogen peroxide. Allow 5 min for oxidation of sulphite to sulphate. Add 30 mL of concentrated nitric acid and put the loosely stoppered flask in a heated water-bath for 4 hr. The dithionate is quantitatively oxidized to sulphate. Evaporate the solution almost to dryness to remove the nitric acid, add 20 mL of water and adjust the pH to 4. Add 25 mL of methanol and continue as in (a).

**Calculations**  Plot the e.m.f. against volume of titrant and take the point of inflection as the end point. Calculate the sulphite content from the difference between titrations (a) and (b). Calculate the dithionate content from titrations (a) and (c), remembering that 1 mole of dithionate is converted into 2 moles of sulphate . . . .

We have compared the gravimetric [assay] with procedure (a), after decomposition of the sample [of fuel derived from refuse] by bomb combustion. The precision of the two methods was checked by burning five samples of the same material and analysing the resulting solutions by both methods. The means and standard deviations [$N = 5$] were $(0.224 \pm 0.012)$ and $(0.226 \pm 0.008)$%, for the gravimetric and titrimetric methods, respectively. The agreement is good, but the titrimetric method is more precise and takes only 30 min, whereas the gravimetric method needs 3 hr.
[Ref: Siskos, P. A.; Diamandis, E. P.; Gillieron, E.; Colbert, J. C. *Talanta,* **1983,** *30,* 980–982.]

## 9.3 The Chemistry of Titration

Multiple equilibria (discussed in Chapter 2) are always involved in titrations. Let us delve into this point in more detail. To understand the chemistry underlying titration, it may be helpful to describe a simple titration and analyze the results in detail.

Before continuing, please realize that titration is *not* a method used for trace analysis. In general, the lower limit for the use of titration to do precise determinations is for solutions containing approximately a tenth of a milligram of the analyte. This limit arises mostly because of two requirements.

1. The reagent–analyte reaction must go to completion. If the reagents get too dilute, the equilibrium shifts back toward unreacted forms. So smaller amounts of analyte must be contained in smaller volumes. Thus:
2. The volume of the analyte solution must be large enough to handle—on the order of a mL.

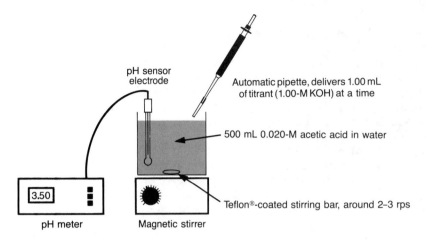

pH sensor electrode

Automatic pipette, delivers 1.00 mL of titrant (1.00-M KOH) at a time

500 mL 0.020-M acetic acid in water

3.50

Teflon®-coated stirring bar, around 2–3 rps

pH meter     Magnetic stirrer

**Figure 9.1**  Diagram of the experimental equipment for a neutralization titration demonstration. A weak acid, acetic acid, is titrated with a strong base, potassium hydroxide. The solution was stirred continuously so that it was homogeneous. The data points were determined by adding sequential 1.00-mL-volumes of KOH to the solution and measuring the pH. The results are shown in Table 9.1.

The pH was determined with an electrode connected to a pH meter, which displayed the measured pH on a digital display. More details about measuring pH this way are described in Chapter 12. The purpose of this demonstration is to illustrate the changes in pH that occur during a neutralization titration. In most neutralization titrations, we are interested only in finding the volume of titrant needed to neutralize the analyte completely.

The amount and precision of titrant addition is not a problem. Contemporary automatic titrators can add tens of microliters with precisions of less than a microliter. The limitations do not prevent titration from being a highly useful assay method for many different analytes.

### Demonstration of a Simple Titration

The demonstration experiment, which was done one day for a class, is illustrated in Figure 9.1. A beaker contained 500 mL of a 0.020-M solution of acetic acid. This had been made by mixing pure[2] acetic acid with distilled water. Since the acetic acid could be weighed accurately (0.600 ± 0.001 g; f.w. 60.05) and the water volume measured accurately (500.0 ± 0.2 mL), the acid concentration was known to 0.2%. This preparation satisfied requirements C and D of the list in Section 9.1. In this demonstration, the concentration of the sample solution is *known.*

Immersed in the solution was a sensor, an electrode, that can be used to measure the hydrogen-ion activity (*not* concentration) in the presence of other ions and molecules. Its presence satisfied requirement E. The electrode's response was calibrated immediately before the experiment by using standard solutions of accurately known pH.

The solution was stirred constantly with a magnetic stirrer. This ensured that the solution was homogeneous and that the measured pH represented the entire solution.

---

[2] According to the batch assay, the acetic acid was minimally 99.7% $CH_3COOH$.

**Table 9.1**   Data for Titration of Acetic Acid with KOH*

| Added Titrant (mL) | pH (expl) | [H$^+$] (expl) | [OH$^-$] (calc) |
|:---:|:---:|:---:|:---:|
| 0.0 | 3.5 | $3.2 \times 10^{-4}$ | $3.2 \times 10^{-11}$ |
| 1.0 | 4.0 | $1.0 \times 10^{-4}$ | $1.0 \times 10^{-10}$ |
| 2.0 | 4.2 | $6.3 \times 10^{-5}$ | $1.6 \times 10^{-10}$ |
| 3.0 | 4.4 | $4.0 \times 10^{-5}$ | $2.5 \times 10^{-10}$ |
| 4.0 | 4.5 | $3.2 \times 10^{-5}$ | $3.2 \times 10^{-10}$ |
| 5.0 | 4.7 | $2.0 \times 10^{-5}$ | $5.0 \times 10^{-10}$ |
| 6.0 | 4.9 | $1.3 \times 10^{-5}$ | $7.9 \times 10^{-10}$ |
| 7.0 | 5.1 | $8.0 \times 10^{-6}$ | $1.25 \times 10^{-9}$ |
| 8.0 | 5.4 | $4.0 \times 10^{-6}$ | $2.5 \times 10^{-9}$ |
| 9.0 | 6.1 | $8.0 \times 10^{-7}$ | $1.25 \times 10^{-8}$ |
| 10.0 | 8.5 | $3.2 \times 10^{-9}$ | $3.2 \times 10^{-6}$ |
| 11.0 | 11.3 | $5.0 \times 10^{-12}$ | $2.0 \times 10^{-3}$ |
| 12.0 | 11.5 | $3.1 \times 10^{-12}$ | $3.2 \times 10^{-3}$ |
| 13.0 | 11.7 | $2.0 \times 10^{-12}$ | $5.0 \times 10^{-3}$ |
| 14.0 | 11.8 | $1.3 \times 10^{-12}$ | $6.3 \times 10^{-3}$ |

* Acetic acid: 500 mL of 0.020 M; KOH: 1.00 M.

To this solution was added 1.00-M KOH solution, 1.00 mL at a time. The concentration of the KOH had been calibrated by titrations against replicate samples of a National Bureau of Standards standard acid, potassium acid phthalate.

The stoichiometry of the potassium hydroxide reaction with acetic acid is

$$K^+ + OH^- + HOAc \rightleftharpoons K^+ + H_2O + {}^-OAc$$

or

$$OH^- + HOAc \rightleftharpoons H_2O + {}^-OAc$$

omitting the K$^+$ spectator ion. The choice of KOH, a strong base, as the reagent, and the calibration of the concentration of the KOH solution satisfied requirement A. Thus far, requirements A, C, D, and E had been fulfilled. Only B was left.

The base was added to the sample solution 1.00 mL at a time with a pipette. A pipette can deliver specific volumes of liquids precisely.[3] (The pipette illustrated in Figure 9.1 is commonly used with water solutions.) The use of the pipette to measure the volume of added KOH satisfied requirement B. Thus all the requirements for a titration were present.

The pH of the initial acetic acid solution was measured as pH 3.5. Next, 1 mL of the KOH titrant was added to the solution, followed by a delay of at least 10 s to allow

---

[3] The multiple addition using the automatic pipette will introduce more cumulative error than a burette over the volume range of this titration. This explains why the end point was not at 10.00 mL. However, 1.00-mL additions are *much* easier to demonstrate.

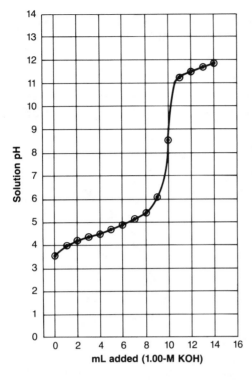

**Figure 9.2** Results of the titration of 500 mL of 0.020-M acetic acid with 1.00-M KOH as illustrated in Figure 9.1. Here are graphed the measured pH-values of the solution versus the total volume of KOH solution that had been added.

the stirred solution to become homogeneous. Only then was the pH reading on the meter recorded. Further 1.00-mL additions were done the same way 15 times, and the pH values were noted. These are listed in Table 9.1. The data are also plotted as circles on a graph shown in Figure 9.2, with a continuous line drawn to join the data points. This is a **titration curve.** The table and titration curve show the same results in two different ways.

The titration curve in Figure 9.2 shows the pH behavior clearly. What we see is an initial slow rise in pH with the addition of base. However, only after 7 or 8 mL of the KOH titrant have been added does the rise become rapid. The pH rises very rapidly between 9 and 11 mL of added KOH, and then the curve that is drawn flattens again. This flattening means, of course, that the pH does not change much with addition of more base. Note that the slope of the graph looks even flatter on the top (above pH 11) than in the early flat part (pH 4 to 5). Let us now analyze this curve in detail using the law of mass action and the descriptive chemistry of the reactions. This information is reflected in the equilibrium constants and stoichiometries for the reactions that occur: the chemical reactions between acetic acid, KOH, and water.

## 9.4 Acetic Acid–KOH Titration: Calculations and Equilibria

Before any base was added, the pH of the acetic acid solution was measured to be 3.5. Then, upon addition of KOH, the solution pH changed as plotted in Figure 9.2. Can we explain these **empirical** pH values? (*Empirical* means based solely on an experi-

**Table 9.2** Result of Calculations of Species Concentrations

| Added Titrant (mL) | [HOAc] (M) | [⁻OAc] (M) | $\dfrac{[^-\text{OAc}]}{[\text{HOAc}]}$ | $K_a$ | $pK_a$ |
|---|---|---|---|---|---|
| 0.0 | 0.02 | $(3.2 \times 10^{-4})$* | 0.016 | $5 \times 10^{-6}$ | 5.3 |
| 1.0 | 0.018 | 0.002 | 0.111 | $1 \times 10^{-5}$ | 5.0 |
| 4.0 | 0.012 | 0.008 | 0.667 | $2 \times 10^{-5}$ | 4.7 |
| 5.0 | 0.010 | 0.010 | 1.00 | $2 \times 10^{-5}$ | 4.7 |
| 6.0 | 0.008 | 0.012 | 1.50 | $2 \times 10^{-5}$ | 4.7 |
| 8.0 | 0.002 | 0.018 | 9.0 | $4 \times 10^{-5}$ | 4.4 |
| 9.0 | 0.001 | 0.019 | 19.0 | $2 \times 10^{-5}$ | 4.7 |

* The value is set equal to the proton concentration.

ment, without reference to theory.) We seek an explanation that tells us *how* the pH values are related to the amount of added KOH.

The precedure involves using our data (or someone else's) to calculate an equilibrium constant for the acid. Then, from the law of mass action, the entire titration curve can be *explained*. At the same time, the calculation tests whether the law of mass action applies to the titrated solution and the ions that we believe are present.

It is important to realize that we can only explain the experiments with theories. Remember, if the data do not agree with the theory, then the theory is either wrong or incomplete. *Incomplete* means that some chemical reaction or chemical species that affects the measured behavior has not been included in the calculation. Put another way, the experiment is primary, and the explanation using chemical theory follows.

In the remainder of this section, we first ascertain whether the equilibrium of acetic acid/acetate can explain the results, which were experimentally obtained. Following that, the behavior of various regions of the titration curve are explained. These regions are the initial pH, the range around pH 4–5, the end point, and the region beyond the end point.

### Determining $K_a$

Here, we shall see how the equilibrium constant $K_a$ for acetic acid is determined from the data. You will see that we do the same equilibrium calculation for each data point (see Table 9.2). What we seek especially to know about the titration is how the concentrations of $H^+$, HOAc, and $^-$OAc change with the addition of $OH^-$. This is the key to understanding the titration curve.

In order to calculate $K_a$, we need to determine three concentrations—$[H^+]$, [HOAc], and [⁻OAc]—to put into the mass-action expression

$$K_a = \frac{[H^+][^-\text{OAc}]}{[\text{HOAc}]}$$

During the titration, with the electrode we measured $a_{H^+}$, the $H^+$ activity. We need to find [HOAc] and [⁻OAc]. We know the original value of [HOAc] = 0.020 M. From this value and the amount of $OH^-$ added, we can find the two needed concentrations

[HOAc] and [⁻OAc] at any point. The method of this calculation is important and will be done step by step.

Recall that the reaction

$$OH^- + HOAc \rightleftharpoons H_2O + {}^-OAc \qquad \text{(9-1) and (3-27)}$$

goes essentially to completion; the equilibrium lies far to the right. Therefore, for every mole of $OH^-$ added, there is a *decrease* of one mole of HOAc and an *increase* of one mole of ⁻OAc. Through this stoichiometric relationship we can find the concentration of acetate and acetic acid. (The reaction's effect on the water activity is negligible.)

Equation 9-1 says

$$\text{moles } OH^- \text{ added} = \text{moles } {}^-OAc \text{ formed} = \text{moles HOAc decreased}$$

It should be recognized that this information is the same as contained in a conservation equation. From the data of the titration, the conservation equation that applies here is

$$0.020 \text{ M} = [HOAc] + [{}^-OAc] \qquad \text{(9-2)}$$

The same type of conservation equations were used in Sections 2.5 and 2.7 when we were solving simultaneous equilibrium problems.

Notice that the units of the quantities in Eq. 9-2 are mol $L^{-1}$. Since the initial volume of the solution was 500 mL, Eq. 9-2 could also be written with units of moles.

$$0.010 \text{ moles} = \text{moles HOAc} + \text{moles } {}^-OAc \qquad \text{(9-3)}$$

---

**EXAMPLE**

Calculate the $K_a$ and p$K_a$ for the solution after 4.00 mL of 1.00-M KOH titrant has been added to 500 mL of a solution 0.020 M in acetic acid. This is the calculation done for the data of line 3 in Table 9.2.

**SOLUTION**

From the experimental data in Table 9.1, we find that the pH of the solution was 4.5. Equating the activity of $H^+$ with its concentration (an approximation) gives

$$[H^+] = 10^{-pH} = 10^{-4.5} = 3.2 \times 10^{-5} \text{ M}$$

We now must find the concentrations of acetic acid and acetate. This is done by calculating the changes in the numbers of moles and then converting to concentrations of the species. That is, addition of 4.00 mL of 1.00-M KOH means that

$$\text{moles KOH} = 0.00400 \text{ L} \times 1.00 \text{ mol } L^{-1} = 0.00400 \text{ moles KOH added}$$

Then, using the information in the stoichiometric equation 9-1, we find that

$$\text{moles } {}^-OAc = 0.00400$$

after 4.00 mL of KOH titrant was added. From the conservation equation 9-3

$$\text{moles HOAc} = 0.0100 \text{ moles} - 0.0040 \text{ moles} = 0.0060 \text{ moles}$$

Since the mass-action equation requires that the quantities of the species be in concentration units, mole quantities are now converted to mol $L^{-1}$. This requires knowing the solution volume. But what is the solution volume? The original volume was 500 mL, and to this were added 4.00 mL, giving a total volume of 504 mL. The final concentrations (to within $\pm 1\%$ relative error) are

$$[HOAc] = \frac{0.0060 \text{ mol}}{0.504 \text{ L}} = 0.0120 \text{ molar}$$

and

$$[^-OAc] = \frac{0.0040 \text{ mol}}{0.504 \text{ L}} = 0.0080 \text{ molar}$$

Substituting into the mass-action expression, we obtain

$$K_a = \frac{[H^+][^-OAc]}{[HOAc]} = \frac{(3.2 \times 10^{-5})(0.0080)}{0.0120} = 2 \times 10^{-5}$$

and

$$pK_a = 4.7$$

Note: Approximations made include assuming that $a_{H^+} = [H^+]$ and that the ionic strength effects are irrelevant throughout the addition of 4 mL of 1-M KOH. (In this particular case, in which both the acid and base concentrations are raised to the same power, any volume change cancels out in the calculation of $K_a$.)

As shown in Table 9.2, from the data that were collected, the equilibrium constant is $2 \times 10^{-5}$ in the concentration range and at the temperature of the solution. (An explanation for the discrepancy of the point at 7.0 mL added will be made later.) This compares with a "literature value" of $1.74 \times 10^{-5}$ at 25 °C ($pK_a = 4.76$). The equilibrium constant calculated from the demonstration experiment data agrees with the $K_a$ value for acetic acid, which others have experimentally determined. More important, though, is the confirmation that the description of the system using the stoichiometric equation and the equilibrium constant expression is self-consistent. We can say we have a correct chemical model of the system.

The disagreement of about 10% between the average $K_a$ value obtained in this experiment and the literature value can arise from a number of sources:

1.  Experimental errors may exist. Among these are a loss of titrant or solution (by splashing), a surplus of titrant (by dripping), and nonlinearity of the pH electrode, which causes the reading to be in error at pH values between calibration points.
2.  The temperature of the solution used in the demonstration may not have been 25 °C.
3.  The $K_a$ values were calculated using a mixed equilibrium constant (Section 2.2) with concentrations of acetic acid and acetate but with the activity of hydrogen ion as measured by the electrode. That is,

$$K_a = \frac{a_{H^+}[^-OAc]}{[HOAc]}$$

This approximation will become worse the further the concentrations differ from activities, and this difference increases with ionic strength.[4]

4.  The ionic strength changes throughout the titration.[5]

To review, the equilibrium constant $K_a$ for acetic acid was calculated from the data. The data were obtained using calibrated solutions and measuring equipment to find the pH of the acid solution as a function of the amount of added base. Together with the descriptive chemistry of the base-neutralization reaction (which has been confirmed by other methods), the data were used to find the equilibrium constant for the dissociation of acetic acid.

### The pH of the Initial Sample Solution: 0.020-M Acetic Acid

From the regularity of $K_a$ (as seen in Table 9.2), we have concluded that the equilibrium of acetic acid/acetate controls the pH of the solution. We can now use this information to calculate the pH expected for a 0.020-M acetic acid solution. This is the same problem that was described in detail in Chapter 2, and we state the result only: The original 0.020-M acetic acid solution is expected to have a pH of 3.2. This value is consistent with the rest of the data, but it differs from the pH that was measured, pH 3.5. Possible causes of this discrepancy are the presence of some base in the water before the acid was added and a lack of correction for the consequences of changing ionic strengths. As an example of the latter, the following table indicates the possible magnitude of the ionic strength effect.

| Ionic Strength (M) | Acetic Acid pK$_a$ (25 °C) | $K_a$ | Expected pH 0.020-M Solution |
|---|---|---|---|
| 0 | 4.76 | $1.7 \times 10^{-5}$ | $3.2_4$ |
| 0.2 | 4.64 | $2.3 \times 10^{-5}$ | $3.1_8$ |
| 3 | 4.52 | $3.0 \times 10^{-5}$ | $3.1_2$ |

*Source:* Sillen, L. G.; Martell, A. E. "Stability Constants of Metal-ion Complexes," Supplement No. 1, Special Publication No. 25; The Chemical Society: London, 1971.

### Description of the Region of pH 4–5

So far, we have found that the equilibrium

$$HOAc \rightleftharpoons H^+ + {}^-OAc$$

with its mass-action expression

$$K_a = \frac{[H^+][^-OAc]}{[HOAc]} = 2 \times 10^{-5}$$

can be used to describe the titration curve of acetic acid in the pH range 3 to 6.2. The values of $K_a$ in Table 9.2 show this. Let us now consider the slopes of the curve plotted in Figure 9.2.

---

[4,5] Numerical calculations of the magnitude and effects of ionic strength are left to you as problems 9.10 and 9.11.

Between 0 and 1 mL of base, the pH rises faster per milliliter of added KOH than further along in the titration—after a few milliliters have been put in. The rise is again faster after about 6 mL are added. In between, the pH change for a given amount of added base is the smallest: The slope of the curve is a minimum. In the pH range around 4.5 to 5, the solution is a **buffer** which, by definition, resists a change in pH when base or acid is added, relative to other regions of the titration. (See Figure 9.3 where the region is labeled.) Let us investigate why this behavior occurs. Additional details about buffers are contained in Supplement 10A.

One way to see the explanation of the changes in slope of the titration curve is to change the form of the mass-action expression. Begin with

$$K_a = \frac{[\text{H}^+][^-\text{OAc}]}{[\text{HOAc}]}$$

Next, take the negative logarithm of both sides.

$$\text{p}K_a = \text{pH} - \log \frac{[^-\text{OAc}]}{[\text{HOAc}]}$$

A rearrangement of the equation gives

$$\text{pH} = \text{p}K_a + \log \frac{[^-\text{OAc}]}{[\text{HOAc}]} \tag{9-4}$$

This form of the mass-action expression is often called the **Henderson–Hasselbach equation.** Note that since this is merely the mass-action equation written in a different mathematical form, there is no new information in it. But sometimes this form is convenient to use. The general form of the equation is[6]

$$\text{pH} = \text{p}K_a + \log \frac{[\text{conjugate base}]}{[\text{acid}]} \tag{9-5}$$

An alternate derivation of this equation is shown in Supplement 9A.

The form of Eq. 9-5 shows that the solution pH is describable as the sum of a constant, the $\text{p}K_a$, and a logarithm of a ratio. Thus the pH will change in the same way as the logarithm of the ([base]/[acid]) concentration ratio changes. For instance, the pH will change by one pH unit for every factor-of-ten change of the ratio. The mathematical key to understanding the changes in the slope of the titration curve lies in seeing how this ratio varies when a fixed volume of base is added.

A most effective way to clarify such behavior is to look at extreme regions. Assume that a small, fixed amount of $\text{OH}^-$ is added to the solution, and the pH change observed. Let us answer the following two questions.

1.  At what pH will the change be smallest?
2.  At what pH will the change be largest?

---

[6] Notice that Eq. 9-5 does not make sense if the acid is a strong acid and is fully dissociated. For a strong acid, the concentration of the undissociated acid in the denominator will approach zero; the $K_a$ becomes undefined—approaching infinity. Also, the equation may be only approximate when used with hydrogen-ion activities but with the acid and conjugate base concentrations.

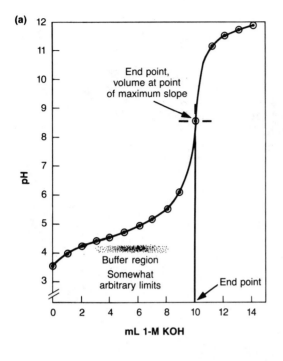

(a)

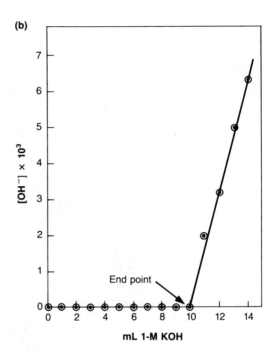

(b)

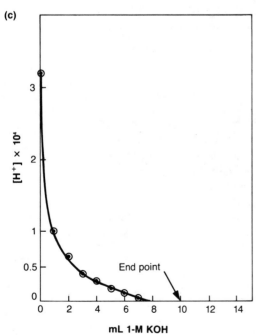

(c)

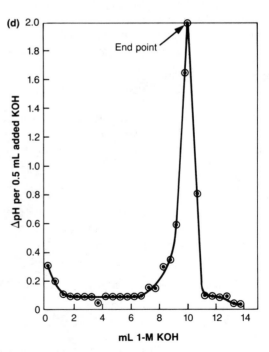

(d)

To answer the questions, we note the following.

> From the stoichiometry of the reaction, the magnitude of the changes in [acid] and [base] are equal and opposite: One increases when the other decreases.
> The result depends on a change in a ratio of two numbers.

Mathematically, with a fixed hydroxide addition, the change in the ratio ([base]/ [acid]) will be the smallest when the *relative* change in *both* concentrations is the smallest. But, recall that the *absolute* changes are the same. So the smallest relative change in *both* concentrations occurs when they are both the same size. In other words, for acetate/acetic acid the smallest changes in the ratio will occur when

$$1 = \frac{[^-OAc]}{[HOAc]}$$

which occurs in the center of the buffer region. The pH at the center of the range is, thus,

$$pH = pK_a + \log 1 = pK_a$$

This is a general result for any acid–base pair: The buffer region is centered around the $pK_a$ of the acid. We have answered question 1.

Conversely, when one of the concentrations—either $[^-OAc]$ or $[HOAc]$—is much smaller than the other, a fixed addition of base will cause a much larger *relative* change in the smaller concentration. The change in the ratio will then be larger than in the buffer region.

To answer question 2, there are two extreme points at which the pH change will be the largest. The first is before any base has been added. (Here, $[^-OAc]$ is at its smallest value.) The second is just before all the acid has reacted with the added base. (There, $[HOAc]$ is at its smallest value.) The volume at which all the acid has reacted and at which the curve has its steepest slope is the end point of the titration. It is labeled in Figure 9.3a. As the end point is approached, the volume of base needed to change the

---

**Figure 9.3**  Results of the titration of 500 mL of 0.020-M acetic acid with 1.00-M KOH titrant plotted in four different ways. The base titrant was added 1.00 mL at a time. The horizontal axis is the same in all four graphs: titrant volume.

a. Typical pH versus titrant volume plot with the buffer region and end point labeled.

b. Solution hydroxide concentration as a function of titrant added. The values used here are shown in Table 9.1. The plot of the concentration of $OH^-$ has an abrupt break at the end point. A general name for this graph might be titrant concentration versus titrant volume. It is the most common type of plot for titrations that are not based on acid–base neutralizations.

c. Plot of the $H^+$ concentration. This plot does not help to visualize the end point and suggests why a pH plot is used for acid–base neutralization titrations.

d. Plot of the slope of the titration curve plotted in 9.3a. The points of the graph were calculated by taking the pH differences over half-milliliter intervals along the titration curve. Their positions on the volume axis are plotted at the mid-points of each half-milliliter interval. A similar plot can be produced electronically if the titrant is added at a constant rate. The end point is then found at the peak value. This is one basis of operation of automatic titrators.

**Table 9.3**   Volume of 1-M KOH Needed to
Change the pH by 1 Unit in a 0.02-M Acetate
Solution

| pH Range | Volume 1-M KOH (mL) |
|----------|---------------------|
| 4.0 to 5.0 | 4.9 |
| 5.0 to 6.0 | 3.0 |
| 6.0 to 7.0 | 0.5 |
| 7.0 to 8.0 | 0.05 |

solution pH by large amounts becomes quite small. Comparisons between different regions of the titration are shown in Table 9.3.

### The pH at the End Point

As the titration progresses, the pH changes have been explainable as resulting from the quantitative transformation of acetic acid to acetate due to its reaction with hydroxide. The pH depends on the ratio of acetate and acetic acid. The result is a titration curve with an initial, relatively rapid rise, then a flattening as the buffer region is titrated, followed by an evermore rapidly rising slope as the end point is approached.

Now, let us answer the following questions.

1.  What species are *in* the solution at the end point?
2.  What is the pH there? From Figure 9.2 you can see that the pH is *not* 7.

If we assume that the end point and equivalence points are the same (a good assumption), answering the first question is easy. At the end point, the acetic acid has all been converted to acetate. Therefore, the solution contains acetate ions, and the concentration of $K^+$ in the solution equals that of acetate.

To answer the second one, note that the solution is *exactly* the same as the one obtained if we had added 0.010 mol of KOAc, potassium acetate, to 510 mL of water. When potassium acetate is added to water, the salt dissociates, and the acetate will hydrolyze. We can explain the pH at the end point in the same way as we explain a conjugate base hydrolysis. It follows that the calculation of the pH at the end point is done as for any other hydrolysis problem. Such a calculation for acetate was done in Section 2.4 and is not reproduced here.

To understand the end-point equilibrium, it is necessary to know some descriptive chemistry of acetate salts. Sodium acetate, potassium acetate, rubidium acetate, and cesium acetate all are completely dissociated in water. This means that we can treat the potassium ion as a spectator and consider only the acetate-ion hydrolysis.

---

**EXAMPLE**

What is the pH at the end point of the titration of 500 mL of 0.020-M acetic acid by 1.00-M KOH in water?

**SOLUTION**

The chemical equation that describes the hydrolysis is

$$^-OAc + H_2O \rightleftharpoons HOAc + OH^-$$

At 25 °C, the hydrolysis constant describing the reaction is

$$K_h = \frac{K_w}{K_a} = \frac{1.00 \times 10^{-14}}{1.76 \times 10^{-5}} = 5.68 \times 10^{-10}$$

For a 0.0196-M solution (0.010 mol in 510 mL),

$$5.68 \times 10^{-10} = \frac{[OH^-][HOAc]}{[^-OAc]} = \frac{X^2}{[0.0196 - X]}$$

$$X = [OH^-] = 3.33 \times 10^{-6} \qquad \text{Thus, pOH} = 5.48$$

At the end point, the pH is expected to be 8.52.

---

## The pH beyond the End Point

We are able to describe the pH at any volume along the titration curve to be a consequence of the acid–base interaction of acetic acid and acetate. The acid–base equilibrium controls the solution pH. The concentrations of acetic acid and acetate ion are determined by the initial concentration of the acid and the amount of KOH that is added and reacts completely with HOAc to produce $^-$OAc.

The pH continues to change with added KOH in the region beyond the end point. Can the acetic acid–acetate pair control the pH here? Let us reason through the answer. The acetic acid has already been converted to acetate; so the acid form can have no effect. It might be that the acetate itself has some influence. However, a 0.02-M acetate solution can contribute, at most, only the hydroxide due to hydrolysis. The pH at the end point, pH 8.53, indicates that the 0.02-M acetate can generate only about $10^{-5}$-M OH$^-$. This is three orders of magnitude too small to bring the solution into the range of the pH measured after the end point, pH 12 ($[OH^-] = 10^{-2}$ M). We must conclude that the acetic acid–acetate equilibrium is no longer determining the pH of the solution. Therefore, after the end point, some other acid–base equilibrium must be controlling the pH changes when KOH is added.

This changeover of controlling equilibrium is an important idea. The end point marks where there is a change in the chemical equilibrium that determines the pH. (This concept is discussed further in Section 9.5.)

The only other possible acid in this simple potassium acetate solution is water itself. The chemical equilibrium that determines the pH beyond the end point must be, then,

$$^*OH^- + HOH \rightleftharpoons H^*OH + OH^-$$

The asterisk is used to label the hydroxide added as titrant. (Of course, in the solution itself, we cannot actually distinguish the hydroxide added from that from the hydroxide in water as is done with asterisks in this equation.) So after the end point, the equilibrium constant describing the pH in relation to the hydroxide concentration is

$$K_w = [H^+][OH^-] = 1.0 \times 10^{-14} \text{ at 25 °C}$$

To see if this idea is correct, let us see whether the calculated pH agrees with that measured at one point (volume) in the titration.

---

**EXAMPLE**

What pH do we expect the solution to have after 3.00 mL of 1.00-M KOH has been added *beyond* the end point?

**SOLUTION**

This addition will produce 513 mL of $5.8_5$-mM KOH solution:

$$5.8_5 \times 10^{-3} \text{ M} = \frac{3.00 \text{ mL} \cdot 1.00 \text{ M}}{513 \text{ mL}} = [OH^-]$$

With

$$K_w = [OH^-][H^+] = 10^{-14} \qquad \text{at 25 °C}$$

then

$$1.0 \times 10^{-14} = [H^+](5.8 \times 10^{-3})$$

and

$$[H^+] = 1.7 \times 10^{-12}, \qquad \text{or the pH} = 11.7$$

This result is in agreement with the experimental data shown in Table 9.1. This agreement gives us confidence that the water is, in fact, the next acid system that is titrated after acetic acid.

---

## 9.5 Alternate Titration Curves

You might ask the question, Why are the data plotted as pH versus the volume of KOH added? The answer is that this is a useful way to graph the data. However, there are many possible ways to do so. Three others are illustrated in Figure 9.3.

When a titration is used as the assay method in an analysis, the main piece of information that is desired is the volume of titrant added to the end point: the quantity from which we calculate the content of analyte in a sample. So any graph that helps us measure the end point is useful. Of the four, it appears that only the plot of $[H^+]$ versus the volume of added KOH is *not* helpful in finding the end point.

The other three graphs show a significant change at the end point even though the same information is plotted in different ways. For acid–base titrations such as acetic acid with KOH, you will see mostly plots of pH versus titrant volume. For most other titrations, such as those covered in Chapter 10, plots of the type shown in Figure 9.3b are used. This graph helps us find the end point by indicating the volume at which excess titrant first appears. Figure 9.3d, the plot of the slope of the pH curve, certainly shows the end point clearly. If the $\Delta$pH data are encoded into an electrical signal, it is relatively easy to build a circuit to detect the end point. This is the basis of operation for many automatic titrators.

## 9.6  Assigning Equilibria to Different Parts of Titration Curves

An interesting and important relationship showed up in the analysis of the experimental titration curve of acetic acid: The pH of the solution is determined by different equilibria in different parts of the titration curve, and the end point marks the point of separation. (See Figure 9.4a.) At the end point, the pH is determined by a hydrolysis equilibrium. In the example of acetic acid titrated with hydroxide, the applicable equilibrium at the end point is the acetate hydrolysis.

How does one tell which equilibrium controls the pH in each region? This is not easy to determine in complicated solutions with many acids and bases present. However, in the simpler acid–base solutions that you will become acquainted with here, the applicable equilibrium can be decided from the numerical values of the $pK_a$ of the acids or the $pK_b$ of the bases present.

Some general descriptive chemistry simplifies the assignment of the controlling equilibrium. In water, the strongest acid that can exist is $H_3O^+$, and the strongest base that can exist is $OH^-$. If any acid is more acidic than $H_3O^+$, it will donate its protons to water to form more $H_3O^+$. If any base is more basic than $OH^-$, it will extract protons from water to form more $OH^-$. This is called the **leveling effect.** As a result of this effect, we did not have to consider such species as $H_2OAc^+$ in the possible equilibria of acetic acid; the descriptive chemistry of acetic acid indicates that the diprotonated species does not exist in water solution. Unfortunately, there are no easy rules for deciding what protonated species fall into such a class, and finding the information for less common acids and bases can be quite frustrating.

As a result of the leveling effect, after the last end point due to reaction of analyte(s), the final plateaus of aqueous-solution titration curves are determined by the water equilibrium,

$$H_2O \rightleftharpoons H^+ + OH^-$$

With a strong base as titrant, the plateau occurs in the base pH range; similarly, with a strong acid as titrant, the plateau occurs in the acid pH range.

The explanation for this assignment is quite straightforward. As you read in Chapter 2, control of pH depends on the concentrations of acid–base pairs. In the limit of low concentration of either an acid or its base, the control of the pH shifts to another pair. Ultimately, this control shifts to the water itself. As noted above, the titration of acetic acid with hydroxide provides a straightforward example.

### Assigning pH-controlling Equilibria to a Carbonate Titration

Let us assign the controlling equilibria in a titration of an aqueous solution of sodium carbonate, $Na_2CO_3$. The compound dissociates completely, and sodium acts as a spectator ion. Thus, the titration is of the carbonate moiety, $CO_3^-$. Figure 9.4b illustrates a titration curve for 0.10-M sodium carbonate in a solution titrated by a strong acid. In the titration curve of carbonate, there are three plateaus and two end points. All five features will be assigned pH-controlling equilibria.

**(a)**

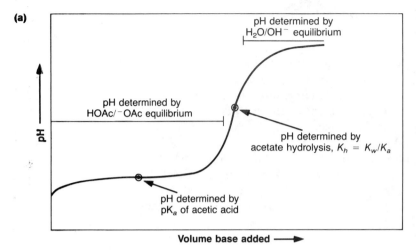

**Figure 9.4** Comparison between a monoprotic neutralization titration (acetic acid) and a sequential diprotic titration ($CO_3^-$). The two sequential titrations are explained in the same manner as two individual monoprotic titrations.

a. Acetic acid titration. The end point occurs at the volume at which all the acetic acid has just been neutralized by the strong base. The pH is determined by hydrolysis of acetate.

b. $CO_3^-$ titration. The first end point is the volume where all the $CO_3^-$ has just been neutralized by the strong acid. The pH at the end point is determined by hydrolysis of $HCO_3^-$. The second end point is the volume at which all the $HCO_3^-$ has just been neutralized by the strong acid. The pH at the end point is determined by hydrolysis of $H_2CO_3$. The $HCO_3^-$ neutralized in the second range may have two origins: (1) any $HCO_3^-$ in the original sample and (2) $HCO_3^-$ from neutralization of $CO_3^-$ in the sample. The equivalents of titrant needed between the first and second end points equals the sum of the two sources of $HCO_3^-$.

**(b)**

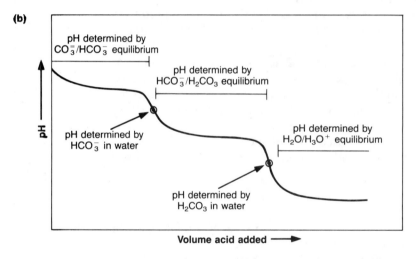

From the discussion above, we expect to find that one of the three plateaus results from the water equilibrium; two plateaus should be buffer regions. We might guess at the possible protonation chemistry of carbonate as being,

$$CO_3^= \underset{-H^+}{\overset{+H^+}{\rightleftarrows}} HCO_3^- \underset{-H^+}{\overset{+H^+}{\rightleftarrows}} H_2CO_3 \underset{-H^+}{\overset{+H^+}{\rightleftarrows}} H_3CO_3^+$$

However, the descriptive chemistry of carbonate reveals that $H_3CO_3^+$ is a stronger acid than water and does not need to be considered. In other words, water will protonate to form $H_3O^+$ before carbonic acid, $H_2CO_3$, will form $H_3CO_3^+$.

We are left with

$$CO_3^= \underset{-H^+}{\overset{+H^+}{\rightleftarrows}} HCO_3^- \underset{-H^+}{\overset{+H^+}{\rightleftarrows}} H_2CO_3$$
$$pK_2 = 10.22 \qquad pK_1 = 6.46$$

Since this is a titration of a base with a strong acid, we expect the final plateau (the one with the greatest volume of titrant added) to be due to the water equilibrium.[7] The pH after the second end point is determined by the concentration of protons from the titrant. This conclusion is illustrated in Figure 9.4.

How do we make the assignments of the other two plateaus, the initial and middle ones? The two carbonate pH-determining equilibria are

$$CO_3^= + H^+ \rightleftharpoons HCO_3^-$$

and

$$HCO_3^- + H^+ \rightleftharpoons H_2CO_3$$

To assign the controlling equilibria, it is necessary only to determine the order of protonation of $CO_3^=$ and $HCO_3^-$. The $CO_3^= - HCO_3^-$ equilibrium has the highest $pK_a$; $pK_a = 10.22$. Therefore, the initial plateau — the one at the highest pH — must be controlled by the carbonate–bicarbonate equilibrium. The middle plateau is then due to the $HCO_3^- - H_2CO_3$ equilibrium; $pK_a = 6.46$. These are the assignments shown in the figure.

### Assigning End Point Equilibria

The equilibria that control the pH of the two end points of the carbonate titration are relatively simple to discover once the equilibria controlling the plateaus have been assigned. The method is the same as in the example of acetic acid. All you need ask is, What are the major species in solution at each end point?

At the first end point (left-most in Figure 9.4b), all the carbonate has been converted to bicarbonate, $HCO_3^-$, so the equilibrium is the same as if a dissociated bicarbonate salt were added to water. Thus the pH is determined by the hydrolysis of bicarbonate,

$$HCO_3^- + H_2O \rightleftharpoons H_2CO_3 + OH^-; \qquad K = \frac{[OH^-][H_2CO_3]}{[HCO_3^-]}$$

At the second end point, all the bicarbonate has been converted to carbonic acid, $H_2CO_3$, so the equilibrium is the same as if carbonic acid were added to water.[8] The

---

[7] One way to see why the water equilibrium applies only at the final, acid plateau is to consider why the water equilibrium does not control the pH of the initial solution. In the solution, $CO_3^=$ is added to 0.10 M, and it hydrolyzes with the stoichiometry

$$CO_3^= + H_2O \rightleftharpoons HCO_3^- + OH^-$$

Because the water alone can contribute only about $10^{-7}$-M hydroxide, the hydroxide concentration in the carbonate solution arises primarily from the hydrolysis of carbonate. The water equilibrium does not control the pH of the solution at the initial point.

[8] This could be done by adding gaseous $CO_2$ to the water since the reaction $H_2O + CO_2(g) \rightleftharpoons H_2CO_3$ occurs readily. The reverse reaction causes the fizz that occurs when a carbonated drink is opened.

controlling equilibrium is then

$$H_2CO_3 + H_2O \rightleftharpoons H_3O^+ + HCO_3^-; \qquad K = \frac{[H^+][HCO_3^-]}{[H_2CO_3]}$$

In the same manner as that just described, any other diprotic or any polyprotic acid titration or base titration can be analyzed. The calculations for any individual region of the curve are done as they are for monoprotic acids. All that is necessary is to separate the *appropriate* or *applicable* equilibrium from those that are possible. In this way, a complicated system can be analyzed logically—by splitting it into a number of simpler problems involving only one or two equilibria.

One final reminder: These calculations are used to match the experimental data. If the results agree with the experiment, then we have some confidence that we understand the underlying chemistry. After all, the chemistry is what we seek to understand first. Once the chemistry is clear, then the technique can be used for analysis with confidence, and problems that arise in the analysis can be approached intelligently.

## 9.7 Titration Volumetric Calculations

In contrast to the moderately involved calculations required to explain the chemical changes during titrations, the calculation for obtaining sample content from the end-point volume is usually quite straightforward. There are three parts to such a calculation:

1. Relate the number of moles of titrant to the number of moles of material titrated.
2. Find the total mass or volume of the analyte.
3. Convert the result to the form desired, for example, milligrams or volume percent or ppm.

The following are a few examples of such calculations.

---

**EXAMPLE**

A pound of dried, white powder was given to you. It is a sample from a carload of low-grade salicylic acid (formula weight 138.12, $pK_{a1} = 2.97$, $pK_{a2} = 13.40$). It is to be acetylated to make aspirin. Find out how pure (in %) the material is if none of the impurities are acidic or basic. Let us consider only one of the replicate samples weighing 0.4208 g. It was titrated with 21.92 mL of a 0.1354-M KOH solution to its first end point.

**SOLUTION**

To solve the problem requires knowing the stoichiometry of the titration reaction. To the first end point, the reaction that occurs can be written

$$OH^- + H_2sal \rightleftharpoons H_2O + Hsal^-$$

The abbreviation sal$^-$ is used for salicylate. From the equation, the stoichiometry is found; each mole of base added reacts with one mole of salicylic acid. This is the first of the three parts needed for the calculation.

For the second part, we must find the total mass of salicylic acid that was titrated. This can be found by multiplying the number of moles times the molecular mass. It can be done in two stages. Initially,

moles salicylic acid $= 1 \times$ moles KOH $= 1 \cdot 0.1354$ M $\cdot 0.02192$ L $= 2.968 \times 10^{-3}$ moles

The factor 1 reflects the stoichiometry found in the first step. Then,

mass of salicylic acid $=$ moles salicylic acid $\times 138.12 = 0.4099$ g

The third stage of the calculation is to convert this number into the desired form. Asking for the percentage purity is the same as asking us to calculate the percentage by weight of the sample that is salicylic acid. Thus,

$$\% \text{ w/w salicylic acid} = \frac{0.4099 \text{ g}}{0.4208 \text{ g}} \cdot 100 = 97.4\%$$

## Comments on the Example

Notice that at no time was it necessary to know the volume of solution in which the salicylic acid was dissolved.

The calculation, which was carried out in three stages, could be compressed into a single equation, that is,

$$\% \text{ w/w} = 100 \cdot \frac{(\text{stoichiometric number})(\text{M KOH})(\text{vol KOH in liters})(\text{f.w. acid})}{\text{mass of sample}}$$

However, this equation would hold only for problems that are the same type: finding the % w/w of a solid sample. If you were running such titrations quite often, this is the formula you might use for your notebook or for a computer program. You would plug in the numbers and calculate the result. However, such equations cover up the similarity of *all* volumetric calculations — the three steps noted above.

In practice, for a new titrimetric assay it is sometimes difficult to find the stoichiometric number, which follows from the stoichiometry of the reactions involved. Nevertheless, it is an important part of the validation of the method.

The calculations of content become even easier if the sample is a liquid. Say it is an acid, and you wish to find its normality. It will be titrated with a base. At the equivalence point of the titration,

equivalents of base used $=$ equivalents of acid (as $H^+$) reacted

But the rest of the stages of a volumetric calculation need not be done because,

equivalents of base used $=$ base normality $\times$ base volume

and

equivalents of acid reacted $=$ acid normality $\times$ acid volume

As a result, we can write the important, simple, general equation

titrant normality $\times$ volume $=$ sample normality $\times$ volume          **(9-6a)** or

$$N_1 \times V_1 = N_2 \times V_2 \qquad \text{(9-6b)}$$

If it is desired to find the molarity of an acid or base, its formula and chemistry must be known so that the molarity can be related to the normality.

---

**EXAMPLE**

Assume that you have 500 L of an aqueous sulfuric acid solution for which you wish to find the molarity. You titrate a 50.00-mL aliquot. It requires 42.20 mL of 0.1354-M KOH. What is the molarity of the sulfuric acid solution?

**SOLUTION**

Since KOH can donate one hydroxide,

$$\text{normality}_{KOH} = \text{molarity}_{KOH}$$

Thus, letting $N_{H_2SO_4}$ = normality of sulfuric acid, from Eq. 9-6b

$$0.1354\,N \cdot 42.20\ \text{mL} = N_{H_2SO_4} \cdot 50.00\ \text{mL}$$

$$N_{H_2SO_4} = 0.1143\,N$$

Then, since

$$\text{molarity}_{H_2SO_4} = \frac{1}{2}\,\text{normality}_{H_2SO_4}$$

$$M_{H_2SO_4} = \frac{0.1143\,N}{2} = 0.0571\ \text{M}$$

## 9.8 Titration Curves and pK$_a$ Values

The titration curve for acetic acid as titrated by the strong base KOH was explained using the equilibrium mass-action expressions and the stoichiometry of the base–acid interaction. The titration curve consists of two plateaus with a pH jump between. The first plateau is centered at a volume half way to the end point where the pH equals the pK$_a$ of the acid. The pH values of the second plateau are determined by the equilibrium of water itself, pK$_w$ = 14.0. (When a monoprotic acid is titrated, the second plateau always is that of the water equilibrium.) It follows that the lower the pK$_a$ of the acid being titrated, the larger the jump of pH at the end point to the water plateau. This is illustrated in Figure 9.5. Since a jump in pH is needed to determine the end point, there is an upper limit to the pK$_a$ of an acid that can be titrated in water.

The same ideas apply to polyprotic titrations such as carbonate. This is seen in Figure 9.6 where the spread between pK$_a$ values and the sharpness of the break are closely related. In general, a larger jump allows a more precise measurement of the end point as well as ensuring a minimum titration error: ensuring that the end point and equivalence point coincide.

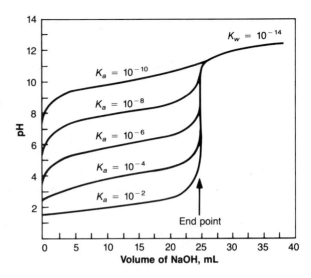

**Figure 9.5** Calculated titration curves for weak acids with differing $pK_a$ values titrated with an aqueous strong base. Each curve is calculated as if 25.00 mL of 0.1000-$N$ acid is titrated with 0.1000-$N$ strong base. Since the curve in the region to the right of the end point is determined by the same equilibrium in all cases, the pH jump at the end point will be less when there are smaller differences between the $pK_a$ and the pH range of the water-equilibrium plateau. The precision will be reduced. The pH jump for the weaker acids can be enhanced by using another solvent or solvent mixture.

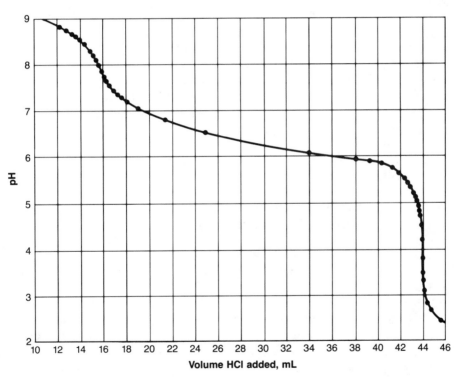

**Figure 9.6** Part of a titration curve for a carbonate/bicarbonate mixture titrated with standardized HCl. The early points are not shown, but both end points are included. Notice that where the pH jump is smaller, the end-point slope is shallower. (The end point is not as sharp.) The first end point occurs where all the $CO_3^-$ has just been neutralized by the strong acid. The second end point occurs where all the $HCO_3^-$ has just been neutralized. The $HCO_3^-$ neutralized in the second range has two origins: (1) the $HCO_3^-$ in the original sample and (2) $HCO_3^-$ from neutralization of the $CO_3^-$ in the sample. The number of data points is small in the buffer region and greater in the end-point region. This saved time since the volumes at the end points comprise the desired data.

### Titrations in Different Solvents

One way to overcome the limitation of water is to use another solvent for the titration. For instance, a solvent with a pK greater than 14 might be used for even weaker acids. As a result, the solvent-determined plateau is shifted further from the plateau determined by the analyte pK-value. The end point then becomes more clear; this means that the jump at the end point is larger upon adding a given amount of titrant. Note that in the precipitation titration described in Section 9.2, a mixed solvent, 1:1 methanol:water, was used for this reason.

## 9.9 Indicators

Titration methods can be useful in seeking answers to questions such as, What is the concentration of acetic acid in this sample of white vinegar?, or How much iodine is there in this pharmaceutical preparation?, or How many times its own weight in stomach acid will this little white pill neutralize? In other words, usually we are interested only in the volume of standardized titrant that is required to reach the end point. From this volume and the volume or weight of the sample, the value of the analyte sought can be calculated. With this viewpoint, the pH values at other points along the titration curve are irrelevant.

As we have seen, near the end point of acid–base titrations there is a rapid change in the pH of the solution as the titrant reacts with the last remaining part of the titrated species. At the end point, a different chemical equilibrium begins to determine the pH. A jump in the pH occurs with a small addition of titrant. This jump in pH can be detected by a change in the color of a dye added to the solution. The dye which changes color is called, reasonably, a **colorimetric indicator.** It indicates the end-point pH jump.

Let us consider a titration of a solution containing two monoprotic acids with different $pK_a$ values. The more acidic component will be a strong acid: It is totally dissociated in the water. The less acidic material has a $pK_a$ of 7.6. Let us do this imaginary titration on 100 mL of solution containing 0.100 M of both acids. We shall titrate the solution with 0.200-M KOH. The titration curve will appear as shown in Figure 9.7a. The equilibria that determine the pH values of the three plateau regions are labeled on the figure.

Next, let us do a second imaginary titration. Everything is the same except that we reduce the concentration of the weak acid ($pK_a$ 7.6). Thus, we now titrate 100 mL of solution that contains 0.100 M in the strong acid and 0.010 M in the weak acid. The titration curve will now appear as shown in Figure 9.7b. The second plateau becomes much narrower; the volume of titrant needed to neutralize the weak acid is smaller by a factor of 10. Nevertheless, the same neutralization reaction is occurring as it did when the concentration of the weak acid was higher, so the same pH range is being traversed from beginning to end.

Finally, let the weak acid be another 100 times more dilute, 0.0001 M, while keeping the strong acid the same as before. In other words, we shall now titrate 100 mL of solution that contains 0.100 M in the strong acid and only 0.0001 M in

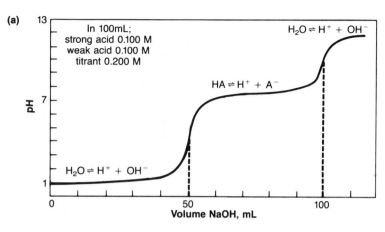

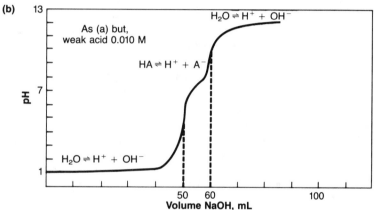

**Figure 9.7** Some properties of neutralization indicators are shown by a titration of a solution composed of a mixture of a strong acid and a weak acid. The middle region of the curve is due to titration of the weak acid. As the quantity of the weak acid is reduced (acid in Figure a > b > c), its influence on the titration curve is lessened. If the weak acid is highly colored and exhibits a color change with pH, then a dilute solution can be used as a colorimetric indicator to detect the end point of the titration of the strong acid. To ensure optimal precision in the assay, the $pK_a$ of a neutralization titration indicator should have a value near to the pH of the solution at the end point of the strong-acid titration.

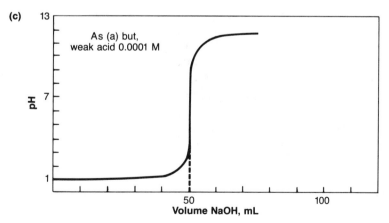

the weak acid. In this case, the titration curve will appear as in Figure 9.7c. Now we cannot see the effect of the weak acid on the pH of the solution although we know that there is a second end point between pH 8 and 9. Also, even though we cannot see the effects, we know that the weak acid must be proceeding through its usual reaction. But the plateau and end-point pH change are occurring over a small relative volume addition: only 1/500th of that needed to titrate the strong acid to its end point. In other words, while 50.0 mL of the KOH are needed to neutralize the strong acid, all the weak acid present is titrated to its end point with only 0.1 mL.

The disappearance of the effects of the weak acid illustrates a limitation of the titration technique; only a relatively narrow range of concentrations can be titrated at one time. However, we can use this limitation beneficially if the weak acid is a highly colored material that changes color from its acid to its base form. First, because the color is intense, the color can be seen even though the concentration is small. Second, the change of color occurs without changing the titration of the stronger acid significantly. In the above example, the relative error was 0.1 mL in 50 mL, or 0.2%. Thus, we can use a highly colored, weak acid as a colorimetric indicator of the end point of the titration of the strong acid.

The above example is consistent with a solution in which the strong acid is HCl and the weak acid, which we will now call an indicator, is phenol red (or phenolsulfonphthalein, $pK_a = 7.6$). In its *acid form,* abbreviated HA or HIn, it is yellow. In its *base form,* $A^-$ or $In^-$, it is red. As seen from the chemical structures in Figure 9.8, this indicator is closely related to both phenolphthalein and bromcresol green. Phenolphthalein is commonly used in manual volumetric acid–base titrations. It is colorless in acid and reddish-purple in base.

## The Properties of Indicators

For an indicator to work as a marker of the end point of a neutralization titration of an acid or a base, ideally its $pK_a$ (often more than one — see below) should be nearly the same as the end-point pH.[9] The reason is as follows. An obvious color change can be seen by eye only when a major proportion of the indicator changes color. As illustrated in Figure 9.8, the color change is due to a change in the chemical form of the indicator, from its acid (protonated) form to its basic form. The chemical transformation for which a color change would be observable is from about 90% in one form to 90% in the other. This means that the ratio of acid-to-base forms of the indicator must change from *approximately*

$$\frac{[\text{HIn}]}{[\text{In}^-]} = 9$$

---

[9] Indicators used in water are weak acids or weak bases (for the color-change reaction) and are used to indicate end points of titrations of acids and bases, respectively.

**Methyl orange**
range pH 3.1–4.4
red–yellow

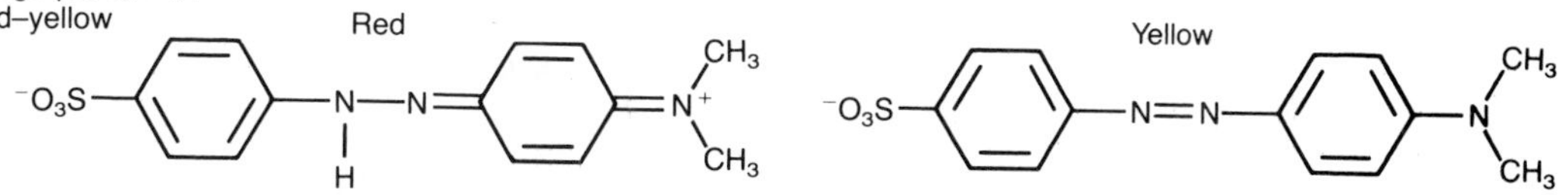

**Bromcresol green**
range pH 3.8–5.4
yellow–blue-green

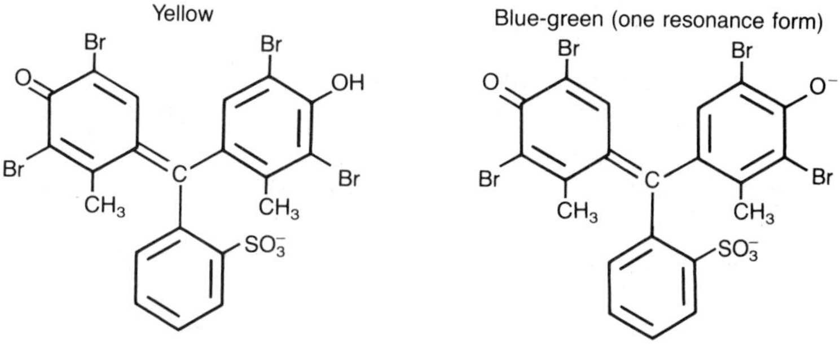

**Methyl red**
range pH 4.2–6.2
red–yellow

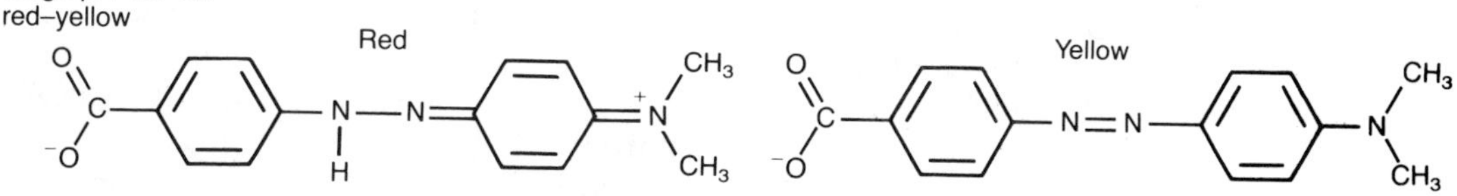

**Phenol red**
range pH 6.8–8.4
yellow–red

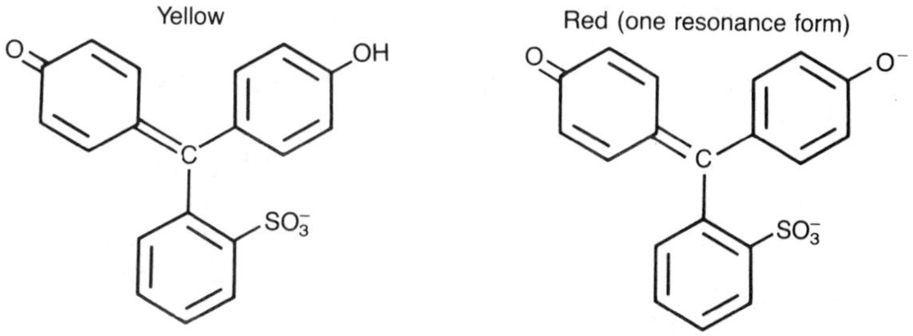

**Phenolphthalein**
range pH 8.0–9.6
colorless–pink

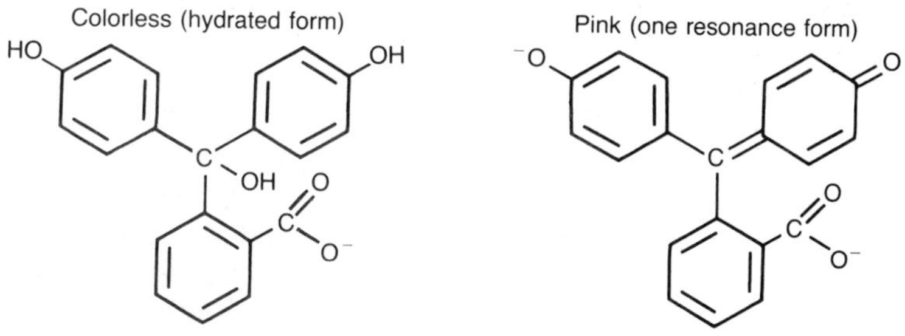

**Figure 9.8** Chemical structures of five commonly used neutralization colorimetric indicators. Each of these indicators has a different useful pH range. More data appear in Appendix VII.

to

$$\frac{[\text{HIn}]}{[\text{In}^-]} = 0.11$$

or vice versa. Compare these ratios to the mass-action expression

$$K_a = \frac{[\text{H}^+][\text{In}^-]}{[\text{HIn}]} \qquad (9\text{-}7)$$

Since $K_a$ is a constant, to change from 90% in one form to 90% in the other, the hydrogen-ion concentration must change by about a factor of 80. Thus, the color change will be observable only over a range of pH values, not just at a single, specific pH. This range is called the **indicator range,** and for *all* color-changing indicators is at least 1.5 to 2 pH units.

As a result of this property of changing form over a pH range, indicators are listed by their useful pH ranges for a color change rather than by their $pK_a$ values. A second reason for doing this is that the changes often involve two or more protons on an indicator molecule. (For example, see phenolphthalein in Figure 9.8.) Thus a listing of the $pK_a$ values are not always easy to interpret in terms of deciding on a useful range. Again, listing an operational range is more useful. A third reason is that the indicators' color changes are all different. For instance, it may be easier to see a change from colorless to pink or from yellow to blue than from purple to green. Again, a useful pH range is a better guide. Even then, some caution may be needed because individuals differ in their abilities to perceive specific color changes.

An indicator is most accurate in determining the end point when its useful pH range lies completely within the steeply rising part of a titration. When an incorrect indicator is used, the results tend to be significately less precise as illustrated in Figure 9.9.

Titrations with two end points require two different indicators. The reason can be seen by considering the titration curve of carbonate shown in Figure 9.6. If the end-point detection is done with color-changing indicators, one indicator is required to change over the range of the first end point that is centered around pH 9. A second is required to change around pH 4. In addition, the colors of the two indicators cannot interfere; otherwise, two separate titrations will be required: one to the first end point with the first indicator and another to the second end point with only the second indicator present. The best situation is to have a colored-to-colorless transition for the first and a colorless-to-colored transition for the second. Then, both indicators can be present together, and both end points can be detected on each sample. As an example, for the titration in Figure 9.6, phenolphthalein is used at the first end point and usually methyl orange for the second one.

Finally, note that there is nothing special about the equilibrium properties of a colorimetric indicator. Therefore, its $pK_a$ value(s) will depend on the ionic strength of the solution and the temperature just as any other equilibrium. In addition, any changes in solvent will change the $pK_a$ value(s). As a result, the useful range in water at temperatures around 20 °C will not necessarily be the useful range in other solvents or under more extreme conditions.

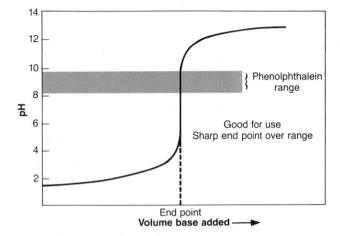

**Figure 9.9**    The use of a good and an inappropriate indicator to determine an end point. The titration curve results from an acid with a $pK_a$ around 2. The shaded regions illustrate the useful ranges of phenolphthalein and methyl orange. Phenolphthalein is a good indicator to use because the entire useful range is encompassed by the pH jump at the end point. On the other hand, the end point for methyl orange will be far less certain because the color change occurs over a relatively large volume of added base.

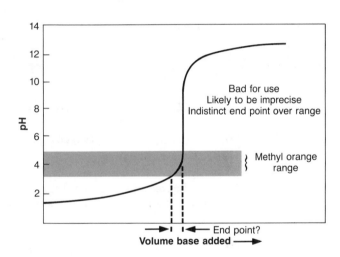

# Suggestions for Further Reading

An advanced textbook on potentiometric titrations: not much about how to do the work but loaded with cautions about the techniques.

Serjeant, E. P. "Potentiometry and Potentiometric Titrations"; Wiley: New York, 1984.

Two books in a series on modern titrimetry.

Jordan, J., Ed. "New Developments in Titrimetry"; Marcel Dekker: New York, 1974.
Wagner, W.; Hull, C. J. "Inorganic Titrimetric Analysis: Contemporary Methods"; Marcel Dekker: New York, 1971.

A comprehensive text on titrations carried out in nonaqueous solvents.

Huber, W. "Titrations in Nonaqueous Solvents"; Academic Press: New York, 1967.

Broadly ranging chapter on the theory and practice of titration of inorganic compounds.

Wilson, C. L. In "Comprehensive Analytical Chemistry"; C. L. Wilson and D. W. Wilson, Eds.; Elsevier: Amsterdam, 1960; Vol. 1B, Chap. 7.

Useful practical guide to various titrimetric techniques, reagents, and indicators.

Barnard, A. J., Jr., et al. In "Handbook of Analytical Chemistry"; L. Meites, Ed.; McGraw–Hill: New York, 1963; Sect. 3.

A monograph on various types of titrations with end points determined photometrically. Somewhat dated.

Headridge, J. B. "Photometric Titrations"; Pergamon: Oxford, 1961.

A comprehensive set of chapters on acid–base titration in aqueous and nonaqueous solutions, complexometric, precipitation, and redox titrations.

Numerous authors. In "Treatise on Analytical Chemistry"; I. M. Kolthoff and P. J. Elving, Eds.; Wiley–Interscience: New York, 1974; Pt. I, Vol. 11, Sect. I–2, Chaps. 114–119.

A carefully developed textbook approach to aspects of equilibria with titrations in some chapters.

Ramette, R. W. "Chemical Equilibrium and Analysis"; Addison–Wesley: Reading, MA, 1981.

Only three pages, but worth reading for the discussion about equilibrium calculations. Written at an intermediate level.

Freiser, H. "Acid–Base Reaction Parameters." *J. Chemical Education* **1970**, *47*, 809–811.

## Problems

9.1    Calculate the pH you would expect to observe in a 0.200-M acetic acid solution in water at 25 °C.

9.2    Using the information in Tables 9.1 and 9.2, calculate the pH expected after 7.00 mL of base is added. At what volume of added base does the expected pH value equal the pH actually measured at 7.00 mL? Use the best value of $K_a$ found experimentally, $2.0 \times 10^{-5}$.

9.3    Calculate and plot the expected titration curve for 100.0 mL of 0.0150-M monoprotic acid with $pK_a = 3.5$ for titration with KOH, 0.0100 M.

9.4    As calculated, the pH of a 0.0200-M acetic acid solution is 3.23. However, in an experiment that was run on 500.0 mL of solution, the pH was measured and found to be 3.5. One hypothesis to explain the discrepancy is that there was some small amount of base in the solution prior to the addition of acetic acid. What would the normality of the base have to be to bring the solution from pH 3.23 to 3.5?

9.5    a.    How much 0.0200-M KOH must be added to change 500.0 mL of a 0.0200-M acetic acid solution from pH 4.00 to pH 5.00? Assume that there is no change of volume for the present. Assume that the activity of $H^+$ equals its concentration, and $K_a = 1.74 \times 10^{-5}$.

       b.    Recalculate part a but include changes in the volume. What do you expect the total volume of the solution to be at pH 4.00 and at pH 5.00?

9.6     50.00 mL of 0.1000-M HCl (a strong acid) is titrated with 0.1000-M NaOH (a strong base). T = 25 °C.
   a.   What equilibrium determines the pH values of the solution before the end point, at the end point, and after the end point?
   b.   What is the volume of the solution and its pH at the midpoint between the beginning and the equivalence point?
   c.   What is the volume of the solution and its pH at the equivalence point?
   d.   Repeat parts b and c for a titration of 50.00 mL of 1.00-mM HCl with 1.00-mM NaOH.

9.7     What are the pH, pOH, and concentrations of $CO_3^-$ and $HCO_3^-$ after 1.000 mL of 0.1500-M HCl is added to 100.0 mL of 0.0100-M sodium carbonate solution? Ignore the added 1% volume. Assume that $a_{H^+} = [H^+]$ and T = 25 °C.

9.8     What do you expect to be the volumes of added acid at the equivalence points in a titration of 2.000 mL of a solution that is 0.4000-M in $Na_2CO_3$ and 0.3000-M in $NaHCO_3$ and is titrated with 0.1024-M HCl? Ignore activity effects. Do the calculation for three cases:
   a.   The original sample is titrated directly.
   b.   The original sample is titrated after having added it to 8.000 mL of water.
   c.   The original sample is titrated after having added it to 98.00 mL of water.
Does the added water make a difference?

9.9     For the above problem, what will be the volumes of the acid added to bring the pH to pH = 10.22 and then to pH = 6.46 when the initial sample was added to
   a.   8.00 mL of water?
   b.   98.00 mL of water?
Does the volume of water make a difference?

9.10     Calculate in parts a–d the ionic strength of 500.0 mL of a solution that is 0.020 M in acetic acid that is titrated with 1.000-M KOH at the following points. Ignore the change in volume due to the titrant addition.
   a.   The initial acetic acid solution
   b.   Where the pH = $pK_a$ of acetic acid
   c.   At the end point
   d.   At a point 5.00 mL past the end point
   e.   By what factor does the ionic strength increase between the beginning and the point at conditions of part d?
   f.   Do the same calculations for a solution that originally has 0.020-M acetic acid and 0.100-M potassium perchlorate, $KClO_4$, a completely dissociated salt.
   g.   By what factor does the ionic strength increase between the beginning and the point 5.00 mL past the end point in this solution with added salt?

9.11     Calculate the ionic strength for the following solutions. (The quantities are in molal units. Sufficient accuracy will be obtained if you assume that the molality and molarity are equal.)
   a.   A solution of 0.0200-$m$ acetic acid in water
   b.   The same solution at the point $pK_a$ = pH after the solution is brought to that point by adding 1.000-$m$ KOH. Ignore volume changes.

c. The following formula gives the activity coefficients for potassium acetate versus ionic strength ($I$).

$$\log \gamma_{\pm} = \frac{-0.82 \sqrt{I}}{1 + \sqrt{I}} + 0.33\, I$$

What is the value of $\gamma_{\pm}$ for the solutions of parts a and b?

d. Since acetic acid is uncharged, its activity will be nearly unchanged by changes in ionic strength. Assume that the activity coefficient is the same for acetate and protons. If

$$K_{measured} = \frac{a_{H^+}[^-OAc]}{[HOAc]}$$

what do you expect to be the value of the ratio of equilibrium constants for parts a and b. In other words, calculate

$$\frac{K_{measured}(0.02 \text{ M in water})}{K_{measured}(pK_a = pH, 0.02 \text{ M})}$$

9.12   You have available to you the following indicators: phenolphthalein, methyl red, methyl orange, and bromocresol green. Assume that for the calculation any indicator present is at a low enough concentration that the actual titration of the indicator can be neglected. Assume that the indicator ranges are applicable at their exact values. 50.00 mL of 0.1000-M HCl is titrated with 0.1000-M NaOH.

a. Calculate the expected titration curve in the region of the equivalence point. The pH at the equivalence point and at $\pm0.05$-mL, $\pm0.10$-mL, and $\pm0.50$-mL NaOH should be sufficient.
b. Which of the indicators would be satisfactory to determine the end point to moderate accuracy (within 1% volume)?
c. Repeat parts a and b for 50.00 mL of 1.00-mM HCl with 1.00-mM NaOH.
d. Which indicators can be used for each of the titrations if 0.1% accuracy is desired?

9.13   At 25 °C an approximately 0.10-M HCl solution is to be calibrated by titration with 0.1000-M NaOH. 40.00 mL of the HCl solution is placed in a flask and titrated. The equivalence point at pH 7.00 is reached when 42.68 mL of base had been added. Also, identical titrations were done using two different indicators: methyl red (range pH 4.2–6.2) and phenol red (range pH 6.8–8.4). Assume that the end points determined by eye were those at the top of each of the ranges of the indicators: that is, pH 6.2 and 8.4, respectively.

a. What volume of base beyond the equivalence point will be needed to reach the pH 8.4 end point?
b. What is the volume shortfall of base at which the pH 6.2 end point is reached?
c. What will be the molarity of the HCl as determined using each of these indicators compared to that found at the pH 7.00 equivalence point?

*9.14   Calculate the titration curve (at enough points to construct a smooth curve) of 100.00 mL of a 0.0100-M $H_2Fe(CN)_6^=$ solution as it is titrated with 0.2000-M KOH. The two different equilibria cannot be separated as well as carbonic and phosphoric acids. The equilibria are

$$H_2Fe(CN)_6^= \rightleftharpoons H^+ + HFe(CN)_6^{3-}, \qquad pK_a = 3.0$$

---

\* Indicates more involved problem.

and

$$HFe(CN)_6^{3-} \rightleftharpoons H^+ + Fe(CN)_6^{4-}, \qquad pK_a = 4.25$$

Assume the volume remains 100mL. You may wish to write a simple computer program to solve this.

**9.15** The seldom-encountered acid, nonsuch acid, abbreviated $H_4Ns$ is one of the (exceedingly) rare tetraprotic acids. It has the following $pK_a$ values:

$$H_4Ns, 2.0; \qquad H_3Ns^-, 4.5; \qquad H_2Ns^=, 7.5; \qquad HNs^{3-}, 10.0$$

A 50.00-mL sample for a test titration is composed of 10.00 mL of five solutions: each of these is 0.0500-M in, respectively, $H_4Ns$, $NaH_3Ns$, $Na_2H_2Ns$, $Na_3HNs$, and $Na_4Ns$.
   a. How many equivalence points will there be in the titration?
   b. What volumes of 0.1000-M NaOH are required to reach these equivalence points for the sample?

**9.16** The following data were obtained by titrating 4.00 mL of an unknown carbonate/bicarbonate mixture with 0.1455-M HCl. What is the concentration of $CO_3^=$ and $HCO_3^-$ in the original mixture?

| mL HCl | pH | mL HCl | pH | mL HCl | pH |
|--------|------|--------|------|--------|------|
| 0.00 | 9.75 | 16.91 | 7.45 | 43.50 | 4.98 |
| 2.02 | 9.61 | 17.19 | 7.39 | 43.57 | 4.98 |
| 4.00 | 9.50 | 17.44 | 7.30 | 43.68 | 4.74 |
| 7.00 | 9.28 | 17.70 | 7.24 | 43.71 | 4.70 |
| 9.96 | 9.03 | 18.20 | 7.16 | 43.83 | 4.38 |
| 12.47 | 8.75 | 19.10 | 7.00 | 43.90 | 4.05 |
| 12.85 | 8.70 | 21.45 | 6.69 | 43.95 | 3.70 |
| 13.40 | 8.60 | 25.02 | 6.48 | 44.00 | 3.48 |
| 13.65 | 8.55 | 29.32 | 6.21 | 44.05 | 3.30 |
| 13.92 | 8.50 | 34.00 | 5.97 | 44.11 | 3.18 |
| 14.34 | 8.40 | 38.12 | 5.80 | 44.21 | 3.00 |
| 14.80 | 8.30 | 39.27 | 5.75 | 44.44 | 2.88 |
| 15.05 | 8.20 | 40.49 | 5.60 | 44.80 | 2.57 |
| 15.29 | 8.12 | 41.49 | 5.60 | 45.05 | 2.45 |
| 15.50 | 8.01 | 42.01 | 5.50 | 45.33 | 2.38 |
| 15.71 | 7.95 | 42.56 | 5.40 | 45.71 | 2.32 |
| 16.00 | 7.80 | 42.90 | 5.30 | 46.45 | 2.32 |
| 16.20 | 7.71 | 43.12 | 5.20 | 47.72 | 2.00 |
| 16.40 | 7.62 | 43.30 | 5.11 | | |
| 16.78 | 7.55 | 43.40 | 5.05 | | |

*Source:* Data courtesy of A. Borchers.

**9.17** [Ref: M. F. Lipton, et al. *J. Organomet. Chem.* **1980**, *186*, 155.]
   A convenient manner to determine concentrations of alkyllithium reagents is by titration of pure 1,3-diphenyl-2-propanone tosylhydrazone (m.w. 378) in tetrahydrofuran (THF). The reactant undergoes the following reactions with alkyllithium reagents in THF.

$$H_2\text{tosylhydrazone} + RLi \rightleftharpoons [\text{H tosylhydrazone}]^- + RH + Li^+$$
$$\text{(colorless)} \qquad\qquad\qquad \text{(colorless)}$$

$$[\text{H tosylhydrazone}]^- + RLi \rightleftharpoons [\text{Li tosylhydrazone}]^- + RH$$
$$\text{(colorless)} \qquad\qquad\qquad\qquad \text{(orange)}$$

R is an alkyl group. A commercially available solution of methyllithium ($CH_3Li$) in an organic solvent was the titrant in this method.

Into an oven-dried 50-mL Erlenmeyer flask is weighed 0.2835 g of the tosylhydrazone solid. The flask is then covered with a serum cap (a tightly fitting rubber cap with a relatively thin section so that syringe needles can be inserted through it). Through an inlet and vent in the cap the flask is purged with pure nitrogen. Then, 10.0 mL of anhydrous THF is added, and the contents are stirred. The flask is cooled in an ice bath (to reduce reaction of RLi with THF), and the reagent liquid is added dropwise with a 1.00-mL syringe until the orange color persists. The color indicates the end point. (The product is a self-indicator.) The syringe volume could be read to the nearest 0.01 mL.

In the titration, 0.452 mL of the $CH_3Li$ solution was required to obtain the orange color.

 a. What is the molarity of the methyllithium solution?
 b. Is the volume of THF in the reaction mixture crucial to obtain a precise result?
 c. Could the same result be obtained with the methyllithium in the flask and with the tosylhydrazone reagent added? How would the color be used as an indicator?

9.18   An assay procedure for the enzyme papain is done by a titrimetric assay of the acid produced by the enzymatic hydrolysis of benzolyl-L-arginine ethyl ester (BAEE). A unit of the enzyme is defined as the amount of enzyme that hydrolyzes 1 $\mu$mol min$^{-1}$ of BAEE at 25 °C and pH 6.2 under the assay conditions. The enzyme is quite unreactive until it is activated by treatment with mild reducing agents such as cyanide or cysteine. An enzyme diluent solution is prepared consisting of

    10 mL    0.01-M edta
    0.1 mL   0.06-M mercaptoethanol
    10 mL    0.05-M cysteine-HCl prepared fresh daily
    70 mL    doubly distilled water

The substrate solution is made fresh daily by mixing the following

    15.0 mL   0.058-M BAEE made fresh daily
    0.8 mL    0.01-M edta
    0.8 mL    0.05-M cysteine-HCl

This solution has the pH adjusted to 6.2 with HCl or NaOH and diluted to 21.0 mL with doubly distilled water.

The assay is done on an automatic titrator that operates by adding titrant NaOH to the reaction solution to keep the pH *constant* at 6.2 as the reaction progresses to form a carboxylic acid by the reaction

$$R-\overset{\overset{\displaystyle O}{\|}}{C}-O-C_2H_5 + H_2O \xrightarrow{\text{enzyme(cat)}} R-\overset{\overset{\displaystyle O}{\|}}{C}-OH + HOC_2H_5$$

An assay was run under the following conditions. The titrant was standardized 0.0164-$N$ NaOH. The enzyme in solution was activated by reaction in the enzyme diluent solution for 30 min. While the enzyme was activating, into the titration vessel were placed 5.00 mL of substrate solution, 5.00 mL of 3.0-M NaCl, and 5.00 mL of doubly distilled water. The solution was allowed to equilibrate to 25 °C. At zero time, 1.000 mL of the enzyme solution was added, and the titrator was turned on to keep the pH adjusted to pH 6.2. After a *constant rate of NaOH addition* was achieved, the titration was allowed to progress for 5.00 min. During the five minutes, 3.811 mL of titrant was used. How many units mL$^{-1}$ are present in the enzyme solution?

## SUPPLEMENT 9A

# An Alternate Derivation of the Henderson–Hasselbach Equation

A titration curve of a weak acid with a strong base is explained by the equilibrium of the acid and water. To solve the algebraic model of the equilibrium requires the usual mass-action expressions, that is,

$$K_a = \frac{[\text{H}^+][\text{A}^-]}{[\text{HA}]} \qquad\qquad \text{(9A-1)}$$

$$K_w = [\text{H}^+][\text{OH}^-] \qquad\qquad \text{(9A-2)}$$

and the conservation equation for the acid

$$C_{\text{HA}} = [\text{HA}] + [\text{A}^-] \qquad\qquad \text{(9A-3)}$$

Another conservation equation is obtained from the total quantity of base added. If the base is NaOH,

$$C_{\text{B}} = [\text{Na}^+] \qquad\qquad \text{(9A-4)}$$

This equation provides information equivalent to the stoichiometric equation for the titration. If the titration reaction goes to completion, that is,

$$\text{HA} + \text{NaOH} \rightleftharpoons \text{A}^- + \text{H}_2\text{O} + \text{Na}^+$$

lies far to the right, then[10]

$$C_{\text{B}} = [\text{Na}^+] = [\text{A}^-] \qquad\qquad \text{(9A-4)}$$

---

[10] Some workers like to use a charge-balance equation, namely,

$$[\text{Na}^+] + [\text{H}^+] = [\text{A}^-] + [\text{OH}^-]$$

in place of Eq. 9A-4.

For moderate concentrations of acid and base, the concentrations of $[\text{H}^+]$ and $[\text{OH}^-]$ become negligibly small. This is an algebraic way to treat the assignment of the controlling equilibrium to the acid–base pair. Because neither H$^+$ nor OH$^-$ are important, Eq. 9A-2 can be ignored in this region.

By rearrangement,

$$[A^-] = C_B \tag{9A-4}$$

$$[HA] = C_{HA} - [A^-] = C_{HA} - C_B \qquad \textbf{(9A-3) + (9A-4)}$$

Substituting for [HA] and $[A^-]$ in Eq. 9A-1 gives

$$K_a = \frac{[H^+]C_B}{C_{HA} - C_B}$$

or

$$[H^+] = K_a\left(\frac{C_{HA} - C_B}{C_B}\right)$$

Taking the negative logarithm of both sides,

$$pH = pK_a + \log\left(\frac{C_B}{C_{HA} - C_B}\right) \tag{9A-5}$$

This form of the equation for the pH is equivalent to the Henderson–Hasselbach equation: Eq. 9.5. What may be clearer from this form is that when the weak acid and its conjugate base are both present in sufficient concentrations to control the pH, the pH should *not* change with dilution of the solution. Both $C_{HA}$ and $C_B$ change by the same relative amount upon dilution. In fact, there are changes upon dilution associated with the change in ionic strength. These are described in more detail in Supplement 10A on buffers.

# 10

# Other Titrimetric Methods

## 10.1  Classifying Other Kinds of Titrations

In Chapter 9 you have read how the procedures of neutralization titrations depend on the equilibrium and stoichiometry of the acid–base reactions. A fundamental requirement for a titrant is that it reacts with the analyte effectively *to completion*. Neutralization titration curves are explained by use of the law of mass action: The solution pH depends on the equilibrium between an acid and its conjugate base. This criterion is the same as saying the reaction is reversible. Acid–base equilibria are not unique in these respects. Exactly the same criteria can be fulfilled by the following kinds of reactions:

1.  Oxidation–reduction reactions
2.  Reactions in which a precipitate is formed (for example, the literature reading of Section 9.2)
3.  Reactions in which a coordination complex is formed
4.  Reactions in which tightly bound ion pairs or stable polymer pairs or colloids are formed

Titrimetric assays are based on all these types of reactions.

The principles and practice of all the methods of titration are so similar that if you understand the details of the acid–base titrations you will be able to comprehend the fundamentals of all the other types. The following needs are exactly the same in all:

A. Reproducible stoichiometry
B. Adequate reaction rates (forward and reverse)
C. Standardized titrants
D. Volumetric measurement equipment
E. End-point detection
F. Careful sample preparation

The greatest differences among the types of titrations exist only in the nature of the underlying chemistry. For instance, for all four types noted above, colorimetric indicators have been discovered, but the indicators for each work in different ways — with different types of chemical changes. Instrumental detection of end points also may depend on different principles for each type. In this chapter are described some of the similarities and differences in the chemistry between neutralization (acid–base) titrations and oxidation–reduction, complexation, and precipitation titrations.

## 10.2 Oxidation–Reduction Titrations

In oxidation–reduction (redox) titrations, instead of monitoring the pH, we can monitor the electrochemical potential of the solution, $E$. Since measurement of the electrochemical potential is often an integral part of the method, oxidation–reduction titrations are often called **potentiometric titrations.**[1]

### Similarities between Redox and Acid–Base Titrations

To carry out a redox titration, a measured volume of an oxidizing agent (or reducing agent) of known concentration is added to oxidize (or reduce) the analyte.[2] Compare this to an acid–base titration during which a measured volume of an acid or base titrant of known concentration is added to neutralize the analyte.

The electrochemical potential $E$ of a redox titration is analogous to the pH in the acid–base titration. The measured $E$-value is controlled by an oxidation–reduction equilibrium. This is similar to the way the pH of a solution is controlled by an acid–base equilibrium.

In both types of titrations, at the end point there is a sharp jump in the quantity being monitored. In a redox titration, the jump is in $E$. This jump occurs where a new redox couple begins to determine the potential of the system. In an acid–base titration, the jump in pH at the end point is where a new acid–base pair begins to determine the pH of the solution.

---

[1] The name *potentiometric titration* applies more broadly. Both the lead-sensitive electrode of the case study of Section 9.2 and the pH electrode produce electrical potentials. Thus, potentiometric titrations are done for all types of underlying chemistries: acid–base, redox, precipitation, complex formation, and colloid formation or ion-pair formation.

[2] The titrant can also be added through an electrochemical reaction in the analyte solution. These are called coulometric titrations and are described in Section 12.8.

In a redox titration, the $E^{\circ\prime}$ value[3] of the indicator should be near the $E$-value at the end point. In an acid–base titration, the $pK_a$ value of the indicator should be near the pH of the end point.

The close similarities in the concepts and practice of neutralization and redox titrations are reflected in the forms of two fundamental equations. For neutralization, the pH is described by the Henderson–Hasselbach equation.

$$pH = pK_a + \log \frac{[\text{conjugate base}]}{[\text{acid}]} \qquad \textbf{(9-5)}$$

For redox potentials, the half-cell potential is described by the Nernst equation (formal-potential form).

$$E = E^{\circ\prime} + \frac{2.303\,RT}{n\mathcal{F}} \log \frac{[\text{oxidized form}]}{[\text{reduced form}]} \qquad \textbf{(2-46)}$$

The descriptive calculations associated with redox titration curves are analogous to those done for neutralization titrations.

---

**EXAMPLE**

A 50.00-mL sample of 0.1000-M $FeSO_4$ solution is titrated with 0.1000-M $Ce(SO_4)_2$. The stoichiometry of the reaction that occurs is

$$Fe^{2+} + Ce^{4+} \rightleftharpoons Fe^{3+} + Ce^{3+}$$

Calculate the concentrations of $Fe^{2+}$, $Fe^{3+}$, $Ce^{3+}$, and $Ce^{4+}$ at the equivalence point. What is the value of $E$ at the equivalence point? Assume that both activity effects and any metal hydrolysis can be ignored and that no metal-sulfate compounds form. This means that $E^{\circ\prime}$ becomes $E^{\circ}$ in Eq. 2-46.

**SOLUTION**

Necessary information:

$$Ce^{4+} + e^- \rightleftharpoons Ce^{3+} \qquad E^{\circ} = 1.61 \text{ V vs. NHE}$$
$$Fe^{3+} + e^- \rightleftharpoons Fe^{2+} \qquad E^{\circ} = 0.771 \text{ V vs. NHE}$$

Volume of the solution at the equivalence point = 100.00 mL

At the equivalence point, two conservation equations hold:

$$[Fe^{3+}] + [Fe^{2+}] = \frac{0.1000 \text{ M} \cdot 50.00 \text{ mL}}{100.00 \text{ mL}} = 0.0500 \text{ M}$$

and

$$[Ce^{4+}] + [Ce^{3+}] = 0.0500 \text{ M}$$

At *any* point during the titration, at equilibrium

$$E_{Fe^{3+}/Fe^{2+}} = E_{Ce^{4+}/Ce^{3+}}$$

---

[3] $E^{\circ\prime}$, the formal potential, is defined in Section 2.10.

Therefore, at 25 °C, since the two Nernst expressions are equal,

$$0.771 + 0.0591 \log \frac{[Fe^{3+}]}{[Fe^{2+}]} = 1.61 + 0.0591 \log \frac{[Ce^{4+}]}{[Ce^{3+}]}$$

Carrying out the appropriate algebra, we obtain the general result

$$\log \frac{[Fe^{3+}][Ce^{3+}]}{[Fe^{2+}][Ce^{4+}]} = \frac{1.61 - 0.771}{0.0591}$$

which yields

$$\frac{[Fe^{3+}][Ce^{3+}]}{[Fe^{2+}][Ce^{4+}]} = 1.57 \times 10^{14}$$

The concentrations of the ions are found as for any competitive equilibrium, by using the mass-action expressions and the conservation relations. From the stoichiometry of the reaction, at the equivalence point

$$[Fe^{2+}] = [Ce^{4+}] = X$$

Then

$$[Fe^{3+}] = [Ce^{3+}] = 0.0500 - X$$

By substituting the algebraic concentrations of the four ions into the equilibrium expression, we find

$$\frac{(0.0500 - X)(0.0500 - X)}{X^2} = 1.57 \times 10^{14}$$

Solving for $X$ (which is made especially easy by taking the square root of both sides of the equation), we find

$$X = [Fe^{2+}] = [Ce^{4+}] = 4.0 \times 10^{-9} \text{ M}$$

Then

$$[Fe^{3+}] = [Ce^{3+}] = 0.0500 \text{ M}$$

The equivalence point is found from the Nernst equation of either metal.

$$E = E^{\circ}_{Fe^{3+}/Fe^{2+}} + 0.0591 \log \frac{[Fe^{3+}]}{[Fe^{2+}]}$$

$$E = 0.771 + 0.0591 \log \frac{0.0500}{4.0 \times 10^{-9}} = 1.19 \text{ V vs. NHE}$$

Note: Numerous assumptions are mentioned in the question, which, if not true, would change the calculated result.

---

In this example, the potential at the equivalence point $1.19 \text{ V} = \frac{1}{2}(1.61 + 0.771)$ V; it is the average of the two $E^{\circ}$ values (both written as reductions). This will be true only when the oxidant and reductant react in an equimolar ratio. It is not a general result.

## Differences between Redox and Acid–Base Titrations

Acid–base and redox titrations are, indeed, nearly identical in the manner in which competitive equilibria cause the behavior displayed by their titration curves. However, there are also major differences in the chemistry and in titration practice.

The one significant chemical difference arises at the initial point on the titration curve — before any titrant is added. In an acid–base equilibrium system, the solvent will almost always act as an acid or base. For instance, in an acid solution the solvent takes up a proton. Thus some acid and its base are both present in the solution. These define the value of $[H^+]$, and the origin of the pH is understandable. However, in general, the solvent used in a redox reaction is not easily oxidized or reduced. Thus, when the species to be titrated is dissolved in the solvent, there is no reaction. But the solution's electrochemical potential can only be established by some redox couple,

$$\text{oxidized form} + n\,e^- \rightleftharpoons \text{reduced form}$$

$$E = E^{\circ\prime} - \frac{0.059}{n} \log \frac{[\text{reduced form}]}{[\text{oxidized form}]}$$

If the solution contains only one of the **electroactive species,** either the reduced form or the oxidized form alone, this redox equilibrium remains undefined. The potential $E$ will not be determined by the analyte redox couple. In practice, it is nearly impossible to figure out the origin of the measured $E$ before any titrant is added.

To run a redox titration sometimes requires an extra step in the sample preparation. The analyte to be titrated must be converted so that it all exists in a single oxidation state. The reason for this extra step is straightforward. During the sample pretreatment, it is difficult or impossible to eliminate all undesirable oxidants and reductants from the sample. For instance, oxygen from the air is a powerful oxidant and can only be eliminated with extra effort. Clearly, the presence of any oxidizing and/or reducing impurities can contribute to an error in the titration. For example, one of the common oxidizing titrants is cerium(IV) ion, $Ce^{4+}$. (In the older literature, this is called ceric ion.) In the redox determination of iron by oxidative titration of its iron(II) form, $Fe^{2+}$, the quantitative reaction that occurs is

$$Ce^{4+} + Fe^{2+} \rightleftharpoons Ce^{3+} + Fe^{3+}$$

But what happens to the accuracy if, before the titration is begun, some oxygen is present and oxidizes some of the iron(II) into the product, iron(III)? The titration to the end point will not be quantitative since the iron(II) has already been partially "titrated" by oxygen. A pretreatment is required which will reduce all the iron to the iron(II) form but, in addition, will leave no *excess* reducing agent in the solution. (Will the assay result be high or low if some excess reducing reagent is present?) Some of the more commonly used reagents and methods for such pretreatments are outlined in Table 10.1.

## Redox Titrants

Just as neutralization titrants are strong acids or strong bases, the titrants for redox titrations are strong oxidizing or strong reducing agents. Two other requirements are

**Table 10.1**   Examples of Reagents Used for Pretreatment of Redox Samples Oxidizing and Reducing Agents*

| Agent | Conditions of Use | Method Used to Destroy Excess |
|---|---|---|
| *Preoxidizing agents* | | |
| Sodium peroxide | Fusion | Dissolve in water, boil |
| Hydrogen peroxide† | Alkaline solution | Boiling: $2 H_2O_2 \rightleftharpoons 2 H_2O + O_2$ |
| Potassium persulfate [peroxydisulfate] | Acid solution | Boiling: $2 S_2O_8^- + 2 H_2O \rightleftharpoons$ $4 SO_4^- + 4 H^+ + O_2$ |
| Ozone, $O_3$ | Acid and alkaline | Boiling or inert gas purge ($N_2$, Ar) |
| Perchloric acid [hazardous with organics] | Acid | Dilution with water |
| *Prereducing agents* | | |
| Metal amalgams of Zn, Cd, Pb, Ag | Solution | Passed through column of solid amalgam granules |
| Powdered metals | Solution | Filtration |
| Sulfur dioxide as gas | Acid | Boil or heat and purge with $CO_2$ |
| Hydrogen peroxide† | Acid | Boiling |

* For further information about preoxidation and prereduction agents, see Blaedel, W. J.; Meloche, V. W. "Elementary Quantitative Analysis"; Harper & Row: New York, 1963; pp. 818–822 and cited references.
† Hydrogen peroxide is both an oxidizing and reducing agent.

desired. First, the stoichiometry of the reactions should be well understood. For redox titrants, this means that the number of electrons transferred in a reaction is known. Second, the reaction should be *clean*. This means that only one reaction occurs, and it goes to completion rapidly. These requirements are relatively stringent. As a result, the great majority of redox titrations are done with only a few different titrants. The identities of some redox titrants along with the number of electrons transferred by each are listed in Table 10.2.

## Redox Colorimetric Indicators

Colorimetric indicators can be convenient reagents to show the end points in neutralization titrations. Similarly, indicators have been discovered that can display the end point of redox titrations. These redox indicators change color with a change in the solution's electrochemical potential.

Just as a pH indicator is characterized by a $pK_a$ value or pH range for its color change, redox indicators are characterized by a half-cell electrochemical potential $E^{\circ\prime}$ or an EMF range. Also, in exact analogy to the pH indicators, which change over a range of pH, the transformation of color of redox indicators progresses over a range of potentials. Some of these colorimetric indicators and their $E^\circ$ values (these are at

**Table 10.2**   Commonly Used Titrants for Redox Titrations

| Name of Titrant | Common Product of Reaction | Electron Equivalents per Mole Reagent |
|---|---|---|
| *Oxidants* | | |
| Permanganate, $MnO_4^-$ | In acid $Mn^{++}$ | 5 |
| | In base $MnO_2$ | 3 |
| Iodine, $I_2$ | $I^-$ | 2 |
| Cerium (IV), $Ce^{4+}$ (ceric ion) | $Ce^{3+}$ | 1 |
| Dichromate, $Cr_2O_7^-$ | In acid $Cr^{3+}$ | 6 |
| *Reductants* | | |
| Thiosulfate, $S_2O_3^-$ | $S_4O_6^-$ | 1 |
| Iron(II), $Fe^{++}$ (ferrous ion) | $Fe^{3+}$ | 1 |

pH $= 0$) are listed in Table 10.3. Most of these values will change with pH, as was discussed in Section 2.9.

There is a special group of redox indicators that are without a parallel in aqueous acid–base titrations: the **self-indicators.** They are highly colored reagents that can act both as the titrant and as an indicator of the end point. As titrants, they are deeply colored materials. But after reacting with the analyte, they become pale in color or even colorless. Thus, during a titration, while the titrant is reacting with the analyte (and becoming pale to colorless), the solution appears colorless. However, after the end point, the excess titrant remains and colors the solution. A straightforward way to see how much of the colored, unreacted titrant appears in solution is to look at Figure 9.5b and substitute "highly colored material" for the vertical axis label $[OH^-]$. Negligible amounts of the colored form remain until the end point, but then its concentration increases linearly with titrant addition. Some self-indicating titrants are listed as a separate set in Table 10.3.

### Redox Volumetric Calculations

Calculating analyte content from the data of redox titrations is done in exactly the same way as for neutralization titrations. Similarly, for developing the method, often the most difficult part is to ascertain the stoichiometric relationship between the titrant and analyte. The reactants and products of the reaction must be sufficiently well known. Data on some commonly used reagents are included in Table 10.2.

A sample calculation of analyte content follows. The method described is a quantitation using an **indirect titration.** In an indirect titration, a quantitative reaction is run to completion between the analyte and an excess of a reagent. The product of the first reaction (the amount is directly proportional to the quantity of analyte) is then titrated to an end point. An indirect titration is less desirable than a direct titration because, due to the extra steps, the precision is inherently lower. Indirect titrations are used when

**Table 10.3**  Examples of Color-Changing Redox Indicators*

| Indicator | Color | | E°(pH = 0)† |
|-----------|-------|-------|-------------|
| | *Reduced* | *Oxidized* | |
| Tris(5-nitro-1,10-phenanthroline) iron(II)sulfate [Nitroferroin] | Red | Pale blue | + 1.25 V |
| Tris(1,10-phenanthroline) iron(II)sulfate [Ferroin] | Red | Pale blue | 1.06 |
| Diphenylaminesulfonic acid | Colorless | Red-violet | 0.84 |
| Methylene blue | Colorless | Blue | 0.34 |
| *Self-indicators* | | | |
| Permanganate [$MnO_4^-/Mn^{++}$] | Very pale pink | Pink | 1.51 |
| Iodine (+ starch) [$I_2/I^-$] | Colorless | Blue | 0.536 |

* All these indicators are used in strongly acid solutions. Further information on redox indicators can be found in "Handbook of Analytical Chemistry", L. Meites, Ed.; McGraw-Hill: New York, 1963; Sect. 3, and references therein.

† The potentials are at unit activity of $H^+$. These potentials will change significantly with pH and less so with other titration conditions such as ionic strength. High-acid concentrations (~ 1M) will keep both the pH and ionic strength nearly constant during a titration.

*Source:* Kolthoff, I. M.; Stenger, V. A. "Volumetric Analysis," 2nd Ed.; Interscience; New York, Vol 1., pg. 140.

1. the analyte cannot be determined by a direct titration, and
2. the reaction of analyte and titrant is so slow that it will not go to completion within a reasonable time.

**EXAMPLE**

Chlorine in water can be determined in the following way. An aliquot of the sample is mixed with a solution containing an excess of KI. This reacts quantitatively with the stoichiometry

$$Cl_2 + 2 I^- \rightleftharpoons 2 Cl^- + I_2$$

The flask must be closed so that no volatile iodine escapes. The iodine formed is then titrated with $Na_2S_2O_3$ in neutral or slightly alkaline solution.

$$I_2 + 2 S_2O_3^- \rightleftharpoons 2 I^- + S_4O_6^-$$

The end point is determined by the disappearance of the iodine color (pale yellow). Starch can be added to the solution as an aid in the colorimetric end-point determination. Starch and iodine form an intensely colored material, so the disappearance of the last color is far easier to detect than for the pale iodine alone.

A 100.0-mL aliquot of chlorinated drinking water was treated with excess KI solution and allowed to react in a closed flask. The pale yellow solution was titrated to an end point with

7.14 mL of 0.0114-$N$ sodium thiosulfate solution. What was the content of $Cl_2$ in molarity and in ppm(w/v) of the water?

**SOLUTION**

From the first chemical equation, we see that the normality of the iodine ($I_2$) titrated equals that of the chlorine ($Cl_2$) in the solution. From the second equation, we see that it requires *two* equivalents of $S_2O_3^-$ per *mole* of iodine (and, thus, two equivalents per mole of chlorine). Thus,

$$N_{Cl_2} \cdot 100 \text{ mL} = (0.0114 \text{ } N)(7.14 \text{ mL})$$

and

$$N_{Cl_2} = 8.14 \times 10^{-4} \text{ } N$$

This normality must now be converted to molarity and ppm(w/v) of $Cl_2$.

$$\text{molarity}_{Cl_2} = \tfrac{1}{2} \text{ normality }_{Cl_2}$$

Thus

$$M_{Cl_2} = \tfrac{1}{2}(8.14 \times 10^{-4}) = 4.07 \times 10^{-4} \text{ M}$$

The molecular mass of chlorine, $Cl_2$, is 70.91. The mass of chlorine in a liter of $4.07 \times 10^{-4}$-M solution is

$$(70.91 \text{ g/mol})(4.07 \times 10^{-4} \text{ mol/L}) = 0.0290 \text{ g/L}$$

With a liter taken to be 1000 g, the relative concentration of $Cl_2$ in a liter is

$$\frac{0.0290}{1000} = 2.90 \times 10^{-5} \quad \text{or} \quad 29.0 \text{ ppm(w/v)}$$

# 10.3 Redox Titration Case Reading

One of the aims of the research quoted in the report that follows was to find out whether a redox titration could, in fact, be done. The method which was developed can be used to quantify analytes belonging to an important group of antibiotics called the *cephalosporins.*

As noted in Section 10.1, one requirement for measuring a potential or determining an end point for a redox titration is that an electrochemical equilibrium exists. A solution potential is not defined unless forward and reverse reactions both "go." If a solution potential is not defined, a useful titration end point will not be observable. The results found and reported in this work illustrate this effect since the reaction of the titrant and analyte is normally *not* reversible. The product of the reaction does not convert back to the reactants during the time it takes to do the titration. When a platinum wire is placed in contact with the solution, the potential measured does not result from the analyte redox couple. (More background about electrochemical potential measurements are presented in Chapter 12.)

This major difficulty was eliminated by the clever use of an oxidation–reduction catalyst. The catalyst has two necessary properties: It somewhat speeds up the attainment of an equilibrium between the reagents, and, in addition, the catalyst's redox

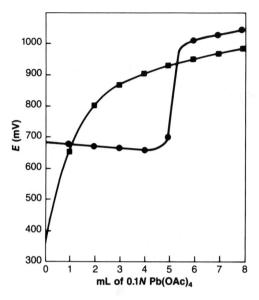

**Figure 10.1**   The titration curves of two potentiometric titrations of diphenyl thioether in 70% acetic acid with 0.1-$N$ Pb(OAc)$_4$ titrant. The squares and circles are markers for the curves and are *not* data points. The automatic titrator produces a continuous output of $E$ versus the volume of titrant, which is added at a steady rate (● ● ●) in the presence of KBr with a mole ratio KBr : analyte of 1 : 100. A small inflection corresponding to the titration of KBr can be seen in the upper part of the curve (■ ■ ■) without KBr in the solution.

reaction is rapid in both directions. As a result, a platinum wire in contact with the solution can register the potential of the catalyst reaction. The potential of the catalyst reaction, in turn, is fixed by the titrant–analyte reaction potential. The titration curve thus becomes measurable. The difference between the titrations measured without and with the catalyst present is shown in Figure 10.1.

Also, note that the titration was found to work best in a solvent of 70% acetic acid, a mixed solvent that is 70% acetic acid and 30% water.

> With the aim of investigating . . . thioethers of pharmaceutical interest, and in order to develop rapid and precise methods for the analysis of products and intermediates containing the R—S—R′ group [R and R′ represent any general organic group bonding through a C–S bond, for example, thioethers], we have continued our study . . . of titrating them [with detection done] by potentiometric techniques. . . .
>
> . . . thioethers . . . take part in addition reactions with [bromine] forming [compounds] of the type:

$$\mathrm{Br_2} + \begin{array}{c} R \\ \diagdown \\ S \\ \diagup \\ R_1 \end{array} \rightleftharpoons \begin{array}{c} R \\ \diagdown \\ S^+\!\!-\!\!Br \cdot Br^- \\ \diagup \\ R_1 \end{array} \tag{1}$$

> In an aqueous acidic environment these derivatives give rise to the corresponding sulfoxides [see chemical equation (4) below]. . . . We . . . hypothesized that the addition of catalytic quantities of bromine water to the medium . . . should render possible . . . their titration. [The experimental results have confirmed that] even the most unreactive thioethers [are] readily oxidized to sulfoxides with an oxidizing agent of appropriate strength, such as . . . [Pb$^{4+}$ salts] provided

that . . . bromine [is] present. . . . We . . . also obtain the same results with KBr, which gives rise to bromine in an oxidizing medium [$2\,Br^- \rightleftharpoons Br_2 + 2\,e^-$].

A Mettler automatic titrator was used. . . . At the beginning, the titrant was added at a rate of 4 mL/min. In the proximity of the potential jump, the titrator adjusted automatically the delivery rate of the titrant to the shape of the potentiometric curves. The reaction rate was comparable to that of the usual neutralization reactions. . . .

Figure [10.1] shows the potentiometric curves for [diphenylthioether].

. . . the titration is only possible in the presence of bromine. . . . The explanation . . . may be as follows. In the presence of an oxidizing agent, the following reaction commences:

$$2\,Br^- \xrightarrow{\text{ox}} Br_2 \qquad (2)$$

Immediately, the following series of reactions is preferred:

$$Br_2 + \underset{R_1}{\overset{R}{\diagdown}}S \rightarrow \underset{R_1}{\overset{R}{\diagdown}}S^+ - Br \cdot Br^- \qquad (3)$$

$$\underset{R_1}{\overset{R}{\diagdown}}S^+ - Br + Br^- \xrightarrow{\text{HOAc/HOH}} \underset{R_1}{\overset{R}{\diagdown}}S=O + 2\,HBr \qquad (4)$$

The hydrobromic acid liberated is continuously oxidized by the [$Pb^{4+}$] added during the titration and is immediately used up in reaction (3). The redox system is thus complex and is represented by the first section of the potentiometric curve [up to the end point].

The system [described] in reaction (2) is [the controlling equilibrium] when all of the thioether has reacted and when an excess of KBr is [present]. [See Figure 10.1 and its caption.]

The redox potential that is measured after the thioether and KBr are titrated is due to the half-reaction

$$Pb^{4+} \rightleftharpoons Pb^{++} + 2\,e^-$$

[Ref.: Reprinted with permission from C. Casalini et al. *Anal. Chem.* **1977**, *49*, 1002–1004. Copyright 1977 American Chemical Society.]

## 10.4 Precipitation Titrations

As described in the case study of Section 9.2, titrimetric assays can be based on the reaction of a titrant and analyte which form a precipitate. In that study was described a precipitation titration in which a lead-selective sensor was used in conjunction with $Pb^{++}$ as titrant. A similar titration curve would appear for pPb versus volume of titrant as is found for pH versus titrant for a base being titrated with acid. An alternate titration curve, shown in Figure 9.3b, could also be used; it would have $[Pb^{++}]$ in place of the titrant $[OH^-]$.

There is an interesting difference between neutralization and precipitation titrations and their equilibrium constants. As you have seen, when a neutralization titration is half-way to the end point the solution's $pH = pK_a$. But consider the precipitation titration reaction

$$Pb^{++} + SO_4^= \rightleftharpoons PbSO_4(s); \qquad K_{sp} = [Pb^{++}][SO_4^=] = 1.7 \times 10^{-8} \text{ at } 25 \text{ °C}$$

The pPb or pSO₄ only equals $pK_{sp}$ when the other species is at unit activity — around 1-M concentration (the exact concentration depends on the activity coefficient).

There is one consistent relationship between the lead and sulfate concentrations and the $K_{sp}$, but this is at the equivalence point where the lead concentration in solution equals that of the sulfate. (The lead concentration will then increase rapidly with titrant being added.) At the equivalence point,

$$[Pb^{++}] = [SO_4^=] = \sqrt{K_{sp}}$$

This is a general result for 1:1 precipitate formation.

However, not all reactions that produce insoluble products are good candidates for use. One reason is that many precipitation reactions are quite slow. As a result, the titration may require a long time to complete, and the end point can become uncertain.

In a precipitation titration, as the end point is approached, the analyte forming the precipitate becomes more and more dilute. Thus, the titrant must be added even more slowly if the reaction is slow to begin with. One of the reasons such reactions tend to be slow is that ongoing precipitation occurs at the surfaces of the precipitate particles. (See Section 8.3.) Considered as a concentration, these surfaces are relatively dilute in comparison with dissolved reagents.

A second drawback is that precipitation reactions tend *not* to be highly specific. Impurities with similar chemical properties can coprecipitate.[4] As a result, there is a tendency for the results to be high. Sample preparation to remove impurities tends to be difficult, and the extra sample handling tends to reduce the assay precision.

A third drawback has been that colorimetric indicators for precipitation titrations tend to be tricky to work with. They require carefully controlled conditions of pH and ionic strength to work moderately well. Another drawback to using such indicators is that the precipitate tends to make the solution opaque and the indicator color difficult to see. Only an especially intense color can be seen at all well.

However, the problems with the indicators have been circumvented with the advent of selective sensors. An example of this is in the case study in Section 9.2 in which a lead-selective electrode was used to determine a titration curve. As a result of such developments, there has been some renewed interest in precipitation titrations.

---

[4] Such coprecipitation is the basis of the chemistry of carriers, as discussed in the literature reading in Section 6.4. Coprecipitation is described in more detail in Section 8.3.

## EXAMPLE

A sample was known to contain only chloride, iodide, and nitrate anions. A potentiometric titration was carried out to determine the chloride and iodide content. Standardized silver nitrate was used as titrant, and a silver wire immersed in the solution was used to determine the solution potential as the precipitates formed. The measured potential began with a plateau around $-0.25$ V, exhibited two end points, and ended with a plateau near $+0.43$ V as the titrant was added. The standardized silver nitrate solution was 0.1103 M. The burette's initial reading was 1.36 mL, and the volumes at the two end points were 12.77 mL and 42.97 mL. Given that the original sample was 0.4332 g, what are the percentages (w/w) of chloride and iodide in the sample? What is the concentration of silver at each of the equivalence points?

## SOLUTION

Necessary information:

$K_{sp}(AgCl) = 1.78 \times 10^{-10}$ at 25 °C
$K_{sp}(AgI) = 9.8 \times 10^{-17}$ at 25 °C
a.w. iodine = 126.90, and a.w. chlorine = 35.45

Stoichiometries for the halides' reactions with silver are

$$Ag^+ + Cl^- \rightleftharpoons AgCl_s \quad \text{and} \quad Ag^+ + I^- \rightleftharpoons AgI_s$$

Of the two halides, iodide is the less soluble. Therefore, the first end point—when less silver has been added to the solution—is due to the iodide, and the second due to chloride. The volume of reagent for each ion type is

for iodide,     $12.77 - 1.36$ mL $= 11.41$ mL
for chloride,   $42.97 - 12.77$ mL $= 30.20$ mL

With a AgNO$_3$ solution $= 0.1103$ M, there are

for iodide,     $0.01141$ L $\cdot$ $0.1103$ M $= 1.258 \times 10^{-3}$ mol
for chloride,   $0.03020$ L $\cdot$ $0.1103$ M $= 3.331 \times 10^{-3}$ mol

With the appropriate atomic weights,

for iodide,     $1.258 \times 10^{-3}$ mol $\cdot$ $126.90 = 0.1596$ g
for chloride,   $3.331 \times 10^{-3}$ mol $\cdot$ $35.45 = 0.1181$ g

For a sample weighing 0.4332 g, the percent (w/w) composition is

for iodide,     $(0.1596 \text{ g}/0.4332 \text{ g}) \cdot 100 = 36.8_4\%$
for chloride,   $(0.1181 \text{ g}/0.4332 \text{ g}) \cdot 100 = 27.2_6\%$

At each equivalence point the metal-ion concentration equals the halide concentration. Thus, letting $X = [Ag^+]$,

for iodide,    $K_{sp} = 9.8 \times 10^{-17} = [Ag^+][I^-] = X^2$ and $[Ag^+] = [I^-] = 9.9 \times 10^{-9}$ M
for chloride,                                                     $[Ag^+] = [Cl^-] = 1.3 \times 10^{-5}$ M

assuming that the titration was done at 25 °C and ignoring ionic strength effects.

## 10.5 Complexometric Titrations (Chelometric Titrations)

Coordinating agents and chelates are molecules (such as ammonia) or ions (such as cyanide) that can exist independently in solutions and that are capable of binding to metal ions, such as zinc. Molecules composed of metals and chelates, or metals and coordinating agents, are called coordination complexes. Some representative coordination complexes are described in Figure 2B.1 in Supplement 2B.

A titration based on the formation of coordination complexes is called a **complexometric** titration. The usual procedure is to use a chelate as titrant to determine metal-ion analytes. Table 10.4 lists some chelates used as titrants for complexometric titrations.

If someone wished to assay for zinc with a complexometric titration, one possible titrant might be ammonia, which reacts to form the zinc–amine complex.

$$Zn^{++} + 4\,NH_3 \rightleftharpoons [Zn(NH_3)_4]^{++}$$

In aqueous solution, the zinc ion actually has water molecules bound around it, and the ammonia replaces the water. Numerous studies have shown that the following reactions occur. The equations may look messy at first, but you will see by inspecting them that each step shows the replacement of 1 water by 1 ammonia molecule.

$$[Zn(H_2O)_4]^{++} + NH_3 = [Zn(H_2O)_3NH_3]^{++} + H_2O \qquad K = 186 \quad \text{(10-1)}$$
$$[Zn(H_2O)_3NH_3]^{++} + NH_3 = [Zn(H_2O)_2(NH_3)_2]^{++} + H_2O \qquad K = 219 \quad \text{(10-2)}$$
$$[Zn(H_2O)_2(NH_3)_2]^{++} + NH_3 = [Zn(H_2O)(NH_3)_3]^{++} + H_2O \qquad K = 251 \quad \text{(10-3)}$$
$$[Zn(H_2O)(NH_3)_3]^{++} + NH_3 = [Zn(NH_3)_4]^{++} + H_2O \qquad K = 11 \quad \text{(10-4)}$$

Incidently, you might ask, Isn't water also a coordinating agent and isn't the aquo-zinc ion $[Zn(H_2O)_4]^{++}$ also a coordination complex? The answer to that is yes — right, on both counts.

For a solution consisting of water, ammonia, and zinc ions (and some counter-ions such as $Cl^-$ or $NO_3^-$), the series of equations above represent a set of simultaneous equilibria of the type with which you are now familiar. However, the ammonia and water also take part in another competitive equilibrium system: the simple acid–base equilibrium with the well-known form

$$NH_3 + H_2O \rightleftharpoons NH_4^+ + OH^- \qquad \text{(10-5)}$$

which you can recognize as a hydrolysis reaction.

These five equations — 10-1 through 10-5 — will be tedious to solve as simultaneous equations. There are some tricks that can be used, and some of these are discussed in Supplement 2B. It is generally easier to use a computer program to do the calculations to find the species that are present under a given set of conditions. But here we are interested only in how coordination complex formation might be used as the basis of a titrimetric method to determine metals in solution.

Two factors affect the equilibria 10-1 to 10-5 which, in turn, influence how a complexometric titration is carried out. These two points are true for all complexometric titrations.

1. The concentration of each of the complexes, $[Zn(NH_3)_n]^{++}$, will depend on the pH of the solution. (This is so because the concentration of $NH_3$ depends on the pH, by Eq. 10-5.) So to have any well-defined equilibria, the pH will have to be buffered. In addition, the protons, $H^+$, compete

**Figure 10.2** The calculated titration curve of 0.1-M $Zn^{++}$ in water with ammonia as titrant. The curve that is plotted is pZn versus the *mole ratio* ammonia/$Zn^{++}$. The pZn is the negative log of the concentration of aquo-zinc ions uncoordinated by ammonia. The pH is assumed to be high enough so that there is negligible $NH_4^+$. In addition, it is assumed that the zinc remains completely in the form $[Zn(H_2O)_n]^{++}$ and does not hydrolyze, that is, does not form species of the type $[Zn(H_2O)_n(OH)_m]^{(2-m)+}$.

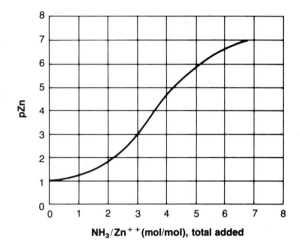

NH₃/Zn⁺⁺(mol/mol), total added

with the $Zn^{++}$ for the ammonia, so the proton concentration must be held rigorously constant.

2. As for any equilibrium involving charged species, the equilibrium constants will depend on the ionic strength. Thus, the ionic strength should be kept approximately constant.

Consider that you have a sample containing an unknown quantity of zinc. The sample is dissolved in water, the solution is strongly buffered for pH, and the ionic strength is brought to a known value. Can you titrate the solution with ammonia and get a good measure of the zinc content? The answer is "not likely." The difficulty is that the stoichiometry involves so many species simultaneously.

Notice that the equilibria represented by Eqs. 10-1 through 10-4 have similar equilibrium constants. As the ammonia is added, there will be four different ammonia complexes formed under approximately the same conditions. To observe the end point of the titration, some method would be needed to observe either the disappearance of the last zinc or the appearance of an excess of free ammonia. However, as illustrated by the titration curve in Figure 10.2, there will not be a simple jump in the amount of free ammonia or free zinc at any point during a titration with ammonia. Thus, ammonia is *not* a good titrant to use. This is true for metals other than zinc as well.

It is necessary to find some titrant that will have a *simple* stoichiometry and will bind to zinc strongly: The coordination reaction must go to completion. There are a number of different reagents that satisfy these criteria. The most commonly used ligands belong to the group of molecules called polyaminocarboxylic acids. Reading this word in parts, it says that there are many *(poly-)* amino groups along with carboxyl groups, (—COOH) on the molecule. The structures and mode of metal bonding of a few of these are shown in Figure 10.3.

All the molecules in Figure 10.3 are classified in the group of molecules that bind as ligands at more than one site. This group is called the chelates or chelating agents. The word *chelate* comes from the Greek word for *claw*. The chelates can be classified by

Ethylenediamine tetraacetic acid
(EDTA or H₄edta). [Hexadentate]

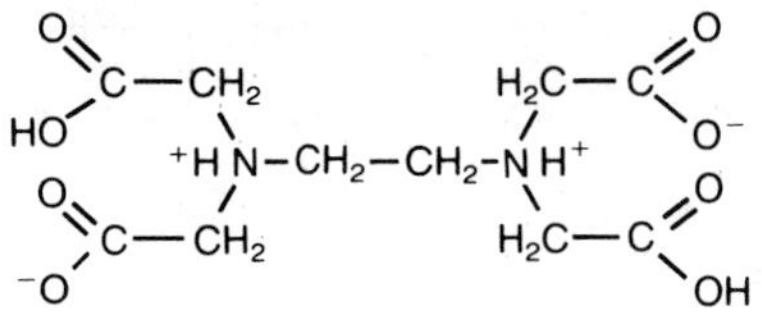

Chemical form of the molecules at pH 7
in the absence of metal ions.

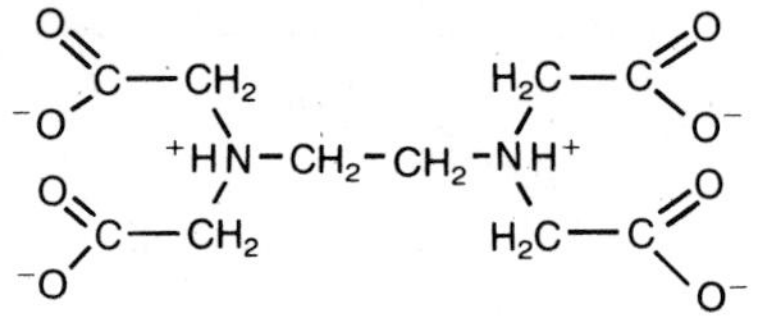

Structure of the edta–metal chelate
for numerous metals.

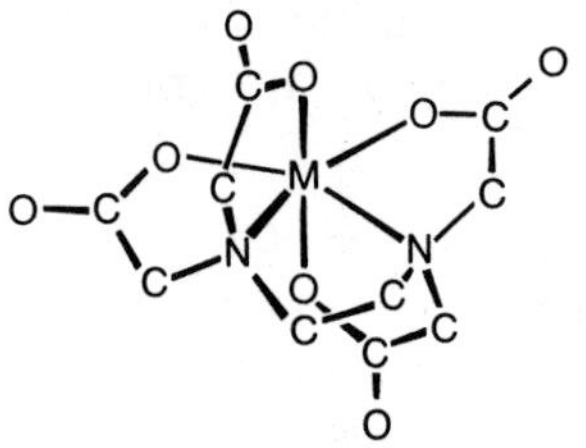

Bis(aminoethyl)glycolether–N,N,N′N′–tetraacetic acid (EGTA or H₄egta).
[Octadentate]

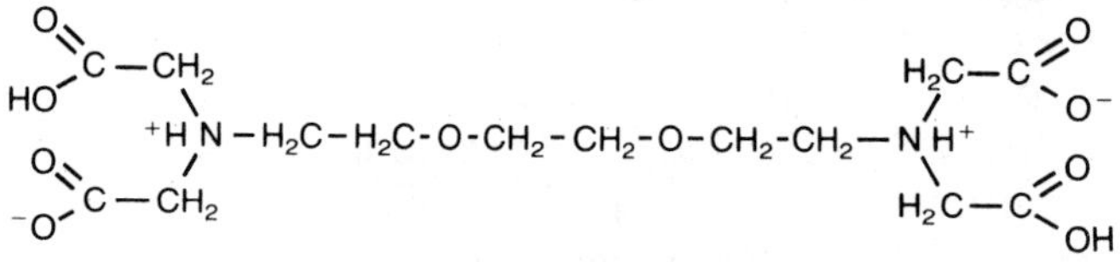

Nitrilotriacetic acid (NTA or H₃nta).
[Tetradentate]

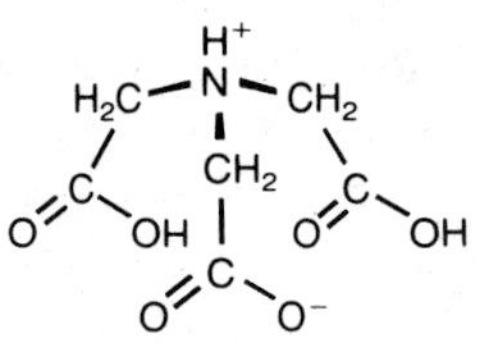

**Figure 10.3**  Three titrants used for complexometric titrations. The *Chemical Abstracts* name for H₄edta is N,N′-1,2-ethanediylbis[N-(carboxylmethyl)-glycine.

their **dentation:** the number of sites that can bind to a metal ion. A chelate such as ethylenediamine,

$$H_2N—CH_2—CH_2—NH_2$$

is *bi*dentate (with two teeth). The two nitrogen sites each bind in a manner similar to ammonia, but they are covalently linked together into a single molecule.[5]

The chelate most commonly used for complexometric titrations is edta (see Figure 10.3), which is an abbreviation for ethylenediaminetetraacetate. Its great advantage is that it reacts with many metals with the simple stoichiometry

$$edta^{4-} + M^{n+} = [Medta]^{n-4} \qquad \text{(10-6)}$$

where $n$ is the charge on the metal ion, M. Edta is also inexpensive and chemically stable.

The reaction 10-6 above describes a reaction that, for zinc and edta in solution, takes the place of all four reactions, 10-1 through 10-4:

$$edta^{4-} + Zn^{++} = [Znedta]^{=}$$

Six sites on the molecule can bind to zinc: two nitrogen sites and four carboxylic acid sites, as seen in the structure in Figure 10.3. However, just as for ammonia and acetic acid, each of these sites can also bind protons. So the titrated solution still must be well buffered for pH and approximately fixed in ionic strength to obtain precise results.

From a theoretical point of view, equilibrium calculations involving edta are just as hard as for four separate ammonia molecules. In the case of zinc, we would substitute one metal–edta chelate equilibrium and six ligand–proton equilibria—a total of seven equilibrium equations—for four metal–ligand (ammonia) equilibria and one ligand–proton equilibrium—a total of five. (Such equilibrium calculations are covered in more detail in Supplement 2B.) However, substituting edta for ammonia in a complexometric titration produces great rewards. Next, you will read why.

### Edta As Titrant

One of the more common tests for which edta titration is used is quantitating calcium in water—part of the characteristic called water hardness. Hardness is customarily expressed as ppm of calcium as calcium carbonate. The hardness is expressed this way even though it is caused by calcium and magnesium sulfates and bicarbonates. Let us use this assay as an example to illustrate the method of complexometric titration using a colorimetric indicator to determine the end point.

The stoichiometry of the reaction of calcium and edta is simply

$$Ca^{++} + edta^{4-} \rightleftharpoons Caedta^{=}$$

---

[5] The nomenclature can be confusing in this area. Recall that a *coordination compound* is the same as a *coordination complex* or a *metal complex*. A *ligand* is also called a *coordinating agent* or *complexing agent*. A *chelate* is also called a *ligand* or, more properly, a *complexing agent*, a *coordinating agent*, or a *chelating agent*.

with an equilibrium constant for *formation* expressed as

$$K_f(\text{Caedta}^=) = \frac{[\text{Caedta}^=]}{[\text{Ca}^{++}][\text{edta}^{4-}]}$$

Note that for complex formation, the equilibrium constants are usually written and tabulated as formation constants. This is in contrast to the equilibrium constants for acids and bases, which are usually written and tabulated as dissociation constants.

The value of the calcium–edta formation constant at pH 9.0 (and *only* at pH 9.0), at 25 °C, and at 0.2-M ionic strength is

$$K_f(\text{Caedta}^=) = 3.8 \times 10^{10}$$

What does this number mean? If we mixed 50.00 mL of a 0.100-M edta solution in water strongly buffered to pH 9.0 together with 50.00 mL of 0.100-M calcium chloride solution, the resulting solution at equilibrium would have a volume of 100.0 mL and would contain the following:

0.05-M [Caedta]$^=$
$4.4 \times 10^{-6}$-M Ca$^{++}$
$4.4 \times 10^{-6}$-M edta not bound to calcium
0.1-M Cl$^-$
$1 \times 10^{-9}$-M H$^+$
Various components of the buffer

These numbers are reported with less precision than the volume and concentrations since they are less certain. Note that of all the calcium and edta only a small part is left unbound—around 1 part per 10,000. As a qualitative description, we can say that the reaction has gone to completion.

However, if any more edta is added above the amount that will bind quantitatively with the calcium, no more free calcium remains to bind with it. So the amount of edta added in excess will be left in solution. But this is just the same behavior as when a strong base was added to the solution of acetic acid. After the acetic acid was used up, the excess base remained in the solution unreacted. A plot of the concentration of free edta in the solution during the titration will be similar to Figure 9.4b, with [edta] replacing [OH$^-$]. There are direct parallels between acid–base titrations and complexometric ones. A number of these are shown in Table 10.4.

### Colorimetric Indicators Used in Complexometric Reactions

The properties of indicators used in complexometric titrations are parallel to those of pH-sensitive indicators. The indicator must be intensely colored in at least one form (bound or unbound to the metal) and must change color when the metal-ion analyte binds with it. Because the color of these indicators changes upon binding (and dissociation) of metal ions, they have been given the name **metallochromic** indicators. The name simply means *metal-colored.*

From the similarities between the acid–base and complexometric titrations, you might guess what chemical properties are required for a metallochromic indicator. You might guess that the indicator must bind the metal ion less strongly than the

**Table 10.4**  Comparison of Neutralization and Complexometric Titrations

| Factor | Caedta Complexometric Titration (buffered near neutral pH) | Acetic Acid–NaOH Neutralization Titration |
|---|---|---|
| Titrant species | $H_2edta^-$ | $OH^-$ |
| Titrant reaction (to completion) | $H_2edta^- + Ca^{++} \rightleftharpoons [Caedta]^- + 2\,H^+$ | $OH^- + HOAc \rightleftharpoons H_2O + {}^-OAc$ |
| Titrated species | $Ca(H_2O)_6^{++}$ | HOAc |
| Titration product | $[Caedta]^-$ | $H_2O$ |
| Titrant competing with | $H_2O$ | $^-OAc$ |
| Titrant binds with | $Ca^{++}$ | $H^+$ |
| Species monitored during titration | edta (using indicator) $Ca^{++}$ (potentiometrically) | $OH^-$ via pH (indicator or potentiometric) |
| End-point jump with change in | edta | $OH^-$ |

titrant, edta. Otherwise the indicator would bind in preference to the titrant and would not be displaced to change its form and color. However, again following the analogy, you might suspect that the indicator must bind more strongly to the metal ion than the water does, otherwise the indicator would not react with the metal ion at all. The water would always win out. Concerning both properties of the indicator, you would be correct.

The metallochromic indicator system, however, has one property that differs in principle from those of acid–base systems. The indicator binding to a metal ion may be dependent on the solution pH; the $H^+$ competes with the metal ion. As indicated before, the complexometric titration must be carried out in pH-buffered solutions — certainly not the conditions of an acid–base titration.

The indicator reaction will be dependent on the ionic strength and perhaps on direct competition of some other metal-ion species in the solution. This causes one of the problems with chelometric titrations: the lack of selectivity. Complicated solutions require significant pretreatment to remove possible interferents. The interferents must be removed or masked (prevented from reacting). One example of such masking is described in the example that follows in which $CN^-$ is the masking agent.

When the sample has been prepared — interferents removed or masked and the pH and ionic strength fixed — calculating the content of a sample from the data is as straightforward as for any other titration.

---

**EXAMPLE**

A 100.0-mL sample of tap water was placed in an Erlenmeyer flask. To this was added 1-g ascorbic acid. (The ascorbic acid reduces any oxygen dissolved in the solution and protects the indicator from air oxidation.) To this was added 1 mL of 50% (w/v) NaOH to bring the pH to about 13 and the ionic strength to about 0.1. Then 0.1 g of NaCN was added to reduce interference from any Fe, Cu, Ni, and the like that may be present. (Their metal–cyano complexes are sufficiently stable under the titration conditions so that edta cannot coordinate with the metal ions.) After the solution was warmed slightly with gentle mixing for a few minutes, it was cooled. Eriochrome Black T indicator was then added. (Dilute solutions of the indicator are pink when complexed with metals and blue when uncomplexed in solutions in the range pH 7 to pH 11.) This solution was titrated to the point where the last tinge of purple disappeared, leaving a blue solution. At that point, 14.11 mL of 0.0237-M $Na_2H_2$edta had been used. What is the hardness of the water as $CaCO_3$ in ppm? Water with a hardness below 100 is generally considered to be soft. Water with a hardness above 300 is considered to be hard.

**SOLUTION**

Necessary information:

> The formula weight for $CaCO_3$ is 100.09.
> The stoichiometry for the reaction at this pH is $Ca^{++} + edta^{4-} \rightleftharpoons Caedta^-$

Magnesium ion also has the same stoichiometry, and the titration includes the sum of Mg and Ca.

There are

$$0.01411 \text{ L} \cdot 0.0237 \text{ M} = 3.34 \times 10^{-4} \text{ mol of } Ca^{++} \text{ and/or } CaCO_3$$

This is

$$3.34 \times 10^{-4} \text{ mol} \cdot 100.09 \text{ g mol}^{-1} \cdot 1000 \text{ mg g}^{-1} = 33.5 \text{ mg of } CaCO_3 \text{ in the } 100.0 \text{ mL}$$

$$\frac{33.5 \text{ mg}}{0.1000 \text{ L}} = 335 \text{ mg L}^{-1} = 335 \text{ ppm hardness (Ca as } CaCO_3)$$

## Suggestions for Further Reading

The readings listed here are in addition to those in Chapter 9.

An advanced textbook on potentiometric titrations. Not much about how to do the work but loaded with cautions about the techniques.
Serjeant, E. P. "Potentiometry and Potentiometric Titrations"; Wiley: New York, 1984.

Best of the breed for theory.
Ringbom, A. J. "Complexation in Analytical Chemistry; A Guide for the Critical Selection of Analytical Methods Based on Complexation Reactions"; Interscience: New York, 1963.

The best place to start if you are actually going to do complexometric titrations.
Pribil, R. "Applied Complexometry"; Pergamon: Oxford and New York, 1982.

Broadly ranging chapter on the theory and practice of titration of inorganic compounds.
Wilson, C. L. In "Comprehensive Analytical Chemistry," Vol. 1B; C. L. Wilson and D. W. Wilson, Eds.; Elsevier: Amsterdam, 1960, Chap. 7.

A monograph on the theory and practice of edta titrations.
Flaschka, H. A. "EDTA Titrations"; Pergamon: Oxford, 1964.

Contains a broad coverage of all aspects or titrations in nonaqueous solvents.
Huber, W. "Titrations in Nonaqueous Solvents"; Academic Press: New York, 1967.

An extensive review article describing oxidizing agents and reducing agents that can be used for nonaqueous redox titrations.

Barrek, J.; Berka, A. "Redox Titrants in Nonaqueous Media"; *CRC Critical Reviews in Analytical Chemistry* **1984**, *15*, 163–221.

Despite its title, mostly titration theory. The chapters on automatic titration (about one-fourth of the book) are quite informative.

Svehla, G. "Automatic Potentiometric Titrations"; Pergamon: Oxford, 1978.

# Problems

### Redox Titrations

10.1    A standardized iodine solution is used to titrate hydrazine sulfate in a buffered sodium bicarbonate solution. The reactions that occur are

$$I_2 + 2\ e^- \rightleftharpoons 2\ I^-$$
$$N_2H_4 \cdot H_2SO_4 \rightarrow N_2 + SO_4^- + 6\ H^+ + 4\ e^-$$

What is the molarity of a solution of hydrazine sulfate when 27.29 mL of 0.1000-$N$ iodine is required to titrate 25.00 mL to the equivalence point?

10.2    $KMnO_4$ is being standardized by titration with $As_2O_3$ (f.w. 197.82) in acid solution. The reaction is

$$2\ MnO_4^- + 5\ H_3AsO_3 + 6\ H^+ \xrightleftharpoons{ICl(cat)} 2\ Mn^{++} + 5\ H_3AsO_4 + 3\ H_2O$$

What is the normality of a $KMnO_4$ solution that requires 45.00 mL to titrate a 0.1500-g sample of $As_2O_3$?

10.3    Standard $KMnO_4$ (f.w. 158.038) was prepared by dissolving about 3 g of reagent grade $KMnO_4$ in 950 mL of distilled water in a beaker. The beaker was covered with a watch glass, heated almost to boiling for 1 h, and set in the dark to age overnight. The resulting solution was filtered through a glass sinter and stored in the dark. The solution was standardized against primary standard sodium oxalate (f.w. 133.9995) that was dried at 110 °C for 1 h. A 1.3653-g portion of the oxalate was weighed out and dissolved in 250.0 mL water. This relatively unstable solution was used immediately. A 50.00-mL aliquot of the $Na_2C_2O_4$ solution was pipetted to a 250-mL beaker, and 50 mL of water containing 5–6 mL of conc. $H_2SO_4$ was added. The solution was heated to 90 °C and slowly titrated with the permanganate to the end point of the first perceptible persistent pink color. (Too fast an addition of permanganate results in unwanted side reactions.) 45.43 mL of the $KMnO_4$ solution were required. The reaction for the titration is

$$5\ H_2C_2O_4 + 2\ MnO_4^- + 6\ H^+ \rightleftharpoons 10\ CO_2(g) + 2\ Mn^{++} + 8\ H_2O$$

Calculate the normality and molarity of the standard $KMnO_4$ solution.

10.4    You are familar with the formula for chemical equivalence,

$$\text{volume}_1 \times \text{normality}_1 = \text{volume}_2 \times \text{normality}_2$$

For titration, however, we sometimes like to put the result into a plug-in equation. Consider the following titration reaction.

$$I_2 + S_2O_3^- \rightleftharpoons 2 I^- + S_4O_6^-$$

a.  Write an equation that expresses the normality of $I_2$ (f.w. 253.82) in the original solution in terms of the volume of the sample solution $V$ and the volume and normality of the thiosulfate ($S_2O_3^-$) only. All known quantities should be collected into a single algebraic factor:

$$\text{normality } (I_2) = \text{factor } x \ . \ . \ .$$

b.  Do the same as in part a but express the concentration of $I_2$ in mg/L.

10.5   An aqueous solution was titrated for alcohol in an acidic solution. The procedure was done accurately, and the solution was reported as containing 0.550% (w/w) ethanol. The solution contained no ethanol, however. It only contained methanol. The reactions under the experimental conditions are

$$C_2H_5OH \rightarrow H_3CCOOH \qquad \text{(unbalanced)}$$
$$CH_3OH \rightarrow CO_2 \qquad \text{(unbalanced)}$$

What is the w/w content of methanol in the solution?

10.6   The reaction of $Cr(VI)$ (as $Cr_2O_7^-$) with $Fe(II)$ is reasonably rapid. As a result, the titration chemistry can be determined from equilibrium considerations. The following reactions may occur in an $FeCl_2$ solution titrated with $K_2Cr_2O_7$.

| | |
|---|---|
| $Cl_2 + 2 e^- \rightleftharpoons 2 Cl^-$ | $E° = 1.36$ V |
| $Cr_2O_7^- + 14 H^+ + 6 e^- \rightleftharpoons 2 Cr^{3+} + 7 H_2O$ | $E(1\text{-}N\ H_2SO_4) = 1.05$ V |
| $Fe^{3+} + e^- \rightleftharpoons Fe^{2+}$ | $E(1\text{-}N\ H_2SO_4) = 0.70$ V |
| $Fe^{3+} + e^- \rightleftharpoons Fe^{2+}$ | $E(1\text{-}N\ H_2SO_4; 0.5\text{-}M\ H_3PO_4) = 0.61$ V |

Possible redox indicators:

| | | |
|---|---|---|
| Ferroin | Transition potential at pH $= 0$ | 1.06 V |
| Sodium diphenylamine sulfate | Transition potential at pH $= 0$ | 0.85 V |

a.  Which of these two indicators is the better one to use?
b.  The 0.5-M phosphate is added to decrease the $E$ of the $Fe(II)/Fe(III)$ couple in order to increase the sharpness of the end point. Does it do so by interacting preferentially with the $Fe(II)$ or the $Fe(III)$?
c.  Will the dichromate oxidize the chloride to any appreciable extent at pH $= 0$?

10.7   A 2.000-g sample is to be assayed for barium as $BaCl_2$. The procedure used includes precipitation of barium as $BaCrO_4$, washing the precipitate, redissolution, reduction of dichromate with $Fe^{++}$, and back-titration of the excess $Fe(II)$ with $KMnO_4$. The equations for the reactions are

$$Ba^{++} + CrO_4^- \rightarrow BaCrO_4(s)$$
$$2 BaCrO_4(s) + 2 H^+ \rightarrow 2 Ba^{++} + Cr_2O_7^- + H_2O$$
$$Cr_2O_7^- + 6 Fe^{++} + 14 H^+ \rightarrow 2 Cr^{3+} + 6 Fe^{3+} + 7 H_2O$$
$$5 Fe^{++} + MnO_4^- + 8 H^+ \rightarrow 5 Fe^{3+} + Mn^{++} + 4 H_2O$$

a. The washed $BaCrO_4$ precipitate was dissolved in HCl, and 50.00 mL of 0.1000-$N$ $FeSO_4$ are added. The excess $Fe^{++}$ requires 6.35 mL of 0.1500-$N$ $KMnO_4$ to titrate to the equivalence point. What is the percentage of Ba as $BaCl_2$ in the sample?
b. The formula weight of $BaCl_2$ is 208.25. What is the equivalent weight of $BaCl_2$ in this titration?

10.8  50.00 mL of 0.1380-M $FeSO_4$ is titrated with 0.0893-M $Ce(SO_4)_2$. The reactions occurring in the titration are

$$Ce^{4+} + e^- \rightleftharpoons Ce^{3+}; \qquad E° = 1.61 \text{ V}$$
$$Fe^{3+} + e^- \rightleftharpoons Fe^{++}; \qquad E° = 0.771 \text{ V}$$

Calculate the expected values, ignoring possible activity effects.
a. At the equivalence point, what is the total concentration of the two iron species?
b. At the equivalence point, what is the total concentration of the two cerium species?
c. At the equivalence point, what is the value of the ratio

$$\frac{[Fe^{3+}][Ce^{3+}]}{[Fe^{2+}][Ce^{++}]}$$

d. At the equivalence point, what is the value of $[Fe^{2+}]$?
e. At the equivalence point, what is the value of $[Ce^{++}]$?
f. At the equivalence point, what is the expected value of $E$?

*10.9  50.00 mL of 0.1000-M $FeSO_4$ is titrated with 0.1000-M $Ce(SO_4)_2$. What is the expected composition of the solution—$[Fe^{2+}]$, $[Fe^{3+}]$, $[Ce^{3+}]$, and $[Ce^{++}]$—and the solution's $E$-value after 40.00 mL of the standard $Ce(SO_4)_2$ is added?

10.10  It is sometimes significantly misleading to take $E°$ as a measure of oxidizing power. For example, dichromate in acid reacts as

$$Cr_2O_7^= + 14 \text{ H}^+ + 6 \text{ e}^- \rightleftharpoons 2 \text{ Cr}^{3+} + 7 \text{ H}_2O; \qquad E° = 1.33$$

a. Calculate the potential you expect for this reaction at pH = 0 for a solution containing 1.33-mM $K_2Cr_2O_7$, 1.33-M $CrCl_3$, and 1.00-M HCl. Assume that activity effects can be ignored. T = 25 °C.
b. Chloride should be oxidizable by this system since

$$Cl_2 + 2 \text{ e}^- \rightleftharpoons 2 \text{ Cl}^-; \qquad E° = 1.36 \text{ V}$$

However, it is not. An experimental measurement of $E$ for the $Cr_2O_7^=/Cr^{3+}$ system shows 1.09 V. Assuming that activity effects are negligible, and assuming that the effect is due entirely to chloride, does the chloride bind more strongly to the $Cr_2O_7^=$ or the $Cr^{3+}$?
*c. If the concentration of $Cr_2O_7^=$ is 1.33 mM, what should the concentration of $Cr^{3+}$ be in order to measure a formal potential at pH0?

10.11  The following half-reactions of nitrate in aqueous acid have been measured.

$$NO_3^- + 4 \text{ H}^+ + 3 \text{ e}^- \rightleftharpoons NO + 2 \text{ H}_2O; \qquad E° = 0.96 \text{ V}$$

---

* Indicates more involved problems.

$$NO_3^- + 3\,H^+ + 2\,e^- \rightleftharpoons HNO_2 + H_2O; \qquad E° = 0.94\,V$$
$$2\,NO_3^- + 4\,H^+ + 2\,e^- \rightleftharpoons N_2O_4 + 2\,H_2O; \qquad E° = 0.80\,V$$

However, a quantitative analysis of nitrate can be made by reaction with excess $FeCl_2$ in the presence of a catalytic amount of molybdenum and back-titrating the excess $Fe^{++}$ in acid with dichromate with diphenylamine as indicator. The reactions are

$$2\,NO_3^- + 6\,Fe^{++} + 8\,H^+ \rightleftharpoons 2\,NO + 6\,Fe^{+++} + 4\,H_2O$$
$$Cr_2O_7^- + 6\,Fe^{++} + 14\,H^+ \rightleftharpoons 2\,Cr^{3+} + 6\,Fe^{+++} + 7\,H_2O$$

a. 0.0100-M $FeCl_2$ and 0.0100-M $Cr_2O_7^-$ are used. If 25.00 mL of the iron solution was reacted with 100.0 mL of a nitrate-containing solution, and 3.47 mL of the chromate solution was needed to titrate the excess Fe(II), what was the original concentration of nitrate in the sample?

b. Can you suggest why the reaction forming NO is stoichiometric when the two other reactions can occur as well? Assume that $a_{H^+} = 1$. Consider both equilibrium and kinetics.

### 10.12   [Ref: Burke, K. E. *Anal. Chem.* **1974**, 46, 882.]

Total sulfur in nickel can be determined in the following way. A sample of nickel is dissolved in a solution 0.9 M in cupric potassium chloride ($CuCl_2 \cdot 2\,KCl \cdot 2\,H_2O$) and hydrochloric acid. All the sulfur is taken up and precipitated as a copper sulfide residue. This process concentrates the sulfur. The residue is collected on a low-sulfur filter paper which is dried and then burned in a combustion furnace at above 1850 °C for two minutes. The sulfur on the paper produces $SO_2$ which is passed into acidified water to form sulfurous acid, $H_2SO_3$. The acid produced is titrated with potassium iodate to a redox end point determined with the aid of thyodene-iodide indicator. The titration stoichiometry in an acid solution is

$$IO_3^- + SO_3^- \rightarrow SO_4^- + I_2 \qquad \text{(unbalanced)}$$

A sample of 9.888 g of nickel alloy was dissolved overnight in the dissolution solution. After drying and combustion, a blank required 0.11 mL to titrate. The potassium iodate solution contained 0.0444 g/L of $KIO_3$ (f.w. 214.00). If the sample required 1.96 mL of titrant, what is the sulfur content of the sample in ppm?

### 10.13   [Refs: EPA-600/2-79-200, Level 2 Sampling and Analysis of Oxidized Inorganic Compounds; EPA-600/4-79-020, Methods for Chemical Analysis of Water and Wastes.]

**Summary of the Method**   The determination of bromide and iodide consists of two separate experiments. The iodide is first determined in the sample, and then a second experiment determines the combined iodide and bromide. The bromide content of the sample is calculated from the difference between the iodide and the combined iodide and bromide determination.

The iodide in the sample is oxidized to iodate, $IO_3^-$, with saturated bromine water in an acid buffer solution. The excess bromide is destroyed by the addition of sodium formate. Potassium iodide is added to the sample solution with the resulting liberated iodine being equivalent to the iodate initially formed in the oxidation step. The liberated iodine is determined by titration with sodium thiosulfate.

In a second sample, iodide and bromide are oxidized to iodate and bromate with calcium hypochlorite. The iodine liberated by the combined reaction products is measured after destruction of the excess hypochlorite and addition of potassium iodide.

**Interferences** Iron, manganese, and organic matter interfere with the above methods. Treatment of the initial samples with calcium oxide removes the interferents.

The following procedure was followed in analysis of wastewater. Sodium thiosulfate titrant was prepared by diluting 100.0 mL of a stock 0.0375 $N$ (including 5 mL/L of chloroform as preservative) solution to 500.0 mL. The solution was standardized by titration with potassium biiodate, $K_2I_2O_6$, and found to be 0.0076 $N$.

100.0 mL of the sample was placed in a beaker and stirred with addition of $H_2SO_4$ to pH approximately 6.5 as determined with a pH meter. The sample was transferred to a 250-mL conical flask, and the beaker was rinsed with distilled water twice, with the rinse being added to the flask. To this solution, 15 mL of sodium acetate solution (275 g/L) and 5 mL acetic acid solution (1:8 acid:water) were added and mixed. Then, 40 mL of bromine water solution (0.2 mL bromine in 800 mL of distilled water) was added. After five minutes, 2 mL of sodium formate solution (500 g/L) was added, and the solution mixed. After five minutes, bromine fumes were removed from over the solution over a period of 30 s using a stream of nitrogen gas. Finally, approximately 1 g of potassium iodide was added to the sample followed by 10 mL of 1:4 $H_2SO_4$ solution. After a five minute reaction period, the sample and a blank were titrated with the sodium thiosulfate solution using a 10 mL burette (markings each 0.1 mL). As the end point was approached, amylose indicator was added. (The end point is reached when the blue color yields to a colorless solution.) A similar procedure was followed, with calcium hypochlorite added as oxidant.

The following data were collected.

| Sample | Titrant Volume Readings (mL) | |
| --- | --- | --- |
| | Beginning | Ending |
| Blank 1 | 0.03 | 0.07 |
| $Br_2$ oxidant | 0.07 | 1.63 |
| Blank 2 | 1.63 | 1.82 |
| $Ca(ClO)_2$ oxidant | 1.82 | 8.34 |

a. What is the sample content in mg/L of $I^-$?
b. What is the sample content in mg/L of $Br^-$?

## Complexometric Titrations

10.14 Water hardness is defined as the combination $Ca^{++}$ plus $Mg^{++}$ reported as $CaCO_3$. In fact, other metal ions can contribute if their chemistries are similar. One method used to determine hardness is to acidify a water sample with HCl, boil the acid solution to remove $CO_2$, and neutralize with NaOH. The solution is buffered at pH 10 with an ammonia buffer. This solution is titrated with edta to the blue end point of Eriochrome Black T. The stoichiometry of the edta–metal interaction is 1 : 1. If a 50.00-mL sample of water requires 31.63 mL of 0.0136-M edta solution to titrate it to the end point, what is the water hardness in ppm as $CaCO_3$?

10.15 [Ref: Fritz, J. S.; Fulda, M. O. *Anal. Chem.* **1954**, *26*, 1206–1208.]
Zirconium in the form Zr(IV) can be determined by direct titration with edta in aqueous solution forming a 1 : 1 complex. However, in order to speed up the equilibration by heating to 70–90 °C, significant hydrolysis occurs with a consequent loss of accuracy. Thus, the procedure is to take a sample solution with approximately 3–5 mM in zirconium, add a slight excess of edta solution, adjust the acidity to pH 1.4 with ammonia solution, heat the solution to near boiling, add Eriochrome Cyanine indicator, and back-titrate with zirconium chloride to the first permanent pink color.

To a sample weighing 0.01070 g dissolved in hydrochloric acid was added 5 g of zinc amalgam to reduce any iron present which might otherwise interfere. To this was added 10.00 mL of 0.0477-M standardized edta solution. After adjusting the pH, adding indicator, and heating, the sample required 5.91 mL of 0.0384-M zirconium oxide in 5% HCl solution to titrate to the end point.
  - a. What percentage of the sample is zirconium?
  - b. Any iron present is reduced from Fe(III) to Fe(II) by pretreatment with zinc amalgam. Given that this eliminates iron interference, how do the Fe(III)–edta and Fe(II)–edta complexes compare in stability with the Zr–edta complex in these conditions?
  - c. Is the pink form of the indicator the protonated form or the zirconium complex?
  - d. With a direct titration, under the conditions of the experiment, which is more stable: the complex of zirconium with the indicator or the complex with edta?
  - e. With back-titration, under the conditions of the experiment, which is more stable: the complex of zirconium with the indicator or the complex with edta?
  - f. Write the equations for the zirconium reaction with the edta and the indicator for the process of back-titration. At this pH, the predominent form of edta is $H_5$edta, and the metal-free indicator is in the form $H_2In^-$. It has been found that the indicator forms a 2 : 1 complex with zirconium. The edta as well as the indicator both lose all their protons upon complexation.

## SUPPLEMENT 10A

# Buffers

A buffered solution resists changes in some property of the solution. Four types of buffers are most commonly found in analyses.

  1. Ionic strength buffers, which reduce the *relative* change in ionic strength with changes in the concentrations of ionic species
  2. pH buffers, which resist the change in pH with added acid or base
  3. EMF buffers, which resist changes in the solution electrochemical potential with added oxidant or reductant

4.  Metal-ion buffers, which resist the change in a specific metal-ion concentration when that ion is added or removed

The action of pH, EMF, and metal-ion buffers results from the equilibrium properties of at least one chemical reaction involving the buffered species. As for any equilibrium, these depend on the following solution properties.

A.  Temperature
B.  Ionic strength
C.  pH (for EMF and metal-ion buffers); metal-ion content (for pH buffers)
D.  Molds, yeast, and/or bacteria growing in the buffer solutions

The first two influences are, perhaps, obvious. Point C reflects the usual competition between protons and metals for ligands. Thus, the components of an *ideal* pH buffer will not coordinate with metals while the components of an *ideal* EMF or metal-ion buffer will not bind protons or hydroxyl groups. Neither of these ideals is obtainable, although in a few cases, the difference in affinity for metals and protons is quite large. Finally, over a wide range of pH and ionic strength, aqueous buffer solutions will support the growth of microorganisms. Over time, the solution components are degraded, and metabolic byproducts are produced. These change the buffer properties. Refrigeration impedes their growth, and growth inhibitors can be added if they do not affect the analysis at hand.

## 10A.1 Ionic Strength Buffers

As was discussed in Chapter 2, the activities of ions in solution depend on the ionic strength of the solution. The ionic strength is defined as a summation over all ions present:

$$\mu(\text{or } I) = \tfrac{1}{2} \sum_i c_i z_i^2 \qquad \text{(2-20)}$$

A number of different equations are used to describe the ionic-strength dependence of ion activities. In a simple equation describing dilute solutions (less than about 1 mM), the single-ion activities change as $10^{\sqrt{I}}$.

The ionic strength of an analyte solution can change for a number of reasons: Adding ionic species as in titration or having different concentrations of ionic species in a series of samples are two examples. In order to minimize the effects of possible changes in ionic strength, it is common to make up the solutions with a constant high-ionic strength: on the order of 0.1–1 M. Solutions possessing ionic strengths cause any *absolute* changes in ionic strength to become small *relative* changes. This reduces significantly the changes in activity effects on ionic equilibria. Thus, the high-ionic strength acts as an ionic-strength buffer.

The high-ionic strength is obtained by adding fully dissociated salts to the solutions: salts such as $NaCl$, $KCl$, $K_2SO_4$, or $KClO_4$. These are used because they can be obtained in highly pure forms at relatively low cost.

## 10A.2 pH Buffers

A wide range of pH buffers are available. There are seven U.S. National Bureau of Standards primary-standard buffers. Their properties are listed in Table 10A.1. Some useful buffering substances with widely varying $pK_a$ values are listed in Table 10A.2. The $pK_a$ values of a number of buffers, especially useful with biological analyses, are listed in Table 10A.3. Many of these are **zwitterionic** buffers. They have both positively and negatively charged sites on an electrostatically neutral molecule. A number of these buffers have been found that buffer well in the physiological range (pH 6–8), and most have few biological effects such as inhibition of enzyme activity. In addition, they are highly water soluble, which allows concentrated stock solutions to be made and minimizes buffer partitioning into biological phases such as membranes.

### The Dilution Value

According to the Henderson–Hasselbach equation, Eqs. 9-5 and 9A-5, upon dilution of a pH-buffered solution in which the weak acid–conjugate base equilibrium controls the pH, the concentrations of both the weak acid and its conjugate base are equally diluted. Thus, according to the equation

$$pH = pK_a + \log \frac{[\text{conjugate base}]}{[\text{acid}]} \tag{9-5}$$

the pH should remain the same. However, the above algebra does not include changes in activities upon dilution. Dilution causes the ionic strength to decrease, and the pH will change. An indication of the effect of dilution is given by a function called the **dilution value** ($\Delta pH_{1/2}$). The numerical value is the change in pH found when a solution of initial concentration $c_i$ is diluted by an equal volume of pure solvent so that the concentration becomes ($c_i/2$).

$$\Delta pH_{1/2} = (pH)_{c_i/2} - (pH)_{c_i} \tag{10A-1}$$

Representative values are shown in Table 10A.1.

Increases in ionic strength also affect the pH-buffer equilibria. The results of these changes are indicated in Table 10A.4. Such effects must be considered if pH changes of this magnitude are unacceptable (for reference, a $\Delta pH$ of 0.04 equals a 10% change in $[H^+]$).

### Buffer Capacity

A pH-buffer region is passed through during an acid–base neutralization titration. As the concentration of analyte (acid or base) increases, the change in pH *with a given amount* of added titrant will decrease; the solution has a higher **buffer capacity**. The buffer capacity for a weak acid HA in the presence of its salt MA is described by the derivative equation 10A-2.

$$\frac{db}{dpH} \approx 2.303 \left( \frac{C_{HA} \cdot K_a \cdot [H^+]}{(K_a + [H^+])^2} + [H^+] + [OH^-] \right) \tag{10A-2}$$

**Table 10A.1** U.S. National Bureau of Standards Primary Standard Buffers (at 25 °C)

Buffer Compositions

| I.D. | Composition | Molality | Molarity | g L$^{-1}$ salt* at 25 °C |
|---|---|---|---|---|
| A | Potassium tetroxalate | 0.05 | 0.04962 | 12.61 |
| B | Potassium hydrogen tartrate | 0.0341 | 0.034 | Saturated at 25° |
| C | Potassium hydrogen phthalate | 0.05 | 0.04958 | 10.12 |
| D | Potassium dihydrogen phosphate<br>Disodium hydrogen phosphate | 0.025<br>0.025 | 0.02490<br>0.02490 | 3.39<br>3.53 |
| E | Potassium dihydrogen phosphate<br>Disodium hydrogen phosphate | 0.008695<br>0.03043 | 0.008665<br>0.03032 | 1.179<br>4.30 |
| F | Borax | 0.01 | 0.009971 | 3.80 |
| G | Calcium hydroxide | 0.0203 | 0.02025 | Saturated at 25° |

Buffer Properties

| I.D. | A | B | C | D | E | F | G |
|---|---|---|---|---|---|---|---|
| Name | Tetraoxalate | Tartrate | Phthalate | Phosphate | Phosphate | Borax | Ca hydroxide |
| Formula | $KH_3(C_2O_4)_2 \cdot 2\ H_2O$ | $KHC_4H_4O_4$ | $KHC_8H_4O_4$ | $KH_2PO_4$<br>$Na_2HPO_4$ | $KH_2PO_4$<br>$Na_2HPO_4$ | $Na_2B_4O_7 \cdot 10\ H_2O$ | $Ca(OH)_2$ |
| pH at 25 °C | 1.679 | 3.557 | 4.008 | 6.865 | 7.413 | 9.180 | 12.454 |
| Dil. value, $\Delta pH_{1/2}$ | +0.186 | +0.049 | +0.052 | +0.080 | +0.07 | +0.01 | −0.28 |
| $db/dpH$ (equiv./unit) | 0.070 | 0.027 | 0.016 | 0.029 | 0.016 | 0.020 | 0.09 |
| $dpH/dT$ (units/°C) | +0.001 | −0.0014 | +0.0012 | −0.0028 | −0.0028 | −0.0082 | −0.033 |

* Weights in air near sea level. Tartrate, phthalate, and phosphates dried at 110 °C for 1–2 h. Potassium tetroxalate dihydrate dried at less than 60 °C. Borax not heated. $CO_2$-free distilled water should be used.

Source: Bates, R. J. Journal of Research of the National Bureau of Standards 1962, 66A, 179–184.

**Table 10A.2** Approximate pK$_a$ Values of Aqueous Buffering Systems (at 25 °C)

| pK$_a$ | Proton* | Compound |
|--------|---------|----------|
| 1.27 | 1 | Oxalic acid |
| 1.97 | 1 | Maleic acid |
| 2.35 | 1 | Glycine |
| 2.95 | 1 | Phthalic acid |
| 3.13 | 1 | Citric acid |
| 3.75 |   | Formic acid |
| 4.21 | 1 | Succinic acid |
| 4.76 |   | Acetic acid |
| 4.76 | 2 | Citric acid |
| 5.41 | 2 | Phthalic acid |
| 6.04 | 2 | Histidine |
| 6.40 | 3 | Citric acid |
| 6.95 |   | Imidazole |
| 7.20 | 2 | Phosphoric acid |
| 7.67 |   | N-Ethylmorpholine |
| 8.08 |   | Tris(hydroxymethyl)aminomethane |
| 8.95 | 4 | Pyrophosphoric acid |
| 9.24 |   | Boric acid |
| 9.25 |   | Ammonia |
| 9.78 |   | Glycine |
| 10.33 | 2 | Carbonic acid |
| 10.72 |   | Triethylammine |
| 11.12 |   | Piperidine |
| 12.38 | 3 | Phosphoric acid |

* For polyprotic acids, this indicates to which proton of the stepwise removal of pK$_a$ refers: pK$_1$, pK$_2$, etc. If only one proton ionizes, no number appears.

where $db$ is the differential amount of strong base added to the buffer solution in equivalents L$^{-1}$. Also, $C_{HA} = [HA] + [A^-]$, and the other quantities have their usual meanings. Adding a strong acid to the solution is expressed as $-db$; the pH decreases with acid, and the change in pH is $-d$pH. As a result, the buffer capacity is always a positive quantity.

Between pH 3 and 11 (that is, where neither $[H^+]$ nor $[OH^-]$ is an important term in Eq. 10A-2) the buffer capacity is well described by the first term in the parentheses. The maximum buffer capacity occurs when pH = pK$_a$. At that point,

$$\left(\frac{db}{d\text{pH}}\right)_{max} \approx \left(\frac{2.303}{4}\right) C_{HA}; \qquad \text{pH} = \text{pK}_a \qquad (10A\text{-}3)$$

**Table 10A.3**    Selected pH Buffers with Special Applicability to Biological Media

| Approximate $pK_a$ Value at 20 °C | Abbreviation | Structure |
|:---:|:---:|:---:|
| 6.1 | MES | O⟩⟨NHCH₂CH₂SO₃⁻ |
| 6.8 | ADA | H₂NCOCH₂N⟨CH₂COO⁻ / CH₂COONa (H) |
| 7.2 | MOPS | O⟩⟨NH—CH₂—CH₂—CH₂SO₃⁻ |
| 7.5 | HEPES | HOCH₂CH₂N⟨ ⟩NCH₂CH₂SO₃⁻ (H) |
| 8.1 | TRIS | HOH₂C, HOH₂C—CNH₂, HOH₂C |
| 8.4 | TAPS | HOH₂C, HOH₂C—C—NH₂—CH₂—CH₂—CH₂—SO₃⁻, HOH₂C |
| 9.3 | CHES | ⬡—NH₂—CH₂—CH₂—SO₃⁻ |
| 10.4 | CAPS | ⬡—NH₂—CH₂—CH₂—CH₂—SO₃⁻ |

*Sources:* Good, N. E., et al. *Biochemistry* **1966**, *5*, 467–477; Ferguson, W. J., et al. *Analytical Biochemistry* **1980**, *104*, 300–310.

With the approximation that ionic-strength effects can be ignored, a buffer's pH remains unchanged upon dilution. However, the *buffer capacity* decreases directly with dilution.

The buffer capacity also changes with pH. If the ratio ($[A^-]/[HA]$) is decreased by a factor of 10—from unity to 1/10—the pH of the solution changes from $pK_a$ to $pK_a - 1$. A similar change to pH + 1 occurs if the ratio is increased by a factor of 10. When the solution pH = $pK_a - 1$, then $[H^+] = 10 \, K_a$. With the approximation $(K_a + [H^+]) \approx [H^+]$, Eq. 10A-2 becomes

$$\frac{db}{dpH} \approx 2.303 \left( \frac{C_{HA} \cdot 0.1 \cdot [H^+][H^+]}{[H^+]^2} \right)$$

**Table 10A.4**   The Effect of Added Salt on the pH of Three NBS Standard Buffers

| I.D. | C | D | F |
|---|---|---|---|
| Buffer | Phthalate | Phosphate | Borax |
| Salt added | KCl | NaCl | NaCl |
| $\Delta$ pH 0.02-$m$ salt | −0.019 | −0.022 | −0.014 |
| $\Delta$ pH 0.05-$m$ salt | −0.044 | −0.051 | −0.035 |

*Source:* Bates, R. G. "Determination of pH," 2nd. ed.; Wiley: New York, 1964.

and

$$\frac{db}{d\text{pH}} \approx \left(\frac{2.303}{10}\right) C_{\text{HA}}; \qquad \text{pH} = \text{pK}_a - 1 \qquad \textbf{(10A-4)}$$

Comparing Eq. 10A-4 with 10A-3, you can see that the buffer capacity falls by a factor of about 2.5 to 40% of the maximum within one pH unit of $\text{pK}_a$. In practice, it is best to use a buffer within $\pm 1$ pH unit of its $\text{pK}_a$.

## 10A.3 EMF Buffers

EMF buffers resist changes in the electrochemical potential upon addition of oxidizing or reducing equivalents. EMF buffers are compounded by mixing the two species of a redox couple, for example, $Fe^{+++}$ and $Fe^{++}$, together. In order for the electrochemical couple to control the EMF, they must dominate all the other couples in the solution. This dominance is a complicated function of concentration and the rate of electron transfer. In other words, a redox couple can be present in the solution, but the electron transfer rate might be so slow that it will not buffer the solution. Such a consideration is not explicitly stated with regard to pH buffers since the $H^+$ transfer rate is fast.[6] One result of the requirement of a fast electron transfer is that only a few organic compounds will function as EMF buffers in aqueous solutions.

In analogy with pH buffers, the maximum buffering capacity for a given total concentration of the redox couple occurs when the concentrations of the oxidized and reduced components are equal. The potential when the concentrations of the two are equal is the formal potential of the half-cell, $E^{\circ\prime}$. The formal potential will change with the following:

1. The identity of the redox couple
2. The solution pH (due to binding with $H^+$ or $OH^-$)
3. Any chemical binding such as coordination that may occur
4. The effects of ionic strength

---

[6] To cite an extreme case, the methyl hydrogens of acetic acid do not participate in the buffering action of an acetic acid–acetate buffer since they do not exchange at an appreciable rate. Only the proton on the oxygen does so.

5. The temperature
6. The solvent

However, since the formal potential is relatively easy to measure, it is relatively easy to adjust it to a chosen value by choice of the ratio [oxidized form]/[reduced form]. As with pH buffers, it is best to work with a buffer near its $E°'$. In this case the equivalent of $\pm 1$ pH unit is $\pm (0.06/n)$ volts, where $n$ is the number of electrons transferred in the redox reaction of the buffer.

## 10A.4 Metal-Ion Buffers

Metal ions in solution can be bound so strongly with a chelate that the bound ions are made inert to many reactions. The ions, which are still present in the solution but unreactive, are said to be **sequestered.** For instance, heavy metals catalyze a number of oxidative reactions in foods. The chelate edta is often added in small amounts to sequester the metal ions, which eliminates the catalytic action and slows the degradation.

When a chelate is present in stoichiometric excess, it will buffer metal-ion concentrations because each free-ion concentration is established primarily by the identity of the chelate and by the pH of the solution. Compare

$$\log [M] = \log K_f + \log \frac{[\text{chelate}]}{[\text{Mchelate}]}$$

for a 1:1 metal-chelate to the Henderson–Hasselbach equation, Eq. 9-5. Secondary contributions to the buffered free-ion concentration are made by the temperature and ionic strength as usual.

### Calculating the Buffered Metal-Ion Concentration Expected

The reason the pH of the solution has a major effect is that the metal ions and hydrogen ions compete for the chelate binding. Some details of the complexity of the proton–metal-ion competition for binding are described in Supplement 2B. Analogous to the pH of a pH buffer, the metal-ion–chelate system can be characterized by a pM. To find the expected pM of a metal-ion buffer involves calculations of the type explained in Supplement 2B.

These calculations rapidly become complicated and can be simplified by use of the concept of an effective formation constant. Briefly, the effective formation constant is found by calculating what *fraction* of the ligand in solution will be in the form $L^{n-}$ to react as

$$L^{n-} + M^{m+} = [LM]^{(m-n)}$$

when the ligand can also bind stepwise with $q$ protons

$$L^{n-} + H^+ = HL^{(1-n)}$$
$$\vdots$$
$$H_{(q-1)}L^{(q-1-n)} + H^+ = H_q L^{(q-n)}$$

## EXAMPLE

A buffer solution for $Cu^{++}(aq)$ is being made using edta as ligand. The pH of the solution must be 4.0. What concentration of $Cu^{++}$ will be present if the solution has $C_{edta} = 0.0020$ M and $C_{Cu} = 0.0010$ M? The stoichiometry of the Cuedta reaction is $1:1$.

## SOLUTION

From Table 2B.2, we find the formation constant for $[Cu-edta]^-$ is

$$Cu^{++} + edta^{4-} = [Cuedta]^-; \qquad K_f = 10^{18.9} = 7.9 \times 10^{18}$$

To use the data given to us, we want to know

$$K_{eff}(\text{pH } 4.0) = \frac{[Cuedta^-]}{[\text{uncomplexed edta(all forms)}][Cu^{++}]}$$

where the following is true.

$$K_{eff} = K_f \cdot \alpha_0$$

The value of $\alpha_0$ is the fraction of the ligand in the form $edta^{4-}$, that is,

$$\alpha_0 = \frac{[edta^{4-}]}{[\text{uncomplexed edta(all forms)}]}$$

Its value can be calculated at pH 4.0 from Eq. 2B-8. The values of $\alpha_0$ for a number of pH values are given in the following tabulation.

| pH | $\alpha_0$ for edta |
|-----|-----|
| 2.0 | $3.7 \times 10^{-14}$ |
| 4.0 | $3.6 \times 10^{-9}$ |
| 6.0 | $2.2 \times 10^{-5}$ |
| 8.0 | $5.4 \times 10^{-3}$ |
| 10.0 | 0.35 |
| 12.0 | 0.98 |

Therefore, $K_{eff}(\text{pH } 4.0) = 7.9 \times 10^{18} \cdot 3.6 \times 10^{-9} = 2.8 \times 10^{10}$.

Given the large formation constant, most of the copper will be bound. As a result,

$$[Cuedta] = 0.0010 \text{ M}$$

Then

$$[\text{uncomplexed edta(all forms)}] = 0.0020 \text{ M} - 0.0010 \text{ M} = 0.0010 \text{ M}$$

The complex dissociates slightly to form $[Cu^{++}] = X$. As a result, the

$$[\text{uncomplexed edta(all forms)}] = (0.0010 + X)$$

Substituting into the equation for $K_{eff}$, we obtain

$$2.8 \times 10^{10} = \frac{[0.0010]}{[0.0010 + X][X]}$$

and

$$X = 3.5 \times 10^{-11} \text{ M} = [\text{Cu}^{++}]$$

The pCu = 10.45 at pH 4.0.

## EXAMPLE
To 100 mL of the above copper buffer solution was added 10 mL of 5.0-mM $\text{Cu}^{++}$ solution. The pH remained fixed. What do you expect the $\text{Cu}^{++}$ concentration to be in this solution?

## SOLUTION
The total volume of the solution will be 110 mL, and

$$C_{\text{Cu}} = \frac{1 \times 10^{-3} \text{ M} \cdot 100 \text{ mL} + 5 \times 10^{-3} \text{ M} \cdot 10 \text{ mL}}{110 \text{ mL}}$$

$$= 1.36 \times 10^{-3} \text{ M}$$

$$C_{\text{edta}} = \frac{2 \times 10^{-3} \text{ M} \cdot 100 \text{ mL}}{110 \text{ mL}} = 1.82 \times 10^{-3} \text{ M}$$

With the same assumptions as above,

$$[\text{uncomplexed edta(all forms)}] = (1.82 - 1.36) \times 10^{-3} \text{ M} = 0.46 \times 10^{-3}$$

$$2.8 \times 10^{10} = \frac{[0.00136]}{[0.00046 + X][X]}$$

and

$$X = 1.05 \times 10^{-10} \text{ M} = [\text{Cu}^{++}]$$

The pCu = 9.97.
Even though enough $\text{Cu}^{++}$ is added to increase its concentration by $0.36 \times 10^{-3}$ M in the solution, as a result of the buffering action, the actual increase was only

$$(1.05 - 0.35) \times 10^{-10} \text{ M} = 0.70 \times 10^{-10} \text{ M}$$

Even though the buffered concentration increased by a factor of about 3, this increase is only about 1 part in 30,000 of the amount of $\text{Cu}^{++}$ added.

# 11

# Kinetic Methods

## 11.1 The Chemistries of Kinetic Analyses

Kinetic methods of analysis are those in which we determine analyte content by measuring the *rate* of a chemical reaction. The analyte can have one of two different chemical actions:

1. The analyte reacts with the reagent and is transformed in the reaction. This is called the **direct kinetic method.**
2. The analyte acts as a catalyst for a reaction in the **catalytic method.**

The practical example for this chapter describes a catalytic method of analysis.

Experimentally, there is little difference between these methods. The rates of reaction must be determined under a set of standard, reproducible conditions. The rates are determined by monitoring the change in the reagent or product concentrations. The results are then related to the initial concentration of the assayed material.

The great majority of kinetic assays are run in solution. If the samples are solids, the concentration of the analyte in solution must be relatable to its mass in the solid. A gas-phase kinetic assay method is described in Supplement 17F.

## 11.2 Why Kinetic Methods?

Precise timing is important in kinetic assays. So why do we bother doing them? Why not use an equilibrium assay method which can be left for a few extra seconds or minutes while we answer the phone?

There are four possible reasons for developing kinetic assays.

1. Kinetic assays can be extraordinarily specific. Even in the relatively nonspecific assay for molybdenum described in the example quoted in this chapter, only a few other metal ions interfere. As a result, most of the other components that remain after acid digestion of the sample do not interfere enough to require extra steps in the sample preparation. You may know about another class of highly specific kinetic assays— those that use the selectivity of enzymes to cause a specific reaction (one of the many, perhaps thousands, that are possible) to occur at a significant rate in an organism. Enzymatic assays often can be run without sample pretreatment: Just add reagents, stir, and measure.

2. Kinetic assays also can be quite sensitive, as the practical example will show. Each molybdenum catalyzes the production of a large number of iodine molecules. A catalytic kinetic assay can amplify the effect of the analyte and be useful for trace analysis.

3. When the rate of a reaction is so slow that waiting for equilibrium to be reached would be impractical, the rate of reaction itself can be observed relatively quickly and used for a kinetic assay.

4. A useful assay reaction might be irreversible: Equilibrium cannot be attained. Nevertheless, the rate of reaction can be observed to quantify the analyte.

Incidentally, all chemical and biochemical methods that use enzymes as reagents are kinetic methods. As a result, in numbers of individual determinations, the majority of all chemical and instrumental analyses are carried out using kinetic methods. This is due, in large part, to the millions of clinical and diagnostic chemical analyses done by kinetic means every day throughout the world.

## 11.3 Practical Example: A Trace Molybdenum Analysis

A demand and a commercial market always exist for less expensive methods to carry out chemical analyses. Inevitably, some manner of automating the sample collection, preparation, and/or assay will be developed to reduce the operator time required for each analysis. If the samples are already in liquid form, later steps of an analysis usually involve adding reagents, waiting for reactions to occur, and heating or cooling the solution. One way to automate these steps is to pump a sample solution through a network of tubes and to add reagents automatically to the sample as it moves along the network toward an assay section. Timing can be regulated by varying the lengths

**(a)**

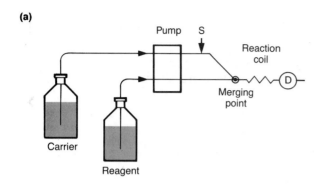

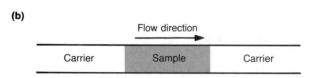

**Figure 11.1**   Flow injection analysis. (a) A schematic diagram for a flow injection analyzer. The box represents a pump. The sample is injected at S into a nonreactive carrier fluid. The reagent to be mixed with the sample is pumped through a separate tube to mix with the sample stream at the merging point. The two solutions are allowed to mix thoroughly and react in the reaction coil, which is symbolized by a jagged line. The liquid sample, now reacted with the reagent, passes through detector D and out of the device. The carrier and reagent streams both are flowing when the sample is added. (b) The sample volume flows as a plug (or bolus) between pure carrier-liquid sectors. The sample plug must be long enough so that only the ends mix and dilute with the carrier. One way of avoiding the intermixing is to place a gas bubble ahead of and behind the sample; this is called segmented flow. (c) The output as it might appear for a series of samples and standards. Either the height of the peaks or the area might be used as quantitative measures of the content of each. [Reprinted from Karlberg, B. I., *American Laboratory* **1983**, *15*, 73. Copyright 1983 by International Scientific Communications, Inc.]

**(b)**

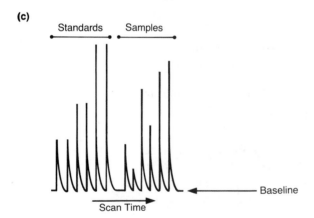

**(c)**

of the tubes and the flow rate. The temperatures can be changed by heating or cooling individual sections of the tube through which the sample passes. This general method is called, reasonably, **flow injection analysis** (FIA). A brief introduction to the technique is illustrated in Figure 11.1. Devices for flow injection analysis fall in the class of instruments called automatic analyzers. Perhaps, because sampling and digestion procedures often come first and most often are done manually, the classification might more correctly be called automatic wet-chemical assayers.

In this example, the authors want to find an improved assay for trace levels of molybdenum in agricultural plant specimens. The chemical basis of the assay is the conversion of iodide to iodine by hydrogen peroxide. Molybdenum acts as a catalyst in the reaction

$$2\ I^- + H_2O_2 + 2\ H^+ \xrightarrow[\text{as catalyst}]{\text{Mo}} I_2 + 2\ H_2O$$

The amount of iodine that is produced depends on the concentration of molybdenum (the catalyst). In the assay that was developed, the color intensity of the iodine is measured after a fixed time of catalyzed oxidation. The methods that the authors describe include the usual sample preparation, elimination of interferences, oxidation–reduction reactions, and statistical testing. Each of these had to be considered within the framework of trace analysis.

Some of the terms in the quote may be unfamiliar to you. *Borosilicate* glass, for example, is the most common type of laboratory glassware (formed of borates and silicates). You may be more familiar with one of its trade names, Pyrex®. Another term, *100 volume* $H_2O_2$, is an old measure of the concentration of hydrogen peroxide solutions and is the same as a 30% by weight $H_2O_2$ solution in water.

When you read this example, notice especially the step in the procedure in which fluoride is added to the solution to reduce interference from iron. Iron ions also catalyze the assay reaction and, thus, act as an interference in a molybdenum assay. The added fluoride coordinates the iron strongly (but not the molybdenum), and the iron coordination complex is not active in the reaction. Because the fluoride acts to mask the effects of iron, fluoride is called a masking agent for iron.

**Preparation of Sample**    For a single determination a mass of sample of 0.25 g in a final volume of 2.5 mL is sufficient. . . . 0.25 g of oven-dried plant material [was weighed accurately and placed] into a silica crucible and ashed overnight in a furnace at 550 °C. [This operation removes the organic components of the sample.] Transfer the cooled ash into a 20 × 150-mm borosilicate test-tube and add 2 mL of 2-$N$ hydrochloric acid, 0.5 mL of 100-volume hydrogen peroxide. . . . Take to dryness . . . add 2.5 mL of 0.125-$N$ hydrochloric acid to the residue, heat to 60° for 15 minutes, mix thoroughly and re-heat at 60 °C for a further 10 minutes. Allow to stand overnight, then determine the molybdenum content by the catalytic method on the automatic analyzer. . . . The incorporation of hydrogen peroxide with the hydrochloric acid ensures that all the iron contained in the sample is oxidized to iron(III) [the importance of which is discussed later].

[The flow diagram for the automated analysis is shown in Figure 11.2.]

**Interferences**    Using the . . . procedure, no interference was found from solutions containing up to 0.5% of potassium, sodium, calcium, or magnesium as their chlorides (equivalent to 5% of each [as a percentage of the weight of the] plant material). [These ions are present in plants at concentrations (w/w) generally less than 4% potassium, 2.4% calcium, and 0.5% magnesium.] Iron also catalyzes the [conversion to] iodine from iodide by hydrogen peroxide. . . . [The catalytic activity of iron(III) was 200 times less than that of molybdenum.] As iron concentrations commonly fall in the range 50–300 ppm, this interference seriously limits the usefulness of the method. [The iron present at 50–300 ppm will produce activity equivalent to 0.4–1.5 ppm of molybdenum.] . . . the addition of ammonium fluoride to complex iron(III) reduced the interference [so the activity of iron(III) was reduced by a factor of 5000. Thus the

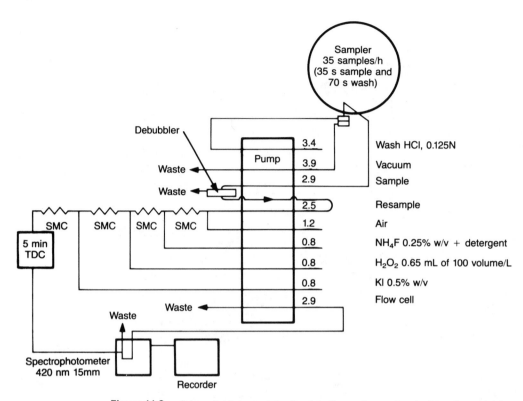

**Figure 11.2**    Schematic diagram of the flow injection analyzer set up to determine molybdenum by catalytic oxidation of iodide and masking of iron with fluoride coordination. The pump simultaneously pumps nine different streams, each adding the material noted on the right side. The numbers on the lines are the pumping rates of each in mL/min. A debubbler removes gas bubbles that might have entered the line up to that point. Single-mixing coils (SMC) allow each reagent to mix thoroughly before the next is added. The time-delay coil (TDC) requires 5 min for the sample to travel through it. This is the fixed time allowed for the oxidation of iodide (the last reagent to be added) to occur. The numbers associated with the spectrophotometer (Chapter 17) are the wavelength of light and length of liquid through which the light passes. Notice that only a part of the total volume is pulled through the spectrophotometer.

iron at 50–300 ppm of the plant mass will have activity equivalent to only 0.00008–0.0003 ppm of molybdenum.]

. . . iron(II) solutions . . . caused a depression in the base line, presumably because of the [lowering of the peroxide concentration due to] oxidizing this [reduced form of] iron. For this reason, it was important to use an oxidizing digestion.

**[Validation]**   The real test of any routine analytical method is how it performs against established methods. We compared results . . . with those from Rua-kura Agricultural Research Centre . . . where in excess of 10,000 plant samples are analyzed for molybdenum annually by [another method]. The results were satisfactory, particularly in view of the purposely chosen range of low concentrations. [See Figure 11.3.]

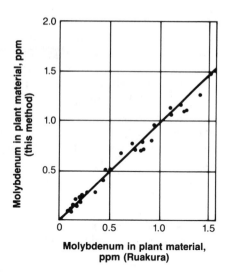

**Figure 11.3** Plot of the results from a series of samples. The results were obtained using two different methods to determine molybdenum: the kinetic method described in this reading and the method used at the Ruakura Agricultural Research Centre. If the two methods agreed perfectly, the points would all lie on the line at 45°.

**Table 11.1** Table of Productivity Comparisons (Man-hours per 100 samples)

| Method | Weighing and Digestion | Mo Determination | Total |
|---|---|---|---|
| This work | 2.5 | 1* | 3.5 |
| Ruakura procedure | 6 | 5 | 11 |

* Includes reagent preparation and chart reading only; automatic analyzer running time does not involve technician time.

> . . . the increased productivity of the automated catalytic method described in this paper is considerable [as shown in Table 11.1].
> [Ref: Quin, B. F.; Woods, P. H. *Analyst* **1979**, *104*, 552–559.]

## 11.4 Assay Kinetics and Assay Types

(You may wish to review Supplement 1B on the mathematics of elementary chemical kinetics before continuing.)

For many kinetic methods, we utilize reactions that can be described by

$$A + R \underset{k_r}{\overset{k_f}{\rightleftharpoons}} P \tag{11-1}$$

where

A is a reactant (it may be the material to be assayed),
R is an added reagent,
P is the reaction product,
$k_f$ is the forward rate constant, and
$k_r$ is the reverse rate constant.

All the reactions are started at zero time ($t = 0$). Concentrations at zero time will be signified as [species name]$_0$; that is, the symbols $[A]_0$, $[R]_0$, and $[P]_0$ are the initial concentrations of A, R, and P. The symbols $[A]_t$, $[R]_t$, and $[P]_t$ represent the concentrations of the species at any time $t$ that comes after $t = 0$.

The kinetic behavior of such a system is more complicated than desirable for a straightforward kinetic assay. But, we can choose the reagent concentrations and the observation period that make the quantitation simpler. How these choices are made and their effects are described in Supplement 11A. Let us assume that such simplification has been made by the choice of conditions. The reactions that we monitor then *appear* to be of the type

$$A \xrightarrow{k} P \qquad (11\text{-}2)$$

This is a first-order reaction[1] that is not reversible. The experimentally observed kinetics for this reaction can be described by

$$\text{rate} = \frac{d[A]_t}{dt} = -k[A]_t \qquad (11\text{-}3)$$

When the reaction's stoichiometry is as shown in Eq. 11-2, the rate described in Eq. 11-3 can also be elucidated from the product of the reaction:

$$\text{rate} = \frac{d[P]_t}{dt} = k[A]_t \qquad (11\text{-}4)$$

As these equations show, for our assay we can measure either the rate of disappearance of the reactant A or the rate of appearance of the product P. Both measures contain the same information. Of course, we shall choose the one that is easier to measure. In the following discussion, let us assume that the concentration of the product of the reaction is monitored.

---

[1] For the reaction with a rate of product production given by the equation

$$\frac{d[P]}{dt} = k[A]^n[B]^m$$

The exponents $n$ and $m$ are generally integers or half-integers. Here, $n$ is the order of the reaction with respect to A, and $m$ is the order with respect to B. If $n = 1$, the reaction is said to be first order with respect to A; if $n = 2$, it is second order with respect to A, and so forth. The sum $(n + m)$ is the overall order of the reaction. These are empirical values and often cannot be predicted from the stoichiometric equation for the reaction. For example, the apparently analogous gas-phase reactions

$$H_2 + I_2 = 2\,HI \qquad \text{and} \qquad H_2 + Br_2 = 2\,HBr$$

have the respective rates of formation given by

$$\frac{-d[H_2]}{dt} = k[H_2][I_2] \qquad \text{and} \qquad \frac{-d[H_2]}{dt} = k'[H_2][Br_2]^{1/2}$$

## The Catalytic Method

When the analyte acts as a catalyst, its effect is to change the rate constant $k$ of the reaction. Therefore, for the catalytic method, the rate measure, as indicated in Eqs. 11-3 and 11-4, can be used directly. The rate, as a function of analyte concentration, is calibrated through the use of standards. From the working curve (rate versus concentration) the unknown catalyst concentrations can be determined. If the assay is well designed (and the chemistry cooperative), the rate $d[P]_t/dt$ and rate constant $k$ are linearly related to the catalyst concentration. The molybdenum determination described in Section 11.3 is a catalytic determination. The changes in rates were ascertained by measuring the changes in the amount of product after a 5-min interval. The molybdenum concentration is directly proportional to that change.

## The Direct Method

The direct method is not quite as straightforward. For the direct method, $k$ is a constant, and the rate varies with [A]. As the reaction progresses, [A]—which is one of the reactants—is used up. However, we seek the amount of analyte present before any reaction occurred; we seek [A]$_0$. Somehow, the measurement(s) must allow us to determine that original concentration.

Although Eqs. 11-3 and 11-4 can be used for catalytic assays directly, they are not adequate for the direct method. We need an equation to find [A]$_0$. Solving the differential equation (Eq. 11-4), we find

$$[P]_t = [A]_0(1 - e^{-kt}) \tag{11-5}$$

Another equation describes the *rate* of the reaction related to [A]$_0$:

$$\frac{d[P]_t}{dt} = k[A]_0 e^{-kt} \tag{11-6}$$

Using either of these equations, we can calculate [A]$_0$ from [P]$_t$ and $t$, because $k$ is known from calibration runs.

## 11.5 Methods for Determining [A]$_0$ or $k_{catalytic}$ from Rates

Assume that you have made a choice of reagent concentrations so that the kinetics of the reaction act as described by Eqs. 11-3 and 11-4. Then, the complete set of data from an assay might appear as shown in Figure 11.4. These data might be recorded directly on a chart as the instrument response versus time. Alternatively, the data might be plotted by hand from kinetic data measured individually at a number of sequential times with the data points then connected by a smooth curve. As will be shown below, only a few data points are needed to determine [A]$_0$ from the assay.

The data that are collected can be related to the concentration of the material to be assayed in three different ways. Each of them has advantages in different circum-

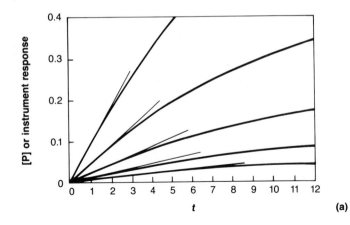

**(a)**

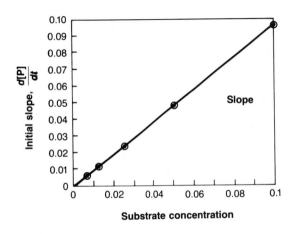

**(b)**

**Figure 11.4**    (a) Demonstration of the derivative method for a direct kinetic assay. A series of kinetic runs with initial analyte concentrations $[A]_0$ of 0.100, 0.050, 0.0250, 0.0125, and 0.0063 in arbitrary units. The time units are arbitrary as well. The tangents to these curves at $t = 0$ are drawn. In (b), the values of the tangent slopes are plotted versus the substrate concentration. The working curve is a straight line. Different experimental methods would be used to obtain the data if the time units were milliseconds to seconds, as opposed to minutes or hours.

stances. These are the **derivative method,** the **fixed-time method,** and the **variable-time method,** which are illustrated in Figures 11.4 through 11.6, respectively.

When using these methods, the experimental measurement for a direct assay and a catalytic assay are the same. However, the experimental conditions are somewhat different. A direct kinetic assay requires conditions such that the rate constant $k$ remains the same for the standards and unknown.

To obtain simple kinetic behavior in a catalytic assay, it is best to have the reactant concentration(s) remain constant, which can be done by having $[A]_0$ in large excess (as well as by keeping all other conditions constant). More details about the experimental conditions that simplify the kinetic behavior are to be found in Supplement 11A.

### The Derivative Method

**Direct Assay**    The derivative method involves measuring the analyte concentration by determining the slope of the concentration-versus-time curve, as illustrated in

Figure 11.4. The slope is quite simply related to the concentration of reactant being assayed. As expressed by Eq. 11-6, the slope of the curve is proportional to the value of $[A]_0$. This link between the slope and $[A]_0$ can, perhaps, be seen more clearly if we rearrange Eq. 11-6 to

$$[A]_0 = \frac{d[P]_t}{dt} \cdot \frac{1}{k} e^{kt} \qquad (11\text{-}7)$$

For a fixed $k$ (direct method), the change in [P] with time, $d[P]_t/dt$, is proportional to $[A]_0$.

Notice that all tangents to the curves are drawn to reflect the slopes of the curves at the beginning of the run. According to Eq. 11-6 or 11-7, a tangent drawn at any other single time works as well. How can this be? The reasons can be seen by studying Eq. 11-7. With the sample and standards run under exactly the same conditions, the value of $k$ is fixed for all runs. As a result, the factor $(1/k)$ is a constant. Again, since $k$ is constant, at any *single* test time, the factor $e^{kt}$ will be the same for all the samples. Thus, the slopes $(d[P]_t/dt)$ are still directly proportional to $[A]_0$, and the assay method still works.

The derivative method for determining analyte concentrations is commonly used. The experiment can be run, plotted, and analyzed in the manner shown in Figures 11.4a and b. On the other hand, it is relatively easy to use electronic means to measure the slope: the change in the instrument's output over time. The slope output can then be related simply to the analyte's concentration.

**Catalytic Assay**    Establishing experimental conditions such that $[A]_0$ remains essentially constant (in large excess), Eq. 11-4 can be written

$$\frac{d[P]_t}{dt} = k[A]_0$$

Therefore, the rate of the reaction is directly proportional to the rate constant. Assuming a linear relationship between $k$ and the analyte (catalyst) concentration, we can write

$$\frac{d[P]_t}{dt} = [\text{analyte}] \times \text{constant}$$

The initial slopes of the kinetic plots are proportional to the analyte concentrations.

### Fixed-Time Method

**Direct Assay**    Equation 11-5 shows the relationship of $P_t$, the concentration of product P at a time $t$, to $[A]_0$, the initial concentration of A. It is useful to ask, What sort of relationship between $[P]_t$ and $[A]_0$ can we find if we consider the concentrations at two different times? Let us call the times $t_1$ and $t_2$. The concentration of P at time $t_1$ is

$$[P]_{t_1} = [A]_0(1 - e^{-kt_1}) \qquad (11\text{-}8)$$

whereas at $t_2$ this is

$$[P]_{t_2} = [A]_0(1 - e^{-kt_2}) \qquad (11\text{-}9)$$

Now, subtract Eq. 11-8, which describes the concentrations at $t_2$, from Eq. 11-9, which describes the concentrations at $t_1$. Then,

$$\Delta P = [P]_{t_2} - [P]_{t_1} = [A]_0(e^{-kt_1} - e^{-kt_2}) \qquad (11\text{-}10)$$

But what happens if the times $t_1$ and $t_2$ are kept fixed throughout the series of experiments in the analysis? (Hence the name *fixed-time method*.) In that case, the quantity in parentheses in Eq. 11-10—the difference between the two exponential functions—remains at a constant value. So with a fixed-time interval, Eq. 11-10 becomes quite simple.

$$\Delta P = [A]_0 \times \text{constant} \qquad (11\text{-}11)$$

The constant can be determined by running standards, and the unknowns are then run with the same fixed-time range. The change in the concentration of P over that time range is directly proportional to $[A]_0$, the initial amount of A, which is what we set out to find. This whole process is described pictorially in Figure 11.5.

**Catalytic Assay**   Establishing experimental conditions such that $[A]_0$ remains essentially constant (in large excess), the product produced at a fixed time $t$ for two different rates $k_1$ and $k_2$ is

$$[P]_{k_1} = k_1[A]_0 t \qquad \text{and} \qquad [P]_{k_2} = k_2[A]_0 t$$

Therefore

$$\Delta P = [P]_{k_1} - [P]_{k_2} = (k_1 - k_2)[A]_0 t \qquad \text{or} \qquad \Delta P = (k_1 - k_2) \times \text{constant}$$

If $k_2 = 0$, then we obtain a relationship similar to Eq. 11-11

$$\Delta P = k_1 \times \text{constant} = [\text{analyte}] \times \text{constant}'$$

If $k_2 \neq 0$ (such as a non-zero blank), then a linear correction is called for.

## Variable-Time Method

**Direct Assay**   In the variable-time method we measure how long the reaction mixture requires to reach a predetermined, *fixed concentration* level of reactant or product. The variable-time method for determining changes in reaction rates is illustrated in Figures 11.5 and 11.6. To see the limitations on the technique, let us look at its algebra.

Assume that we want to analyze two samples that contain different concentrations of analyte. The reactions for both are started at the same time $t_0$. To describe the changes in $[P]_t$ for the two samples, we can write two equations.

The reaction of sample 1 is described by Eq. 11-10 as

$$\Delta[P]_1 = [A]_{01}(e^{-kt_0} - e^{-kt_1}) \qquad (11\text{-}12)$$

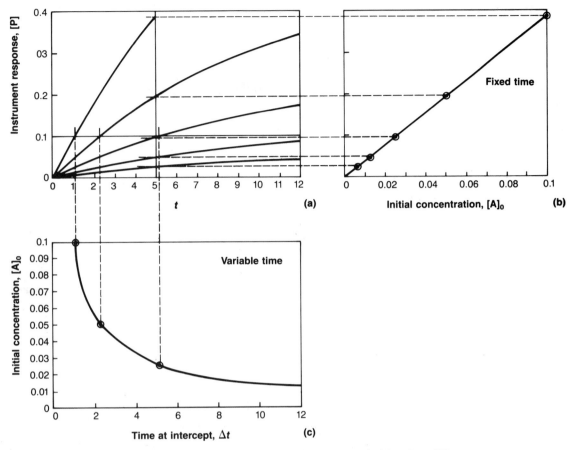

**Figure 11.5**    Demonstration of the fixed-time method (a and b) and variable-time method (a and c and Figure 11.6) for a direct kinetic assay. The kinetic runs are the same as in Figure 11.4a. For the fixed-time method, the concentrations are measured at a single time. Only one concentration measurement is needed for each run. Here, the concentrations are measured after 5 time units—at the vertical line. The points are projected to (b) where the values proportional to $[P]_5$ are plotted versus the value of $[A]_0$. This working curve is a straight line through zero. For the variable-time method, the time required to reach a fixed concentration is measured. The horizontal line at 0.1 indicates the chosen, fixed concentration. However, measurement at only one time point is necessary for each run. These times are projected to a plot of $[A]_0$ versus the times. The working curve is not linear since linearity is expected only for a plot of $1/\Delta t$. A $1/\Delta t$ plot is shown in Figure 11.6.

The reaction of sample 2 is described by

$$\Delta[P]_2 = [A]_{02}(e^{-kt_0} - e^{-kt_2}) \qquad (11\text{-}13)$$

As you can see, the only differences between the two equations are the subscripts on the values of $\Delta P$ and $[A]_0$ and the final time. The differences reflect the changes in the two initial concentrations, $[A]_{01}$ and $[A]_{02}$, that are being assayed.

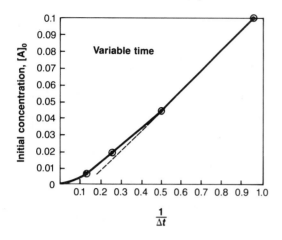

**Figure 11.6** Demonstration of the variable-time method (continued from Figure 11.5). The graph is of $[A]_0$ (the initial concentration of analyte) versus $1/\Delta t$ (the inverse of the time required to reach a fixed concentration of P). Notice that the working curve is not linear nearer zero because the approximations made for the method no longer hold. Linearity is not expected if the reaction proceeds for too long a time when the solutions contain low initial concentrations of analyte.

We now run into a problem: These equations describe quite complicated behavior. For an assay, we would like to observe some simpler form. A more direct measurement tends to be more precise.

It turns out that the behavior can be made quite simple if we measure only the product concentrations near the beginning of the reaction (when less than 1–2% of the reagent A has reacted). Mathematically, this simplification can be expressed by an approximation. The values of the exponential functions are, approximately,

$$e^{-kt} \approx 1 - kt + k^2 t^2 + \cdots$$

For values of $kt$ that are small (this is where the reaction of only 1–2% of the reagent comes in), we can write

$$e^{-kt} \approx 1 - kt \tag{11-14}$$

Now, substitute Eq. 11-14 into Eq. 11-12.

$$\Delta[P]_1 = [A]_{01}(1 - kt_0 - 1 + kt_1)$$

But this becomes, quite simply,

$$\Delta[P]_1 = [A]_{01} k(t_1 - t_0) \tag{11-15}$$

Define

$$\Delta t_1 = (t_1 - t_0)$$

and similarly define $\Delta t_2$ for the other sample. Thus, for the two samples,

$$\Delta[P]_1 = [A]_{01} \cdot k \cdot \Delta t_1 \tag{11-16}$$

and

$$\Delta[P]_2 = [A]_{02} \cdot k \cdot \Delta t_2 \tag{11-17}$$

Now comes the trick of the technique. To characterize the reaction rates, we choose the time when the *same* amount of product, $\Delta P$, is produced by *both* reactions. To quantify the analyte with the assay, the question we ask is, How are the initial concentrations related to the times needed to produce the *same fixed amount* of P? Having the same amount of P means that

$$\Delta[P]_1 = \Delta[P]_2 = \Delta P$$

We can combine Eqs. 11-16 and 11-17 for the same, fixed $\Delta[P]$ to obtain

$$[A]_{01} \cdot \Delta t_1 = [A]_{02} \cdot \Delta t_2 \qquad \textbf{(11-18)}$$

The initial concentrations are *inversely* related to the time needed to reach the fixed concentration. This inverse relation is also *linear* if the reaction is not allowed to proceed too far. Stated algebraically, the analytical relationship is

$$[A]_0 = \text{constant} \cdot \frac{1}{\Delta t} \qquad \textbf{(11-19)}$$

This relation is shown graphically in Figure 11.6.

**Catalytic Assay**    For the case of a catalytic assay with $[A]_0$ effectively constant, Eqs. 11-16 and 11-17 are transformed to

$$\Delta[P]_1 = [A]_0 k_1 \Delta t_1 \qquad \text{and} \qquad \Delta[P]_2 = [A]_0 k_2 \Delta t_2$$

Equation 11-18 then becomes

$$k_1 \Delta t_1 = k_2 \Delta t_2$$

An inverse relationship between $k_i$ and $t_i$ is seen. Assuming a linear relationship between the analyte concentration and the rate, the analytical relationship becomes

$$[\text{analyte}] = \text{constant} \cdot \frac{1}{\Delta t}$$

in parallel with Eq. 11-19.

## Suggestions for Further Reading

A comprehensive but readable work with numerous examples. The examples are from before 1968.
Mark, H. B., Jr.; Rechnitz, G. A. "Kinetics in Analytical Chemistry"; Interscience: New York, 1968.

An excellent general review of elemental determinations using contemporary kinetic methods. Highly useful discussions of drawbacks and numerous useful tables as entries to the literature.
Müller, H. "Catalymetric Methods of Analysis"; *CRC Critical Reviews in Analytical Chemistry* **1982**, *13*, 313–372.

A concise review of some kinetic methods with references to examples.

Guilbault, G. G. In "Treatise on Analytical Chemistry," Pt. I, Vol. 1, 2nd ed.; I. M. Kolthoff and P. J. Elving, Eds.; Wiley–Interscience: New York, 1978; Chap. 11.

Two reviews with numerous references to the literature. Little biochemistry in either.

Malmstadt, H. V.; Delaney, C. J.; Cordos, E. A. "Reaction-rate Methods of Chemical Analysis"; *Critical Reviews in Analytical Chemistry* **1972,** *2,* 559–619.

Mottola, H. A. "Catalytic and Differential Reaction Rate Methods"; *Critical Reviews in Analytical Chemistry* **1974,** *4,* 229–280.

An encyclopedic set of books of techniques for analyses of enzymes involving kinetic analyses.

Bergmeyer, H. U.; Grassl, M. "Methods of Enzymatic Analysis," 3rd. ed.; Verlag Chemie: Weinheim and Deerfield Beach, FL, 1983.

## Problems

11.1   What method of measurement is used in the example in Section 11.3 to this chapter? Derivative, variable-time, fixed-time?

11.2   The enzyme nitrate reductase catalyzes the reduction of nitrate to nitrite. It is assayed by measuring the rate of production of nitrite. The assay reaction is carried out in the presence of a reducing agent, sodium dithionite, to remove oxygen. The reaction is initiated by mixing solutions of nitrate with the enzyme. The reaction is stopped by bubbling the solution with air at the end of the 10.00-min reaction period. (This also removes the sodium dithionite.) The nitrite that has been produced is then reacted with color-producing reagents, and the absorbance at 540 nm is measured versus a blank. The blank consists of an identical enzyme solution that has been treated identically except that instead of being mixed with the sodium nitrate solution, it has been mixed with an equal volume of water. The final assay solutions have volumes of 2.5 mL. The following data were found.

| Solution | Instrument response |
|---|---|
| Blank | 0.004 |
| Blank + 20-$\mu$M NO$_2^-$ | 0.269 |
| Blank + 40-$\mu$M NO$_2^-$ | 0.550 |
| Blank + 60-$\mu$M NO$_2^-$ | 0.803 |
| Blank + 80-$\mu$M NO$_2^-$ | 1.087 |
| Blank + 100-$\mu$M NO$_2^-$ | 1.355 |
| Unknown | 0.664 |

The activity of the enzyme is measured in units. By definition, 1 unit of enzyme reduces 1 $\mu$mol of nitrate to nitrite min$^{-1}$ at 30 °C and pH 7.0.

    a.  Plot the working curve. What is the sensitivity of the assay in instrument response versus enzyme units of activity?

    b.  The unknown solution was made from 0.0196 g of dry powder. What is the activity in units per gram dry weight?

    c.  Assume that the highest purity enzyme obtained to date contained 32 units per gram dry weight and that the material was "pure." How pure is this preparation?

That is, what fraction (w/w) of the unknown is "pure" enzyme relative to that purest material?

11.3   RNA polymerase II(B) cleaves DNA and RNA and is involved in RNA transcription processes. One unit of enzymic activity causes incorporation of 10 pmol of uridine-5'-triphosphate (UTP) into a precipitatable product of denatured DNA in 15 min at 25 °C. This incorporation is found by using radioactive UTP, [($^3$H)UTP]. The standard [($^3$H)UTP] produces $3.7 \times 10^4$ nuclear decays per second in 100 pmol of the compound. Thus, the number of radioactive decays counted is directly related to the amount of UTP that has become incorporated into the precipitated DNA up to the end of the reaction time. The progress of an enzyme purification was monitored with this kinetic method. Complete the accompanying table.

|  | Purification Step | | | | |
|---|---|---|---|---|---|
|  | *1* | *2* | *3* | *4* | *5* |
| Total protein (mg) | 191 000 | 6 830 | 1 480 | 92 | 30 |
| Sample taken (mg) | 20 | 0.20 | 0.10 | 0.05 | 0.01 |
| Radioactivity counts/s above background | 4 650 | 1 430 | 2 550 | 18 100 | 9 300 |
| Total enzyme activity (units) | 12 000 | 13 200 | — | — | — |
| Enzyme activity (units/mg-protein) | 0.063 | — | — | — | — |
| *n*-fold purification | 1 | 31 | — | — | — |
| Extraction yield (%) | 100 | 110 | — | — | — |

11.4   [Figure 11.4.1 and method reproduced from Koupparis, M. A., et al., *Analyst* **1982**, *107*, 1309.]
This exercise involves finding the concentration of nitrite in water. The method consists of injecting solutions of reagents and a sample together into a flowing stream in a carefully controlled manner. After injection, the two solutions automatically are mixed rapidly and thoroughly and pass into a detector. The flow is then stopped, and the progress of the reaction of nitrite in the sample is monitored. First, the nitrite reacts with an aromatic amine. The product of this first reaction then couples to a second aromatic compound to form a dye. In this case, the dye is reddish-purple. The resulting data—the time-dependence of the dye development—are shown in Figure 11.4.1. The curve labels are ppm nitrite in the samples for each. The absorbance scale is linear and directly proportional to the dye (and nitrite) concentration.

  a.  Determine the calibration curve for nitrite with the derivative method. There appear to be some irregularities in the curves at early times. How do you overcome this problem?
  b.  Determine the calibration curve for nitrite with the fixed-time method.
  c.  For the fixed-time method, will your results tend to be more or less precise or have the same level of precision if the time chosen is 2 rather than 5 s?
  d.  Determine the calibration curve for nitrite with the variable-time method. Consider carefully where you will fix the concentration line in this case.
  e.  Is the fixed-time method or the variable-time method superior to the other?
  f.  The following data were obtained from an unknown. What concentration does the sample have (in ppm $NO_2^-$)? Calculate the concentration using each of your three calibration plots.

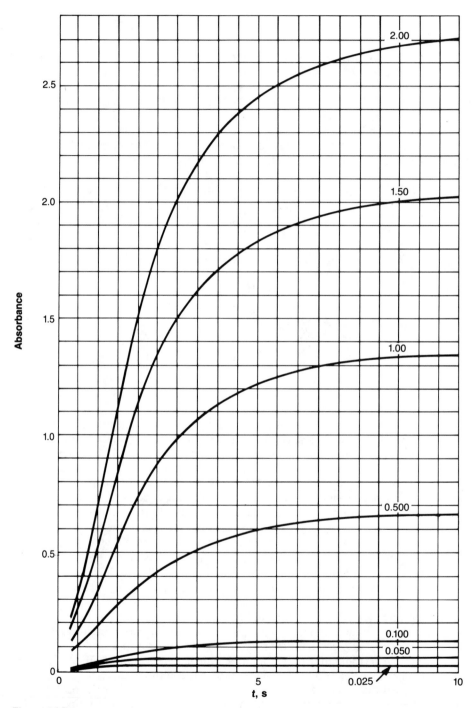

**Figure 11.4.1**

| Time (s) | Absorbance | Time (s) | Absorbance |
|----------|-----------|----------|-----------|
| 0 | 0.0 | 4.0 | 1.36 |
| 0.5 | 0.31 | 5.0 | 1.50 |
| 1.0 | 0.43 | 6.0 | 1.57 |
| 1.5 | 0.66 | 7.0 | 1.60 |
| 2.0 | 0.90 | 8.0 | 1.62 |
| 2.5 | 1.07 | 9.0 | 1.63 |
| 3.0 | 1.20 | 10.0 | 1.64 |

11.5   Another kinetics problem can be found as problem 9.18.

## SUPPLEMENT 11A

# Obtaining Simple Kinetic Behavior

In this section we illustrate how reaction conditions are chosen such that reactions of the type

$$A + R \underset{k_r}{\overset{k_f}{\rightleftharpoons}} P \qquad \qquad \textbf{(11A-1)}$$

can exhibit simpler kinetic behavior. The change with time of the concentrations of the species present can be expressed by three equations. In this case, all three contain the same information: The right-hand sides are the same.

$$\frac{-d[A]_t}{dt} = k_f[A]_t[R]_t - k_r[P]_t \qquad \qquad \textbf{(11A-2)}$$

$$\frac{-d[R]_t}{dt} = k_f[A]_t[R]_t - k_r[P]_t \qquad \qquad \textbf{(11A-3)}$$

$$\frac{d[P]_t}{dt} = k_f[A]_t[R]_t - k_r[P]_t \qquad \qquad \textbf{(11A-4)}$$

Differential equations such as these are mathematical models for the reaction kinetics. With this in mind, we shall not distinguish between the equation and the chemical reaction from now on.

The general kinetic behavior of this reaction is complicated and is not particularly useful for straightforward assays. However, the reactions can usually be run under special conditions in which the kinetic behavior is simpler and more useful. One of the greatest simplifications occurs when the reverse reaction does not occur $P \nrightarrow R + A$. Thus, assay reactions that have slow or effectively zero reverse rates are chosen. If this is not true in actual assays, then corrections for the effect must be made on the assay results.

When no reverse reaction occurs, then $k_r = 0$. Equation 11A-1 becomes

$$A + R \xrightarrow{k_f} P$$

and the behavior of the reaction is described correctly by

$$\frac{-d[A]_t}{dt} = k_f[A]_t[R]_t \qquad \text{(11A-5)}$$

This is Eq. 11A-2 written without the second term. Similarly modified, Eqs. 11A-3 and 11A-4 have the same form.

Even with such simplifications, the behavior of this system over time can still be more complicated than desired for a straightforward analysis, a point that is illustrated in Figure 11A.1. Plotted in Figure 11A.1a are the values of $[A]_t$, $[R]_t$, and $[P]_t$ that would be measured under the specific conditions cited. In Figure 11A.1b are *changes* in concentration over time ($d[A]_t/dt$, and so on). The data of Figure 11A.1 *can* be analyzed to find the analyte concentration changes. But to derive the desired concentration from these data is far more difficult than necessary even when no reverse reaction occurs. As will be shown, further simplifications arise by adjusting the concentrations of the reagents.

### Pseudo First-order Behavior

Note that the stoichiometry of reaction 11A-1 shows that one A and one R combine to form each product molecule. If we add a large excess of R (say a factor of 20) to the solution, then, while all of the A will be used up, only 5% of the R reacts. The value of [R] changes only slightly during the reaction. We may rewrite the equation for the change of P with time as

$$\frac{d[P]_t}{dt} \approx k_f[A]_t[R]_0 \qquad \text{(11A-6a)}$$

or

$$\frac{d[P]_t}{dt} \approx k_f'[A]_t \qquad \text{(11A-6b)}$$

where $k_f' = k_f[R]_0$. These approximations can be made arbitrarily more precise by further increasing [R].

Since $d[A]_t/dt = -d[P]_t/dt$, the form of Eq. 11A-6b is a first-order kinetic reaction like Eq. 11-2. A semilogarithmic plot of [P] versus time will be linear. As noted, this simple behavior results only under the special initial conditions: a large excess of the reagent R. Since the reaction is not truly first-order but only acts as though it were, these conditions are call **pseudo first-order** conditions.

When the large excess of R is present, the concentration of $[A]_0$ can be found from the extrapolation of the data to zero time on a semilogarithmic plot: log[A] or log[P] versus $t$. Any of the other three data-treatment methods shown earlier (derivative, fixed-time, variable time) may also be used.

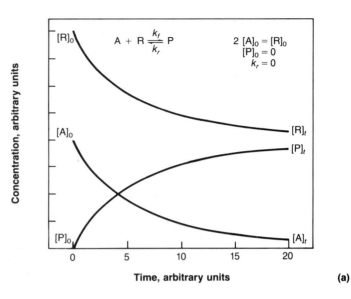

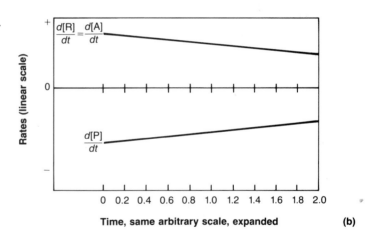

**Figure 11A.1**   (a) Graph of the concentrations of A, R, and P for the reaction

$$A + R \rightarrow P$$

with initial concentrations $2\,[A]_0 = [R]_0$ and $[P]_0 = 0$. The units of both time and concentration are arbitrary. This graph illustrates the behavior of a relatively simple reaction system. Pseudo first-order conditions would require a far larger excess of R over A. (b) Graph of the slopes of [A], [R], and [P] with time (from figure a) versus an expanded time scale. Only at relatively short times will the system act as a pseudo zero-order system. A true zero-order rate system will exhibit fixed values of the derivatives.

## Pseudo Zero-order Behavior

There is another way to eliminate many of the difficulties of complex kinetic behavior. Notice in Figure 11A.1b that the rates of change in the concentrations, $d[A]/dt$, $d[R]/dt$, and $d[P]/dt$ are time-dependent. If we measured a concentration using Eq. 11A-6,

$$[\text{analyte}] \propto \frac{d[P]_t}{dt} \approx k_f'[A]_t \qquad \textbf{(11A-6b)}$$

over the course of the reaction, we would find a changing assay value. However, if we look for only a relatively short time, at the early part of the curve, the derivative plots are nearly constant. In the early time region, the rate alone could be used to determine $[A]_0$. The mathematical description of these conditions is obtained by modifying Eq. 11A-6: substitute $[A]_0$ for $[A]_t$.

$$\frac{d[P]}{dt} \approx k_f[A]_0[R]_0 \tag{11A-7}$$

and

$$\frac{-d[R]}{dt} \approx k_f[A]_0[R]_0 \tag{11A-8}$$

Equations 11A-7 and 11A-8 say that the rate of change of $[P]$ or $[R]$ is directly proportional to the concentration $[A]_0$, the value we want to measure. As a result, the initial slope of the plot of *either* $d[P]/dt$ or $d[R]/dt$ can be used to determine $[A]_0$. Therefore, if we use only the initial times of the reaction, the behavior appears to be simpler. This results in a more straightforward approach to the measurement of $[A]_0$.

Note, peripherally, that the rate is proportional to $[R]_0$, not $[R]_t$. But $[R]_0$ is a constant; it does not change. As a result, by having a large excess of $[R]_0$ and measuring the generation of product for a short time, the rate of the reaction does not appear to depend on *either* $[A]$ or $[R]$. These conditions are called **pseudo zero-order** conditions.

# 12

# Electrochemical Methods

## 12.1 The Variety of Electrochemical Methods

The relationships of the concentrations, charges, and energies in chemical reactions involving charged species have been reviewed in Section 2.9. These relationships apply in two general types of reactions: those in which oxidation and reduction occur and those in which dilution or diffusion of charged (and related uncharged) species occurs. The electrical currents and potentials that result from these kinds of reactions can be determined precisely to ascertain chemical content. In all cases, the techniques are used to measure analytes in a solution.

In this chapter are presented some of the instrumental methods used to measure electrical currents, charge, and/or potentials for analytical uses. As is usual, the most important part of the instrument is the transducer, the part of the instrument that interacts with the sample. In many cases, the details of the chemistry that occurs at the site where the measured species interact with the transducer are not well understood. Yet, the methods can still be exceedingly useful. Given the sophistication of contemporary electronics, most of the problems with the electrochemical methods lie in controlling the chemistries of the samples and the transducers. We shall, therefore, concentrate on this aspect of the measurement.

Since you are already familiar with the theory of electrochemical cells as described by the Nernst equation, experimental methods to which it is applicable are presented first. These are the techniques of **potentiometry,** the measuring of potentials.

Following these, the techniques of measuring the changes in resistance of electrolyte solutions are described. You already have worked with Ohm's law, which relates the current to the potential and resistance. The inverse of resistance, $1/R$, is called **conductance.** This property is measured in **conductimetry** and is related to a concentration.

As was noted in Section 2.9, the faraday constant $\mathscr{F}$ is the proportionality constant that relates equivalents to coulombs. Because a current of 1 coulomb/second is an ampere, we can relate the number of moles of a substance oxidized or reduced to the current and to the time period over which the current flows. This is the basis of a number of different techniques. They are called, reasonably, **coulometric** methods. Contemporary forms of coulometry require an accurate and precise measurement of electrical current and time. Coulometric techniques are presented in the third portion of this chapter.

Finally, a group of techniques combine the measurement of potential, current, and time. The methods use the changes in the current (over time) while the voltage applied to the electrochemical cell is changed in a precisely controlled manner. Because the voltage and current are measured simultaneously, these are called the **voltammetric** methods, of which there are several. They are highly sophisticated in concept and are, perhaps, experimentally the most elegant yet simple methods for trace analysis. A discussion of these techniques concludes the chapter.

## 12.2 Potentiometry

### Reference Half-cells

In the potentiometric methods, we measure the potential of an electrochemical cell in order to find the concentration of some component of a sample solution. The concentration is related to the potential through the Nernst equation or a modification of it. The unknown solution will form a half-cell, that is, one-half of the electrochemical cell. In order to make the measurement, we need to use another half-cell. If we want a precise measurement of the unknown quantity, we need a reproducible and stable half-cell. If we want a measurement that can be compared to those made in other laboratories or on other days, we need a half-cell that possesses an accurately known potential.

**Table 12.1**  Temperature Dependence of a Saturated Calomel
Electrode (SCE)*

| Temperature °C | Potential $V$ vs. NHE = 0 V (experimental) |
|:---:|:---:|
| 10 | 0.2507 |
| 25 | 0.2412 |
| 40 | 0.2308 |
| 70 | 0.2078 |

\* The potentials can be calculated at other temperatures using the following
empirical formula, with $E$ in volts vs. NHE and $T$ in °C.

$$E_{SCE} = 0.2412 - 6.61 \times 10^{-4}(T - 25) - 1.75 \times 10^{-6}(T - 25)^2 - 9.0 \times 10^{-10}(T - 25)^3$$

Probably the two half-cells most commonly used as reference voltages are the
**calomel** and the silver–silver chloride half-cells. Calomel is an old name for mer-
curous chloride, $Hg_2Cl_2$; the redox reaction in the half-cell is

$$Hg_2Cl_2(s) + 2 e^- \rightleftharpoons 2 Hg^0 + 2 Cl^-$$

The reaction is established in a half-cell with the configuration

$$\| KCl(saturated), Hg_2Cl_2(s) \,|\, Hg$$
$$E = +0.241 \text{ V vs. the normal hydrogen electrode (NHE) at 25 °C}$$

This potential varies with temperature, as can be seen in Table 12.1. Recall that the
half-cell potential of the normal hydrogen electrode is defined to be zero volts exactly.
When the aqueous KCl solution is saturated and excess KCl crystals are present, the
electrode is called a **saturated calomel electrode** (SCE). The electrodes can have any
one of literally scores of different shapes without affecting the potential of the half-cell
configuration.

Sometimes, instead of using a saturated solution of KCl, a 1-M solution is used.
The cell is then referred to as a **normal calomel electrode.** The half-cell is

$$\| KCl(1 \text{ M}), Hg_2Cl_2(s) \,|\, Hg \qquad E = +0.280 \text{ V vs. NHE at 25 °C}$$

The silver–silver chloride half-cell utilizes the redox couple

$$K^+ + Cl^- + Ag^0 \rightleftharpoons AgCl(s) + K^+ + e^-$$

The cell can be as simple as a silver wire coated with silver chloride precipitate
immersed in a KCl solution. The cell configuration is

$$\| KCl(saturated), AgCl(s) \,|\, Ag \qquad E = 0.197 \text{ V vs. NHE,}$$
$$-0.045 \text{ V vs. SCE at 25 °C}$$

This may also be constructed with nonaqueous solvents that will dissolve a chloride
salt.

Like the calomel reference, the Ag–AgCl cell potential is strongly temperature
dependent. The Ag–AgCl cell has the advantage of greater stability at higher temperatures but has the drawback that the cell potential is sensitive to light.

## Potentiometric Precision

It is difficult to measure a cell potential more accurately than about $\pm 1$ mV $=$ $\pm 0.001$ V. (Greater precision is possible; greater accuracy is difficult to obtain. See later discussion of **junction potentials**.) Because the potential is related to the concentrations we seek, an error of $\pm 1$ mV in potential leads to an error in analyte concentration. We can ask, What concentration change does 1 mV represent for monovalent, divalent, and trivalent charge transfers ($n = 1, 2$, and 3, respectively)? To answer this question, you need only refer back to Section 2.9. From the Nernst equation, we found that a ten-fold (decade) change in concentration of a monovalent, divalent, and trivalent species was equivalent to a potential change of a concentration cell of 59 mV, 29 mV, and 20 mV, respectively, at 25 °C. The effect of a 1-mV error can be calculated in the following way.

$$\text{Let } E = E^0 - \frac{0.059}{n} \log K_{act} \quad \text{and} \quad E^* = E + \Delta E = E^0 - \frac{0.059}{n} \log K_{act}^*$$

Then, from subtraction of the above equations and the properties of logarithms,

$$-\frac{\Delta E \cdot n}{0.059} = \log \frac{K_{act}^*}{K_{act}}$$

and

$$\frac{10^{-n\Delta E/0.059}}{0.059} = \frac{K_{act}^*}{K_{act}} = K_{rel}$$

The results are that for an error $\Delta E = 1$ mV at 25 °C,

$$K_{rel}(\text{monovalent}) = 0.96$$
$$K_{rel}(\text{divalent}) = 0.92$$
$$K_{rel}(\text{trivalent}) = 0.89$$

Thus, an uncertainty of 1 mV produces an uncertainty[1] of 4%, 8%, and 11%, respectively, in the extent of dilution (or concentration) involving a mono-, di-, or trivalent charge transfer. These percentages represent the approximate limits of accuracy for the potentiometric measurement of concentration.

But what if these limits of accuracy are too wide for potentiometry to be of use in a specific assay? Does this mean that the great benefits of potentiometric analysis must be given up? The answer is no. The electrodes (that is, ion-selective electrodes, Section 12.4) can be used as detectors to ascertain the progress and end points of titrations. An example is the use of a pH electrode in a neutralization titration, and the lead-sensitive electrode mentioned in the literature reading in Section 10.3. If proper precautions are taken, the limits of precision of a potentiometric titration can be up to two orders of magnitude better than the precision of the electrode alone.

---

[1] % uncertainty $= 100\,(1 - K_{rel}) = 100\,(K_{act} - K_{act}^*)/K_{act}$

## How an Inert Electrode Senses the Potential of a Redox Couple

In the description of some half-cells in Section 2.9, you read that the element composing the electrode need not be the same as the ions in solution—for instance, a platinum wire can be used with a $Fe^{+++}/Fe^{++}$ couple. Other materials could be employed in addition to platinum, among them carbon, in the form of graphite or glassy carbon, and gold. Why can these so-called inert electrodes be used to measure the potential of the redox reaction in solution? This question is answered in this section.

In the solution, the electrochemical equilibrium, as any equilibrium, is a continuously active process. For example, in a solution that contains a mixture of $Fe^{+++}$ and $Fe^{++}$, electrons are constantly transferred from the iron(II)—$Fe^{++}$—ions to the iron(III)—$Fe^{+++}$—ions, causing the pairs to exchange oxidation states. The reaction can be written

$$Fe^{++} + {}^*Fe^{+++} \rightleftharpoons Fe^{+++} + {}^*Fe^{++}$$

The asterisk is used here to label a specific iron atom regardless of its ionic form. The equilibrium concentrations of both Fe(II) and Fe(III) ions are maintained, but electron transfers of this type are occurring between them all the time. Interpretation of experimental data has produced a picture of the **electron transfer reaction,** in which two separate processes are occurring.

1.  The two ions must approach sufficiently close to one another. This happens by random collisions. The name given to this motion is **mass transfer.**
2.  During a collision between the components of a redox pair, such as $Fe^{++}/Fe^{+++}$, the electron is transferred from the reduced species to the oxidized species: the **electron transfer** step. It is characterized by a rate constant, just like any other reaction.

In all the examples that we shall consider, the electron transfer reaction is very fast once the two ions are sufficiently close together. As a result, the rate of the oxidation–reduction reaction will be controlled by the rate of mass transfer—by how often the ions collide.

The potential of the redox couple, here $Fe^{+++}/Fe^{++}$, is determined by the ratio of the relative concentrations of the Fe(III) and Fe(II) ions. As you know, the potential's dependence on the concentrations can be described by the Nernst equation. But what happens when the wire is placed into the solution? How does the wire measure the solution potential?

The answer involves finding how the potential of the wire itself is determined. From electrostatic theory, we explain that the potential of a metal is determined by the excess or defect of electron charges. Thus, to change the potential of a wire, some electrons must be added to or removed from it: Some current must flow. Let us assume for now that the wire is not connected to any measuring device. Then, the current to change the wire's potential must result from some electrons being transferred between the ions in the solution and the wire. The molecular events leading to this electron transfer can be depicted as shown in Figure 12.1.

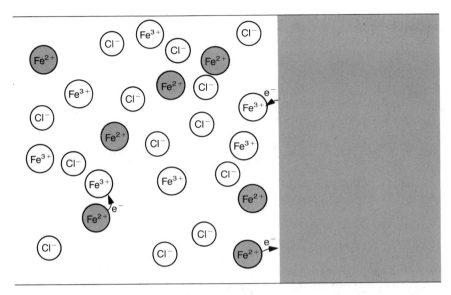

**Figure 12.1**    Diagram of how electrochemists explain the manner in which an inert electrode responds to the electrochemical potential of a redox couple in solution. In this example of a mixed Fe(II)/Fe(III) chloride solution, transfer of electrons from Fe(II) to Fe(III) ions constantly occurs when the ion pairs collide. Similarly, when the electrode and ions are sufficiently close, electron transfer can occur: either transfer from Fe(II) ions to the electrode or from the electrode to Fe(III) ions. Thus, the inert electrode can act as an intermediary in the electron exchange between Fe(III) and Fe(II). The electrode will not vary from the potential of the redox couple, because if it did, an excess of electrons flowing to or from the electrode would cause the potential to return to that of the redox couple.

Let us assume that when the wire is placed into the solution, its potential is more negative than the redox couple in solution: The wire has a sufficiently great excess of electrons. As a result, electrons will be transferred from the wire to any $Fe^{+++}$ ions that are in the *immediate vicinity* of the wire surface. The fast electron transfer, indicated in the figure by the arrow from the wire to an $Fe^{+++}$ ion next to the surface, will not occur unless the wire's surface and the ion are sufficiently close together. As a result of this transfer of an electron, the potential on the wire will become less negative, because it has lost a negative charge to the solution. Such electron transfers will occur until the potential of the wire is the same as that of the solution.

What happens if the wire initially is more positive than the solution potential? Then electrons will be transferred to the wire from any $Fe^{++}$ ions that are sufficiently close. This is indicated by the arrow from an adjacent $Fe^{++}$ to the wire. Such an addition of negative charge to the wire will cause the wire's potential to become more negative. Electrons will flow into the wire until its potential is the same as that of the solution. Thus, at equilibrium, a wire that can exchange electrons with the ions in solution will have the same potential as the solution. If we sample the potential carefully so that very little current flows, the inert electrode's measured potential is equal to that of the solution. With contemporary instruments, this is quite easy to do.

## The Experimental Measurement

As illustrated in Figure 12.2, an electrochemical cell can be compared with an electrical circuit. The circuit's voltage source is equivalent to the potential difference of the two electrochemical half-cells. The circuit's resistor represents the sum of the resistances of the cell's ionic solutions, the salt bridge, and the wires connecting the electrodes to the voltmeter. In most real cells, the resistance of the wires is negligible compared with those of the solution and salt bridge.

Ideally, the measurement of electrochemical potentials should be done without allowing any current to flow. The reason is clarified by using the equivalent electrical circuit. We know from Chapter 3 that for the simple circuit shown in the figure,

$$V_{meas} = V_{cell} - V_{resist}$$

However, from Ohm's law, we can write

$$V_{meas} = V_{cell} - IR_{cell}$$

The $IR_{cell}$ term becomes zero only when there is no current flow ($I = 0$). Only then will the measured potential be the potential of the electrochemical cell.

A question that naturally follows from this is, Isn't there some current flowing into the voltmeter? Otherwise, how could you get a response? It is, in fact, true that a voltmeter draws some current. However, if the current is small, the potential can be measured as accurately as needed. This point is illustrated in Figure 12.3.

From inspection of the circuit, we note that the resistances of the meter and the cell are in series. What will the measured voltage be in this case? The meter measures the potential drop across its resistor, $R_{meter}$. Thus,

$$V_{meas} = IR_{meter}$$

According to Ohm's law, the current that flows through the circuit depends on the resistances of both the cell and the meter. This is described algebraically by

$$V_{cell} = I(R_{cell} + R_{meter})$$

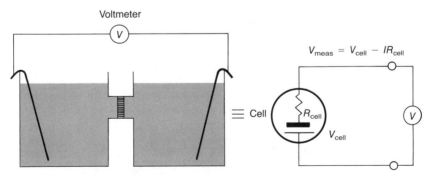

**Figure 12.2**  A diagram of an electrochemical cell shown with an electrical circuit equivalent to the chemical system. The cell acts like a battery in series with a resistance. The resistance arises from the limited ability of the ionic solution to conduct charge.

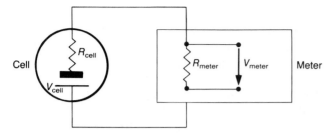

**Figure 12.3**  Circuit schematic for an electrochemical cell connected to a measuring meter. If the resistance of the meter is far larger than the resistance of the cell, the electrochemical potential can be measured accurately and precisely. The *IR* drop of the cell will not cause any error. This result can be shown by a calculation similar to the voltage divider formula. (See the calculation in the text.) Typical values of the resistances are

$$R_{cell} = 10^3 \Omega \quad \text{and} \quad R_{meter} = 10^{12} \Omega$$

The measured potential is expected to be within about 1 part in $10^9$ of $V_{cell}$.

The ratio of the measured potential to the true potential is

$$\frac{V_{meas}}{V_{cell}} = \frac{R_{meter}}{R_{cell} + R_{meter}} \tag{12-1}$$

The behavior of this equation shows that the larger the meter resistance, $R_{meter}$, the closer the measurement will be to the true value of the cell's potential.

---

**EXAMPLE**

Find the error in a typical potentiometric experiment, given that typical values for electrochemical cells and simple, contemporary voltmeters are

$$V_{cell} = 0.5 \text{ V}$$
$$R_{cell} = 10^3 \Omega = 1 \ k\Omega$$
$$R_{meter} = 10^{12} \Omega = 1 \ T\Omega$$

**SOLUTION**

From Eq. 12-1, the ratio of the measured potential to that of the cell is

$$\frac{V_{meas}}{V_{cell}} = \frac{10^{12}}{10^3 + 10^{12}} = 1 - (1 \times 10^{-9})$$

and the absolute error in the measured potential is expected to be

$$V_{meas} - V_{cell} = (1 \times 10^{-9}) \times 0.5 = 5 \times 10^{-10} \text{ V}$$

This is only 0.5 nV, not too much to worry about since the error is less than the errors due to interface potentials (discussed below), which have values above a millivolt or so.

---

## 12.3  Liquid Junction Potentials: An Interface Potential

The interface potentials create difficulties in making accurate measurements of cell potentials. However, they are also the basis of the measurements made with **ion-selective electrodes.** (An example is the ion-selective pH electrode with which we can

**Figure 12.4**  (a) Example of a cell that will have an interface potential between the two half-cells. (b) Schematic diagram illustrating the imaginary boundary (somewhere in the neck between the two reservoirs) across which the rate of transfer of ions is different in each direction. This difference in diffusion rates causes a separation of charge, which is the interface potential. Since the interface potential occurs between two liquid phases, it is called a liquid junction potential. There is a contribution to the interface potential from each ion. Only the part contributed by chloride is shown specifically.

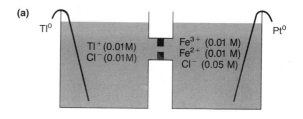

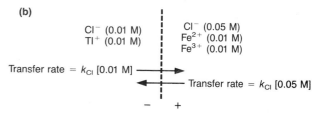

measure the activity of $H^+$ in the presence of any number of other species.) The interface potential that arises between liquids with different compositions is discussed next.

Until now, we have discussed the generation and measurement of electrochemical potentials as if they were entirely a result of redox reactions that take place at electrodes. That is an idealization. Actually, all of the ions and solvent molecules in the solutions and their interactions must be considered. Let us return to the original set of two half-cells that were used as an example in calculating cell potentials in Chapter 2: the $Tl/Tl^+$ and $Fe^{++}/Fe^{+++}$ cells. Figure 12.4a is a diagram that includes all the ions in the two solutions, assuming that chloride salts of the metals are used.

However, there is a significant difference in the concentrations of chloride ion between the two cells, and we know that such a concentration difference is equivalent to an electrochemical potential. But, the voltage due to the chloride concentration difference is not measurable at the electrode, since the chloride is neither oxidized nor reduced. To understand this point, recall that the reason we were able to measure the voltage equivalent to a $Cu^{++}$ concentration was through *two* reactions:

$$Cu^{++}(0.1 \text{ M}) \rightarrow Cu^0 \rightarrow Cu^{++}(1 \text{ M})$$

(See Section 2.9 and Figure 2.3.) But since the chloride ion is not reactive with the electrode, the diffusion potential will not originate *in the same way*.

In fact, the chloride diffusion potential appears at the interface between the two solutions. This potential, called a **liquid junction potential,** arises as illustrated in Figure 12.4b. Think of an imaginary plane separating the two half-cells. This is indicated in the figure by a dashed line. The chloride ions can move back and forth across the position of the plane. We can associate a rate constant, $k_{Cl}$, with the rate of transfer of chloride ions across the plane. The rate at which the chloride ions will pass across the plane to the left or to the right will depend on the the concentrations of chloride ion on each side. In the language of elementary kinetics,

$$\text{transfer rate}_{Cl}(\text{to left}) = k_{Cl}[0.05 \text{ M}]$$

and

$$\text{transfer rate}_{Cl}(\text{to right}) = k_{Cl}[0.01 \text{ M}]$$

There is a significant imbalance in the number of chloride ions that will transfer across the imaginary plane to the left: five times more. Since the ions are negatively charged, and since more of them will move to the left than to the right, an excess of negative charge will build up on the left. This excess of charge causes the left side to be more negative than the right: An electrochemical potential forms. The liquid junction potential gets its name because the potential appears at a junction between two liquid solutions.

You probably noticed that this picture is oversimplified. After all, what happened to the positive ions? In fact, all the positive ions have to be considered, and each species of positive and negative ion has a different rate constant associated with it. These rate constants for transfer are called **ionic mobilities** and are experimentally determined. More will be said of this topic in the section on conductimetry (Section 12.8).

This picture, which we use to explain the origin of liquid junction potentials, does bring up an important, basic point. The actual generation of an electrochemical potential results from a separation of charges on the molecular level. The location of such a separation is at an interface between bulk phases containing mobile, charged species.

These junction potentials can be quite large. In the cell

$$\text{Ag} \,|\, \text{AgCl}, \text{HCl}(0.1 \text{ M}) \,\|\, \text{KCl}(0.1 \text{ M}), \text{AgCl} \,|\, \text{Ag}$$

the entire diffusion potential arises from $H^+$ diffusing to the right and $K^+$ to the left. All the other parts of the cells and solutions are the same. The potential for this has been measured. It is[2]

$$E_{\text{junction}} = 28 \text{ mV}$$

As described above, the formation of a junction potential depends on there being different ionic concentrations and/or ions with different ionic mobilities on opposite sides of an interface. The formation of these potentials can be suppressed. The interface potential will be minimized if we can minimize the differences between the two sides of the interface. This can be done by adding an overwhelming concentration of the same, electrochemically inert, strong electrolyte to both sides. Commonly, this is done by making up all solutions with more than 0.1-M potassium sulfate or other strong electrolyte in them. In such a solution, the potassium sulfate is called the

---

[2] Ref: MacInnes, D. A. "The Principles of Electrochemistry"; Dover: New York, 1961; Chap. 13. This result is due entirely to the different rates of diffusion—the different ionic mobilities—of protons and potassium ions.

**background electrolyte.** Without a background electrolyte, most electrochemical quantitative analysis cannot be done accurately or precisely.

## 12.4 Ion-selective Interfaces and Ion-selective Electrodes

In 1875, Lord Kelvin, while studying electrolyte conduction in solids, suggested that glass could conduct electricity as do electrolyte solutions by means of ion movement. This is called **electrolytic conduction.** Around 1905, M. Cremer discovered a "glass electrode potential." This was an electrochemical potential that formed when a thin glass membrane was placed between two aqueous solutions of different pH. The quantitative dependence on pH was measured in 1909 by the German chemists Fritz Haber and Z. Klemensiewitz. They used a bulb of soft glass containing an acid with a platinum wire immersed in the acid solution. When the bulb was placed into an external solution, they found that the measured potential for their electrochemical cell depended directly on the pH difference between the two solutions. The development of the pH electrodes and pH meters has continued since.

This brief history presents the two main points you should understand about ion-selective electrodes. The first is that the origin of the electrode potential is at the surfaces of a membrane separating two electrolyte solutions. The second is that electrical conduction through the membrane is necessary. If there were no conduction, there would be an open circuit, and no measurement would be possible. A peripheral point, only implied in this brief account, is that until recently the production of ion-selective electrodes was done by trial and error.

As mentioned above, the membrane for the pH electrode is composed of soft glass. For the measurement of other ions, membranes are constructed from thin liquid layers or crystals of various kinds or from chemically modified polymers (special plastics). These materials will be described briefly after presenting the general principles of ion-selective electrodes as well as some of the difficulties associated with them. A list of various selective electrodes is presented in Table 12.2. The majority of electrodes that are currently available commercially are specific for ionic species. The nonionic species for which electrodes are used are gases. However, the electrodes operate as ion-selective electrodes (ISE) after the gas that is separated from the matrix by passing through a selective membrane then reacts to form ionic species. Thus, as presented below, a similar mechanism of operation applies to all the electrodes listed in the table.

In the description of the origin of liquid junction potentials, an assertion was made without explanation that all electrochemical potentials arise at an interface (not in the bulk) and that they are due to a separation of charge at the molecular level. The details of the proof of these statements are well beyond the level of this text.[3] To

---

[3] A good place to start looking further is in Koryta's book *Ion Selective Electrodes,* which is cited in the references.

**Table 12.2**   Ion-Selective Electrodes and Their Uses

| Ion for Which the Electrode is Selective | Useful Concentration Range (M and ppm) | Uses |
| --- | --- | --- |
| Ammonia, ammonium $NH_3$, $NH_4^+$ | $10^{-6}$–0.1 M<br>0.02–2000 ppm | The electrode measures dissolved ammonia. Other substances can be measured by chemical pretreatment including ammonium nitrate and organic nitrogen. Ammonia is commonly measured in foods, effluents, soil, and biological systems. |
| Bromide $Br^-$ | $10^{-6}$–0.1 M<br>0.08–8000 ppm | Water pollution analyses, photographic processing |
| Cadmium $Cd^{++}$ | $10^{-7}$–0.1 M<br>0.01–11,000 ppm | $Cd^{++}$ activity can be measured directly or through EDTA titrations. Used in plating bath analysis and as end-point detector in titration of sulfide in wood-pulping liquors. |
| Calcium $Ca^{++}$ | $10^{-5}$–1 M<br>0.4–40,000 ppm | Direct measurement of $Ca^{++}$ in biological fluids and total calcium by EDTA titrations. Also soil, food, and water. |
| Carbon dioxide $CO_2$ | $5 \times 10^{-5}$–0.01 M<br>2–400 ppm | Dissolved $CO_2$ and carbonate can be measured directly. Samples analyzed include serum, wine, water, and soft drinks. |
| Chloride $Cl^-$ | $3 \times 10^{-5}$–0.1 M<br>1.0–3500 ppm | Chloride activity is critical in processing of foods, resins, and insecticides. Also used for biological material and in medical research of cystic fibrosis. |
| Copper $Cu^{++}$ | $10^{-6}$–0.1 M<br>0.06–6000 ppm | Plating and etching baths, soil, sewage, pharmaceuticals, food, and beverage products |
| Cyanide $CN^-$ | $10^{-6}$–$10^{-3}$ M<br>0.03–250 ppm | Analysis of plating baths, mineral extractions, and waste water |
| Fluoride $F^-$ | $10^{-6}$–0.1 M<br>0.02–2000 ppm | Widely used in analysis of fluoridated water supplies, toothpaste, bone, plating baths, water discharge, and fluoride metabolism |
| Fluoroborate $BF_4^-$ | $3 \times 10^{-6}$–0.1 M<br>0.3–9000 ppm | Quantitating fluoroboric acid in plating baths and analysis of boron in soil, plant tissue, and glass upon conversion to fluoroborate |
| Hydrogen $H^+$ | 0–14 pH | Measures $a_{H^+}$ of solutions |

*(continued)*

**Table 12.2** (continued)

| Ion for Which the Electrode is Selective | Useful Concentration Range (M and ppm) | Uses |
|---|---|---|
| Iodide I⁻ | $10^{-7}$–0.1 0.01–10,000 ppm | Commonly used in analysis of salt brine, industrial waste and in pharmaceutical research |
| Lead $Pb^{++}$ | $10^{-6}$–0.1 M 0.2–20,000 ppm | Major uses in analyses of electro-plating baths and to sense lead ti-trant in the indirect determina-tion of sulfate |
| Nitrate $NO_3^-$ | $6 \times 10^{-6}$–1 M 0.4–60,000 ppm | Primarily in analyses of soil and fertilizer, food, water treatment biologicals, and in pharmaceuti-cal and photographic manufac-turing |
| Nitrogen oxide $NO_x$ | $4 \times 10^{-6}$– $5 \times 10^{-3}$ M 0.2–200 ppm | Gas-sensing electrode used pri-marily for determining nitrite, $NO_2^-$, in foods, ground water, plating baths and atmospheric $NO_x$ |
| Perchlorate $ClO_4^-$ | $3 \times 10^{-6}$–0.1 M 0.3–10,000 ppm | Primary use has been in explo-sives and solid-propellant in-dustry |
| Potassium $K^+$ | $10^{-5}$–1 M 0.4–40,000 ppm | Analyses of soil, fertilizers, bio-logical fluids, and wines |
| Silver/sulfide $Ag^+$, $S^=$ | $10^{-7}$–0.1 M $Ag^+$ | Applications in the pulp and paper industry and in silver re-covery from photographic solu-tions |
| Sodium $Na^+$ | $10^{-5}$–0.1 M 0.2–2000 ppm | Sodium analyses of dietetic foods, high-purity water sources, tissues, glass, pulping liquor, and biologicals |
| Thiocyanate $SCN^-$ | $5 \times 10^{-6}$–1 M 0.3–60,000 ppm | Most commonly used in the plat-ing industry |

*Source:* Primary reference: "The Beckman Handbook of Applied Electrochemistry"; Beckman Instruments, Inc.: Fullerton, CA, 1980.

proceed, let us take these ideas as given and see how they can be used to explain the behavior of ion-selective electrodes.

The model that is described below is oversimplified. The reasons that such a simple model is presented are the following:

A.  The analytical determinations using ion-selective electrodes are carried out under reproducible conditions of ionic strength, temperature, and sample pretreatment. Thus, knowledge of the exact nature of the error, if conditions change, is not necessary in order to use ISEs in analysis.

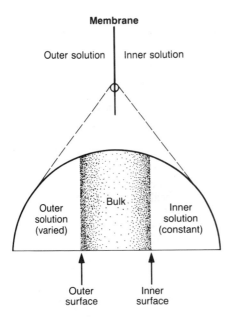

**Membrane**

Outer solution | Inner solution

**Figure 12.5**    Illustration of how the structure of a membrane is conceptualized as consisting of separate regions having surface and bulk properties. The experimentally measured properties are attributable to all the regions—the surfaces, the bulk, and the solutions. However, the details of the explanation are still debated.

B.    The interferences that result from these factors can be approached empirically by the use of appropriate standards.
C.    Not all the points of detail are generally agreed upon.

Consider a thin membrane separating two liquid volumes (Figure 12.5). For simplicity, we can think of the membrane as having two surfaces and a region of bulk material between them. In many cases, these surfaces are not well understood in their chemical details. Accordingly, in the figure they are shaded gray without specific features. The two surfaces are in contact with the solutions on either side of the membrane.

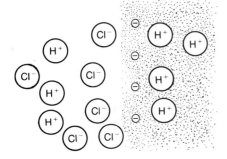

Solution          Surface          Membrane

**Figure 12.6**    Illustration of how an ion-selective surface of a proton ion-selective electrode (pH electrode) is conceptualized. The protons are able to enter the surface region, but the chlorides are mostly excluded. The excess positive ions in the surface region and the excess negative ions in the solution region produce a voltage at the interface. At least part of the selectivity can be attributed to negative charges fixed at the surface. They produce a barrier to the passage of negative charges.

Now, consider that the membrane surface is special; it will allow only certain specific ions to enter (or permeate) into this surface layer and will exclude any others that may be present. More properly, the permeation occurs across the interface between the membrane and the solution (Figure 12.6). When a membrane exhibits this type of behavior, we say that the membrane is **perm-selective.**

As noted above, this diagram is certainly overly simple, given the complexities of ionic charge, activity coefficients, and the different diffusion rates of each species of ion. All of these factors affect the potential at the surface.

The solution in contact with the surface is hydrochloric acid in water. The protons, with their positive charges, penetrate into the glass leaving the chloride with its negative charges excluded outside. This creates a voltage across the interface.

How does this voltage change with changes in the concentration of protons in the solution? In the presence of a higher proton concentration, a greater number of protons will diffuse into the interface. This is similar to the effects that cause the liquid junction potential between two solutions. In this case, however, the interface consists of one solution and one solid phase. The higher the concentration of protons diffused into the solid, the greater the excess of anions left outside. As a result, a larger potential develops across the interface. This potential is described by the Nernst equation.

$$E = E^\circ - \frac{RT}{\mathscr{F}} \ln a_{\mathrm{H}^+} = E^\circ + 0.059\, \mathrm{pH} \qquad \text{at 25 °C}$$

Thus, the potential that develops across the glass membrane can be used to determine the pH of the solution.

There are two more points to notice in this figure. First, there are two surfaces. There are potentials across the inner surface and the outer surface. The potential that is measured from one side of the membrane to the other is the sum of these, as illustrated in Figure 12.7. Because of this structure, we must keep a standard solution in contact with the inside surface of the electrode. The membrane is then exactly analogous to the two half-cells in a concentration electrochemical cell.

Second, the circuit of the electrochemical cell must be complete. Two connections must be made: one of them through the membrane and the other from the inside of the electrode back to the solution. In the case of the glass pH electrode, the connection through the membrane is made by ions such as sodium and lithium moving through the solid glass. However, they are not highly mobile, so to get a good connection—a lower resistance—the glass membrane must be as thin as possible. Almost every practicing analytical chemist has been reminded of this after inadvertently breaking a glass pH electrode.

The connection needed to complete the circuit from the electrode through the electrolyte solution is made through a salt bridge from the reference electrode. (A number of commonly used salt bridges are described in Supplement 12.A.) An illustration of a glass pH and reference electrode pair is shown in Figure 12.8.

If the two surfaces of the membrane act together like a concentration cell, you

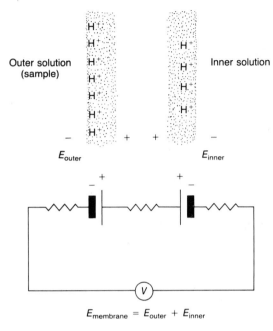

Surfaces not to scale (expanded)

Outer solution (sample)

Inner solution

$E_{outer}$     $E_{inner}$

$E_{membrane} = E_{outer} + E_{inner}$

**Figure 12.7**    Diagram illustrating the consensus of the operation and structure of a $H^+$ ion-selective electrode. The potential measured across the membrane is the sum of two opposing potentials on either surface. Also shown is the schematic of an electrical circuit equivalent to the electrode in solution. The resistance between the surfaces is much larger than the resistances of the ionic solutions.

might reasonably ask, Why do you need a reference electrode? The answer is that the reference electrode eliminates problems caused by changes in junction potentials when making measurements on solutions with different compositions. With the junction potentials throughout the system at *steady* values and the internal pH of the electrode interior fixed, the response of the glass electrode to protons becomes simply

$$E_{membrane} = \text{constant} + \frac{RT}{\mathscr{F}} \ln a_{H^+}(\text{solution}) \qquad (12\text{-}2)$$

The constant results from the junction potentials. The value of the constant is unimportant for practical use. The electrode potential that corresponds to a specific pH is found by using standard buffers. Then, changes in pH follow the regular Nernstian form of the second term of the right-hand side. Notice that the response of the electrode will be temperature dependent.

### Range of Response of Ion-selective Electrodes

Ion-selective electrodes, like all transducers, have a limited range of response. These ranges of ion concentrations are listed for some representative electrodes in Table 12.2. Note that although the electrodes respond to ion activity levels, the limits are stated in molar concentrations. For most ISE, the concentration range extends down

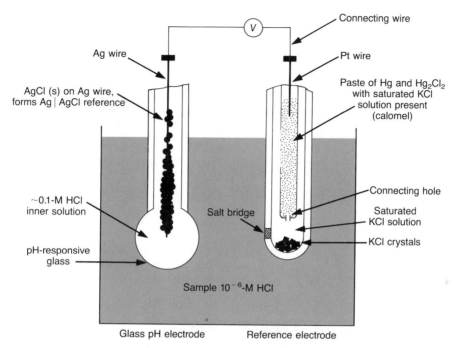

**Figure 12.8**  Illustration of the most important parts of a pH electrode and a calomel reference electrode, as used to determine [H$^+$]. The KCl salt bridge minimizes differences in the interface potential between the reference electrode and different test solutions. The design provides two reference potentials in series with the pH-sensitive glass membrane.

to about $10^{-6}$ M. However, in practice, the usable range also depends on the other components in the solutions. As an example, see Figure 12.9 in which the response of an ISE to lithium is shown at different values of sodium activity. From a study of the figure, you should see that the linear range for lithium decreases as the sodium concentration is raised. At the extreme, the potential being measured does not respond to changes of lithium concentration, and the electrode becomes useless for analytical purposes. This lower, unresponsive, range occurs at higher concentrations of lithium as the sodium concentration is raised. The effects of interference, such as sodium on a lithium ISE, can be treated quantitatively in a manner described next.

### Interferences in Ion-selective Electrodes

Look back at Figure 12.6, which shows an illustration of how the charge separation of chloride (and other anions) from hydrogen occurs across the interface of the glass electrode. The electrode is selective for hydrogen because of the special chemistry of the protons with the glass surface. At least part of this special interaction is due to the negative charges at the surface, which attract the adjacent hydrogen cations and repel the adjacent chloride anions.

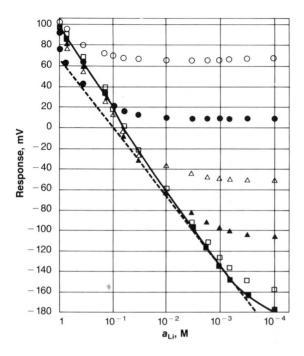

**Figure 12.9**   Plot of the lithium response of an ISE as it varies with the concentration of sodium also in the solution. The plot shows the measured potential in mV versus activity of lithium. The dashed line is the Nernstian response: 59 mV per decade at 25 °C. The electrode was a Beckman sodium ion electrode (Serial No. 39278). The temperature was 25 °C, and the pH was buffered by keeping the solution saturated with $Ca(OH)_2$. Sodium concentrations are $\bigcirc$, 1 M; $\bullet$, 0.1 M; $\triangle$, 0.01 M; $\blacktriangle$, 0.001 M; $\square$, 0.0001 M; $\blacksquare$, pure LiCl response. Note that the electrode could be used for either sodium or lithium ion–selective detection. [Ref: Reprinted with permission from Buck, R. P. *Anal. Chem.* **1974**, *46*, 261. Copyright 1974 American Chemical Society.]

Now consider what will happen at the surface of the glass membrane in sodium hydroxide solution with pH = 13. The activities of the ions in the solution are[4]

$$a_{H^+} = 10^{-13} \quad \text{and} \quad a_{Na^+} \approx 0.1$$

The sodium is about $10^{12}$ times more concentrated than the protons. Such an overwhelming excess of sodium ions might be expected to interfere to some degree with the protons: Both are cations. Some of the sodium ions pass across the interface and contribute to the potential measured by the pH electrode.

Interference such as by $Na^+$ on the pH measurement is expressed quantitatively using the **selectivity coefficient.** The selectivity coefficient is usually written as $k_{i,j}$ or $K_{i,j}$. The equations in which it appears, Eqs. 12-3 and 12-4, will be used without proof of their validity.

For the case of sodium interfering with protons, the potential determined experimentally with the electrode is expressed algebraically,

$$E_{ISE} = E_0 + \frac{RT}{\mathscr{F}} \ln(a_{H^+} + k_{H,Na} a_{Na}) \tag{12-3}$$

The value of $E_0$ (note, *not* $E°$) is a potential that consists of the reference electrode potential and the junction potentials that are present. We do not need to know $E_0$,

---

[4] Since the standard state in solutions of ions is 1 M, the activities are approximately equal to molarities.

but it must have a constant value. Equation 12-3 says that the sodium activity will contribute to the measured potential. The $Na^+$ contribution to the potential is proportional to the selectivity coefficient: A smaller selectivity coefficient indicates less interference. As an example, values for $k_{H,Na}$ can be as low as $10^{-15}$ but vary depending on the type of glass composing the electrode. If $k_{H,Na} = 10^{-15}$, then Eq. 12-3 indicates that the $Na^+$ contribution will equal that from $H^+$ when $a_{Na^+} = 10^{15}a_{H^+}$.

The equation including the selectivity coefficient is slightly more complicated if the ionic charge of the analyte $z_i$ and the interferences $z_j$ are not the same. Then the equation for each interference is,

$$E_{ISE} = E_0 + \frac{RT}{\mathscr{F}z_i} \ln(a_i + k_{i,j}a_j^{z_i/z_j}) \tag{12-4}$$

This equation reduces to the simpler form of Eq. 12-3 when $z_i = z_j$.

### How Selectivity Coefficients Are Determined

Equations such as 12-4 and tables of measured selectivity coefficients comprise a technique for bookkeeping: for storing the information derived from an experiment. Selectivity coefficients will be dependent on the temperature and the specific electrode. Here, the word *specific* alludes to the type of electrode and the manufacturer and even the production batch if production of the surface material is difficult to control. From such selectivity coefficients we can reconstruct the changes in potential due to different interferents. However, the calculation will match the measured behavior only if the specifics are the same.

In Eqs. 12-3 and 12-4, the value of $E_0$ includes a number of unknown junction potentials. An experiment to determine the selectivity coefficients must be organized so that the junction potentials are as constant as possible. Then, the selectivity coefficients can be determined by measuring the cell potential with a number of different concentrations of each interferent present. First, the potential is measured without the interferent. The potential can be expressed as

$$E_{ISE}^{(1)} - E_0 = \frac{0.059}{z_i} \log a_i \qquad \text{at 25 °C} \tag{12-5a}$$

The potential is then measured again after some interferent is added. The potential will be characterizable by

$$E_{ISE}^{(2)} - E_0 = \frac{0.059}{z_i} \log(a_i + k_{i,j}a_j^{z_i/z_j}) \qquad \text{at 25 °C} \tag{12-5b}$$

---

**EXAMPLE**

Given the data in Table 12.3, determine $k_{Ca,Mg}$ for a $Ca^{++}$ ISE. The experiment was carried out by measuring the potential of the electrode versus a standard half-cell and then adding aliquots of concentrated magnesium chloride to the solution to change the magnesium concentration while leaving the volume nearly constant. The solution was stirred, and the potential at the new equilibrium measured. The solution used was at 25 °C and contained 0.25-M KCl as background electrolyte.

**Table 12.3**  Data to Determine $k_{Ca,Mg}$ for a $Ca^{++}$ ISE

| Concentrations $Ca^{++}$, $Mg^{++}$ (mMolar) | Potential vs. Reference (millivolts) |
|---|---|
| 0.10,   0 | −353.0 |
| 0.10,   10 | −341.8 |
| 0.10,   50 | −326.4 |
| 0.10, 100 | −318.3 |

## SOLUTION

To carry out the calculation, we first find the difference between expressions 12-5a and 12-5b. For clarity, the *subscript ISE will be dropped.*

$$E^{(2)} - E^{(1)} = 0.0295 \log\left[\frac{a_{Ca}^{(2)}}{a_{Ca}^{(1)}} + k_{Ca,Mg}\left(\frac{a_{Mg}}{a_{Ca}}\right)\right]$$

The ISE measurement is sensitive to the activities of the ions. However, since in day-to-day use concentrations will be measured, we do the calculations with the ion concentrations. As long as the solution is constantly kept 0.25 M in KCl, the *ratios* of ion activities that are measured will be close to the *ratios* of the respective ion concentrations. Thus, for the experiment here,

$$\Delta E = E^{(2)} - E^{(1)} = 0.0295 \log\left\{1 + k_{Ca,Mg} \cdot \frac{[Mg^{++}]}{[Ca^{++}]}\right\}$$

Rearranging and taking the antilog of each side, we obtain

$$10^{(\Delta E/0.0295)} = 1 + k_{Ca,Mg} \cdot \frac{[Mg^{++}]}{[Ca^{++}]}$$

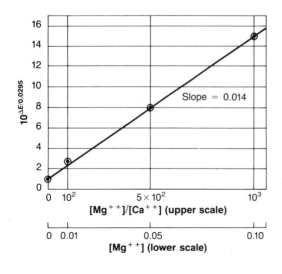

**Figure 12.10**    A graph to find the selectivity coefficient $k_{Ca,Mg}$. This is a plot of

$$10^{\Delta E/0.0295} \text{ versus } [Mg^{++}]/[Ca^{++}]$$

The data are in Table 12.3. The slope, equal to 0.014, is the selectivity coefficient.

Using the data in the table, we plot this function in Figure 12.10 for each of the four data sets. The slope of the graph is $k_{Ca,Mg}$ and equals 0.014 in 0.25-M KCl at 25 °C.

You might wonder about our determining the selectivity coefficient in 0.25-M KCl. After all, might not the $K^+$ ion interfere too? Yes, it might; but, from appropriate tables, the interference by potassium is found to be small. More importantly, however, is that any $K^+$ interference will be constant with a constant $K^+$ concentration. Notice that the answer to the question was that the selectivity coefficient is 0.014 *in 0.25-M KCl.* If more than one interfering ion is present, a sum of correction terms is needed.[5] However, this subject will not be pursued further here.

### Types of Ion-selective Electrodes

You have read that ion-selective electrodes operate as transducers of ion activity because the potentials across their surfaces change with changes in the activities of ions adjacent to a surface. The electrode's selectivity arises from the interface's selectivity in permitting only a specific ion or ions to enter into the membrane itself. Separation of charges causes the potential. As noted earlier, another property of these transducers is that the interior part of a membrane must be able to conduct charge, so that there is a complete electrical circuit.

A number of different kinds of membrane materials satisfy these criteria. Often, ion-selective electrodes are rather arbitrarily divided into categories depending on the composition of their membranes. Besides the glass membranes, there are solid-state membranes; stabilized liquid membranes; and solid, chemically modified polymer membranes. However, all of them work on the same general principle of selective separation of charge at the interface between the membrane and the analyte solution.

## 12.5 Electrochemical Methods Using Current Flow

The response of ion-selective electrodes depends only on the motions of charged species and not on the occurrence of oxidation–reduction reactions. The remainder of the electrochemical methods discussed in this chapter depend on oxidation and/or reduction of analytes.

Quantitative analyses using oxidation–reduction chemistry can be done in a number of different ways. The most direct of them, perhaps, is simply to reduce an ionic metal species out of solution by producing its metallic form; the metal deposits on the electrode where the reduction occurs. It is then straightforward to dry the metal-coated electrode and weigh (by difference) the metal that is plated out. This is the method of **electrogravimetry.** It can be highly precise (better than $\pm0.02\%$) and accurate if done under the optimum conditions.

---

[5] Entry to the literature: Buck, R. P. *Anal. Chem.* **1978,** *50,* April 17R–29R.

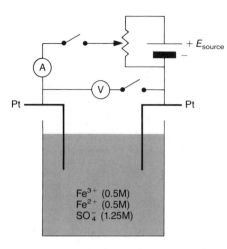

**Figure 12.11**    Diagram of the experimental setup and circuit diagram used to illustrate the response of an electrochemical cell with $E_{cell} = 0$ when it has an external potential applied through nominally inert electrodes.

V is a voltmeter

A is an ammeter

Alternatively, as noted at the beginning of this chapter, we could measure for a relatively short time (seconds) the electric current that oxidizes or reduces the species of interest: the *amperometric methods*. A third possibility is to run the oxidation or reduction for a time long enough to oxidize or reduce *all* the analyte and measure the total charge: the *coulometric methods*. All three of these methods require a moderate ($\mu$A and greater) current during at least some part of the analysis. This is in contrast to the very small currents (less than picoamps) allowed to flow when measuring potentials in potentiometry.

It is found experimentally that when a $\mu$A or greater current flows through an analytical electrochemical cell, the current is not related to the potential by Ohm's law, $I = V/R$. This is non-Ohmic behavior, and the explanation for it will be described next.

## Behavior of an Electrochemical Cell with an Applied Potential

Let us consider the apparatus illustrated in Figure 12.11. It has two electrodes which are initially at the same potential; if the voltmeter is connected by closing the appropriate switch, the measured potential will be zero. After making this reading, the voltmeter is disconnected.

Any voltage between zero and $E_{source}$ can be applied to the electrodes by using the battery and the voltage-dividing circuit. (The circuit and its effects were described in Section 3.4.) In an experiment, the external voltage is applied to the cell for a few seconds by connecting the battery circuit. Before disconnecting the battery, the ammeter is read to find the steady current flowing through the whole circuit including

**Figure 12.12** (a) Graph of the current that flows (vertical axis) when an external potential (horizontal axis) is applied to the cell illustrated in Figure 12.11. Initially the current rises linearly with potential—the cell exhibits ohmic behavior. Then, the current decreases relative to that expected if the ohmic behavior were to continue. Finally, the current levels off at the limiting current value. The limiting current depends on the identity of the electrolyzed species and its concentration. The leveling off is due to concentration polarization which is characterized by the potential $E_{cp}$. (b) The equivalent circuit for the cell. In the ohmic part of the curve, $E_{cell}$ = constant for the cell illustrated in Figure 12.11. The departure from ohmic behavior is equivalent to increasing $E_{cell}$. (Do not confuse this equivalent circuit with the control circuit of Figure 12.11.)

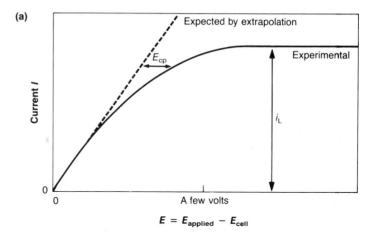

(a)

$$E = E_{applied} - E_{cell}$$

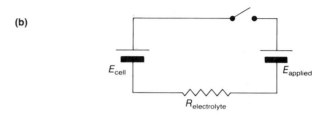

(b)

the cell. This last step is repeated at a number of known voltages, and the results are plotted as a graph of current versus applied potential.

A representative graph of the current that flows at different applied potentials is shown in Figure 12.12a. Let us look at the behavior of the current that is plotted there. No current flows when the applied potential is zero. The current then exhibits a region in which its increase is a linear function of the applied potential. At the end of this range there is a departure from linearity; less current flows than we would have expected had the linear behavior continued. This departure becomes quite large, and finally a plateau is reached where there is no further increase in current as the potential is increased. The value of the current at this highest level is called the **limiting current** and is written in equations as $i_L$.

We have been quite vague about the numerical values of the current and voltage in Figure 12.12a, because the values can depend on the surface areas and surface qualities of the electrodes, both of which may be highly variable. The surface area may not be simply the area you would measure with a ruler. It is determined by the interface structures and compositions at the atomic level, both those of the electrode surface and the adjacent solution. Let us now seek to explain the behavior illustrated in the figure.

### The Linear Region

When the difference between the applied potential $E_{app}$ and the electrochemical cell potential $E_{cell}$ is zero, no current flows. We can write this fact algebraically as[6]

$$E_{app} - E_{cell} = 0; \quad \text{no current flow}$$

Given that

$$E_{cell} = E_1 - E_2$$

we can also write the applied potential as

$$E_{app} = E_1 - E_2; \quad \text{no current flow} \tag{12-6}$$

However, once the current begins to flow, this simple equation, 12-6, no longer describes the relationship correctly. The reason is that part of the circuit through which the current flows when the switch is closed is the electrolyte solution. And, like all electrolyte solutions, this one has a finite resistance. As a simple number to remember, a 1-M KCl solution has a resistance of about 1 ohm if the areas of the electrodes are 1 $cm^2$, and the electrodes are about 1 cm apart. All the numbers are ones, and the electrodes form opposite sides of a cube 1 cm on each edge.

One way to analyze the behavior is to draw an electrical circuit that will behave the same way as the circuit with the electrochemical cell in it. This is called an **equivalent circuit**. The circuit equivalent to the apparatus shown in Figure 12.11 appears in Figure 12.12b. When $E_{app}$ differs from $E_{cell}$, then a current will flow in the circuit. There will be a voltage drop across the resistor which mimics the voltage drop across the electrolyte solution in the cell. The current that flows will be described by Ohm's law, $I = V/R$. For the cell with electrolyte,

$$I = \frac{E_{app} - E_{cell}}{R_{electrolyte}}$$

assuming the resistance of the apparatus is negligible. If this relation of current to potential is put into the form of Eq. 12-6, then

$$E_{app} = E_1 - E_2 + IR_{electrolyte} \tag{12-7}$$

This equation describes the linear portion of the curve. It is a region of Ohmic behavior.

### The Nonlinear Region

As the voltage applied to the cell is increased further, the current does not increase as Ohm's law suggests. To understand how this behavior is explained, let us return for a time to Figure 12.1 and Section 12.2, describing the reason a platinum wire can be

---

[6] In the cell of Figure 12.11,

$$E_{cell} = E_1 - E_2 = 0$$

since the electrochemical potentials of the two electrodes are the same: $E_1 = E_2$.

used to measure the potential of a redox couple in solution. Recall that there are two different kinetic processes occurring. These are

1. the transport of material to the surface of the wire, and
2. the transfer of an electron to or from the wire to the ions.

The cause of the decrease in the current from the expected value for the $Fe^{+++}/Fe^{++}$ couple is generally agreed to be limited by the rate of the first process—the rate of transfer of ions to the surface of the electrode. The following reasoning is given.

A.   The electrons from the wire will be transferred into the solution by reducing the $Fe^{+++}$ and will be removed from the solution by oxidation of $Fe^{++}$. The conduction in the solution will, however, involve all the ions in the solution.
B.   If the ions cannot reach the electrodes, they cannot be reduced or oxidized.
C.   As the applied potential increases, the transport of ions to the wire electrode surfaces cannot keep enough Fe ions arriving at the surface fast enough to take up all the charges. This slow transport will begin to limit the current that can be pushed through the solution.
D.   When the absolute limit of ion transport is reached, a further increase of applied potential will not increase the current flow. A limiting current is reached.

Let us construct a picture at the molecular level that can explain this behavior. We will look at the cathode, where reduction of $Fe^{+++}$ occurs.

The potential *at the surface* of the electrode defines the relative amount of $Fe^{+++}$ and $Fe^{++}$ ions that are incident at the interface of the electrode and the solution. Therefore, before any external potential is applied, the rate of electron exchange between the electrode and the iron ions is fast enough that the potential of the redox couple at the interface matches the electrode potential.

Another way to look at this process is that at the electrode surface, the relative concentrations of $Fe^{++}$ and $Fe^{+++}$ will be described by the Nernst equation for the redox couple at the potential of the electrode. At the surface of the electrode,

$$E_{electrode} = E^{\circ} + \frac{RT}{n\mathscr{F}} \ln \frac{[Ox]_{surface}}{[Red]_{surface}} \qquad (12\text{-}8)$$

When no current flows, the concentration ratio at the interface equals that of the bulk solution.

If an applied potential is put across the cell, the voltage dependence of the current is described by Eq. 12-7. At the cathode, electrons flow from the electrode to the $Fe^{+++}$ ions. The $Fe^{+++}$ ions there are converted to $Fe^{++}$ ions, which diffuse away from the electrode surface into the solution, which has fewer $Fe^{++}$ ions and more $Fe^{+++}$ ions than are *now* near the cathode. This is illustrated in Figures 12.13a and b. One figure is an illustration of the molecular distribution, and the other is a graph of the concentrations, which is the way such information is usually presented.

At the electrode surface, the relative concentrations of $Fe^{++}$ and $Fe^{+++}$ will be described by the Nernst equation at the electrode potential. Meanwhile, unlike the situation when no current is flowing, the potential away from the electrode (the bulk

**(a)**     $[Fe^{2+}] = [Fe^{3+}]$

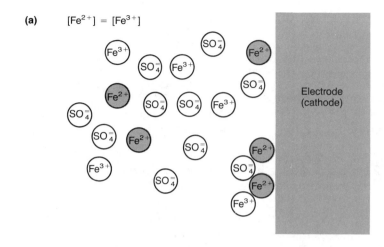

Electrode (cathode)

**(b)**

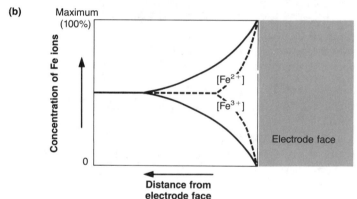

Maximum (100%)

Concentration of Fe ions

$[Fe^{2+}]$

$[Fe^{3+}]$

Electrode face

0

**Distance from electrode face**

**Figure 12.13**   (a) Illustration of the conceptualization of the events occurring at a cathode/solution interface. The solution is a mixture equally of Fe(III) and Fe(II) sulfate. As the $Fe^{+++}$ ions (open circles) arrive in proximity to the electrode surface, an electron may be transferred to form an $Fe^{++}$ ion (shaded circles). The ratio $[Fe^{+++}]/[Fe^{++}]$ at the surface of the electrode is described by the Nernst equation with $E$ being the electrode potential. (b) The $Fe^{++}$ ions formed at the interface are in excess at the electrode in comparison with the bulk. There is, thus, a net diffusion of $Fe^{++}$ away from the interface. However, since the $Fe^{++}$ ions were formed from $Fe^{+++}$, there is a dearth of $Fe^{+++}$ at the electrode surface. Thus, there is a net diffusion of $Fe^{+++}$ toward the electrode. After a time, the concentration distribution in space of $Fe^{+++}$ and $Fe^{++}$ occurs as is plotted in the graph here. The *bulk solution* in this case begins where both $Fe^{+++}$ and $Fe^{++}$ are equal. The distance at which the concentrations are equal to the bulk-solution values moves with time: ----- early time; —— later time.

solution) has a different value of $[Fe^{+++}]/[Fe^{++}]$. This concentration ratio is tending to change (by diffusion) from the bulk value to the ratio defined at the electrode surface by the electrode's potential.

As the potential at the cathode becomes more negative, a point is reached where essentially all the iron ions near the surface are going to be $Fe^{++}$ ions. Then, the only way to get more electrons into the solution is to wait for more $Fe^{+++}$ ions to get near enough to the electrode surface to pick up more electrons. At this point, the flow of current will not depend on the potential at the electrode but only on the rate that $Fe^{+++}$ ions diffuse to and contact the cathode surface. The current reaches its limiting value, the limiting current, $i_L$.

Focus now at the change in concentration that occurs between the surface of the electrode and the bulk solution. As illustrated in Figure 12.13b, at the surface of the electrode in the presence of a sufficiently negative potential, all the ions are $Fe^{++}$. At the same time, if the current has flowed for only a short period, in the bulk solution the ratio $[Fe^{+++}]/[Fe^{++}]$ is still near unity, and $E_{\text{bulk soln}} = E^{\circ}_{Fe^{+++}/Fe^{++}}$. The differences in ionic concentrations between the surface and the bulk is *equivalent to an electro-*

*chemical concentration cell.* Thus, as you know, there is a voltage that is equivalent to this concentration change. This voltage is called the **concentration polarization potential** or **concentration polarization.** Concentration polarization occurs at both electrodes and tends to limit the current that can flow. We shall call the total concentration polarization from both electrodes $E_{CP}$. Remember that $E_{CP}$ can appear only when a current flows through the cell.

Equation 12-7, which describes the current flow with applied potential, must be amended to

$$E_{app} = (E_1 - E_2 + E_{CP}) + IR \qquad \text{(12-9a)}$$

We can rearrange this equation to express the current as a function of potential.

$$I = \frac{[E_{app} - E_1 + E_2 - E_{CP}]}{R} \qquad \text{(12-9b)}$$

or

$$I = \frac{[E_{app} - E_{cell} - E_{CP}]}{R} \qquad \text{(12-9c)}$$

### Kinetic Overpotential

Using the same apparatus as is drawn in Figure 12.11, we can electrolyze water, with the reactions yielding hydrogen gas at the cathode and oxygen gas at the anode. It is likely that you have seen a demonstration of this experiment in a previous chemistry course. The energy needed to convert water to $H_2$ and $O_2$ has an equivalent electro-

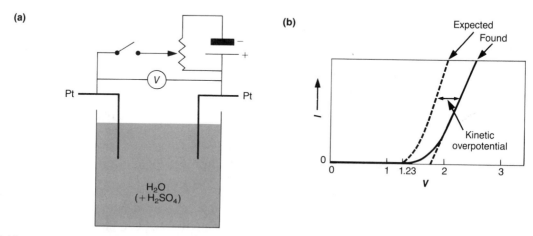

**Figure 12.14**    (a) Line drawing of the experimental setup that might be used to measure the effects of the kinetic overpotential on the electrolysis of water to hydrogen and oxygen. (b) The graph shows the current expected if no effects of overpotential occurred (dashed line). The equilibrium potential is 1.23 V. The current measured as the applied potential is changed is shown as the solid line. The electrolysis does not begin until a potential greater than 1.23 V is applied across the cell. The difference between the lines is the kinetic overpotential; it is slightly current-dependent. As indicated by the total current $I$, the electrolysis occurs at a faster rate as the potential is raised.

chemical potential. It is 1.23 V and is independent of pH. We expect to see the decomposition of the water begin if this potential is applied, since the electrochemical energy is then just equal to the decomposition energy. However, the electrolysis of water does not begin to any extent until about 1.7 V is applied. These facts are illustrated in Figure 12.14. The difference in these two voltages, 1.23 V and 1.7 V, is called an **overpotential.**

The cause of the overpotential is not the concentration polarization, as can be shown from two pieces of evidence.

1. The overpotential does not change with the rate of diffusion to the surface of the electrode.
2. The overpotential even shows up in the limit of zero current. Compare Figures 12.11 and 12.14 to see the difference graphically.

This overpotential is attributed to the process of electron transfer from the electrode to the adjacent ions in the solution, but the mechanism is not fully understood. In addition, it depends significantly on the material composing the electrode and the detailed nature of the electrode surface. This is one reason that clean liquid mercury is used as an electrode in many electrochemical amperometric analyses. Compared with other electrode materials, its surface is significantly more reproducible.

In review: When a current is allowed to flow through an electrochemical cell, electrochemical reactions will not take place at the potentials described by the simple Nernst equation. There are three causes for this behavior. The first is the resistance of the electrolyte solution in the cell. The second is the concentration polarization that developes at the electrodes. And, in some cases, there is the third, the kinetic overpotential. These fundamental effects must be considered when using any of the following electrochemical methods.

## 12.6 Electrogravimetry

The method of electrogravimetry involves the deposit and subsequent weighing of a solid product of an electrochemical oxidation or reduction. It can be used with great precision and accuracy when only one platable species is in the solution. However, for routine determinations when somewhat lower precision is needed (in the range $\pm 0.2 - 1\%$), the method has been significantly supplanted by spectroscopic methods such as atomic absorption (Chapter 18).

The simplest manner of carrying out the electrodeposition is shown in Figure 12.15. A source that provides a constant voltage across the electrodes is connected across the electrodeposition cell. A clean, preweighed, platinum grid shaped into a cylinder is used as a cathode. The anode is a platinum wire immersed in the same solution. The voltage applied across the cell must be large enough to plate out the metal at a reasonable rate while overcoming the effects of the solution resistance, concentration polarization, and any overpotential. However, the rate must be slow enough to produce a good, solid plating of material, so that pieces do not break off when the grid is washed, dried, and weighed after the plating is done.

**Figure 12.15** Diagram of the essential parts of an electrodeposition apparatus using a two-electrode system. The negative electrode is a Pt mesh. The plating of $Cd^{++}$ is illustrated. A cyanide salt might be added to the solution to aid in obtaining a good, solid plate of Cd metal, which helps decrease losses of mass in washing, drying, and weighing. Electrodeposition can be quite precise. However, when a large number of less precise and faster analyses are needed, other spectroscopic and electrochemical methods could be used. In the figure A = ammeter, V = voltmeter.

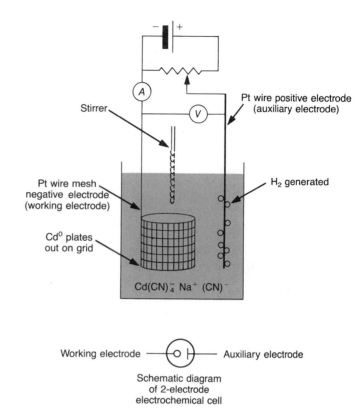

With an apparatus like this, the conditions are determined empirically because the current, solution resistance, and overpotentials change as the plating progresses. In addition, the analyte concentration (cadmium in Figure 12.15) continually decreases while the plating progresses, and so the equilibrium (reversible) potential of the electrode couple is changing with time as well.

To eliminate some of these problems, a **three-electrode potentiostat** is used. Because this is an important piece of equipment in quantitative electrochemistry, some of the details of operation are worth learning.

## 12.7 Three-Electrode Potentiostat

As its name implies, a three-electrode potentiostat consists of three electrodes and holds a potential constant. The three electrodes include the two that are used in the electrodeposition cell—namely, a working electrode (such as the platinum mesh) and an auxiliary electrode (such as the platinum wire). The third electrode is a **reference** electrode, such as an SCE or Ag/AgCl electrode. Such a cell is illustrated in Figure 12.16. The circuit and electrodes operate so as to define the potential at the

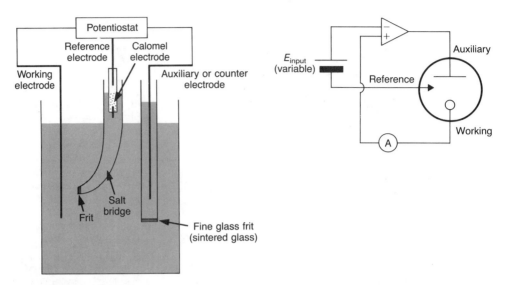

**Figure 12.16**    Line drawing of the components of a three-electrode potentiostat and the schematic diagram for the system. The symbols for the types of electrodes and the cell are as illustrated. The current at the working electrode is measured at potentials maintained automatically by the potentiostat. In order for the system to operate properly, the reference electrode should be closer to the working electrode than is the auxiliary electrode. The electronic circuits of a contemporary three-electrode potentiostat are considerably more sophisticated than the one shown here for illustration. In the figure A = ammeter.

working electrode at a preset value regardless of the changing resistance or concentration polarization of the solution.

The schematic circuit of the potentiostat is shown in Figure 12.16. Inside the circle representing the cell, an arrow is used to represent the reference electrode. The symbols for the other two electrodes are noted in the figure.

Earlier in this chapter you read that if an accurate measure of the potential of an electrochemical cell is wanted, it should be made with a minimum of current flowing. Such a low-current measurement is made between the working electrode and the reference electrode. The potential of the working electrode is then set relative to the reference electrode half-cell potential. The variable voltage source $E_{\mathrm{input}}$ sets the level.

The key to understanding how the potentiostat works is to realize that the amplifier will generate a current at the output until the voltage on the inverting input (defined in Section 3.5 and labeled with a −) is equal and opposite to the voltage on the noninverting input (labeled +). Let us look more closely at the values of the potentials on the two inputs.

We have set the potential at the inverting input to a value $E_{\mathrm{input}}$ *relative to the reference.* The amplifier will generate a potential of the opposite sign at its output. (This is a primary characteristic of the amplifier.) The amplifier will keep increasing its output current through the solution until the voltage on the *non*inverting input equals $-E_{\mathrm{input}}$. The voltage will reach $-E_{\mathrm{input}}$ regardless of the conditions in the solution between the surface of the auxiliary electrode and the surface of the working electrode. But the noninverting input and reference electrode are directly connected.

Therefore, the potential at the surface of the working electrode is *forced* to $-E_{input}$ relative to the reference electrode potential. This means that we can set the potential to the value we want at the surface of the working electrode, and the electronic circuit will keep it at that value regardless of the changing conditions (for instance, conductance) of the solution. The solution conductance has a lower usable limit, though. (After all, the potentiostat will not work across an open circuit.) The solution must be able to conduct sufficient current between the auxiliary and working electrodes.

### Electroseparation

With a plating apparatus based on the three-electrode potentiostat, sometimes more than one metal can be plated out sequentially — one at a time. We say that the metals can be **electroseparated** if they have formal potentials $E^{\circ\prime}$ that differ by more than about 0.2 V from each other. This property can be used in electrogravimetry.

The separation method can also be used to remove a number of interferents, such as heavy metals, from alkali metals in sample preparation. In aqueous solution, the heavy metals such as $Cu^{++}$, $Ni^{++}$, and $Cd^{++}$ plate out whereas the alkali metals do not.

Another possible use is to remove electrochemically active organic impurities (those that may be relatively easily oxidized or reduced) from solutions. An apparatus such as that shown in Figure 12.17 can be used to "clean up" solutions used in electrochemical trace analysis. The metal ions that are reduced to $M^0$ either dissolve in the liquid mercury or deposit on it. Organic reactive species are reduced and rendered inactive. The mercury can later be cleared of the dissolved metals by putting an electrolyte solution in the apparatus and allowing time for the metal ions to be

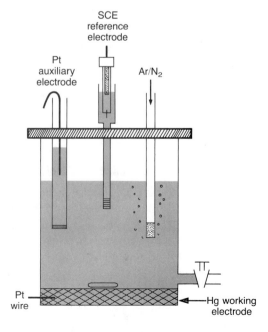

**Figure 12.17** Diagram of an electrolytic cell used for separations or purification of solutions by electrochemical means. One use could be to remove metals or interfering organic compounds from solutions to be used with electrochemical trace analysis. The cell is set up so that solution can be drawn off while the electrolysis still is going on. In that way, any metals that have been removed will not reoxidize and redissolve in the solvent. The three electrodes are connected to a three-electrode potentiostat. The working electrode is a mercury pool which provides a large surface area for reaction. The auxiliary electrode is a platinum wire isolated in a tube with a frit that allows electrical contact but prevents solution mixing. In this way, products of the electrolysis at the auxiliary electrode are prevented from contaminating the main solution. The tubes isolating the reference and auxiliary electrodes should contain the same electrolyte solution as in the main reservoir. [Ref: Rubinson K. A.; Mark, H. B., Jr. *Anal. Lett.* **1982,** *15,* 989–993, by courtesy of Marcel Dekker, Inc.]

oxidized by oxygen from the air and to return to the solution as ions. Rarely will organic compounds reoxidize to chemical forms that will be reactive at potentials less negative than the "clean up" potential.

## 12.8 Conductimetry

As you have learned, an ionic solution conducts electrical current by means of the motions of ions in the solution. One way of measuring how well the solution conducts would be to place two electrodes into the solution and connect an ohmmeter to them. The resistance that would be measured will depend on

1.  the surface area of the electrodes,
2.  the shape of the electrodes,
3.  the positions of the electrodes relative to each other in solution,
4.  the identities of the species in the solution,
5.  the concentrations of the species, and
6.  the temperature.

As noted in this list, the resistance that is measured depends not only on the solution properties but also on the instrument geometry — numbers 1 – 3 of the list. Within a single laboratory and with the same electrodes, this measurement could be useful for analysis. But the information could not be usefully communicated between laboratories. As a result, the apparent resistance — or its reciprocal, conductance — is not used to describe the properties of ionic solutions.

The units that we use for analytical measurements account for the differences in electrode area and geometry. A resistance is converted to a unit *as if* the measurement had been carried out in a cell of volume 1 cm$^3$ between two electrode plates, each 1 cm$^2$, that are placed parallel to each other and 1 cm apart. This measure of resistance in a specific geometry is called **resistivity.** The units for this quantity are, for instance, ohm-cm. [(1 ohm/1 cm length) · 1 cm$^2$ area = 1 ohm-cm.] Other units are shown in Table 12.4. The reciprocal of this quantity is called the **conductivity** or **specific conductance,** with units as shown in Table 12.4. Typical values of ionic specific conductance are those for KCl and are listed in Table 12.5.

### Conductivity and Ionic Concentration

As noted in the preceding list, the solution conductivity depends on the identities of the ions present and their concentrations. All the ions present in the solution participate in the conduction process. To divide up these contributions, it is helpful to consider the conductivity per equivalent of each ion (per ionic charge). This is called the **equivalent conductivity.** It is designated by a Greek $\Lambda$. The equivalent conductivity attributable to a specific ion is labeled $\lambda_i$. It is always true that

$$\Lambda = \sum_{\text{all ionic species}} \lambda_i$$

**Table 12.4**  Units Used in Conductimetry

| Quantity | Name of Unit | Unit Symbol |
|---|---|---|
| Resistance | ohm | $\Omega$ |
| Conductance | $ohm^{-1} = mho = siemens$* | $\Omega^{-1} = \Omega^{-1} = S$ |
| Resistivity ($\rho$) | ohm $\cdot$ cm <br> siemens $\cdot$ meter | $\Omega \cdot cm$ <br> $S \cdot m$ |
| Conductivity ($\kappa$) | mho $\cdot$ cm$^{-1}$ = siemens $\cdot$ cm$^{-1}$ <br> siemens $\cdot$ m$^{-1}$ | $\Omega^{-1} \cdot cm^{-1} = S \cdot cm^{-1}$ <br> $S \cdot m^{-1}$ |

* Siemens is the SI unit.

**Table 12.5**  Specific Conductances of KCl Solutions

| Concentration (M) | Conductivity (S $\cdot$ cm$^{-1}$) | | |
|---|---|---|---|
| | 19 °C | 20 °C | 21 °C |
| 1 | 0.1001 | 0.1020 | 0.1040 |
| 0.1 | 0.01143 | 0.01167 | 0.01191 |
| 0.01 | 0.00125 | 0.00128 | 0.00130 |

This equation is a quantitative statement of how the total conduction always consists of contributions from all the ions present. Representative values of equivalent conductivities for solutions with *only one salt present* are listed in Table 12.6.

You will note from inspection of the table that the equivalent conductivities are dependent on concentration. This is due to ion–ion interactions—the cause of varying ionic activities as well (Section 2.2).

One way to see some systematic properties of conductivities is to decrease the concentrations of the salts until only the interactions of the ions with the solvent occur. This can be done by extrapolating the equivalent conductivity to zero ion concentration. As the concentration approaches zero, the equivalent conductivity approaches a maximum value called the **limiting equivalent conductivity.** In Table 12.6, the limiting equivalent conductivities are those listed in the second column under its symbol, $\Lambda^0$.

A benefit of determining the limiting ionic conductivities is that—unlike at higher concentrations—the single-ion contributions can be added together such that

$$\Lambda^0 = \lambda^0_+ + \lambda^0_-$$

where $\lambda^0_+$ and $\lambda^0_-$ are the contributions of the cation and anion, respectively. A number of single-ion limiting conductivities appear in Table 12.7.

An important point to note from the values in the table is that H$^+$ is about five times more efficient in conducting charge in water than are the other cations, and OH$^-$ is about three times as efficient as other anions. This means that a conductimetry experiment will be about five times as sensitive to changes in proton concentration as to other cations. Similarly, a threefold factor is expected for hydroxide.

**Table 12.6**  Equivalent Conductivity in Water at 25°C
(S cm² equiv⁻¹)*

| Electrolyte | Limit $\tilde{c} \to 0$ $\Lambda°$ | Concentration (equiv · L⁻¹) | | |
|---|---|---|---|---|
| | | *0.001* | *0.01* | *0.1* |
| KCl | 149.86 | 146.95 | 141.27 | 128.96 |
| NaCl | 126.45 | 123.74 | 118.51 | 106.74 |
| HCl | 426.16 | 421.36 | 412.00 | 391.32 |
| AgNO₃ | 133.36 | 130.51 | 124.76 | 109.14 |
| KNO₃ | 144.96 | 141.84 | 132.82 | 120.40 |
| NH₄Cl | 149.7 | — | 141.28 | 128.75 |
| LiCl | 115.03 | 112.40 | 107.32 | 95.86 |

* The units are (S cm² equiv⁻¹) since $\Lambda = \dfrac{\kappa}{\tilde{c}}$ where $\tilde{c}$ is the equivalent concentration in equiv cm⁻³.

**Table 12.7**  Limiting Ionic Conductivity ($\lambda°$) in Water
(S cm² equiv⁻¹) at 25 °C

| Cations | $\lambda°_+$ | Anions | $\lambda°_-$ |
|---|---|---|---|
| H⁺ | 349.8 | OH⁻ | 197.6 |
| Li⁺ | 38.69 | F⁻ | 55.4 |
| Na⁺ | 50.11 | Cl⁻ | 76.34 |
| K⁺ | 73.50 | Br⁻ | 78.14 |
| Rb⁺ | 77.8 | I⁻ | 76.97 |
| Cs⁺ | 77.3 | CN⁻ | 82 |
| NH₄⁺ | 73.4 | NO₃⁻ | 71.44 |
| Ag⁺ | 61.92 | ClO₄⁻ | 67.4 |
| Tl⁺ | 74.7 | $\frac{1}{2}$SO₄²⁻ | 80 |
| $\frac{1}{2}$Mg²⁺ | 53.06 | $\frac{1}{3}$Fe(CN)₆³⁻ | 99.1 |
| $\frac{1}{2}$Ca²⁺ | 59.50 | $\frac{1}{4}$Fe(CN)₆⁴⁻ | 111 |
| $\frac{1}{2}$Sr²⁺ | 59.46 | Formate⁻ | 54.6 |
| $\frac{1}{2}$Zn²⁺ | 52.8 | Acetate⁻ | 40.9 |
| $\frac{1}{2}$Cd²⁺ | 54 | Chloroacetate⁻ | 39.8 |
| $\frac{1}{2}$Pb²⁺ | 70 | Dichloroacetate⁻ | 38 |
| $\frac{1}{2}$Mn²⁺ | 53.5 | Trichloroacetate⁻ | 35 |
| $\frac{1}{3}$Al³⁺ | 63 | *n*-Propionate⁻ | 35.8 |
| $\frac{1}{3}$Co(NH₃)₆³⁺ | 99.2 | Benzoate⁻ | 32.3 |
| N(CH₃)₄⁺ | 44.92 | Picrate⁻ | 31.39 |
| N(C₂H₅)₄⁺ | 32.66 | | |
| N(C₃H₇)₄⁺ | 23 | | |

## Conductimetry in Practice

Experimental values of conductivities are almost always found by using a cell with rigidly fixed electrodes. Three types of such conductivity cells are illustrated in Figure 12.18. With such cells, quantitation using conductance measurements can be done simply by calibrating the instrument with known concentrations of the ion-forming analyte and then measuring unknowns. The calibration curves are slightly nonlinear, as can be seen from the values in Table 12.5 for different concentrations of KCl.

Alternatively, the cell can be calibrated with a material of known specific conductance. Sets of solutions having a range of concentrations of KCl in highly purified water are common calibrants. Then the specific conductance of the unknown material can be determined. Not only can analyses be done in this way, but fundamental properties of ionic materials can be determined, such as the ionization constants of acids.

Conductance measurements adequate for numerous applications can be carried out simply with a battery, an ammeter, and two electrodes immersed in the solution to be tested. Such simple conductance meters are used to determine the purity of water coming from a still. (So-called conductivity water has a conductivity of around $10^{-6}$ S $\cdot$ cm$^{-1}$.) However, we encounter some major problems with such a simple system if more precise measurements are needed.

You have read about these problems in Section 12.5. The difficulty lies in the influence of overpotentials and concentration polarization on the currents that pass through an electrolytic cell when an external voltage is connected to the electrodes. The true voltage across the bulk solution will be uncertain. These effects will certainly

(a)  (b)  (c)

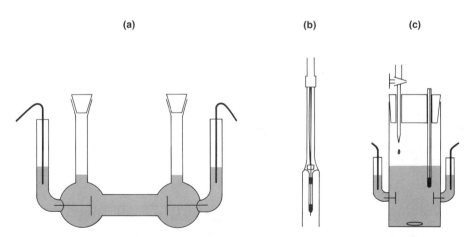

**Figure 12.18**  Illustrations of three different types of conductivity cells. In all three, the electrodes are constructed from platinum foil covered with platinum black. Samples fill the cells so that the geometry is the same for all runs. The temperature should be held precisely constant. Control of the temperature to 0.1 °C allows precision in the measurements to be in the range $\pm 1$–2%. (a) The sample cell is filled with the solution. (b) A dipping cell with an open bottom that is immersed in the sample. (c) A cell used for titration with conductometric monitoring.

cause errors in the measurement of the conductance of an electrolyte solution. This is because

$$V = IR = \frac{I}{\text{conductance}}$$

and if the true voltage is unknown, so is the conductance.

Recall that the concentration polarization can be explained as arising from a concentration difference induced between the bulk solution and the volume at the electrode surface. The kinetic overpotential is due to properties of the surface of the electrode and the species in solution. Both of these effects must be minimized to obtain accurate and precise conductivity changes.

The problems due to the kinetic overpotential can be overcome by choosing the right electrodes. These are usually made of platinum foil that has more platinum plated onto the surface. The plated metal is not shiny. The surface appears to be black, and the electrodes are said to be coated with *platinum black.*

The possible problems due to concentration polarization are overcome by doing the measurement so quickly that no appreciable concentration polarization can build up near the surfaces of the electrodes. In practice, two identical electrodes with opposite polarities (one positive, one negative) are rapidly switched between positive and negative at a rate of about 1000 times per second. This is done by means of a rapidly oscillating voltage applied to the electrodes. The process is illustrated in Figure 12.19. In this way, conductances are measured far more precisely than can be done with DC measurements.

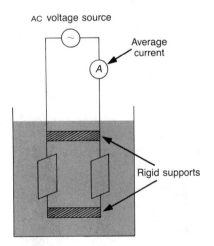

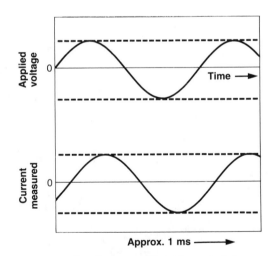

(Amplitude proportional to conductivity)

**Figure 12.19**    Illustration of a conductivity experiment using an oscillating voltage to minimize effects of concentration polarization. The applied potential is graphed on the top right as a function of time. It is a sine wave. The current that flows through the cell is plotted below. The current will follow the voltage, reversing direction twice each cycle. The output of the instrument is the average current as related to the average applied voltage.

The utility of conductance measurements is broad since concentration changes in any ionic solution can be monitored. Some of the uses for conductimetric techniques are for monitoring

contamination in streams and rivers,
salt content in boiler systems,
acid concentrations of solutions used in industrial processes,
the concentration of liquid fertilizer as the fertilizer is applied, and
the end points in titrations where an ion concentration of the analyte or
    titrant changes (for example, neutralization titrations by $[H^+]$ or $[OH^-]$).

With proper standards and temperature control, high-precision conductimetry can be done with errors in the range of 0.1%. However, such precision is not routine. One of the problems is the strong temperature dependence of conductivity. As seen from Table 12.5, ionic conductance typically changes by about 2% per °C. Quantitative conductivities can be determined to about ±1–2% if the resistance between the electrodes is in the range of 1000 ohm, and the temperature is thermostated to within ±0.1°.

## 12.9 Coulometry

Earlier, the technique of electrogravimetry was described. The technique involved passing a current at a constant applied potential through an electrochemical cell. The metal being assayed plates out on a preweighed cathode. To yield precise results, all of the metal must be plated out on the electrode. However, numerous materials can be reduced or oxidized that do not plate out onto an electrode. In addition, there is a lower limit to the method: when the amount of material becomes too small to weigh. The method of coulometry can be used in both these cases.

To carry out a coulometric assay, we measure the total charge that is needed to complete an electrochemical oxidation or reduction. This charge is relatable directly to the total amount of material electrolyzed. Recall the simple relationship between a mole of charge, the faraday, and the number of coulombs:

$$96,491 \text{ coulombs} = 1 \text{ faraday}$$

Further, recall that

$$1 \text{ coulomb/s} = 1 \text{ amp (ampere)}$$

Thus, the current in amps passing for a given time can be related to the number of moles. The fundamental equation in coulometry needed for analysis is

$$\text{number of moles} = \frac{\text{coulombs for total electrolysis}}{96,491 \cdot n}$$

or

$$\text{number of moles} = \frac{[\text{current(amps)} \cdot \text{time(s)}]}{96,491 \cdot n}$$

The letter $n$ represents the number of electrons transferred per mole, as in the Nernst equation.

As mentioned above, coulometry is especially useful when the total amount of electroactive material (meaning oxidizable or reducible at an electrode) is smaller than can be weighed precisely. Let us see how much material is *not* small by coulometric methods.

It is relatively easy to measure a total charge of millicoulombs (mC) over periods of minutes. (See, for instance, Figure 12.22.) And,

$$1 \text{ coulomb} = (96{,}491)^{-1} \text{ mol of e}^- = 1.03637 \times 10^{-5} \text{ mol of e}^-$$

Thus, 1 mC is equivalent to about $1 \times 10^{-8}$ mol of electrons ($1 \times 10^{-8}$ faraday). For a one-electron reduction, this is 10 nanomoles (nmol) of material. For a species with a formula weight of 100, 10 nmol is equal to 1 microgram ($\mu$g) of material—truly a weight difficult to determine directly with precision or accuracy.

In order to do a coulometric determination, an analyte is exhaustively oxidized or reduced, usually in a stirred solution. The redox reaction is driven by an external voltage source such as a three-electrode potentiostat. The apparatus is similar in design to that in Figure 12.16. The working electrode can be composed of any of a number of electrically conducting materials.

As noted in Section 12.5, at the surface of an electrode, the concentrations of the electroactive species are described by the Nernst equation with a potential equal to that of the electrode. However, through diffusion and stirring, eventually the concentrations *set* at the electrode surface are reached throughout the solution volume, and the current ceases. As illustrated in Figure 12.20, the current flowing decreases with time as the electroactive species is depleted.

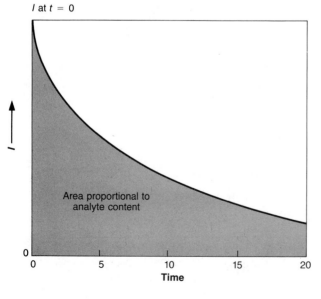

*I* at *t* = 0

Area proportional to analyte content

0    5    10    15    20
**Time**

**Figure 12.20**   If an electrode at a fixed potential were electrolyzing some species in a vigorously stirred solution, the electrolysis current would vary with time as shown in this graph. The current is proportional to the amount of electrolyzed species left in the solution. After all of the material is electrolyzed, the current will stop flowing, assuming no other species could be electrolyzed. The curve has an exponential form since the rate of electrolysis is proportional to the amount of the material that is left in solution—analogous to a reaction with first-order kinetics. The electrolysis rate depends on the stirring rate, electrode area, electrode material, electrode potential, initial concentration of the electrolyzed species, and, perhaps, concentrations of other species in the solution.

If the solution is not stirred, the time needed to complete the reaction can be quite long. A one-millimeter diameter wire in the middle of an *unstirred* solution in a 100-mL beaker would take up to a month to electrolyze the analyte completely. With stirring, this is reduced to tens of minutes. An alternate to stirring is to use only a thin layer of solution, so that diffusion alone is sufficient to homogenize the concentrations in a few minutes.

---

**EXAMPLE**

A sample of L-dopa, a drug used for treatment of the symptoms of Parkinson's disease, is electrolyzed exhaustively with 42.0 $\mu$C. The process requires two electrons per molecule. The formula weight of L-dopa is 197.2. What is the mass of L-dopa in the sample?

L-dopa

**SOLUTION**

The conversion from coulombs to moles is

$$\frac{42.0 \times 10^{-6}}{96,491 \cdot 2} = 2.17_6 \times 10^{-10} \text{ mol}$$

The total mass of L-dopa is then

$$2.17_6 \times 10^{-10} \text{ mol} \cdot 197.2 \text{ g/mol} = 4.29 \times 10^{-8} \text{ g or } 42.9 \text{ ng}$$

---

### Background Currents: Competitive Electrolysis

As you have read, the mechanism of electrolysis at an electrode is the transfer of electrons to or from the electroactive species in solution. This electron transfer continues until the relative concentrations of the redox couple are equal to the equilibrium constant—$K = [Ox]/[Red]$—for the appropriate half-cell. The products of the electrode reaction then migrate into the bulk solution by diffusion, and new material arrives at the electrode from the bulk, again by diffusion.

Let us ask, What happens when two redox reactions have the same formal potential, and all four electroactive species are present in the solution? Quite simply, both reactions will "go" at the electrode. The rate at which each of the reaction products appear depends on the concentrations of the reactants and the overpotentials of the respective reactions.

When a typical electrochemical determination is done, we do not expect to have two redox reactions with the same formal potential occurring at the same time. However, there are always materials other than the species of interest that *can* be oxidized or reduced. These include the electrodes themselves, the background electrolyte, dissolved gases, as well as reagent impurities. At some potential, even the

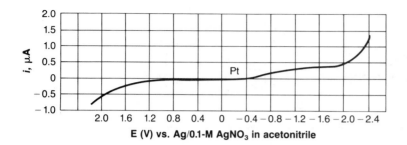

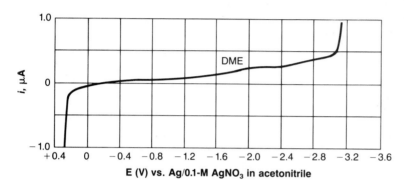

**Figure 12.21**   Background currents for two different electrode materials as they vary over a potential range. The solvent for both is purified acetonitrile with background electrolyte of 0.1-M tetraethylammonium perchlorate, $Et_4N^+ClO_4^-$. The upper curve was obtained with a platinum electrode, the lower curve with a dropping mercury electrode. The horizontal scale is in volts relative to a Ag/AgCl reference electrode. The reference electrode was made with a AgCl-coated silver wire immersed in a 0.1-M $AgNO_3$ solution in acetonitrile, a nonaqueous electrode. As a result, the potential of the voltage scale cannot be compared directly with an experiment that employs an aqueous Ag/AgCl reference. Note the relatively flat baseline until the ends, where solvent breakdown occurs or the electrode itself electrolyzes. These two steep portions of the curve limit the usable voltage range. This range varies depending on the specific conditions. [Ref: Reprinted with permission from Sherman, E. O.; Olson, D. C. *Anal. Chem.* **1968,** *40*, 1174–1175. Copyright 1968 American Chemical Society.]

solvent itself will electrolyze. As an example, at large enough potentials, water will be electrolyzed to gaseous oxygen and hydrogen at the positive and negative electrodes, respectively. Chloride can be oxidized to chlorine or even higher oxidation states. Platinum or gold electrodes can be oxidized to their respective ions and go into solution. Thus, components of the electrochemical cells and solutions must be chosen with care to enable a meaningful determination to be carried out. The currents from these interfering reactions are called **background currents.**

Examples of such currents are seen in Figure 12.21. Here are shown the background currents as they change with potential in the nonaqueous solvent acetonitrile ($CH_3CN$) with a background electrolyte of tetraethylammonium perchlorate [$(C_2H_5)_4N^+ClO_4^-$]. Currents are shown for electrodes composed of two different materials: platinum and mercury. Note first, the negative-potential region (toward more-reducing conditions). With the mercury electrode (DME), the background current does not rise sharply until potentials are more negative than they are for the platinum electrode. This is due to a difference in the hydrogen overpotentials on the two metals. This fact means that the mercury electrode should be chosen for work at potentials more negative than $-2.0$ V on this scale. This sharp rise in current at the extreme negative potential is called the point of solvent **breakdown;** the solvent electrolyzes.

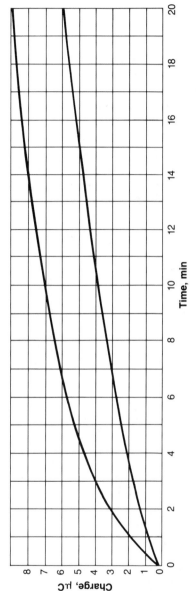

**Figure 12.22** Data of a coulometric experiment for lead. The lower curve is the background run. The upper curve results with lead present. The graph, which was plotted as the experiment was progressing, is a plot of microcoulombs versus time in minutes. As is usual for coulometric experiments, the solution was stirred. The conditions were as follows: sample volume, 3.0 mL; potential, −700 mV versus SCE; working electrode, Hg pool. The background electrolyte is a buffer made from 0.1-M sodium citrate with 0.005-M acetic acid added. [Data courtesy of Matthew Doyle.]

On the other end of the range, at the left side of the figure, the mercury electrode shows a sharp increase in negative current (electrons are flowing into the electrode; oxidation is occurring). In this region, the mercury electrode surface is being oxidized. This fact means that when potentials more positive than about zero volts on this scale are needed, of the two metals platinum is the necessary choice for the electrode.

For coulometric assays, ideally, the background currents from all the other materials in the cell should be zero. But this ideal state is not found in practice, although the currents can be relatively small. The current from the sample must somehow be separable from that of the background. To assess the situation, a blank can be run as shown in Figure 12.22.

The data of Figure 12.22 are from a coulometric determination of lead in a complex biochemical mixture. The lower trace shows the change in charge due to the background current alone as run on a blank sample. The top trace is with the lead-containing species. The difference between the two traces is due to the lead. The potential used was −0.700 V versus SCE, low enough to reduce the lead ions to metallic lead while keeping the background current as small as possible. In problem 12.11 you are asked to calculate the quantity of lead present from the measured charge.

The total charge measured for a coulometric assay is the sum for the two processes: the number of coulombs for the analyte and the number attributable to background processes. This relationship is commonly expressed quantitatively using the ratio called the **electrochemical efficiency,** or just **efficiency.** By definition,

$$\text{efficiency} = \frac{\text{coulombs used in reaction of interest}}{\text{total number of coulombs used}}$$

A similar expression can be written for the electrochemical efficiency in terms of the current passed into the cell.

$$\text{efficiency} = \frac{\text{current electrolyzing analyte}}{\text{total current}}$$

The efficiency can be expressed as a percentage. Optimum conditions for a coulometric assay occur when the efficiency is the greatest. In a bad case, a slight variation in background current will swamp the current from the material to be assayed, and no determination is possible at all. The efficiency depends on the **current density,** the electrode material, and the solution components. The current density is the current per unit area of the electrode, for instance, A cm$^{-2}$.

An assumption made in coulometry is that the analyte being measured is completely electrolyzed by the time the experiment is finished. Otherwise, the relation between the charge transferred and the quantity of analyte will be uncertain or even meaningless. Let us *define* 99.99% reduction of the material to be considered quantitative. Assuming that neither overpotentials nor concentration polarization need be considered, what potential must be used to obtain quantitative results? (This same calculation can be made for "complete" plating in electrogravimetry.) For the conditions in which 99.99% of the material is reduced, only 0.01% is left in the oxidized form. For a one-electron reduction, the Nernst equation describing this condition is

$$E = E^{\circ\prime} + 0.059 \log \frac{[0.001]}{[0.9999]}$$

or

$$E = E^{\circ\prime} - 0.059 \times 3 = E^{\circ\prime} - 0.177 \text{ V}$$

Therefore, the potential must be nearly 0.2 V more negative than the formal potential in order to have a quantitative (99.99%) reduction.

### Coulometric Titrations

Coulometry, the method used to obtain the data in Figure 12.22, involves holding a constant potential on the working electrode and measuring the total charge needed for complete electrolysis of the solution. A relatively sophisticated electronic circuit was used to integrate the total charge. However, if the electrochemical efficiency of an oxidation or reduction is nearly 100%, then a second, simpler, coulometric method can be used in analyses.

This simpler method involves use of a constant-current source and an accurate timer. The constant-current source delivers, as its name implies, a constant *current* regardless of the potentials or resistances of the electrolysis cell. In order to measure the total number of coulombs, all that has to be done is to measure the time from the beginning of the experiment to the time when the electrolysis is finished. This method is unlike coulometry done at a constant potential, where the current inherently falls to the level of the background (recall Figure 12.20).

As a consequence, we are required to find some other method to discover the point at which the analyte's redox reaction is finished and some other redox reaction (perhaps the background electrolysis) begins. This description sounds like a titration. In fact, the constant-current coulometric method is called **coulometric titration** because of its similarity to regular titration. Professor G. W. Ewing has called it "titrating with electrons,"[7] a most concise descriptive phrase. In general, the analyte itself is not directly electrolyzed. Instead, a titrant is generated quantitatively by electrolysis at an electrode rather than by adding a reagent solution.

The reagent used is generated electrolytically from a precursor that is in excess over the analyte in the solution. In Table 12.8 are listed a number of reagents that can be generated coulometrically. These reagents include $H^+$ and $OH^-$ generated for neutralization titrations. Redox reagents are among the group. In addition, chelates can be generated by reducing a strongly binding metal ion (usually mercury) which releases the chelate to bind with another species. Precipitation reactions can be done by generation of precipitant metal ions such as silver.

---

[7] Ewing, G. W. *American Laboratory* **1981**, *13*, 16–22.

**Table 12.8**  Examples of Coulometric Titrations

| Assayed Substance | Reagent Generated | Precursor | Titration Type |
|---|---|---|---|
| $Br^-$ | $Ag^+$ | $Ag°$ anode | Precipitation |
| $Fe^{++}$ | $Cl_2$ | HCl | Redox |
| $H_2O$ | $I_2, I_3^-$ | KI (pH < 9) | Karl Fisher reagent |
| Organic acids | $OH^-$ | $H_2O$ | Neutralization |
| Bases | $H^+$ | $H_2O$ | Neutralization |
| $Ca^{++}, Zn^{++}$ | $Hedta^{3-}$* | $HgNH_3edta^-$ | Complexometric |
| Olefins | $Br_2$ | KBr (pH < 5) | Olefin addition (redox) |

* edta = ethylenediaminetetraacetate. See Table 10.4 for the structure.
*Further references:* Stock, J. T. biennial reviews in *Anal. Chem.* 1986, 1984, 1982, 1980, etc. Farrington, P. S. In "Handbook of Analytical Chemistry" Meites, L., Ed. McGraw–Hill: New York, 1963, Table 5.55.

One of the primary advantages of coulometric titrations over volumetric ones is that standardized solutions need not be prepared. Thus, there are no problems with primary standards and, more importantly, with storing unstable reagent solutions.

Another benefit of coulometric titrations is the relative ease of handling smaller quantities of titrant and sample. To remind yourself why, recall the relation between charge, current, and mass. The charge corresponding to a microgram mass can be measured quite simply with contemporary equipment. In fact, the generation of enough reagent for larger samples can take so long that coulometric titrations become inconvenient.

As for any titration, some method is needed to determine the end point. You are aware of a number of such methods. We could use redox indicators or a potentiometric method. For the latter, a voltmeter is used to sense the potential jump of an electrode, in the same way that a pH electrode is used to detect the pH jump at the end points in acid–base titration.

As noted above, one requirement of this method is that the electrochemical efficiency in generating the reagent must be near 100% to achieve the desired precision and accuracy. Maximizing the efficiency requires two conditions. First, the coulometric titration requires a careful choice of reagent precursor to use under the needed conditions. Second, it requires that a relatively low current be kept passing into the solution across a given area of the electrode (a low current density). However, finding the best reactions and conditions can be worthwhile because of the previously stated advantages.

Coulometric titrations are especially amenable to automation, since not only can the end point be determined directly from the signal of a transducer, but the quantity of titrant generated can be controlled precisely, and the electrochemical reagent addition can be stopped at the end point — all by the same electronic circuit. There is an added advantage in eliminating mechanical devices for adding reagent: The instrumentation becomes far simpler. A few specific examples of coulometric titrations will be described in the following section on amperometry, the measurement of current, which can be used conveniently to detect the end points of coulometric titrations.

## 12.10 Amperometry

In amperometry, as the name implies, we measure the current that passes through a solution and causes oxidation or reduction of analyte. Amperometric assays depend on quantitative measurements of the limiting current (illustrated in Figure 12.12). The reason that quantitative analysis can be done with the method is that the limiting current is directly proportional to the concentration of the species being oxidized or reduced at the electrode.

The fundamental relationship used for amperometric assays concerns the relationship of the diffusion-limited current and the concentration of electroactive species. It is, simply,

$$i_L = \text{constant} \times \text{concentration of electroactive analyte} \qquad \textbf{(12-10)}$$

The constant is determined through the use of standards.

The types of materials that can be quantitated include metals and organic species. Usually, for the organics, the determination is done by reducing a bond or bonds of the molecule. Some of the organic-bond types that can be reduced in this way are listed in Table 12.9.

Two general types of amperometric methods are commonly used. In the first type, we measure the change of current at a *fixed potential*. This is usually used to monitor the end point during a titration. When end points are measured this way, the titrations are often called **amperometric titrations.** (This is somewhat a misnomer since the reagent is being added either coulometrically or volumetrically. A parallel misnomer is to call neutralization titrations indicator titrations or potentiometric titrations.)

**Table 12.9**  Bond Types That Can Be Reduced for an Amperometric Determination

| | | | | |
|---|---|---|---|---|
| C—C | C—N | C—O | C—S | C—X |
| C=C | C=N | C=O | C=S | |
| C≡C | C≡N | | | |
| | N—N | N—O | N—S | |
| | N=N | N=O | | |
| | | O—O | O—S | O—X |
| | | | S—S | S—X |

Condensed benzenoid rings                                    X = halogen
Some heterocyclic rings

*Source: Chemical and Engineering News, March 18, 1968, p. 96.*

The second general method involves measuring the changes in the current as a function of a *changing potential* applied to the working electrode of an electrochemical cell. There are a number of these **voltammetric** methods, which include **polarography, cyclic voltammetry,** and **anodic stripping voltammetry.** These three methods are used to measure the concentrations of electroactive species in solutions. The four amperometric methods are covered in the remainder of this chapter.

### Fixed Potential Amperometry: Amperometric Titrations

In order to determine the end points of titrations, two different amperometric measurement techniques are common. In the first, we monitor the current flowing between two identical electrodes immersed in the titrated solution. In the second, we monitor the current flowing at one electrode.

### The Two-electrode System

The first end-point detection technique involves the use of two identical inert electrodes (such as platinum wires) inserted in the test solution. Since two electrodes are used in this amperometric method, it is called a **biamperometric** detection of an end point. A diagram of the experimental apparatus for a volumetric titration is shown in Figure 12.23a. The experimental setup for this method with coulometric addition of titrant is illustrated in Figure 12.23b.

A potential at some value between a few millivolts to a volt is placed across the two elecrodes. During the titration, the current is monitored; it is in the range of microamps, as seen in Figure 12.24. No potentiostat, reference electrode, or salt bridge is needed. Thus, the technique is experimentally quite simple and inexpensive. It is also quite effective.

Let us consider a specific example of biamperometric detection in detail: the titration of a 0.1-M iodine solution in water with KI present. This is titrated with 0.1-M sodium thiosulfate.

The descriptive chemistry of the reaction of iodine and thiosulfate is shown by the equation

$$2\ S_2O_3^= + I_2 \rightleftharpoons S_4O_6^= + 2\ I^-$$

The sulfur–oxygen species on the right side, called tetrathionate, forms specifically when iodine is the oxidizing agent.

The results of this titration, as monitored by biamperometric apparatus, are illustrated in Figure 12.24a. At the end point there is a sharp break in the curve of the measured current. This is just what we want in order to determine a titration end point. What is the chemistry that underlies this simple and useful method?

In Section 12.5, you read how *inert* electrodes transmit current from the metal to the solution. This involves simultaneously oxidizing some species at the more positive electrode and reducing some species at the more negative electrode. Electrons must be transferred between the electrodes and the solution for the circuit to be complete. In other words, electrolysis must occur in order that the conduction by

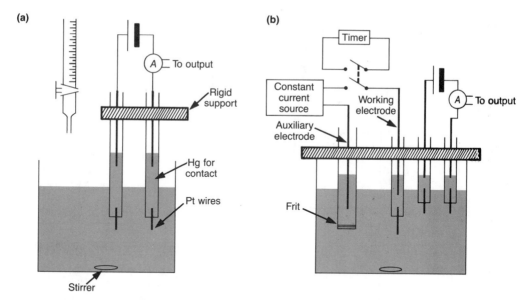

**Figure 12.23**  Line drawings of the essential parts needed to do a titration with biamperometric detection: (a) shows the equipment for use when the titrant is added volumetrically; (b) shows the equipment for use with coulometric generation of titrant. The biamperometric detection electrodes and electronics are the same for both. For the coulometric generation of titrant, the exact time needed to generate the reagent is recorded. With a highly stable current source, the number of coulombs required to titrate the sample is calculated from the mathematical product of the current and time. If the electrolysis efficiency is 100%, then no calibration of the titrant is needed.

electrons in the wires be converted to conduction by ions in the solution. Thus, with the use of inert electrodes, the potential applied to the electrodes must be large enough to electrolyze *some* species in solution at *each* electrode. If there is no electrolysis at *either* of the electrodes, then no current will flow. This straightforward idea is the basis for biamperometric amperometry.

In our example, the two platinum electrodes were placed into the iodine/iodide solution, and a fixed voltage of approximately 0.2 V was placed across them. As seen in Figure 12.24a, a current appears and is quantified with an ammeter. In order to have a current, there *must* be some electrolysis at both electrodes. Let us see what the reactions are at each.

At the more negative electrode, where reduction must be occurring, the reaction is

$$I_2 + 2\,e^- \rightarrow 2\,I^-$$

At the more positive electrode, where oxidation must be occurring, the reaction is the reverse

$$2\,I^- \rightarrow I_2 + 2\,e^-$$

As the titrant thiosulfate is added, the amount of iodine in the solution is decreased. But recall that since the rate of the reactants colliding with the electrode surface is

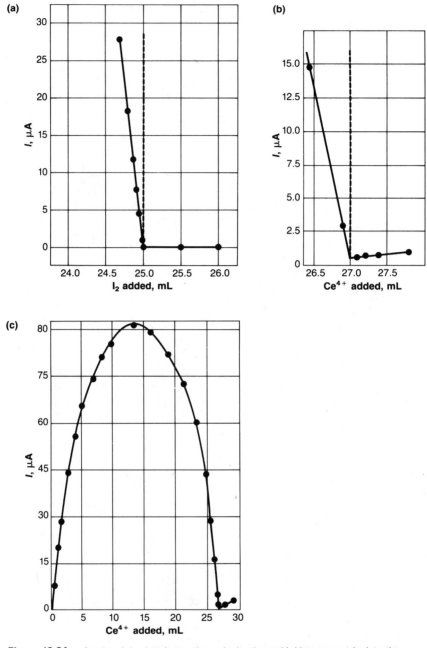

**Figure 12.24**   Graphs of the data from volumetric titrations with biamperometric detection. The graphs are microamperes versus volume of reagent added. (a) The region near the end point for a titration of 0.1-*N* thiosulfate with a solution of 0.1-*N* $I_2$ in aqueous KI. The electrodes were two platinum wires with a 50-mV potential between them. This is a "dead-stop" end point. (b) The region near the end point for a titration of $K_4Fe(CN)_6$ with $Ce^{4+}$. The products are $Fe(CN)_6^{3-}$ and $Ce^{3+}$. The applied potential was 50 mV. Current flows after the end point due to electrolysis of the titrant. (c) The same as Figure b except the entire titration curve is shown, not just the region of the end point. Note that no current flows initially. [Redrawn with permission from Stone, K. G.; Scholten, H. G. *Anal. Chem.* **1952,** *24,* 671–674. Copyright 1952 American Chemical Society.]

proportional to their concentrations, the current flowing is proportional to the amount of electroactive species in solution. Thus, as the amount of iodine drops, the current measured by the detecting electrodes drops. At the end point, there is no iodine left in the solution to be reduced, and no current flows. This is true even though there is plenty of iodide, $I^-$, available for oxidation; but to observe a current, *both* oxidation and reduction must be possible.

You might reasonably ask, What about the thiosulfate and tetrathionate? After all, there is also tetrathionate in solution at the end point. The answer to the question is a matter of descriptive chemistry. The tetrathionate will not be reduced at the platinum electrode under these conditions. Also, the thiosulfate will not be oxidized even if it is added in excess. So beyond the end point, the current in the detection circuit remains almost zero.

This behavior—having a current flow that "stops dead" at the end point—led to the naming of this titration behavior. It is called the **dead-stop end point,** even though in other situations this behavior is not followed. For instance, if the thiosulfate is titrated by iodine, the exact reverse of the above reaction, the behavior would be just opposite that in Figure 12.24a. However, this is *not* called a "dead-start" end point.

Different behavior is observed when both anodic and cathodic electrolysis can occur with *both* the reactant and the titrant. This situation is illustrated in Figure 12.24b for the titration of $Fe(CN)_6^{4-}$. The current does not drop to zero but begins to rise. However, the end point is still easy to determine.

The total titration curve detected with biamperometric detection is shown in Figure 12.24c. Both $Fe(CN)_6^{3-}/Fe(CN)_6^{4-}$ and $Ce(III)/Ce(IV)$ couples contribute to the currents. Writing the equations for the detection reactions is left to you for an exercise in problem 12.22.

### One-electrode System

The second amperometric titration monitoring technique utilizes a three-electrode potentiostat. The experimental setup is illustrated in Figure 12.25a. Only the working electrode is in direct contact with the titrant solution. The auxiliary and reference electrodes are separated from the test solution by salt bridges. The electrode potential is set at a value such that the limiting current $i_L$, due to either the titrant or the analyte, is measured. The limiting current will be directly proportional to the concentration of the electrolyzed species. This idea is expressed in Eq. 12-11.

$$i_L = \text{constant} \times \text{concentration of electroactive analyte} \qquad \textbf{(12-11)}$$

As illustrated in Figure 12.25b, the species concentration can be monitored and graphed as a function of added titrant. The end point of the titration will appear as a junction between two straight lines. (Compare Figure 9.3b.) As illustrated, the three possible types of behavior of the measured currents are:

1. The analyte is electrolyzed at the applied potential but the titrant is not.
2. The titrant is electrolyzed at the applied potenial but the analyte is not.
3. Both the titrant and analyte are electrolyzable at the potential of the working electrode.

**(a)**

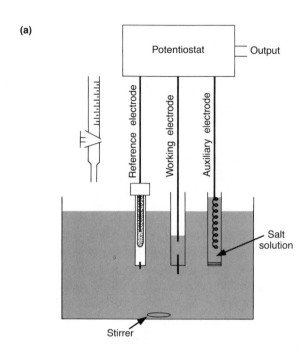

**(b)**

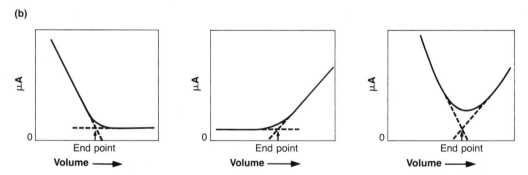

**Figure 12.25** (a) Line diagram of the experimental apparatus required to do a volumetric titration with single-electrode amperometric monitoring. The reaction is followed by measuring the current at the working electrode of a three-electrode potentiostat with the auxiliary electrode isolated. (b) Three types of behavior that might be seen plotted as the amperometric output versus the volume of titrant added. Compare these with Figure 9.3b. The end point is the point of the break in the output response. The only difference between them is the slopes of the lines on either side. Three cases are illustrated. (Left) Analyte electrolyzed at the electrode, titrant is not. (Middle) Titrant is electrolyzed at the working electrode, analyte is not. (Right) Both analyte and titrant are electrolyzed at the working electrode. The break is usually not as sharp as desired for a precise measure of the end point. The intersection of line segments can give a more precise answer. These are the dashed lines.

The end point is at the junction of the extrapolation of the two line segments of the graph. Since the current is proportional to concentration, the lines in the graph may be curved if the volume of the titrated solution changes significantly during the titration. There are two ways to overcome this problem. Either use a small amount of high-concentration titrant or correct the measured current for changes in volume of the solution.

All that you have learned about titrations holds for amperometric titrations. The techniques of amperometric detection are simply flexible, alternative methods to determine the progress of the titration and its end point.

Amperometry at a fixed potential tends to be a nonspecific assay method. As long as the electrode potential is adequate, all species coming into contact with the electrode will be electrolyzed. However, with adequate separations prior to electrolysis, highly specific amperometric electrodes have been developed. One such electrode, for the specific determination of a sugar, is described in Supplement 12B.

Another use for fixed-potential amperometry is in liquid chromatography (Chapter 14). A working electrode is placed so that electroactive analytes pass over it after they are separated. The electrolysis currents can be used to quantify the amounts of a number of analytes sequentially. This is called *liquid chromatography with electrochemical detection* (LC/EC). For a number of electroactive species, LC/EC is a powerful combination for trace analysis. Nanogram amounts of electroactive compounds can be determined directly.

## 12.11 Voltammetry

As you have read, amperometry can be used to quantitate the *changes* in concentration of numerous different electrolyzable species. These can be both metal ions and organic species. One problem with amperometric measurements is that the origin of the currents is not always obvious: How much of the current is from background interferents, and how much is from the electroactive analytes? One way to determine the background currents is to run a blank, *if* every possible electroactive species that might contribute to the background is known. The blank then consists of all the interferents in the right proportion; such a blank can get exceedingly complicated. However, there is an easier way.

In a single solution, both the background current and the limiting current due to the electroactive analyte can be measured and separated. This can be done by scanning the potential over a range of voltages. Refer to Figure 12.26. Shown is a graph of the current for reduction at a working electrode versus the scanned potential. As shown in the lower graph, the potential is changed at a constant rate from zero volts to more negative potentials. Such a time-dependent change in voltage is called a **linear voltage scan.**

Initially, the electrode potential is set so no electrolysis of the analyte species occurs although there may be some background currents. The voltage scan is then begun. As the potential is brought toward the formal potential of the species being investigated $E^{\circ\prime}$, some additional electrolysis occurs. The current continues to increase as the potential is scanned further, and eventually the region of the diffusion-limited current is reached. As noted in the principal equation of analytical amperometry, Eq. 12-11, the diffusion-limited current is proportional to the concentration in solution of the material being reduced.

The current that is due to the background electrolysis is extrapolated into the region where the analyte limiting current is present. The difference between the baseline extrapolation and the limiting current is taken to be due to the analyte alone. This extrapolation is shown in Figure 12.26.

There is further information in the plot of current. The position of the curve on the voltage axis is characteristic of the species being assayed. This position can be useful in identifying an unknown material that is electrolyzed. The voltage used to characterize the horizontal location of the curve is called $E_{1/2}$, and it is the midpoint of the rise in current.

$E_{1/2}$ is equal to $E^{\circ\prime}$ if the redox reaction of the species being measured is **reversible**. *Reversible* in the electrochemical sense is slightly different from a *reversible reaction*. In most chemistry, reversible means that a reaction *can* proceed in both forward and

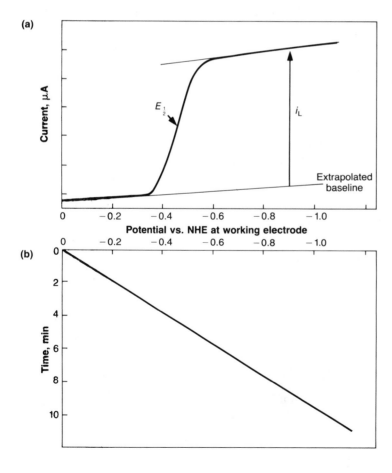

**Figure 12.26**   Illustration of the output of a voltammetric experiment using a working electrode consisting of a few-mm diameter flat disk, one side of which contacts the solution. It is rotated about its axis at a moderate rate to keep bulk solution passing across its surface. (This is called either a **rotating disk** or **rotating button** electrode.) (a) The current flow is plotted versus potential. At the left is the background current. As the potential is scanned, the electrolysis of analyte begins, and the current rises rapidly until it reaches the limiting current. The limiting current is defined as the difference between the extension of the background current and the plateau region. The midpoint of the rise (and point of maximum slope) is $E_{1/2}$. The potential is scanned at a fixed rate of 100 mV/min for these experiments. (b) Graph of the scan voltage versus time. Note that a and b share the same voltage axis.

reverse directions. In electrochemistry, it means that the redox reaction *does* go easily both forward and backward. One requirement for electrochemical reversibility is that the rates of electron transfer for *both* oxidation and reduction, as was illustrated in Figure 12.1, are extremely fast.

In this sense, relatively few electrochemical reactions are reversible. In aqueous solutions, most electrochemical reactions that are reversible at the electrodes involve metal ions. Organic aromatic groups such as those containing benzene or naphthalene rings are also reversibly electrolyzed at electrodes, but only in certain solvents. The solvents must be nonaqueous and cannot donate a proton to the reactant(s). Representative of these are the organic aprotic solvents such as acetonitrile ($CH_3CN$) and methylene chloride ($CH_2Cl_2$), which is also known as dichloromethane. In this book, only analyses involving reversible reactions will be considered.

## DC Polarography and the Dropping Mercury Electrode

The scanned potentials in voltammetry are usually produced using a three-electrode potentiostat so that we can know accurately the working electrode's potential. However, problems can arise from adsorption of organic species or some "gunk" on the electrode surface during electrolysis: An example of the latter is a salt precipitate. Another possibility is an organic "tar" or polymer that coats the electrode when some organic species is electrolyzed. If the voltammetry is done in organic solvents, this is quite likely to occur at the regions of solvent breakdown, when the solvent itself electrolyzes. (The region of solvent breakdown was discussed in association with Figure 12.21.)

In order to overcome this problem, one approach is to use an electrode that continually renews its surface. This apparently strange requirement can be satisfied by using a **dropping mercury electrode (DME)** as the working electrode in an electrochemical cell. A pictorial view is shown in Figure 12.27. In the illustration, the auxiliary and reference electrodes are the same as usual: a calomel reference with a KCl salt bridge and a platinum wire auxiliary in a compartment separated from the bulk solution. When the working electrode is a DME, the voltammetric apparatus is called a **polarograph,** for historical reasons.

The working electrode consists of a drop of mercury. The drops are formed from a reservoir of mercury, which is held high enough so that the pressure of mercury forces the liquid slowly through a capillary tube. The mercury droplets continually grow and, when they become large enough, fall off the capillary tip to the bottom of the container. The system is designed so that the largest droplets are less than a millimeter in diameter. The mercury flow rate is constant and is regulated so that a new drop forms about every 4 to 6 seconds. In addition, most contemporary polarographs have a small plunger attached to the capillary tube. At regular intervals, the plunger is driven to strike the capillary and dislodge the drop that has formed at the end. In this way, the time between drops (the **drop time**) and drop size can be controlled and synchronized with electronic controls. The synchronization is essential for more sophisticated voltammetric determinations, which will be discussed soon.

The data from a voltammetric scan with a dropping mercury electrode are shown in Figure 12.28. They are labeled DC in the figure, since the technique just described is

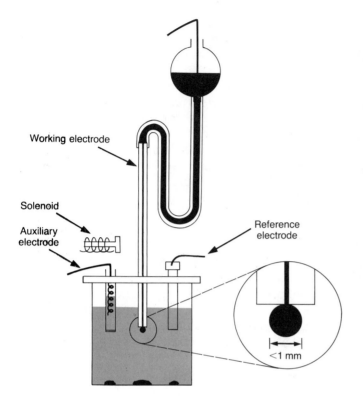

**Figure 12.27** A line drawing of the apparatus used for forming a working electrode from a constantly flowing mercury stream. This is a dropping mercury electrode, DME. The reference and auxiliary electrodes are set up in chemical isolation and connected either with a frit (common for the auxiliary) or KCl salt bridge (common for the reference). The working electrode itself is a small drop of mercury at the tip of a thin-bore capillary tube. The drop falls off when it gets larger than about one millimeter in diameter. It can be knocked off at a constant size by activating the solenoid at regular intervals. This apparatus can be used for all the forms of voltammetry. Often the experiments are conducted in deoxygenated solutions under an inert atmosphere such as $N_2$ or Ar.

called DC **polarography**. The "hash" of the scan is due to the changing size and surface area of the mercury drop. The current is smaller when a drop is small than when the drop becomes large. The larger drop has a larger area in contact with the solution, and so more electrolysis occurs, resulting in a larger current. Note that the currents are in the μA range.[8]

As the data for an assay, the heights of the peaks of the oscillating current are most useful. The peaks of the current curve occur just before the mercury drops fall off the capillary. We can, if we want, keep the useful information—the magnitudes of the peaks of the current—and eliminate recording the distracting "hash" current. This can be done by electronic circuits. The current is not recorded until just before the mercury drop is knocked off by the solenoid. At that time, the current is measured for a short time and recorded on the graph. This technique is called **sampled DC polarography**. An example of a sampled DC polarogram is shown in Figure 12.29. This voltammogram (sampled DC polarogram) was scanned from more-positive to more-negative potentials: from left to right in the figure. In the solution were Cu, Zn, Pb,

---

[8] The Ilkovic equation discussed in Chapter 1 applies to polarographic limiting currents. It is

$$i_L = (7.082 \times 10^4)nv^{2/3}t^{1/6}D_A^{1/2}C^0$$

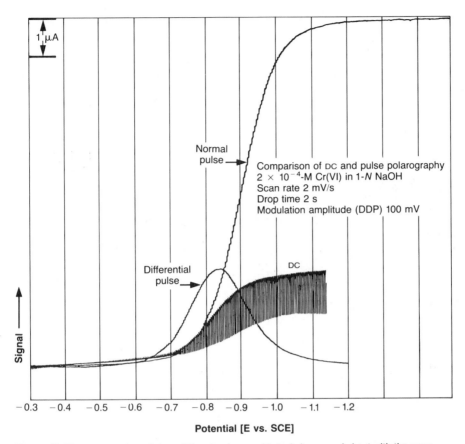

Figure inner labels:

1 μA

Normal pulse →

Comparison of DC and pulse polarography
$2 \times 10^{-4}$-M Cr(VI) in 1-*N* NaOH
Scan rate 2 mV/s
Drop time 2 s
Modulation amplitude (DDP) 100 mV

Differential pulse →

DC

Signal →

$-0.3$　$-0.4$　$-0.5$　$-0.6$　$-0.7$　$-0.8$　$-0.9$　$-1.0$　$-1.1$　$-1.2$

**Potential [E vs. SCE]**

**Figure 12.28**　Comparison of three different polarographic techniques carried out with the same apparatus on the same sample. The sample is a strongly basic solution (NaOH) with $2 \times 10^{-4}$-M Cr(IV) in it. The scan rate of the potential was 2 mV/s, with a drop interval of 2 s. The magnitude of the potential jump for the differential pulse polarogram was 100 mV. This is called the modulation amplitude (see Figure 12.32). Notice that the peak of the differential pulse polarogram appears at the point of maximum rise of the DC polarogram. However, the midpoint of the normal pulse output is not at the same potential. Quantitation for the DC is done using the limiting current on the top of the "hash." Quantitation by normal pulse also is based on the limiting current. For the differential pulse curve, the quantitation is done using the peak height or area above the baseline.

and Cd in the concentrations noted in the figure caption. Notice how the current that is recorded is the *sum* of currents from *all* the ions that can be reduced at the electrode's potential. Each limiting-current plateau serves as a baseline for the metal reduced at a more-negative potential. Thus, as the working electrode's potential is made more negative, the electrolysis current arises first from the most easily reduced ion. As the voltage scan continues, the potential passes through the formal potential (here it is equal to $E_{1/2}$) and into the region of diffusion-controlled current. The same sequence occurs for the second most easily reduced ion. The limiting current then becomes the sum of those for both ions. At the end, the total limiting current is the

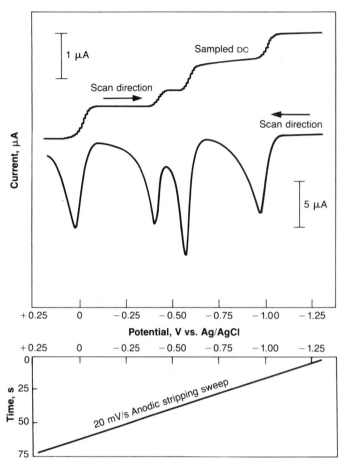

**Figure 12.29**    Illustration of a sampled-DC polarogram and the anodic stripping voltammogram for the same solution containing four different metals: 2.5 ppm in Cu and Zn and 5 ppm in Pb and Cd. The solution was 0.1-M sodium acetate with 0.01-M $HNO_3$ added, and the pH was 5.5. For the sampled-DC polarogram (upper curve), the scan rate was 10 mV/s beginning near +0.2 V. The anodic stripping determination (middle curve) was done by depositing at −1.300 V versus Ag/AgCl for 90 s, followed by a rest period. The positive scan was then made at a rate of 20 mV/s. The voltage scan protocol is plotted on the lower graph. Notice the difference in scales for the currents for these two methods. [Figure courtesy of John Wise and William Heinemann. Reprinted from "Laboratory Techniques in Electroanalytical Chemistry," P. T. Kissinger and W. R. Heineman, Eds.; Dekker: New York, 1984, by courtesy of Marcel Dekker.]

sum of all four. The assignment of the ions to each of the polarographic steps in Figure 12.29 is problem 12.16.

Note that if the solution were stirred, and the time allowed were long enough, the bulk of the solution would be depleted of the reducible ions. This would then be the same as a coulometric experiment but with a very small mercury electrode. For example, compare the apparatus sketched in Figure 12.27 with that in 12.17 and notice how similar they are.

There are differences, however. In the coulometric experiment, we want all the ions to contact the surface as quickly as possible. A large electrode surface area and vigorous stirring helps this. On the other hand, for a voltammetric assay, we want the current to be as reproducible as possible. The most common way to achieve this is to use a smaller electrode immersed in a solution that is completely still. The currents in a stirred solution are generally not as reproducible as those in a completely still one.

There is one other effect that must be considered in obtaining reproducible limiting currents which can be used for a polarographic assay. It should be clear that whatever species is electrolyzed changes its charge. Reduced species will become more negatively charged; oxidized species will become more positive. (An exception to this rule might be caused by further chemical reactions occurring away from the surface of the electrode.) Thus, near an electrode where electrolysis is occurring, a number of ions with different charges and perhaps some neutral species will be present. Ions of different charge will be attracted to or repelled from the charged electrode surface at different rates. The effective migration rates will vary with the potential. As a result, the limiting currents will depend on voltage and also on ionic charge. In order to obtain reproducible limiting currents, these complicated voltage-dependent effects must be eliminated. But you have already read about how to reduce the effects of ionic interactions in solution and at interfaces: Raise the ionic strength with inert electrolyte. Thus, quantitative voltammetric assays are inevitably done with a high concentration of background electrolyte present. One commonly used solution is $0.1-1$ M $Na_2SO_4$. Under these conditions, the current depends only on the rate of diffusion. The resulting diffusion-limited current is then proportional to the bulk-solution concentration of the species being reduced or oxidized at the working electrode.

### Pulsed Voltammetry

Electrochemists call the part of the total current that is due to electrolysis a **faradaic current.** However, whenever an electrode has a *changing* potential applied to it, there is another contribution to the observed current besides electrolysis. This other contribution, called a **nonfaradaic current,** is due to a capacitance at the surface of the electrode. This capacitance causes the electrode to act like a capacitor, some of the properties of which were described in Section 3.6. One characteristic is that a current arises upon changing the voltage across a capacitor. For a step jump in voltage, the currents rise or decay with a simple, exponential time dependence.

How can the surface of an electrode be like a capacitor? A somewhat oversimplified answer is illustrated in Figure 12.30. The system described is a mercury electrode in contact with an aqueous ionic solution. (There is general agreement on the major features of the electrode–solution interface. However, disagreement about details still exists.) Immediately adjacent to the mercury surface is a layer of adsorbed water molecules. If there are any ions in this first layer, they tend to be anions, such as $I^-$, rather than cations, such as $Na^+$. In either case their numbers are relatively small compared with the water. When the metal electrode is uncharged, the anions and cations outside the first water layer are randomly mixed. However, when the electrode becomes charged, the ions of the opposite charge are attracted to it, and those with the same charge as the surface are repelled. Say the electrode is negatively charged. As a result, an excess of positive charges will locate nearer to the surface, outside of the surface water layer. The negative charges that are expelled from this region cause an excess of negative charge to be located further away. This region of separated charges is called the electrical **double layer.**

What do we see in our experiment as a result of the formation of the double layer from randomly distributed ions? The motion of the anions and cations as they rearrange *is* an electric current in the solution at the interface. This current of double-layer formation is called the **capacitive current.** Since no electrons are trans-

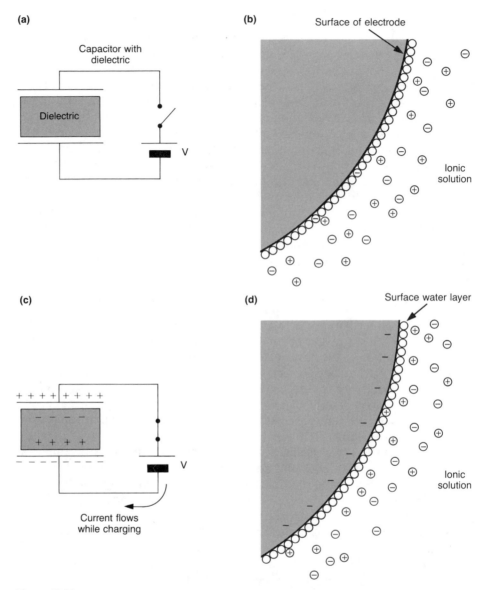

**Figure 12.30** Illustration of the similarity of behavior for the surface of a metallic electrode in an ionic solution and a capacitor with a dielectric between its plates. (a) The capacitor is uncharged, and the charges within the dielectric are random. (b) The electrode is uncharged, and the ions are randomly distributed in the region of the interface. If ions are adsorbed, they tend to be anions. (c) The voltage is applied to capacitor plates causing an electric field to appear between the plates. Charges appear at the surface of the dielectric. For the charges to appear, electric currents must flow within the dielectric. When the capacitor plates are discharged, the charges in the dielectric revert to the original random distribution. (d) The electrode is negatively charged, causing the nearby positive ions to tend to move toward the surface and the accompanying negative ions to tend to move away. This charge rearrangement is a current and produces a less random distribution of charge at the interface—the double layer. When the charge of the electrode is removed, the ions return to the more random distribution, and the double layer "collapses" again producing a current.

ferred in the process, the capacitive current is a nonfaradaic current, and it is transient. In a 1-M salt solution, the double layer forms in less than 100 microseconds after the electrode is charged. The capacitive current flows no longer than that. When the electrode surface is discharged, the double layer disappears within the same length of time. A transient capacitive current in the opposite direction accompanies the disappearance.

A double layer forms at an electrode interface upon charging and disappears when the electrode is discharged, accompanied by the associated nonfaradaic currents. A nonfaradaic current will also be generated at the surface of an electrode composed of a growing drop of mercury held at a fixed potential. To make this clear, consider the following picture shown in Figure 12.31. Assume that we have a small uncharged drop of mercury as a working electrode. Now instantaneously put a fixed potential on the drop. A nonfaradaic current flows until the double layer is established.

In a second step, let us consider that we can instantaneously enlarge the drop so that the surface area becomes twice as large but retains the same potential. However, since the surface area is larger, more charges must move at the interface to establish

**Figure 12.31**    Illustration of a thought experiment to show how a growing drop of mercury continually generates capacitive (nonfaradaic) currents. First the mercury is rapidly charged with a step potential jump, as illustrated in the lower graph. The middle graph shows the capacitive current that rises and rapidly decays (in about 100 $\mu$s in a 1-M solution). Then an *imaginary* instantaneous jump in drop size is made. The area of the drop increases instantaneously, but the potential stays the same, as shown on the lower graph. However, more double layer forms, which produces more capacitive current with the same decay time. A continuously growing drop produces a continuous capacitive current.

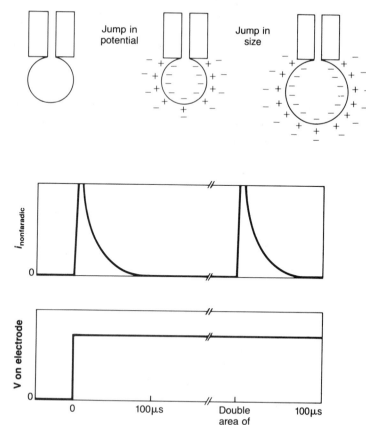

the double layer corresponding to the one that surrounded the smaller drop at the same potential. A current flows. In other words, the surface area of the double layer increases in proportion to the increase in drop area at constant potential. The capacitive current flows for a short time when the drop is instantaneously enlarged. Thus, when the drop is continually increasing in size, there is a continual capacitive current. This current will add to the faradaic currents producing the total observed current.

In order to assay many materials voltammetrically at concentrations in the range $10^{-4}$ to $10^{-5}$ M, we need not worry about nonfaradaic current; the faradaic currents from the electrolysis of the analyte will be much larger. However, when we want to determine concentrations of more dilute solutions, in the range of $10^{-6}$ M and below, special efforts must be made to separate out the faradaic currents that are used for the assay from the nonfaradaic currents and background currents that are simultaneously present. To do this, two different **pulsed voltammetry** methods are used. (With a DME working electrode, the term **pulsed polarography** is applied.) Understand before you proceed that the experimental cell and DME or other working electrodes look the same for all the voltammetric experiments. However, the potentiostat is used in a different way.

The purpose of using pulsed voltammetric methods is to decrease the influence of the nonfaradaic currents on the measurements. For clarity, let us consider pulsed polarography and the DME. Based on the discussion above, we can state two properties of the nonfaradaic currents:

A.  The faster the *relative* growth of the mercury drop surface area, the larger the nonfaradaic currents.
B.  After a voltage change, the nonfaradaic currents decrease very rapidly relative to the faradaic currents (surface area fixed).

To minimize the capacitive currents, we take advantage of these two properties. (The basis of the techniques that have been developed are illustrated in Figure 12.32 and are described in more detail in Supplement 12D.)

A.  The sampling of the total current is delayed until the mercury drop is as large as possible, at which time the relative area of the drop is changing most slowly. The measurement is made with the drop area nearly constant.
B.  The potential is pulsed to the desired level, and the resulting current is not sampled until a time (after hundreds of microseconds) when the nonfaradaic currents have decreased significantly.

This manipulation of the potential on the working electrode and the collection of data is done electronically. An example of the results of a **normal pulse** polarogram is shown in Figure 12.28. Each step of the polarogram represents a single pulse from an initial potential at around $-0.3$ V to the potential at which the step appears. In a normal-pulse polarogram, the point of inflection of the curve is not at the same potential as the $E_{1/2}$ of a DC polarogram.

There are two more benefits of waiting until the droplet is large and then pulsing the potential to the test value. We can obtain a relative increase of the analyte's faradaic current compared with that in a DC or sampled DC polarographic assay. The reasons are the following:

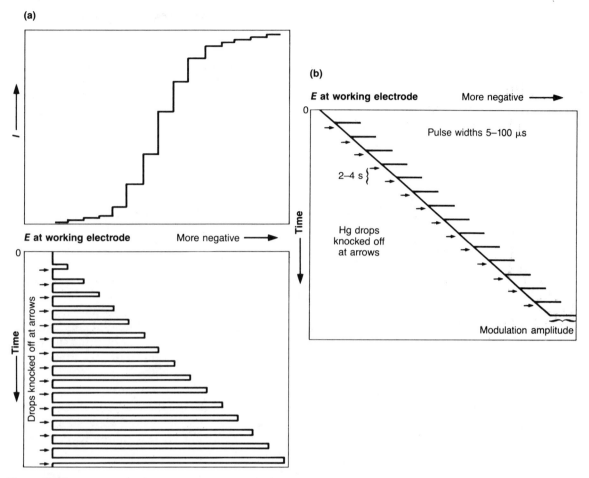

**Figure 12.32**    Graphs of the voltages on the working electrode when doing normal pulse and differential pulse polarography. (a) In the normal pulse system, pulses are applied after the drop has had time to grow. Immediately after the pulse, the drops are mechanically knocked off the capillary so that the drop size is reproducible for each pulse. The voltage pulses continually increase in magnitude from a fixed baseline potential. This protocol enhances sensitivity but distorts the polarographic-current shape somewhat. The general relations between pulse size, time, potential, and measured current are shown. (b) The differential pulse polarography pulse protocol differs in that the pulses are all the same size (called a modulation amplitude). The base potential is changed at a fixed rate as is done in DC polarography. The pulses are superimposed on the changing potential with a pulse height in the range of 12–100 mV. The instrument output has the shape of a derivative of a sampled-DC polarogram. The pulse- and current-sampling protocol (see Supplement 12C) minimizes the effects of nonfaradaic currents and subtracts background currents. Differential pulse polarograms are shown in Figures 12.28 and 12.33.

1.  The larger drop has a larger surface area, so the faradaic current is larger.
2.  Since the pulse is from a voltage where no electrolysis occurs to a region of electrolysis, the *bulk* concentration of the electrolyzable species is at the surface of the mercury drop when the voltage pulse is applied. This is not true of DC or sampled-DC polarograms.

The results due to the second point can be seen from the experimental data shown in Figure 12.28 by comparing the level of the limiting currents of the DC and normal pulse polarograms shown there.

To review: In a pulsed polarogram, the plateau region of the diffusion current is still proportional to the concentration of the electroactive species, and the midpoint of the current rise is still dependent on the species being reduced even though it is not equal to $E_{1/2}$. Pulsed-polarographic methods enable measurements of less concentrated solutions than DC or sampled DC polarography permit. Solutions from $10^{-6}$ to $10^{-7}$ M can be analyzed with precisions as good as $\pm 3\%$. Further details of the voltage pulses and data collection are discussed in Supplement 12C.

### Differential Pulse Polarography

You have read that improvement in the sensitivity of polarography can be obtained by using voltage pulses instead of a linear sweep of the working electrode potential and that, with careful sampling of the currents, we can separate the interfering capacitive currents from the analytically useful faradaic currents. However, the faradaic background currents from the solution itself will not be removed in this way. We could run blanks, if they could be made exactly equivalent to the assayed solution but without its analyte. However, an experimentally easier way has been devised to minimize the background currents. The method is called **differential pulse polarography.** In it, the background currents in the same solution are sampled and subtracted from the total currents due to the sum of the faradaic currents: those of the sample plus those of the background.

The experimental apparatus is the same as before. However, the pulsing voltages placed on the working electrode (see Figure 12.32b) and the sampling of the currents are different. With contemporary potentiostats, the pulse train generation is done automatically. Some details of the pulse sequence and current-sampling protocol are presented in Supplement 12C.

As can be seen in Figure 12.28, the resulting output is a measure of the *slope* of current versus voltage of a DC or sampled DC polarogram. The steepest slope of a standard polarogram is at the $E_{1/2}$ of the material being electrolyzed. In a differential pulse experiment, the *peak* appears at the $E_{1/2}$. The area of the peak above the baseline is directly proportional to the concentration of the electroactive species being assayed. Alternatively, the peak height can be used to measure concentration if the polarograms are the same shape.

An example of how the differential pulse polarography background suppression allows more precise analyses at lower concentrations is shown in Figure 12.33. The data shown in that figure result from 0.36 ppm of tetracycline. This is equivalent to a $8 \times 10^{-7}$-M solution. With a baseline correction, precision in the range of a few percent is possible.

### Scanning Voltammetry – Cyclic Voltammetry

All the polarographic methods described earlier are run with the voltage scan rate around 100 mV per minute. This allows time for the diffusion-limited currents to be established at reproducible values. The subsequent limiting currents of the analytes

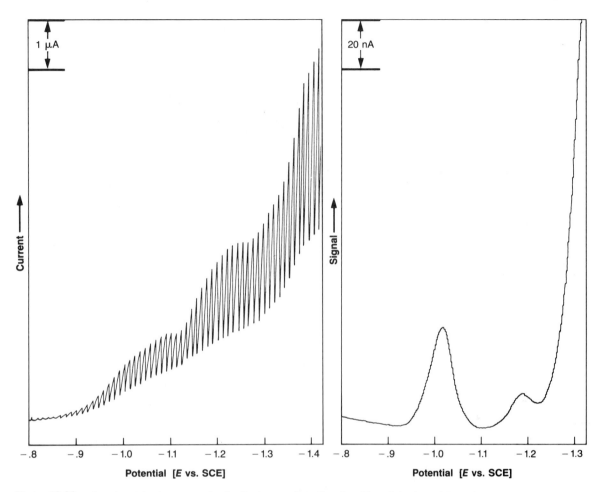

**Figure 12.33**   Experimental polarograms showing the improved results using differential pulse polarography compared with DC polarography. The polarograms are of tetracycline hydrochloride (an antibiotic). Both polarograms were obtained from samples in a 0.1-M acetate buffer, pH 4, with a dropping mercury electrode as the working electrode. A drop was generated each second. Notice that the current scales of the polarograms are different. Also, the concentration for the DC polarogram is 180 ppm while the concentration for the differential pulse determination was 0.36 ppm—a factor of 500 more dilute.

are the principal data that we use to quantify the contents of numerous, diverse materials.

If, however, we wish to find formal potentials $E°'$, an alternative method is often faster and easier to use. This is **cyclic voltammetry.** In such experiments, the electrodes and other components of the sample cell can be the same as have been illustrated throughout this chapter. The working electrode can be gold, platinum, carbon, mercury, or any other conducting material. If mercury is used, it is usually not a dropping mercury electrode. Instead, a small, fixed drop of mercury is used. This is called, reasonably, a **hanging mercury drop electrode.**

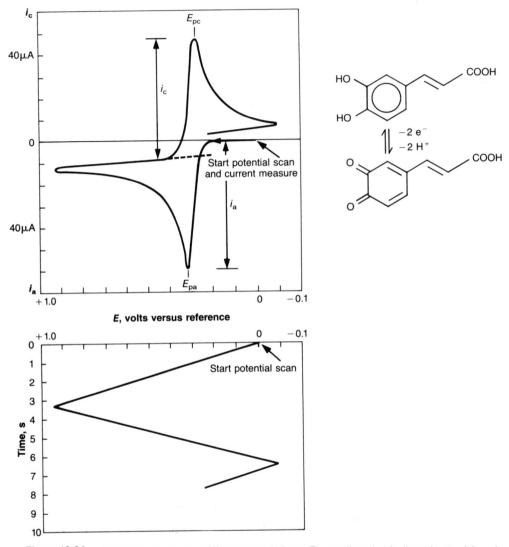

**Figure 12.34** A cyclic voltammogram of the compound shown. The anodic and cathodic peak potentials and peak currents are defined. The currents are measured from the baseline extended from a region before the electrolysis begins. The potential is scanned as shown in the lower graph. The voltage axes of both graphs are the same. The rise in the currents of the cyclic voltammogram is similar to that for a polarogram. However, instead of reaching a limiting current, the current falls off due to depletion of the analyte in the region near the electrode. The cathodic peak is due to reduction of the material oxidized earlier in the scan. The original oxidation produces the anodic peak. The "cyclic" shown here begins at a potential of zero volts relative to the reference potential, and the direction of the first scan is toward more-positive potential. The electrode used for this experiment could not be mercury, since mercury itself will oxidize in the voltage range and would produce a large background current.

Another difference in this technique is in the way the potential on the working electrode is changed over time. As for all voltammetric techniques, the voltage is scanned, and the resulting current is measured simultaneously. However, the potential is not scanned at 100 mV/min, but much faster—up to 100 V s$^{-1}$. However, in normal use, the rate of voltage scan is nearer 0.01–1 V s$^{-1}$. Also, there are no pulsed-potential or special current-sampling protocols.

The voltage scan is not unidirectional as in polarography but is scanned in both forward and reverse directions. An example of a cyclic voltammogram and a graph of the values of time and voltage associated with the potential scan are shown in Figure 12.34.

In this voltammogram, the linear scan of the potential starts at 0.0 V. It is taken to slightly over +0.9 V, then reversed to about −0.1 V, and finally brought back to about +0.2 V. This is slightly more than one **cycle,** which was completed upon returning to 0.0 V.

Before the underlying causes of this strangely shaped curve are described, we need some definitions and an explanation of the conventions of the plot.

A. The horizontal axis, the voltage axis, is written with more negative potentials to the right and more positive to the left. This convention arose from polarography. However, an international scientific body has suggested that this convention be reversed. So read the axis labels carefully.

B. The currents for reduction are plotted increasing upward as they are in the polarograms in Figure 12.24. These are called cathodic currents and are labeled $i_c$. Thus, the oxidative currents or anodic currents, $i_a$, are downward.

C. The voltage scan started at zero and went positive. A voltage scan toward more-positive potentials is called a **positive scan.** A voltage scan toward more-negative potentials is a **negative scan.**[9]

D. There are peaks in the currents, as can be seen in Figure 12.34. The potential at which the peak is reached in the cathodic current is called the **cathodic peak potential** and is abbreviated $E_{pc}$ or $(E_p)_c$. The former notation is used here. The **anodic peak potential** is, as shown, abbreviated $E_{pa}$.

E. The currents at the peaks are abbreviated $i_c$ and $i_a$ for cathodic and anodic peak currents, respectively.

F. The peak currents are measured as the current above background levels. The background extrapolation is indicated by the dashed lines in the figure. Notice how similar this measurement is to finding the limiting current above background for a polarogram (illustrated in Figure 12.26). This should be no surprise since much is the same except the rate of the voltage change on the working electrode. The heights of

---

[9] In place of positive scan and negative scan, you may find the terms *anodic scan* and *cathodic scan*. Their use is discouraged.

the peak currents (all other conditions remaining the same) are proportional to the concentrations of the electroactive species involved.

Now that the conventions and definitions have been explained, two important relationships of cyclic voltammograms are stated without proof.[10] They are true for electrochemically reversible reactions.

The first is that the formal potential of the electrolyzed species is the mean of the peak potentials. Algebraically,

$$E^{\circ\prime} = \frac{E_{pc} + E_{pa}}{2} \tag{12-12}$$

This simple relationship makes it easy to find a formal potential. Problem 12.24 will give you some experience in calculating the formal potentials from cyclic voltammograms or cyclics, as they are often called.

The second relationship is that the difference in voltage between the peaks is inversely proportional to the number of electrons transferred. Algebraically,

$$E_{pa} - E_{pc} = \frac{0.059}{n} \qquad \text{at 25 °C} \tag{12-13}$$

This relationship holds for electrochemically reversible couples. If the reaction is not reversible at the electrode, then the peaks will be further apart. Thus, the relationship holds only in the limit of reversibility. Otherwise, Eq. 12-13 describes a lower limit to the peak separation. With this caveat, the value of $n$, the number of electrons transferred in the reaction, can be determined from cyclic voltammograms.

As you can see, the positions of the observed peaks are quite useful. Redox potentials can be found in a few minutes. Each redox couple produces a set of peaks around its $E^{\circ\prime}$. Thus a cyclic voltammogram can quickly show the presence of all species that undergo oxidation–reduction reactions at the electrode—within the limits set by solvent electrolysis. A limitation is that the formal potentials must be sufficiently different to resolve the individual reactions occurring at different potentials.

But why are there peaks in the current at all? A simplified answer is that the voltage scan is so fast that an experiment "isolates" the region around the electrode from the bulk solution. There is not enough time for diffusion to transport material between the bulk solution and the region adjacent to the electrode surface.

For each peak, the rise in the current from the baseline through the region of the formal potential is caused in the same way as in polarography: by an electrolysis rate that depends on the electrode potential as compared to the formal potential of the redox couple. The rise can be compared to the increases in current of the sampled DC polarogram in Figure 12.29.

The decrease in the current is similar to the decreases seen in coulometry (as illustrated in Figure 12.20) in which the current decreases as the electroactive species becomes exhausted. During the relatively fast voltage scan, the small, "isolated"

---

[10] The rigorous derivations can be found in the Bard and Faulkner reference listed at the end of this chapter.

volume near the electrode becomes exhaustively electrolyzed. In effect, the diffusion-limited current decreases as the region around the working electrode becomes depleted of reactant. A more detailed answer is quite complex mathematically, and you are referred to the general references for details.

## 12.12 A Special Application of Cyclic Voltammetry

Of all the polarographic methods, cyclic voltammetry alone has all of these four special properties.

1. The peak currents can be used to quantify electroactive species.
2. The voltage at which the peaks occur can be used to aid identification of the species being electrolyzed.
3. It allows a number of different species to be determined in a single scan if the $E^{\circ\prime}$ values are far enough apart.
4. It is fast. Seconds to minutes are all that are needed to run a cyclic.

These four properties of cyclic voltammetry allow it to be used to determine the change in concentration of some neurotransmitters inside the brain cavities of animals.

Neurotransmitters are molecules that act as messengers between nerve cells. When a traveling nerve pulse reaches the end of a nerve cell, a number of neurotransmitter molecules are released and diffuse to the surface of the adjacent cell where they bind and cause the adjacent cell to activate. (This is only one of the mechanisms by which nerve cells transmit information among themselves.) Two of these transmitters are electroactive: dopamine and 5-hydroxytryptamine. Their structures are

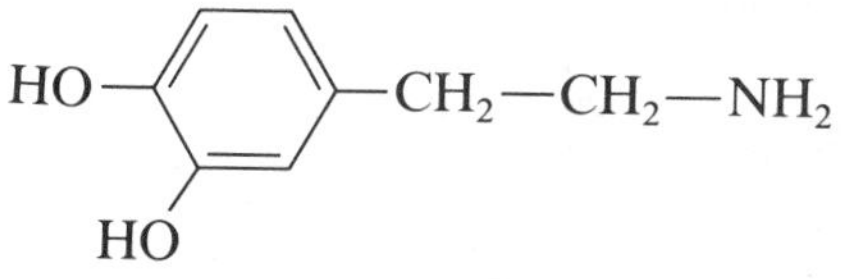

dopamine 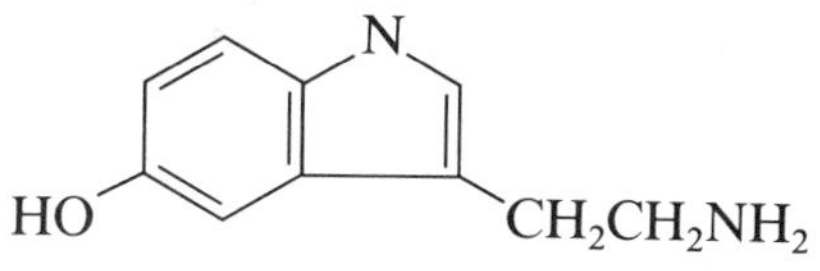                      5-hydroxytryptamine (5HT)

Acetylcholine and $\gamma$-aminobutyric acid (GABA), shown below, are also neurotransmitters but are not electroactive.

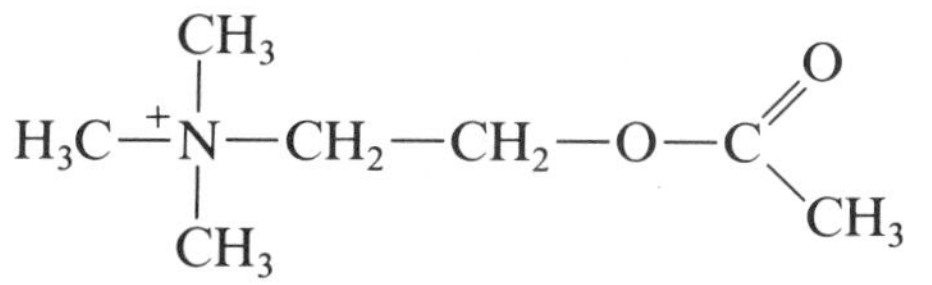

acetylcholine 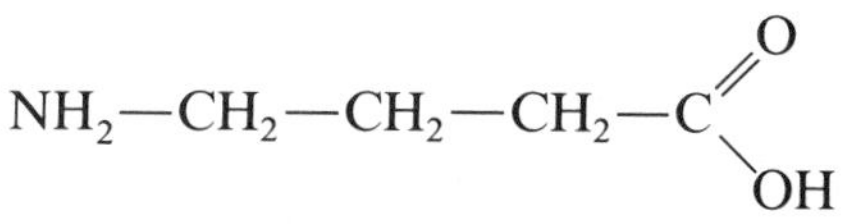                      $\gamma$-aminobutyric acid (GABA)

One further definition might be useful before you read the following brief excerpt from two papers describing this technique: Cerebrospinal fluid is the fluid that bathes the brain and spinal cord. The authors also mention homovanillic acid (HVA), which is a metabolite of dopamine. (The structure is shown on the following page.)

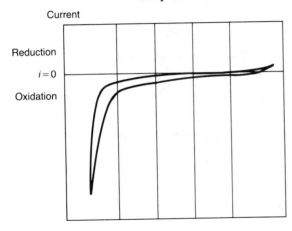

homovanillic acid (HVA)

An in vivo voltammetry technique has been developed which enables direct and continuous measurement to be made of the release of neurotransmitter metabolites . . . . The measurement is simple in principle. A potential is applied to a micro carbon electrode [placed into specific brain regions]. One then measures the minute current which results from the oxidation or reduction of small molecular weight constituents near the sensor electrode. The potential at which the electrolysis occurs serves as a qualitative indication of the constituent being monitored, and the current itself is directly proportional to the concentration of this component . . . .

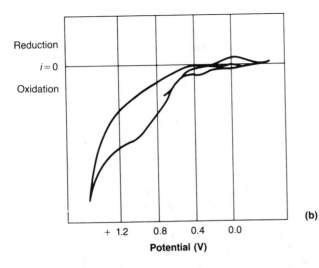

**Figure 12.35**   Cyclic voltammograms in cerebrospinal fluid using a working electrode made of a carbon fiber. (a) The voltammogram of the background fluid. (b) The voltammogram after nerve stimulation. The baseline has changed. If it had not, the current as indicated by the upper trace of b at around +1.2 V would lie at least as far toward the reduction region (upward) as the comparable current trace before stimulation. [Reprinted by permission from Wightman, R. M., et al., *Nature* **1976**, *262*, 145–146. Copyright © 1976 Macmillan Journals Limited.]

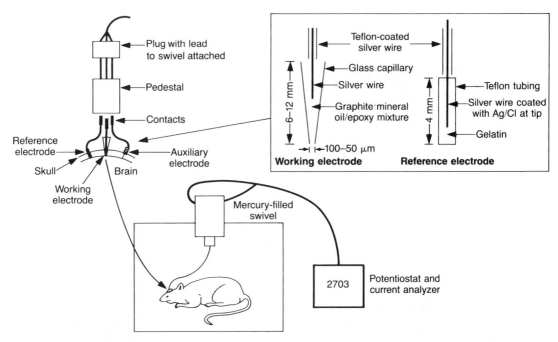

**Figure 12.36**    Diagram showing the basic setup for in vivo voltammetric measurement of dopamine and 5HT release from the brain. The pedestal with the electrode contacts inserted into it is cemented onto the skull with dental cement. The working and reference electrodes are constructed as shown at top right. The working electrode is inserted into the specific brain regions. The reference electrode is placed on the brain surface. The auxiliary electrode is a steel screw fixed into the skull with a Teflon-coated silver wire attached to it. [Ref: Adams, R. N., et al. *British Journal of Pharmacology* **1978**, *64*, 470P–471P.]

The catecholamines and their metabolites as well as 5-hydroxytryptamine and its metabolites are readily oxidized. [Thus the technique is applicable to nerve synapses which use these compounds as transmitters.] Acetylcholine, GABA, and various other aliphatic amines and acids which have suspected neurotransmitter roles are not electroactive and are not sensed by the electrode in the conditions used . . . .

Figure [12.35] illustrates the qualitative aspects of the electrochemical experiment. The cyclic current–voltage curves (voltammograms) . . . were initiated by a triangular wave potential sweep from 0.0 to 1.4 V and then returned to −0.2 V. (All potentials are measured against a micro SCE normally placed in contact with brain tissue over the frontal sinus area.) Figure a shows the baseline level of components . . . which can be oxidized within this potential range (primarily ascorbate and micromolar levels of HVA). . . . [As seen in Figure b], following stimulation . . . a large current [arises] with a plateau at around +0.8 V which is a result of HVA oxidation. (Although this is a large increase, all electrochemical currents referred to herein are at most several hundred nA.) The small peaks around +0.1 V . . . come from oxidation and reduction of dihydroxy-phenylacetic acid . . . .

[Figure 12.36 illustrates the apparatus used to obtain the electrochemical information in freely moving unanesthetized rats.]

[Refs: Wightman, R. M., et al. *Nature* **1976,** *262,* 145–146. Adams, R. N., et al. *British Journal of Pharmacology* **1978,** *64,* 470P–471P.]

## 12.13 Anodic (and Cathodic) Stripping Voltammetry

Anodic stripping voltammetry, a method that combines coulometric and voltammetric techniques, is used for trace analysis for metals in the range of $10^{-6}$- to $10^{-9}$-M solutions. This is in the range of ppb and below. The steps of this technique are illustrated in Figure 12.37. The output from an anodic stripping experiment is shown in Figure 12.29.

The lower limit of concentration for which differential pulse polarography is able to yield good results is in the range $10^{-6}$ to $10^{-7}$ M in the electroactive species. However, if the material being assayed can be concentrated, the limit of the working range *as a fraction of the sample* can be lowered significantly. In other words, if the analyte is first concentrated by a factor of 100, then the assay method can be used to obtain results on samples that are 100 times more dilute. This is the principle behind stripping voltammetry. A hanging mercury drop electrode is commonly used as the working electrode, although a thin film of mercury deposited on platinum or carbon (graphite) can be used as well. We shall discuss only a mercury drop electrode.

Initially, a solution with the sample in it is electrolyzed for 0.5 to 5 min with stirring. This is similar to a coulometric determination of material, only in this case the total charge might be too small to measure in the presence of faradaic background from the solution. In the example shown in Figure 12.29, the electrolysis was done in a stirred solution for 90 s at $-1.3$ V vs. SCE.

At this negative potential, ions of metals that form amalgams with mercury will be reduced and will dissolve in the mercury drop. The metals that will do so are listed in Table 12.10. In this way, the elements are concentrated from the large solution volume into the small mercury drop, less than 1 mm in diameter.

Next, the drop is isolated electrically and given time to equilibrate. This equilibration time and isolation are necessary for three reasons. If the circuit to the working electrode remained connected, more ions would be reduced, or some of the amalgamated (dissolved) metals might reoxidize. (Whether oxidation or further reduction occurred would depend on the potential.) Second, the metals that have been reduced have entered the drop through its outer surface. It takes some time for the metal atoms to diffuse and become equally distributed inside. And third, the external solution, which was stirred during the reduction–concentration step, needs time to become still. In other words, after the concentration step, some time is needed to obtain an equilibrated, unstirred solution both inside the mercury drop and in the aqueous solution surrounding it.

Following the rest period, the potential is swept linearly in the positive direction starting from the deposition voltage. In turn, the metals are reoxidized. A peak is produced for each, and the peak appears at a characteristic potential for its metal. In addition, the height of the peak is proportional to the concentration in the sample solution. The peak heights are also proportional to the deposition (concentration)

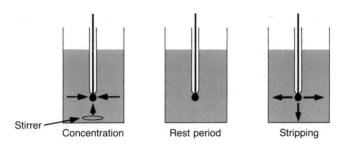

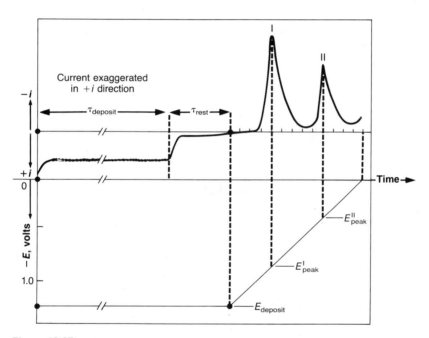

**Figure 12.37**    Illustration of the steps of an anodic stripping determination using a mercury-drop working electrode. The solution is stirred while the metals are reduced and amalgamated in the electrode. Then there is a rest period when the solution is allowed to become still, and the amalgam is allowed to become homogeneous. Following this, the potential is scanned in the positive direction. The metals reoxidize at their characteristic potentials, producing current peaks that rise when each metal begins to oxidize and fall off due to depletion of the metal in the mercury drop. The area under each peak and above the baseline is proportional to the concentration of each metal. For well-behaved (reproducible) systems, the peak heights can be used as a measure of the concentration with equal precision. The overall precision and accuracy of analyses depend on the artistry of the person doing the analysis and sample preparation. Another example of a stripping voltammogram is shown in Figure 12.29. The stripping peaks are quite near the $E_{1/2}$ of each metal.

**Table 12.10**   Elements Determinable by Anodic and Cathodic Stripping

*By anodic stripping*
Ag*, As, Au*, Ba, Bi, Cd, Cu, Ga, Ge, Hg*, In, K, Mn, Ni, Pb, Pt, Sb, Sn, Tl, Zn

*By cathodic stripping*†
Br⁻, Cl⁻, I⁻, S⁻, thio compounds

\* Determined on solid electrodes, such as carbon or gold.
† Cathodic-stripped compounds form mercury precipitates at the electrode which are subsequently stripped off in a negative scan.

time; more ions deposit during longer deposition times. As a result, standards must be used with the technique.

Why is the current peaked as opposed to having a series of plateaus as with polarography? The answer to that question is similar to the answer given for cyclic voltammetry. The rise is a result of the formal potential being reached for the reaction

$$\text{metal(amalgam)} \rightleftharpoons M^{n+}\text{(aqueous)} + n\,e^-$$

The falloff is due to the exhaustion of the metal atoms in or on the mercury drop. The material has been completely electrolyzed. To see this similarity more clearly, look at the shapes of the anodic currents in Figure 12.29 and compare them with the anodic current in Figure 12.34. Both have a sharp rise on the leading (time) edge with a slower falloff on the trailing side.

Even greater sensitivity can be obtained if the anodic scan is done with differential pulse voltammetry. Concentrations in the range of $10^{-9}$ M can be determined with precisions of 3–5%. However, before that level of metal is able to be determined, all the problems of sample preparation for ultratrace analysis will be encountered. Anodic stripping is one of the most effective trace analytical methods for the amalgamating metals, and the equipment needed is relatively inexpensive compared with the spectroscopic instruments that can be used at the same concentration levels.

## Suggestions for Further Reading

The best next place to go for more electrochemistry if you are going to do experiments.
Sawyer, D. T.; Roberts, J. L., Jr. "Experimental Electrochemistry for Chemists"; Wiley: New York, 1974.

A practical approach to ion-selective electrodes with manufacturers' data and numerous examples of applications.
Covington, A. K. "Ion-Selective Electrode Methodology"; CRC Press: Boca Raton, FL, 1979; Vols. I and II.

A physical–chemical and descriptive-chemistry approach to ion-selective electrodes. The text covers the factors responsible for ISE response. There is not a list of uses.
Koryta, J.; Stulik, K. "Ion-selective Electrodes," 2nd. ed.; Cambridge Univ. Press: Cambridge and New York, 1983.

A two-volume set of books on the general properties and various types of ion-selective electrodes. Volume 2 contains an extensive literature bibliography.
Freiser, H., Ed. "Ion-selective Electrodes in Analytical Chemistry"; Plenum: New York, Vol. 1, 1978 and Vol. 2, 1980.

Still the classic work on the subject of oxidation potentials.
Latimer, W. M. "The Oxidation States of the Elements and Their Potentials in Aqueous Solutions"; Prentice–Hall: New York, 1952.

A textbook balanced between theory and practical information with mostly pictorial explanations of the details of electroanalytical chemistry. It is encyclopedic, so know what you are looking for first.
Kissinger, P. T.; Heineman, W. R., Eds. "Laboratory Techniques in Electroanalytical Chemistry"; Dekker: New York, 1984.

A short textbook with mathematical descriptions of the various electrochemical techniques. The equations are not derived, however.
Vassos, B. H.; Ewing, G. W. "Electroanalytical Chemistry"; Wiley: New York, 1983.

An advanced textbook treatment of the topics in analytical electrochemistry. Significant mathematical and thermochemical sophistication required.
Bard, A. J.; Faulkner, L. R. "Electrochemical Methods"; Wiley: New York, 1980.

A complete coverage of the physical chemistry underlying electroanalytical chemistry. Extremely well written and highly recommended.
Bockris, J. O'M.; Reddy, A. K. N. "Modern Electrochemistry"; Plenum: New York, 1970; in two volumes.

A highly readable, advanced text on electrochemistry at solid electrodes. A useful book for finding out why electroanalytical chemistry assays do what they do.
Adams, R. N. "Electrochemistry at Solid Electrodes"; Marcel Dekker: New York, 1969.

A complete book of amperometric titrations.
Stock, J. T. "Amperometric Titrations"; Interscience: New York, 1965.

A brief, clear description at an intermediate level of the capabilities of cyclic voltammetry. A good entry into the primary literature. Emphasis on the electrochemical mechanisms rather than the analytical capabilities.
Heinze, J. "Cyclic Voltammetry"; *Angewandte Chemie International Edn.* **1984,** *23*, 831–847, in English.

An advanced textbook on potentiometric titrations. Not much about how to do the work but loaded with cautions about the techniques.
Serjeant, E. P. "Potentiometry and Potentiometric Titrations"; Wiley: New York, 1984.

A nonmathematical, practical treatment of stripping analysis approximately equally divided between the principles, descriptions of the instrumentation, and applications. A useful next place to read if you plan to do stripping analysis.
Wang, J. "Stripping Analysis, Principles, Instrumentation, and Applications"; VCH Publishers: Deerfield Beach, FL, 1985.

Sets of multivolume compendia of electrochemistry data for various systems listed by element and/or compound.
Meites, L., et al. "CRC Handbook Series in Organic Electrochemistry"; CRC Press: Boca Raton, FL, 1977.

Meites, L., et al. "CRC Handbook Series in Inorganic Electrochemistry"; CRC Press: Boca Raton, FL, 1980.

A multivolume compendium of the electrochemical information, both on EMF and kinetic data, of all the elements. Later volumes contain similar information for organic compounds.

Bard, A. J., Ed. "Encyclopedia of Electrochemistry of the Elements"; Marcel Dekker: New York, 1973.

This book contains fascinating reports of applications of micro-ISE electrodes (small enough to be placed inside a single cell) and enzyme electrodes in biology. It is written at a moderately advanced level.

Kessler, M., et al., Eds. "Ion and Enzyme Electrodes in Biology and Medicine"; University Park Press: Baltimore, 1976.

## Problems

### Potentiometry

12.1    At 25 °C the value of the standard potential for a Ag/AgCl reference electrode is

$$\| Cl(a_{Cl^-} = 1), AgCl(s) \,|\, Ag \qquad E^0 = +0.2222 \text{ V}$$

The value of the potential for the normal silver–silver chloride electrode is

$$\| KCl(1 \text{ M}), AgCl(s) \,|\, Ag \qquad E = 0.237 \text{ V}$$

What is the activity coefficient for chloride ion in 1-M KCl solution at 25 °C?

12.2    A beaker has a pH $= 0.7$ sulfuric acid solution in it. A few volts are applied to two platinum electrodes in the solution for a long enough time that bubbles of oxygen and hydrogen form and adhere on the opposite electrodes. The potential is subsequently measured with a sensitive voltmeter. Note that the pressure of oxygen and hydrogen is 1.0 atm for each. Answer the following questions.
   a.   Assign the positive and negative electrode for the electrolysis of water.
   b.   Assign the positive and negative electrode for the subsequent galvanic cell formed with the gases present on the electrodes.
   c.   For the galvanic cell, from the data in Appendix III, calculate the potentials of the anode and the cathode half-reactions. Assume the pressure is 1.0 atm over the whole depth.
   d.   Calculate the equilibrium potential of the cell. What is the pH dependence of the reaction?

12.3    A fluoride ion-selective electrode responds to both $F^-$ and $OH^-$ but not to HF. A measurement is made and interpreted assuming that all the fluoride ion is unassociated and no other species interferes with the ISE. Data: $K_a(HF) = 7.2 \times 10^{-4}$ at 25 °C; $k_{F,OH} = 0.06$ under the conditions of the experiments. Answer the following questions.
   a.   Will the formation of HF tend to make the measured result low or high?
   b.   Will the presence of $OH^-$ tend to make the measured result low or high?
   c.   Given a true fluoride concentration of 1 ppm, at what pH will the presence of HF create an error of 3% in the measured fluoride?
   d.   Given a fluoride concentration of 1 ppm, at what pH will the presence of $OH^-$ create an error of 3% in the measured fluoride?

12.4    For the method of standard additions in potentiometry, the potential developed in a known volume of sample $V(s)$ is recorded. An aliquot of standard of volume $V(a)$ is added, and the new potential recorded. Show that the concentration of the sample $C(s)$ is related to the concentration of the aliquot $C(a)$ by

$$C(s) = \frac{C(a)V(a)}{(V(s) + V(a))\ 10^{\Delta E/S} - V(s)}$$

$\Delta E$ is the difference in potential between the (sample + aliquot) and the sample alone. The letter $S$ represents the slope of the electrode response in mV per decade (ten-fold) change in activity. The slope of the electrode response must be known under the conditions of the experiment to obtain accurate values with the method. Compare this equation with that derived in Problem 6.5.

12.5    Using a calcium ion-selective electrode in a determination, what is the maximum concentration level of magnesium that can be tolerated in a $10^{-4}$-M $Ca^{++}$ solution and still produce less than a 10% error due to magnesium interference? Under the conditions, $k_{Ca,Mg} = 0.014$. Assume that the activity coefficients of $Ca^{++}$ and $Mg^{++}$ are equal.

12.6    A measurement of the potassium concentration difference between the interior and exterior of kidney tubules (microscopic tubular structures involved with excretion) were made using a double microelectrode. Two microelectrodes are fused together side-by-side. One electrode is a reference electrode. The other is a $K^+$ ion-selective electrode with a liquid membrane. The response of the ISE to changes in potassium activity is $+50$ mV per decade increase in $a_{K^+}$. The activity coefficient for potassium at the solution ionic strength is 0.77. With the double electrode, the measurement inside the kidney tubule produces a potassium ISE potential of $+68$ mV relative to the potential measured with the double electrode in the solution surrounding the outside of the tubule. A spectrometric method was used to show the external potassium concentration to be 2.5 mM.
    a.   What is the external potassium activity?
    b.   What is the internal potassium activity?
    c.   What is the internal potassium concentration?

12.7    [Ref: Light, T. S.; Capuccino, C. C. *J. Chem. Ed.* **1975**, *52*, 247.]
One way to enable accurate determinations of ion concentrations is to keep the ionic strength of the test solution high and relatively constant. In this way, the activity coefficients do not change appreciably, and concentrations of standards and samples can be related directly to each other. To this end, it is useful to add a total ionic strength adjustment buffer (TISAB) to a solution for an ISE analysis. For fluoride, the TISAB has the following compostion: NaCl 1.0 M, acetic acid 0.25 M, sodium acetate 0.75 M, sodium citrate 0.001 M at pH 5.0. The total ionic strength is 1.75. Equal volumes of sample and TISAB are added in one fluoride ISE assay. The purpose of the citrate is to coordinate metal ions such as $Fe^{+++}$ and $Al^{+++}$ which might bind $F^-$ and lower its concentration. Other chelating agents may be used as well.
    A sample of toothpaste weighing 188.0 mg was placed in a 250-mL beaker containing 50 mL of TISAB. The mixture is boiled for 2 min and cooled. The solution is transferred with two washes into a 100-mL volumetric flask and diluted to the mark with distilled water. This solution had a reading of $+175$ mV with a $F^-$ ion-selective electrode which had a response of $-60.0$ mV/decade increase. Then, separately, two 0.010-mL spikes of 10 mg/mL $F^-$ were added to the sample, mixed, and measured. The readings were 73.6 and 55.3 mV, respectively. What was the concentration in ppm of $F^-$ in the original toothpaste?

*12.8    This exercise leads to a derivation of one form of a **Gran plot** for potentiometric titrations. [Ref: Gran, G. *Acta Chim. Scand.* **1950,** *4,* 559; *The Analyst.* **1952,** *77,* 661.] When a potentiometric method is used to determine concentrations, the logarithm of the concentration is proportional to the voltage output. For the concentration of free ions, this can be written

$$E_0 = \text{constant} \pm S \log \gamma_0 C_0$$

Here $E_0$ is the potential versus the reference, $\gamma_0$ is the activity coefficient, and $C_0$ is the analyte concentration. The slope (in mV per decade) of the logarithmic response is $S$.

The volume of the solution initially is $V_0$. If, after the addition of a volume $V_s$ of standardized reagent with concentration $C_s$, then the activity coefficient becomes $\gamma_a$, and

$$E_a = \text{constant} \pm S \log \gamma_a \frac{V_0 C_0 - V_s C_s}{V_0 + V_s}$$

The sign used depends on the charge of the ion being measured.

    a.   If $\Delta E = E_a - E_0$, derive the expression for $C_0$ in terms of $V_0$, $V_s$, $C_s$, $\Delta E$, $\gamma_a$, and $\gamma_0$. The constant has the same value in both expressions. Compare this equation with that derived in Problem 12.4 for standard addition.

    b.   Using the equation for $E_a$, rearrange it so as to complete the form below. (Let the constant be $k$.)

$$(V_0 + V_s) \cdot 10^{\pm (E_a - k)/S} = \ \cdots$$

    c.   If the left-hand side of this equation is plotted versus $V_s$, derive the relationship between $C_0$, $V_0$, and $C_s$ at the intercept (left-hand side $= 0$) with the $V_s$-axis, $V_{s\text{-intercept}}$. Such a plot is one form of a Gran plot. Among other properties, it produces a straight-line plot compensated for added volumes. Thus, longer extrapolations are possible when volumes change appreciably such as for some standard additions and potentiometric titrations. (The titrations are, in essence, treated as determinations with standard additions.)

*12.9    [Ref: Hershcovitz, H., et al. *J. Chromatog.* **1982,** *252,* 113.]

A silver electrode coated with a sparingly soluble salt can be used as a transducer for detecting different negative ions in a flowing stream. If $X^-$ is the anion being detected, the reaction that occurs is

$$AgX(s) + e^- \rightleftharpoons Ag^0 + X^-$$

for which the potential is

$$E = E^{\circ}_{Ag/AgX} - \frac{RT}{\mathscr{F}} \cdot \ln a_{X^-}$$

The transducer works by keeping a concentration of $Ag^+$ at the interface. This local concentration changes depending on the solubility of the salt formed on the silver wire. Thus, if $Br^-$ or $SCN^-$ appear in solution, their salts will form, and the potential will change.

    a.   What term should be written in the blank in the following equation to make it a correct relation?

$$E^{\circ}_{Ag/AgX} = E^{\circ}_{Ag/Ag^+} + (RT/\mathscr{F}) \underline{\hspace{2cm}}$$

    b.   If the silver wire is used as an electrochemical detector, what function would you

---

\* An asterisk denotes problems that are more complex.

plot to obtain a straight-line calibration with concentrations (activities) of anions? Assume that $T = 25\ °C$ to calculate numerical values.

c.  If $K_{sp}$ for silver bromide is $5.2 \times 10^{-13}$, what potential do you expect to measure in the cell shown below.

$$Ag\,|\,AgBr,\ Br^{-}(a = 1)\,\|\,H^{+}(a = 1),\ H_2(atm)\,|\,Pt$$

d.  Under the conditions of part c, what potential do you expect to measure against a reference electrode with potential $+0.224\ V$ versus NHE?

e.  With the same setup as in part c, what potential do you expect to measure if the bromide concentration drops to $10^{-5}\ M$?

f.  With the same conditions as in part e, what difference in potential do you expect to measure if $10^{-5}$-M bromide is substituted by $10^{-5}$-M chloride? Assume that $K_{sp}(AgCl) = 1.82 \times 10^{-10}$ under the conditions here.

g.  What concentration of $Ag^{+}$ do you expect to be near the surface of the electrode under the conditions of part f?

## Coulometry

12.10   For its reduction reaction, hydrogen has a kinetic overpotential $\eta$ (see Figure 12.14) of $-0.09\ V$ on platinum and $-1.04\ V$ on mercury at low-current densities. For its reduction, cobalt has no kinetic overpotential on either material. If a solution contains $0.1$-M $Co^{++}$ at pH 1, will cobalt metal or hydrogen gas be the primary product (more easily reduced) when the cathode is

a.  mercury, and
b.  platinum

12.11   The following problems refer to Figure 12.22.

a.  At what time is the lead reduction–amalgamation complete?
b.  What is the concentration of lead in the solution in ppb?

12.12   A coulometric titration of Fe(II) by electrochemically generated Ce(IV) was carried out in the following manner. A stock solution of Ce(III) sulfate was prepared. This was composed of 100 g of $Ce_2(SO_4)_3 \cdot 8\ H_2O$ and 100 mL of concentrated sulfuric acid brought to 1.000-L volume. Of the solution, 30 mL was placed in a 100-mL beaker, and a platinum electrode was placed in the solution as the generating anode. A second platinum wire, isolated by a fritted separator, was used as the cathode. Two additional platinum wires were used in a biamperometric end-point circuit. The constant current source supplied 0.1000 mA. A 5.00-$\mu$L aliquot of the $Fe^{++}$ sample was added, and the titration completed in 132 s. How many micrograms of iron were in the aliquot?

12.13   The following questions refer to Table 12.8.

a.  Write the half-reactions for the generation of reagents (2nd column) from their precursors (3rd column) for the examples of coulometric titrations shown in the table.

b.  Which of the six reactions occur at the more positive electrode and which at the more negative electrode?

12.14   One important method to determine water in materials is the Karl Fisher titration in a dry, nonaqueous solvent. Karl Fischer reagent is composed of iodine, sulfur dioxide, pyridine, and methanol. The sulfur dioxide, pyridine, and methanol are in excess. The reactions that occur in a titration with water are

$$C_5H_5N \cdot I_2 + C_5H_5N \cdot SO_2 + C_5H_5N + H_2O \rightleftharpoons 2\ C_5H_5N \cdot HI + C_5H_5N \cdot SO_3$$

and $SO_3$ is *irreversibly* scavenged by the excess methanol through the reaction

$$C_5H_5N \cdot SO_3 + CH_3OH \rightarrow C_5H_5N(H)SO_4CH_3$$

    a.   The Karl Fisher reaction can also be run coulometrically. Write the equation for the electrochemical regeneration of the reagent $C_5H_5N \cdot I_2$. You will need to use the information that the second reaction is irreversible.

    b.   How many electrons are required per $H_2O$ molecule in the coulometric titration?

12.15   A Karl Fischer reaction with coulometric generation of reagent was run to determine the water content of a sample. A sample of 1.000 mL was run to determine its water content. The reaction end point was reached after 158 s with a current of 0.1000 mA. What was the water content on a ppm weight/volume basis? (Use correct significant figures.)

## Amperometry

12.16   The following refer to Figure 12.29.

    a.   Assign the jumps in the sampled DC polarogram to each of the four metals present in solution (from left to right).

    b.   Assign the peaks in the anodic stripping polarogram to each of the four metals present in solution (from right to left).

12.17   [Ref: Last, T. A. *Anal. Chim. Acta* **1983**, *155*, 287.]
Figure 12.17.1 shows the currents due to material arriving in a flowing stream at the working electrode after the three components present have been chemically separated. The sample consists of 90 ng of 6-hydroxydopa, 87 ng of L-dopa, and 100 ng of tyrosine. This is the order of arrival in time from first to last. The three different response curves are determined with the

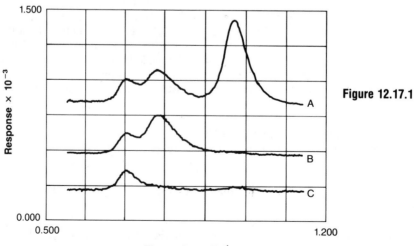

**Figure 12.17.1**

working electrode held at three different potentials: 1.0 V (curve a), 0.6 V (curve b), and 0.3 V (curve c) versus Ag/AgCl.

    a.   List the three different substances in order of ease of oxidation.

    b.   If the experiment were run with a reference electrode of SCE, would the order of oxidation change?

12.18   In the Figure 12.18.1 are differential pulse polarograms showing a blank run (lower curve) and with an analyte present (upper curve). The computer in this instrument, using its usual program, automatically drew the straight-line baseline shown in the figure. It then calculated the height at the peak relative to this baseline.

    a.   What percentage error will the computer result produce in comparison with a correct measurement of the height above the blank?

**Figure 12.18.1**

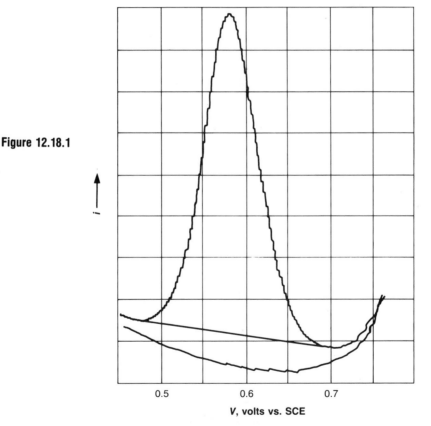

V, volts vs. SCE

    b.   Given the true baseline, what percentage error will the computer result produce in comparison with a correct measurement if the area under the curve is used instead of the height? (Extrapolate the scans until they meet.)

12.19   In Figure 12.19.1 are classical DC polarograms of a solution with $Cd^{++}$ present in 0.2-M HCl. There are three standards and an unknown. Given a solution volume of 50.00 mL, what was the original $Cd^{++}$ concentration? [Data courtesy of Dan McMullen.]

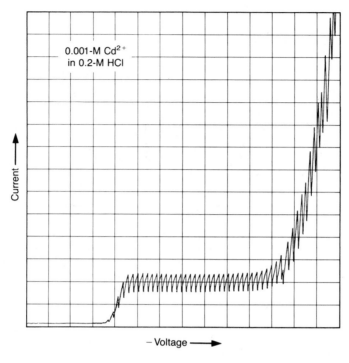

**Figure 12.19.1**

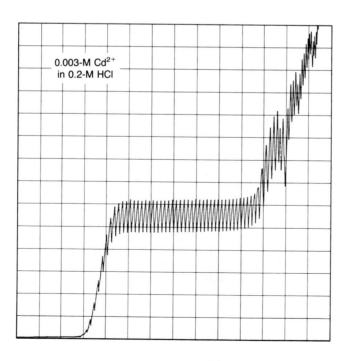

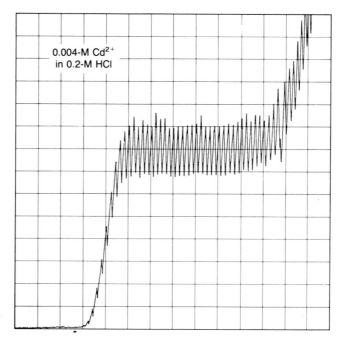

0.004-M Cd$^{2+}$
in 0.2-M HCl

**Figure 12.19.1** *(continued)*

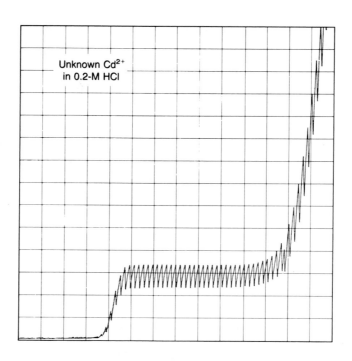

Unknown Cd$^{2+}$
in 0.2-M HCl

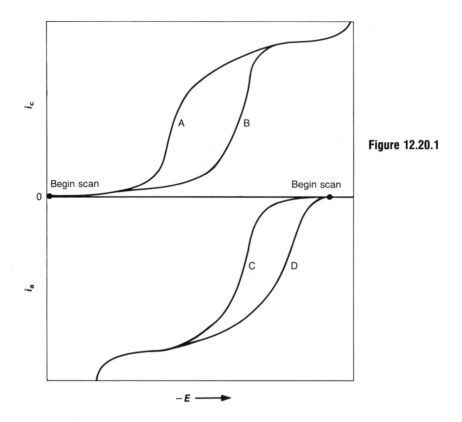

**Figure 12.20.1**

12.20   Figure 12.20.1 shows four voltammetric scans. Scans A and B begin at zero volts versus the standard half-cell scanning toward more negative potentials. Scans C and D begin at some negative potential and scan toward zero.

    a.   For which compound, A or B, is reduction more difficult?

    b.   For which compound, C or D, is oxidation easier?

*12.21   [Ref: Barnes, R. M. *J. Chem. Ed.* **1972**, *49*, 272.]

Figure 12.21.1 shows a continuous current recording of an amperometric titration of cyclamate in 1-M $HNO_3$ with the electrode at an applied potential of $+1.1$ V versus SCE. Each peak begins with an addition of 0.250 mL of 1.00-M $NaNO_2$. The decrease of each peak represents the course of the titrant's reaction as the system progresses to equilibrium. The reaction of nitrite and cyclamate is

$$\text{cyclohexyl–NHSO}_4^- + HNO_2 \longrightarrow \text{cyclohexene} + N_2\ (g) + H^+ + SO_4^= + H_2O + \text{other organic products}$$

    a.   Plot the steady (equilibrium) currents against the total volume of titrant. Some of the curves may have to be extrapolated to their steady values.

    b.   What is the total weight of cyclamate reported as calcium cyclamate, $Ca(cyclam)_2$? F.w. calcium cyclamate is 396.54.

    c.   Is the current that is measured due to cyclamate, $HNO_2$, or some unknown species in the solution? It may be a combination.

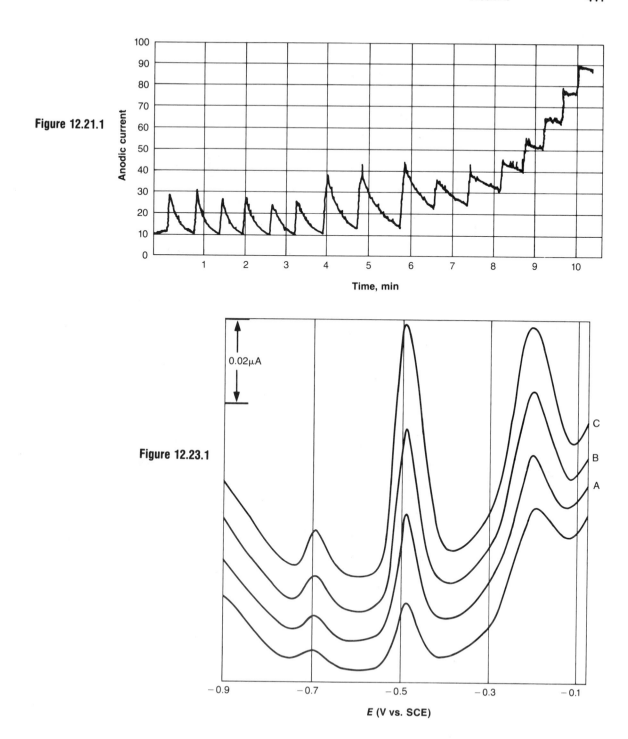

Figure 12.21.1

Figure 12.23.1

12.22   This exercise refers to a biamperometric titration of $Fe(CN)_6^{4-}$ with $Ce^{4+}$ ion, as shown in Figure 12.24. The potential difference between the indicator electrodes is 50 mV.

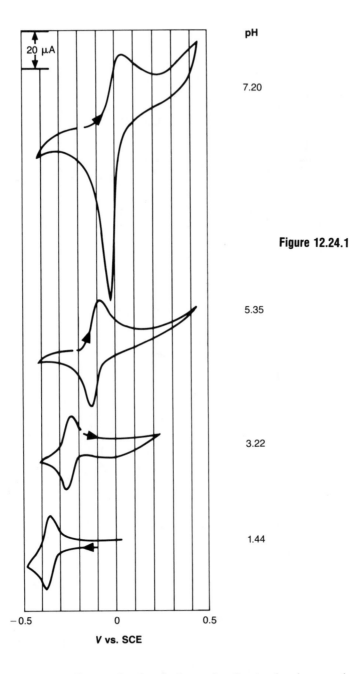

pH

7.20

**Figure 12.24.1**

5.35

3.22

1.44

−0.5                    0                    0.5

*V* vs. SCE

a.  Write out the chemical equation for the titration reaction.
b.  Write the equations for *all* the reactions involving the metal-containing species that occur at both the more positive and the more negative electrodes *during* the titration.
c.  What metal-containing species are in the solution at the end point when there is no current flowing?

*12.23   [Ref: Adeloju, S. B., et al. *Anal. Chim. Acta* **1983,** *148*, 59.]
Figure 12.23.1 shows the differential pulse stripping voltammogram for the simultaneous determination of copper, lead, and cadmium in a sample of shark meat. Deposition time was 90 s at −0.9 V versus SCE. The unlabeled scan is the sample. The scans labeled a, b, and c are with spikes as follows (all concentrations are in $\mu$g/L): (a) +0.05 Cd, 0.5 Pb, 0.5 Cu; (b) +0.10, 1.0, 1.0, respectively; (c) +0.20, 1.5, 1.5, respectively. The baselines of the scans are displaced to separate them. In other words, the left-hand sides of the scans would coincide if they were not shifted. Draw in the appropriate baselines, assign the peaks of the voltammogram, and determine the concentrations of each of the three metals.

12.24   Shown in Figure 12.24.1 is a set of cyclic voltammograms of a compound in aqueous solution at various pH values.
   a.   What are the values of $E^{\circ\prime}$ at the pH values shown?
   b.   What is the standard potential $E^{\circ}$ of the compound versus SCE?
   c.   What is the standard potential $E^{\circ}$ of the compound versus NHE?

# SUPPLEMENT  12A

# Salt Bridges

Salt bridges are constructed to make electrical contact between ionic solution reservoirs of different composition without generating excessive extraneous interface potentials. In order to work effectively the region of contact should be kept clean. To stop clogging, most salt bridges have the KCl solution flowing out into the sample solution. The rate of flow is different for different types of bridges. Shown in the accompanying figure, 12A.1, are some commonly used bridges arranged in order of decreasing rate of flow.

The two conflicting problems in choosing which to use are the following:

1.   How easily the sample will clog the junction. A sample of ketchup will be worse than a pure phosphate buffer, for instance.

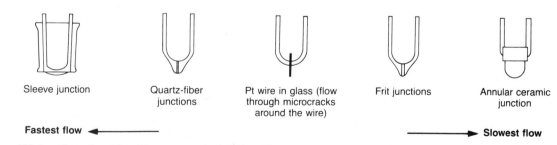

Sleeve junction          Quartz-fiber junctions          Pt wire in glass (flow through microcracks around the wire)          Frit junctions          Annular ceramic junction

**Fastest flow** ◄————————                                   ————————► **Slowest flow**

**Figure 12A.1**   Illustration of five different types of salt bridges. They are arranged in order of flow rate of electrolyte, with the fastest at the left. [Ref: Beckman Handbook of Applied Electrochemistry, Bulletin 7386, © Beckman Instruments, 1980.]

2. The size of the sample. For instance, KCl solution from a fast-flowing junction will change the volume and ionic strength of a small sample while the measurement is being made.

Another problem arises if $K^+$ or $Cl^-$ cannot be tolerated in a sample, such as in a determination using a $Cl^-$-selective electrode. In that case, a **double junction** can be used. In a double junction, an electrode containing a KCl bridge (for instance, a calomel electrode) is placed into another solution, such as $KNO_3$, which comprises the second salt bridge into the test solution. Since saturated KCl produces the smallest interface potentials (the ion mobilities of $K^+$ and $Cl^-$ are nearly equal), some of the benefits are lost with a double bridge using other salt solutions. A double junction–reference electrode is illustrated in Figure 12.17.

## SUPPLEMENT 12B

# A Biochemical Amperometric Detector for Glucose

The greater selectivity that an assay method possesses, the simpler can be the sample preparation methods. Some of the most selective chemical systems that exist are enzymes. This great selectivity can be used in an electrochemical method to determine glucose in blood samples injected into the instrument without any preparation at all. Without sample preparation, the assay is speeded up significantly; it needs only a few minutes from blood sample to result. Thus, the method can be used in emergency rooms to determine if, for instance, an entering comatose patient is suffering from diabetic shock.

The enzyme reacts specifically with glucose from the solution. The reaction catalyzed is oxidation of glucose by oxygen. The name of the enzyme is glucose oxidase.

water +     D-glucose     + oxygen $\xrightarrow{\text{glucose oxidase}}$ gluconic acid   + hydrogen peroxide

**Figure 12B.1**  Diagram of the layers of an enzyme electrode that can be used to determine glucose without sample pretreatment. Only small molecules can penetrate the outer membrane. Of these, only glucose reacts with the enzyme to produce $H_2O_2$. Only $H_2O_2$ can penetrate to the electrode, where it is reduced at a platinum electrode. Two of the three electrodes of a three-electrode potentiostat are on the head of the sensor. The sensor is about 5 mm in diameter. [Modified after a figure of Yellow Springs Instruments. Courtesy J. Johnson.]

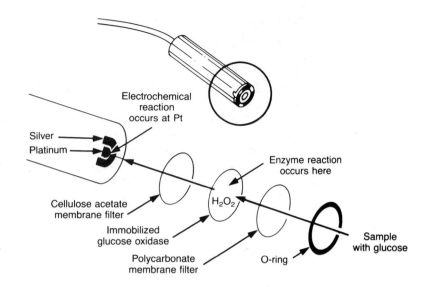

The hydrogen peroxide then reacts at a platinum anode with the reaction

$$H_2O_2 \rightarrow 2\,H^+ + O_2 + 2\,e^-$$

The current resulting from this reaction is directly proportional to the concentration of hydrogen peroxide which, in turn, is proportional to the amount of glucose in the sample solution. The platinum electrode is held at $+0.7$ V versus a Ag/AgCl reference electrode.

Thus, the glucose oxidase contributes its specificity to the assay. The electrode is then used as a transducer to detect the product of the enzyme reaction, $H_2O_2$.

In operation, a 25-$\mu$L sample is injected directly into a volume of buffer solution, which is in contact with a probe containing two of three electrodes. The probe is constructed as shown in Figure 12B.1. Let us follow the sequence of reactions and separations from the solution to the electrode.

The outside membrane eliminates large proteins of molecular mass greater than about 300,000. This protects the electrode from fouling with precipitated proteins. On the inside of the membrane is the enzyme, which is chemically bound. Of all the molecules that penetrate to this layer, only the glucose reacts with the enzyme. The $H_2O_2$ product diffuses out of this layer both into the solution and toward the electrode. The second membrane filter allows only $H_2O_2$ to pass through; it prevents other molecules, which might be oxidized, from reaching the electrode and contrib-

**Figure 12B.2**  The structure of acetaminophen, a possible interferent that would react at the platinum electrode. However, because of its size, the inner membrane prevents it from reaching the electrode surface. The inner membrane is a molecular filter, but it is difficult to classify the separation of $H_2O_2$ from acetaminophen as either a chemical or physical separation.

uting to the assay results. An example of such interferents is acetaminophen (see Figure 12B.2), which may also cause the same symptoms as diabetes in large overdoses.

The probe has two of the three electrodes of a three-electrode potentiostat on its surface. The central electrode is a platinum disk, which is the working electrode. A ring around this disk is composed of silver and, with the chloride ions in solution, is the Ag/AgCl reference electrode. The auxiliary electrode is anchored in the sample chamber.

Thus, within a thickness of about 0.1 mm, two separations, an enzyme-catalyzed reaction, and a electrochemical reduction, are done. And, as a result of this elegant method, the assay can be done without sample preparation.

## SUPPLEMENT 12C

# Voltages, Data Acquisition, and Time in Pulse and Differential Pulse Polarography

In Section 12.11 you read how the nonfaradaic (capacitance) currents have exponential decays. Experience has shown electrochemists that the nonfaradaic currents are shorter lived than the currents due to oxidation or reduction of an electroactive species — the faradaic currents. In a solution approximately molar in salts, the nonfaradaic currents decrease with a half-time of less than 100 $\mu$s. The faradaic currents decline much more slowly. In Figure 12C.1 this behavior is shown pictorially.

The pulse polarographic methods allow us to sample the faradaic currents (those we want to measure) after the nonfaradaic ones have decreased significantly. Illustrated are the currents that arise during a "square-wave" change in the potential on the working electrode. The time course and voltage protocol of the square-wave pulses are shown in Figure 12.32. Pulses are applied during the 5–100 $\mu$s when the mercury drop is the largest and just before it is knocked off. However, normal pulse polarography and differential pulse polarography have different pulse protocols, as can be seen in the figure. For normal pulse polarography, all the pulses begin at the same base voltage and are different in magnitude. On the other hand, for differential pulse polarography, all the pulses are the same magnitude but arise from a changing base voltage. The sampling of currents is also different.

In normal pulse polarography, the sampling of the currents is done electronically at the end of the voltage pulse. This is illustrated in Figure 12C.2a. The current which is measured in the short end-period is then amplified. The output appears as shown for the normal pulse polarogram in Figure 12.28.

As you can see from Figure 12C.2a, this sequence of pulses and delayed sampling reduces the relative contribution of the capacitive currents to the total current. However, the relatively constant (with voltage) faradaic background currents still

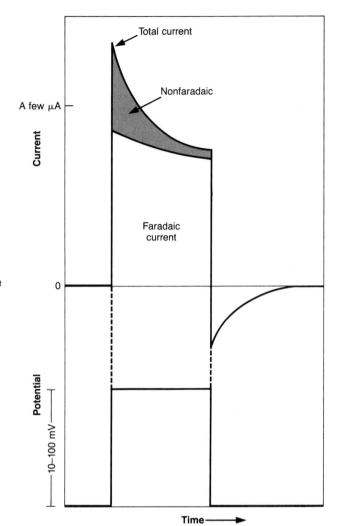

**Figure 12C.1**    Illustration of the basis of pulse amperometry of all types. (a) Plot of the current as it appears during and after the pulse. The total current is the sum of two processes: the faradaic (e⁻-transfer) and nonfaradaic (capacitive) ones. The nonfaradaic current decreases more rapidly so that after waiting an appropriate time (on the order of 0.1 $\mu$s) a larger fraction of the total current arises from faradaic processes. After the end of the pulse, the resulting current in the opposite direction is primarily capacitive current. (b) A plot of the time dependence of a voltage pulse applied to the working electrode of a three-electrode potentiostat.

contribute to the results. The pulse and sampling protocols of differential pulse polarography enables us to minimize the contribution of the background faradaic currents as well.

As shown in Figure 12.32, the pulse protocol for differential pulse polarography consists of pulses of the same magnitude applied on top of a linear voltage sweep. Unlike normal pulse polarography, the resulting currents are sampled twice instead of once. This is illustrated in Figure 12C.2b. The first current sample is measured immediately before the voltage pulse is applied. The currents measured in this time span are due to the background current. The background current's magnitude is then stored electronically. Then the pulse is applied, and the current is measured again after the majority of the capacitive current has decayed. This current sampling is the same as in normal pulsed polarography. This second current sample is composed of

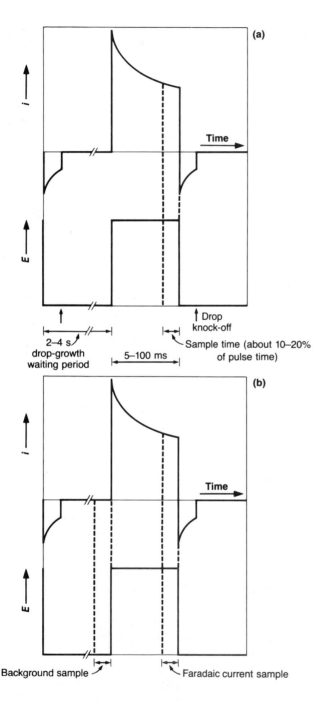

**(a)**

i

E

Time

↑ Drop
knock-off

2–4 s
drop-growth
waiting period

Sample time (about 10–20%
of pulse time)

5–100 ms

**(b)**

i

E

Time

Background sample ⤴

Faradaic current sample ⤴

**Figure 12C.2** Pulse and sampling protocols for normal pulse polarography (a) and differential pulse polarography (b). For normal pulse, the current is sampled at the ends of pulses of fixed length (5–100 $\mu$s) but with constantly increasing magnitude. (See Figure 12.32.) The pulses are applied at fixed intervals (2–4 s) when the mercury drop is large and slowly changing in area. Immediately after the pulse, the drop is knocked off the capillary.

For DPP, the pulse length is fixed in the same range, and all the pulses have the same magnitude on top of a scanning potential. (See Figure 12.32.) Associated with each pulse, currents are measured at two different times. The first current is measured immediately before the pulse and is the background current. The second current measurement is done at the end of the pulse and measures mostly faradaic currents. Both the measurements of current are stored as they are taken. Following this, the value of the background current is subtracted from that of the faradaic current to produce the output response. The output, as far as possible, reflects a minimization of the nonfaradaic currents as well as a correction for the background faradaic currents. (The latter are only changing slowly as the potential is scanned.)

the faradaic currents of both the background electrolyte and the assayed species. Next, the first current sample is electronically subtracted from the second current sample. The first sample is from the background current. The second current sample is the sum of the background current (at the slightly higher potential) and the current from the assayed species. Thus the output of this subtraction is a voltage proportional to

$$output = (background + sample) - background = sample\ alone$$

This subtraction is quite effective because of electrochemical behavior, as illustrated in Figure 12.21. Usually, only at the ends of the working range of a solvent does the background current change rapidly with the potential. Thus, we can subtract directly the background current measured before the voltage pulse from the total current measured in the pulse and get a signal due, for the most part, to the electroactive analyte(s) alone.

# Part IV
# SEPARATIONS

# 13

# General Introduction to Chromatography

## 13.1 The Chromatography Experiment

Could a chemical standard be made without ever having a pure substance? Could we ascertain whether or not an interference was present unless a pure substance were available? The answer to the first question is no. If a material is not pure, then whatever properties are measured will be a sum or average of all the different species present. In order to develop an assay for a specific material, at *some* time we *must* be able to ascertain the properties of every analyte alone. Thus the answer to the second question is also no.

There are a few qualifications to the above negative answers, however. First, as pointed out in the sections in which trace analysis is discussed, *pure* might be considered to be in quotes. The meaning of the word is changing and will continue to change, depending on our increasing capabilities in separations and abilities to detect ever smaller amounts of each component of a material. Second, the selectivity of certain biological reactions (such as antibody binding) is so great that pure molecular species can be monitored without obtaining an isolatable, pure sample.

The reason that we must obtain pure components is straightforward. With only a few exceptions, we are unable yet to measure atomic or molecular properties with less than about 1 ng of a substance. (If the species of interest has m.w. 100, this is $10^{-10}$ mol $= 6 \times 10^{13}$ particles or formula units.) So in order to study the properties of any pure substance, a large number of identical molecules or atoms of that substance must be obtained. To isolate quantities of pure substances involves the science and art of separations. Contemporary practice in analytical separations usually involves simultaneous separation and identification of the components of a sample.

While studying this chapter, you should note that the separations of components of materials are achieved using a wide variety of techniques. The molecular differences upon which the separations are based are quite diverse as well. For example, molecules can be separated by their differences in molecular charge, molecular size, molecular mass, bond polarities, redox potential, and ionization constants.

The principal method for separations based on differing chemical properties is called **chromatography.** We shall let the originator of the contemporary method, Michael Tswett, introduce it himself. In this introductory quote, he uses the word **adsorbent.** This is a solid material to which molecules bind on the surface. A **chromatographic column** is illustrated in Figure 13.1. In this case, it is a tube packed with adsorbent over which flows a liquid which is called the **eluent.** The analyte is placed at the top of the column and washed through with the liquid eluent. The components are said to be **eluted:** They are carried through the packed column by the flowing liquid. In this example, the eluent is petroleum ether, which is a low-boiling-point, organic solvent. Tswett's column was a vertical tube open at the top with the eluent driven under its own weight through the adsorbent.

> If a petroleum ether solution of chlorophyll is filtered through a column of an adsorbent (I use mainly calcium carbonate which is stamped firmly into a narrow glass tube), then the pigments, according to the adsorption sequence, are resolved from top to bottom into various colored zones. . . . Like light rays in the spectrum, so the different components of a pigment mixture are resolved on the calcium carbonate column according to a law and can be estimated on it qualitatively and quantitatively. Such a preparation I term a *chromatogram,* and the corresponding method, the *chromatographic method.*
>
> It is self-evident that the adsorption phenomena described are not restricted to the chlorophyll pigments, and one must assume that all kinds of colored and colorless chemical compounds are subject to the same laws.
> [Ref: Tswett, Michael, 1906, as translated and quoted in *J. Chem. Ed.* **1959,** *36,* 144 and *Ibid* **1967,** *44,* 235.]

Tswett used his eyes to detect the **bands** or **zones** of different materials by their colors. As the compounds separated, there were bands of color on his column from the numerous components in the sample. This experiment showed that chlorophyll is only one of many pigments found in plant leaves.

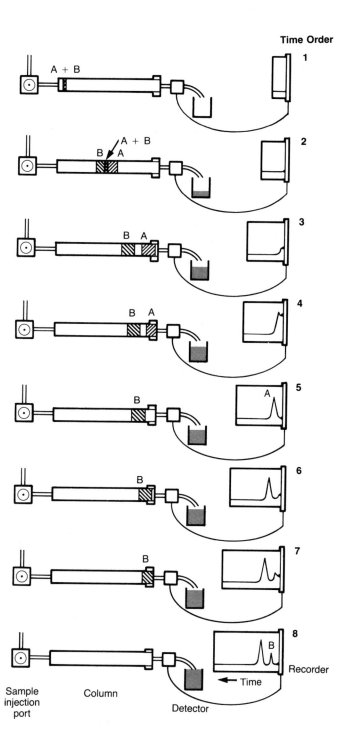

**Figure 13.1** Illustration of a contemporary column chromatographic experiment. The system is shown at different times (not at regular intervals) after the sample enters the column. The equipment shown includes a column packed with a solid adsorbent. The mobile phase passes through this column—here from left to right. Connected at the head end of the column is an injection port which allows the sample to be introduced into the flowing stream without leakage and without stopping the flow. The sample, composed of mixed components A and B, is carried through the column. The materials that elute off the column pass through a detector, and its response is printed on a strip-chart recorder with the paper moving at a constant speed. This is the **chromatogram.** Note that the concentrations of A and B in the eluted volume are always less than in the injected volume. Chromatographic separations always cause dilution of samples.

## 13.2 Nomenclature of Chromatographic Separations

Some of the details of the various techniques of chromatography are presented in Chapters 14 and 15. In this chapter, the characterization of chromatograms will be described first. Then some of the explanations of chromatographic separations are presented.

Chromatographic separations can be carried out in the liquid and gas phases. In both **liquid chromatography (LC)** and **gas chromatography (GC),** the sample is injected into a moving fluid phase, a liquid or gas, respectively. A more general term for the moving liquid or gas is the **mobile phase.** In general use, the term *eluent* applies to liquids.

The components of the sample are carried by the mobile phase through a volume or layer of particles of solid material. The solid material has a number of different generic names. Among these are **solid support, sorbent, static phase, packing material,** as well as adsorbent. Probably the most widely used term is **stationary phase.**

In contemporary analysis, in addition to using our eyes, we can detect the components that have been separated by measuring changes in a number of different chemical or physical properties. Three examples are electrical conduction, light absorption, and evidence for the presence of nitrogen. These are only a few. Essentially every instrumental and wet analytical chemical method has been used to detect the components separated chromatographically.

An eluent that has passed through a column is often also called the **effluent.** The results of chromatographic separations can be obtained by collecting the effluent in a series of **fractions** and carrying out further tests on them. For instance, we might let the effluent drip into a test tube and change the tube after 1 mL has been collected. Then, a chemical test could be run on each tube to see when the desired fraction "came off" the column.

On the other hand, we can make the measurements continuously. In the example of chlorophyll, instead of simply ascertaining the color of the effluent, we might pass the liquid through an instrumental detector. The transducers of the detectors are placed at the end of the column, as seen in Figure 13.1.

The detector registers the changes in some property of the eluent passing through or by it. An example is the change in the amount of red light absorbed by material in the eluent, as the liquid from the column passes through. The change occurs as each component arrives, flows through, and leaves the detector. The change is a rise and fall (or fall and rise) in the detected property, from the baseline level. As plotted on a moving chart, this response produces a series of peaks over time. It is this chart of detector response versus time that is now called a **chromatogram** (see Figure 13.1) in contrast to Tswett's nomenclature.

The reason for the wide applicability of chromatographic methods is the variety of conditions that can be used to separate components of materials. For instance, we can use different mobile phases (gases or liquids) and different solid supports (minerals, polymers, or liquid-coated solids). In addition, we can change geometries of the stationary phase (inside columns or on plates) or the amount of it (large for preparative and smaller for analytical). The techniques and specific names of these chromatographies will not be considered in detail until later chapters. But remember, all of them have a solid support with a mobile phase flowing past it as well as some way to introduce the sample and detect the presence of separated components.

The remainder of this chapter follows in this order: A practical example is quoted; then the ways to characterize chromatograms and separations are explained; the factors affecting chromatograms are shown; and finally, a widely accepted description of the causes of chromatographic separations is explained.

## 13.3  A Practical Example

Chromatographic methods are pervasive in chemical analysis. The following example illustrates an application in the area of microbiology. A chromatographic assay can help identify different species of bacteria. This is done by analyzing their metabolic products; each species has a different set of products.

There are a few terms in the quoted example that may need definition. The chromatographic separation is of molecules in the gas phase, in which the mobile phase is called the **carrier gas.** In order to keep molecules from condensing and to ensure reproducible separations, the column is heated. The sample is injected into the column through an injection port with the use of a syringe.

The column has packed in it a solid support called Chromosorb W/AW. This is a trade name. As is a common practice, the company supplying the support material is written in parentheses. The Chromosorb name is prefixed with the numbers 100/120. These numbers define the hole size in standard screens for sifting powders. The higher the number, the more lines in the mesh, and the smaller the particles must be to pass through. The powder used to pack these can pass through a 100 screen but not through a 120 screen. Thus, the set of screen numbers describes the range of particle size of the stationary phase.

Each of the two powder packings is coated with a high-boiling-point liquid. These are the numbers prefixed with SP-. The percentage is a (w/w) percentage of the liquid to solid. Since the sample is in the gas phase and the stationary phase is the liquid on the solid support, the technique is called **gas–liquid** chromatography (GLC). The solid support is ignored in the name — it provides a large surface area for the liquid to spread on.

Notice the care with which the blanks and controls are done and the need for pure materials to avoid background interferences. The analytes are dilute, and the sample is relatively small. These precautions even enter into the manner of growing the bacteria that are being used. Finally, an emulsion is a suspension of small oil drops in another liquid; for example, mayonnaise is an emulsion.

> Gas–liquid chromatography (GLC) techniques are now well established as being extremely useful for characterizing and identifying microorganisms. A variety of GLC techniques have been developed which allow analysis of various cellular constituents and metabolic products of microorganisms in vitro and in vivo. . . .

458

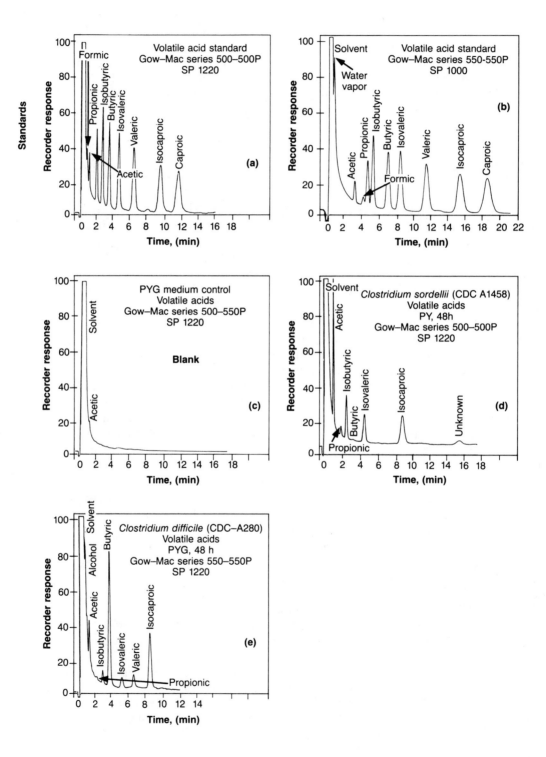

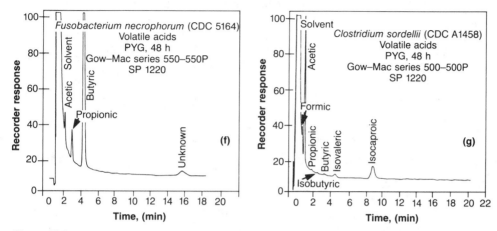

**Figure 13.2**  Chromatograms of volatile acids from (a,b) standard solutions, (c) a blank, and (d–g) solutions containing metabolic products from anaerobic bacteria. The presence of a specific *set* of analytes is used to identify the bacterial species that is growing in the medium. Chromatograms of standards are run with different packing materials in each of two columns. Not only do the peaks for each acid elute at different times on the two different adsorbants, but even the order of elution of acetic and formic acids is reversed. Only the analyses on SP1220 are shown here.

Metabolic product analysis has been used for many years in characterizing anaerobic bacteria, and the taxonomic classification of these microorganisms now relies heavily on metabolic product profiles. . . .

Chromatographs manufactured by various companies have been used satisfactorily for analysis of volatile and nonvolatile acid products of anaerobic bacteria. . . . all [are] equipped with thermal conductivity detectors and two 6-foot × ¼-inch stainless steel columns.

The chromatograph is fitted with 1 column packed with 15% SP-1220/1% $H_3PO_4$ on 100/120 Chromosorb W/AW (Supelco) for analysis of volatile acids and a 2nd column containing 10% SP-1000/1% $H_3PO_4$ on 100/120 Chromosorb W/AW (Supelco) for nonvolatile acids. . . . Helium is used as the carrier gas.

The following conditions are usually used for analysis of volatile and nonvolatile acids.

| | |
|---|---|
| Column temperature | 135–140 °C |
| Injection port temperature | 200 °C |
| Detector temperature | 180–185 °C |
| Carrier gas flow rate | 100–120 cc/min |

**Performance Testing**  Uninoculated PYG medium [this is the growth medium used to cultivate the bacterial cultures—PYG abbreviates peptone–yeast extract–glucose], when tested by GLC as described in this paper, should show only trace amounts, if any, of volatile and nonvolatile acids. Appropriate strains of control microorganisms should show adequate growth in PYG and characteristic acid products when tested by GLC after 48 hours of incubation at 35 °C. Strains of *Bacteroides fragilis, Clostridium difficile,* and *Fusobacterium necrophorum* . . . are suitable control microorganisms.

**Table 13.1**  Volatile and Nonvolatile Acid Profiles of Various *Clostridia* in PYG Cultures

The identification of the specific species depends on the presence or absence of the specific products. The *C.* in each name is an abbreviation for *Clostridium.*

| Species | Acid Products* | |
|---|---|---|
| | *Volatile* | *Nonvolatile* |
| *C. bifermentans* | A,IC,(P),(IB),(IV) | HCA |
| *C. clostridiiforme* | A | PAA |
| *C. difficile* | A,IB,B,IV,V,IC,(P) | HCA |
| *C. paraperfringens* | A,B | L |
| *C. septicum* | A,B | — |
| *C. subterminiale* | A,IB,B,IV,(P),(V),(IC) | PAA |
| *C. tetani* | A,P,B | (HCA) |

\*  A = acetic acid              IV = isovaleric acid
   B = butyric acid             P = propionic acid
   IB = isobutyric acid        L = lactic acid
   C = caproic acid            HCA = hydrocinnamic acid
   IC = isocaproic acid        PAA = phenylacetic acid
   V = valeric acid             PY = pyruvic acid
   ( ) = variable; if produced, there is usually only a trace amount.

**Analysis of Volatile Acid Products**   This procedure is used for analysis of short-chain, volatile acids (including formic, acetic, propionic, isobutyric, butyric, iso-valeric, valeric, isocaproic, caproic, and heptanoic) in PYG broth cultures.

1. Using a capillary pipette, inoculate a tube of PYG medium with growth from an active . . . culture of the microorganism to be tested, and incubate in an anaerobic jar or glove box with the cap loose at 35 °C for 48 hours or until good growth is obtained.
2. Transfer 2 mL of the PYG culture to a clean, screw-cap tube (e.g., 13 × 100 mm).
3. Add 0.2 mL of 50% (v/v) aqueous sulfuric acid to acidify the culture to pH 2 or less.
4. Add 1 mL of ethyl ether, tighten cap, and mix by gently inverting the tube about 20 times.
5. Centrifuge briefly (1500–2000 RPM) to break the ether–culture emulsion layer.
6. Place the tube in a freezer or a dry ice–alcohol mixture, and hold until the aqueous phase is frozen.
7. Rapidly pour off the ether layer into a clean, screw-cap tube.
8. Fill a 20-microliter syringe . . . with ether extract, adjust the volume of extract in the syringe to 14 microliters, and inject it into the chromatograph. . . .
9. Identify the volatile acids by comparing the elution times of products in the bacterial extract with those of known acids in a standard volatile acid mixture chromatographed as described for the bacterial extract on

the same day. [Sample] chromatograms obtained with the standard volatile acid mixture and extracts of representative cultures of anaerobic bacterial are shown in [Figure 13.2]. . . .

**Quality Controls**   Proper control of all steps involved in analyzing volatile and nonvolatile acid products by GLC is essential to avoid erroneous results. Glassware, media, reagents, and solvents that are free of extraneous acids and appropriate standard acid mixtures must be used. It is also important to use the same medium for growing cultures to be analyzed for acid products as that which was used in generating data for the identification tables used. [These are tables such as Table 13.1.] The composition of the culture medium can dramatically influence the metabolic products elaborated by a microorganism grown in it.
[Ref: Lombard, G. L.; Dowell, V. R., Jr. "Gas Liquid Chromatography Analysis of the Acid Products of Bacteria"; U. S. Dept. of Health and Human Services, 1982, Atlanta, Ga.]

## 13.4   Descriptions of Experimental Chromatograms

Figure 13.3 shows chromatograms obtained using liquid chromatography. The first two show individual solution samples. These samples were mixed, and the third chromatogram was run on the mixture. The vertical axis is proportional to the response from a detector as the liquid effluent from the chromatographic support passed by.

In the chromatogram, the horizontal axis is volume or, equivalently, time. The volume measured is more fundamental in the theory of chromatography. However, in much of modern chromatography the fluid **flow rate** is kept constant, and time is used to label the horizontal axis. The relationship is simply,

$$\text{flow rate(volume/time)} \times \text{time} = \text{volume} \qquad \textbf{(13-1)}$$

The plot of time was made by using a strip chart recorder which passes paper at a constant rate under a pen that follows the response of the detector. Thus, the horizon-

**Figure 13.3**   Chromatograms showing the independence of components in chromatography. Chromatograms (a) and (b) are from two different samples. When these two samples are mixed and run (c), all the peaks appear and remain in the same positions in time as they did in the two original samples. The peaks project both above and below the baseline. This is a property of the type of detector used. The sample A is from a sample of nerve toxin that sometimes occurs in shellfish. Sample B is a chemically modified form of A. The components were separated using liquid chromatography. [Ref: Rubinson, K. A. *Biochim. Biophys. Acta* **1982**, *687*, 315–320.]

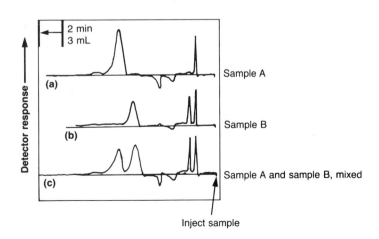

tal plot is easily calibrated for volume by measuring the flow rate of the effluent for a fixed time and comparing it with a length on the paper.

The chromatogram that results from a blank sample is called the *baseline.* In the figures, it is the straight line drawn in each graph. Notice that some of the chromatographic peaks fall below the baseline. This is merely a characteristic of the type of detector being used.

What do the peaks in the chromatogram mean? Since, in this case, the detector response is linearly proportional to the concentration of materials in the mobile phase, the shape of each peak is indicative of the concentration distribution of the associated component in the effluent.

From inspection of the chromatogram in Figure 13.3, we can discern some of the characteristics of a chromatographic separation. Notice that the third chromatogram, that of the mixture of samples, is a sum of the separate chromatograms of the two samples. It appears that whatever is the cause of the separation, we shall be able to treat each component independently.

Notice that some of the peaks reach the baseline before the next peak begins. However, in the chromatogram of the mixtures, the graph does not reach the baseline between the two largest peaks. These two components are *not* **baseline resolved.** Having every component baseline resolved is desirable. On the other hand, we say that the components are **unresolved** only if there are two or more components that elute together so that only one peak appears. For the intermediate case, we say the components are partially resolved.

The samples were injected at a time when the pen was on the right-hand side of the graph. The paper moved to the right, and the chromatogram was plotted. Thus, elapsed time is plotted increasing to the left. Note that the peaks are broader the later that they pass through the detector. This is a general property of chromatograms. In addition, notice that the first peak that appears is not at the time of injection. There is a delay. This delay is also a general property of chromatographic separations. The volume of mobile phase that passes through the column between the time the sample was injected and the top of the first peak is called the **void volume,** or **dead volume.** More will be said of this later.

There is one more set of descriptive terms that commonly appear in publications concerning chromatography. Notice the direction in which time is plotted in the chromatograms shown in Figures 13.2 and 13.3. They are opposite. If we wish to describe the earlier eluting portion of a band, referring to the directions right or left can thus be confusing. Instead, we refer to the **leading edge** and **trailing edge** of the band. The leading edge refers to the material of a band that elutes at an earlier time (or volume). The trailing edge results from the material eluting at a later time (or volume).

## 13.5 Parameters of Chromatography

As chemists use the term, a *parameter* is a quantity that can take on different values and characterizes some process, operation, or result. A well-known example in mathematics is the radius of a circle. The radius can change in value from circle to circle,

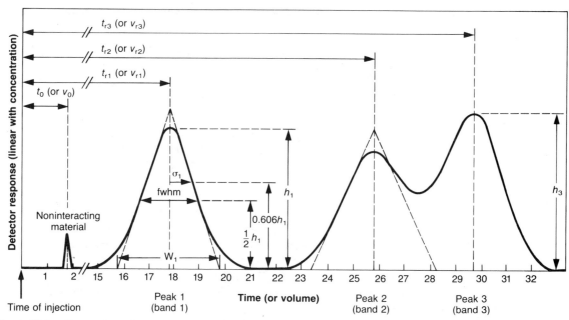

**Figure 13.4**   Hypothetical chromatogram showing the parameters that are used to characterize a chromatogram. Each band can be described by a peak position and a peak width. Pairs of bands are characterized by a separation factor or resolution of the corresponding peaks.

but it always characterizes a circle. Parameterizing data in chromatography, as in other methods, allows for ease of tabulation and communication of that data.

The shapes, positions, and resolution of characteristic bands can be parameterized. The labels used to do so are illustrated in Figure 13.4. Parameters may also be correlated successfully with other chemical factors. In the case of chromatography, parameters describing the chromatograms can be successfully correlated with descriptions of the molecular processes underlying the separations. However, the data is primary, and the descriptions are models that follow from the data and descriptive chemistry.

### Parameters for Individual Bands

The first parameter that is used to characterize a chromatogram is the volume that elutes off the column between the injection of the sample and the first peak that elutes, or comes off. As noted this volume is called the void volume, or dead volume. Its value, written $v_0$, corresponds to the liquid (or gas) volume surrounding the solid support. This is the minimum volume of eluent that can carry any component of the sample from the point of injection to the detector.

The first peak of a chromatogram that appears may not appear at the void volume of the column. There are two reasons for this possibility to occur.

   1.   The detector is not sensitive to the material in the band and, thus, will not respond.

2.   All the components in the mixture interact with the solid support, and thus, none elute in the void volume.

In addition to the void volume, each peak requires a number of other parameters to describe it. One is the volume (or time) at which the maximum of the peak appears. This is called the **retention volume,** or **elution volume** (or **elution time**). The quantity is abbreviated $v_r$ or $t_r$ in this text and is illustrated in Figure 13.4. The retention time alone can be used to identify components of a sample, as you have seen in the quoted reading. The acids produced by bacteria were identified this way.

The width of the peak — its **bandwidth** — is also parameterized. (The same term, bandwidth, is used in spectrometry for a different quantity. Do not confuse them.) One parameter is the *full width at half maximum,* abbreviated FWHM or $W_{1/2}$. For a band having height $h_i$, the FWHM is the volume eluted between the edges of the band at a position one-half of the total height. This volume is labeled on band 1 in Figure 13.4 along with $h_1$. Another parameter that can be used to characterize the width derives from assuming that the band is gaussian in shape. It is the standard deviation, abbreviated $\sigma_1$. A third possible parameter (also shown in Figure 13.4) conveys the same information about the distribution of the material in the effluent volume; it is the peak width at the base, $W_1$. This is found by drawing tangent lines at the inflection points on both sides of the band outline and extending them down to the baseline. The time between these points on the baseline is a useful measure of the width. A gaussian band shape is often a good approximation to the shape observed experimentally. In that case, for any band $i$,

$$W_i = 1.70 \text{ FWHM} = 4\sigma_i \quad \text{(gaussian band shape)} \qquad \textbf{(13-2)}$$

The relationship is proven in Supplement 13A.

You saw that the peaks in the chromatograms of Figure 13.3 are independent of each other. As a result, descriptive parameters of each band must be independent as well. To describe a chromatogram the following parameters will be sufficient:

A.   The dead volume for the run
B.   For each peak, the effluent volume when the peak occurs
C.   For each peak, the peak width

These quantities are sufficient to characterize the quality of an experimental separation (baseline or partial) and the possibility that a good separation can be done. In addition, we use a few other quantities derived from them. These are defined next.

The net elution volume or net retention volume is defined as the difference in the eluent volume of a peak and the void volume.

$$\text{net elution volume (peak } i) = v_{ri} - v_0 \qquad \textbf{(13-3)}$$

A parameter, the **capacity factor** $k_i'$, can be defined for each band. It is

$$k_i' = \frac{v_{ri} - v_0}{v_0} \qquad \textbf{(13-4a)}$$

**Table 13.2** Relative Impurity of Two Adjacent Chromatographic Bands with Equal Concentrations of Material in Each If Cut Midway between Peaks*

| $R_s$ | Relative Impurity Content |
|-----|-----|
| 1.5 | 0.001 |
| 1.0 | 0.023 |
| 0.8 | 0.045 |
| 0.5 | 0.16 |

* The bands are assumed to be gaussian in shape, symmetric, and of the same area. The results are different if the concentrations of the two separated species are unequal.

For a constant eluent flow rate, the respective times may be used instead of volumes in the equation.

$$k_i' = \frac{t_{ri} - t_0}{t_0} \qquad \text{(13-4b)}$$

### Parameters Describing Pairs of Bands

The parameters describing each band of a chromatogram do not express information about the relationships between the bands. For example, in the fictitous but representative chromatogram in Figure 13.4, at around 21 time units, no material of any band is eluting. If a fraction that contained all the solution eluting between 15 and 21 time units were collected, it would contain none of the material that makes up bands 2 and 3. (I hope by now you have become suspicious of the term *none* when it is used. Let me define it here as being less than one part-per-thousand in number, 0.1 mol %, of the component.)

If a fraction of the eluent from time 22 to about $27\frac{1}{2}$ were collected, you might guess from the appearance of the two bands that some of the material composing peak 3 would be contained in the fraction. The fractions are not completely separated. Two parameters are used to quantitate the amount of mixing of the materials comprising the various eluted bands: the **resolution** and the **separation factor.**

The resolution of bands 1 and 2 is defined by

$$\text{resolution} = R_s = \frac{v_{r2} - v_{r1}}{\frac{1}{2}(W_2 + W_1)} = \frac{\text{peak separation}}{\text{average peak width}} \qquad \text{(13-5)}$$

The equation applies to any pair of bands. The denominator in this equation is the average of the two baseline widths, and the numerator is the separation of the peaks. $R_s$ is a measure of how much mixing of material there is between two adjacent bands. Some representative values are shown in Table 13.2. It is helpful to remember that when $R_s = 0.5$, the bands are quite mixed, but at $R_s = 1.5$, the impurity is about 0.1%.

Note that resolution is a dimensionless quantity, so the units of the peak widths must be the same (volume or time) as the numerator units. All calculations of $R_s$ are made so that its value is positive.

Recall that we have assumed that the detector has a linear response. However, the detector response to each component still must be calibrated. For instance, in Figure 13.4, peak 3 might contain 50 times the mass of peak 2, but the detector is not as sensitive to the material eluting in peak 3 as to that in peak 2. So be careful to differentiate the *plot* resolution and the *mass,* or number, resolution, which the plot represents. They cannot be related without calibration.

The separation factor is defined as

$$\alpha_{1,2} = \frac{v_{r2} - v_0}{v_{r1} - v_0} = \frac{k_2'}{k_1'} \tag{13-6a}$$

The calculation is made with the larger volume (more slowly eluted band) as numerator. Under the conditions of constant eluent flow, we may also write

$$\alpha_{1,2} = \frac{t_{r2} - t_0}{t_{r1} - t_0}; \text{ constant flow of mobile phase} \tag{13-6b}$$

### Efficiency

The major goal of chromatography is to separate the components of a sample. The separation of components depends on the widths of the peaks and the spacing (in time or volume) between the peaks. As a result, an effective separation of pairs of components can be achieved with either narrow or broad bands. The separations are equally good if the $R_s$ values are the same. However, if the bands are broad, the separation may take more time. All other effects being equal, we prefer narrow bands of material.

We quantify this preference for narrow bands with the parameter called the system **efficiency** ($N$). This property, which is a property of the entire chromatographic separation *system,* is defined from the characteristics of a single band. The defining equation is

$$N = \left(\frac{v_{ri}}{\sigma_i}\right)^2 = \left(\frac{t_{ri}}{\sigma_i}\right)^2 = 16\left(\frac{t_{ri}}{W_i}\right)^2 \tag{13-7}$$

This equation is also written

$$N = \left(\frac{2.35\, t_{ri}}{\text{FWHM}}\right)^2 = 5.52\left(\frac{t_{ri}}{\text{FWHM}}\right)^2$$

All the terms in the equations have been defined in Figure 13.4. $N$ is a dimensionless number; the values of $\sigma_i$ and $W_i$ are measured in the same units as the numerators. The difference between resolution $R_s$ and efficiency $N$ is illustrated in Figure 13.5.

The reason that $N$ is used as an abbreviation for efficiency is that the parameter is also called the **number of theoretical plates.** Some details of this concept and more about the importance of Eq. 13-7 will be presented in Section 13.7.

The resolution — or degree of separation achieved — is determined by the choice of stationary phase, mobile phase, temperature, and length of the stationary phase

(a)

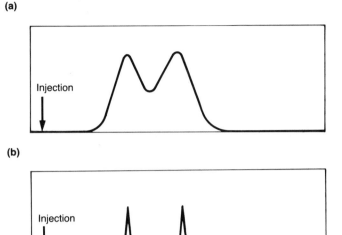

**Figure 13.5** Illustration of two ways to improve a separation. Assume the flow rate is the same for all three chromatograms and the samples are identical: (a) shows two partially resolved components; (b) illustrates the chromatogram after improving the efficiency; the components have the same $k'$ values; (c) illustrates the effect of improving the separation by altering the $k'$ values. The conditions of (c) are not as good as those in (b); it takes longer to obtain a separation of the same quality.

The sensitivity of the vertical scale in the middle chromatogram is decreased. If the vertical scales were all the same, the peak heights of (b) would be significantly larger than those of (a) or (c). However, the peak areas in all three would be equal since the quantities of analytes are equal.

(b)

(c)

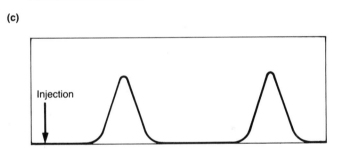

through which the separation occurs. This is in contrast to the efficiency, which is a measure of the band spreading. The efficiency is determined by factors of column construction and packing, and the velocity of the mobile phase.

---

**EXAMPLE**

The values of the chromatographic parameters are calculated below for bands 1 and 2 of Figure 13.4. The values for band 3 and its relationships with the others are left to you as a problem. For the artificial example of Figure 13.4, we assume a fluid flow of 3 mL/time unit.

The elution volumes are $v_{r1} = 17.8 \times 3 = 53.4$ mL

$$v_{r2} = 25.7 \times 3 = 77.1 \text{ mL}$$

The capacity factors are $k_1' = \dfrac{17.8 - 1.7}{1.7} = 9.5$

$$k_2' = \dfrac{25.7 - 1.7}{1.7} = 14.1$$

The widths (in time units) are $W_1 = 4.0$ units

$$W_2 = 5.0 \text{ units (uncertain due to overlap)}$$

The resolution of 1 and 2 is $R_{1,2} = \dfrac{t_{r2} - t_{r1}}{\frac{1}{2}(W_2 + W_1)}$

$$= \dfrac{25.7 - 17.8}{\frac{1}{2}(5.0 + 4.0)}$$

$$= 1.76$$

The separation factor is $\alpha_{1,2} = \dfrac{k_2'}{k_1'} = 1.49$

The column efficiency is $N = 16\left(\dfrac{17.8}{4}\right)^2 = 317$

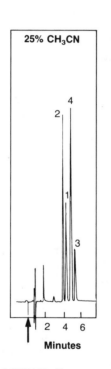

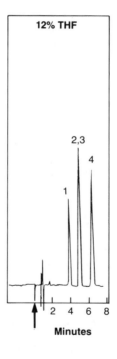

**Figure 13.6a**　Chromatographic separations can be changed by changes in mobile phase composition. In these column LC chromatograms of some antidepressant drugs, all conditions are the same except the solvent composition. The conditions are noted in the figure. Not only are the elution times different, but the order of elution is changed. Note the loss in resolution in the chromatogram on the right. The arrow indicates the injection time. The term **isocratic** literally means *equal power*. In chromatography, it means that the composition of the mobile phase is constant throughout the time of the separation. [Ref: Reprinted from *American Laboratory* **1979**, *11* (August), 9. Copyright 1979 by International Scientific Communications, Inc.]

Instrument:　　Altex Model 322 MP with
　　　　　　　Model 155 Detector
Column:　　　Ultrasphere-IP, 150 × 4.6 mm
Mobile Phase:　(Isocratic)
　　　　　　　A = 0.01 M KH$_2$PO$_4$
　　　　　　　　　0.01 M Nonyl Amine
　　　　　　　　　pH = 3.0
　　　　　　　B = Organic modifier as noted (v/v)
Flow Rate:　　　2.0 mL/min
Peak identification
1. Imipramine
2. Desipramine
3. Amitriptyline
4. Nortriptyline

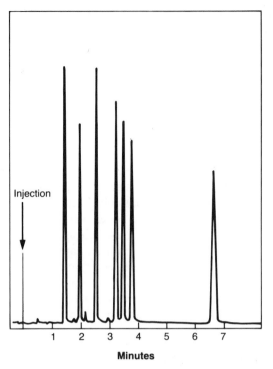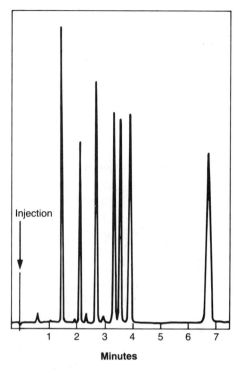

**Figure 13.6b**　Even small changes in the stationary phase cause changes in the separation. The stationary phases of these are *in name* the *same material*. However, the packings are from two different manufacturers. Their slight differences cause the shifts in peak positions (most easily seen in the second and sixth major peaks) as well as relative peak heights (especially clear in the fourth through sixth peaks).
Conditions: Column 100 × 4.6 mm; column temperature 30 °C; mobile phase 60/40 methanol/acetonitrile; flow rate 1.0 mL/min; detector absorbance 0.05 absorbance units full scale; left graph, solid support Ultrasphere ODS-3 $\mu$m, $N = 15{,}500$, $A_s = 0.97$ (see Section 13.11 for definition); right graph, solid support Perkin Elmer ODS-3 $\mu$m, $N = 14{,}000$, $A_s = 0.80$. [Ref: Courtesy B. G. Archer, Altex (subsidiary of Beckman), 1716 Fourth St., Berkeley, CA, 94710.]

It is important to realize that for any two specific substances these parameters can change. They all depend on the exact nature of the solid support and the eluting fluid as well as on temperature. The ways some factors affect chromatograms are shown in Figure 13.6.

## 13.6 Quantitation in Chromatography

Under fixed conditions, each species that passes through a chromatographic column elutes with a characteristic time for its maximum concentration, and each elutes over a range of volume. You have seen how to characterize each band with a number of different parameters. From the conditions (type of static phase, identity of mobile phase, flow rate, temperature, and detector type), the parameters alone can be used to identify the presence of specific components in a material. As a result, the chromatogram (along with the set of fixed conditions) can be used as a method of *qual*itative analysis. This was demonstrated in the practical example earlier in the chapter. However, we have not considered how to determine the quantity of each component. Quantitation can be done quite precisely: With internal standards, precisions can approach ±0.1%. Thus, contemporary chromatographic equipment is arranged both to separate and allow a *quan*titative analysis of the components nearly simultaneously.

### Percent Recovery

Regardless of how the materials composing the bands are detected, there can be a more fundamental problem. In an extreme case, it is possible that some materials in a mixture will not elute off the solid support at all. In other words, the material is effectively lost. In this extreme case, even the qualitative analysis will be incorrect.

However, it is also possible that only some fraction of the material is lost on the support. In that case, the qualitative analysis could be correct; we could know that a substance is present. But the quantitation may have low precision if the loss is erratic. The loss of material is described by the parameter *percent recovery* in a manner similar to that associated with sample preparation (Section 6.3 and Supplement 6B).

$$\% \text{ recovery} = 100 \times \frac{\text{quantity injected into mobile phase}}{\text{quantity recovered from mobile phase}} \tag{13-8}$$

The percent recovery must be determined experimentally. Its value depends on the identity of the individual component, the mobile phase, the solid phase, and the temperature. It may also depend on the other components of the sample in the mobile phase. However, let us assume that there is 100% recovery for all the components in the examples and problems of this chapter.

### Detector Response

How can we quantitate the components with the necessary precision? You might expect that the precision is closely allied with the characteristics of the detector. (The mechanisms of operation of a number of detectors are described in Supplements 14A and 15A.)

The importance of the detectors in quantitative chromatography must be emphasized. The choice of detector depends not only on the chromatographic conditions but also on the chemical and physical characteristics of the analytes present, their concentrations, and the sample size that can be used. All of these factors contribute to the precision. Precise quantitation can most easily be accomplished if the output from the detector is on a scale that is an exact linear function of the concentration of component. Algebraically, this situation can be described by a simple equation.

$$\text{value on output scale} = S_{\text{component}} \, [\text{component}] \tag{13-9}$$

$S$ represents a calibration constant that relates the concentration in the gas or liquid effluent and the output scale. As noted earlier, we have assumed that any detector

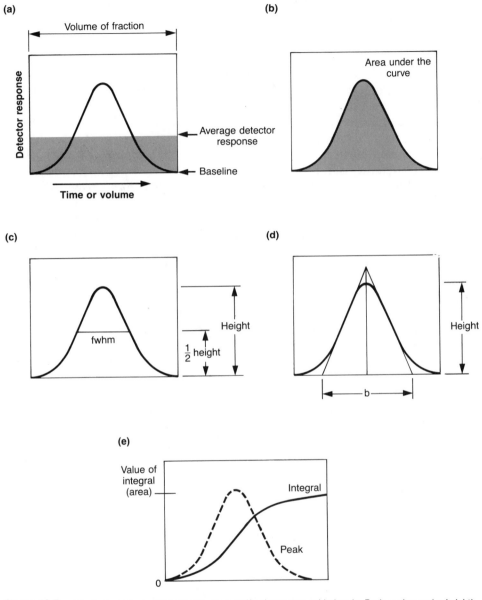

**Figure 13.7**   Illustration of a number of ways to quantify chromatographic bands. Each peak area (or height) must be calibrated using standards to be able to relate it to a specific concentration of the species it represents. A constant mobile-phase flow rate is assumed. (a) The content of a zone is the product of the concentration and volume. A fraction may be collected that contains all the material of a single band, and its content can be measured in a separate determination. (b) The area under the peak's curve can be measured and calibrated with standards. (c) The area can be approximated as a triangle. The area is calculated as FWHM × height or ½ base × height. (d) The peak is integrated electronically — the magnitude of the integral is proportional to content.

referred to in the examples and problems exhibits a linear response in the concentration range of the analyte.

## Quantitation Techniques

If all components of a material are baseline separated, then there are a number of ways to determine the quantity of each component. The components could be collected as separate fractions, and each of these fractions quantitated. In order to quantitate them, we could use titrimetric methods or some instrumental detection method such as an electrochemical method. The total quantity of a component would equal the mathematical product of concentration times the volume of the fraction. The quantity is illustrated in Figure 13.7a.

However, this average concentration is usually measured more conveniently without collecting fractions. Recall how the plot of the detector response versus time was obtained. The measurement is made on a small volume of the effluent as it passes through the detector. When the output response is linear, the concentration of a species is proportional to the area under the graph of response versus time. Those of you who have studied calculus will recognize that the area is the integral of the graphed output curve.

A number of techniques are used to measure the areas under chromatogram peaks. Some are quite exact, and some are approximations which are simpler to measure while still allowing satisfactory precision. The more commonly used techniques are illustrated in Figures 13.7b to 13.7d.

When the plot is on graph paper, there are two ways to determine the area shown in Figure 13.6b. One is simply to count the number of squares of graph paper under the curve and above the baseline. If the squares are large, the partial squares can be summed carefully. Alternately, the squares with more than half their area under the curve can be counted as one and those with less than half can be ignored. This averages the squares randomly. Another direct method is to weigh the area under the curve cut out of the graph paper (or, better, a photocopy since the original data should not be destroyed). This can be weighed without much effort to the nearest 0.1 mg. While such a method may seem strange at first glance, it is highly precise if your hand is steady when you cut out the curve (another good reason to use a photocopy — if you slip . . .).

Figure 13.7c illustrates two methods to approximate the area under the curve. Measure the height of the curve and the FWHM. Then the area is approximated by

$$\text{area} \approx \text{FWHM} \cdot \text{height}.$$

The alternate method is to draw straight lines tangent to the sides of the output plot. The area is then approximated by the area of the triangle.

$$\text{area} = \tfrac{1}{2} \text{ base} \cdot \text{height}$$

Instead of doing these manual operations, which can become tedious, the area measurements can be done electronically. An output corresponding to a chromatographic band and the corresponding output from an **integrator** are shown in Figure 13.7d. The integrator puts out a signal directly proportional to the area. Thus, instead of measuring the area, one needs only to read off the height of the curve above the integrator baseline. More typical now, however, are instruments with built-in computers that can calculate the areas numerically after "deciding" on the location of the baseline. On such a computer-controlled instrument, the areas of peaks are usually supplied in a numerical list along with the peak times.

The above methods all account for the areas under chromatographic peaks even if the widths of the bands change. However, when the experimental conditions are carefully controlled, the widths will remain the same from run to run. In this case, an excellent measure of the amount of material present may be made from the height of the peak alone. The relative precision of a measurement made this way can be better than 1%. However, if there is any doubt about the constancy of the chromatographic conditions, then one of the area measurements should be used.

The precision of all of the measurements made from a chromatogram — as opposed to being made directly on the fractions — depends on having a detector with a linear response to the concentration. In addition, the mobile-phase flow must continue at a constant rate for the method to work. If a chart recorder is used, as opposed to a computerized unit, the chart movement also must be constant if area measurements are used to quantitate analytes. If either of these conditions does not exist, then the precision of the analysis is certain to decrease.

## 13.7 Explaining Chromatographic Separations

### Extractions

As we shall see in this section, chromatographic separations can be thought of as similar to carrying out a large number of **extractions,** one after the other. Thus, to understand something about the underlying chemistry of chromatography, it is worth some time to study about extractions.

Illustrated in Figure 13.8 is a simple liquid–liquid extraction. The technique entails using two liquids that are not too soluble in each other — they are immiscible. Let us say the two liquids in the illustration are water and carbon tetrachloride, $CCl_4$. Since the $CCl_4$ is more dense, it composes the bottom layer. Dissolved in the water is a

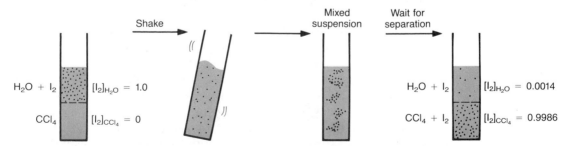

**Figure 13.8**    A simple extraction experiment. Two immiscible liquids are shaken and allowed to separate. At equilibrium, some components in the solution might be soluble in both phases. The relative concentrations of the solute in the two phases are described by a distribution coefficient $K_D$. The value of $K_D$ depends on the identities of the solute and the two solvents, the temperature, and may depend on the other components present as well.

solute: In this example, it is iodine. The mixture is shaken vigorously and then allowed to sit while the two liquids again form separate layers. If the iodine concentration is high enough, we can see its purple color in the organic phase.

We now determine the concentrations of iodine in both the aqueous and organic phases. (For instance, the iodine that is in each phase can be determined by a redox titration.) If we did this experiment at a fixed temperature but using a number of different initial concentrations of iodine, we would find that the results could be described by the following expression.

$$K_D = \frac{[I_2]_{CCl_4}}{[I_2]_{H_2O}} \approx 7 \times 10^2$$

$K_D$ is an equilibrium constant called the **partition coefficient,** or **distribution coefficient.**

The partition coefficient is a ratio of concentrations. Finding the *amounts* of material in each phase requires knowing the volumes of both phases. Let us call the volume of phase $i$, $V_i$. For the example of iodine partitioning between water and carbon tetrachloride,

$$\begin{aligned} \text{total moles of molecules} &= \text{moles in } CCl_4 + \text{moles in } H_2O \\ &= [I_2]_{CCl_4} V_{CCl_4} + [I_2]_{H_2O} V_{H_2O} \end{aligned} \tag{13-10}$$

And for the general case of partitioning between two phases,

$$\text{total moles of molecules} = C_1 V_1 + C_2 V_2 \tag{13-11}$$

where the $C_i$ are the concentrations (mol/unit volume) in the respective phases.

### $K_D$ and Elution Times

Now let us see how the elution time of a chromatographic band is related to the partition coefficient of the compound in the band. The important ideas are illustrated in Figure 13.9. After finding the relationship between $K_{Di}$ and $t_{ri}$, we shall derive the relationship between $K_{Di}$ and the capacity factor $k'_i$.

These relationships will be derived in some detail. The final results are in Eqs. 13-15 and 13-17. It is easy to misuse these equations unless you understand the assumptions in the derivations.

We begin by asking the question, What fraction of the time is a solute molecule in the mobile phase? When the solute molecules are in the mobile phase, they are being swept along at the velocity of the mobile phase. When a molecule is associated with the stationary phase, it does not move. Thus, the *average* velocity of a single molecule will depend on the average time it is stuck on the solid compared with the time it is dissolved in the mobile phase. An equation can be written expressing this idea. The equation includes the quantity **linear velocity.** This is the average velocity with which the mobile phase is flowing along the long direction of the column.

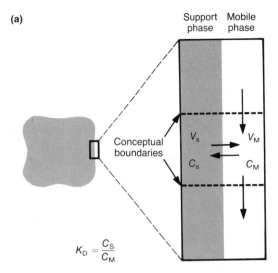

**(a)**

$$K_\mathrm{D} = \frac{C_\mathrm{S}}{C_\mathrm{M}}$$

**Figure 13.9**    (a) Illustration of the idea of chromatography as an extraction. Here, an extraction occurs as the mobile phase moves past the stationary phase. Included in the model is the idea that a separation, which is *equivalent* to a single *equilibrium* extraction, occurs over a specific length between two "conceptual boundaries." The length (or height) between these conceptual boundaries is a theoretical plate. (b) If we considered only the *average* time spent in the mobile phase relative to the *average* time anchored on the solid phase, a chromatogram would appear as shown here. The separation of the components is explained, but the finite widths of the bands are not.

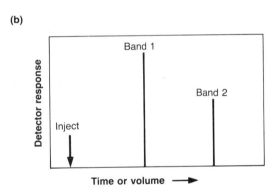

**(b)**

average rate of travel =
    linear velocity of mobile phase × fraction of time molecule is in mobile phase
    + linear velocity of stationary phase × fraction of time molecule is in stationary phase

Because the linear velocity of the stationary phase is zero, the second term contributes zero to the average. If we call the average linear velocity of the mobile phase $\bar{u}$, then

average rate of travel = $\bar{u}$ × fraction of time molecule is in mobile phase

The key to relating the distribution coefficient $K_{\mathrm{D}i}$ to the elution time $t_{ri}$ is the following idea: The fraction of *time* a molecule is in the mobile phase equals the fractional *number* of the same species that are in the mobile phase. Written as an equation for species A, this idea becomes

fraction of time molecule A is in mobile phase =

$$\frac{\text{number of molecules of A in mobile phase}}{\text{total number of molecules of A in mobile and stationary phases}}$$

The numerator and denominator are calculable from the terms of Eq. 13-11. The number of moles of molecules is equal to the algebraic product of the molar concentrations and volumes. Thus, with the subscripts M and S referring to the mobile and stationary phases, respectively,

$$\text{fraction of time molecule is in mobile phase} = \frac{C_M V_M}{C_M V_M + C_S V_S} \quad \textbf{(13-12)}$$

Equation 13-12 can be expressed in terms of the distribution coefficient, $K_{Di} = C_S/C_M$. The algebraic steps to convert Eq. 13-12 are

$$\text{fraction of time in mobile phase} = \frac{1}{(C_M V_M + C_S V_S)/C_M V_M} = \frac{1}{1 + (C_S V_S/C_M V_M)}$$

By substituting $(C_S/C_M) = K_{Di}$, the final form is

$$\text{fraction of time molecule is in mobile phase} = \frac{1}{1 + K_{Di}(V_S/V_M)} \quad \textbf{(13-13)}$$

Let us call the average rate of travel for a component $\bar{u}_i$. Then

$$\bar{u}_i = \text{average rate of travel} = \bar{u} \times \frac{1}{1 + K_{Di}(V_S/V_M)} \quad \textbf{(13-14)}$$

Notice in the equation that the volumes of the stationary and mobile phases and the linear velocity of the mobile phase remain the same for every component. The average velocity of the component varies only through differences in its distribution coefficient.

In experiments, we seldom measure a true velocity of the mobile phase. In column chromatography, what is measured is the time that a component takes to go through a fixed length column (as illustrated in Figure 13.1). But the path through the column is longer than the column length. The path is tortuous—it weaves back and forth as it goes around solid particles of the packing. So we really measure an *effective* velocity and an *effective* path length.

The time to traverse the length of the column at the effective velocity is the elution time $t_{ri}$. Algebraically, this is a simple distance, rate, and time equation.

$$t_{ri} = \frac{\text{effective path length traversed}}{\text{average rate of travel}}$$

We abbreviate effective path length by $L_{eff}$. The average rate of travel is given in Eq. 13-14. Thus, we find what we sought—the direct relationship between the elution time of a band and the distribution coefficient associated with the material.

$$t_{ri} = \frac{L_{eff}}{\bar{u} \times 1/[1 + K_{Di}(V_S/V_M)]} \quad \textbf{(13-15)}$$

The second goal was to relate the capacity factor to $K_{Di}$. Recall the definition of the capacity factor:

$$k_i' = \frac{t_{ri} - t_0}{t_0} \qquad (13\text{-}4b)$$

We need only find the expression for $t_0$, the time for the void volume to elute. Recall that a solute will elute in the void volume when it is not adsorbed at all. This is the fastest possible rate of travel through the column. However, if there is no adsorption, $K_{Di} = 0 = C_S/C_M$. The average rate of travel will, thus, be $\bar{u}$, and the void volume elution time will be

$$t_0 = \frac{L_{eff}}{\bar{u}} \qquad (13\text{-}16)$$

The expression for $k_i'$ as it relates to $K_{Di}$ can be found by substituting Eq. 13-15 and 13-16 into Eq. 13-4.

$$k_i' = K_{Di} \cdot \left(\frac{V_S}{V_M}\right) \qquad (13\text{-}17)$$

This equation relates experimental results, the $k_i'$ values, with a chemical distribution coefficient for that species. Note that the relative volumes of the stationary and mobile phases are required. This is *not* an easy number to obtain except for gas–liquid chromatography, which is the technique used by the authors of the reading in Section 13.3. The total volume of the liquid coating can be measured.

Numerous other useful relationships follow from Eq. 13-17. Especially interesting is the dependence of the separation factor $\alpha_{i,j}$ on the distribution coefficient. Deriving this relationship is left to you as problem 13.6.

What is the purpose of working in detail through this model? Why bother relating the retention times to distribution coefficients? The reason is that you can obtain a clearer view of where chemistry applies to chromatography. One great benefit to be derived from this interrelationship is that a large amount of descriptive chemistry of adsorption and the distribution coefficients of a wide range of molecules already exists. In addition, there are numerous correlations of the strengths of adsorption on specific surfaces and their variation with chemical structure. For instance, the adsorption may depend on the presence of functional groups on the analyte molecules — for instance, alcohols or acids. As a result of these chemical data, it is possible to design better experiments with less wasted trials. "Designing better experiments" means making an intelligent choice of mobile phase, temperature, and solid support to effect a better and/or faster separation. More details of the range of these choices will be presented in the chapters on gas chromatography and liquid chromatography. An outline of possible factors is presented in Table 13.3.

After studying this section, you may be interested to go back and read the quote from Tswett at the beginning of the chapter. You will see that he already understood the basis for the relative retention times and the elution order of the components being separated.

**Table 13.3** Factors of Choice for Separation and Quantitation by Column Chromatography

1. Stationary phase
   a. Cross-section shape and size
   b. Length
   c. Particle size*
   d. Material composition

2. Mobile phase
   a. Composition
   b. Flow rate

3. Temperature

4. Detector

5. Sample size (volume, concentration)

* An alternative in both GC and LC is to use an open narrow capillary tube without packing. The tube wall serves as the stationary support. This is a rapidly developing area.

### The Concept of a Theoretical Plate

The relationship of the elution time to an equilibrium constant $K_{Di}$ suggests an alternate way to visualize the chromatographic process. This is as a number of sequential extractions. In Figure 13.9 there are some lines labeled *conceptual boundaries*. We can imagine that within the volume contained between the boundaries, a single extraction equivalent to that shown in Figure 13.8 occurs. We then consider that the mobile phase moves onward, and another single extraction is done. This continues until the material elutes off the stationary phase.

For historical reasons, the individual extractions are called **theoretical plates.** The separations of components can then be labeled as *equivalent* to a number of theoretical plates. The number of theoretical plates is denoted by $N$, and, as you read,

$$N = \left(\frac{4\,t_{ri}}{W_i}\right)^2 = 16\left(\frac{t_{ri}}{W_i}\right)^2 \tag{13-7}$$

The variables $t_{ri}$ and $W_i$ were defined in Figure 13.4. Recall that the efficiency is a measure of the relative retention of the solute compared with the width of the peak. Figure 13.5 illustrates the relationship between efficiency and the quality of a chromatographic separation. The efficiency parameter is useful when comparing chromatographic separations under different conditions.

Another useful measure is the number of theoretical plates *per unit length* of support material. Assume that the support material packing has an externally measured length $L$ (this is *not* $L_{\text{eff}}$, the eluent internal path length) and that the entire column length can be used to separate components. If the column is found to have $N$ theoretical plate equivalents (using Eq. 13-7 and data from a chromatogram), then the **height equivalent to a theoretical plate,** $H$, is

$$H = \frac{L}{N} = \frac{L}{16}\left(\frac{W_i}{t_{ri}}\right)^2 \qquad \textbf{(13-18)}$$

Height equivalent to a theoretical plate is usually abbreviated as the capitalized initials HETP. This name results from its history—you might observe that length equivalent to a theoretical plate might be more accurate. A chromatographic system is better when the HETP is smaller. The system becomes more "perfect" as the HETP approaches zero.

---

**EXAMPLE**

If the chromatogram of Figure 13.4 was obtained on a column 10 cm long, what is the HETP of the system?

**SOLUTION**

At the end of Section 13.5, the efficiency of the separation was calculated; $N = 317$. Therefore,

$$H = \frac{L}{N} = \frac{10.0 \text{ cm}}{317} = 0.032 \text{ cm}$$

---

## 13.8  A More Complete Model

In the previous sections, we have considered a picture, or model, that is used to explain chromatographic retention times and, hence, the underlying cause of separations. However, the simple model does not tell the whole story. If it did, actual chromatograms would appear as shown in Figure 13.9b. Each peak would come out at a single time that is characteristic for the eluted material. But the bands would not elute over a span of time as they actually do.

The idea of a sequential series of extractions is quite useful and can be used to explain the widths of chromatographic peaks. However, by concensus, the model presented next is a better viewpoint, although disagreement about various points still exists, as will be pointed out.

Thus far, the widths of peaks have been considered in defining sets of variables describing chromatograms. The model that will be described in the following sections has been used to understand the chemical factors involved in this spreading. It is important to realize that the details of the model are *not* generally used on a daily basis. However, the concepts in it *are* used when thinking about how to run better separations and what conditions or materials should be varied to do so.

The model explains variation in $H$, the HETP, as it depends on the average linear velocity of the mobile phase, $\bar{u}$. Recall that $H$, defined in Eq. 13-18, is a measure of the ratio between a band's width and its retention time. (This result arises because $H$ depends on $N$, the efficiency. A change in efficiency is seen in Figure 13.5.)

In a simplified form, the model equation can be written as

$$H = A + \frac{B}{\bar{u}} + C\bar{u} \qquad \text{(13-19a)}$$

This equation is read: The HETP is determined by some constant value, a factor that is directly proportional to the mobile phase velocity, and some factor that is inversely proportional to it. Equation 13-19a is called the **van Deemter equation.** Each of the parameters $A$, $B$, and $C$ are constants determined from certain physical properties of the stationary and mobile phases. All are positive numbers.

As we shall discuss, the $A$-term is not needed for well-packed GC columns and for GC capillary columns. In other words,

$$H = \frac{B'}{\bar{u}} + C'\bar{u}, \qquad \text{for GC}$$

The superscript primes indicate that the $B$ and $C$ values differ from those when an $A$-term is included.[1]

A slightly more complex form of the equation is more helpful. The second term, $C\bar{u}$, can be divided into two parts, ascribing part of the effect to the solid support and part to the mobile phase. Thus, $C$ is describable as a sum of $C_M$ and $C_S$ from the mobile phase and stationary phase, respectively. (These are not the concentrations in each phase. The two different uses of $C$ is an historical accident.) With this separation, the model equation becomes

$$H = A + \frac{B}{\bar{u}} + (C_M + C_S)\bar{u} \qquad \text{(13-19b)}$$

Let us investigate the general algebraic properties of this equation. In the following sections, more details of the origins of the parameters $A$, $B$, $C_S$, and $C_M$ will be presented.

Recall that $\bar{u}$ is the actual velocity (for instance, in cm/s) of the mobile phase inside the column. It is *not* the volume flow rate of solvent (in, for instance, mL/min) through the column.

Equations 13-19 describe the *relative broadening* (not the retention times) of chromatographic bands. If you were given a set of $C = (C_S + C_M)$, $B$, and $A$ values, then using Eq. 13-19 alone, you could recreate the experimentally found HETP as it

---

[1] One other form of the van Deemter equation has been used to describe the HETP changes with $\bar{u}$.

$$H = A''\bar{u}^{1/3} + \frac{B''}{\bar{u}} + C''\bar{u}$$

Here, the double prime superscripts indicate that the constants differ from those in Eq. 13-19a. The applicability of the $A$-term and various equations for $H$ are discussed in Katz, E.; Ogan, K. L.; Scott, R. P. W. *J. Chromatog.* **1983**, *270*, 51–75 and in Poole, C. F.; Schuette, S. A. "Contemporary Practice of Chromatography"; Elsevier: Amsterdam, 1984. The short reference by S. J. Hawkes listed at the end of the chapter is highly recommended reading for anyone interested in the inapplicability of the $A$-term in GC.

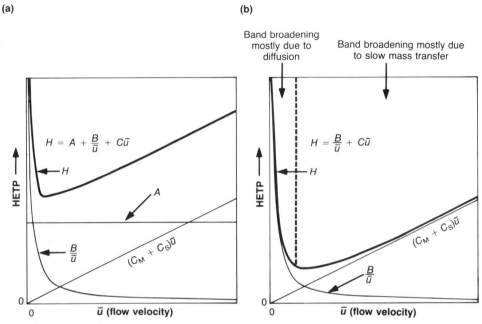

**(a)**     **(b)**

Band broadening mostly due to diffusion

Band broadening mostly due to slow mass transfer

$$H = A + \frac{B}{\bar{u}} + C\bar{u}$$

$$H = \frac{B}{\bar{u}} + C\bar{u}$$

HETP

$\bar{u}$ (flow velocity)

**Figure 13.10** Van Deemter plots of the behavior of the height equivalent to a theoretical plate (HETP) as the mobile-phase flow velocity changes for (a) HPLC and (b) for capillary GC and GC with well-packed columns. The curves labeled $H$ show the experimental behavior. We can attribute the flow-dependent behavior to two independent causes. The first is longitudinal diffusion (along the column length) characterized by $B$. The second is the broading effect of transverse diffusion (perpendicular to the long axis) and rates of reaction at the support, characterized by $C$. A third factor, characterized by $A$, is needed for LC and for less perfect GC systems. The contributions from each are calculated from the terms in Eq. 13-19. The best experimental situation occurs at the minimum. Graphs a and b are plotted on different scales as can be seen from the quantitative van Deemter plots for specific gas and liquid chromatography experiments shown in Figure 13.11.

depends on $\bar{u}$. A plot of such values for both HPLC and GC is illustrated in Figure 13.10.

At $\bar{u}$ near zero, the $(B/\bar{u})$-term is the largest. It decreases in value with increasing $\bar{u}$ and tends toward zero. In an opposite trend, the $C\bar{u}$-term is zero at $\bar{u} = 0$, and it increases with increasing $\bar{u}$. The $C\bar{u}$-term increases without limit as $\bar{u}$ increases. The sum of these terms, the HETP, has a value that is large and decreasing at very low flow velocities and large and increasing at very high mobile-phase velocities. There must be a minimum value of $H$ at some value of $\bar{u}$ in between. The exact value of $H$ at the minimum and the $\bar{u}$-value at which it occurs depend on the values of $B$ and $C = (C_S + C_M)$. Contributions from processes described by $A$ raise and lower the entire curve.

Figure 13.11 illustrates representative experimentally determined plots of the HETP as it depends on $\bar{u}$ for both gas chromatography and liquid chromatography. As you see, there is a significant difference between gas chromatography and liquid chromatography, both in the values of $H$ and in the optimum $\bar{u}$-range. GC separa-

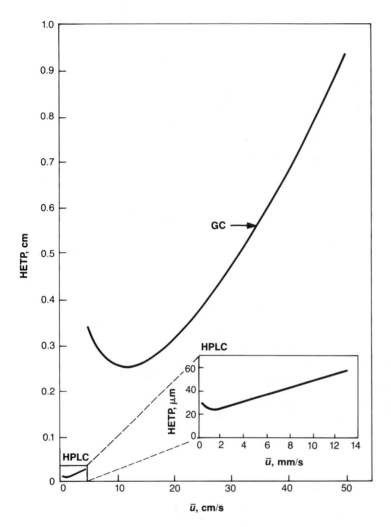

**Figure 13.11**    Plots of HETP versus flow velocity for specific gas chromatography and specific liquid chromatography systems. The gas chromatography experimental data are calculated from a peak with $k' = 7.90$ on a chromatogram run with nitrogen as the mobile phase carrier gas. The temperature was 175 °C. The liquid chromatography plot is from data using a 4.6-mm inside diameter column packed with particles 10 $\mu$m in diameter. The liquid was pumped at high pressure through the column. HPLC stands for high pressure (or high performance) liquid chromatography. Gas and liquid chromatographies are significantly different in practice as indicated, perhaps, by the significant differences in these two plots.

tions are often done near the optimum $H$ versus $\bar{u}$ point. However, the flow rates at which the optimum occurs for LC are so low that LC separations are usually done somewhat above the optimum value.

## 13.9 The Zone-broadening Equation in More Detail

In a typical chromatographic analysis, a sample is injected into the flowing stream which carries the material through the region where the stationary phase is located. The effluent from the stationary phase carries the separated materials through a detector.

In chromatographic separations, the individual components *always* elute more diluted than when introduced at the beginning of the solid support. Dilution of the material is equivalent to band spreading. Each factor of Eq. 13-19 represents a different process contributing to this broadening.

Explaining the mathematical behavior of Eq. 13-19 is not very helpful in deciding how to modify experimental conditions to obtain a better or faster separation. In this section, the physical origins of $B$, $C_M$, and $C_S$ are explained in more detail. This description is presented primarily in a pictorial way. The purpose is to help you understand how and why chromatographic conditions can be varied to obtain better separations.

Let us consider that the injection of the sample is "instantaneous." This means that the time for the injection to take place is very short compared with the width (in time) of any effluent band. Thus, if no band broadening occurred at all in the elution, each band in the chromatogram would appear to be instantaneously narrow, as are those in Figure 13.9b.

## The Broadening Process Accounted for by $B/\bar{u}$

Diffusion is the random motion that tends to spread molecules uniformly throughout space. In the mobile phase, the molecules diffuse randomly along the flow direction—both forward and backward. This process is illustrated in Figure 13.12. The broadening caused by diffusion in these directions is called **longitudinal diffusion** and is characterized by $B/\bar{u}$.

Why is the term inversely proportional to $\bar{u}$? The reason is that when the flow is faster, the time between injection and elution/detection is shorter. So there is less time for the material to spread out by longitudinal diffusion.

## The Broadening Process Accounted for by $C\bar{u}$

Not only does diffusion occur along the flow direction, as illustrated in Figure 13.12, the molecules in the mobile phase diffuse across the flow direction as well. In this way, the solute molecules reach the surface of the stationary phase and reenter the mobile phase. In addition, once the analytes are near the stationary surface, some time is required for the chemical interaction to occur. Adsorption and desorption have kinetic rates just as any other chemical reaction.

We describe the two processes separately.

1. A diffusion across the flow direction—the $C_M$ part
2. A reaction rate of the analyte and the stationary phase—the $C_S$ part

These are illustrated in Figure 13.13. Either or both of these processes can be "fast" or "slow" relative to the mobile phase linear velocity $\bar{u}$. If both processes are fast relative to the flow velocity, there will be little zone spreading: The value of $C = (C_M + C_S)$ will be small. If *either* process is slow, it will contribute to the zone broadening. The value of $C$ will be larger. The line between fast and slow, as you may guess, is not sharp. Fast and slow are extremes.

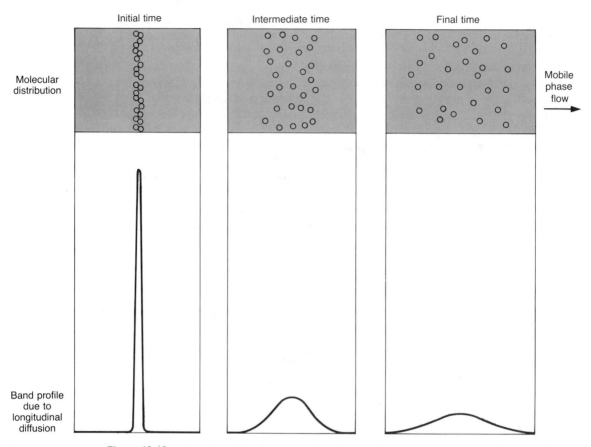

**Figure 13.12**    Illustration of the mechanism of longitudinal diffusion in chromatography. This is the *B*-term in Eq. 13-19. The molecules of the sample initially are together in a plane. As they pass through the particle bed, the molecules tend to diffuse randomly. The diffusion takes place both along and against the direction of mobile phase flow, as indicated in the figure. The longer time the sample is in the column, the wider will be the distribution of the component molecules. Thus, the contribution to the bandwidth increases with a slower flow rate. This process contributes little to *liquid* chromatography bandwidths.

As illustrated in Figure 13.13, if the mobile phase flow rate is slow relative to the diffusion across the flow direction, each solute remains within the same small volume as it is carried along. However, if the mobile phase flows faster than the diffusion, then the molecules that must diffuse over a range of distances to reach the support will be carried different lengths before adsorbing. The band thus spreads out. This spreading will be directly proportional to the velocity of the mobile phase. Since the entire broadening process occurs in the mobile phase, we express the effect as $C_M\bar{u}$.

The same sort of effect will occur if the actual adsorbing or desorbing process is relatively slow. The mobile phase will carry the unbound material beyond the point where the same solute remains bound because of slow adsorption/desorption kinetics. The band will broaden, and the spread will be directly proportional to the flow velocity. This effect is expressed as $C_S\bar{u}$.

Another process causes the same effect as slow adsorption/desorption kinetics and is included in the $C_S\bar{u}$ term. The mechanism is illustrated in Figure 13.14. The solid support particles usually are not "solid" on the molecular level. There are uneven pores and channels, smaller and larger, filled with the mobile phase into which the molecules that are being separated can enter. The presence of these pores also causes band broadening by changing the path length a molecule travels. The molecules can

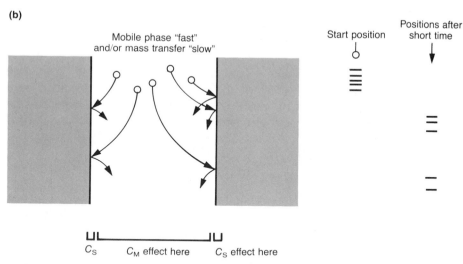

**Figure 13.13**    Illustration of the mechanisms of "mass transfer resistance" broadening in chromatography. The $C$-term in Eq. 13-19 depends on these effects. The molecules of the sample initially are together. As they pass through the particle bed, the molecules tend to diffuse randomly. The diffusion takes place perpendicular to the solvent flow. (a) shows this perpendicular diffusion at equilibrium and *ignores* diffusion in other directions. However, when the mobile phase is started, the molecules follow the paths shown (b). The positions of the particles are spread out as shown on the scale to the right. This spreading can be considered to be due to the system not being at equilibrium. The amount of spreading is proportional to the rate of flow and contributes the term $C_M\bar{u}$ in Eq. 13-19b. In addition, slow interactions at the surface will hold back the molecules more than we would expect from $K_D$ alone. The effect is to cause more spreading in the zone as the solute in the mobile phase continues moving past. This process contributes to the HETP as $C_S\bar{u}$.

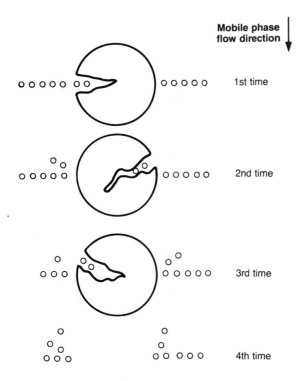

**Mobile phase flow direction**

1st time

2nd time

3rd time

4th time

**Figure 13.14** Illustration of how the inner structure of packing materials can cause band broadening. A number of molecules all start at the same position in the mobile phase. These molecules interact with a heterogeneous particle of the solid support. Some of the molecules migrate into a crevice and then migrate out again. However, they are now behind the original line. This process continues throughout the particle bed. The deeper the pores, the greater the spreading will be. This is because a deeper pore will allow a greater migration length and, thus, a greater delay. Also, the faster the mobile phase moves, the greater the spreading will be. The process also contributes to the term $C_s\bar{u}$ in Eq. 13-19b. One way to reduce this broadening is to make the particles smaller. That way the longest possible pore length is reduced.

be trapped inside a pore only to emerge later into the mobile phase. The amount of band spreading depends on the depth of these pores, since the deeper the pore, the longer the unwanted excursion can take. Therefore, one more benefit of having smaller particles is that the maximum length that these pores and channels can have is limited by the size of the particles. The contribution to $H$ by the particle size is proportional approximately to the square of the diameter. That is,

$$H \propto d_p^z; \qquad z = 1.4 \text{ to } 2.0$$

These two effects, slow exchange kinetics and diffusion into heterogeneous pores, both cause spreading of the band proportional to the flow velocity. The faster mobile phase simply moves the unassociated molecules further before those associated with the solid reemerge into the mobile phase. These two effects cannot be separated and collectively contribute the term $C_s\bar{u}$ to Eq. 13.19b.

### Packing of the Stationary Phase and the $A$-term

All the molecular processes described above assume a homogeneously packed stationary phase. One of the arts of chromatography is creating such a well-packed column from a small pile of particles. For if there are any significant inhomogeneities, broadened bands will result — the chromatographic efficiency will be decreased. Inhomogenities consist of larger channels through which the mobile phase passes more rapidly than through the rest of the bed. This spreads the analyte zones, as illustrated in Figure 13.15. The process is called **eddy diffusion.**

The problems with such channels tend to be diminished in three ways.

A.  Having highly regular particles. Spheres of equal size seem to be particularly good. Irregular channels have less tendency to form.

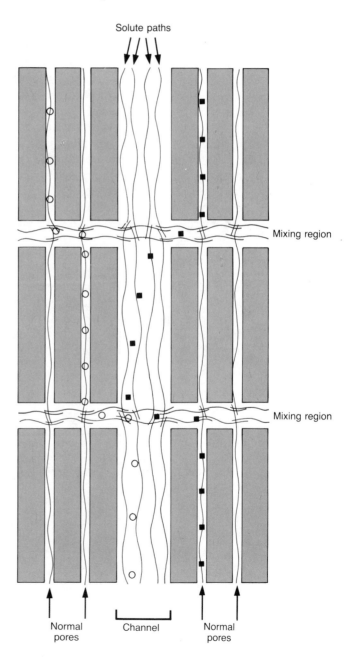

**Figure 13.15** Diagrammatic representation of how a larger channel affects the quality of a chromatogram. The solutes travel faster in the channel. However, if there is easy transfer between the channel and normal packing regions, as shown here, all the particles will experience the same average rate of travel. Only if the channel is isolated will there be zone spreading due to its presence. *Isolated* in this situation means that not all the solute molecules will pass through the channel for some length of time as they pass through the region. Isolation is reduced with smaller particles and smaller column cross sections. If some isolation remains, it is generally accounted for by the $A$ in the van Deemter equation. The process is called eddy diffusion.

B.  Having smaller particles. The larger channels that do form tend to be closer to the size of the average. Also, there are more openings for analyte to diffuse back into the main bed.
C.  Having a thinner (smaller diameter) bed. Any channels formed will be less isolated. The term *isolated* is described below.

If channels are *not* isolated, they do not create as much zone broadening. The reason is that the analyte molecules diffuse across the flow direction, which, in effect, causes them to jump randomly back and forth between the larger channels and the normal regions. (See Figure 13.15.) If the mixing between regions is rapid, it tends to cause all the molecules to experience the same *average* mobile phase velocity. Thus, the effect of minor channels becomes simply one of speeding up the average flow rate somewhat.

The *A*-term of the van Deemter equation reflects the interaction between the quality of the packed bed and the isolation of the irregularities. Regardless of the quality of the packing of the stationary phase, it appears that an *A*-term is necessary to describe the HETP of the columns in LC. However, for capillary GC columns and highly homogeneous packed GC columns, the *A*-term apparently is not needed to describe the HETP results. The larger channels are small enough and the transfer among the regions so rapid that the effect described by the constant *A* value disappears.

Nevertheless, if a column packing is poor (with numerous large channels), then the HETP versus $\bar{u}$ plot may require an *A*-term to fit it. A similar result may arise if, for example, the sample injector volume is too large, even though the column is excellently packed. Then an *A*-term again may be required to fit the GC results, but the effect does not arise from a process on the column. It was from such effects of injectors and detectors in early GC experiments that caused eddy diffusion (the *A*-term) to be considered an intrinsic part of the gas chromatography separation process.

## 13.10 Improving Separations

To develop an improved separation means improving the resolution of the bands of components and/or shortening the time required to obtain the separation. We should like some guidelines as to the sorts of conditions to change to improve a separation. This guidance is obtained from Eq. 13-20, which can be derived from other equations in this chapter (as you are asked to show in problem 13.13). In the equation, the resolution of two *adjacent* bands can be closely approximated by three factors. And, importantly, each of the factors can be calculated from chromatographic parameters that you know how to find from chromatograms.

The equation is

$$R_s = \frac{1}{2} \sqrt{N} \left( \frac{\alpha - 1}{\alpha + 1} \right) \left( \frac{\bar{k}'}{1 + \bar{k}'} \right) \qquad \text{(13-20a)}$$

or

$$R_s = \frac{1}{2} \sqrt{\frac{L}{H}} \left( \frac{\alpha - 1}{\alpha + 1} \right) \left( \frac{\bar{k}'}{1 + \bar{k}'} \right) \qquad \textbf{(13-20b)}$$

In the equation, $\bar{k}'$ is an average capacity factor for two bands. Call the bands 1 and 2.

$$\bar{k}' = \frac{k_1' + k_2'}{2}$$

In deriving the equation, it was assumed that the efficiencies measured for each band are equal: $N_1 = N_2$.

The three terms of Eq. 13-20 have the following significance.

> The first term depends only on the column efficiency, a function of $L$ and $H$.
> The second term depends only on the separation factor. It is a measure of the *difference* between $K_{D1}$ and $K_{D2}$.
> The third term is an average function of $k_1'$ and $k_2'$. It thus is a measure of the average $K_D$ for the components. However, as Eq. 13-17 shows, this factor also depends on the amount of stationary phase ($V_S$ is proportional to the surface area).

What does this equation tell us about choosing the chromatographic conditions? Let us first consider what choices we have. These are listed in Table 13.3. The aim of our chromatography is to have an $R_s$ value greater than about 1.5 for all sets of adjacent bands. As noted before, an $R_s$ greater than this value means that the mutual overlap of the bands on the chromatogram will be less than 0.1%. This is only true when the detector is equally sensitive to both components. If the sensitivity of the detector is not equal for two adjacent eluting components, the impurity levels in each component are not equal; the impurity of component one in band two is not equal to the impurity of component two in band one. For the discussion below, the detector sensitivity is assumed to be equal for both components.

Let us look at the behavior of $R_s$ with variation in $\alpha$ and $\bar{k}'$ with a fixed $N$-value. The results are illustrated in Figure 13.16. In Eq. 13-20, the $\bar{k}'$ term is a measure of how long a separation takes. (Recall that $k' = (t_r - t_0)/t_0$.) As you can see from the graphs of Figure 13.16, lengthening the time of a separation while keeping $\alpha$ and $N$ constant is not much use. It appears that efforts to improve (increase) $\alpha$ may be rewarding. (Recall that $\alpha$ is a measure of the difference in interaction of the compounds with the stationary phase.)

Consider an extreme example: $\alpha = 1$ which means that the two compounds have exactly the same average interactions with the adsorbant. The second term of Eq. 13-20 becomes zero, indicating that a separation is impossible.

But what happens when $(\alpha - 1)$ is very small? As indicated in the graphs of Figure 13.16, just letting the time increase is of little help. As a result, there are only a few possible paths to pursue. The first is to change the mobile phase, stationary phase, or temperature to increase $\alpha$. By understanding the possible chemistries of retention, a decent guess might be made as to what changes to try to improve the relative retention factor. This is where descriptive chemistry comes in, and the art of analysis is applied.

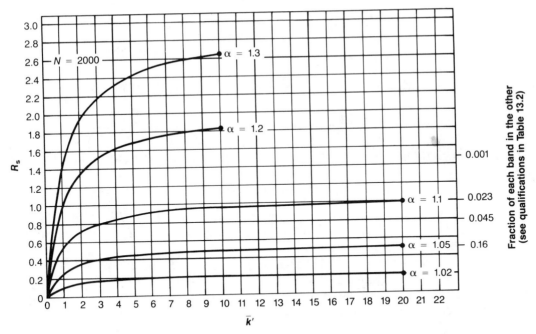

**Figure 13.16**    Plot of the calculated values of $R_s$ for a column with $N = 2000$ for various values of $\alpha$ and $\bar{k}'$. It is much more effective to improve the separation factor $\alpha$ than to run the separation for a longer time. In other words, it is more effective to change the relative retentions than the absolute retentions.

The second way to increase the efficiency is to change the column length. As the first term of Eq. 13-20b indicates, if the HETP remains the same, then doubling the length will double the efficiency. However, this increases $R_s$ by only $\sqrt{2}$. Another way to increase the efficiency is to decrease $H$, the HETP. Some general ways to do this were presented in the previous section; for example, use smaller particles or better packing technique.

However, as illustrated in Figure 13.16, the best single approach is to improve the separation factor $\alpha$. This means increasing the differences in $K_D$ between the different components.

---

**EXAMPLE**

In a gas chromatographic assay, two unidentified components had $k'$ values of 10 and 12. The value of $v_0$ was 2.5 mL. With a column 200 cm long, the resolution of the two components was calculated to be 1.2. If the chemical conditions were already optimized, how long a column should be used to obtain a resolution of 1.5? If a new column of that length were packed exactly the same way with the same stationary phase, what would you expect the value of $v_0$ to be?

**SOLUTION**

With the same stationary-phase conditions (and cross section), the HETP will remain the same. From Eq. 13-20,

$$\frac{R_{new}}{R_{old}} = \frac{\sqrt{L_{new}}}{\sqrt{L_{old}}}$$

$$\sqrt{L_{new}} = \frac{R_{new}\sqrt{L_{old}}}{R_{old}} = 1.5 \cdot \frac{\sqrt{200}}{1.2}$$

$$L_{new} = 312 \text{ cm}$$

From the properties of packings, the void volume is proportional to the total packing volume assuming identical packing conditions and materials. Since the volume of packing material is proportional to the length of the column (if the diameter is the same),

$$\frac{v_0(new)}{v_0(old)} = \frac{L_{new}}{L_{old}}$$

$$v_0(new) = 2.5 \text{ mL } (312/200) = 3.9 \text{ mL}$$

## 13.11 Asymmetrical Peaks

So far, the band shapes that we have considered have been symmetrical. This means that the distribution of mass in the band is the same ahead of the midpoint peak as it is following the midpoint. However, notice that the peaks in the chromatograms in Figures 13.2 and 13.3 are not symmetrical. The peaks are extended at the trailing edges.

This band asymmetry can be described by a single parameter called, reasonably, the **asymmetry.** It is abbreviated $A_s$ and is defined as shown in Figure 13.17. As shown in Figure 13.18, there are three classifications of the shape of chromatographic bands:

**Figure 13.17**    Definition of the peak asymmetry factor $A_s$. The leading edge of the band in the figure is to the left. If $A_s = 1$, the band is symmetric. A good value for an experimental chromatogram is about 1.1. The zone is, then, slightly tailing.

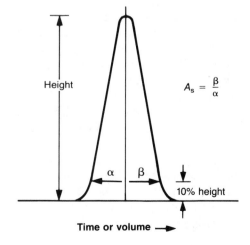

Height

$$A_s = \frac{\beta}{\alpha}$$

$\alpha$ | $\beta$

10% height

Time or volume ⟶

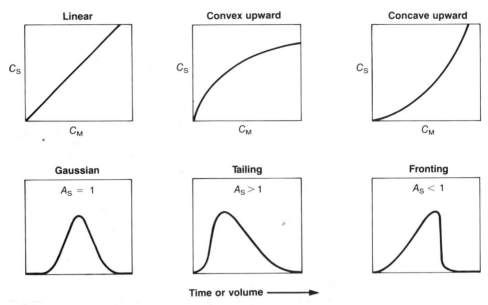

**Figure 13.18**     Each band shape is associated with an isotherm type. The isotherms at the top are plots of the concentration of analyte "in" the solid phase $C_S$ versus the concentration in the mobile phase $C_M$. A straight-line isotherm indicates these two concentrations are directly proportional. A curve that is convex upward means that as $C_S$ increases, a smaller proportion associates with the solid phase. The opposite is the case for a curve that is concave upward. The band shapes associated with each isotherm are shown below. Fronting is not commonly seen.

symmetric or gaussian, **tailing,** and **fronting.** The asymmetries are, respectively,

$$\begin{array}{ll} \text{Symmetric} & A_s = 1.0 \\ \text{Tailing} & A_s > 1.0 \\ \text{Fronting} & A_s < 1.0 \end{array}$$

The asymmetry results from the distribution coefficient $K_D$ being dependent on the concentration of the material in the band. This dependence on concentration is illustrated as a graph called an **isotherm.** Since the word means *constant temperature,* you might expect that these properties are dependent on temperature, and you would be right. The isotherms related to the three classifications of peak shapes are shown in Figure 13.18. For optimal separation of peaks, the minimum asymmetry is desired. If the bands are asymmetric, for a given resolution (by the strict definition for symmetric peaks) the zones will overlap more.

As you can see from Figure 13.18, isotherms are plots of the concentration (or amount) of material on the solid support versus the concentration in the mobile phase. The explanation of the shapes of the isotherms follows from the interactions that arise between molecules on the solid support. The linear isotherm indicates that there is *no* interaction between molecules.

Tailing is a common type of behavior. It occurs when a smaller *proportion* of the material in the mobile phase can be held by the stationary phase when the concentration is increased. Put another way, the interaction of some material with the stationary phase somehow prevents further material from associating as strongly. Chemists usually explain this as being due to a fixed number of sites on the solid phase that "fill up" with molecules.

Fronting is a less common behavior. Here, as material interacts with the stationary phase, it becomes even easier for more molecules to be taken up. The explanation for this is that the molecules that bind at a site on the solid support will bind with other molecules on sites next to them, enhancing the ability of these others to bind. It is behavior seen most often when samples that are too large are put on the column. The surfaces of the stationary phase fill up, and the solute molecules bind with each other in addition to the stationary phase.

The best separation behavior occurs when bands are symmetric. If either fronting or tailing occurs, the resolution decreases. Larger amounts of each component mutually mix. As a result, either more time and/or a higher efficiency will be required to obtain an equally good separation of asymmetric peaks compared with peaks that are symmetric.

## Suggestions for Further Reading

A useful guide (although not comprehensive) to the practice of LC and GC.
Walker, J. Q.; Jackson, M. T.; Maynard, J. B. "Chromatographic Systems: Maintenance and Troubleshooting," 2nd ed.; Academic Press: New York, 1977.

A somewhat out-of-date but solid textbook.
Karger, B. L.; Snyder, L. R.; Horvath, C. "An Introduction to Separation Science"; Wiley: New York, 1973.

A complete, useful, and readable guide to HPLC.
Hadden, N., et al. "Basic Liquid Chromatography"; Varian, Inc.: Walnut Creek, CA, 1971.

A readable and useful guide with experimental details for GC, LC, and TLC.
Zweig, G., and Sherma, J., Eds. "Handbook of Chromatography," Vol. 2; CRC Press: Cleveland, OH, 1972.

All the equations of this chapter that are related to the plate theory of chromatography are carefully and completely derived, as well as those for chromatographic peaks that are not symmetric. It is relatively sophisticated mathematically. Not for the faint of heart.
Said, A. S. "Theory and Mathematics of Chromatography"; Verlag: Heidelberg, 1981.

An excellent book containing the nitty gritty of thin-layer chromatography practice. Includes manufacturers and extensive references. Highly recommended.
Touchstone, J. C.; Dobbins, M. F. "Practice of Thin Layer Chromatography"; Wiley: New York, 1978.

An recent discussion of the theory of zone dispersion. Highly recommended.
Hawkes, S. J. *J. Chem. Ed.* **1983,** *60,* 393.

A short but interesting overview of TLC.
Stahl, E. "A Quarter Century of Thin-Layer Chromatography—An Interim Report"; *Angewandte Chemie, International Edition in English* **1983,** *22,* 507–516.

An up-to-date review of capillary column chromatography—including GC, HPLC, and supercritical fluid chromatography (the pressure keeps a normally gaseous compound liquified as the mobile phase).

Novotny, M. "Capillary Separation Methods: A Key to High Efficiency and Improved Detection Capabilities"; *Analyst* **1984**, *109*, 199–206.

## Problems

13.1    Assume that the solvent flow accidently stopped for 20 s just as the peak of zone 1 of Figure 13.4 reached the detector.
   a.  Assuming the horizontal axis is volume in mL, draw a diagram of the appearance of peak 1 as it would appear.
   b.  Will the peak height appear to increase, decrease, or remain the same?
   c.  Will the peak area appear to increase, decrease, or remain the same?

13.2    Assume that by accident the sample that was used for the chromatogram of Figure 13.4 was injected over a period of 10 s. The solvent flow is precisely controlled by a pump.
   a.  Will the capacity factor $k_1'$ increase, decrease, or remain the same?
   b.  Will the peak heights appear to increase, decrease, or remain the same?
   c.  Will the peak areas appear to increase, decrease, or remain the same?

13.3    In the chromatogram of Figure 13.4, for band 3, calculate the elution volume, capacity factor, bandwidth (volume), and FWHM (volume). Assume that the flow rate is 3 mL/time unit. Assume that the band is exactly symmetrical so that the triangle approximation may be made by reflecting the tangent from the trailing to leading edge of band 3.

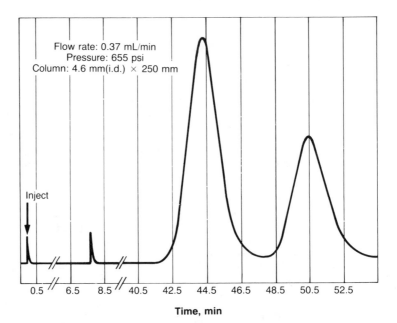

**Figure 13.5.1**

13.4　Calculate for Figure 13.4 the separation factor and resolution for band 3 with both bands 1 and 2.

13.5　Use Figure 13.5.1 to do the following:
　　a.　Calculate the elution volumes of the two peaks.
　　b.　Calculate the capacity factors of the two peaks.
　　c.　Calculate the widths (in time units) of the two peaks.
　　d.　Calculate the resolution of the two peaks.
　　e.　Calculate the separation factor.
　　f.　What is the column efficiency?
　　g.　What is the HETP of the column?
　　h.　If the column were doubled in length, that would the resolution then be?
　　i.　If the column were doubled in length, what would the capacity factors then be?

13.6　Derive the simple relationship between the measured separation factor $\alpha$ and the distribution coefficients characteristic of the two bands involved.

13.7　An experiment was run to determine percent recoveries. The peak heights were used for the assay. After the standards were run and then the samples, all the recoveries were found to be significantly greater than 100%. The operator decided that there was some change in the experimental conditions between the time the standards were run and the time the samples were run.
　　a.　Could the results occur if all the $k_i$ values were constant?
　　b.　Could the results occur if all the $t_i$ values were constant?
　　c.　Would there be a simple way to change the assay to avoid the problem in the future?

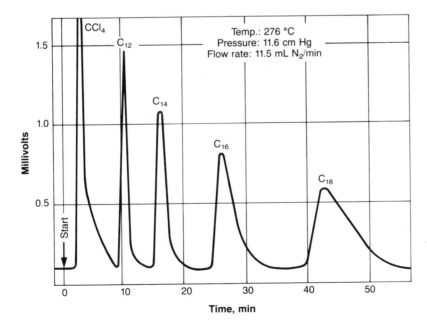

**Figure 13.8.1**

**13.8**   [Ref: Beerthuis, R. K., et al. *Ann. N.Y. Acad. Sci.* **1959,** *72*, 616.]
Shown in Figure 13.8.1 is a chromatogram of normal fatty acids in $CCl_4$. This was done with a packed column with i.d. 0.4 cm., length 150 cm, filled with Apiezon L on Celite (50 to 100 $\mu$m) in 1 : 4 ratio. Assume that the response of the detector is the same per mole for all the acids, and the sample contained 240 $\mu$g of the $C_{18}$ acid, what masses of $C_{12}$, $C_{14}$, and $C_{16}$ acids are present?

**13.9**   Calculate the asymmetry factor for all four bands of the chromatogram of the fatty acids in Figure 13.8.1.

**13.10**   From the last band of the fatty acid chromatogram in Figure 13.8.1, calculate the efficiency of the column.

**13.11**   Figure 13.11.1 shows a gas chromatogram collected by a computerized system. The computer drew baselines as it was programmed to do. If peak heights are used to quantify the components, what percentage of the true quantity would be reported for the peak at 23.5 min?

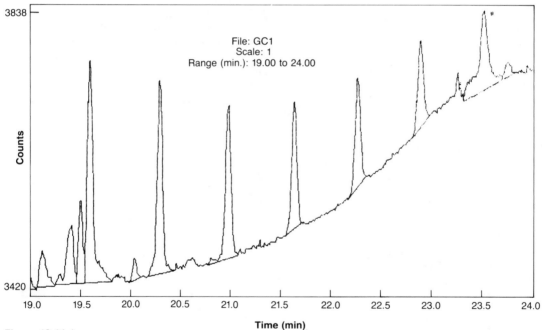

**Figure 13.11.1**

**13.12**   If two identical, symmetric, gaussian peaks, $A$ and $B$, have an $R_s$ value of 1.5, and the detector response for $B$ per unit mass is exactly twice the response for $A$,
     a.   What is the percentage of $B$ in $A$?
     b.   What is the percentage of $A$ in $B$?

**13.13**   Show by proper substitution and algebra that Eq. 13-20 follows from the definitions of $R_s, \bar{k}', k_i', N$, and $\alpha$. First make the assumption that $N$ is equal for both peaks. Following that, calculate the values inside each bracket. Then multiply the values in each of the brackets.

**SUPPLEMENT 13A**

# Proof That $W_i = 4\sigma_i$ for Gaussian Peaks

Ignoring subscripts, we describe a gaussian curve by

$$f(x) = e^{-x^2/2\sigma^2}$$

The derivative of the gaussian is

$$\frac{df(x)}{dx} = \frac{-x}{\sigma^2}\,e^{-x^2/2\sigma^2}$$

As shown in Figure 13A.1, tangents are drawn through the points of inflection to the base of the curve. The points of inflection occur where the half-width is $\sigma$. As shown in Figure 13A.1, $W = 2\sigma + 2B$. We seek the value of $B$.

At the point of inflection,

$$x = -\sigma$$

Therefore,

$$Y = f(-\sigma) = e^{-1/2}$$

The slope at the point of inflection may be expressed two ways. They are equivalent.

$$\left(\frac{df(x)}{dx}\right)_{x=-\sigma} = \frac{Y}{B}$$

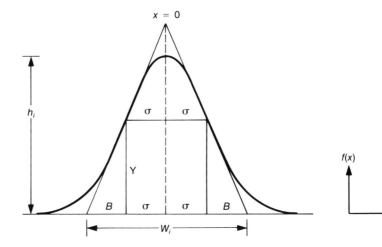

**Figure 13A.1**    A gaussian curve used to approximate the mass distribution of a zone in chromatography. The straight lines are tangents to the points of inflection.

but

$$\frac{Y}{B} = \left(\frac{df(x)}{dx}\right)_{x=-\sigma} = \frac{1}{\sigma} e^{-1/2}$$

At $x = -\sigma$,

$$\frac{Y}{B} = \frac{e^{-1/2}}{B} = \frac{1}{\sigma} e^{-1/2}$$

Therefore,

$$\frac{1}{B} = \frac{1}{\sigma} \quad \text{or } B = \sigma$$

and the value of $W$ is $4\sigma$.

## Supplement Problem

13.14   Consider a gaussian peak (as shown in Figure 13A.1) that has a peak area of 1.00 and a peak height of 0.638. A triangle approximation is made to the area by drawing two lines tangent at the point of inflection of the curve. The point of inflection is where the half-width is $\sigma$. What percentage of the area of the gaussian is the area of the triangle?

# 14

# Liquid Chromatography

## 14.1 Types of Liquid Chromatography

In liquid chromatography (LC) the mobile phase is a liquid. Within that broad definition are a myriad of techniques, and many of them are associated with more than one name. The names focus on many different aspects of LC methods. For instance, one of the primary classification schemes of LC is by the physical shape of the solid support such as *column* chromatography (as initially done by Tswett and described in Chapter 13), *thin-layer* chromatography (with the solid support forming a layer on a plate), and *capillary* liquid chromatography (with the inner surface of the capillary as the solid support). Other names arise based on the direction of flow of the mobile phase: *ascending* chromatography, *descending* chromatography, and *flat-bed* chromatography. Classification is also based on the efficiency of the separations, such as *high-performance* liquid chromatography or *high-performance* thin-layer chromatography. (This last classification is, of course, relative. Look for "very high performance chromatography" in the future followed by "ultrahigh performance.") Sometimes names of LC methods identify solutes that are separated and detected, such as *ion* chromatography and *amino acid* analysis (both done on columns). It is useful to be aware of all these special names. However, the underlying principles of all are those you read about in Chapter 13.

**Table 14.1**  Characteristics of Selected Liquid Chromatography Detectors

| Type | Approximate Limit of Detection* | Approximate Linear Range† | Comments |
|---|---|---|---|
| Ultraviolet and visible absorption | $10^{-11}$ g | $10^4$ | Specific for light-absorbing compounds (see Chapter 17). |
| Differential refractive index | $10^{-9}$–$10^{-10}$ g | $10^3$ | Universal detector. Measures changes in refractive index. |
| Electrochemical: amperometric | $10^{-10}$–$10^{-11}$ g | $10^5$ | Specific detector. Compound must be electroactive (see Chapter 12). |
| Electrochemical: conductometric | $10^{-8}$ g/mL‡ | $10^5$ | Specific detector. Compound must be ionic (see Chapter 12). |
| Fluorescence | $10^{-14}$ g§ | $10^5$ | Specific detector. Compound must be fluorescent (see Chapter 17). |

\* Depends on the peak width. Value for a measurement without preconcentration of the sample. These are constantly being improved.
† Of a detector (transducer + electronics) without chemical interferences present. These are constantly being improved.
‡ Limited by the inherent conductance of the eluent.
§ This value is for strongly fluorescing compounds without interferents.
*Source:* Varadi, M. *Pure and Appl. Chem.* **1979,** *51,* 1175–1182, and references therein.

In this chapter, we focus on the special characteristics of liquid chromatography. These include the benefits of having a wide range of possible properties for the liquid mobile phase and having the choice of a number of significantly different kinds of stationary-phase chemical interactions on which to base the separations.

One of the reasons that LC has become ubiquitous in analytical laboratories is the development of high-performance (or high-pressure) liquid chromatography (HPLC). HPLC is a column chromatography using apparatus similar to that shown in Figure 13.1 in the previous chapter. The solid phase is carefully packed with particles that are less than 10 $\mu$m in diameter, providing unprecedented resolution and efficiency. The size of the particles makes it necessary to pump the mobile phase through the column at pressures up to thousands of pounds per square inch (the commonly used, nonmetric pressure measurement). HPLC pumps keep a highly precise eluent flow rate regardless of the conditions in the column. Thus, as you saw in examples in the last chapter, the positions of the peaks in time can be used for identification of analyte species. This characteristic eliminates overly frequent calibration, a benefit that shortens the average time needed for analyses.

An example of an HPLC assay is quoted below. It was written by an analyst with the Division of Drug Product Quality, Bureau of Drugs, Food and Drug Administration, USA. At one point the author uses the phrase "sensitivity and accuracy of LC." What is meant is that the *detector* is sensitive and precise, and the simplicity of the analyses enhances their accuracy and precision. Some of the capabilities of detectors

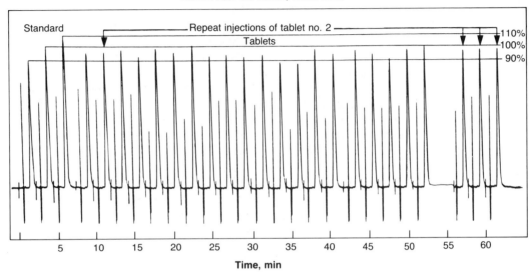

**Tablet Uniformity**
**Prednisolone—Trade name—5 mg tablets**

Packing: Bondapak $C_{18}$/Corasil
Solvent: Acetonitrile/Water (22/78)
Solvent flow: 0.8 cc/min, Chart 12″/hr

**Figure 14.1** Repetitive and rapid assays of steroid tablets done with HPLC. The order of samples is (L to R) three standards, one each of 90, 100, and 110% of the dose; samples numbered 1 through 20; three additional replicates of tablet number 2 to test the reproducibility of the assay. All this was done in slightly over one hour. The quantitation was done by peak height analysis as opposed to peak area. Each chromatogram consists of a narrow mark to show the time the sample was injected, the void volume peak, and the peak for the steroid. Further information on the assay conditions is shown in the figure. [Figure courtesy of Waters Associates, Milford, MA.]

used for LC are listed in Table 14.1. Their mechanisms of operation are described in Supplement 14A.

One of the parameters by which the quality of pharmaceutical formulations can be gauged accurately is by performing chemical analysis of the unit dose recommended on the manufacturer's label. The sensitivity and accuracy of LC, as well as the simplicity of sample preparation needed for the separations, allow it to perform single-tablet assays with relative ease. Assays of prednisolone [$\Delta^1$-hydrocortisone, an anti-inflammatory steroid] . . . is a good example [of a determination] to assure unit dose uniformity and chemical equivalency of drug products. [A reproduction of a uniformity test is shown in Figure 14-1.]
[Ref: Reprinted from Shroff, A. P. *International Laboratory,* May/June 1976, pp. 9–25. From original Waters Associates Bulletin An-124. "Assay of Steroids— Quantitative Assay of Final Dosage Form"; July 1973. Courtesy of Waters Associates, Milford, MA]

## 14.2 Solid Phases and Solute Interactions

Four basic types of chromatographic supports are used in liquid chromatography. The associated chromatography is classified by the general type of chemical interaction that occurs between the solid support and the solutes. The classifications are denoted **normal-phase** (or **adsorption**), **reversed-phase, ion exchange,** and **gel filtration** (or **exclusion** or **gel permeation**) chromatographies. The interactions are illustrated in Figure 14.2 and described in some detail below.

In addition, some solid supports are designed to interact with specific chemical functional groups, such as thiols (—SH) or diols (—COH—COH—), that may be on the solutes being separated. When the specific functional groups are in proteins, the solid phase can be made with molecules that bind to one specific protein. Chromatography with such site-specific groups on the stationary phase is called **affinity chromatography.**

It is imperative to understand that *no* LC separation occurs by means of only one type of interaction. There are *always* interactions besides the named, predominant, type. So if you read "the separation was done by adsorption chromatography," it means that adsorption is the predominant mechanism of interaction with the solid support. However, there also will be some contribution to the separation from at least one of the other mechanisms, such as ion exchange or gel filtration.

### Normal-phase (Adsorption) Liquid Chromatography

In adsorption chromatography, the solid support is more polar than the mobile phase. *Polar,* in this context, is used the same way that we speak of polar bonds in molecules. There is a larger dipole moment associated with the bonds of the solid support than with the bonds in the solvent molecules. The solid supports that are used for normal-phase chromatography are mostly inorganic polymers. The two that are most commonly used are hydrated silicon-oxygen and aluminium-oxygen polymers. These are called, respectively, *silica* (or silica gel) and *alumina.* Their chemistries are quite complicated. For instance, when these inorganic polymers are more hydrated—contain more water in their structure—solute molecules tend to bind onto the surfaces less strongly. The silica and alumina are then said to be less "active."

Recall that the elution time of a given solute depends on the competition between the solid and mobile phases. In normal-phase chromatography, the more polar solvents compete for solutes more effectively than the less polar solvents. The solutes move more rapidly and elute earlier with more polar solvents.

The polarity of the solvent is not the only factor that will cause better solvation and faster elution. Another could be the hydrogen bonding ability of the solvent. You should realize that the chemistry of solvation can be quite involved. As a result, there are as yet no easy, general schemes to use to predict which solvent–solid pair will produce optimal results for a given separation/assay.

Nevertheless, for normal-phase chromatography, the solvent polarity is the primary factor. For convenience, solvents have been classified into an **eluotropic series.** This classification places various solvents into a semiquantitative order based on their abilities to elute solutes from a specific solid support. Table 14.2 lists the

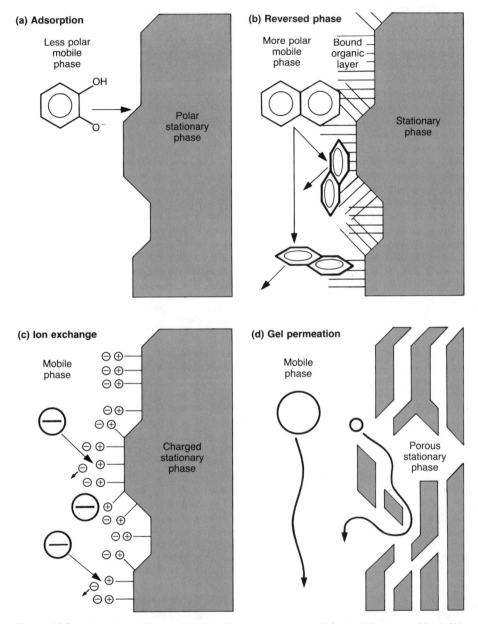

**Figure 14.2**   Illustration of the way that four different mechanisms of LC retardation are explained. Although one of these four mechanisms may be predominant, there is always some contribution from at least one other. (a) Adsorption, in which a polar stationary phase is in equilibrium with a less polar mobile phase. (b) Reversed phase, in which the stationary phase is less polar than the mobile phase (60–70% of HPLC assays are done with reversed-phase techniques). (c) Ion exchange, in which the stationary phase has co-valently bound ionic groups. (d) Gel permeation, in which the smaller molecules, which are able to enter pores in the stationary phase, are held up longer than larger molecules, which are unable to penetrate the gel matrix.

**Table 14.2**   Solvent Strength Parameter $\epsilon°$ for Alumina Supports: The
Eluotropic Series

| In Alphabetical Order | | In Numerical Order Low to High | |
|---|---|---|---|
| *Solvent* | $\epsilon°$ | *Solvent* | $\epsilon°$ |
| Acetic acid | 1.0 | Pentane | 0.00 |
| Acetone | 0.56 | Petroleum ether | 0.01 |
| Acetonitrile | 0.65 | Hexane | 0.01 |
| Benzene | 0.32 | Cyclohexane | 0.04 |
| Carbon tetrachloride | 0.18 | Carbon tetrachloride | 0.18 |
| Chlorobenzene | 0.30 | Xylene | 0.26 |
| Chloroform | 0.40 | Toluene | 0.29 |
| Cyclohexane | 0.04 | Chlorobenzene | 0.30 |
| Dimethylsulfoxide | 0.62 | Benzene | 0.32 |
| Dioxane | 0.56 | Ethyl ether | 0.38 |
| Ethyl acetate | 0.58 | Chloroform | 0.40 |
| Ethylene dichloride | 0.49 | Methylene chloride | 0.42 |
| Ethyl ether | 0.38 | Tetrahydrofuran | 0.45 |
| Hexane | 0.01 | Ethylene dichloride | 0.49 |
| *iso*-Propanol | 0.82 | Methylethylketone | 0.51 |
| Methanol | 0.95 | Dioxane | 0.56 |
| Methylene chloride | 0.42 | Acetone | 0.56 |
| Methylethylketone | 0.51 | Ethyl acetate | 0.58 |
| Pentane | 0.00 | Dimethylsulfoxide | 0.62 |
| Petroleum ether | 0.01 | Acetonitrile | 0.65 |
| *n*-Propanol | 0.82 | Pyridine | 0.71 |
| Pyridine | 0.71 | *iso*-Propanol | 0.82 |
| Tetrahydrofuran | 0.45 | *n*-Propanol | 0.82 |
| Toluene | 0.29 | Methanol | 0.95 |
| Water | Large | Acetic acid | 1.0 |
| Xylene | 0.26 | Water | Large |

eluotropic series for alumina. The order is mostly in the order of the polarity (dipole moment) of the solvent molecules. Hexane is low, ether in the middle range, and water on the high end.

### Reversed-phase Liquid Chromatography

In reversed-phase chromatography, the solid support is *less* polar than the mobile phase. Two fundamental types of solid supports are used, the most common being composed of silica particles with hydrocarbon molecules chemically bonded to the

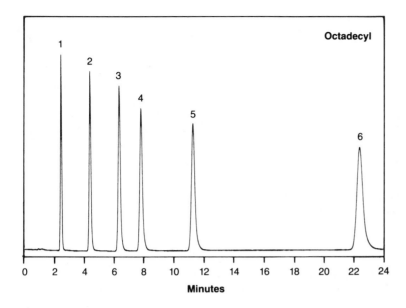

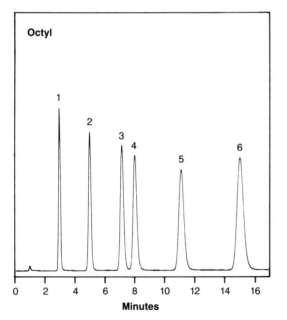

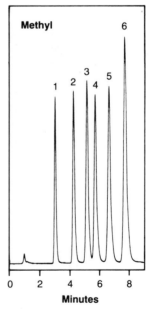

**Figure 14.3** Comparison of reversed-phase chromatographic materials with different carbon chain lengths. All three columns are the same except that the bonded phases have 18 (octadecyl), 8 (octyl), or 1 (methyl) carbon chains bonded to the surface of silica. The elution times become progressively shorter with shorter carbon chains for the organic solutes listed below. In addition, their capacity factors change somewhat. Peak identification: 1, uracil; 2, phenol; 3, acetophenone; 4, nitrobenzene; 5, methyl benzoate; 6, toluene. Operating conditions: Columns, IBM columns 4.5 × 250 mm; mobile phase, 50/50 methanol/water isocratic elution; flow rate 1.0 mL/min; detection at 254 nm. [Figures courtesy of IBM Instruments.]

surface. Of these, the most often used organic groups are $-CH_3$, $-C_8H_{17}$, and $-C_{18}H_{37}$. The last, an 18-carbon chain, is the most common. The organic coating of the particles is formed from a silicon compound containing the $C_{18}H_{37}$ group, the octadecyl group. The abbreviation ODS is sometimes used for this type of support. These bonded organic groups have an effect similar to that which would be produced by an extremely thin organic solvent layer on the surface of the silica particles. The interactions of the solutes with the support and the mobile phase are, thus, much like those in a liquid–liquid extraction. The longer the carbon chain length, the more "organic" these bonded layers become. Therefore, longer chains interact more strongly with solutes that prefer the organic phase. The effects of this difference are illustrated in Figure 14.3.

One disadvantage of these chemically bonded supports is that they decompose — the organic groups slowly hydrolyze and fall off, leaving a normal-phase silica surface exposed. Thus the retention indices such as $k'$ can change over time under the same experimental conditions.

The second type of solid support used for reversed-phase chromatography is organic polymer beads. A typical polymer is a resin composed of polystyrene and divinylbenzene. The divinylbenzene component forms bonds between the polystyrene polymer chains — so-called crosslinks. The crosslinks cause a physical stiffening of the polymer. Stiffness is necessary to resist deforming under high pressure.

Reversed-phase chromatography is quite popular since the chemical reactions of adsorption/desorption tend to be fast. Thus, the chromatographic peaks tend to have less spreading due to slow surface reactions — $C_s\bar{u}$ effects — than do normal-phase supports. Also, there is usually less tailing with reversed phase, which is indicative of more linear adsorption isotherms. (See Section 13.11.)

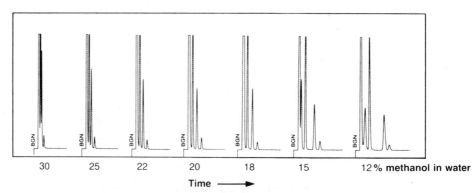

30    25    22    20    18    15    12% methanol in water

Time ⟶

**Figure 14.4**   Illustration of the effects of changing solvent conditions on a reversed-phase column. The sample is an analgesic preparation. Each run is isocratic, and the eluents range from high (30%) methanol to low (12%) methanol : water mixtures. The methanol is a better solvent for these materials than is the water. Conditions: ODS column 0.26 × 25 cm; eluent methanol (% shown)/water + 0.5% $H_3PO_4$; flow rate 1.0 mL/min; detection ultraviolet light absorption at 250 nm. [Data courtesy The Perkin–Elmer Corporation.]

**Table 14.3** Ion Exchange Groups

| Type | Active Group | pH Range of Operation | Application Example |
|---|---|---|---|
| Strongly acidic cation exchanger | $-SO_3^-$ | $1-14$ | Amino acids, inorganic separations |
| Weakly acidic cation exchanger | $-COO^-$ | $5-14$ | Transition elements, organic bases |
| Strongly basic anion exchanger | For example, $-N(CH_3)_3^+$ | $1-12$ | Alkaloids, fatty acids |
| Weakly basic anion exchanger | For example, DEAE, $-C_2H_4N(C_2H_5)_2$ | $1-9$ | Organic acids, amino acids |

In contrast with the eluotropic series for adsorbants such as silica and alumina, less polar solvents are *more* powerful eluents in reversed-phase separations. Figure 14.4 shows an example of the changes in elution times with changes in eluting solvent for a reversed-phase separation. As can be seen from the chromatograms, when the eluting solvent contains more methanol in the water, it becomes more strongly eluting. The components of the mixture elute faster in that case. In general, the reversed phase eluting power is the reverse of Table 14.2.

There is a similar mechanism of adsorption in **partition chromatography.** For partition chromatography, the solid support is coated with an organic solvent such as *n*-butanol, benzene, or chloroform. The mobile phase is water and/or a polar solvent. This is, thus, a liquid–liquid extraction chromatography. You will see the name partition chromatography in the older literature. It has largely been supplanted by chromatography using bonded organic stationary phases since the unbonded coating solvent tends to wash off too easily.

### Ion Exchange Liquid Chromatography

In solution, one of the causes for two ions to bind is the attraction of unlike charges. As an example, we explain the strengths of acids by the magnitude of the negative charge at the site of proton binding. High-charge sites tend to hold the protons more strongly. In the same way, if, for example, positively charged groups are anchored to a stationary phase, then ions of the opposite charge will be attracted to them. These ionic interactions can be used to separate eluting species by differences in their average charge.

Stationary-phase materials with bound charges are called ion exchange resins. The reason they are called ion *exchange* resins is because ionic solutions always contain equal numbers of positive and negative charges—that is, they are electrically neutral. As a result, the ions bound on the resin support are always associated with *some* ion of the opposite charge from the mobile phase. Thus the associated ion can leave the bound site only when it is exchanged with a new one, as illustrated in Figure 14.2.

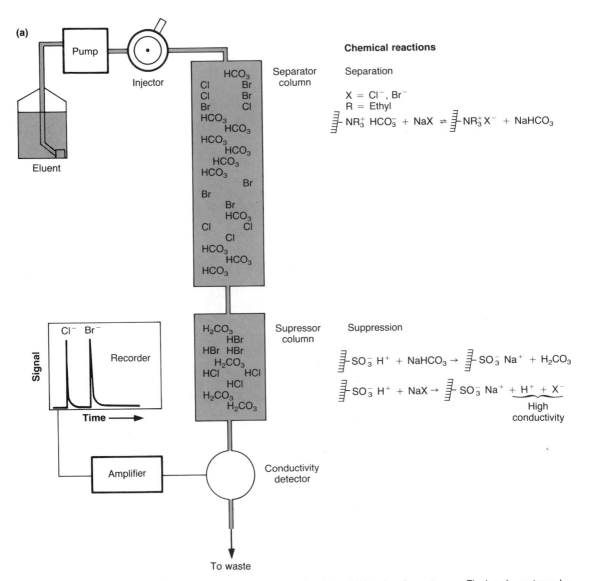

**Figure 14.5** (a) Ion chromatography equipment and (b) an ion chromatogram. The ion chromatography apparatus employs a pump to keep the eluent flow rate constant, so that the retention times alone may be used to identify the component peaks. The separation occurs as a result of differences in the relative affinities that ions have for an ion exchange resin. For the anions used in this illustration, the separation is done using a quaternary ammonium resin (with —$NR_3^+$ groups). The eluent is aqueous sodium bicarbonate solution. The reversible reaction shown in the figure is the reaction that determines the $k'$ for each ion. If there were no bicarbonate in the eluent solution, the anions would merely bind with and remain on the resin sites with a simultaneous release of the ionically bound bicarbonate which they displace. The anions would not elute. (In the language of ion exchange, when chloride exchanges for bicarbonate, the resin is said to be converted from the bicarbonate *form* to the chloride form.)

The suppressor column changes the nature of the effluent to make it more suitable for the conductivity detector. The detection is done by measuring the electrical conductivity of the effluent. When the zones of ions pass through the detector, the conductivity rises. (See also Supplement 14A and Section 12.8.) However, to

Solid supports composed of cross-linked polymers with fixed negative charges bound to them exchange positive ions (cations) and are called **cation exchange resins.** Cation exchange resins are further divided into two groups, *weak cation* exchangers and *strong cation* exchangers. The terms weak and strong have to do with the $pK_a$ of the groups bound to the resin and *not* to the properties of the solutes being separated.

In analogy, polymers with covalently linked positive charges on them exchange negative ions (anions). These resins are called **anion exchange resins.** Both weak and strong anion exchangers are made and used for separations. Table 14.3 lists common cation and anion exchange groups.

However, the ion exchange is not quite as straightforward as it may sound since competitive chemical reactions occur simultaneously with all the charged components of the solution. In other words, the bases and conjugate acids of the solutes as well as the base and conjugate acid of the groups on the solid support and of any

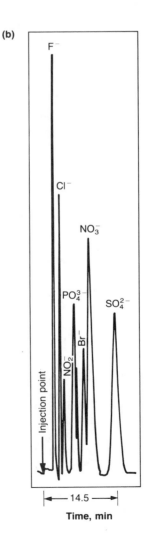

**(b)**

**Figure 14.5** *(continued)*    obtain better sensitivity to changes in ion concentrations, any background ions from the eluent should be removed. This is done by converting the bicarbonate ions to the weak acid carbonic acid in the suppressor ion exchange column. (The conductivity of aqueous carbonic acid is significantly less than the equivalent bicarbonate solution.) The resin in the suppressor is initially in the acid form, with $H^+$ ionically bound. These protons are exchanged for the $Na^+$ in the effluent through ion exchange. As a result, all the anions flow through in their acid forms. The $HCO_3^-$ becomes a weak acid, $H_2CO_3$, but the other anions such as $Cl^-$ and $NO_3^-$ form strong acids with high conductivities and their peaks stand out. Note that the suppressor column is used to do an ion exchange and not a separation. This method of removing the background ions is only one of many that are now being used.

Eventually, all the protons of the suppressor resin are exchanged in reactions that will leave the resin in the sodium form predominantly. The resin must be **regenerated** periodically by passing a solution high in $H^+$ through the column, reversing the supression ion-exchange reaction and regenerating the acid form.

The sample chromatogram was run under the following conditions: Columns, Dionex (trade name) anion separator 3 × 500 mm and Dionex anion suppressor 6 × 250 mm; eluent, 0.003-M $NaHCO_3$/0.0024-M $N_2CO_3$ at 30% of a preset maximum flow rate; detection, conductometric. [Chromatogram is from EPA-600/2-79-200, "Procedures Manual: Level 2 Sampling and Analysis of Oxidized Inorganic Compounds," USEPA, Washington, D.C., 1979.]

buffers that may be needed are all present. We need not consider all the details of this complicated simultaneous equilibrium system, however, since the strengths of the binding interactions are proportional to the *average* charge of each acid–base pair involved. This concept is discussed next.

Figure 14.5b illustrates the chromatogram for a separation of a number of anions. What causes the ions to be separated? To understand the underlying chemistry, let us consider an extreme case. For both anion and cation exchangers, neutral molecules will pass through the resin bed without interacting with the fixed, charged groups. Thus a neutral molecule will elute in the dead volume $v_0$.

Let us now consider only a positively charged support — an anion exchanger. If no interaction occurs between neutral solutes and the support, then the solutes that possess a neutral charge on the largest fraction of the material will pass by the stationary phase with the least retardation. The solutes that have the largest fraction in its charged form will be most retarded. The solutes with fractions between these extremes will elute between them.

The average charge of an anion such as the conjugate base of an acid depends on the $pK_a$ of the acid and the pH of the mobile phase. Since the percentage of a solute in its acid form is so crucial in its retention, the pH of the mobile phase is almost inevitably buffered to a fixed and carefully chosen pH. This buffering also keeps the bands from spreading due to variations in pH in different regions of the resin bed. There is one more complication to consider. The stationary-phase ionic charge is also dependent on the pH. For instance, if the concentration of $OH^-$ gets very high, it may neutralize the anion exchange resin itself, and ion exchange will effectively cease. Thus, strong anion and strong cation exchangers have a wider pH range in which they operate.

A relatively recent development using ion exchange chromatography is becoming so widely used that it is worth understanding how it works. This is **ion chromatography.** It is used to quantify, in a single experiment, a number of anions or cations down to the ppm level. In essence, it works by using ion exchange resins for separating the ions chromatographically. The band positions are used for identification, and the areas under the bands are used to quantify the amounts of each ion in the mix.

The general setup of the equipment for the determination, the reactions that occur in each section, and a sample chromatogram are shown in Figure 14.5. The figure caption describes the process in more detail. Ion chromatography in the form described involves the use of two ion exchangers, one to separate the ions and one to effect a total ion exchange of $Na^+$ for $H^+$.

### Gel Filtration Liquid Chromatography

Gel filtration separations differs from the other three types since the molecules are not separated through differences in chemical interactions. The separation is made on the basis of the effective *sizes* of molecules. Again, though, let me remind you that no separation is accomplished due to a single pure mechanism. The mechanism of gel filtration, illustrated in Figure 14.2, is perhaps best described using the alternate name for the process, **exclusion chromatography.** The support used in gel permeation chromatography is a solid that has pores with controlled (as carefully and precisely as possible) cross sections. These pores permeate the volumes of the small beads packed in the columns. The separation occurs because only smaller molecules can enter the pores and be retained away from the path of the mobile phase. Thus, the larger molecules, which do not enter the pores of the gel, elute first followed by the smaller molecules. A gel permeation chromatogram of a series of oligomers is shown in Figure 14.6. (Oligomers are molecules with sizes approximately in the range between tetramers and polymers.)

The retention times of molecules follow an interesting behavior, which is illustrated in Figure 14.7. Plotted on the graph is the logarithm of the molecular weight versus the volume of eluent. Notice that at low volumes all the polymers with masses greater than about $10^5$ daltons elute together; at higher volumes all the particles with

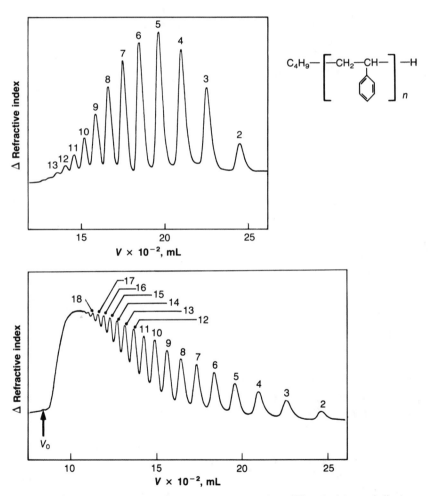

**Figure 14.6**    Example of gel permeation chromatogram of two different mixtures of oligomers (small polymers) of styrene. The general formula is shown. The peaks are labeled with the value of $n$ in the formula. Eluent, tetrahydrofuran (THF); detection, refractive index. [Ref: Heitz, W.; Boemer, B.; Ullner, H. *Makromol. Chem.* **1969**, *121*, 106.]

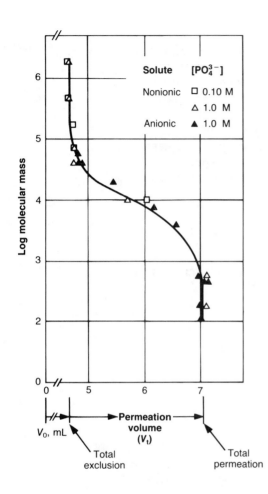

**Figure 14.7**   Example of a calibration curve for gel permeation chromatography (GPC). Note that the molecular mass scale is logarithmic versus the elution volume. The calibration was done with both ionic and nonionic solute standards. The gel exclusion packing, called *controlled-pore glass,* is a hard silica glass that has been prepared under special conditions so that it is permeated with molecular-sized pores. The diameter of the pores of this material average 75 Å. Total exclusion occurs only for polymers over about m.w. 100,000. [Ref: Cooper, A. R.; Matzinger, D. P. *J. Appl. Poly. Sci.* **1979,** *23,* 419. Copyright © 1979. Reprinted by permission of John Wiley & Sons, Inc.]

masses less than $10^3$ daltons elute together. In between, the elution volume varies nearly linearly with the logarithm of the molecular mass. How can we explain these characteristics?

As illustrated in Figure 14.2, if a molecule is larger than the largest pore size, then it will not enter into the pore throughout its passage through the column. It does not matter whether the molecules are 10% larger than the pore or two times the size. All molecules over a certain size limit will flow continuously with the mobile phase and elute together in the void volume. This explains the lack of resolution at the void volume of the column. On the other hand, all molecules below a certain size will spend an equal time tortuously making their way through the pores of the static phase. Thus there is a lower limit to the molecular size resolution. These low molecular mass particles elute off the column in a volume called the **total permeation volume,** $v_t$. All molecules that interact exclusively through an exclusion mechanism will elute from the column between these limits, $v_0$ and $v_t$.

The relatively linear part of the calibration curve can be explained qualitatively through Figure 14.8. Stated simply, the *accessible* volume of pores that can impede

**Figure 14.8**    Diagram of one model that explains in a simple way the behavior observed in gel filtration. The volume of pores accessible to the molecules depends on the molecular size; the retardation of a molecular species is proportional to its accessible volume. Consider a spherical molecule in a regularly shaped pore, as shown. The accessible volume is the volume of the pore inside a surface that lies at a molecular radius inward from the true pore surface. Three different-sized molecules are shown along with the cross-hatched accessible volume. When the pore radius equals the molecular radius, the accessible volume becomes zero. No entry to the pore is possible if the molecular radius is larger.

smaller molecules is greater than the volume available to larger molecules. Therefore, smaller molecules spend more time impeded inside the pores than do the larger ones, the smaller ones elute later.

With careful calibration, gel permeation chromatography can be used to determine the molecular size (and approximate molecular weight) of both biochemical and industrial polymers. To do so, the molecular sizes of interest must lie on the linear portion of the logarithmic calibration curve. What can you do if the range of molecular sizes of interest is not correct for the column being used? The answer is to change the column packing to one for which the expected molecular sizes are in the log-linear range. Most manufacturers produce packings with different **nominal pore size.** The pore size of the packing used in the experiment giving the results shown in Figure 14.6 was 75 Å. Often the calibration is given as a molecular weight exclusion limit, which is the molecular mass that is slightly too large to penetrate into the pores. These exclusion limits range from around m.w. 1000 to m.w. 300,000. The useful range in which the fractionation can be calibrated is from about $\frac{1}{3}$ of the exclusion limit upward. For example, a packing with an exclusion limit of 100,000 daltons can be calibrated for masses between 30,000 and 100,000 daltons.

## 14.3  Filters and Batch Separations

The differences between the molecular properties of the components of a sample can be so great that the components can be separated without using chromatography. The one or two components of interest can be adsorbed on a solid support while the rest of the matrix passes through. After the support is rinsed to be sure that the unwanted components are fully removed, the adsorbed species can be eluted with a more strongly eluting solvent for further processing or assaying. For such separations, which include affinity chromatography, stationary phases consisting of short columns of adsorbant (range of 1–2 cm) can be used. Special filter papers also work for some separations. The filter paper acts like an extremely short column, with the paper as the solid phase. Different filter papers are available: direct phase, reversed phase, and ion exchange as well as papers with special bonded groups such as —SH. The name *batch separation* can be used for all these techniques. Such separations are commonly used in sample preparation to clean up the analyte for an assay.

Another batch separation technique involves plastic disk filters that act like exclusion gels. The holes in these filters are precisely controlled to be homogeneous and of molecular size. Such filtration, often called **molecular filtration,** is used to separate macromolecules from small molecules and ions such as salts, sugars, and solvents. Because of the small pore size, molecular filtration is usually done by forcing the liquid phase through the disk filter under a pressure difference. Either gas pressure is applied above or a vacuum is applied below.

A similar, nonchromatographic method that is used to separate molecules by size is **dialysis.** In a dialysis, again a membrane with molecule-size pores is employed. Usually the membrane is a tubular sack which is immersed in a relatively large reservoir. The separation is done by allowing adequate time (up to days) for the small molecules to diffuse through the membrane into the external reservoir. Those molecules that are too large to pass through are retained inside the sack. In general, dialysis is much slower than gel filtration and molecular filtration.

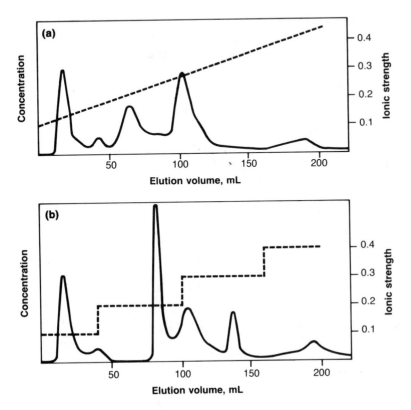

**Figure 14.9** Illustration of the differences in a separation that can be made by employing a linear or stepwise gradient. Here, the gradient consists of an increase in ionic strength using NaCl solution; the concentrations are shown as the dashed lines and refer to the scale on the right-hand side. The biochemical sample is bovine blood serum. The fourth peak in chromatogram A results from serum albumin, a common protein. Stepwise gradient elution causes the albumin to be eluted in two distinct peaks (4th and 5th) and also changes the shapes of the others. None of the materials is baseline resolved.

Conditions: Column, quarternary ammonium QAE-Sephadex (trade name) A-50; bed dimensions, 1.5 × 25 cm; sample, 4 mL of 3% (w/v) freeze-dried bovine serum; eluent 0.1-M tris(hydroxymethyl)aminomethane-HCL buffer pH 6.5 + NaCl. The buffer name is usually abbreviated as Tris-HCl. The eluent is gravity fed (no pump is used). [Chromatograms courtesy of Pharmacia Fine Chemicals.]

## 14.4 Gradients

So far, we have discussed liquid chromatography as if the optimal conditions are found by a single selection of three things: the solid-phase, the mobile-phase composition, and the flow rate. However, we are not limited to having a single mobile-phase composition throughout the separation; we are not limited to isocratic conditions. If the eluting power of the solvent is changed during a separation, we say that we are running a **gradient elution.** The gradient is created by mixing solvents with different eluting capabilities, beginning with the poorest eluting solvent mixture and proceeding to a better eluting solvent mixture, while the separation is taking place. This change can be done by jumps (stepwise gradient elution) or by a constantly changing solvent composition (continuous gradients). Why take the trouble? Because resolution can be improved. Figures 14.9 and 14.10 illustrate some of the improvements in separations obtained by the use of solvent gradient elutions.

**Figure 14.10**   Illustration of the differences in a separation that can be made by changing the steepness of the gradient. The gradient is made by increasing the ionic strength using increasing amounts of NaCl. As can be seen, a significant change in the resolution occurs with a slight change in the gradient steepness. The separation is done on an ion exchange gel.

Conditions: Sample, $O_2$-hemoglobin, CO-hemoglobin, albumin monomer, and albumin dimer; column, Diethyla-minoethyl-Sephadex A-50; eluent, 0.1-M Tris-HCl at pH 8.3 + NaCl; flow rate 8 mL cm$^{-2}$ h$^{-1}$. (Note that the use of flow rates of this type means that the column diameter is redundant information.) [Chromatograms courtesy of Pharmacia Fine Chemicals.]

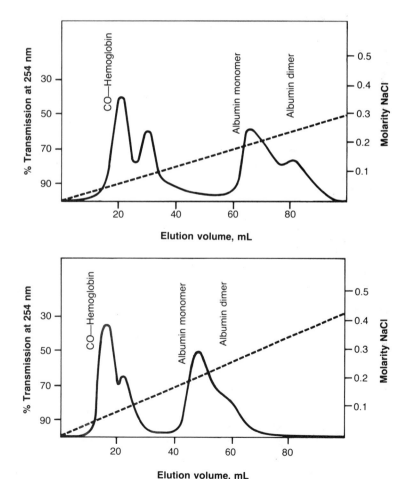

1 ASP
2 THR
3 SER
4 GLU
5 PRO
6 GLY
7 ALA
8 CYSTINE
9 VAL
10 MET
11 ILE
12 LEU
13 TYR
14 PHE
15 LYS
16 HIS
17 NH$_3$
18 ARG

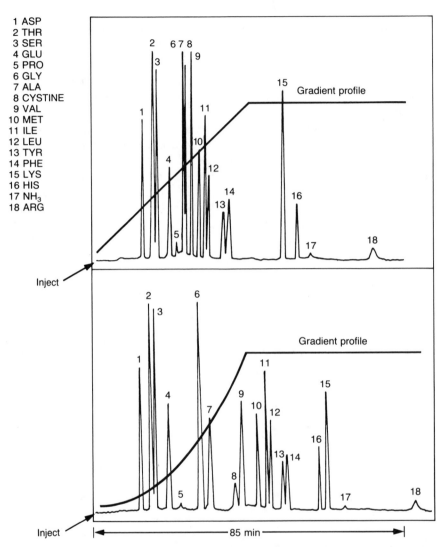

**Figure 14.11**    Illustration of the differences in a separation that can be made by changing the shape of a gradient from linear to nonlinear. Separation of the amino acids is done by ion exchange chromatography with an ionic strength and pH gradient elution. Buffer I: 0.2-*N* Na$^+$, pH 3.08. Buffer II: 1.0-*N* Na$^+$, pH 7.4. The primary improvement is in separating peaks 6–8. Note the reversal of the positions of peaks 15 and 16. [Figures courtesy of Waters Associates, Milford, MA.]

Continuous gradients can be *linear,* which means that the solvent composition changes at a fixed rate with the volume. The gradients can also be varied so that they are *nonlinear:* the composition of the eluent is made to change nonlinearly with volume. One example of the effect of changing the shape of a gradient is shown in Figure 14.11. The general rule for gradients is, If it works to produce a better separa-

**Table 14.4**   Properties of Eluents That May Be
Varied to Produce Gradient Elutions

Ionic strength

pH

Hydrogen bonding ability

Strength in eluotropic series

Dielectric constant

Concentration of reagents that compete for
the stationary-phase binding site

tion, use it. A number of different properties of mobile phases can be changed to effect a gradient elution. Some of these are listed in Table 14.4.

## 14.5 Particle Size, Pressure, and HETP

Another factor of choice in chromatography, as noted in Table 13.3, is the particle size of the stationary phase. As the particle size is reduced, the HETP is reduced as well. As a result, the same resolution should be obtainable on shorter lengths of stationary phase. The benefit of reducing the particle size is illustrated in Figure 14.12, which shows column-liquid-chromatography van Deemter plots for three different particle sizes with other conditions remaining the same.

However, changing the particle size cannot be done independently, without adjusting at least one other experimental variable: either the flow rate or the pressure required to force the mobile phase through the support. Figure 14.13 illustrates this interdependence, which can be expressed in two empirical equations. (The terms are defined below.)

$$\text{HETP} = D_0 \bar{u}^n d_p^z, \tag{14-1}$$

with the variables in the ranges $n = 0$ to $1$ and $z = 1.4$ to $2.0$, and

$$\text{pressure drop} = \Delta P = \frac{\bar{u} L \eta}{\Theta d_p^y}, \tag{14-2}$$

with the variables in the ranges $y = 1.8$ to $2.0$ and $\Theta \approx 600$. In these equations,

$d_p$ is the particle diameter in cm,
$D_0$ is a solute diffusion coefficient ($\approx 10^{-5}$ cm$^2$ s$^{-1}$ in water),
$\Delta P$ is the pressure drop in atmospheres (1 atmosphere = 14.696 psi),
$\bar{u}$ is the average linear velocity of the mobile phase in cm/s,
$\eta$ is the viscosity in centipoises (cP) of the mobile phase ($\eta_{H_2O} = 1$ cP at 20 °C),
$L$ is the support length, and
$n, y, z$, and $\Theta$ are empirically determined constants.

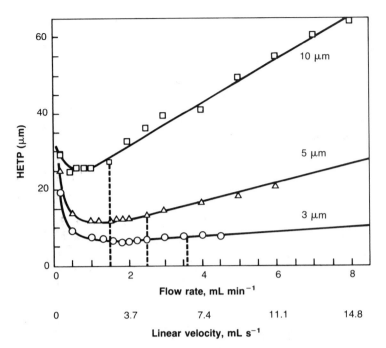

**Figure 14.12**    Graph of experimentally determined HETP versus flow rate for different particle diameters in HPLC. The flow rate in mm/s is related to the flow rate in mL/min for columns with 4.6-mm inside diameter. The dashed lines indicate the optimum flow rates for each column. The reason that the optimum is not at the minimum of the curves is that the time required for an assay has been taken into account. For instance, because there is such a small variation with flow rate in the HETP of the column with the 3-$\mu$m packing, the flow can be increased (and analysis time decreased) by a factor of about two with little loss of efficiency. [Courtesy The Perkin-Elmer Corporation.]

What is important to see in Eq. 14-1 is that the HETP decreases (desirable) as $d_p$, the particle diameter, decreases. Thus, to obtain the same efficiency, shorter columns can be used if the particle size is smaller, as can be seen in Figure 14.13. Both $\bar{u}$ and $D_0$ are constant, and the column lengths are different for each of the chromatograms shown.

However, as can be seen from Eq. 14-2, the trade-off for a smaller particle size comes with a necessary increase in pressure on the mobile phase, increasing approximately as the inverse square of the particle diameter. In Figure 14.13, note that the pressure required to obtain a 1.0 mL/min flow rate with a shortened column for the smaller particle size is greater than the pressure needed for a longer column packed with larger particles.

Not all column chromatography is HPLC. When larger particles are used, the pressures can be lower. **Open column** chromatography is done with particle diameters $\geq 100$ $\mu$m. The columns are usually mounted upright, and the mobile phase is driven through by gravity as was Tswett's early chromatogram. To speed separations when the columns get longer, a few atmospheres of gas pressure can be applied to the solvent. This is called **flash chromatography.** Alternatively, a low-pressure, fixed flow rate pump may be used. However, when the particles are in the 3- to 10-$\mu$m range — the HPLC range — a pump must be used directly on the solvent to achieve the pressures needed. Typical HPLC pressures are in the range of a few thousand psi ($\sim 100$ atmospheres).

**Figure 14.13**    Illustration of the effect of particle size on analysis time. As the particle size is reduced, the analysis time can be shortened. In this example the efficiency is kept approximately constant by reducing the column length as the particle size is reduced. The trade-off is the higher pressure required to keep the same flow rate for the three columns.

Peak identification: 1, phenol; 2, acetophenone; 3, nitrobenzene; 4, methylbenzoate; 5, anisole; 6, benzene; 7, toluene. Conditions: Octyldecylsilane packing of diameters noted with chromatograms; column A, 300 × 4.6 mm; column B, 150 × 4.6 mm; column C, 75 × 4.6 mm; mobile phase methanol/water 60/40; flow rate 1.0 mL/min; temperature 21 °C; pressure on A = 640 psi, on B = 1770 psi, and on C = 2540 psi. [Ref: Cook, N. H. C.; Olsen, K. *J. Chromatog. Sci.* **1980,** *18,* 512–524. Reproduced from the *Journal of Chromatog. Science* by permission of Preston Publications, Inc.]

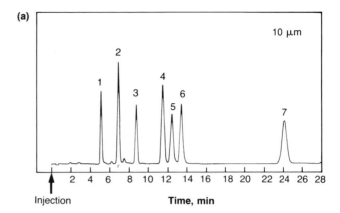

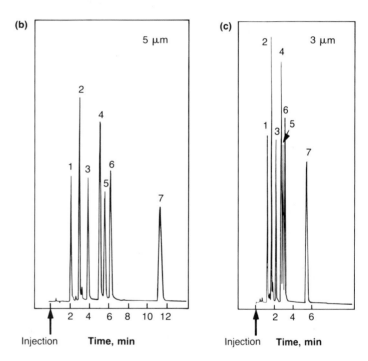

## 14.6 Planar Solid Supports

So far, in both this chapter and the last, chromatography has been presented in the context of packed columns of stationary phases. These stationary phases are equilibrated with the mobile phase, and then a sample is introduced into the "head" of the

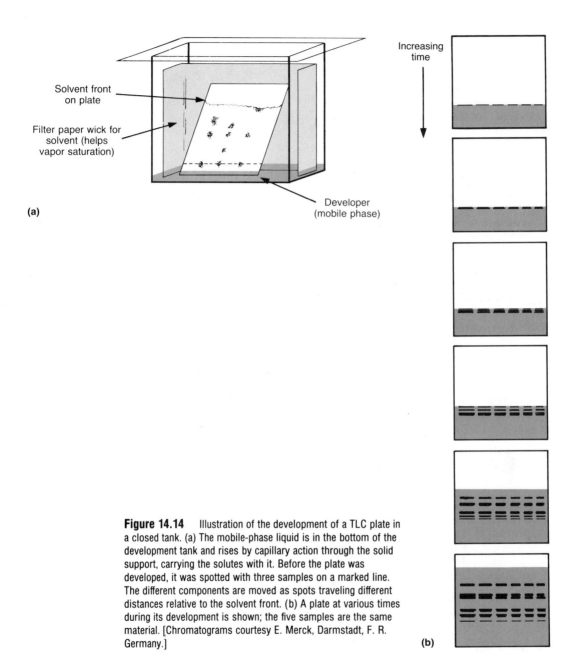

**Figure 14.14** Illustration of the development of a TLC plate in a closed tank. (a) The mobile-phase liquid is in the bottom of the development tank and rises by capillary action through the solid support, carrying the solutes with it. Before the plate was developed, it was spotted with three samples on a marked line. The different components are moved as spots traveling different distances relative to the solvent front. (b) A plate at various times during its development is shown; the five samples are the same material. [Chromatograms courtesy E. Merck, Darmstadt, F. R. Germany.]

Solvent front on plate

Filter paper wick for solvent (helps vapor saturation)

Developer (mobile phase)

(a)

Increasing time

(b)

column and carried through it until eluted and detected at the end. However, this is not the only method that is used for chromatographic analysis.

In another geometry, the stationary phase is a layer of adsorbant particles a fraction of a millimeter thick attached to a solid backing plate of aluminum, plastic, or glass. A separation carried out on these stationary phases is called **thin-layer chromatography (TLC).** (Unfortunately, sometimes *thick-layer chromatography,* with layers 0.5 – 2 mm thick and primarily used for preparations, is also abbreviated TLC.)

TLC separations are usually done as shown in Figure 14.14. A few microliters of the sample dissolved in a minimum volume of solvent is placed onto the *dry,* solid support above the edge of the plate. The solvent from the sample is then allowed to dry. To *develop* the chromatogram, one edge of the dry plate is placed into a pool of developing solvent. The solvent, drawn up by capillary action, is allowed to rise to a predetermined height on the plate. The plate is then removed from the solvent tank and dried. The sample(s) are quantitated in one of a number of ways, which are described in Supplement 14B.

Figure 14.14 shows a closed container, which keeps the atmosphere saturated with the vapor of the solvent. This saturated vapor prevents solvent from evaporating off the face of the plate. Alternatively, depending on which conditions work best, the plate may be developed in an open tank. The atmosphere is not saturated with solvent.

The primary advantage of TLC is that a number of samples and standards can be analyzed simultaneously; with column chromatography, individual samples must be analyzed sequentially. One other advantage is that samples that are difficult to resolve may be developed in two different solvents run in perpendicular directions. An example of the results of such a two-dimensional development scheme is shown in Figure 14.15.

The way that we explain the cause of TLC separations is the same as for column chromatography. The general chromatographic theory is the same. As is true for column chromatography, the retardation on a TLC plate depends on the strengths of interactions between the solute and stationary phase. However, as can be seen in Figures 14.14 and 14.16, each of the individual components has traveled a different distance through the stationary phase. In column chromatography, all components must travel through the same length of stationary phase to elute. Thus the values of $k'$ (which depend on the time spent passing through the entire length of the stationary phase) cannot be determined in exactly the same way for TLC as for column chromatography.

### Characterizing the Retention of TLC Spots

For *qual*itative analysis, spots on a TLC plate are characterized by the distance they are moved relative to the distance that the mobile phase traverses. These distances are shown in Figure 14.16. The parameter used to characterize this ratio is called the **retention factor** or **retention index,** abbreviated $R_f$. (Do not confuse $R_f$ with the resolution $R_s$.) For TLC,

$$R_f = \frac{\text{distance from origin to spot}}{\text{distance from origin to solvent front}} \quad \text{(14-3)}$$

522

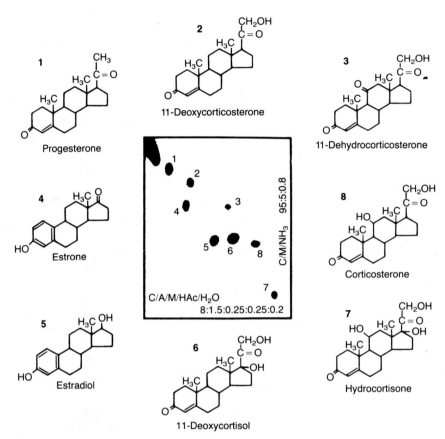

**Figure 14.15** Results of a TLC plate developed in two dimensions in order to achieve separations that could not be done with a single development. The spots are numbered with the various steroids that were separated. The development was first done in the bottom-to-top direction. Compounds 3 and 4 travel together as do compounds 5, 6, and 8. After the plate was dried, these compounds were then separated by development from right to left with a different developing solvent system (see below). The sample was originally spotted at the point of compound 7, which was not moved by either solvent.

Conditions: Solid-phase adsorbant silica gel H (Merck [the manufacturer]) plus 10% by weight magnesium silicate; development, 1st, chloroform/methanol/28%-aqueous ammonia 95/5/0.5; 2nd, chloroform/acetone/methanol/acetic acid/water 8/1.5/0.25/0.25/0.2. The steroids are normally colorless, so the spots were made visible after the plate was dried by spraying on a solution 98% sulfuric acid + 37% formaldehyde solution (97/2 water/formaldehyde). The plate was then heated at 180 °C for 30 min. The steroids become charred and produce dark colored spots for identification. [Ref: Rouser, G. *J. Chromatog. Sci.* **1973,** *11,* 63. Reproduced from the *Journal of Chromatographic Science* by permission of Preston Publications, Inc.]

**Figure 14.16** Measuring the $R_f$ value from a thin-layer chromatography plate. The point of application is marked, and the plate developed. The final position of the solvent front is marked with a pencil while the plate is still wet. For the first and second spots, $R_f = a/b$. There is a problem with spot 3 since the spot is significantly tailed. This is equivalent to tailing in column chromatography. The point chosen to represent the spot is where the peak would be if it were a column chromatogram. This is in the center of the head circle, as if the shape were considered a circle with a tail.

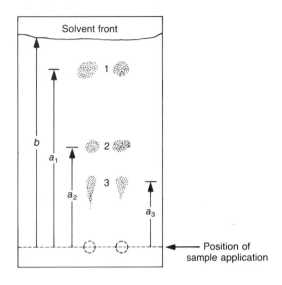

For each component, the retention factor is related to a capacity factor $k'_i$ in the following way.[1]

$$R_{fi} = 1/(1 + k'_i) \qquad \text{(14-4)}$$

(Deriving Eq. 14-4 from the definitions of $k'$ and $R_f$ is problem 14.1.) The $R_f$ values depend on the same experimental conditions as do the $k'$ values of column chromatography: the composition of the mobile phase, the identity of the stationary phase, the temperature, and the identities of the compounds separated. Usually, standards in TLC are run simultaneously with a number of samples on the same plate.

The materials used for the stationary phase of a thin-layer plate can be the same as those packed into a column: alumina, silica, bonded alkylsilanes, and cellulose. However, even if exactly the same material is used both for column packings and for TLC adsorbants, the behavior of a given solute may still not be exactly comparable. The main reason for this is that in TLC the solvent and solutes interact with dry surfaces that were not initially pre-equilibrated with the solvent, whereas the column stationary phase is continuously immersed in solvent. An example of this difference can be seen in Figure 14.17.

### High-Performance Thin-Layer Chromatography

Since separations by TLC and column chromatography are based on the same phenomena, it is expected that improvements in TLC performance are obtainable in the same general ways that conventional liquid chromatography was improved. The

---

[1] This equation is not exactly correct since the solvent front travels somewhat faster than the mobile phase as a whole.

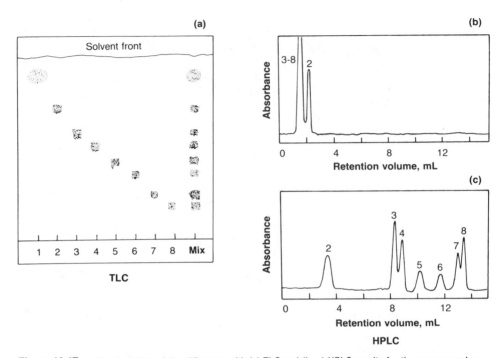

**Figure 14.17**    Demonstration of the difference with (a) TLC and (b, c) HPLC results for the same sample on approximately the same chemical surface. Both the TLC and HPLC adsorbants are octaldecylsilane modified silica gel (reversed phase). The samples are polystyrene polymers with different molecular masses from the monomer (m.w. 105) to a large polymer (m.w. $1.0 \times 10^7$). The separation is primarily due to a reversed-phase adsorption. The TLC (a) and chromatogram (b) of the HPLC are directly comparable in experimental conditions but significantly different in result. The mobile phases were nearly the same: TLC, methylene chloride–methanol 79:21 (v:v); HPLC (chromatogram b), the same solvents but 77.5:22.5. Only when a gradient was run for the HPLC (linear from 53:47 $CH_2Cl_2$:$CH_3OH$ to pure $CH_2Cl_2$ in 18.8 min) did a moderately similar chromatographic separation occur, chromatogram c. The authors explain this by the fact that the solvent composition varies over the development distance of the TLC. The mixed solvent system becomes partly unmixed. [Ref: Reprinted with permission from Armstrong, D. W.; Bui, K. H. *Anal. Chem.* **1982**, *54*, 706–708. Copyright 1982 American Chemical Society.]

HETP should be decreased by reducing imperfections in packing and by using smaller particles for the solid phase. These expectations have been fulfilled in high-performance TLC (HPTLC) plates. (Do not confuse flat TLC *plates* with the concept of theoretical *plates*.)

By use of particles having a more homogeneous-size distribution than on conventional TLC plates, the packing is improved (in part by reducing channel formation). Also, average particle diameters are reduced from about 20 $\mu$m on conventional TLC plates to 5 $\mu$m on HPTLC plates. The final result of these changes is to reduce the HETP value from around 30 $\mu$m (conventional) to around 12 $\mu$m (HPTLC). The variation in HETP, which is less than a factor of three, may not seem like a very significant amount. However, in addition, less sample is applied to the HPTLC plates, which produces smaller spots on the HPTLC plate coating. (This is equivalent to a more rapid injection of sample into a chromatographic column.) As a result, the total number of usable theoretical plates is in the range of 600 for conventional TLC and 5000 in HPTLC. An equivalent separation can be obtained with a shorter migration distance in HPTLC, which, in turn, means that a shorter time is required for equivalent analyses.

One drawback is that the sample size for HPTLC must be significantly smaller (in the nanogram range) than for conventional TLC (in the microgram range). Either commercially available mechanical sample applicators or carefully supported microsyringes are needed in order not to damage the surface of an HPTLC plate when samples are applied. In general, HPTLC requires more equipment and more expertise on the part of the analyst than conventional TLC.

## 14.7 Paper Chromatography

Historically, a wide range of materials has been separated by elution along the length of a paper sheet. The mobile phase is usually a water–organic mixture. The solid phase is the main component of paper—cellulose (or chemically modified forms of it). It is generally accepted that the cellulose adsorbs water onto its surface. The interaction of the stationary phase with the solutes is then a combination of two types: a liquid–liquid extraction between the mobile phase and the water adsorbed on the cellulose surface as well as some contribution from direct adsorption on the underlying cellulose. Paper chromatography is relatively slow and has been replaced in large measure by reversed-phase HPLC and reversed-phase TLC.

## Suggestions for Further Reading

An encyclopedic textbook/reference on gas, liquid, and thin-layer chromatography including equipment, sample preparation, and theory. It is exhaustive and contains large numbers of literature references. However, you will need to know what you are looking for since the organization is complex. It should be easily readable after studying the topics in this chapter.
Poole, C. F.; Schuette, S. A. "Contemporary Practice of Chromatography"; Elsevier: Amsterdam, 1984.

A monograph on HPLC.
Snyder, L. R.; Kirkland, J. J. "Introduction to Modern Liquid Chromatography," 2nd. ed.; Wiley: New York, 1979.

A complete, useful, and readable guide to HPLC.
Johnson, E. L.; Stevenson, R. "Basic Liquid Chromatography"; Varian Assoc.: Walnut Creek, CA, 1978.

A moderate length review of HPLC with numerous examples. The balance is similar to this chapter.
Pietrzyk, D. J. In "Physical Methods in Modern Chemical Analysis," Vol. 2; T. Kuwana, Ed.; Academic Press: New York, 1980.

Contains mostly explanations of the theory and equipment for gel-permeation chromatography with some experimental detail.
Yau, W. W.; Kirkland, J. J.; Bly, D. D. "Modern Size-exclusion Liquid Chromatography"; Wiley–Interscience: New York, 1979.

A collection of chapters by different authors on gel-permeation chromatography. They cover the details and problems of GPC. Especially recommended is Chapter 7 on precision and accuracy.
Janca, J., Ed. "Steric Exclusion Liquid Chromatography of Polymers"; Dekker: New York, 1984.

A monograph on ion chromatography including what can be done and what happens rather than practical instruction. However, it is useful and contains lots of references.
Smith, F. C., Jr.; Chang, R. C. "The Practice of Ion Chromatography"; Wiley: New York, 1983.

More of a how-to-do-it book than Smith and Chang. It has straightforward descriptions of experimental ion chromatography with numerous examples.
Fritz, J. S.; Gjerde, D. T.; Pohlandt, C. "Ion Chromatography"; Huthig: Heidelberg and New York, 1982.

A short review article on developments in ion chromatography by its inventor.
Small, H. "Modern Inorganic Chromatography"; *Anal. Chem.* **1983,** *55,* 235A–242A.

A well-written set of review articles on HPLC detectors.
White, P. C., "Recent Developments in Detection Techniques for High-performance Liquid Chromatography. Part I. Spectroscopic and Electrochemical Detectors"; *Analyst* **1984,** *109,* 677–697.
White, P. C., "Recent Developments in Detection Techniques for High-performance Liquid Chromatography. Part II. Other Detectors"; *Analyst* **1984,** *109,* 973–984.

An intermediate level article (less mathematical than most) on LC optimization.
Glajch, J. L.; Kirkland, J. J. "Optimization of Selectivity in Liquid Chromatography"; *Anal. Chem.* **1983,** *55,* 319A–336A.

Everything you wanted to know about the practice and past uses of thin-layer chromatography.
Kirchner, J. G., "Thin Layer Chromatography"; Wiley–Interscience: New York, 1978.

An excellent book containing the nitty gritty of thin-layer chromatography practice. Includes manufacturers and extensive references. Highly recommended.
Touchstone, J. C.; Dobbins, M. F. "Practice of Thin Layer Chromatography," 2nd ed.; Wiley: New York, 1978.

Relatively up-to-date book on the detailed theory and practice of thin-layer chromatography.
Zlatkis, A.; Kaiser, R. E. "HPTLC: High Performance Thin-layer Chromatography"; Elsevier: Amsterdam, 1977.

## Problems

14.1   Show from the definitions of $R_f$ and $k'$ that Eq. 14-4 is correct if the motion of the solvent front indicates the motion of the mobile phase.

14.2  What are the $R_f$ values of spots 1, 2, and 3 in Figure 14.16?

14.3  From the data supplied in Figure 14.13, use band 7 to determine the exponent for $d_p$ in Eqs. 14-1 and 14-2. Assume that $\bar{u}$ does not depend on particle size. Are $y$ and $z$ approximately equal?

14.4  [Ref: Fuller, E. N. et al. *J. Chromatog. Sci.* **1982,** *20,* 120. Reproduced from *Journal of Chromatographic Science* by permission of Preston Publications, Inc.]
On the gel filtration chromatogram in Figure 14.4.1 four of the five peaks are labeled with the molecular weights of the fractions (278,000, etc.). What is the average molecular weight of the middle fraction?

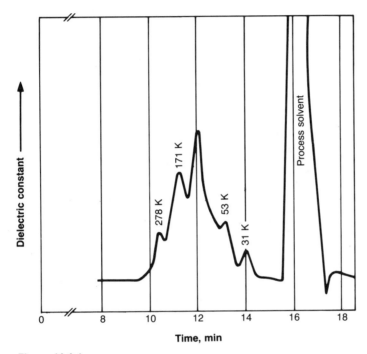

**Figure 14.4.1**

14.5  [Ref: Jackson, M. A.; Fisher, J. E. *Ind. Res. & Devel.* **1983,** February, 130. © 1983, Industrial Research & Development.]
The graph in Figure 14.5.1 shows the changes in retention times for a number of amino acids as they change with acetate concentration in the eluent. This is predominately an ionic strength effect. Which of the amino acids are neutral at pH 5.15, the pH of the experiments?

14.6  This problem refers to Figure 13.5.1, p. 494. Since the bands are somewhat asymmetric, the operator decided that the $R_s$ value should be increased to 2.0. Since the conditions were already optimized, a longer column was needed.
    a.  What column length is needed to achieve the desired separation?
    b.  What pressure will be needed to keep the flow rate constant?
    c.  What pressure will be needed to keep the same retention time?

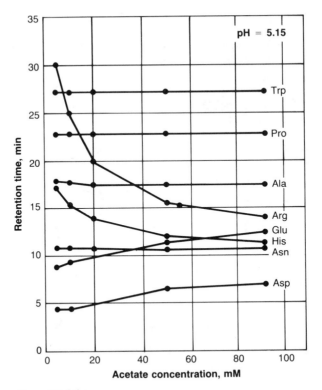

**Figure 14.5.1**

14.7   From band 6 of the chromatograms in Figure 14.3,

   a.   estimate the ratio of partition constants of methyl : octyl : octadecyl with toluene
        under the conditions of the separation. Assume $V_S$ remains constant. (You may
        wish to make a photocopy enlargement.)

   b.   Let us assume that the reversed-phase adsorbant surfaces are homogeneously cov-
        ered with the chains sticking out like bristles on a brush. The $V_S$ would vary with
        chain length (1 : 8 : 18). What are the ratios of $K_D$ for toluene with $V_S$ varying this
        way? Does this simple picture produce results consistent with the ideas of reversed-
        phase separations?

14.8   [Ref: Jenke, D. R. et al. *Anal. Chim. Acta* **1983,** *155,* 279.]
Figure 14.8.1 shows an ion chromatogram run *without* a suppressor column. There are three
different detectors in the effluent stream one after the other. They are 1, photometric at
280 nm; 2, refractive index; and 3, conductivity. Which of these would yield the most precise
results under the same conditions but with a tenfold reduction in ion concentrations?

14.9   [Data from Kroeff, E. P. and Pietrzyk, D. J., *Anal. Chem.* **1978,** *50,* 502.]
The capacity factor of DL-tyrosine changes with pH on a reversed-phase support. With a
phosphate buffer 0.02 M and solvent of ethanol : water 5 : 95, the pH values and the associated

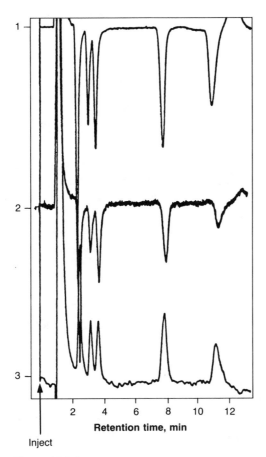

**Figure 14.8.1**

$k'$ values are:

| pH | $k'$ |
|------|------|
| 2.3 | 3.33 |
| 4.8 | 1.75 |
| 6.5 | 1.56 |
| 8.9 | 1.94 |
| 11.4 | 7.28 |

The $pK_a$ values of tyrosine are $pK_1 = 2.20$, $pK_2 = 9.11$, and $pK_3 = 10.07$. At low pH, the molecule is a monocation. At high pH values, it is a dianion.

    a.  If the mechanism of retention is a pure reversed-phase one, which form of the molecule would have the highest $k'$?

    b.  What type(s) of interaction(s) is/are occurring on this reversed-phase column: H-bonding (adsorption), reversed-phase, ion exchange, exclusion?

## SUPPLEMENT 14A

# Detectors for Column LC

Table 14.1 lists the sensitivities and linear ranges of selected LC detectors. Their mechanisms of operation are described here. The sample volumes of most contemporary detectors are in the range of $5-20 \mu L$ for use with HPLC. There are good reasons for keeping the volume small. Since the volume of eluent within the cell is mixed by the flow of liquid, the earlier and later eluting volumes intermix. The final result is that each band is broadened due to this mixing, and the larger the cell the larger the eluent volume that is mixed. For a cell volume of a few microliters, the effect of mixing produces only a minor broadening of the narrowest bands. Larger cell volumes cause quite obvious decreases in the resolution that is achieved on the column and even cause nearly complete mixing of adjacent bands.

In this supplement are shown examples of the construction of five of the most commonly used types of detectors: UV–visible light absorption, refractive index, conductivity, amperometric, and fluorescence. These examples are only representative structures since the details of construction differ among units from different manufacturers. Conductivity and amperometric techniques are discussed in Chapter 12. UV–visible and fluorescence spectrometries are included in Chapter 17.

### UV–Visible Detectors

Figure 14A.1 illustrates the construction and operation of a UV detector which uses filters to produce the wavelength used. Although this wavelength is fixed, any of a large number of wavelengths can be selected. The volume of sample needed is kept low by using narrow-bore tubing where possible, and sensitivity is increased by constructing the longest possible path length within the constraints of the optics. Other parts of the detector are similar to a normal spectrophotometer in this wavelength region (Chapter 17).

### Fluorescence Detectors

The general principle of fluorescence detectors is similar to that of the UV–visible units, but the emitted light intensity is measured 90° to the exciting beam. (Further description of the measurement of fluorescence can be found in Chapter 17.) A diagram of a representative sample cell and optics is shown in Figure 14A.2. The flow of the sample is perpendicular to the page.

Shown in Figure 14A.3 is a chromatogram with simultaneous UV and fluorescence detection. Note that the compounds do not respond uniformly in both detection modes. This is especially obvious in the range of bands 1–10 and 53–63. Such differences in response and sensitivity must be taken into account in the choice of detector(s) for LC.

Eluent

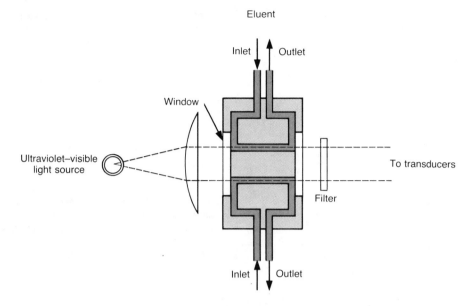

**Figure 14A.1** Diagram of the light path and liquid flow path for a dual-beam LC sample cell, one design of a light absorption LC detector. Light from the source on the left passes through a window and two tubes filled with liquid. One of the tubes contains a sample of the eluent. It is not flowing so as not to waste it. The other tube has the effluent from the column passing through. The difference in light absorption between the two is put out as a signal to plot as the chromatogram. The wavelength of the light is selected by the filter. The volume of the cell through which the light passes can be as low as a few microliters. Often, when the volumes become that small there can be problems caused by changing temperatures of the effluent and/or reference solvent. This is a major cause of baseline drift. Thus, there tends to be a trade-off between stability and zone broadening due to mixing in the cell.

**Figure 14A.2** Diagram of the light path and liquid flow path for one possible design of a fluorescence LC detector. Light from the source on the left passes through a filter that selects a wavelength to excite fluorescence in the eluent stream flowing perpendicular to the paper. The light is collected and passed through a second color filter, which selects the desired assay wavelength. The light energy is converted to an electrical signal by the transducer.

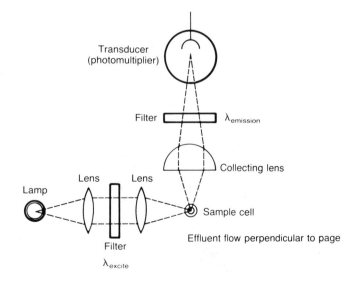

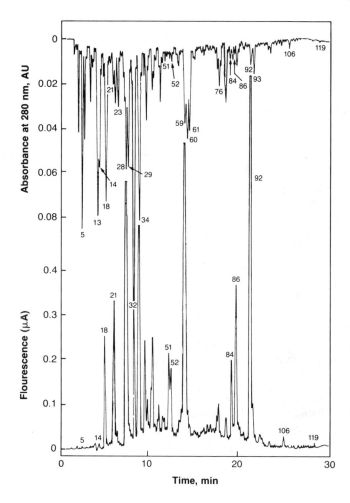

**Figure 14A.3**   Illustration of the difference in response between absorption and fluorescence detectors in an HPLC chromatogram. Note, for instance, the differences in bands 59–61. Two of the three compounds do not show up in the fluorescence detection while one appears to be relatively quite large. Conditions: Sample 10 mm³ of urine extract with about 100 ng of a number of aromatic organic acids; column, 5-μm LiChrosorb RP-18 (a reversed-phase C₁₈ column) 25 cm long; flow rate, 2 mL/min gradient elution from 0.1-M phosphate buffer pH 2.1 mixed to form gradient with acetonitrile; temperature 70 °C. The fluorescence detector had an excitation wavelength of 238 nm and emission wavelength at 340 nm. The UV absorbance was measured at 280 nm. The output scales are (top) absorbance units and (bottom) microamps from the fluorescence detector. [Ref: Reprinted with permission from Molnar, I.; Horvath, C.; Jatlow, P. *Chromatographia* **1978**, *11*, 260. Pergamon Press, Ltd.]

## Differential Refractive Index Detector

Only one of a number of different designs of refractive index (RI) detectors is illustrated in Figure 14A.4.[2] The design shown depends on the refracting (light-bending) properties of prisms. The angle through which a prism will refract an incoming beam depends on the refractive index of the material comprising the prism. If two prisms of identical shape are placed edge to edge, as shown in Figure 14A.4, and if both prisms have the same refractive index, then the light will pass straight through

---

[2] The construction and operation of other types of RI detectors can be found, for example, in Snyder, L. R.; Kirkland, J. J. "Introduction to Modern Liquid Chromatography," 2nd ed.; Wiley–Interscience: New York, 1979.

**(a) Top view**

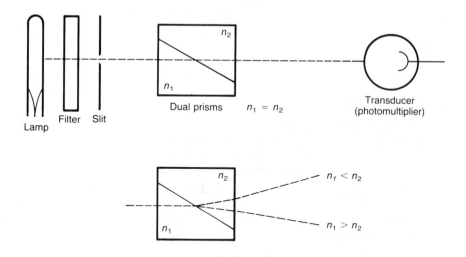

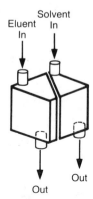

**(b) Perspective view of prism**

**Figure 14A.4** Diagram of the light path and solvent flow for one design of a refractive index detector. The light is passed through a filter to select a single wavelength after which the light passes through two hollow prisms that are filled with liquid. One prism contains the solvent as reference; the chromatographic eluent passes through the other. The light is bent when the refractive indices of the two liquids differ. This change in refractive index appears as a change in the amount of light that falls on the transducer. The refractive index of the eluent changes with any added solute. Thus, the RI detector is a universal detector. However, it is relatively less sensitive than the absorbance or fluorescence detectors if the solutes absorb light or are fluorescent.

to the transducer. If, however, one prism has a different refractive index from the other, the light path will shift. The active area of the transducer is arranged so that any shift of the difference in refractive index changes the amount of light falling on it. This response is amplified to produce a signal proportional to the differential index of refraction of the prisms.

This principle is implemented for liquid samples by using hollow prisms that are filled with the liquid of the sample and references. When the index of refraction of the liquid changes, the light path shifts, and the output of the transducer changes. Sensitive instruments require that the prisms and their liquid contents be carefully thermostated.

### Amperometric Detector

A representative structure for an amperometric LC detector appears in Figure 14A.5. It consists of a three-electrode potentiostat (described in Section 12.7) connected to the working, reference, and auxiliary electrodes as shown. The reference and auxiliary electrodes are downstream from the working electrode. The effluent enters the detector head and passes in a thin channel over the working electrode. The output signal arises from the current generated at a fixed, preset potential. If the analytes have different (order of 100 mV) formal potentials, further discrimination between analytes can be made by careful choice of the electrode potential. This point is illustrated in Figure 12.17.1 in the problem set of Chapter 12.

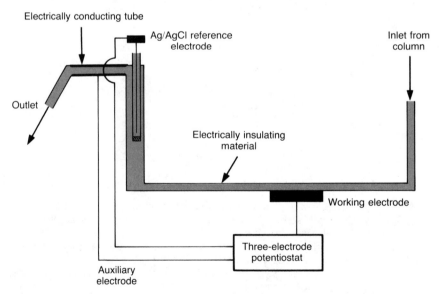

**Figure 14A.5**   Diagrammatic representation of a general design for an amperometric HPLC detector. A three-electrode potentiostat (Section 12.7) is connected to a working electrode in contact with the effluent stream. The working electrode may be constructed of a material such as platinum, gold, glassy carbon, or graphite. Any reference electrode can be used. Ag/AgCl wire in a KCl-filled capillary is shown here. The auxiliary electrode here is a metal capillary tube in contact with the effluent. Other designs have the reference and auxiliary directly across from the working electrode, a design that allows less concentrated ionic solutions to be used as eluents.

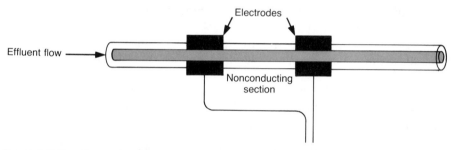

**Figure 14A.6** Diagram (cross section) of one design for a LC conductance detector. The two electrodes are stainless steel and are in contact with the eluent but separated by a nonconducting region.

## Conductivity Detector

The conductivity of a solution is measured using common AC bridge equipment (described in Section 12.8). The cell for a flowing effluent stream can be made as illustrated in Figure 14A.6. The electrodes are metallic tubes at either end of a nonconducting tube. As long as the geometry is constant, precise relative conductivities can be measured.

## SUPPLEMENT 14B

# Detection and Quantitation for TLC[3]

The detection of the location of spots on a TLC plate is primarily done by observing one or more of the following characteristics:

a. Inherent colors of the compounds
b. Inherent fluorescence of the compounds
c. Changes in fluorescence of a fluorescent indicator that is incorporated in the stationary phase. The whole plate fluoresces under UV light when *no* solutes are present. The spots where solutes reside appear as dark patches on the plate.
d. Colors of compounds after reaction with gaseous reagents such as ammonia, sulfuric acid, or iodine
e. Colors of compounds after reaction with group-specific or metal-specific reagents sprayed evenly onto the plate: for example, dimethylglyoxime for nickel and fluorescamine for amines

---

[3] Primary ref: Issaq, H. J.; Barr, E. W. *Anal. Chem.* **1977**, *49*, 83A–96A.

Quantitative analysis for TLC is, even potentially, not as precise ($\pm 5-10\%$) in general as is possible for HPLC or GC ($\pm 1\%$ and, with care, $\pm 0.1\%$). For quantitative analysis, the sample must be quantitatively applied onto the plate. This can be done by using syringes that can place microliter volumes onto the plate with $\pm 1\%$ volume error.

The quantitative measurement of a separated component can be done either by removing the chromatographed material from the stationary phase or by making the measurement in situ. Removing the material from the support is done in two ways.

1. Quantitatively scrape from the plate the area of the stationary phase that contains the desired component. The analyte is then eluted with a highly eluting solvent, and an assay is run in any of the usual ways. Automatic scrapers are available commercially.
2. Use a spot eluter, which works something like a wet vacuum. Eluting solvent is sprayed on the spot while, at the same time, the wash is sucked up into a container.

Running the assay while the component is still on the plate can also be done in a number of ways. The simplest is to run a series of standards along with the unknown and visually compare the color or fluorescence intensity. Precision is $\pm 15-50\%$. Instruments are available to determine directly the fluorescence and/or the color of spots on the plates. In order to be as precise as possible, the instruments must be able to quantitate the measurement *over the whole area of the spot*. The material is generally more concentrated at the center of the spot, with decreasing concentrations toward the perimeter. As a result, the most precise measurements are made by scanning the analyte and reference spots in a series of narrow regions (analogous to the scanning of a television screen), recording the output as a function of position, and integrating the total intensity associated with each spot. Precisions in the range of $\pm 1\%$ are possible under the best conditions of sample application, development, and quantitation, but $\pm 5\%$ is more common.

# 15

# Gas Chromatography

## 15.1 Comparison between Gas and Liquid Chromatography

Gas chromatography (GC) is an analytical method that has been studied and used intensively for more than 25 years in solving a wide variety of analytical problems. Thus, for most kinds of compounds for which GC is an appropriate separation method, good separation conditions have been found.

The parameters describing the chromatograms for both gas and column liquid chromatography are calculated in the same manner. With consistent conditions, the retention times or volumes are used in the same way to ascertain the chemical identity of the component in each band. The techniques used to measure the areas of the printed chromatographic peaks are also the same.

The same questions also arise in both qualitative and quantitative analysis. Do two overlapping bands appear as one? Is the detector response in the linear range? What is the best way to choose the correct baseline when it is drifting? How precisely measured are the volumes of injections of samples? Is the manner (rate) of injection precisely the same for all samples? What is the best technique for measuring the peaks: heights or areas?

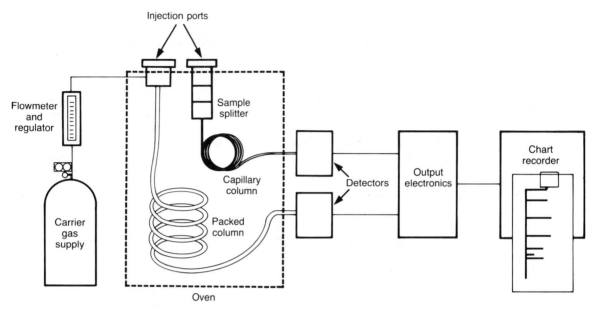

**Figure 15.1**    Diagram (not to scale) of the basic parts of a GC/GLC chromatograph. The carrier gas eluent, stored in a tank under high pressure, has its pressure precisely regulated to produce a fixed rate of gas flow through the column. The sample is injected into a heated injection port and passes into the chromatographic column. If the column is a capillary column (Section 15.2), the sample is usually split, and only part passes into the column, with the rest going to waste. At the end of the column is the detector, of which there are more than twenty types.

Gas chromatography equipment is quite different from that used for HPLC. GC requires precise control of the flow of gases instead of liquids. The column is almost always significantly longer and narrower. Packing materials and stationary phases differ. The temperature range is usually different. The detectors differ both in construction and in operation. A general GC instrument is illustrated in Figure 15.1.

## 15.2 Nomenclature

Gas chromatography may be divided and named according to the type of columns and column packings used. Two primary types of packings are dry adsorbants, which bind with the analytes, and relatively inert packing materials coated with one of a number of high-boiling liquids; the analytes dissolve in and evaporate from the liquids. When gas chromatography is done with the latter, it is often called gas–liquid chromatography (GLC).

A second set of names is based on a division of the diameters of the columns used. If the column diameter is 2 mm and above, this is simply gas chromatography with a packed column. If the internal diameter of the column is in the range of 0.2–0.5 mm, this is called **capillary column** gas chromatography, or simply *capillary* GC. The trend in GC seems to be toward capillary GC for reasons that will be described in more detail later. One value that suggests the reason for this trend is the efficiency, which can be over 100,000 theoretical plates in columns 30–100 m long. Capillary GC is the most powerful separatory technique that now exists to resolve the components of a complicated mixture. Capillary GC has made it relatively straightforward to view the real complexity of the chemical world. The chromatogram in Figure 15.2 illustrates this point. Each peak represents *at least* one compound in an extract of oak leaves, and these are only the volatile components!

## 15.3 Literature Example

The analytical methodologies needed to solve specific problems are not, in general, made effective and reproducible without significant effort. Through give and take between analysts (not always calmly, either) difficulties are overcome and standard methods are developed. To add another dimension, seldom does only one assay methodology work.

In the example that follows, the problem was to determine the artificial sweetener sodium cyclamate in juices and soft drinks. Two separate papers were published about a year apart by analysts in two different industrial laboratories. The first author suggests that a gas chromatographic procedure be used instead of a gravimetric one,

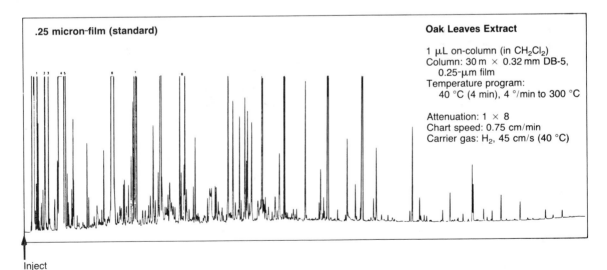

.25 micron-film (standard)

**Oak Leaves Extract**

1 μL on-column (in $CH_2Cl_2$)
Column: 30 m × 0.32 mm DB-5,
  0.25-μm film
Temperature program:
  40 °C (4 min), 4 °/min to 300 °C

Attenuation: 1 × 8
Chart speed: 0.75 cm/min
Carrier gas: $H_2$, 45 cm/s (40 °C)

Inject

**Figure 15.2**  Chromatogram of the volatile components of oak leaves. The conditions of the separation are listed on the figure. This was done on a capillary column with hydrogen as the carrier gas and a linear temperature gradient. Total time for the chromatogram was about one hour. Even though a large number of components have obviously been separated, a statistical calculation suggests that a significant fraction of the peaks that appear actually are due to coeluting species. This calculation is outlined in Supplement 15B.

in which one measures the weight of a precipitate (Chapter 8). A second set of analysts report a problem in the sample preparation method for the gas-chromatographic assay. In addition, they recommend that a "simpler direct procedure" be used instead of GC.

Terms mentioned that might be unfamiliar are *light petroleum,* a solvent of mixed hydrocarbons that boils in a specific temperature range; *Apiezon L* and *siliconized Embacel,* a high-boiling organic liquid and a solid support respectively, which together form the stationary phase in the GLC chromatographic column; and *100 to 115 mesh,* a measure of the size of the Embacel particles, which are coated with the Apiezon L. The mobile phase (or **carrier gas**) is argon, and benzene is used as an internal standard. This GC work was done in 1965 on what now would be dated equipment. However, it is more the sample preparation that is debated.

The present recommended method for determining cyclamates in soft drinks is based on the fact that they react with nitrous acid in the following manner:

the sulfate produced being precipitated and weighed as barium sulfate. This method is, however, very tedious when applied to citrus drinks due to the large amount of fibrous matter present. Another modification is the direct titration of the acidified cyclamate solution with standard sodium nitrite solution with starch–iodide used as an external indicator. This method has only proved suitable in our hands as a very rough check on the cyclamate content for quality-control purposes.

The method described [here] is based on the analysis of the cyclohexene produced, after reaction with nitrous acid, with a gas–liquid chromatographic technique. The method, which is both rapid and very specific, is suitable for analyzing cyclamates in the range of 0 to 1 mg per mL of fruit drink.

In addition to cyclohexene, gas-chromatographic evidence indicated the presence of at least two other products of reaction, probably cyclohexanone and cyclohexanol, which had retention times of about 50 and 58 minutes under the conditions described. The presence of these compounds prevented the extension of the range about 1 mg per mL since they would have interfered unduly with subsequent chromatographic analyses [run on the same column]. . . .

**Method**

Apparatus:
   Pye [the manufacturer] Argon Chromatograph
   Column. 4 ft $\times$ 0.25 in.; 10 percent Apiezon L on acid-washed, siliconized
      Embacel, 100 to 115 mesh, at 50 °C.
   ⋮

   Argon flow-rate, 40 mL per minute.
   ⋮
Reagents:
   Sodium cyclamate solution, 0.20 percent, aqueous.

Zinc acetate solution. Add 21.9 g zinc acetate dihydrate and 3 mL of glacial acetic acid to water and dilute to 100 mL.

Potassium ferrocyanide solution, 10.6 percent, aqueous.

Light petroleum. Boiling range, 30° to 40 °C. Analysis by the conditions above should give no peaks corresponding to benzene or cyclohexene. If peaks are obtained, re-distil the light petroleum and collect the fraction distilled at 30° to 35 °C.

Light petroleum solution. Add enough benzene (about 2 drops) to 50 mL of the light petroleum such that on analysis of a 1 $\mu$L sample by the conditions given above, a benzene peak is obtained 2 to 2½ inches high [on the chart].

Sodium nitrite, 0.5 M.

⋮

**Procedure** . . . Transfer by pipettes, an aliquot of . . . orange drink containing . . . preferably about 25 mg [of sodium cyclamate], 5 mL of concentrated hydrochloric acid, 1 mL of zinc acetate solution, and 1 mL of potassium ferrocyanide solution into a 50-mL calibrated flask, shake well, and dilute to the mark with water. Filter the solution through a fluted Whatman No. 41 filter-paper, and extract a portion of the filtrate with three 50-mL aliquots of chloroform and then twice with 25-mL aliquots of light petroleum. Transfer by pipette, 20 mL of the final aqueous solution, 1 mL of light-petroleum solution, and 1 mL of sodium nitrite solution into a 25-mL calibrated flask, shake the contents, and analyse the light petroleum solution [by transferring a 1.00 $\mu$L sample by microsyringe into the chromatograph and recording the chromatogram]. Calculate the height of the cyclohexene peak relative to that of the benzene peak, and by means of the calibration curve, determine the concentration of sodium cyclamate in the original orange drink.

[Ref: Rees, D. I. *Analyst* **1965**, *90*, 568–569.]

A problem was found with this analytical methodology. This was reported in the second paper.

A gas-chromatographic procedure has been published by Rees for the determination of sodium cyclamate in soft drinks, but we have had difficulty in applying this method to soft drinks obtained from several sources.

According to Beck, if the acid content of a solution is greater than that corresponding to pH 0.65, cyclamic acid is formed which, unlike sodium cyclamate, is soluble in the [organic] solvents involved in the procedure and hence is discarded [in] the solvent extraction steps.

We have confirmed Beck's observations, and have also shown that as the organic acid content of these samples varies (often considerably), this must be taken into account, otherwise erroneously low cyclamate values are obtained.

In the procedure described by Rees, the cyclohexene produced by the nitrite reaction is determined as a basis for calculating the sodium cyclamate content of the sample. Other products, such as monochlorocyclohexene, cyclohexanone, and cyclohexanol are also formed, and this stresses the importance of rigid standardisation in preparing solutions for gas chromatographic evaluation, to ensure that the amount of cyclohexene produced is reproducible. In our modified method, the test solution is acidified with sulphuric acid in preference to hydrochloric acid before it is reacted with sodium nitrite, and this precludes the formation of monochlorocyclohexane. . . .

We, however, consider this procedure to be lengthy, and it involves the use of expensive instrumentation; in view of this we recommend the much simpler direct procedure described in the following paper (*Analyst* **1966**, *91*, 522). [Title: Titration of cyclamate with sodium nitrite with electrometric end-point detection, a potentiometric titration method.]

[Ref: Richardson, M. L.; Luton, P. E. *Analyst* **1966**, *91*, 520–521.]

## 15.4 Samples Analyzed

A few types of samples can be analyzed chromatographically using either LC or GC. Examples are moderately high-boiling organics (<200 °C) and amino acids. The amino acids can be analyzed directly with LC, using a number of possible techniques such as reversed phase or ion exchange. For GC, which requires volatile samples, the samples must be **derivatized.** Derivatization is a chemical treatment used to modify the chemical groups that cause low vapor pressure (high boiling points). A simple example of derivatization is the conversion of a high-boiling organic acid into an ester with a lower boiling point: for example, $C_5H_{11}COOH$ (b.p. 187°C), $C_5H_{11}COOCH_3$ (b.p. 127.3°C).

GC is routinely used mostly for analyses of organic compounds with boiling points below about 450 °C and for gas analysis. These have vapor pressures that are high enough so that they will be carried in the gaseous mobile phase. However, a more common limitation than an overly low vapor pressure is decomposition. Since the column of a GC chromatograph must be heated above ambient temperature when higher-boiling materials are being separated, the compounds being separated may decompose at the temperatures required to obtain a chromatogram in a reasonable time.

On the other hand, the GC column may be cooled during a separation of low-boiling materials such as methane or $H_2$. For such materials, gas chromatography can be run conveniently under conditions ranging from room temperature down to dry ice temperature, −80 °C.

## 15.5 GC Equipment and Chromatographic Conditions

### Columns

Recall that in Section 13.8 you read about the spreading of chromatographic bands when isolated channels occur in the packing. One way to eliminate this problem is to have only one channel, through which all the material passes. This is the idea behind capillary GC columns. The surface upon which the sample interacts is spread on the wall of the tube. One drawback to the capillary column is that to achieve optimum separations the amount of sample must be kept small (less surface area per unit length

| Type | WCOT | SCOT | Micropacked | Packed |
|---|---|---|---|---|
| Graphic Description |  ⭕ |  ◎ |  ⬤ |  ⬤ |
| Typical inside diameter | 0.25 mm 0.50 mm | 0.50 mm | 1 mm | 2 mm 4 mm |
| Typical length | 10–100 m | 10–100 m | 1–6 m | 1–4 m |
| Typical efficiency | 1000–3000 plates/meter | 600–1200 plates/meter | 1000–3000 plates/meter | 500–1000 plates/meter |
| Sample size | 10–100 ng | 10 ng–1 μg | 10 ng–10 μg | 10 ng–1 mg |
| Linear flow rate (optimum) | 20–30 cm/s $H_2$, He 10–15 cm/s $N_2$, Ar | 20–30 cm/s $H_2$, He 10–15 cm/s $N_2$, Ar | 8–15 cm/s $H_2$, He 3–10 cm/s $N_2$, Ar | 4–6 cm/s $H_2$, He 2–5 cm/s $N_2$, Ar |
| Volume flow rate (optimum) | 1.5 mL/min, $H_2$, He 0.5–4 mL/min, $N_2$, Ar | 2–8 mL/min, $H_2$, He 1–4 mL/min, $N_2$, Ar | 2–6 mL/min, $H_2$, He 1–3 mL/min, $N_2$, Ar | 20–60 mL/min, $H_2$, He 15–50 mL/min, $N_2$, Ar |
| Pressures required | Low | Low | Very high | High |

**Figure 15.3**   Typical properties and characteristics of GC/GLC chromatographic columns. [Ref: Courtesy: Alltech Associates, Inc.]

is available to interact with it than in a packed column). The column **capacity** is smaller. One compromise between packed columns and capillary columns is to produce a capillary column with a surface increased by attaching packing material to it. Some characteristics of packed and capillary columns are presented in Figure 15.3.

You should be aware of two abbreviations that are so common that the original full names are hard to find. These are acronyms for two types of capillary columns used for gas–liquid chromatography: **WCOT** (wall-coated open tubular) and **SCOT** (support-coated open tubular). In the WCOT column the polymer liquid of choice is coated on the inside of the column wall. The SCOT column has support particles on its wall; these are, in turn, coated with the liquid phase. The SCOT has a higher surface area and can handle somewhat larger samples.

The stationary phase — either a solid for GSC or liquid-coated solid for GLC — is chosen so as to achieve optimum resolution and the highest efficiency: to achieve the desired analysis in the shortest possible time. A few of the hundreds of materials used for stationary phases are listed in Table 15.1. Extensive lists of sorbants and the types of compounds for which they are used can be found in commercial catalogs of vendors and handbooks of GC/GLC. The rational selection of sorbants is discussed in more detail later in the chapter.

# Table 15.1 Packing Materials for GC and GLC

*Representative solid adsorbants*

Alumina  
Silica gel  
Carbon  
Molecular sieves (zeolites)  
Porous organic polymers  

} Used for permanent gases and organic materials below mass 150

*Representative liquid stationary phases* (for coating porous packings or capillary surfaces)

In order of increasing approximate polarity

Squalane $[(CH_3)_2-CH-(CH_2)_3-CH-(CH_2)_3-CH-(CH_2)_3-CH-CH_2-CH_2-]_2$ (with $CH_3$ branches)

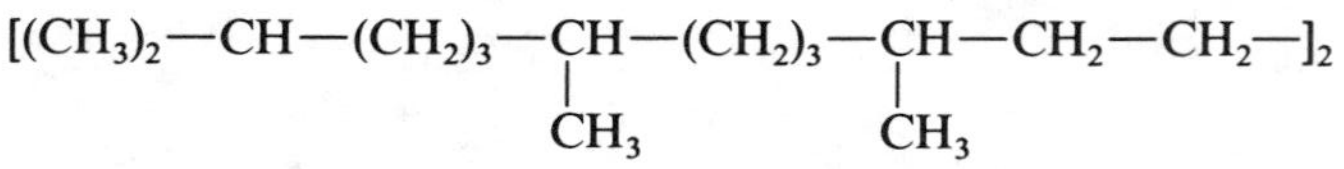

SE-30     poly(dimethylsiloxane)

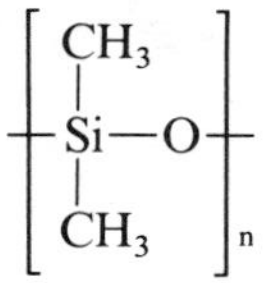

$$\left[ \begin{array}{c} CH_3 \\ | \\ -Si-O- \\ | \\ CH_3 \end{array} \right]_n$$

OV-3     poly(phenylmethyldimethylsiloxane) 10% phenyl  
OV-7     poly(phenylmethyldimethylsiloxane) 30% phenyl

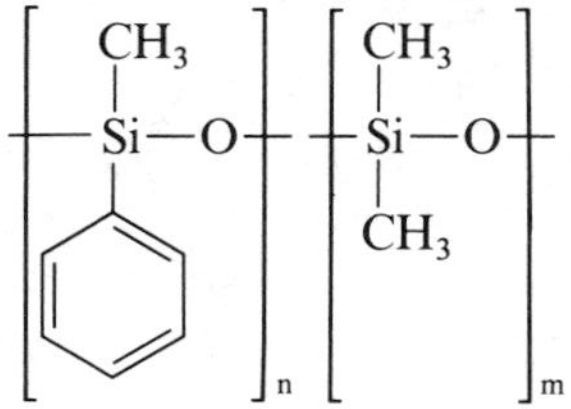

$$\left[ \begin{array}{c} CH_3 \\ | \\ -Si-C_6H_5 \\ | \\ CH_3 \end{array} \right]_n \left[ \begin{array}{c} CH_3 \\ | \\ -Si-O- \\ | \\ CH_3 \end{array} \right]_m$$

DC-710     poly(phenylmethylsiloxane)

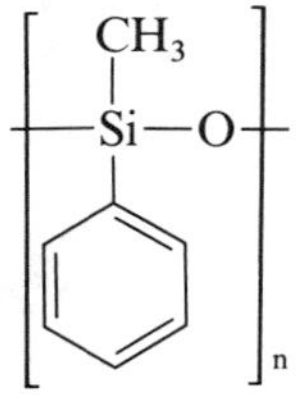

$$\left[ \begin{array}{c} CH_3 \\ | \\ -Si-C_6H_5 \\ | \\ O \end{array} \right]_n$$

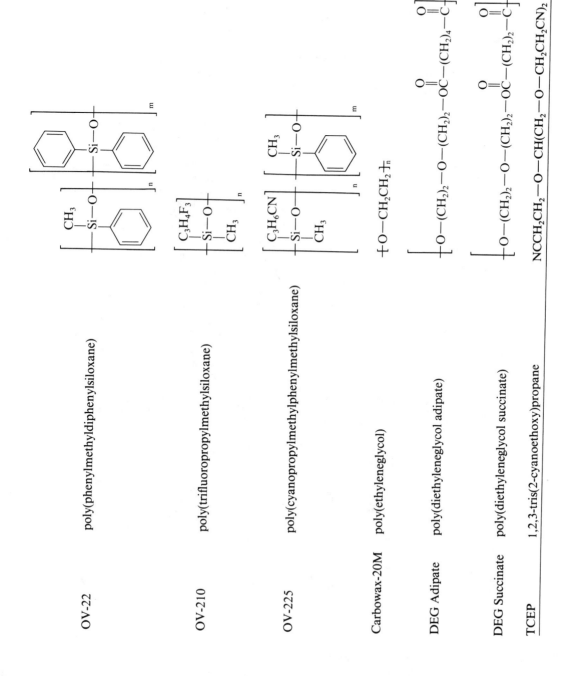

OV-22    poly(phenylmethyldiphenylsiloxane)

OV-210    poly(trifluoropropylmethylsiloxane)

OV-225    poly(cyanopropylmethylphenylmethylsiloxane)

Carbowax-20M    poly(ethyleneglycol)

DEG Adipate    poly(diethyleneglycol adipate)

DEG Succinate    poly(diethyleneglycol succinate)

TCEP    1,2,3-tris(2-cyanoethoxy)propane

All the stationary phases have limits on their usable temperature ranges. Few liquid phases are usable above 300 °C. The inorganic solid phases generally are stable to higher temperatures, although they tend to catalyze decomposition of analytes at these temperatures. As a result, high-boiling materials are, perhaps, better separated by HPLC.

## Temperature in GC

The temperature at which a separation is run is crucial. The choice of column temperature in GC is equivalent to the choice of the mobile phase in LC. A control of temperature to ±0.5° is worthwhile.

**(a)**

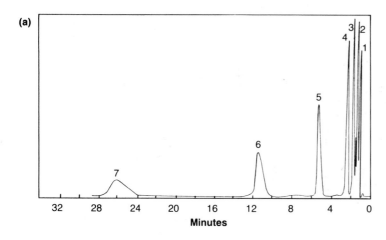

**(b)**

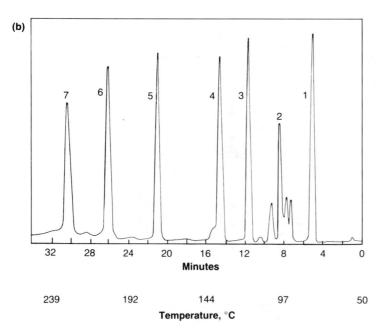

**Figure 15.4**    Illustration of the beneficial effect of running a temperature gradient in gas chromatography: (a) at a constant temperature of 168 °C; (b) with a temperature gradient from 50 to 235 °C with a linear increase of 6 °C/min.

Conditions: Column 5 mm i.d. × 122 cm; Packing 25% DC-200 (viscosity grade 500) silicone oil on 35–80 mesh Chromosorb (brand name). Carrier gas helium with a flow rate at the exit of the column of 35 cm³/min.

Sample: 1, pentane, 2, hexane, 3, heptane, 4, 1-octene, 5, decane, 6, 1-dodecene, and 7, tetradecene. [Ref: Reprinted with permission from Nogare, S. D.; Bennett, C. E. *Anal. Chem.* **1958,** *30,* 1157. Copyright 1958 American Chemical Society.]

In analogy with a gradient in the eluent of LC, a gradient in GC is done using a temperature gradient; *gradient elution* means a change of eluent composition in LC but a change in column temperature in GC. The benefits of temperature gradients in GC are illustrated in Figure 15.4. (Also, note that a gradient of 4 °C min⁻¹ was used to obtain the chromatogram of Figure 15.2.)

However, there are limits at the high end of the temperature range for a gradient. In addition to the sample decomposition mentioned earlier, the stability and vapor pressure of liquid stationary phases also must be considered. When the temperature rises high enough, the liquid phase itself can be transported by the carrier gas through the column. This is called column **bleeding**—a graphic description. Two problems arise if bleeding occurs. First, the column efficiency and retention change, causing changes in the chromatograms. Second, the high-boiling liquid will either plug or "gunk" up every kind of detector.

The maximum operating temperatures for liquid phases are usually listed as one of their characteristics. In addition, to alleviate the problem of clogging detectors, a newly coated column is generally operated for some time at the highest operating temperature without the detector attached. This operation is called **conditioning** the column. In this way, any high-boiling components are vented harmlessly.

The low end of the temperature range that can be used for a specific separation is determined by the properties of both the stationary phase and the chromatographed compounds. In the extreme case, at excessively low temperatures, all the compounds injected will merely stay on the stationary phase, never to move again. At higher, but still too low, temperatures, the adsorption/desorption process is sluggish, and the peaks are broadened as described by the $C_S$-term in Eq. 13-19b. The efficiency can be regained by slowing the carrier flow, but then the analysis time will be longer than necessary.

Raising the column temperature speeds the elution but also tends to reduce the resolution. Thus, an optimum temperature program is found between the constraints of needed resolution and the desire to minimize the time for the separation. The quality of the best separation of complex mixtures may be deceptive, however, with many more components present than the number of peaks indicate. This idea is covered in more detail in Supplement 15B.

### Sample Injection

Samples for GC can be in either gaseous or liquid forms. Solid samples must be vaporized in some manner. Generally, solids are pyrolyzed, and the decomposition products analyzed to characterize the sample. (An example of pyrolysis GC was given in Figure 7.4.)

To obtain the best separations, the injection of the samples should be done over the shortest possible time, because the bands will be broadened by a time equal to the length of the injection period. A band also could be significantly broadened if an injected liquid sample only slowly evaporates. Similar problems occur if some of the sample rests or condenses anywhere on a cooler section of the wall or packing of the column. This condensation could also cause cross contamination between samples: Some part of an earlier sample that only slowly desorbs might be mixed into a later sample. Both of these problems — the peak broadening and the cross contamination — are alleviated by injecting samples into a compartment that is hotter than the column and hot enough to vaporize liquid samples rapidly. (On the other hand, the temperature should not be so high that the components of the sample will be thermally decomposed.) The optimum temperature of this compartment, the injector (see Figure 15.1), is determined experimentally.

As noted earlier, a special problem with capillary GLC involves the necessity for small samples with upper limits in the 500-$\mu$g range. This prevents overloading of the column, which would cause distortion of the band shapes (mostly fronting). It is difficult to handle quantitative measurements and transfers of volumes of the required size — 100 nL or so. As a result, it is common to use a **sample splitter** at the input to a capillary column. The design and general operation of such a splitter are illustrated in Figure 15.5. Its position in the instrument is shown in Figure 15.1. Basically, a larger sample is injected, vaporized, diluted with carrier gas, and the majority of it is thrown away. Typically only 1 part in 100 to 1 part in 1000 enters the

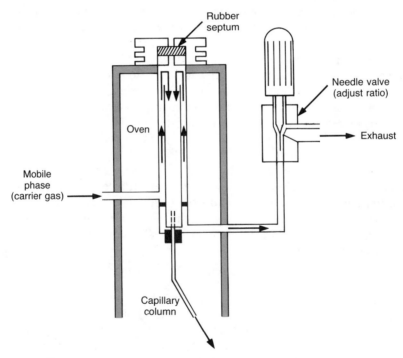

**Figure 15.5**    Schematic illustration of the structure and operation of a sample splitter as used with capillary columns. The apparatus is enclosed in an oven at a temperature higher than that of the column. The sample is injected through a rubber septum, vaporized, and carried toward the column by the carrier gas. The path to the column is long enough so that all components in the sample can be completely vaporized. Part of the sample enters the column while the larger part is driven out to an exhaust. The needle valve allows a precise adjustment of the ratio of sample entering the column to that thrown away. With the valve closed, the injector acts as a typical injector for a packed column, in which all of the sample is applied to the column.

capillary column. One requirement for good precision in quantitation is that the split be linear for each sample and for each component of each sample. In other words, the amount of each component reaching the column should be *equally* directly proportional to its content in the injected sample. This is not a trivial problem when the components boil over a wide temperature range and the amounts of the compounds range from trace to major.

## GC Detectors

A wide range of detectors is available for determining the changes in eluent composition in gas chromatography. A few of these and some of their characteristics are listed in Table 15.2, along with their linear ranges and detection limits. Some of these detectors are among the most sensitive probes of content that exist in chemical analysis. It is their development that allowed trace and ultratrace organic analysis to be developed. Even with capillary GLC, it is possible to do ppm determinations, which requires quantifying $10^{-10}$ g of components.

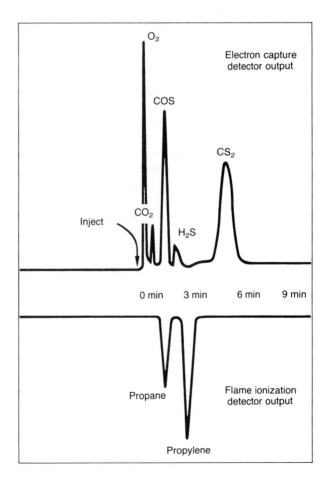

**Figure 15.6** Chromatograms of the same eluent using two detectors with different sensitivities. Top is with an electron capture detector, and the bottom is with a flame ionization detector (see Supplement 15A). Notice that COS and propane elute at the same time. Thus, the two detectors can compensate for an imperfect separation by *their* selectivities. [Ref. Walker, J. Q. *Petroleum Refiner* **1965**, *44*, 171. Reproduced with permission from *Petroleum Refiner* (now *Hydrocarbon Processing)* October, 1965.]

**Table 15.2**   Properties of Selected Gas Chromatography Detectors

| Type | Approximate Limit of Detection* | Approximate Linear Range | Comments |
|---|---|---|---|
| Thermal conductivity | $10^{-5}$–$10^{-6}$ g | $10^3$–$10^4$ | Universal detector. Measures changes in heat conduction. |
| Flame ionization (FID) | $10^{-12}$ g/s | $10^6$–$10^7$ | Universal detector. Measures ion currents from pyrolysis. |
| Electron capture (EC)(ECD) | $10^{-13}$ g/s | $10^2$–$10^3$ | Specific detector. Compound must contain halogen. |
| Conductivity | $10^{-11}$ g | $10^5$ | Specific detector. Compound must contain Cl, S, N, and so on. |

\* Depends on the peak width. Value for a direct measurement without preconcentration of the sample. These are constantly being improved.

*Source:* Varadi, M. *Pure and Appl. Chem.* **1979,** *51,* 1175–1182, and references therein. Drushel, H. V. *J. Chromatog. Sci.* **1983,** *21,* 375–384.

As you may now have come to expect of all transducers, the detectors are not equally sensitive to all compounds. For example, two of the most sensitive, **flame ionization** and **electron capture** detectors, have significantly different selectivities. This property is illustrated in Figure 15.6. As is the normal perverse situation, the most universal detector — the one responding to the widest range of compounds — is the least sensitive of those in common use. This is the **thermal conductivity** detector. More details of the construction and mode of operation of these detectors are presented in Supplement 15A.

A highly sophisticated detector is described in the next chapter. It is a **mass spectrometer,** which can be used to measure the molecular mass of each component as it elutes from the column. If a component is a known compound and has had its mass spectrum recorded, it is highly probable that the component can be identified. This combination technique is called gas-chromatography/mass-spectrometry and is abbreviated GC/MS; it is the method of choice for analysis of complex organic mixtures. GC/MS is described in more detail in Section 16.7.

## 15.6 Relationships between GC Parameters and Molecular Structure

Gas chromatography is a mature field, although innovations are always being made in column design, adsorbant composition, detectors, and injectors, as well as mobile phases. However, finding GC conditions usable for a desired analysis should not require much experimental searching if the general *types* of compounds to be separated are known. Good guides to doing GC separations can be found in catalogs of

GC equipment and supplies. If the answer is not there, then literature references for the conditions usually can be found.

But, as can be seen in Figure 15.2, the mixtures to be analyzed can be highly complex. A number of schemes have been proposed to enable the gas chromatographic retention parameters to be cataloged in order to allow identification of a good proportion of these numerous components. One desire of the originators of these schemes was to create some index number for cataloging that would be transferable: be independent of the carrier gas flow rate, column length, and volume of the stationary phase. The index number still would be dependent on the identity of the solid phase and the column temperature.

In the following sections, we describe briefly some of the techniques used to transmute experimental data into transferable parameters. However, the unreliability of the parameters has caused these techniques to fall into disuse for their original purpose. The index methods have largely been replaced by the use of gas chromatography/mass spectrometry. With GC/MS, the identification of the component in the eluent can be made far more certain and far less dependent on trying to reproduce experimental GC conditions. However, the index methods have become the basis for cataloging the properties of liquid stationary phases for GLC. The basis for the classification will be discussed.

### Homologous Series

The separations in gas chromatography can be explained as functions of chemical interactions with the stationary phase as well as the boiling points of the compounds being separated. If the chemical interactions remain of the same type, then regularities are seen in a **homologous series.** Members of a homologous series of compounds are of the same type, such as *n*-alkanes, carboxylic acids, and ketones. However, the **adjacent members** of the series differ by an incremental —$CH_2$— group. Members of a homologous series are called **homologues.** Examples of parts of two homologous series are listed below.

| *n*-Alkanes | | Secondary Amines | |
|---|---|---|---|
| *Formula* | *Name* | *Formula* | *Name* |
| $CH_4$ | methane | $CH_3(CH_2)_6CHNH_2CH_3$ | 2 amino-nonane |
| $C_2H_6$ | ethane | $CH_3(CH_2)_7CHNH_2CH_3$ | 2 amino-decane |
| $C_3H_8$ | propane | $CH_3(CH_2)_8CHNH_2CH_3$ | 2 amino-undecane |
| $C_4H_{10}$ | butane | $CH_3(CH_2)_9CHNH_2CH_3$ | 2 amino-dodecane |
| $C_5H_{12}$ | pentane | $CH_3(CH_2)_{10}CHNH_2CH_3$ | 2 amino-tridecane |
| ⋮ | | ⋮ | |

The elution behavior of the members tends to be quite regular. An example of this order is shown in the semilogarithmic graph in Figure 15.7. This regularity makes adjacent homologues useful as standards for parametrization schemes used to identify eluting compounds, as is demonstrated next.

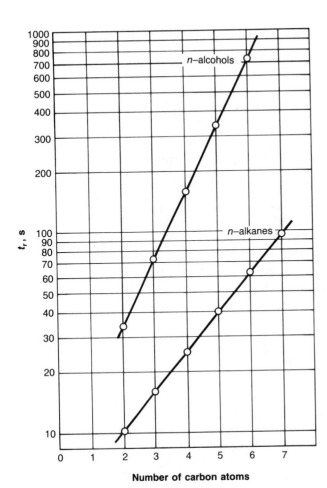

**Figure 15.7**    Illustration of the log–linear relationship of elution volume and structure within a homologous series of compounds. The logarithm of the elution volume (or time) is proportional to the chain length of the compounds. Each series of compounds falls along a different line.

### Retention Index[1]

This simple log–linear relationship between the elution volume and carbon number of the homologues (Figure 15.7) has been used to devise an indexing method for elution times of compounds separated by gas chromatography. The **retention index** is used to quantify the retention of a test compound by comparing it with a pair of adjacent homologue n-alkanes as standards. The homologous alkanes, such as the pair butane/n-pentane, are chosen so that under the specific chromatographic conditions one homologue elutes before the test compound and the other elutes after the test compound. The gas chromatogram of Figure 15.8 illustrates this bracketing. By convention, the number of carbon atoms in the first eluting standard is $z$. The adjacent homologue then has $z + 1$ carbons. The retention index is a ratio of relative retention values $\alpha$. The Kovats index $I$ is defined by the equation

[1] Ref: Kovats, E. *Adv. Chrom.* **1965**, *113,* 303–356.

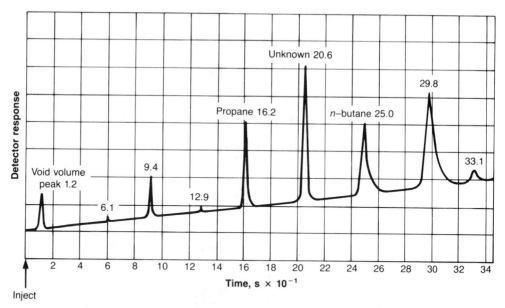

**Figure 15.8** Sample chromatogram used to illustrate the retention index calculation. The retention index was used to allow interlaboratory identification of compounds from the elution time of gas chromatography alone. This method is now mostly out of date, with identification now being done using mass spectrometry in GC/MS. The retention index is still in use as the basis for characterizing liquid phases for GLC.

$$I = 100z$$

$$+ 100 \frac{\log(\text{relative retention of test compound and standard with } z \text{ carbons})}{\log(\text{relative retention of standards with } z \text{ and } z + 1 \text{ carbons})}$$

$$(15\text{-}1a)$$

The basis of the index rests on the following definition: The value of $I$ for all $n$-alkanes under *any* set of fixed conditions on *all* columns is

$$I_{C_zH_{z+2}} = 100z \qquad (15\text{-}2)$$

The alkane index values are 100, 200, 300, 400, 500, and so on. This definition eliminates information about the column type and temperature.

Equation 15-1 can be rewritten using the relative retention parameters $\alpha_{i,z}$ (the separation factor), with subscripts $i$ for the unknown and $z$ for the standard $n$-alkane. Equation 15-1a is rewritten as

$$I = 100z + 100 \frac{\log \alpha_{z,i}}{\log \alpha_{z,z+1}} \qquad (15\text{-}1b)$$

In reading this equation, notice that the possible values of the second term cause the retention index to have a value between $100z$ and $100(z + 1)$, the bracketing alkane retention indices.

---

**EXAMPLE**

The data from Figure 15.8 are used for the following example. Find the retention index of the unknown compound.

**SOLUTION**

The following data can be listed for the chromatogram:

| Peak | $t_r$ (s) | $t_r' = t_r - t_0$ |
|---|---|---|
| Void volume | 1.2 | 0 |
| Propane | 16.2 | 15.0 |
| Unknown | 20.6 | 19.4 |
| n-Butane | 25.0 | 23.8 |

The value $t_r'$ is called the net retention time or adjusted retention time. The same result would be calculated if the values of $k_i'$ were used instead. Now recall that all $\alpha_{i,j}$ are calculated so that they are always greater than 1. (In this nomenclature, $\alpha_{i,j} = t_{rj}'/t_{ri}'$.) So from the definition of $I$ and the chemical formulas for propane and n-butane, we calculate

$$I_{\text{unknown}} = 100 \cdot 3 + 100 \, \frac{\log(19.4/15.0)}{\log(23.8/15.0)}$$

$$I_{\text{unknown}} = 300 + 56 = 356$$

---

## 15.7 Schemes for Characterizing GLC Liquid Phases[2]

---

The separations of GLC are due mostly to the chemical properties of the stationary phases. The carrier gas is inert, and the differences in retention depend on the relative solubilities of the components in the liquid stationary phase. If a component does not dissolve in the liquid, it will elute with the void volume. The simplest rule to use when choosing a GLC stationary phase is *like dissolves like*. A polar compound will dissolve in a polar phase; a nonpolar one in a nonpolar phase. (Examples of stationary phases are listed in Table 15.1.)

However, in addition, there are specific functional group interactions that could be especially useful. For instance, if you wanted to separate some alcohols from some ketones, a liquid phase that is **selective** for either group would help in the separation since ketones and alcohols tend to be approximately equally polar. The term *selective* in chromatography has its colloquial meaning—the stationary phase discriminates between compounds. This merely reflects significant differences in the solubilities of the compounds being separated.

The polarities and selectivities of stationary phases for specific chemical groups can be quantified through comparisons of standard compounds containing the func-

---

[2] Ref: McReynolds, W. O. *J. Chromatog. Sci.* **1970,** *8,* 685.

**Table 15.3**   Standards for Rohrschneider/McReynolds Indices

| Symbols | Index Compound | Basis of Interaction: Classes of Compounds |
|---|---|---|
| *Most used* | | |
| 1  X′ | Benzene | Intermolecular interactions: aromatics and olefins |
| 2  Y′ | *n*-butanol | Electrophilic interactions: alcohols, acids, nitriles, alkyl halides |
| 3  Z′ | 2-pentanone | Electrophobic interactions: ketones, ethers, aldehydes, esters, $(CH_3)_2N-$ |
| 4  U′ | Nitropropane | Specific interactions: nitro and nitrile groups |
| 5  S′ | Pyridine | Specific |
| *Less often cited* | | |
| 6  H | 2-methyl-2-pentanol | |
| 7  J | 1-iodobutane | |

tional groups of interest. A scheme commonly employed is that of the Rohrschneider–McReynolds constants or, simply, McReynolds constants. A description of the classification scheme follows. The overall view to keep in mind is that the larger the McReynolds constant, the greater the retention time.

The basis of this classification scheme is to compare the Kovats index of a standard compound with the Kovats index of the same compound run on a standard stationary phase called *squalane*. (The structure of squalane is shown in Table 15.1.) McReynolds numbers ($\Delta I$) are *differences* between the Kovats index on squalane and the Kovats index of the stationary phase of interest.

The standard compounds that originally were run to determine the McReynolds numbers were benzene, butanol, 2-pentanone, 1-nitropropane, pyridine, 2-methyl-2-pentanol, 1-iodobutane, 2-octyne, 1,4-dioxane, and *cis*-hydrindane. Subsequently, the first five — benzene through pyridine — have been considered sufficient to use in most cases. Since the $\Delta I$ values for the index compounds are so often used, they have been given the abbreviations shown in Table 15.3. McReynolds numbers for 12 stationary phases over a range of polarities are listed in Table 15.4.

Listed in Table 15.4 are the sums of the first five $\Delta I$ values,

$$\text{sum} = X' + Y' + Z' + U' + S'$$

Each sum is, arguably, a measure of the "polarity" of the liquid phase. The higher the sum is, the higher the polarity. In addition, the higher the sum, the longer the retention times for the five standard compounds.

Such tables help us to choose stationary phases when different functional groups are involved. For instance, if a stationary phase is required that will retain ketones more strongly than alcohols, a liquid with a larger $Z'$ value and smaller $Y'$ value should be selected. Thus a look at the table suggests that OV-210 would be a reasonable choice but OV-225 would not.

**Table 15.4**  McReynolds Constants for Selected GLC Liquid Phases

| Liquid† | McReynolds Constants* | | | | | |
|---|---|---|---|---|---|---|
| | $X'$ | $Y'$ | $Z'$ | $U'$ | $S'$ | Sum |
| Squalane | 0 | 0 | 0 | 0 | 0 | 0 |
| SE-30 | 15 | 53 | 44 | 64 | 41 | 217 |
| OV-3 | 44 | 86 | 81 | 124 | 88 | 423 |
| OV-7 | 69 | 113 | 111 | 171 | 128 | 592 |
| DC-710 | 107 | 149 | 153 | 228 | 190 | 827 |
| OV-22 | 160 | 188 | 191 | 283 | 253 | 1075 |
| OV-210 | 146 | 238 | 358 | 468 | 310 | 1520 |
| OV-225 | 228 | 369 | 338 | 492 | 386 | 1813 |
| Carbowax-20M | 322 | 536 | 368 | 572 | 510 | 2308 |
| DEG Adipate | 378 | 603 | 460 | 665 | 658 | 2764 |
| DEG Succinate | 496 | 746 | 590 | 837 | 835 | 3504 |
| TCEP | 594 | 857 | 759 | 1031 | 917 | 4158 |
| Kovats indices on squalane for reference compounds | 653 | 590 | 627 | 652 | 699 | |

| * Symbols | Equals $\Delta I$ for these compounds |
|---|---|
| $X'$ | benzene |
| $Y'$ | *n*-butanol |
| $Z'$ | 2-pentanone |
| $U'$ | nitropropane |
| $S'$ | pyridine |

† Structures and chemical names appear in Table 15.1.

Another consideration in choosing a GLC stationary phase is to get the best efficiency. Higher polarity phases tend to retain compounds longer. The separation time is increased. One can counteract the increased time by increasing the column temperature. However, as the higher temperature lessens the retentions, it also tends to reduce the selectivity of the stationary phase. In other words, faster separations tend to be obtained on less polar phases at lower temperatures. Whether such conditions produce a separation adequate for the problem at hand is for the analyst to decide. But attempting a separation at lower temperatures (20–50 °C) on a less polar phase is a good place to start.

A number of other indices have been developed to aid in selecting the liquid phase used in GLC. They are based on the same kinds of comparisons with standard compounds as Rohrschneider–McReynolds numbers. Often the parameters are related to specific types of interactions such as hydrogen bonding or dipole moments of the liquid phase. These schemes have apparently not replaced McReynolds numbers in popularity. This may be because the other forms are less intuitively clear or are less directly connected to the daily choices of stationary phases.

## 15.8 Optimization of Gas Chromatography Separations[3]

Finding the optimum conditions to carry out an analytical procedure is often a matter of enlightened trial and error, using intuition and the descriptive chemistry of the systems. Numerous methods have been devised to aid the choice of conditions for analyses. Often, a computer program can be written to help evaluate the factors involved. The ability to use these relatively automatic methods depends on how clearly the effective factors can be related mathematically to changes in the procedures that improve the results. Chromatographic separations are especially amenable to such methodical treatment. Aids to optimization of separations are especially important with complicated mixtures, because as the number of components gets large, trial-and-error searches become almost impossible.

To illustrate one method, let us use a hypothetical and relatively simple example of finding optimum GLC conditions. The experiments that are carried out are a number of chromatographic separations run at the same temperature, same gas flow rate, same column size, and same support material, but with different liquid phases coating the support. The liquid phases that are used can be selected from tables or literature references based on the type of sample (such as a crude oil, lemon peel, red wine, or automobile exhaust). Let these two different liquids be called phase A and phase B.

The capacity factors, $k_i'$, of the components of the chromatograms are plotted on the vertical axis of the graph shown in Figure 15.9. Other parameters such as the retention times or retention volumes could be plotted as well.

In this hypothetical example, compounds W and Y are not resolved on pure phase B, and compounds X and Y are not resolved on pure phase A. Stick diagrams of the chromatograms are shown. Approximate envelopes of overlapping bands are shown where two components are close together. After the test chromatograms are run, straight lines are drawn to connect the points for the $k_i'$ for each compound as they vary with column composition. The points on these lines represent the capacity factors for a column composed of a mixture of stationary phases A and B. This linear behavior with packing composition is seen experimentally — it is not an assumption.

The mixed stationary phase is not made by mixing the liquids A and B and then coating the solid support. Instead, it is made by a physical mixing of particles coated with A and particles coated with B. The scale on the composition axis (the $x$-axis) is the fraction of particles coated with A in the mixture of A and B. The support composition and the average coating thickness are assumed to be identical for both A and B.

The next step is straightforward although it takes some time to calculate. Calculate the values of *all* the possible separation factors, the $\alpha_{i,j}$, versus the composition of the stationary phase. This plot is shown in Figure 15.10. It is called a **window diagram** by the originators (see footnote 3). The lowest possible value of $\alpha$ is 1, when no separation of two components is possible.

---

[3] Ref: Laub, R. J.; Purnell, J. H. *J. Chromatog.* **1975**, *71*, 112.

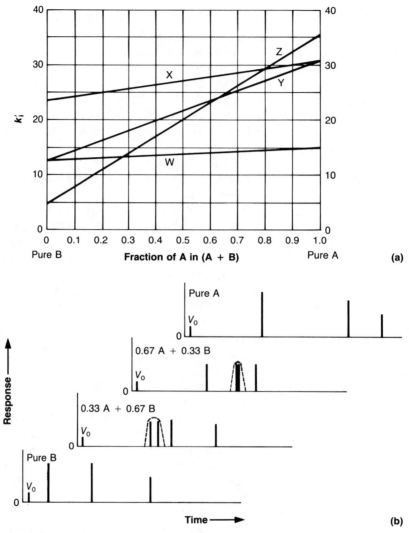

**Figure 15.9** (a) Plot of gas chromatography capacity factors for four compounds chromatographed on combinations of two different column materials, A and B, under the same chromatographic conditions. It is found for GC in general, that if column materials are mixed together in a column, the retention of each species closely follows a linear behavior as a function of column composition as shown in the graph. (b) Representative stick chromatograms are shown with approximate zone shapes for unresolved bands.

As you know, $\alpha_{i,j}$ describes the ease of separation of two components. The shaded areas in Figure 15.10 are the regions that fall below the smallest values of $\alpha_{i,j}$ at each concentration along the horizontal axis. At each column composition, the tops of these shaded sectors indicate the *best* separation of the *most poorly* separated pair of peaks. The highest $\alpha$-value of all these *windows* indicates the optimum separation

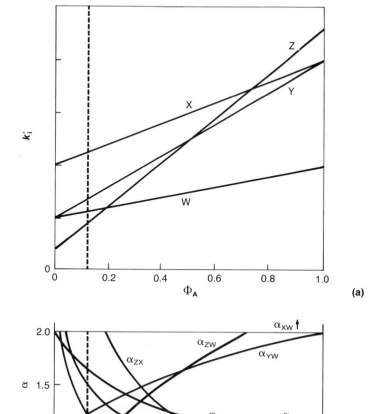

**Figure 15.10**  From the data shown in Figure 15.9, a plot of the separation factors $\alpha_{i,j}$ for all pairs of bands is made as a function of the column composition. This is called a **window diagram** by the developers of this optimization method. The darkened regions are below the lowest values of $\alpha$. The highest value among these regions is the optimum composition. The optimum is indicated by the vertical line. [Ref: Laub, R. J.; Purnell, J. H. *J. Chromatog.* **1975,** *71,* 112.]

conditions. In this case, the maximum point of all the windows is at $\phi_A = 0.12$ with $\alpha_{Y,W} = 1.23$.

This calculation yields the best composition of the two test phases. Next is to find out how long a column of this composition is needed to effect the desired resolution. This is found from Eq. 13-20.

$$R_s = \frac{1}{2} \sqrt{\frac{L}{H}} \left( \frac{\alpha - 1}{\alpha + 1} \right) \left( \frac{\bar{k}'}{1 + \bar{k}'} \right) \tag{13-20b}$$

At $\phi_A = 0.12$, the two bands with the closest $k_i'$ values are bands Y and W. It is the resolution of these two that we desire. The following values are read off of Figure 15.9a at $\phi_A = 0.12$ and are then substituted into Eq. 13-20a.

$$\alpha_{YW} = 1.23 \qquad k_Y' = 14.7 \qquad k_W' = 12.9 \qquad \bar{k}' = \frac{14.7 + 12.9}{2} = 13.8$$

From this substitution, the following numerical result is obtained:

$$R_s = 0.0481 \sqrt{\frac{L}{H}}$$

We next need to decide how much resolution is desired. Let us assume that the separation of Y and W should have an $R_s$ value of 1.5. Thus,

$$\frac{L}{H} = 976$$

Further assume that we found that the HETP = 0.5 cm on the experimental columns. Thus,

$$976 = \frac{L}{0.5} \quad \text{and} \quad L = 488 \text{ cm} = 4.9 \text{ m}$$

Thus, from the data of a limited series of runs, we can decide the length and optimum composition of the column to carry out a separation to the desired degree of resolution.

## Suggestions for Further Reading

An outstanding short book at an intermediate level. Since it includes GC/MS, quantitative GC is included. Highly recommended.
Ettre, L. E. "Introduction to Open Tubular Columns"; Perkin Elmer: Norwalk, CT, 1979.
Freeman, R. R. "High Resolution Gas Chromatography," 2nd ed.; Hewlett–Packard, 1979.
Millard, B. J. "Quantitative Mass Spectrometry"; Heydon: Philadelphia, 1978.

Monographs balanced toward experimental GC.
Novak, J. "Quantitative Analysis by Gas Chromatography"; Marcel Dekker: New York, 1975.
Thompson, B. "Fundamentals of Gas Analysis by Gas Chromatography"; Varian: Walnut Creek, CA, 1977.

Half this book describes the nuts and bolts of capillary column GC.
Jennings, W., Ed. "Gas Chromatography with Glass Capillary Columns," 2nd ed.; Academic Press: New York, 1980.

This book has some theory and relates what can be done but with little practical direction. Half of it describes applications listed by compound type (for instance, amines) or area of study (for instance, food analysis).
Lee, M. L.; Yang, F. J.; Bartle, K. D. "Open Tubular Column Gas Chromatography: Theory and Practice"; Wiley: New York, 1984.

An encyclopedic textbook/reference on gas, liquid, and thin-layer chromatography including equipment, sample preparation, and theory. It is exhaustive and contains large numbers of literature references. However, you will need to know what you are looking for because the organization is complex. It should be easily readable after studying the topics in this chapter.
Poole, C. F.; Schuette, S. A. "Contemporary Practice of Chromatography"; Elsevier: Amsterdam, 1984.

A worthwhile, short review about the problems of capillary GC.
Jennings, W. "Some Aspects of Troubleshooting in Capillary GC"; *J. Chromatog. Sci.* **1983,** *21,* 337–340.

A book with chapters by different authors. It relates how electron capture detectors work. Contains a little theory and lots of applications examples.
Zlatkis, A; Poole, C. F., Eds. "Electron Capture: Theory and Practice in Chromatography"; Elsevier: Amsterdam and New York, 1981.

A long review about a selective detector not covered in this text.
Hall, R. C. "The Nitrogren Detector in Gas Chromatography"; *Critical Rev. Analy. Chem.* **1978,** *7,* 323–380.

A short review article that ages well. It outlines the specifications of more than 20 different detectors that have been used for GC.
Hartmann, C. H. *Anal. Chem.* **1971,** *43*(2), 113A–125A.

In addition, in this mature and widely used technique, it is worthwhile to peruse the catalogs of GC suppliers.
Varian, Alltech, and Supelco are examples.

A book describing equipment, with a qualitative discussion of techniques and data handling. It contains a good description of GC/MS interface problems.
Message, G. M. "Practical Aspects of Gas Chromatography/Mass Spectrometry"; Wiley: New York, 1984.

## Problems

15.1 Assume the *n*-alkanes follow the behavior of a homologous series in Figure 15.8.
    a. Under these conditions, what would you expect for the *n*-pentane retention time?
    b. What is the retention index of *n*-propane?

15.2 The following question refers to the data shown in Figure 15.8. Within the sensitivity of the experiment, is there any ethane in the sample?

15.3 Do the major compounds of the chromatogram shown in Figure 13.11.1 form a homologous series?

15.4 [Figure courtesy of J&W Scientific, Inc., Rancho Cordova, CA.]
Figure 15.4.1 shows the van Deemter curves for three different carrier gases. These were determined from a $C_{17}$ hydrocarbon peak with $k' = 7.90$ in an open tubular column.
    a. In what carrier gas is the diffusion of the hydrocarbon the fastest?
    b. Which carrier gas should be used to obtain the best separation both as to efficiency and speed?

15.5 A mixture of liquids was made of the following composition:

| Material | Amount (g) | Peak Areas (cm²) |
|---|---|---|
| *n*-Hexane | 1.59 | 16.55 |
| Cyclohexane | 1.87 | 18.00 |
| Benzene | 2.28 | 21.20 |
| Toluene | 1.65 | 15.23 |

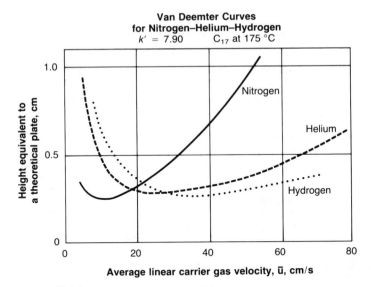

**Figure 15.4.1**

A sample of this mixture was injected into a GC with a thermal conductivity detector. The values found for the areas of the peaks (in cm²) are shown in the table in the third column. What is the thermal conductivity detector's relative response per unit weight of the three other components compared with benzene? Take benzene to be 1.00 exactly.

15.6  [Chromatogram courtesy of Hach Company, Loveland, CO.]
Figure 15.6.1 shows the chromatogram of a gas mixture. There are no components other than those appearing here. The numbers such as ×8 mean that the peak height is reduced eightfold by changing the attenuation of the thermal conductivity detector. The TC detector has a relative response to each component (per unit volume), as shown in the table below. Also included are the values of the heat content for each component. Assume ideal-gas behavior, that is, the volumes of the gases are additive.

| Component | Relative Response of TC Detector | Heat Content (BTU/ft³ at 60 °F) |
| --- | --- | --- |
| Methane | 0.0300 | 1009.7 |
| Ethane | 0.0253 | 1768.7 |
| Propane | 0.0194 | 2517.2 |
| Isobutane | 0.0186 | 3252.6 |
| Butane | 0.0179 | 3267.0 |
| Isopentane | 0.0179 | 3999.7 |
| Pentane | 0.0173 | 4008.7 |

a.  What component of the sample forms the greatest volume fraction, and what is the volume-fraction value?

b.  What component of the sample forms the smallest volume fraction, and what is the volume-fraction value?

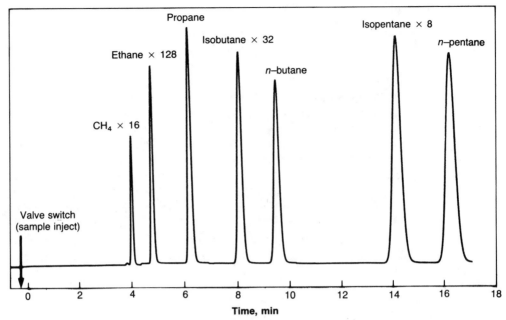

**Figure 15.6.1**

c. What component contributes the largest fraction of the heat content of the sample?
d. What component contributes the smallest fraction of the heat content of the sample?
e. What is the total heat content in BTU/ft³ at 60 °F for the mixture?

## SUPPLEMENT 15A

# Detectors for Gas Chromatography

In this supplement, the structures and basis of operation of three common GC detectors are presented: thermal conductivity, flame ionization, and electron capture detectors. The Hartmann reference describes more details for these and almost twenty others. A short list of some properties of GC detectors is presented in Table 15.2.

All GC detectors have requirements for certain flow rates of the effluent. If a detector is constructed to be used with a packed column, then it may be necessary to augment the much lower effluent flow from capillary columns. Similarly, if the detector requires a low effluent-flow rate, then an effluent splitter will be required. It is similar in operation to the sample splitter described in Figure 15.5.

## Thermal Conductivity Detector (TCD)

A gas flowing across a hot wire will lower the temperature of the wire by carrying away some of the heat. The amount of heat that is conducted away depends on the flow rate and the identity of the flowing gas. In a mixture of gases, the thermal conductivity depends on the inherent thermal conductivity of each of the components and the partial pressure of each. The temperature of the wire depends on the rate that heat is put in as well as the rate it is removed. Thus, if the heating rate is kept constant (by resistance heating from a constant electric current), the wire will reach an equilibrium temperature that depends on the mixture of gases flowing over it. This effect is the basis for the operation of the transducer of a thermal conductivity detector. The detector responds to all gases: It is a universal detector.

If the wire is quite thin, the response will correspond closely to the mixture in contact with the wire at the time. A thin wire has a large surface in contact with the gas relative to its total mass. The geometry of the detector is shown in Figure 15A.1. In the type of thermal conductivity system described here, two separate cells are used. One, the reference cell, has only the carrier gas passing through. The effluent from the

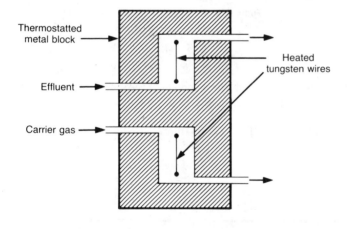

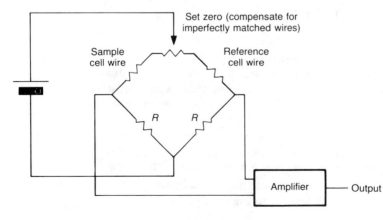

**Figure 15A.1**  A two-cell *thermal conductivity detector* (TCD). The platinum wires change resistance with temperature, acting as two edges of a Wheatstone bridge. The error voltage is amplified to produce the instrument response.

column passes through the other cell. The flow rate in both cells is the same. The temperature difference between the filaments of the reference and the effluent streams is measured.

The temperature difference can be monitored by measuring the ohmic resistances of the wires as they change with temperature. However, the changes in resistance are quite small. A Wheatstone bridge circuit (Section 3.4), which is illustrated in Figure 15A.1, is used to measure the change. The difference in temperature between the reference cell and sample cell filaments shows up as a current imbalance, which is amplified as necessary to produce the output.

### Flame Ionization Detector (FID)

The flame ionization detector is used more than any other type of GC detector. The FID is nearly a universal detector, which means that it responds nearly equally per unit mass of analyte *regardless of chemical structure.* As illustrated in Figure 15.6, the FID is relatively insensitive to a few small molecules: $N_2$, $NO_x$, $H_2S$, $CS_2$, $CO$, $CO_2$, COS, HCOOH, and $H_2O$. As noted in Table 15.2, the detection limit is significantly lower than that of the TCD. A schematic diagram of the construction of a FID detector is shown in Figure 15A.2.

The FID works by burning the effluent in a hydrogen/air flame with a stoichiometric excess of oxygen over hydrogen. These conditions produce an *oxidative flame.* The flame produces ionized fragments of the organic molecules in the effluent stream. The negatively charged fragment ions are driven by the electric field (200–

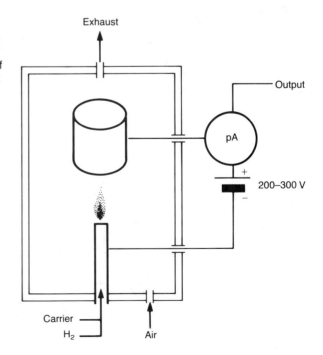

**Figure 15A.2**   Illustration of the operation of a *flame ionization detector* (FID). The oxidative flame causes gas-phase ions to form from the eluting components. The ions are collected, and the resulting current is amplified as the output of the detector. The sensitivity of the detector to each component depends on the flame chemistry of the individual component. [Ref: Reprinted with permission from Hartmann, C. H. *Anal. Chem.* **1971,** *43*(2), 113A–125A. Copyright 1971 American Chemical Society.]

300 V between the electrodes) into the collector. This ion-produced current is amplified to produce the output.

When a FID is being used, there is usually a clutter of gas tanks around the GC apparatus, because in addition to a tank of carrier gas, tanks of hydrogen and compressed air are required for the detector. All the flow rates must be carefully regulated for optimum results.

### Electron Capture Detector (ECD)

The response of an electron capture detector results from the changes in conductivity of a plasma (a gaseous mix of ions and free electrons) in the presence of molecules that have high affinities for free electrons. The plasma is generated by a radioactive source in the transducer cavity, as illustrated in Figure 15A.3. One common radioactive source is $^{63}Ni$ which is chosen for its low-energy beta emission. Low-energy beta (electron) emission is still relatively high-energy in chemical terms, and electrons of this energy (0.067 MeV) each produce up to 1000 plasma electrons through collisions with the carrier gas $N_2$ or $Ar + 10\%$ $CH_4$. The plasma electrons have energies in the range 0.02–0.05 eV. Within the detector head, the conductivity of the plasma under the influence of a voltage produces a **background current** of around 10 nA.

When molecules that have high electron affinities pass through the space containing the plasma, they capture the low-energy plasma electrons to form negative molecule ions. Two general reactions can occur.

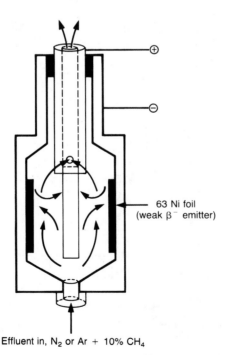

Effluent in, $N_2$ or $Ar + 10\%$ $CH_4$

63 Ni foil
(weak $\beta^-$ emitter)

**Figure 15A.3**   Schematic illustration of the transducer for one type of electron capture detector for gas chromatography. This is called a *coaxial transducer.* The center cylindrical electrode is surrounded by a hollow cylinder covered on the inside with a thin radioactive nickel foil. The beta (electron) radiation from the radioisotope ionizes the gas between the inner and outer electrodes rendering it conductive. Reduction of the conduction indicates the presence of compounds with high electron affinities.

$$AB + e^- \rightarrow AB^- \qquad \text{nondissociative attachment}$$
$$AB + e^- \rightarrow A{\cdot} + B^- \qquad \text{dissociative attachment}$$

In either case, the negative ions that are formed have two properties that cause a diminution in the background current (the output signal). The first is that the negative molecule ions tend to recombine with the positive ions more rapidly than do the plasma electrons. As a result, the total concentration of ions in the plasma is reduced. The second is that the molecule ions are larger and, because of collisions with the other particles in the plasma, move more slowly in the electric field than do the plasma electrons. In other words, the molecule ions have lower *drift velocities,* and the current is proportionally reduced.

The conducting properties of plasmas cause the ECD to have a small linear range (around a factor of 50 in mass). The reason is as follows. If a DC potential is held on the electrodes, the positive plasma ions drift toward the negative electrode, and the negative ions and electrons drift toward the positive electrode. However, the separation and movement of the plasma charge (the space charge) partially neutralizes the applied potential. Therefore, when the plasma ion concentration is diminished by the presence of an eluted species, the space charge decreases as well. As a result, the *effective* DC voltage increases, which leads to a nonlinear response in the output (the same plasma conductivity produces a larger current). The linear range can be increased to about $10^4$ by using pulsed DC detection. By the use of short voltage pulses, the space charge is not allowed to build up to any extent. (Compare this with AC detection of conductance in solutions, Figure 12.19.)

In addition to the problems with linearity, the ECD is significantly species specific, principally to halogenated hydrocarbons such as the insecticide DDT (dichlorodiphenyltrichloroethane). In fact, it is this specificity that has given the impetus to the widespread use of ECDs, despite their drawbacks. The response of the detector increases nonlinearly with the number and type of halogens in a molecule as can be seen from the following list.

| Molecular Formula | Relative Response ECD |
|---|---|
| $CF_3Cl$ | 1.0 |
| $CHCl_3$ | $1 \times 10^4$ |
| $CFCl_3$ | $4 \times 10^5$ |

These drawbacks cannot detract, however, from the outstanding sensitivity of the detector. For instance, hexachlorocyclohexane can be detected at the 0.1 pg level.

## SUPPLEMENT 15B

# How Good is the Resolution in High-Resolution Chromatograms?[4]

When one sees chromatograms with hundreds of peaks, such as that shown in Figure 15.2, the feeling may arise that most of the components are separated. However, the statistical argument outlined below indicates that, in fact, only a minority of the peaks may arise from a single component. The majority of the peaks probably indicate two or more coeluting compounds.

The question we want to answer is, Of the total number of peaks in a complicated chromatogram, how much overlap of components can we expect? This question cannot be answered in any absolute sense, because the total number of components is not previously known. (This difficulty is analogous to determining the "true value" of a standard and requires similar statistical arguments.)

We make a number of assumptions in order to allow a calculation to be tractable. First, the effects of peak-height variations are ignored. Second, the statistical probability that a peak will appear in a given elution volume is assumed to be random. This latter assumption is nearly correct, even though we know that a number of homologous series are likely to be present with their highly regular order. When the arrival of a component at the detector (or a person in the checkout line of a supermarket) is random in time, we can write down a probability that a second component will arrive after a specific time interval, call it $t^0$. Algebraically, we write

$$P(t \geq t^0) = e^{-\alpha} \tag{15B-1}$$

The probability is exponential to some negative power $\alpha$, which will be defined shortly.

Since it is certain (the probability is unity) that there will be a subsequent peak (we do not treat the first and last components because they lack one adjacent $t^0$), then the probability that the next peak will fall *within* the time interval $t^0$ is

$$P(t < t^0) = 1 - e^{-\alpha} \tag{15B-2}$$

One can define a variable called the **peak capacity** $n_c$. This is an integer that indicates how many regularly spaced peaks can just fit into a given retention time with all pairs of peaks resolved by $R_s$. ($R_s$ is defined in Section 13.5.) However, it is clear from Figure 15.2 that the peaks are not regularly spaced in a real chromatogram. Therefore, the numerical value of the peak capacity is the maximum possible number of peaks that can appear having resolution $R_s$ over the time of the experiment.

---

[4] *Sources:* Davis, J. M.; Giddings, J. C. *Anal. Chem.* **1983,** *55,* 418–424. Giddings, J. C.; Davis, J. M.; Schure, M. R. In "Ultrahigh Resolution Chromatography"; S. Ahuja, Ed.; American Chemical Society: Washington, D. C., 1984.

Let us call the average number of components that elute per unit time $\overline{m}$. Then, for the statistics so far described (stated without proof),

$$\alpha = \frac{\overline{m}}{n_c}; \quad 0 < \alpha \leq 1 \tag{15B-3}$$

The value of $\alpha$ is called by the originators of this calculation the *saturation factor*. It is the ratio of the average number of components to the hypothetical maximum number of components separable over the same time interval.

What, then, is the probability that a chromatographic peak is due to a single component in the sample? We require (a) that the component eluting ahead of the one of interest be separated by at least $t^0$ and (b) that the component following also be at least that far away.

In probability theory, the probability that two events happen together equals the product of the probabilities for each (here $P(t \geq t^0) = e^{-\alpha}$). Therefore, the probability that a band is due to a single component is

$$P_1 = (e^{-\alpha})(e^{-\alpha}) = e^{-2\alpha}$$

This equation says that the preceding and following peaks both appear further than $t^0$ away.

Following from the same type of argument, the probability that a band consists of two (and only two) unresolved components is a product of the probabilities of three events: that one other component will elute within the $t^0$ interval, and that the component preceeding and the component following do not fall within the interval $t^0$. Algebraically,

$$P_2 = (1 - e^{-\alpha})(e^{-\alpha})(e^{-\alpha}) = (1 - e^{-\alpha})(e^{-2\alpha})$$

From a continuation of this argument, we find the general formula for $n$ unresolved components composing one band is

$$P_n = (1 - e^{-\alpha})^{n-1}(e^{-2\alpha}) \tag{15B-4}$$

Over the total time of the chromatogram, the *number* of bands with $n$ unresolved components $(N_n)$ is the product of the average number of bands per unit time and the probability that a band has $n$ components. In algebraic terms,

$$N_1 = \overline{m}P_1$$
$$N_2 = \overline{m}P_2$$
$$\vdots \tag{15B-5}$$
$$N_n = \overline{m}P_n$$
$$\vdots$$

The relevant experimentally observable quantity in a chromatogram is the total number of peaks that appear over the total time. Let us call this integer $p$; it is the sum of all the bands composed of a single component, all those with two components, all those with three components, and so forth. In other words, $p$ equals the sum of all the values of Eqs. 15B-5.

$$p = N_1 + N_2 + N_3 + \cdots \tag{15B-6}$$

Factoring $\overline{m}$ from this equation,

$$\frac{p}{\overline{m}} = P_1 + P_2 + P_3 + \cdots \tag{15B-7}$$

We now note that all expressions for $P_n$ contain the factor $(e^{-2\alpha})$. Therefore, the expression for $(p/\overline{m})$ becomes a well-known arithmetic series

$$\frac{p}{\overline{m}} = e^{-2\alpha} \sum_{l=0}^{\infty} (1 - e^{-\alpha})^l \tag{15B-8}$$

where $l = n - 1$. The sum of the series has a value

$$\frac{p}{\overline{m}} = e^{-\alpha} \tag{15B-9}$$

Substituting for the value of $\overline{m}$ from Eq. 15B-3, we find

$$p = n_c \alpha e^{-\alpha} \tag{15B-10}$$

In a word-equation, Eq. 15B-10 says:

Total number of observable peaks =
maximum total resolvable peaks $\times$ function of $\alpha$

If more than $n_c$ peaks appear, the separation cannot be as good as the optimum case with equally spaced peaks. Some peaks must be less resolved than $R_s$.

To determine the best possible case for randomly occurring peaks, we can find the value of $\alpha$ that produces a maximum for $p$ in Eq. 15B-10. After some straightforward calculation, we find the maximum value occurs at $\alpha = 1$. Therefore, the best possible case (compared to the ideal) occurs when

$$p = n_c e^{-1} = 0.37 n_c, \qquad \text{best case}$$

Only 37% of the number of peaks that are possible for the ideal, uniformly spaced chromatogram appear. Clearly, when running a chromatographic separation, the overlap of components will be high unless a significantly larger number of theoretical plates are provided than one would calculate are needed for an ideal chromatogram with the expected number of components.

A similar calculation shows that if a chromatographic system is set up that can separate 100 components distributed equally in time ($n_c = 100$), but a sample with only 50 components (randomly distributed in time) is injected, only 30 peaks are expected to be seen ($p = 30$), and only 18 of these are expected to result from single components. From the statistics, we expect the other 12 peaks to result from two or more unresolved species.

This calculation is approximate, and a number of assumptions have been mentioned; but the gist of the argument suggests the importance of specific detectors and GC/MS in particular to separate and analyze complicated mixtures.

A number that indicates the difficulty of achieving a complete separation of sample components is the number of theoretical plates needed to isolate 90 of a total of 100 components as single-component bands. A calculation analogous to the one above suggests that we would need a chromatograph with 20 million theoretical plates. This is well beyond current technology: A 100-m capillary GC column may have 300,000 theoretical plates.

# Part V

# MASS SPECTROMETRY

# 16

# Mass Spectrometry

## 16.1 Introduction

Mass spectrometry is a name for a collection of techniques that are used to measure the abundances and masses of ions in the gas phase. The term *mass spectrometry* is now a misnomer. It came about because the early data were collected as images on photographic plates and looked like spectra, with a number of separate parallel lines on the film. The separation of molecules, molecular fragments, and atoms by their masses can now be done in a number of ways, some of which are described in this chapter. To carry out the separation, all the species must first be ionized and in gaseous form.

Mass spectrometry is one of the most versatile and powerful tools of chemical analysis. Its broad usefulness is due in part to the wide variety of capabilities possible for each of the three sections of a mass spectrometer. As shown in Figure 16.1, these three basic sections are the mass source, the mass discriminator, and the transducer/detector, which is used to quantitate the selected masses. The designs of the instruments will be discussed after the analysis of simple mass spectral data is described.

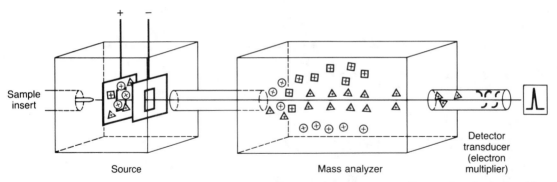

**Figure 16.1**    Diagram of the major features of a mass spectrometer. The sample molecules travel from the left, where they are introduced into the instrument, toward the right. The sample is inserted either as a gas, liquid, or solid. If it is not a gas, the sample must be vaporized by appropriate techniques. The gaseous molecules are ionized in the space between two charged plates. The region where the molecules are introduced, rendered into a vapor, ionized, and accelerated is called the *source*. The ions are accelerated by an electric field, formed into a column, and focused through slits. As shown, the positive ions are repelled by the positive electrode and attracted by the negative one. The acceleration potential is in the range of kilovolts. Shown in the source are three types of gaseous, positive ions with different masses, indicated by circles, triangles, and squares. The accelerated ions pass from the source into one of a number of types of *mass analyzers* (described in Supplement 16B). In the mass analyzer the ions of different masses are separated in space. The ion current from each mass impinges on the transducer at a different time. This small ion current is amplified by the transducer and associated electronics to produce the output signal. The output of the ion current versus time can be calibrated as ion current versus mass. This is the mass spectrum, such as those shown in Figure 16.2.

The presentation of data for mass spectrometry is usually done in one of three ways, which are illustrated in Figure 16.2. All show the **relative abundance** and the masses of the sample molecules (or atoms) and, simultaneously, fragments from decomposition of the molecules. The relative abundance is determined by the relative peak areas or, more commonly, peak heights, just as in chromatography. In mass spectrometry, *relative abundance* is another name for *relative concentration*.

In a plotted mass spectrum, the masses are measured by their positions on the horizontal axis. The numerical mass number can be found by counting based on standard, ubiquitous molecules with known masses. For instance, there is always some dinitrogen $N_2$, mass 28, around. Note that the axis label is not mass, however. Mass spectrometry works *only* on charged species: gas-phase ions. What is always determined is the ratio of the mass to the electric charge on the fragment. Thus the mass label is written correctly as $m/z$, the **mass-to-charge ratio**. In the past, the mass-to-charge ratio has been abbreviated as $m/e$, but $m/z$ is now preferred.

As mentioned, the material to be assayed by mass spectrometry must be in the gas phase. If the sample itself is not gaseous, it must somehow be vaporized. It is in this area that the development and application of new methods are progressing so quickly. Mass spectrometry (MS) was once considered to be mostly a technique for organic chemical analysis; now its applications are reaching into biochemistry and inorganic analysis. Some of the methods for vaporizing samples are described in Section 16.6 and Supplement 16A.

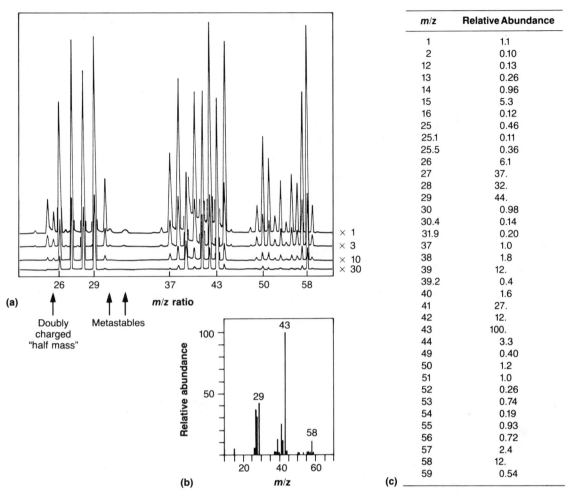

| m/z | Relative Abundance |
|---|---|
| 1 | 1.1 |
| 2 | 0.10 |
| 12 | 0.13 |
| 13 | 0.26 |
| 14 | 0.96 |
| 15 | 5.3 |
| 16 | 0.12 |
| 25 | 0.46 |
| 25.1 | 0.11 |
| 25.5 | 0.36 |
| 26 | 6.1 |
| 27 | 37. |
| 28 | 32. |
| 29 | 44. |
| 30 | 0.98 |
| 30.4 | 0.14 |
| 31.9 | 0.20 |
| 37 | 1.0 |
| 38 | 1.8 |
| 39 | 12. |
| 39.2 | 0.4 |
| 40 | 1.6 |
| 41 | 27. |
| 42 | 12. |
| 43 | 100. |
| 44 | 3.3 |
| 49 | 0.40 |
| 50 | 1.2 |
| 51 | 1.0 |
| 52 | 0.26 |
| 53 | 0.74 |
| 54 | 0.19 |
| 55 | 0.93 |
| 56 | 0.72 |
| 57 | 2.4 |
| 58 | 12. |
| 59 | 0.54 |

**Figure 16.2**   Three different presentations of the mass spectrum of a compound. (a) A plot of the ion current with four different scales plotted at the same time. Note two special features: There is a peak at "mass" $25\frac{1}{2}$. This is due to a mass-51 particle with two charges: a doubly charged ion with $m/z$ $25\frac{1}{2}$. Also, there are broad, low peaks between $m/z$ 30 and 37. These are **metastable ions** — ions that have decomposed while in flight after they have left the source region. (b) A computerized histogram output. The major peak $m/z$ values are labeled, and the largest peak is scaled to be 100. (c) A list of the peak $m/z$ values and relative abundances. Computerized systems generally list both (b) and (c). [Ref: (a) Courtesy of DuPont, Wilmington, DE. (b) and (c) from McLafferty, F. W. "Introduction to Mass Spectroscopy," 2nd ed.; Benjamin: Reading, MA, 1973.]

## 16.2 Case Study

The following example of the use of mass spectrometry is an application in the area of inorganic analysis. The materials to be analyzed are highly radioactive, and so the extreme sensitivity of mass spectral analysis is welcomed. Nanogram samples can be analyzed. The actual sample in this analysis is a single anion-exchange resin bead (~0.2 mm diameter), which contains about 1–3 ng of the metal ions to be assayed. The metals are vaporized by heating the sample to a high temperature.

The method involves the use of an internal standard. In this case, since mass spectrometers discriminate by mass, the internal standard spike can be composed of a less common isotope of each of the atomic species being assayed: Here the standards are $^{233}$U and $^{242}$Pu, both man-made isotopes. Quantitation is done by comparing the magnitudes of mass spectral peaks from the known cencentrations in the spike with unknown concentrations of the common isotopes of U and Pu in the sample. This is called the **isotope dilution method.** The isotope dilution method as used in mass spectrometry is described in Figure 16.3.

One of the most difficult stages of the nuclear fuel cycle to sample is spent reactor fuel dissolver solutions. These solutions are highly radioactive, containing fission products and actinides generated in the operation of the reactor. It is nevertheless essential that such materials be sampled for uranium and plutonium to allow establishing a material balance for accountability and safeguards purposes. A technique of using single anion resin beads selectively to adsorb plutonium and uranium from highly radioactive solutions was developed at this laboratory . . . .

Good separation was achieved from rare-earth and other fission products and most other actinides. In addition to plutonium and uranium, only neptunium

**(a) U isotopic**

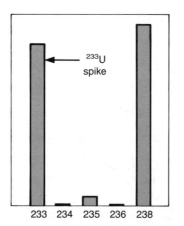

233  234  235  236  238

**(b) Pu isotopic**

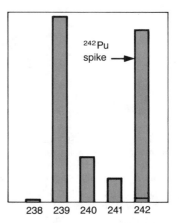

238  239  240  241  242

**Figure 16.3**    Histogram illustrating the mass spectra as used to determine uranium (a) and plutonium (b) using the isotope dilution method. A known amount of an isotopic spike is added to the analyte solution, and they are equilibrated; both isotopes must be in the same chemical form and homogeneously mixed. In (a), the spike for uranium is the man-made isotope $^{233}$U, as indicated. The other four peaks are the naturally occurring uranium isotopes. The spike comprises an internal standard, and the uranium concentration can be determined from the ratio of the $m/z$ 233 and 238 peaks and the known concentration of the spike. (b) illustrates the mass spectrum with a $^{242}$Pu internal standard spike. The horizontal line in the right bar indicates the amount of $^{242}$Pu that is present before the spike is added. A correction must be made for the $^{242}$Pu *not* in the spike, in order to achieve precise results. The use of an internal standard in this manner allows precise quantitation without 100% recovery of the analyte since the ratio of the spike and analyte peaks is used. [Figure courtesy of R. L. Walker, Oak Ridge National Laboratory.]

and thorium adsorb appreciably from 8 M $HNO_3$ solutions. Neither of these elements interferes with mass spectrometric analysis of U and Pu, and the quantities present on a [single ion-exchange] bead do not constitute a radiation hazard . . . .

**Experimental**  Dowex 1, 2% cross-linked anion resin beads in the nitrate form were used. Bead diameters of $150-250\ \mu m$ proved suitable for our needs, being large enough to manipulate individually with relative ease . . . .

For isotopic analysis [in the mass spectrometer] a single bead is loaded into a rhenium canoe-shaped filament. The filament is lightly crimped to hold the bead in place and installed in the mass spectrometer. The filament is carefully heated under high vacuum conditions ($\sim 10^{-5}$ Pa) until a pressure burst [measured on a pressure gauge] signals decomposition of the bead. Plutonium is then analyzed at about 1450 °C; any excess is burned off and uranium analyzed at about 1750 °C . . . .

**Results**  . . . beads were introduced to a sample solution and agitated varying lengths of time . . . . Ten minutes was enough to obtain sufficient adsorption for good analyses to be obtained . . . .

The amount of original input solution taken for isotopic dilution analysis must be chosen through consideration of three primary factors: 1) the permissible level of radioactivity; 2) the cost and availability of isotopic spikes; and 3) the desired accuracy and precision of the isotopic dilution measurements. For this type of analysis, the optimum ratio of the spike isotope to the most abundant sample isotope is 1 . . . .

For isotopic dilution measurement to be accurate, equilibrium between sample and spike must be established. This presents little problem for uranium, but plutonium requires careful adjustment of the valence state and destruction of [its inorganic] polymers and complexes in . . . solutions . . . .

Destruction of polymers and complexes is achieved by evaporating the spiked . . . [samples of the input solution] to near dryness with $HClO_4$ and HF . . . followed by reduction with Fe(II) and sulfamic acid followed by oxidation with $NaNO_2$ . . . .

Extreme care must be taken to avoid contamination of the sample with uranium either from the environment or from other samples. This involves using triple distilled water and redistilled $HNO_3$; distillations were carried out in a quartz still. Containers must be leached in $1:1\ HNO_3:H_2O$ before use. A clean, preferably isolated, laboratory hood should be set aside for resin bead manipulation . . . .

The table [16.1] presents results obtained from a sample synthesized to simulate spent reactor fuel . . . . Twelve analyses were made over a period of several months. The data show that the overall accuracy of the method is $\pm 0.5\%$, with a relative standard deviation of 0.6% and 0.9% for U and Pu respectively.

A secondary benefit of this technique is the ease of storing samples for archival purposes. More than 100 beads [from the same analysis] can be set aside on a single microscope slide. Each bead is isolated from air by collodion [a polymer soluble in organic solvents like ether] to prevent its oxidation. We have analyzed samples stored for more than two years in this manner with no difficulty.
[Ref: Walker, R. L.; Carter, J. A.; Smith, D. H. *Anal. Lett.* **1981,** *14*(A19), 1603–1612.]

**Table 16.1**   Results on Simulated Spent Nuclear Fuel

| Element | Concentration of Standard Solution (mg/g) | Measured Concentration (mg/g) | Bias (%) |
|---------|---------|---------|---------|
| Pu | 2.356 | $2.361 \pm 0.021$ | $+0.21$ |
| U | 231.5 | $232.5 \quad \pm 1.4$ | $+0.43$ |

## 16.3  Analysis of Organic-Molecule Mass Spectra

In the above example, you have read of the use of isotope abundance measurements in determining the amounts of inorganic ions in a sample. Similar measurements are used to help determine the identity of organic molecules and molecular fragments. In this section, we shall present one of the ways that this identification is done.

After data such as those shown in Figure 16.2 are made available, the order of operations is as follows:

1.  *Identify the **molecular ion**. The molecular ion is the ion with its mass equal to that of the molecular formula composed of atoms of the highest natural abundance (that is, $^{12}C$, $^{14}N$, $^{35}Cl$, and see Table 16.2). For example, in Figure 16.2, the peak at m/z 58 is the molecular ion of n-butane.*
2.  *Study the **isotope distribution pattern**. This is the pattern of intensities of peaks that are due to combinations of the different atomic isotopes in a specific molecule or fragment. In Figure 16.2, peaks with m/z of 58 and 59 form an isotope distribution pattern, as do 43 and 44. This point is explained in more detail later.*
3.  *Explain the **fragmentation pattern**. When the sample molecules are ionized in the instrument, some of the energy introduced to cause ionization serves to break up the molecule. The fragments from the breakup help confirm the identification of the molecule. In Figure 16.2, all the peaks below m/z 58 are part of the fragmentation pattern. However, masses 43 and 29 are the most useful.*

We emphasize that this is not the only way to approach the problem of identification. Also, computer-aided search routines do *not* work this way; they use methods of pattern identification which, in essence, look at all the peaks as a group.

Determining the identity of a molecule from its mass spectrum is a skill that can be obtained only with practice. Usually, other spectral information is exceedingly helpful, such as that from ultraviolet, infrared, or nuclear magnetic resonance spectra (see the respective chapters later in the text) or other stoichiometric data such as the wt% C,H,S, and Cl. Let us now look in more detail at the three steps of assigning mass spectra to the molecule from which it comes.

**Table 16.2A**  Natural Isotopic Abundances of Common Elements in Organic Compounds

| Element | Mass | Relative Abundance | Mass + 1 | Relative Abundance | Mass + 2 | Relative Abundance |
|---------|------|--------------------|----------|--------------------|----------|--------------------|
| *Elements with one isotope* | | | | | | |
| F | 19 | 100 | | | | |
| P | 31 | 100 | | | | |
| I | 127 | 100 | | | | |
| *Elements with two major isotopes* | | | | | | |
| H | 1 | 100 | 2 | 0.016 | | |
| C | 12 | 100 | 13 | 1.08 | | |
| N | 14 | 100 | 15 | 0.36 | | |
| Cl | 35 | 100 | | | 37 | 32.5 |
| Br | 79 | 100 | | | 81 | 98.0 |
| *Elements with three major isotopes* | | | | | | |
| O | 16 | 100 | 17 | 0.04 | 18 | 0.20 |
| Si | 28 | 100 | 29 | 5.1 | 30 | 3.4 |
| S | 32 | 100 | 33 | 0.80 | 34 | 4.4 |

**Table 16.2B**  Natural Isotopic Abundances of Some Inorganics

The most abundant isotope of each element is assigned an abundance of 100. The abundances of the less common isotopes are given relative to that, rounded to the nearest integer. Isotopes that are less than 1 on this scale are not included.

| Element | Atomic Masses (Relative Abundances) | | | | |
|---------|------------|------------|------------|------------|------------|
| Be | 9 (100) | | | | |
| B | 10 (23) | 11 (100) | | | |
| Na | 23 (100) | | | | |
| Mg | 24 (100) | 25 (13) | 26 (14) | | |
| Al | 27 (100) | | | | |
| K | 39 (100) | 41 (7) | | | |
| Ca | 40 (100) | 44 (2) | | | |
| Sc | 45 (100) | | | | |
| Ti | 46 (11) | 47 (11) | 48 (100) | 49 (8) | 50 (7) |
| V | 51 (100) | | | | |
| Cr | 50 (5) | 52 (100) | 53 (11) | 54 (3) | |
| Mn | 55 (100) | | | | |
| Fe | 54 (7) | 56 (100) | 57 (2) | | |
| Co | 59 (100) | | | | |
| Ni | 58 (100) | 60 (39) | 61 (2) | 62 (5) | 64 (2) |

*(continued)*

**Table 16.2B**  *(continued)*

| Element | Atomic Masses (Relative Abundances) | | | | | | |
|---|---|---|---|---|---|---|---|
| Cu | 63 (100) | 65 (45) | | | | | |
| Zn | 64 (100) | 66 (58) | 67 (8) | 68 (38) | | | |
| Ga | 69 (100) | 71 (67) | | | | | |
| Ge | 70 (56) | 72 (75) | 73 (21) | 74 (100) | 76 (21) | | |
| As | 75 (100) | | | | | | |
| Se | 74 (1) | 76 (18) | 77 (15) | 78 (47) | 80 (100) | 82 (18) | |
| Rb | 85 (100) | 87 (39) | | | | | |
| Sr | 86 (12) | 87 (8) | 88 (100) | | | | |
| Y | 89 (100) | | | | | | |
| Zr | 90 (100) | 91 (22) | 92 (33) | 94 (34) | 96 (5) | | |
| Nb | 93 (100) | | | | | | |
| Mo | 92 (61) | 94 (38) | 95 (64) | 96 (67) | 97 (39) | 98 (100) | 100 (39) |
| Ru | 96 (18) | 98 (7) | 99 (41) | 100 (41) | 101 (53) | 102 (100) | 104 (58) |
| Rh | 103 (100) | | | | | | |
| Pd | 102 (3) | 104 (34) | 105 (83) | 106 (100) | 108 (98) | 110 (50) | |
| Ag | 107 (100) | 109 (95) | | | | | |
| Cd | 106 (4) | 108 (3) | 110 (43) | 111 (45) | 112 (85) | 113 (43) | 114 (100) | 116 (7) |
| In | 113 (4) | 115 (100) | | | | | |
| Sn | 112 (3) | 114 (2) | 115 (1) | 116 (43) | 117 (23) | 118 (73) | 119 (26) | 120 (100) |
| | 122 (14) | 124 (18) | | | | | |
| Sb | 121 (100) | 123 (74) | | | | | |
| Te | 122 (7) | 123 (3) | 124 (14) | 125 (20) | 126 (55) | 128 (93) | 130 (100) |
| Cs | 133 (100) | | | | | | |
| Ba | 135 (9) | 136 (11) | 137 (16) | 138 (100) | | | |
| W | 182 (86) | 183 (47) | 184 (100) | 186 (93) | | | |
| Re | 185 (59) | 187 (100) | | | | | |
| Os | 186 (4) | 187 (4) | 188 (32) | 189 (39) | 190 (64) | 192 (100) | |
| Ir | 191 (63) | 193 (100) | | | | | |
| Pt | 192 (82) | 194 (97) | 195 (100) | 196 (75) | 198 (21) | | |
| Au | 197 (100) | | | | | | |
| Hg | 198·(34) | 199 (57) | 200 (77) | 201 (44) | 202 (100) | 204 (23) | |
| Tl | 203 (41) | 205 (100) | | | | | |
| Pb | 204 (3) | 206 (51) | 207 (40) | 208 (100) | | | |
| Bi | 209 (100) | | | | | | |

## The Molecular Ion

Assigning the molecular ion peak has the same function as determining the molecular mass of a molecule. This is the most important part of identification in mass spectrometry. The task is not necessarily easy. Notice that for all the organic-compound elements listed in Table 16.2, the most abundant stable atomic isotope is that with the lowest mass. (This pattern is not often found for the inorganics.) The molecular ion peak is often the largest (tallest) peak among the highest $m/z$ group. But again, this is not always true. Let us look at a few examples. Incidentally, the most intense ion peak is called the **base peak.**

In Figure 16.2, the heaviest mass observed, within the range of the spectrum, is at $m/z$ 59. Of the group of lines in this region, that at mass 58 is the base peak and is the molecular ion of $n$-butane.

In Figure 16.4 is the mass spectrum of methanol, $CH_3OH$. We expect the molecular ion to be at $m/z$ 32 from $^{12}C^1H_3^{16}O^1H$. Indeed, there is a large peak at $m/z$ 32. But the peak at 31 is larger. The gas-phase ion $CH_3O^{•+}$ is more abundant than the molecular ion. The reason is that this molecule loses one hydrogen quite easily.

One further complication is that the method of sample ionization can also cause a change in the mass of the molecular ion as well. Most of the mass spectra you will see are run with the ionization done by knocking an electron off the molecule using a beam of accelerated electrons. This is called **electron impact ionization.** The gas-phase interaction yields a positive radical ion, $M^{•+}$. (Capital $M$ is often used to represent mass of the molecular ion.) Mass spectra can also be obtained using **chemical ionization** (more of this later), and the ionic species of the molecule is formed by adding a proton. This is called a *quasimolecular ion,* $MH^+$, and has a mass of $M + 1$. The difference can be seen in the diagrams of the mass spectra in Figure 16.5a and b. Notice that when chemical ionization is used, the masses are shifted up one mass unit due to the addition of a proton to the parent molecule. However, the isotope distribution pattern of the shifted $m/z$ lines is the same in both. It is this unchanging pattern that we shall discuss next.

**Figure 16.4**    The mass spectrum of methanol. [Ref: McLafferty, F. W. "Introduction to Mass Spectroscopy," 2nd ed.; Benjamin: Reading, MA, 1973.]

| $m/z$ | Relative Abundance |
|---|---|
| 12 | 0.33 |
| 13 | 0.72 |
| 14 | 2.4 |
| 15 | 13. |
| 16 | 0.21 |
| 17 | 1.0 |
| 28 | 6.3 |
| 29 | 64. |
| 30 | 3.8 |
| 31 | 100. |
| 32 | 66. |
| 33 | 0.98 |
| 34 | 0.14 |

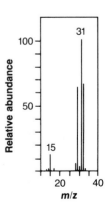

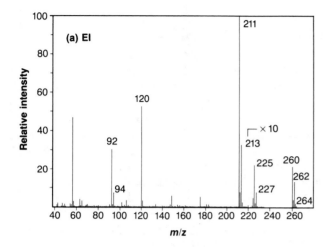

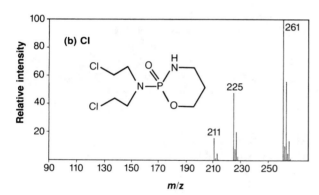

**Figure 16.5** Mass spectra of the same compound with three different types of sample vaporization/ionization. The sample is cyclophosphamide. Its structure is illustrated in (b). Spectrum (a) results from electron impact ionization. Spectrum (b) arises from chemical ionization and (c) from field desorption of the compound. (See Section 16.6 for descriptions.) Notice the significantly different fragmentation patterns. Also note the similar isotope patterns of the molecular ion (or quasimolecular ion in b). [Ref: Fenselau, C. In "Physical Methods in Modern Chemical Analysis," Vol. I; T.Kuwana, Ed.; Academic Press: New York, 1978, p. 103.]

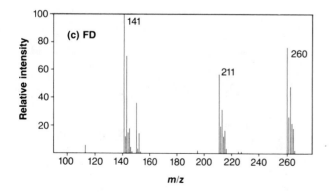

## Isotope Distribution Patterns

The isotope distribution pattern is determined by two properties of the molecules or fragments producing them.

  a.  The identity of the atoms in the molecule
  b.  The number of each of these atoms

The phrase *isotope distribution pattern* means the appearance of a set of peaks with fixed *relative* intensities. For example, isotope patterns of -Cl, -Cl$_2$, -Cl$_3$, and -Cl$_4$ are shown in Figure 16.6. These patterns arise in the following way. The identity of the atoms, chlorine, means that there are two isotopes in the ratio (Table 16.2)

$$^{35}Cl : {}^{37}Cl \quad \text{of} \quad 100 : 32.5$$

If we obtained a mass spectrum of (ionized) atomic chlorine alone, it would appear as shown in Figure 16.6: a peak at $m/z$ 35 about three times the height of $m/z$ 37.

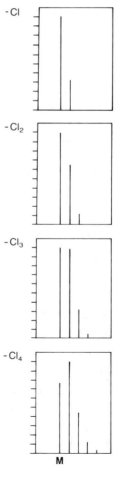

**Figure 16.6**    A histogram and list of the isotope patterns for molecules with one, two, three, or four chlorines. Table 16.3 illustrates the origins of the isotope abundance patterns. The largest peak of each set is scaled to 100.

| | | Relative Abundances | | | |
|---|---|---|---|---|---|
| | **M** | **M + 2** | **M + 4** | **M + 6** | **M + 8** |
| - **Cl** | 100 | 32.5 | | | |
| - **Cl₂** | 100 | 65 | 10.5 | | |
| - **Cl₃** | 100 | 98 | 32 | 3.4 | |
| - **Cl₄** | 77 | 100 | 44 | 10.5 | 0.9 |

**Table 16.3**   Possible Chlorine Isotope Compositions in a Molecule

| -Cl$_2$ | $^{35}$Cl$^{35}$Cl | $^{35}$Cl$^{37}$Cl | $^{37}$Cl$^{37}$Cl | | |
|---|---|---|---|---|---|
| Mass | $M$ | $M+2$ | $M+4$ | | |
| Isomers | 1 | 2 | 1 | | |
| -Cl$_3$ | $^{35}$Cl$^{35}$Cl$^{35}$Cl | $^{35}$Cl$^{35}$Cl$^{37}$Cl | $^{35}$Cl$^{37}$Cl$^{37}$Cl | $^{37}$Cl$^{37}$Cl$^{37}$Cl | |
| Mass | $M$ | $M+2$ | $M+4$ | $M+6$ | |
| Isomers | 1 | 3 | 3 | 1 | |
| -Cl$_4$ | $^{35}$Cl$^{35}$Cl$^{35}$Cl$^{35}$Cl | $^{35}$Cl$^{35}$Cl$^{35}$Cl$^{37}$Cl | $^{35}$Cl$^{35}$Cl$^{37}$Cl$^{37}$Cl | $^{35}$Cl$^{37}$Cl$^{37}$Cl$^{37}$Cl | $^{37}$Cl$^{37}$Cl$^{37}$Cl$^{37}$Cl |
| Mass | $M$ | $M+2$ | $M+4$ | $M+6$ | $M+8$ |
| Isomers | 1 | 4 | 6 | 4 | 1 |

The results for 2, 3, and 4 chlorine atoms in the same molecule are less straightforward to explain, since the reasoning is based on statistics and the existence of the two chlorine isotopes in the ratio 100 : 32.5.

When two or more chlorine atoms are in one molecule, there are a number of *possible* isotope compositions: This means that the structure remains the same, but different isotopes can lie at each position of the molecule. (For two chlorines you might consider the general molecule $^{37}$Cl—R—C—$^{35}$Cl and $^{35}$Cl—R—C—$^{37}$Cl.) In Table 16.3 are the isotopic compositions, along with the masses of each combination. Also included is a number that indicates how many *distinct* isotope isomers there are: In other words, how many different ways the given chlorine isotopes can be arranged on the chlorine positions of the molecule. After some study, you can see why the isotope signature will consist of three different masses for -Cl$_2$, four different masses for -Cl$_3$, and five different masses for -Cl$_4$. In addition, these masses will each be separated by *two* mass units.

This two-mass-unit separation of peaks with all the peaks relatively abundant is a useful signature for bromine- and chlorine-containing compounds. Look at the examples of the mass spectra of cyclophosphamide shown in Figure 16.5. The isotope pattern due to the two chlorines is constant regardless of the method of sample introduction and ionization.

Further calculations and comparisons with experiments are necessary to see whether we can also understand and explain the relative abundances of the sets of lines when more than two chlorines are present on a molecule. Let us carry out the calculation for -Cl$_2$ in detail. The -Cl$_3$ and -Cl$_4$ calculations are included as chapter problems.

**EXAMPLE**

The question to be answered is, If $^{35}$Cl : $^{37}$Cl occur in the ratio 1.00 : 0.32, with what probability will each of the combinations shown on the top of Table 16.3 occur?

**SOLUTION**

This is a calculation of probabilities. The relative probability for the occurrence of each of the possible sets of chlorine isotopes is the mathematical product of the probabilities for each isotope to be at random in each chlorine position. These mathematical products are shown below.

$$^{35}Cl^{35}Cl \qquad\qquad ^{35}Cl^{37}Cl \qquad\qquad ^{37}Cl^{35}Cl \qquad\qquad ^{37}Cl^{37}Cl$$

$$1.0 \times 1.0 = 1.0 \quad 1.0 \times 0.32 = 0.32 \quad 0.32 \times 1.0 = 0.32 \quad 0.32 \times 0.32 = 0.10$$

Notice that we must account for each distinctive isotope isomer. When both isotopes are $^{35}Cl$ or both are $^{37}Cl$, there is only one possible isomer. However, when there is one $^{35}Cl$ and one $^{37}Cl$, two different isomers are possible. (Recall $^{37}Cl—R—C—^{35}Cl$ and $^{35}Cl—R—C—^{37}Cl$; they are distinctive.)

If the relative abundance of the molecule containing $^{35}Cl^{35}Cl$ is called 100, then the relative peak heights will be as tabulated.

| Mass of peak | $M$ | $M + 2$ | $M + 4$ |
|---|---|---|---|
| Relative abundance | 100 | 64 | 10 |
| Number of distinct isotopes $\times$ relative peak height | $1 \times 100$ | $2 \times 32$ | $1 \times 10$ |

As shown above, the relative abundance of each peak is found by multiplying the relative probability of each isomeric mass times the number of distinct isomers having that mass.

The three peaks will be in the ratio $100:64:10$ for $M$, $M + 2$, and $M + 4$, respectively.

It is interesting to note in Figure 16.6 that when molecules contain four chlorines, because of the effect of having a large number of different isomers, the molecular ion does not have the highest relative abundance.

Exactly the same considerations are required in calculating the isotope pattern when carbon $^{12}C/^{13}C$ and/or hydrogen $^{1}H/^{2}H$ are present. However, these calculations of isotope patterns quickly get more complicated as the number of atoms with two or more isotopes increases in a molecule or fragment. As a result, it is convenient to use tables of relative abundances. Table 16.4 lists the isotope patterns expected from various numbers of carbons in a molecule. (The protons present have little effect.) The closeness of agreement with experimentally determined isotope patterns improves along with the quality of the mass spectrometer.

## Fragmentation Patterns

As you have observed in the illustrations of mass spectra, there are numerous lines at $m/z$ values less than that of the molecular ion. As noted earlier, these lines are due to the fragmentation of the molecules. The sets of fragments that appear and their relative abundances are called *fragmentation patterns*. The patterns depend on the parent molecule and the techniques used to vaporize and ionize the sample. Such differences in fragmentation patterns can be seen in Figure 16.5. Notice especially the small amount of fragmentation in the mass spectrum following chemical ionization. Also note that whatever type of source is used, many of the peaks in the three mass spectra are the same.

In fact, there is a well-studied descriptive chemistry of organic molecules in the gas phase, and with extensive practice, specialists can identify significant parts of the

**Table 16.4**   Contribution to Isotope Patterns by Carbon and Hydrogen for the Formula $(CH_{1.5})_n$*

$M$ relative abundance $= 100$

| | Relative Abundance | |
| --- | --- | --- |
| | $(M + 1)$ | $(M + 2)$ |
| $C_1$ | 1.1 | 0.000 |
| $C_2$ | 2.2 | 0.01 |
| $C_3$ | 3.3 | 0.04 |
| $C_4$ | 4.4 | 0.07 |
| $C_5$ | 5.5 | 0.12 |
| $C_6$ | 6.6 | 0.18 |
| $C_7$ | 7.7 | 0.25 |
| $C_8$ | 8.8 | 0.34 |
| $C_9$ | 9.9 | 0.44 |
| $C_{10}$ | 11.0 | 0.54 |
| $C_{11}$ | 12.1 | 0.67 |
| $C_{12}$ | 13.2 | 0.80 |
| $C_{13}$ | 14.3 | 0.94 |
| $C_{14}$ | 15.4 | 1.10 |
| $C_{15}$ | 16.5 | 1.27 |

* The maximum relative error in the listed values is less than 1.5% between the values shown and those calculated for $(CH)_n$ and $(CH_2)_n$.

structures of unknown molecules. This skill is far beyond the level of expertise that we shall need here. However, even at an introductory level, structural information can be obtained from the *separations* of the peaks in the fragmentation pattern. We can identify the masses of some common **neutral fragments** which split off from the molecular ion. The name *neutral fragments* simply describes their electric charge. Neutral fragments break off from a **parent ion** that may or may not be the molecular ion. *Parent ion* is a generic term for an ion that decomposes to generate a neutral fragment and an ion with a lower molecular mass. Two types of simple fragmentation reactions are

$$M^{\bullet+} \rightarrow [M - \text{fragment}]^+ + \text{fragment}^{\bullet}$$

and

$$M^{\bullet+} \rightarrow [M - \text{fragment}]^{\bullet+} + \text{fragment}$$

Table 16.5 lists some neutral fragments that are commonly lost from organic molecules. Let us identify some of them in the examples of *n*-butane (Figure 16.2)

**Table 16.5**  Common Neutral Fragments Lost by Gas-Phase Ions

Mass losses of 4 to 14, 21 to 23, 33, 37, and 38 are unlikely.

In addition to the above, the homologous series of most of the fragments also exist. These fragments differ by an additional $C_nH_{2n}$: masses 14, 28, 42, and so forth.

| $m/z$ Lost | Formula | Particular Sources of Loss |
|---|---|---|
| 1 | H | Labile H, aldehydes |
| 15 | $CH_3$ | Favored from branched site |
| 16 | O | Sulfoxides, nitro compounds |
| 16 | $NH_2$ | Amides, aromatic amines |
| 17 | OH | Acids, oximes |
| 18 | $H_2O$ | Alcohols (primary), aldehydes, ketones, ethers |
| 19 | F | Fluoroalkanes |
| 26 | CN | Alkylcyanides |
| 27 | HCN | Cyanides, aryl—$NH_2$ |
| 28 | $C_2H_4$ | Alkanes, ethyl esters |
| 28 | CO | Aromatic carbonyls, phenols |
| 29 | CHO | Alcohols |
| 29 | $C_2H_5$ | Favored from branched site, ethyl esters |
| 30 | NO | Aromatic nitro compounds |
| 30 | $CH_2O$ | Aromatic methoxy compounds |
| 31 | $CH_3O$ | Methyl esters |
| 32 | S | Sulfides, aromatic thiols |
| 34 | $H_2S$ | Thiols |
| 35,37 | Cl | Labile chloride |
| 41 | $C_3H_5$ | Propylesters, propylamides |
| 42 | $CH_2CO$ | Methylacetates |
| 43 | $C_3H_7$ | Propyl groups |
| 45 | $OC_2H_5$ | Ethyl esters |
| 46 | $NO_2$ | Aromatic nitro groups |
| 79,81 | Br | Alkyl bromides |
| 127 | I | Alkyl iodides |

and methanol (Figure 16.4) and then in the mass spectrum of cyclophosphamide (Figure 16.5).

---

**EXAMPLE: *n*-butane**

We have already identified the molecular ion as $m/z$ 58. The two other important peaks are 43 and 29, which are due to mass losses 15 and 29, respectively. We recognize these losses as likely to be $CH_3^{\bullet}$ and $C_2H_5^{\bullet}$, respectively. The remaining masses, 43 and 29, are consistent with $C_3H_7^+$ and $C_2H_5^+$. Masses 28 and 27 correspond to $C_2H_4^{\bullet+}$ and $C_2H_3^+$.

These last two species are illustrative of a general rule. Ion radicals usually have even-valued masses, except if the radical contains 1, 3, 5, . . . nitrogen atoms. Even-electron (nonradical) species such as $C_2H_3^+$ usually possess odd-valued masses.

---

## EXAMPLE: Methanol

Here the neutral loss of an extremely labile proton from the alcohol causes the peak at $m/z$ 31 to be larger than that of the molecular ion. Mass 15 is $CH_3^+$ and results from the loss of $O^\bullet$. If we thought that $m/z$ 31 were the molecular ion, then the formula of the compound would be $CH_3O$, which is not a stable compound. The fragmentation pattern helps us make the correct assignment of the molecular ion.

---

## EXAMPLE: Cyclophosphamide

The structure of cyclophosphamide is shown in Figure 16.5. As pointed out earlier, the mass spectra contain the splitting pattern indicating that two chlorines are associated with masses 260 and 141. However, masses 225, 211, and 92 appear to show the signature for only one chlorine. Mass 120 does not appear to have any chlorines. The mass 261 in the chemical ionization mass spectrum is the expected $(M + 1)^+$.

Next, note the following mass differences and the conclusions that we can draw from the chlorine isotope patterns:

$$260 - 225 = 35 \qquad \text{loss of one chlorine}$$
$$260 - 211 = 49 \qquad \text{includes loss of one chlorine}$$
$$260 - 141 = 119 \qquad \text{no chlorines lost}$$
$$260 - 120 = 140 \qquad \text{includes loss of two chlorines}$$

The following assignments of the peaks can be made. The conclusions derive from calculations of fragment masses, the chlorine content, and some descriptive chemistry: In the gas phase, it is possible for protons to transfer to the nitrogen.

| $m/z$ | Assignment |
|---|---|
| 260 | Molecular ion $= M^{\bullet+} = C_7H_{15}Cl_2N_2O_2P^{\bullet+}$ |
| 225 | $M^{\bullet+}$ less Cl |
| 211 | $M^{\bullet+}$ less $CH_2Cl$ |
| 141 | $[HN(CH_2CH_2Cl)_2]^+$ |
| 120 | $M^{\bullet+}$ less $N(CH_2CH_2Cl)_2$ |
| 92 | $C_3H_7ClN$ (probably $CH_2NHCH_2CH_2Cl^{\bullet+}$) |

Confirm for yourself that the isotope patterns and the formulas for each assignment are consistent.

---

The mass spectrum of cyclophosphamide also illustrates a general rule of mass spectra: the **nitrogen rule**. The nitrogen rule states that the molecular ion (but not the quasimolecular ion $MH^+$ obtained with chemical ionization) will appear at an even-valued mass if the molecule has an even number of nitrogen atoms: 0, 2, 4, . . . . Conversely, if the molecule has an odd number of nitrogens, the molecular ion will appear at an odd-valued mass.

**Table 16.6** Exact Nuclidic Masses*

| Isotope | Atomic Mass | Isotope | Atomic Mass |
|---------|-------------|---------|-------------|
| $^1H$ | 1.00782522 | $^{19}F$ | 18.9984046 |
| $^2H$ | 2.01410222 | $^{28}Si$ | 27.9769286 |
| $^{12}C$ | 12.00000000 | $^{31}P$ | 30.9737633 |
| $^{13}C$ | 13.00335508 | $^{32}S$ | 31.9720728 |
| $^{14}N$ | 14.00307440 | $^{35}Cl$ | 34.96885359 |
| $^{16}O$ | 15.99491502 | $^{79}Br$ | 78.9183320 |
| $^{18}O$ | 17.99915996 | $^{127}I$ | 126.9044755 |

* Superscripts are nominal masses.

## 16.4 Exact Masses and Chemical Formula Determination

The mass spectra shown in this chapter up to now show the peaks as if each were characterized by masses with unit values such as 48 or 70. In fact, there may be two or more unresolved, distinct molecules contributing to a single peak. A simple example is $(^{14}N)_2$ and $^{12}C^{16}O$; both appear at mass 28, their **nominal mass.** But the **exact masses** of atoms are not integer values. The exact masses of some isotopes are listed in Table 16.6. From these isotope masses, more exact molecular masses can be calculated for $(^{14}N)_2$ and $^{12}C^{16}O$: namely, 28.00061 and 27.99491 atomic mass units (a.m.u.) based on the $^{12}C$ scale. (This scale is composed assuming that $^{12}C$ has an atomic mass of exactly 12.)

The difference in mass between the $^{12}C^{16}O$ and $(^{14}N)_2$ is 0.00570 a.m.u. Thus, if a mass spectrometer can separate molecules with mass differences of 0.00570 a.m.u., we can ascertain which of the two are present in the sample and quantify each species separately.

Since each isotope has a noninteger exact mass, a similar analysis can be carried out for more complex molecules. Molecules with the same nominal mass but with different empirical formulas have slightly different masses. This is illustrated in Table 16.7 in which all the compound formulas in the table have a nominal molecular mass of 203. (Note that the entries of the table are computer-generated possibilities. Not all the entries correspond to actual molecules.)

Therefore, if a measurement of the molecular mass is made accurately enough, we can determine the molecular formula from the mass measurement alone. This is the basis of *high-resolution mass spectrometry.* Experimentally, a high-quality instrument must be used.

**EXAMPLE**

A molecule is found to have an exact molecular mass of 203.1522. If the molecule can contain only C, H, N, and/or O, what is its formula?

**Table 16.7**  Molecular Ions with Exact Masses between 203.0000 and 203.2000

Limited to molecules containing C, H, and less than 4 N and 4 O only.

| (Mass − 203.0000) | Empirical Formula | (Mass − 203.0000) | Empirical Formula |
|---|---|---|---|
| 0.0007 | $C_{13}HNO_2$ | 0.1158 | $C_9H_{17}NO_4$ |
| 0.0093 | $C_9H_3N_2O_4$ | 0.1185 | $C_{12}H_{15}N_2O$ |
| 0.0120 | $C_{12}HN_3O$ | 0.1271 | $C_8H_{17}N_3O_2$ |
| 0.0133 | $C_{14}H_3O_2$ | 0.1284 | $C_{10}H_{19}O_4$ |
| 0.0218 | $C_{10}H_5NO_4$ | 0.1298 | $C_{11}H_{15}N_4$ |
| 0.0246 | $C_{13}H_3N_2O$ | 0.1311 | $C_{13}H_{17}NO$ |
| 0.0331 | $C_9H_5N_3O_3$ | 0.1396 | $C_9H_{19}N_2O_3$ |
| 0.0344 | $C_{11}H_7O_4$ | 0.1424 | $C_{12}H_{17}N_3$ |
| 0.0359 | $C_{12}H_3N_4$ | 0.1436 | $C_{14}H_{19}O$ |
| 0.0371 | $C_{14}H_5NO$ | 0.1509 | $C_8H_{19}N_4O_2$ |
| 0.0457 | $C_{10}H_7N_2O_3$ | 0.1522 | $C_{10}H_{21}NO_3$ |
| 0.0484 | $C_{13}H_5N_3$ | 0.1549 | $C_{13}H_{19}N_2$ |
| 0.0497 | $C_{15}H_7O$ | 0.1635 | $C_9H_{21}N_3O_2$ |
| 0.0570 | $C_9H_7N_4O_2$ | 0.1648 | $C_{11}H_{23}O_3$ |
| 0.0583 | $C_{11}H_9NO_3$ | 0.1675 | $C_{14}H_{21}N$ |
| 0.0610 | $C_{14}H_7N_2$ | 0.1761 | $C_{10}N_{23}N_2O_2$ |
| 0.0695 | $C_{10}H_9N_3O_2$ | 0.1801 | $C_{15}H_{23}$ |
| 0.0708 | $C_{12}H_{11}O_3$ | 0.1873 | $C_9H_{23}N_4O$ |
| 0.0736 | $C_{15}H_9N$ | 0.1886 | $C_{11}H_{25}NO_2$ |
| 0.0821 | $C_{11}H_{11}N_2O_2$ | 0.1999 | $C_{10}H_{25}N_3O$ |
| 0.0861 | $C_{16}H_{11}$ | | |
| 0.0934 | $C_{10}H_{11}N_4O$ | | |
| 0.0947 | $C_{12}H_{13}NO_2$ | | |
| 0.1032 | $C_8H_{15}N_2O_4$ | | |
| 0.1060 | $C_{11}H_{13}N_3O$ | | |
| 0.1072 | $C_{13}H_{15}O_2$ | | |

*Source:* Benyon, J. H. "Mass Spectrometry and Its Applications to Organic Chemistry"; Elsevier: Amsterdam, 1960.

**SOLUTION**

By looking in tables such as Table 16.7, we find that the formula is $C_{10}H_{21}NO_3$. The molecule is most likely *n*-decyl nitrate. Incidently, the molecular weight listed in a handbook is 203.28. The difference is due to the natural isotope distribution: The exact molecular mass applies to the isotopically pure species.

**Figure 16.7**  Illustration of a definition of resolution in mass spectrometry. The resolution is defined experimentally as $M/\Delta M$ when the two identical experimental peaks overlap at 5% of their height producing a 90% valley between the peaks. Reported values may be defined with different peak overlaps.

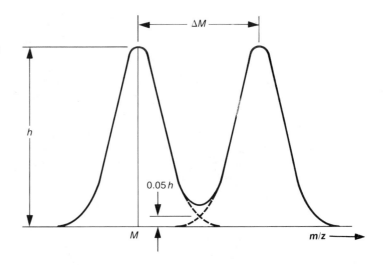

## 16.5 Mass Spectral Resolution

Two parameters that are used to describe the experimental capabilities of mass spectrometers are the **resolving power** and the **resolution.** In the literature, these two terms have been used almost interchangeably. To add to the confusion, resolution also has been defined as the reciprocal of resolving power. In any case, the number that is reported will imply the definition. In current common usage, a larger number refers to a better ability to resolve two components. Thus, for mass spectrometry,

$$\text{resolution} = \frac{M}{\Delta M} \qquad (16\text{-}1)$$

In the equation, $\Delta M$ is the difference between the masses of two species, and $M$ is the mass of the lighter one. With this definition of resolution, a larger value means that smaller mass differences can be resolved.

Sometimes the resolution for an instrument will be reported as *resolution xxx with 90% valley.* This means that the peaks are considered resolved when they appear as in Figure 16.7; the valley is 90% of the peak heights of equal peaks. However, it is more common to leave this information out. This results in ambiguity, which is discussed below.

---

**EXAMPLE**

Find the resolution needed to determine both $(^{14}N)_2$ and $^{12}C^{16}O$ in a gas mixture.

**SOLUTION**

Since the mass difference between $(^{14}N)_2$ and $^{12}C^{16}O$ is 0.0057 a.m.u., and the lighter species has mass 27.99,

$$\text{resolution} = \frac{27.99}{0.0057} = 4900$$

The number we have just calculated says nothing about how far two experimental peaks of equal height are separated in a mass spectrum.

How is the resolving power determined experimentally? First, a mass spectrum is run on the instrument being tested. The peak (a plot of ion intensity versus $m/z$) that results from a single species is scanned. A plot of this single-line envelope is then duplicated. As illustrated in Figure 16.7, the two identical peaks are located so that they cross at 5% of the peak height. The resolution of the instrument is then calculated as shown in the figure. In this case, the definition of crossing at 5% of the peak height — the sum defines a 90% valley — is the link between experimental resolution and the definition in Eq. 16-1. Unfortunately, you cannot *assume* that this definition has been used by everyone.

As you can see from the figure, it will be possible to resolve two different masses even when the peaks are not separated as far as illustrated. So the determination of $(^{14}N)_2/^{12}C^{16}O$ could be done with an instrument with less than 5000 resolving power (90% valley).

The resolving capability of a mass spectrometer is a function of its design and the quality of all of its parts. In general, the better the resolution, the higher the price. Table 16.8 lists the names of some types of mass spectrometers and the resolving power of each. The names of the spectrometers will have more meaning if you read Supplement 16B.

## 16.6 Mass Spectrometers: Principles of Operation

Mass spectrometers, regardless of the design of the various parts, are constructed to separate ions of gas-phase molecules and atoms according to their masses. All mass spectrometers are constructed to carry out four operations:

1. Introduction of the sample as a gas
2. Ionization of the molecules in the gas
3. Separation by mass
4. Detection of separated species

With a few exceptions, the four operations are done under high vacuum, generally around $10^{-7}$ torr (1 torr = 133 Pa = 1 mm Hg). A "high pressure" in mass spectrometry is around $10^{-3}$ torr. The first two steps of the above list occur in the ion source. The third step is done by mass analyzers. The fourth step involves a transducer, usually an **electron multiplier,** which works in a way similar to the photomultipliers described in Supplement 17A.

High-resolution determinations are limited by the quality of the mass separation. Otherwise, the ion source determines which analytical problems can be approached

**Table 16.8** Typical Characteristics of Commercial Mass Spectrometers*

| Type | Extremes of Mass Range | Resolution |
|---|---|---|
| Double focusing | 2–5000 | 20,000–100,000 |
| Single focusing | 1–1400 | 1500 |
| Time of flight | 1–700 (special to 10,000) | 1000 |
| Quadrupole | 2–1000 | 1000 |

* This technology is fast-changing. Special modifications of commercial instruments increase the top of the mass range and the typical resolution.

by mass spectrometry. The ion source determines which sample introduction technique can be used and, thus, which molecules can be brought into the gaseous state and ionized. Some of the more common ion sources are described next.

### Ion Sources for Gaseous Samples

A mass spectrum is obtainable only from samples that can be converted into gaseous ions. If the original sample material is a gas, then all that need be done is to feed it into the spectrometer without losing the vacuum of the instrument and to ionize the molecules. The ionization can be done in a number of different ways. The two most common methods are electron impact (bombardment) (EI), and chemical ionization (CI). You have already seen the differences in the results of using these two methods; there is generally far less fragmentation and a more abundant molecular ion (at $M + 1$) with chemical ionization.

Electron impact ionizers are constructed so that a collimated beam of electrons, which have been accelerated to about 70 eV, crosses through a concentration of neutral molecules from the sample. The electrons interact with the molecules, knocking an electron off each molecule to form positively charged ions. The energy required for this process is usually in the range of 7 to 12 eV. The remaining energy is available for fragmentation. These parent and fragment ions are then accelerated by an electric field into the mass analyzer. A more detailed look at the EI source appears in Supplement 16A.

The chemical ionizers use a reactant gas as a mediator in order to "soften" the ionization conditions of the sample. As you have seen, there is less fragmentation since less "excess" energy (above that needed for ionization) is available to break and rearrange chemical bonds. The reactant gas (often methane or another low-molecular mass hydrocarbon) is introduced into the ionization chamber at about $10^{-3}$ torr; its concentration is far greater than that of the sample. The reagent gas is ionized by the electron beam and, through subsequent gas-phase reactions, adds a proton to form the reagent-gas molecular ion. If methane is used, the reagent ion is $CH_5^+$, and it transfers a proton to analyte molecules with greater proton affinities in a reaction such as

$$CH_5^+ + M \rightarrow MH^+ + CH_4$$

The $MH^+$ ion is then accelerated through an electric field into the mass analyzer. The same ion source can function for both EI and CI methods, which gives added flexibility to the instruments so equipped.

### Ion Sources for Solid and Liquid Samples

Solid and liquid samples can be ionized by EI and CI if they can be vaporized. In many cases this is quite easy. The solid or liquid is placed on the tip of an inert probe which is placed in or near the evacuated ionization chamber. Then the probe is heated until the sample vaporizes into the path of the electron beam.

However, this method is limited to samples that will volatilize before they decompose. Many materials do not fall into this category. Different methods of volatilization are required for such compounds to be analyzed by mass spectrometry. Three of many volatilization methods are described below: **field desorption, fast-atom bombardment,** and **inductively coupled plasma.** All these techniques volatilize and ionize the compounds at the same time.

### Field Desorption (FD)

A field desorption mass spectrum is shown in Figure 16.5c. Note that its fragmentation pattern is different from that of the other two ionization methods. The basis of field desorption is the removal of an electron from a molecule by exposing it to an extremely high electrical potential: around $10^8$ V/cm. This is possible because the potential near a very sharp point is proportional to its sharpness. The points used for field desorption are at the tips of microscopic needles grown on a metal filament. The sample solution is placed on these needles, the solvent is allowed to evaporate, and the probe is then placed in the evacuated sample chamber. A potential of a few kV is turned on between the sample electrode (+) and an external electrode (−). The sample is heated, and at the high field at the tips of the needles, sample molecules lose electrons to the positively charged needle. Since the sample molecules and needle are then both positively charged, the molecules are driven off into the vacuum: The ionized molecules "field desorb." The ions are then accelerated into the mass analyzer.

### Fast-atom Bombardment (FAB)

The process of evaporating molecules really is a matter of giving them enough energy to break the bonds holding them together in a condensed phase without breaking the intramolecular bonds. Heating is not the only way to do this. The energy can be obtained from accelerated atoms hitting the sample and, in effect, knocking the sample molecules and fragments into the gas phase. In the FAB process, the sample is dissolved in a liquid (often glycerol). Then the solution is placed where a beam of accelerated atoms can impinge on its surface and eject the analyte molecules. The volatilized particles are of the types $(M + H)^+$, M, and $(M - H)^-$. The neutral molecules are unaffected by the electric field, but the ions can be accelerated into the mass analyzer. Either positive or negative ions can be selected. FAB is described more fully in Supplement 16A.

### Inductively Coupled Plasma (ICP)

A comparatively new mass spectrometry source for quantitative trace and ultratrace elemental analysis is an inductively coupled plasma torch, which is described in Figure 18.10. The torch operates at temperatures as high as 10,000 K, and the molecules in the solution samples are broken down into their elemental forms. Some fraction of each elemental species is simultaneously ionized [the monocationic $(+1)$ form is preferred for the assay]. The plasma is interfaced with a quadrupole mass filter, and the ions are separated by mass and detected. Because the transducer is relatively insensitive to mass, most ions are relatively equally detectable. The combination of an inductively coupled plasma and mass spectrometer is abbreviated ICP/MS. (The limits of detection of ICP/MS are listed in Table 7.2.)

Quantitation with ICP/MS has interferences analogous to those encountered with any mixture, except that fragmentation is not a problem. Among these interferences are the following:

> The percentage of each element that is in its monocationic form in the torch depends on the identity of the element.
> A significant fraction of some ions are doubly charged under the conditions, producing responses at one-half the expected $m/z$ while lowering the response at the expected $m/z$ region.
> Mass spectral peak overlap of isotopes of different elements occurs. The name given to this overlap is *isobaric* (equal weight) interference. An example of an isobaric interference is $^{87}Rb^+$ and $^{87}Sr^+$.
> The ions of metals forming exceptionally stable oxides tend to form diatomic molecular oxides, which reduces the expected peak intensity at the metal-ion mass. For example, uranium forms both $U^+$ and $UO^+$.

Sample preparation must be carefully considered for two reasons.

> The total amount of salt that can be tolerated is 1%, and 0.1% is preferred. If high-acid concentrations are needed, the sample must be diluted with corresponding dilution of trace analytes.
> Some molecular ions from the acids of digestion can interfere if their concentrations are high enough. For example, $ClO^+$ (masses 51 and 53) from perchloric or HCl interferes with $^{51}V$.

## 16.7 Chromatography/Mass Spectrometry

Molecular fragmentation interferes seriously in the use of mass spectrometry for the analysis of mixtures. A major difficulty involves determining which peaks represent the different molecular ions among the large number of peaks that arise from the fragmentation of higher mass molecules. This problem is only somewhat reduced by the use of chemical ionization, which causes less fragmentation of the molecules in a mixture.

As you can appreciate, one of the great strengths of mass spectrometry is our ability to identify molecular and atomic species from the MS data. On the other hand, the strength of chromatographic methods is the ability to separate materials into their molecular components. However, one of the weaknesses of chromatographic methods is in identifying all the components that elute. In view of the comparative strengths and weaknesses of mass spectrometry and chromatography, it is not surprising that the use of a mass spectrometer as a detector to identify chromatographic bands of unknown mixtures has become probably the most powerful analytical method of chemical analysis.

Mass spectrometry has been coupled with both gas chromatography and liquid chromatography. The gas chromatography/mass spectrometry (GC/MS) techniques are now better developed than those of liquid chromatography/mass spectrometry (LC/MS), partly because it is easier to get rid of excess carrier gas than to eliminate liquid mobile phases. However, the techniques of LC/MS are being developed quickly.

We shall concentrate here on only a few characteristics of GC/MS. First, consider the problem of the total amount of information that must be handled. A simple "real" mixture to be analyzed by GC might consist of 50 components. A list of 50 names of compounds is a comprehensible amount of information. However, it is not the names that are collected, but the mass-spectral data from peaks. Assume that after you optimize the GC separation, on the average one peak elutes from the column every 30 seconds (a relatively long time), and a mass spectrum is obtained to identify each component. Perhaps, on average, there are only 50 important peaks in each mass spectrum. Thus, the total number of MS peaks to be recorded and analyzed will be 50 peaks × 50 compounds = 2500 in half an hour. If 40 samples are run in a day, then each day there will be more than 100,000 pieces of data ($m/z$, relative abundances, and retention indices). A computer is *essential* to deal with this avalanche of information from GC/MS.

A second problem is how to get rid of the excess carrier gas from the GC column. A number of ingenious ways have been invented to do this. Only one of them is explained here — a **jet separator.** A schematic diagram is shown in Figure 16.8. The separator works because the carrier gas molecules, usually helium, diffuse into a region of reduced pressure more quickly than do analyte molecules with far heavier masses.

As illustrated in the figure, the effluent stream from a gas chromatograph passes through an orifice and then across a narrow, evacuated section into another orifice. There are three effects of this passage: (1) Some carrier gas is removed; thus (2), the effluent is enriched in sample molecules; but (3), some of the sample molecules are lost. For better results, the effluent is passed through two stages, as shown in Figure 16.8. The result of the total process is that a large fraction of the carrier gas is removed, but the enriched gas stream does not contain all the sample. In addition, the sample is delayed in reaching the mass spectrometer.

Naturally, the above effects have been named for ease of communication. The relative amount of carrier gas lost compared with the relative amount of sample lost is called the **enrichment factor.** The total percentage of the sample that arrives at the mass spectrometer is the **separation yield.** And the delay time is called the **lag time.**

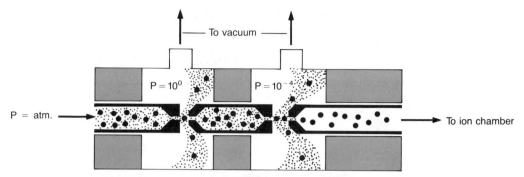

**Figure 16.8**    Diagram of one type of molecule separator used as a GC/MS interface. The idea is to remove as much of the lighter-mass carrier gas as possible while keeping as much of the sample as possible. Here, in two stages, the stream of gas from the GC column expands into a vacuum and impinges on a hole, heading toward the mass spectrometer. The lighter molecules diffuse away more easily; thus, the central train of molecules is enriched in the heavier analyte molecules. The pressure in the second stage is lower than that in the first stage. The enrichment depends on the carrier gas pressure, gas flow rates, and vacuum pumping rate in both stages. The separator yield — equivalent to percent recovery of the sample reaching the mass spectrometer — can be variable from 20% downward. The enrichment and yield can both depend on the identity of each component of the sample. Thus, quantitative analysis from the mass spectra alone is always suspect. Standards for each compound should be run. [Ref: Ryhage, R. *Arkiv. Kemi* **1966,** *26*, 305.]

These three parameters are used to characterize the performances of GC/MS molecular separators.

An important factor is the rate at which the entire mass spectrum can be scanned. As an example, if a GC peak elutes over a period of 10 seconds, it is rather useless to use a mass analyzer that can obtain a spectrum only over a period of one minute. It is now common for a number of different types of mass analyzers to be able to scan the mass range 10–1000 in 100 ms or less. In the extreme, a **time-of-flight (TOF)** mass spectrometer can obtain over a thousand complete spectra a second. Only the mass spectrometers with the largest magnets are too slow to be used with chromatographs. More details about these mass analyzers and their capabilities are presented in Supplement 16B.

## Suggestions for Further Reading

An outstanding short book on quantitation in mass spectrometry written at an intermediate level. It includes GC/MS. Highly recommended.
Millard, B. J. "Quantitative Mass Spectrometry"; Heydon: Philadelphia, 1978.

This book contains information on sample handling and describes in depth how to interpret mass spectra from organic compounds.
Hamming, M. C.; Foster, N. G. "Interpretation of Mass Spectra of Organic Compounds"; Academic Press: New York, 1972.

The construction and performance of all the parts of mass spectrometric instruments are clearly discussed.

Roboz, J.; Chait, E. In "Physical Methods of Chemistry," Pt. 6, Vol. 1; A. Weissberger and B. W. Rossiter, Eds.; Wiley–Interscience: New York, 1977; Chap. 3.

Monograph describing the gas-phase chemistry and equipment of mass spectrometry with chemical ionization. Numerous examples of the mass spectra.

Harrison, A. G. "Chemical Ionization Mass Spectrometry"; CRC Press: Boca Raton, FL, 1983.

Clearly written introductions to organic-molecule mass spectrometry with electron ionization.

Pavia, D. L., et al. "Introduction to Spectroscopy: A Guide for Students of Organic Chemistry"; Saunders: Philadelphia, 1979.

Silverstein, R. M.; Bassler, G. C.; Morrill, T. C. "Spectrometric Identification of Organic Compounds," 4th ed.; Wiley: New York, 1981; Chaps. 4 and 5.

This book contains the possible stoichiometries of molecules containing only C, H, N, and O with nominal masses between 12 and 500. It is used for identification by exact mass measurements. This information is now usually filed in a computer attached to the spectrometer.

Beynon, J. H.; Williams, A. E. "Mass and Abundance Tables for Use in Mass Spectrometry"; Elsevier: Amsterdam and New York , 1963.

The following journal articles review clearly the topics listed in their titles.

Anbar, M.; Aberth, W. H. "Field Ionization Mass Spectrometry"; *Anal. Chem.* **1974**, *46*, 59A–64A.

Fenselau, C. "The Mass Spectrometer As a Gas Chromatograph Detector"; *Anal. Chem.* **1977**, *49*, 563A.

Milne, G. W. A.; Lacey, M. J. "Modern Ionization Techniques in Mass Spectrometry"; *Crit. Rev. Anal. Chem.* **1974**, *5*, 45–105.

Munson, B. "Chemical Ionization Mass Spectrometry"; *Anal. Chem.* **1977**, *49*, 772A.

Simpson, C. F. "Gas Chromatography–Mass Spectroscopy Interfacial Systems"; *Crit. Rev. Anal. Chem.* **1972**, *3*, 1–40.

## Problems

16.1   Explain by way of a calculation the relative intensities of the lines for -$Cl_3$ and -$Cl_4$ as shown in Figure 16.4.

16.2   A mass spectrometer has a resolving power of 5000. To what value of $\Delta M$ is this equivalent if $M$ is

    a.   300?
    b.   900?
    c.   3000?

---

* Denotes more involved problems.

*16.3   An isotope dilution analysis was done on a rock sample in order to assay for Rb. A sample of rock weighing 0.350 g was digested and to the sample was added a 29.45 $\mu$g spike of rubidium. The isotopic composition (number percent or mole percent) of the spike was

$$^{87}Rb = 95.4\%, \qquad ^{85}Rb = 4.6\%$$

The mass spectral results showed that the peak for $^{87}Rb$ was 1.12 times the peak for $^{85}Rb$. Assuming that both isotopes have exactly the same chemistries, what is the Rb content of the rock in ppm (assuming both isotopes with normal isotope distribution)? The natural abundances and atomic masses of the rubidium isotopes are shown in the following table.

| Isotope | % Natural Abundance | Atomic Mass (a.m.u.) |
|---------|---------------------|----------------------|
| $^{87}Rb$ | 27.83 | 86.909 |
| $^{85}Rb$ | 72.17 | 84.912 |

16.4   [Ref: Middleditch, B. S., et al. "Mass Spectrometry of Priority Pollutants"; Plenum: New York, 1981.]

The mass spectra in Figure 16.4.1 — I, II, III, IV, and V — are of important industrial chemicals that are regulated in the workplace. Identify the molecular formula of each and the structure if possible.

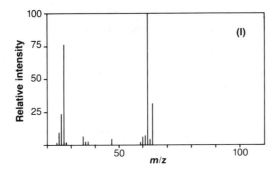

**SPECTRAL DATA (I)**

| Mass | Abundance | Mass | Abundance |
|------|-----------|------|-----------|
| 24 | 2.4 | 48 | 2.0 |
| 25 | 10.3 | 49 | 1.7 |
| 26 | 24.2 | 59 | 2.0 |
| 27 | 76.4 | 60 | 6.5 |
| 28 | 2.3 | 61 | 8.4 |
| 35 | 7.3 | 62 | 100.0 |
| 36 | 2.7 | 63 | 5.0 |
| 37 | 2.6 | 64 | 30.8 |
| 47 | 4.8 | | |

**SPECTRAL DATA (II)**

| Mass | Abundance | Mass | Abundance |
|------|-----------|------|-----------|
| 50 | 16.0 | 78 | 6.7 |
| 51 | 43.8 | 93 | 14.8 |
| 52 | 2.5 | 123 | 75.6 |
| 63 | 2.1 | 124 | 5.4 |
| 65 | 14.0 | | |
| 74 | 7.9 | | |
| 75 | 4.6 | | |
| 76 | 3.7 | | |
| 77 | 100.0 | | |

**Figure 16.4.1**   *(continued on following page)*

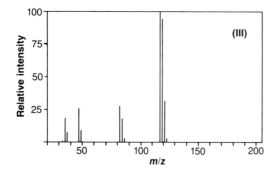

| SPECTRAL DATA | | | |
|---|---|---|---|
| **Mass** | **Abundance** | **Mass** | **Abundance** |
| 35 | 18.7 | 121 | 30.7 |
| 37 | 6.6 | 123 | 3.2 |
| 47 | 25.6 | | |
| 49 | 8.9 | | |
| 82 | 28.3 | | |
| 84 | 18.3 | | |
| 86 | 2.9 | | |
| 117 | 100.0 | | |
| 119 | 93.7 | | |

(III)

| SPECTRAL DATA | | | |
|---|---|---|---|
| **Mass** | **Abundance** | **Mass** | **Abundance** |
| 25 | 3.0 | 51 | 7.8 |
| 26 | 13.1 | 60 | 3.2 |
| 27 | 39.8 | 61 | 8.8 |
| 28 | 3.3 | 62 | 100.0 |
| 35 | 4.2 | 63 | 12.5 |
| 36 | 2.6 | 64 | 31.1 |
| 47 | 2.1 | 65 | 3.2 |
| 48 | 2.2 | 98 | 7.3 |
| 49 | 24.9 | 100 | 4.5 |

(IV)

| SPECTRAL DATA | | | |
|---|---|---|---|
| **Mass** | **Abundance** | **Mass** | **Abundance** |
| 25 | 3.1 | 61 | 8.9 |
| 26 | 13.2 | 62 | 8.4 |
| 27 | 42.8 | 63 | 100.0 |
| 28 | 3.7 | 64 | 4.2 |
| 35 | 5.8 | 65 | 31.4 |
| 36 | 2.9 | 83 | 12.9 |
| 37 | 2.4 | 85 | 8.3 |
| 47 | 2.9 | 98 | 6.8 |
| 60 | 3.5 | 100 | 3.6 |

(V)

**Figure 16.4.1**    (continued)

16.5    [Reprinted from Schulten, H.-R., et al. *J. Trace Microprobe Techn.* **1982,** *1*, 115 by courtesy of Marcel Dekker, Inc.]

Figure 16.5.1 shows the partial mass spectrum of a coal sample being analyzed for metal content. The spectra amplified by different amounts are illustrated. All the peaks are due to positive ions of metals which were obtained by field desorption MS with a laser-assisted vaporization. To separate the organic and anion components, the metals were deposited on the emitter element by electrodeposition. Identify the elements present and the isotopes contributing to the peaks, taking into account the isotope distributions of the elements present.

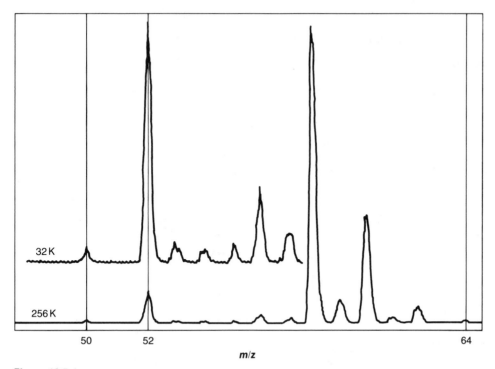

**Figure 16.5.1**

16.6 [Reprinted with permission from Harrison, A. G.; "Chemical Ionization Mass Spectrometry"; CRC Press: Boca Raton, FL, 1983, p. 103. Copyright CRC Press, Inc., Boca Raton, FL.]
Figure 16.6.1 shows mass spectra of four closely related compounds. The mass spectra were obtained by chemical ionization using $H_2$ as the proton donor. What are the compounds?

16.7 Below are illustrated the structures of four compounds with nominal masses of 194. The accompanying table (p. 604) lists the mass spectra I, II, and III obtained with electron bombardment ionization. Identify the compound from which each of the three spectra came.

(a)  (b)  (c)  (d)

| I | | II | | III | |
|---|---|---|---|---|---|
| Mass | Relative Intensity | Mass | Relative Intensity | Mass | Relative Intensity |
| 15 | 14 | 41 | 1 | 40 | 1 |
| 16 | 2 | 42 | 2 | 41 | 7 |
| 18 | 11 | 43 | 42 | 42 | 1 |
| 27 | 3 | 44 | 1 | 43 | 4 |
| 28 | 5 | 50 | 2 | 45 | 1 |
| 39 | 1 | 51 | 2 | 50 | 6 |
| 40 | 4 | 52 | 1 | 51 | 4 |
| 41 | 5 | 53 | 3 | 52 | 1 |
| 42 | 10 | 59 | 1 | 53 | 3 |
| 44 | 2 | 62 | 2 | 62 | 2 |
| 52 | 2 | 63 | 12 | 63 | 11 |
| 53 | 3 | 64 | 9 | 64 | 16 |
| 54 | 3 | 65 | 11 | 65 | 4 |
| 55 | 37 | 66 | 1 | 66 | 1 |
| 56 | 3 | 74 | 1 | 73 | 1 |
| 66 | 3 | 75 | 1 | 74 | 2 |
| 67 | 23 | 76 | 1 | 75 | 3 |
| 68 | 3 | 77 | 3 | 76 | 5 |
| 69 | 1 | 80 | 1 | 77 | 25 |
| 70 | 2 | 91 | 1 | 78 | 3 |
| 81 | 4 | 92 | 39 | 79 | 3 |
| 82 | 18 | 93 | 7 | 81 | 2 |
| 83 | 2 | 94 | 1 | 92 | 24 |
| 94 | 2 | 95 | 2 | 93 | 2 |
| 97 | 2 | 119 | 1 | 104 | 1 |
| 108 | 2 | 120 | 100 | 105 | 1 |
| 109 | 59 | 121 | 41 | 106 | 1 |
| 110 | 6 | 122 | 4 | 107 | 9 |
| 136 | 4 | 123 | 1 | 108 | 2 |
| 137 | 6 | 135 | 1 | 109 | 1 |
| 138 | 2 | 151 | 3 | 120 | 1 |
| 165 | 4 | 152 | 81 | 135 | 100 |
| 193 | 9 | 153 | 7 | 136 | 10 |
| 194 | 100 | 154 | 1 | 137 | 1 |
| 195 | 9 | 163 | 6 | 152 | 56 |
| | | 194 | 1 | 153 | 5 |
| | | | | 165 | 1 |
| | | | | 194 | 9 |
| | | | | 195 | 1 |

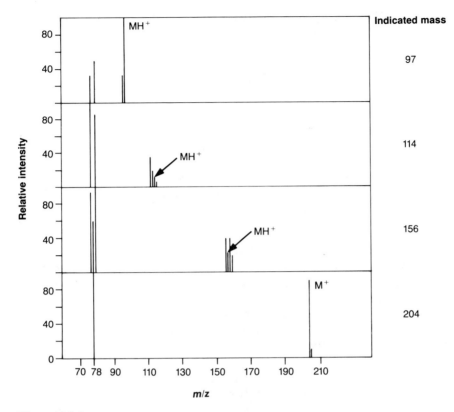

**Figure 16.6.1**

16.8   For the four compounds shown in problem 16.7:
   a.   What are their exact masses?
   b.   What resolution would be necessary to separate them in a mass spectrum?

*16.9   [Combination Radiochemical Kinetics–Mass Spectrometry Problem]
$^{87}$Rb has a radioactive decay (first-order kinetics)

$$^{87}Rb \rightarrow \beta^- + {}^{87}Sr, \quad \text{half-life } 4.8 \times 10^{10} \text{ years}$$

A rock was dated by assaying the product of this reaction. The mass spectrum of a homogeneous sample of the rock showed the $^{87}$Sr/$^{86}$Sr mole ratio to be 2.25. Assume that the original $^{87}$Sr/$^{86}$Sr ratio was 0.700 when the rock cooled. Chemical analysis of the rock gave 15.5 ppm Sr and 265.4 ppm of Rb. [Note: These are for the average atomic mass for the isotopic ratios found by mass spectrometry.] The other isotope ratios in the rock were $^{86}$Sr/$^{88}$Sr = 0.119 and $^{84}$Sr/$^{88}$Sr = 0.007. The isotope masses are as follows:

| Isotope | Mass (a.m.u.) |
|---------|---------------|
| $^{88}$Sr | 87.906 |
| $^{87}$Sr | 86.909 |
| $^{86}$Sr | 85.909 |
| $^{84}$Sr | 83.913 |

The data for Rb are included with problem 16.3.

    a. From the isotope ratios, what is the average atomic mass for Sr in the rock?

    b. What was the original concentration of Rb in the rock in ppm?

    c. What percentage of the original $^{87}$Rb was converted?

    d. What is the age of the rock, that is, the time since it cooled?

## SUPPLEMENT 16A

# Source Ionizers for Mass Spectrometry

Whatever ion source is used, one property that is desired is stability in ion concentration. If the source is not stable over the time needed to run a spectrum, then the peak abundances cannot be compared. The ideal is to have a technique for highly stable sample introduction with a constant fraction of all the gas-phase molecules ionized. The time required for a scan in a mass spectrometer depends on the specific instrument, and times can range from a millisecond to nearly a minute.

All the sources for mass spectrometry must have ion optics to accelerate and focus the ions produced by ionization. The manipulation of the gas-phase ions can be done with suitable electric fields to accelerate and focus and with slits to collimate the ion beam. These are shown in Figure 16A.1 for the electron ionization source.

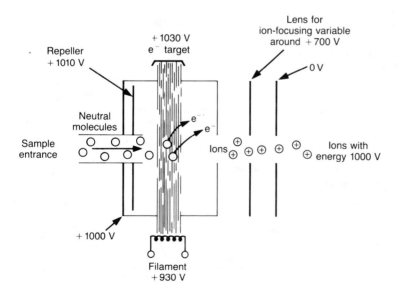

**Figure 16A.1**  Diagram of the operation of an electron impact mass spectrometry ion source. Neutral molecules are vaporized into the chamber. A beam of electrons passes through the vapor, causing positive ions to form. The ions are repelled out of the chamber, focused, and admitted into the mass analyzer. The potentials shown in the figure are such that the ions will enter the mass analyzer distributed around an average kinetic energy of 1000 eV. Notice that although the electron beam is accelerated by 100 V, both electrodes are held at a high enough potential so that the electrons will be driven into the ionization chamber. Some sources have magnets placed so that the electrons of the beam will cross the chamber with helical paths. This makes the effective ionization path much longer.

## Electron Ionization (EI)

Electron impact ionization sources and chemical ionization sources are similar in design. Commercial units that can be used for both are available. However, provision must be made for the relatively high pressure of reactant gas in the CI source being next to the high vacuum of the rest of the system. One way to eliminate this potential difficulty is to have a high-capacity vacuum pump to evacuate the region between the source and the mass separator. In this case, the slits also serve to separate regions of the instrument that operate at different pressures. They allow only a slow leak from the high-pressure region.

The electrons to ionize the sample are emitted from a heated filament and accelerated through electrodes to produce the 70-eV electrons (a commonly used energy). They pass through an aperture to produce a collimated beam. The electrons that do *not* react (the great majority) are collected. The resulting current of unreacted electrons is measured, and a feedback circuit is used to keep the ionizing beam current constant.

The ions that are formed from electron impacts are swept out of the electron beam, accelerated, and collimated into an ion beam in which all the ions have nearly the same kinetic energy: a monoenergetic ion beam. The acceleration potentials and the subsequent kinetic energies for the ions are in the range of kilovolts.

## Fast-atom Bombardment (FAB)

A schematic figure of a FAB source is shown in Figure 16A.2. A beam of atoms, commonly argon, with energies in the range of 3 to 10 keV impinge on the sample at an angle about 20° from a parallel to the surface. As noted previously, the interaction of the atoms with the sample ejects positive and negative ions and fragments from the sample. Depending on the accelerating voltage polarity, ions of one charge or the other are driven into the mass analyzer. The sample can be a solid, but a greater stability of the source is obtained when the solid sample can be dissolved in a solvent with a low vapor pressure. Glycerol has been found to be excellent for the purpose. The sample molecules apparently are ejected from the surface layer of the liquid.

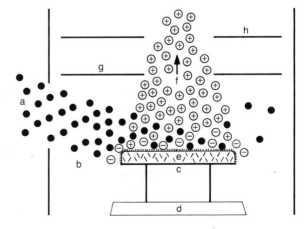

**Figure 16A.2**  Diagram of the operation of a fast-atom bombardment ion source. The atoms have kinetic energies in the range of 3 to 10 keV and arrive at the sample at an angle of about 20° from a parallel with the surface. The labeled features are (a) the atom gun; (b) the atom beam; (c) the metal sample holder; (d) the end of the probe arm used to insert the sample into the chamber; (e) the sample in a low-volatility solvent; (f) the sample's ions driven from the surface; (g) ion-extraction plate, which is shown charged so as to select the positive ions for mass analysis; and (h) ion-lens system leading to the mass analyzer. [Ref: Reprinted with permission from Barber, M.; Bordoli, R. S.; Elliott, G. J.; Sedgwick, R. D.; Tyler, A. N. *Anal. Chem.* **1982,** *54,* 645A–657A. Copyright 1982 American Chemical Society.]

The source of accelerated argon atoms operates by first ionizing argon atoms. The ions are accelerated through an electric field of 3 to 10 kV. Subsequently, the high-speed ions are neutralized to reform argon atoms which impact onto the analyte target.

### Ion Bombardment (SIMS)

A technique closely related to FAB involves ionizing and vaporizing the analyte by bombarding the sample not with atoms but with ions. In general, the results are similar to mass spectra obtained with atom bombardment. The bombarding particles are ions such as $Cs^+$ in the energy range 10–100 keV. The $Cs^+$ is called the **primary ion.** The ions that are ejected from the surface are called the **secondary ions.** Mass spectrometry of the products resulting from the ion bombardment source is called *secondary ion mass spectrometry* or *SIMS*. The masses of the secondary ions are measured. The samples on which the technique is used are generally solids. However, since the material that desorbs comes only from the top few molecular layers, the technique is used as a method of surface analysis. Qualitative elemental analysis can be done with the mass spectra alone. With carefully formulated standards, quantification of the elements at a surface can be done. If molecular species are involved, SIMS exhibits all the problems of analyzing mixtures with mass spectrometry (Section 16.7).

### SUPPLEMENT 16B

# Mass Analyzers for Mass Spectrometry

The purpose of the mass analyzer is to limit the accelerated ions that arrive at the detector to those within a narrow range of mass. Since the detector does not discriminate by mass, the resolving power of a mass spectrometer is, to a large degree, a measure of the performance of its mass analyzer. In order to increase sensitivity, the instruments are designed so that a large fraction of the ions emerging from the source slit will focus onto the detector. Ions are focused in the same sense that light is focused onto a small area. The science of designing to optimize the quality and sensitivity of mass spectra (such as by ion focusing) is called ion optics.

Most mass spectrometers contain one of four types of mass analyzers: single focusing, double focusing, quadrupole, or time of flight. These are described in order.

### Single Focusing

When a moving ion passes through a magnetic field, it is deflected from its original path. If an ion of mass $m$ and charge $z$ accelerates through a potential $V$ and passes through a homogeneous magnetic field of strength $B$, the path that the particle follows while in the magnetic field traces part of a circle with the radius

$$r = \left(\frac{1}{B}\right)\left(2V\frac{m}{z}\right)^{1/2} \tag{16B-1}$$

This property is exploited in a single-focusing mass analyzer. The detector and exit slit are fixed in position as illustrated in Figure 16B.1. Most commonly, the accelerating voltage $V$ is held constant. Therefore, for a fixed $r$ (required by the fixed geometry), scanning the masses of the ions reaching the detector depends on varying the magnetic field $B$. A mass spectrum is obtained, then, by scanning $B$. Ions with masses other than those focused on the detector are lost by collision with the surfaces of the instrument and the plates of the slit.

## Double Focusing

The voltage through which an ion is accelerated depends somewhat on its location in space in the ion beam as it is accelerated. This variation occurs because the charges of the surrounding ions also contribute to the local electric fields. The resulting spread of energies degrades the resolution of the spectrometer, because each mass is detected over a larger range of $B$ than if all were accelerated equally. (For a fixed $r$ in Eq. 16B-1,

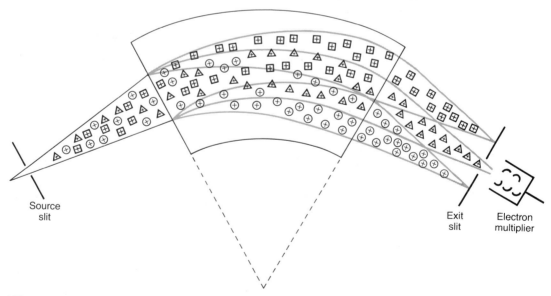

**Figure 16B.1**    Diagram of the operation of a sector-magnet mass analyzer. The ions speed through the magnetic field between the metal plates of the magnet and are deflected depending on their masses. Only a narrow range of masses can impinge on the transducer, having been selected by a precision slit. The theory and practice of ion optics shows us that if the sector angle is near 60° (the shape of the edges have some effect), and if the ions approach perpendicular to the leading edge, then ions of the same $m/z$ will focus together at a point. Thus, in addition to separation of masses, the magnet is an ion lens.

the range of $V$ means that the ions with exactly the same $m/z$ impact on the detector over a range of $B$.) To limit this spread of ion energies, the ion beam can be passed through an energy filter and a second slit. The energy filter is constructed of two electrodes that are sectors of two concentric cylinders, as illustrated in Figure 16B.2. Only ions within a narrow range of energy are passed through the filter and subsequently pass into the magnetic sector. This energy focusing constitutes the second focus in a double-focusing mass spectrometer: The ions are focused first electrostatically (for energy) and second for mass (magnetically).

### Quadrupole Mass Filter

The quadrupole mass filter operates without a magnet. The separation is done by using electric-field effects only. As shown in Figure 16B.3, the mass filter is constructed of four precisely ground metallic rods aligned parallel to each other. A DC potential is applied to the rods in the manner labeled in the figure: Diagonally opposite rods have the same potential on them. The electric field set up by this geometry is called a quadrupole (four poles), from which the mass filter gets its name.

The collimated beam of ions from the source passes through the center of the array of rods. Since the voltage along the rods is constant, there is no effect on the motion of the ions *along* the rods. Further, if only the DC-quadrupole field were present, the ions would be driven out of the center until they collided with the rods of the opposite charge.

However, the inventors of this method were clever. A radiofrequency AC potential is applied to the rods as well. When the ions are about to collide with the rod, an AC field appears that drives them away. This is done in such a way that the ions are driven in a helical path down the length of the rods. This helical path is quite special and depends on the $m/z$ ratio. Thus, with each ion's charge $z = 1$, only one mass at a time can pass through the filter to the transducer.

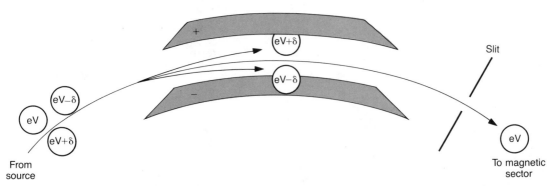

**Figure 16B.2**    Illustration of the operation of an electric energy filter. The two plates of the system are highly polished and precisely shaped to form a sector of two coaxial cylinders. Ions moving through the system will be filtered out of the beam (and impinge on some surface of the instrument) unless they fall in a narrow range of energy. The selectivity depends on the voltage across the plates, their geometry, and the placement and width of the slit. The reason that this filter improves resolution is that it narrows the range of energies of the ions that are accelerated from the source.

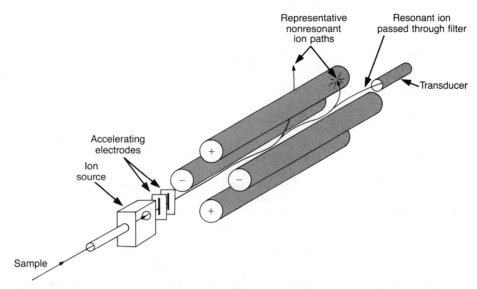

**Figure 16B.3**   Diagram of the operation of a quadrupole mass filter. The combination of AC and DC fields allows only molecular ions with a specific $m/z$ ratio to pass through to the transducer/detector. The rest are filtered out of the beam and impinge on some part of the instrument.

The helical path is $m/z$-dependent for the following reasons. In order for an ion to pass through the filter, every time it comes near an oppositely charged electrode, there must be an AC field arising to drive it away. Otherwise, it will impact and stick on the rod. Thus, there must be a coordination between the time that the AC field appears and the time when an ion arrives (traveling a circular path) near a rod surface. But the time of the ion's arrival from the region of the previous rod depends on the velocity of the ion over the fixed distance between rods. As imparted by the field, the driving *energy* of all the (monovalent) ions is the same. But since $E = \frac{1}{2}mv^2$, the velocity depends on the ion mass: $v = (2E/m)^{1/2}$. Only a small range of masses can pass through the filter at one time along the helical path. The rest collide with a rod surface or are driven outside the region of the rods.

The AC frequency is kept constant throughout the scan. How, then, is a mass spectrum scanned in a quadrupole instrument? This is done by changing the DC and AC potentials together while keeping the ratio of the potentials the same.

## Time of Flight

Time-of-flight mass spectrometers also separate masses without a magnetic field. The operation is illustrated in Figure 16B.4. It works, in essence, by measuring the differing velocities of ions accelerated through the same voltage $V$. Since $E = zV = \frac{1}{2}mv^2$, the velocity depends on the ion mass only when all the ions possess the same charge $z$:

$$v = \left(2V\frac{z}{m}\right)^{1/2} \tag{16B-2}$$

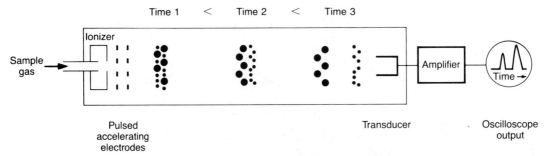

**Figure 16B.4**    Diagram of the operation of a time-of-flight mass spectrometer. The molecules or atoms are ionized in the ionizer at left and drawn out and accelerated into the delay tube by pulsing a voltage onto the acceleration grid, through which the ions pass into the delay tube. The ions (assumed to be singly charged and with different masses, as indicated by the size of the dots) separate as they pass down the evacuated tube. The detector and amplifier must operate at a high rate (and short time constant) to detect the individual masses arriving at the right end of the tube.

In operation, the ionized analytes are accelerated in short pulses about a microsecond long. After passing through a highly evacuated tube about a meter long, the ions in this pulsed sample impinge on a detector at a time that depends on the $m/z$ ratio of the ions.

The physics of this separation is straightforward. The velocity of an ion is given by Eq. 16B-2. If ions of different mass are accelerated through the potential and then allowed to pass through a highly evacuated tube of length $L$, they will not all arrive at the opposite end at the same time. After all, when $L = vt$, then

$$t = \frac{L}{(2V\,z/m)^{1/2}} \tag{16B-3}$$

A typical value of $L$ is 1 m, $V$ around 100 V, and $t$ in the range of microseconds. Separations of adjacent masses are in the range of tens of nanoseconds. The resolution of the instrument (Table 16.8) follows directly and depends critically on the length of the accelerating pulse. The reason is as follows. The minimum time a specific mass can impinge on the detector is determined by the accelerating pulse. A longer pulse produces a longer train of ions. Since the path length and transit times are fixed, the ions of different masses, which overlap more in space and time, are not separated as well. The resolution is reduced.

## Supplement Problem

S16.1    Assume that in the normal running of an analysis, there are two different masses of interest. One of the masses focuses on the detector at exactly half the magnetic field as the other in a sector instrument. One day the scanning circuitry of the magnet breaks down. The operator sets the magnetic field at a fixed value such that the lower of the two masses is detected. If the accelerating voltage is 2000 V, to what voltage should it be changed to enable detection of the higher of the two assayed masses without changing the magnetic field?

# Part VI
# SPECTROMETRY

# 17

# Introduction to Spectrometry

## 17.1 Some Similarities of Spectrometries, from Radiofrequency to $\gamma$-rays

The measurement of the **absorption** and **emission** of light by materials is called **spectrophotometry,** often shortened to **spectrometry,** which is the term used here. The terms *absorption* and *emission* have the same meaning that they have in everyday use: Absorption means *to take up,* and emission means *to give off.*

One reason to use the word *spectrometry* is to eliminate the *photo* (from the Greek *phōs,* light) since, when we use the word *light* colloquially, we usually mean light visible to our eyes. As is reviewed in Supplement 17G, visible light is only a small part of the electromagnetic spectrum, which includes (from long to short wavelengths) radio, microwave, infrared, visible, ultraviolet, X-ray, and gamma-ray radiation. The ultraviolet–visible range is also called the **optical** region of the spectrum. Sometimes, the infrared is also included as part of the optical range. You should recognize the similarities of all the spectrometric methods (Chapters 18–20) that utilize radiation in different regions of the electromagnetic spectrum. Thus, we shall use the word *light* as a general term for electromagnetic radiation simply because it is less cumbersome than *electromagnetic radiation.*

In a parallel vein, depending on their structures, specific instruments used in spectrometry are referred to as **spectrophotometers, spectroradiometers,** or **spectrometers.** The last will be used here to refer to all types.

Spectrometry can be used to determine both the elemental and the molecular content of materials. However, it is important to realize that each spectrometric method has its own benefits, drawbacks, specificities, and interferences. After you have studied each method, you should be able to recognize these four qualities.

Over the entire electromagnetic spectrum, production and detection of the radiation require different techniques. However, there are two fundamental variables that are measured in all spectrometric methods:

1. The wavelength of the radiation
2. The intensity of the radiation at that wavelength

When light energy is absorbed, it is absorbed only at certain wavelengths by a specific sample; at other wavelengths, the material is transparent. The wavelengths at which the energy is absorbed is a characteristic property of the molecules or atoms of the material that is being investigated.

When light is emitted by an atom or molecule, it does so only at its characteristic emission wavelengths. At other wavelengths, no emission occurs.

In uncomplicated cases, changes in intensity, or brightness, are proportional to changes in the amount of the material that absorbs or emits the light.

The distinction between these two properties is most important to understand. The wavelengths that interact depend on the *identity* of the compound, no matter how much there is. The intensity of the light at the wavelengths depend on the *amount* of the compound present.

In the application example in this chapter, the quoted authors discuss the use of spectrometric measurements in two regions of the electromagnetic spectrum, the visible region and the X-ray region, to answer a long-standing question. The problem is to determine whether the "blood stains" on the Shroud of Turin are, in fact, from blood, and its relationship to the history of the cloth. One of the terms used, **fluorescence,** means to emit light as a fluorescent lamp does. A more specific definition will be given later in Section 17.7 and Supplement 17E. *X-ray fluorescence,* as the name of the technique implies, uses the generation and detection of radiation in the X-ray region of the spectrum.

Hemoglobin, the oxygen-carrying molecule of blood, fluoresces with a bright red light only when iron is *not* chemically bound to it. When iron is present, there is no fluorescence. The iron is said to **quench** the fluorescence, in analogy to eliminating the glow of a fire with water.

In this example, the limitation in providing a satisfactory answer to the question lies neither in the analytical spectrometric techniques nor in the sampling. The problem lies in tracing the history of the cloth.

> The chemical composition of the "blood" provides only circumstantial evidence bearing on the question of the origin of the Shroud. A forger could have used blood to produce the stains, or a genuine shroud, containing an image only, could have been touched up with red paint by someone who was impressed but dissatisfied with the overall effect. . . . one important question that might be answered is whether the "blood" or the image was applied first . . . .
>
> Heller and Adler examined one of the . . . samples [removed with a piece of sticky tape] which contained a small, brown crystallite in addition to the "blood"-stained fibrils . . . . They reported that the crystallite and stained fibrils showed intense . . . absorption which suggest[ed] the presence of . . . hemoglobin. To test this hypothesis, they treated the samples . . . to displace the iron [that is normally in the hemoglobin]. Under . . . [ultraviolet] radiation, the treated sample[s] . . . were seen to [emit light] in the red, confirming the suspected presence of . . . blood in the "blood" areas.
>
> . . . X-ray fluorescence studies indicated iron concentrations 20–40 $\mu g$ cm$^{-2}$ above . . . background levels in the "blood" areas. Morris et al. found these numbers to be generally consistent with the expected quantities of iron in comparable blood stains; however, their measurements could not differentiate between actual blood and iron-based pigments because X-ray fluorescence detects only element concentrations without regard to molecular arrangement . . . .
>
> The evidence seems to be sufficient to conclude that the Shroud "blood" areas are blood . . . . This contradicts earlier tentative conclusions that were drawn mainly from the negative results of less sensitive tests.
>
> At this time, the most interesting unanswered question is whether the blood or the image was applied first. So far, the results of most tests have provided little information pertaining to this problem.
>
> [Ref: Schwalbe, L. A.; Rogers, R. N. *Anal. Chim. Acta* **1982**, *135,* 3–49.]

## 17.2 Energy, Wavelength, Frequency, and Temperature

As is shown in Supplement 17G, light, or more generally electromagnetic radiation, can be thought of as wavelike. This is true when we explain phenomena like the operation of a grating or prism in dispersing the visible light into its component colors (described in Supplement 17H). However, when light interacts with atoms and molecules, its description as wavelike no longer explains the interaction. The light must be considered to have the nature of particles of energy, **photons.** In this situation, we find that the frequency of electromagnetic radiation is proportional to the

**Table 17.1**   Commonly Used Measures of Wavelength and Energy

| Name of Unit | Symbol | Unit for | Used in Range of* |
|---|---|---|---|
| nanometer | nm | $\lambda$ | Ultraviolet, visible, near-infrared |
| Ångstrom† | Å | $\lambda$ | X-ray, ultraviolet-visible (in older work) |
| millimicron† | m$\mu$ | $\lambda$ | Visible (in older work) |
| micron, micrometer | $\mu$m | $\lambda$ | Infrared (common in physics literature) |
| wave number (reciprocal cm) | cm$^{-1}$, $\bar{\nu}$ | $\nu/c$ | Infrared, UV-visible (less common) |
| electron volt† | eV or ev | $E$ | X-ray, $\gamma$-ray |
| hertz | Hz | $\nu$ | Radiofrequency, microwave |
| = cycles per s | cps | $\nu$ | Radiofrequency (less common in current literature) |

\* See Figure 17G.4 for the wavelength ranges.
† 1 Å = 0.1 nm, 1 m$\mu$ = 1 nm, 1 eV = $1.602 \times 10^{-19}$ joules

energy of the photon. Quantitatively the relationship is

$$E = h\nu \tag{17-1}$$

The energy also is related to the wavelength of the photon as will be shown below. Again, do not confuse the *energy* of the photons with the *intensity* of the radiation (or how many photons s$^{-1}$) at a specific wavelength.

In practice, numerous measures of energy are used. One measure is the **erg**. If $E$ in Eq. 17-1 is measured in ergs and the frequency $\nu$ in s$^{-1}$, then the constant $h$, Planck's constant, is $6.62 \times 10^{-27}$ erg. For $E$ in joules, $h = 6.62 \times 10^{-34}$ joule s. In chemical analysis, other measures are usually used. A selection of these are listed in Table 17.1.

Since the frequency and wavelength of light are related by

$$\text{speed of light} = c = \lambda\nu,$$

then the energy of the light and the wavelength are inversely related.

$$E = \frac{hc}{\lambda} \tag{17-2a}$$

Sometimes it is convenient to label the radiation by use of the inverse of the wavelength. An example is cm$^{-1}$ (called the **wavenumber** instead of *inverse centimeters*) used in infrared spectrometry. Wavenumber is often denoted by $\bar{\nu}$. The energy is the product of $hc$ and the wavenumber, that is,

$$E = hc\bar{\nu} \tag{17-2b}$$

It is important to recognize that, regardless of the manner of expression, a unique energy can be assigned to each wavelength and frequency of light. The longer the wavelength, the lower the energy and frequency.

Another measure of energy that you know about is the temperature of an object. Temperature is a measure of the average kinetic energy. Since wavelength and frequency are related to energy, then they must be relatable to temperature as well. The

**Table 17.2** Conversion Factors of Wavelengths *in vacuo*, Frequency, and Energy

■ Shaded regions indicate reciprocal relationships. When left column units are multiplied by 10, the units across top are divided by 10.

*Examples*
1. An energy of 1 cm⁻¹ is equivalent to light with *in vacuo* wavelength of $10^7$ nm.
   An energy of 10 cm⁻¹ is equivalent to light with *in vacuo* wavelength of $10^6$ nm.
2. Light with *in vacuo* wavelength of 1 nm has an energy of $1.24 \times 10^3$ eV.
   Light with *in vacuo* wavelength of 10 nm has an energy of $1.24 \times 10^2$ eV.

| One ↓ \ Equals → | nm | Å | cm⁻¹ | eV | MHz | joule*/mol |
|---|---|---|---|---|---|---|
| nm | 1 | 10 | $10^7$ | $1.2398 \times 10^3$ | $2.9979 \times 10^{11}$ | $1.196 \times 10^8$ |
| Å | 0.1 | 1 | $10^8$ | $1.2398 \times 10^4$ | $2.9979 \times 10^{12}$ | $1.196 \times 10^9$ |
| cm⁻¹ | $10^7$ | $10^8$ | 1 | $1.2398 \times 10^{-4}$ | $2.9979 \times 10^4$ | 11.962 |
| eV | $1.2398 \times 10^3$ | $1.2398 \times 10^4$ | $8.0658 \times 10^3$ | 1 | $2.418 \times 10^8$ | $9.648 \times 10^4$ |
| MHz | $2.9979 \times 10^{11}$ | $2.9979 \times 10^{12}$ | $3.3356 \times 10^{-5}$ | $4.1355 \times 10^{-9}$ | 1 | $3.915 \times 10^{-4}$ |
| joule*/mol | $1.196 \times 10^8$ | $1.196 \times 10^9$ | $8.3591 \times 10^{-2}$ | $1.036 \times 10^{-5}$ | $2.505 \times 10^3$ | 1 |
| Spectrometric Region | UV–vis | X-ray UV | Infrared | X-ray γ-ray | Radiofrequency (e.g., NMR) | |

* 4.184 joule = 1 calorie
1 joule = 0.2390 calorie

639

relationship between energy and temperature is usually expressed as either the energy per atom,

$$E = kT \tag{17-3}$$
$$k = \text{Boltzmann's constant} = 1.380 \times 10^{-16} \text{ erg K}^{-1} \text{ atom}^{-1}$$
$$= 1.380 \times 10^{-23} \text{ joule K}^{-1} \text{ atom}^{-1}$$

or as the energy per mole of material,

$$E = RT \tag{17-4}$$
$$R = 8.3170 \times 10^7 \text{ erg K}^{-1} \text{ mol}^{-1}$$
$$= 8.3170 \text{ joule K}^{-1} \text{ mol}^{-1}$$

A list of conversion factors is shown in Table 17.2 (previous page).

---

**EXAMPLE**

Ambient temperature is around 300 K. To what energy, wavelength, and wavenumber is this equivalent?

**SOLUTION**

$$E = (1.380 \times 10^{-23} \text{ joule K}^{-1} \text{ atom}^{-1})(300 \text{ K}) = 4.14 \times 10^{-21} \text{ joule atom}^{-1}$$

$$E = \frac{hc}{\lambda} = \frac{(6.63 \times 10^{-34} \text{ joule s})(3.00 \times 10^8 \text{ m s}^{-1})}{\lambda} = 1.989 \times 10^{-25}$$

$$\lambda = \frac{1.989 \times 10^{-25}}{4.14 \times 10^{-21}} = 4.80 \times 10^{-5} \text{ m} = 0.00480 \text{ cm}$$

$$\bar{v} = \frac{1}{\lambda \text{ (cm)}} = 208 \text{ cm}^{-1}$$

---

## 17.3 The Transformations of Light Energy

In this section and most of the remainder of the chapter, we shall deal with some of the general fundamentals that apply to all spectrometries, regardless of wavelength region — namely, measuring the intensity and wavelength of light.

A short digression on energy and power is probably worthwhile here. Power is measured in quantitative spectrometry, and the power at a specific wavelength (a specific energy) is given by the product

$$P = \text{number of photons s}^{-1} \times \text{photon energy}$$

Thus power = energy/time.

In this context, let us consider the conservation of energy: Energy can neither be created nor destroyed, only changed in form. Then, over a suitable length of time, we might write a similar conservation expression for the power. This is correct in terms of measurements in spectrometry. However, it is more common to use the language

**Figure 17.1** Diagram illustrating the transformations of radiant energy by an atom, molecule, or ion. It is common nomenclature to illustrate heat **(nonradiative transitions)** as wavy arrows and radiation absorbed or emitted as straight arrows, with the sense of energy flow shown by their directions.

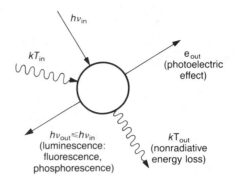

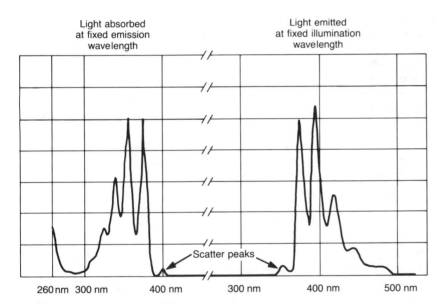

**Figure 17.2** Illustration of the absorption and reemission of radiation by molecules. The energy is absorbed with a wavelength dependence illustrated at the left. This was obtained by measuring the fluorescence reemission at a fixed wavelength while scanning the source wavelength. Absorbed energy is reemitted with a characteristic spectrum as shown on the right. This spectrum was obtained by exciting the molecules with light at a fixed wavelength and scanning the output wavelength to determine the emission spectrum. Note three points. First, the emission spectrum is at lower energy than the absorption spectrum; some energy is lost as heat. Second, spectra are approximately mirror images of each other. This is often the case. And third, the scatter peaks are the points where the wavelengths of the two monochromators coincide. In that case, light, which may be scattered from particles of dust in the liquid sample, passes directly into the detector from the source. The spectra are in the UV–visible range. Conditions: sample, 1 ppm anthracene in methanol; spectral resolution, 2.5 nm, both curves (see Figure 17.11 and Section 17.8); excitation wavelength for emission spectrum, 355.6 nm; sample monochromator setting for excitation measurement, 398.8 nm. [Redrawn from Phillips, R. E. *Spectroscopy,* Series I, Vol. II; International Scientific Communications, Inc.: Green Farms, CO, p. 338. Copyright 1974 by International Scientific Communications, Inc.]

**Table 17.3**   The Transformations of Light Energy Interacting with Atoms

| Energy In | Energy Out | Spectrometry |
|-----------|------------|--------------|
| Heat | Light | Emission |
| Light | Heat | Absorption |
| Light | Light | Phosphorescence, fluorescence |
| Light | Moving electrons | Photoelectron spectroscopies |
| Bonding energy | Light | Chemiluminescence |

of energy and describe the interactions as *per photon.* The ideas are applicable as long as we realize that the power we measure in spectrometry has time included. So let us use the language of energy per photon below.

As conservation of energy dictates, the sum of all the forms of energy entering the sample must equal the sum of all the forms of energy leaving and remaining in the material. The result of the interaction of an atom (or molecule) with its surroundings and with light can be drawn as shown in Figure 17.1 (previous page). The energy is carried into and out from the atom as light, as heat, and as the kinetic energy of any emitted particles such as electrons. Because light can be transformed into heat, the wavelength of the light that is emitted can be longer (lower energy) than the wavelength that excites the atom. This property is illustrated in Figure 17.2 (previous page).

Various names are given to the spectrometries depending on the form of energy that impinges on the atom and the form(s) that leave. These general names are listed in Table 17.3. The first two entries are *emission* and *absorption.* The third, *luminescence,* can be divided into two types: *fluorescence* and *phosphorescence;* on an experimental basis, these can be distinguished by their different light-emission behaviors. After a phosphorescent material is illuminated, the emission may persist for an appreciable time after the illumination is removed. Fluorescence emission ceases virtually immediately. Phosphorescence is usually observed in the visible range. Fluorescence is used analytically in the visible, ultraviolet, and X-ray regions.

The next to last entry of Table 17.3 is *electron emission.* This is called the *photoelectric effect* and is the basis for a number of **electron spectroscopies,** which are most important in the analysis of surfaces.

## 17.4 Quantifying the Amount of Electromagnetic Radiation

The wavelength (or energy) of electromagnetic radiation that is absorbed or emitted by the atoms or molecules of a substance is a characteristic of the atomic and molecular structure of that substance. A number of different measures (Table 17.1) are used to characterize these wavelengths. Some of these measures apply only to one type of spectrometric determination, as noted at the bottom of Table 17.2.

The amount of radiant power that is taken up or given off by the atoms or molecules is proportional to the amount of the substance present. The measures associated with the quantity of radiation are listed in Table 17.4.

**Table 17.4**   Commonly Used Measures of Light

The units $\% \ T$ and $A$ are common in chemical analysis.

| Name of Unit | Symbol | Definition |
|---|---|---|
| Photon | | Quantum of light |
| Photon energy | $E$ | $E = h\nu$ (see Eq. 17-1) |
| Radiant power | $\Phi$ | Rate of flow of radiant energy per second from a source (in watts) |
| Irradiance* | $I$ | Flow of energy per unit area per unit time |
| Transmittance† | $T$ | $P/P_0$; the ratio of the radiant power passed through a sample, $P$, to the power incident on it, $P_0$ |
| Percent transmittance† | $\% \ T$ | $100 \times T$ |
| Absorbance† | $A$ | $-\log T = -\log(P/P_0) = \log(P_0/P)$   [or $I_0/I$] |

\* Also called *intensity* (no longer recommended)
† Further discussed in Section 17.6

There are other similarities among the spectrometries. They all require the following properties.

    A.  A *source* of electromagnetic radiation must be present.

    B.  A *single wavelength* of electromagnetic radiation must be obtained from either the source or the emission of the analyte. (This will be discussed further in Section 17.8.)

    C.  A *transducer* must be present to measure changes in the amounts of electromagnetic radiation.

With these requirements met, there are three *general* types of spectrometric techniques, which differ in practice. These are *emission, absorption,* and *fluorescence* spectrometries. The differences among them are covered in Sections 17.5 – 17.7.

## 17.5 Emission Spectrometry

A generalized representation of the equipment used in emission spectrometry is shown in Figure 17.3. The emission of the sample itself provides the light, and the intensity of the light emitted is a function of the analyte concentration. The illustration shows the light originating from a luminous flame, but gas discharges, such as are seen in a neon sign, can be used as well. Some specialized emission sources are discussed in Chapter 18.

The transducer converts the light into a signal that we can relate to the concentration of the analyte. A number of the transducers for use in different wavelength regions are described in Supplement 17A. A common property of such transducers is that they respond to a range of wavelengths. For example, a transducer for the visible region responds to light of all colors.

As with all quantitative spectrometric methods, we want to isolate a "single wavelength" of light (quotes to be explained in Section 17.8) and measure the changes in intensity at that wavelength. Such light is called **monochromatic,** and the apparatus

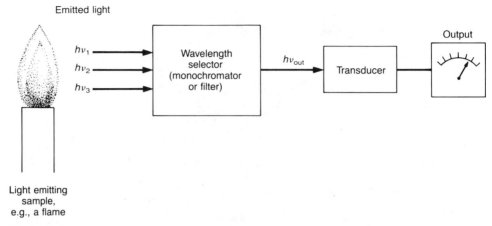

Emitted light

Light emitting
sample,
e.g., a flame

**Figure 17.3**    Generalized experimental equipment design to measure light emission in any wavelength region. The light from the source is collected and passed through a monochromator where a specific wavelength is allowed through. The light power at that wavelength is converted to an electrical signal in the transducer. The signal is amplified and displayed in any of a number of ways as the instrument output. A filter can be used in place of a monochromator if only a single wavelength is needed. The arrows show the direction of light propagation.

used to isolate a single wavelength of the spectrum is a **monochromator**.[1] Its position between the emitting sample and the transducer is shown in Figure 17.3. We then record the emission from a single wavelength at a time. The construction of one type of monochromator is shown in Supplement 17B.

Any transducer, as discussed in Chapter 3, is a converter of power to an electrical signal. A spectrophotometric transducer is no exception. The power that is being converted is that of the impinging photons. The current generated is amplified to provide the output that we relate to the amount of material assayed.

Often, as at the beginning of this chapter, the power that is measured is referred to as the "light intensity." This nomenclature is not exactly correct. As noted in Table 17.4, *intensity* more properly refers to the output of the light source. Power refers to the light that arrives at a certain place, such as at the transducer in Figure 17.3. You might expect that the intensity and power are directly proportional to each other. This is, in general, true. However, there may be problems, such as imperfections in monochromators, that can cause major errors if an exact proportion between $I$ and $P$ is assumed to be true at all wavelengths for a spectrometer. This is spectral interference, which is discussed further in Section 17.14.

## 17.6 Absorption Spectrometry

Absorption spectrometry involves measuring the proportion of light of a given wavelength that passes through a sample. The geometry of the equipment is illustrated in

---

[1] Notice that the monochro*mator* does not measure (*-meter*) the light; it selects one color.

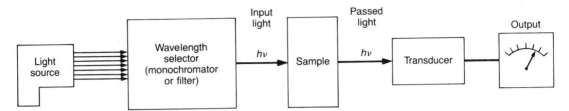

**Figure 17.4**    Generalized experimental equipment design to measure sample absorption or transmittance in any wavelength region. The source light is passed into a monochromator where a specific wavelength is allowed to pass through. The light at that wavelength passes through a sample, which absorbs a part of the power. The remaining light is converted to an electrical signal in the transducer. The signal is amplified and displayed as the instrument output. The monochromator is not needed if the light source is monochromatic (such as a specific radiofrequency or a laser light emission). An alternate design has the monochromator placed between the sample and transducer. There are benefits to each layout.

Figure 17.4. The sample (such as a colored solution) does not emit light itself, so a separate radiant source must be included. Some of these electromagnetic radiation sources are described in Supplement 17C.

Most sources produce light having unwanted wavelengths, in addition to the ones that are desired. (The exceptions to this general characteristic are radiofrequency sources and lasers.) Passing the light through either a monochromator or a filter selects the wavelength that is desired for the assay. The figure shows the monochromator between the source and the sample; however, in some instruments the monochromator is placed on the opposite side of the sample, between the sample and the transducer.

Two measurements of the amount of light absorbed are usually made. In the first, we measure the amount of light (at the chosen wavelength) that falls on the transducer when a blank is in place. Call the power falling on the transducer $P_0$. $P_0$ measures the power that falls on the transducer when the concentration of the assayed material is zero.

Measurements made when the analyte or calibration (standard) samples are in place are compared with the blank measurement. Let us call the power that is measured with the samples or standards present $P$. The comparison that we always make involves the ratio $P/P_0$, with both measured at the same wavelength. There are three terms commonly used to express this ratio (all are listed in Table 17.4).

The first is simply the ratio $P/P_0$, which is called the **transmittance**. The usual abbreviation for transmittance is an upper-case $T$.

$$T = \frac{P}{P_0} \tag{17-5}$$

The second is the **percent transmittance,**

$$\%T = T \times 100 \tag{17-6}$$

The third is the negative logarithm of $T$, which is called the **absorbance** and is abbreviated by an upper-case $A$. The algebraic expression for absorbance is

$$A = -\log \frac{P}{P_0} = \log \frac{P_0}{P} \qquad (17\text{-}7)$$

Another name, **optical density,** is also used for this quantity, but its use is now not recommended.

---

**EXAMPLE**

What values of the absorbance correspond to 100% $T$, 10% $T$, and 1% $T$?

**SOLUTION**

100% $T$, 10% $T$, and 1% $T$ correspond to transmittances of 1.00, 0.10, and 0.010. From the definition of $A$, 100% $T$ has $A = -\log 1 = 0$; 10% $T$ has $A = -\log 0.10 = 1.0$; and 1% $T$ has $A = -\log 0.010 = 2.0$

---

## Absorbance and Concentration

The absorbance of a sample is proportional to the total amount of material that absorbs the incident light. Experimentally, it can be shown that

$$A = \epsilon bc \qquad (17\text{-}8)$$

In other words, the absorbance is directly proportional to the following:

> $\epsilon$, a constant that is a property of the material itself as well as the wavelength at which the measurement is being made;
> $b$, the length of the path through which the light travels in the sample; and
> $c$, the concentration of the material that absorbs the light.

Equation 17-8 describes **Beer's law.** When the concentration $c$ is in (mol/L), then $\epsilon$ will have the units (L/mol · cm) and is called the **molar extinction coefficient,** or **molar absorptivity.** However, various scientific groups now prefer that $\epsilon$ be called only the molar absorptivity. The value of $b$, the path length, is in centimeters. Labeling the units, we obtain the equation

$$A \text{ (unitless)} = \epsilon \text{ (L/mol · cm)} \; b \text{ (cm)} \; c \text{ (mol/L)} \qquad (17\text{-}8)$$

A derivation of Eq. 17-8 is shown in Supplement 17D. Figure 17.5 shows the results of an experiment that illustrates Beer's law.

Note that $\epsilon$ depends on the measurement wavelength. Thus, the value of $\epsilon$ is often written

$$\epsilon_{\lambda(nm)}, \quad \text{such as} \quad \epsilon_{530}$$

If no units are noted, then the use of $\epsilon$ generally implies molar units. Other units that might be used for the absorptivity are, for instance, $(\mu g/L)^{-1} cm^{-1}$. The accompanying concentration unit $c$ then must be in $(\mu g/L)$.

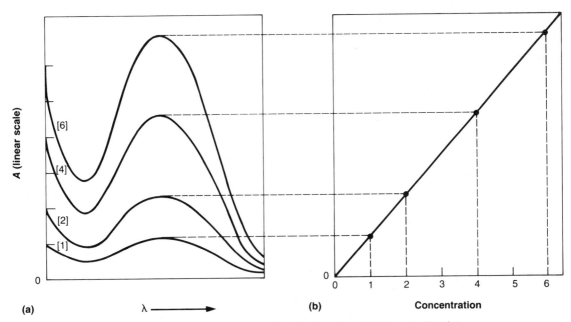

**Figure 17.5**    Illustration of the experimental basis for the Beer–Lambert law. At any fixed wavelength point on the absorbance curves (a), the absorbance is linearly proportional to the concentration (b). If a sample follows Beer's law, the working curve is linear with concentration.

## Absorbance, Concentration, and Precision

Because of the logarithmic relationship between $T$ and $A$, the precision of the measurement of concentration is not the same over the entire range of absorbance. Recall that the ratio of radiant power, the transmittance, is the directly measured quantity. Absorbance is a quantity derived from transmittance. We use absorbance in our analyses because its value is directly proportional to the concentration of the light-absorbing species. To see how the precision changes, let us *assume* that the error in measuring $T$ is 0.5% for all values of $T$ from 0 to 1. This is a relative error of 0.005 for $T$.

Refer to Figure 17.6. On the graph are noted five specific points at which are calculated the resulting error in $A$ from the fixed 0.5% error in $T$. The error values also are listed in the first two columns in Table 17.5. Perhaps the most useful way to think of these five points is as measurements taken for five different concentration standards of the same material. (Assume the relative errors in the concentrations are 0.1%: one part per thousand. As a result, any error in the concentrations can be neglected in comparison with the 0.5% $T$ instrument error.)

As you can see from the graph and the numbers in the third column of the table, the absolute error in absorbance changes with a *constant* relative error in $T$. The absolute error in $A$ decreases continuously from high to low values of $A$.

However, the error that will concern us is the *relative* error in absorbance; it is the measure of our analytical precision. The relative error in $A$ (and thus in the concen-

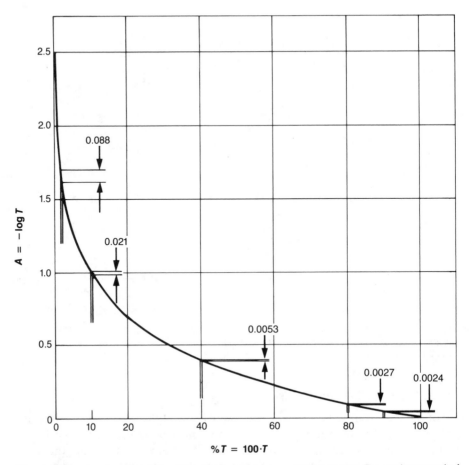

**Figure 17.6**    Graph of the relationship between absorbance and transmittance. The varying errors in *A* (labeled) result from a constant $\frac{1}{2}$% *T* error.

tration) is shown in the right-hand column of Table 17.5. These values are calculated in the usual way:

$$\text{relative error in } A = \frac{\text{absolute error}}{A}$$

The results listed in column 4 of the table are plotted versus both *A* and % *T* in Figures 17.7a and b. Note especially that at the low concentrations, 90% *T* ($A = 0.045$), and at relatively high concentrations, $<2\%\ T\ (A > 1.70)$, the relative error is greater than 5%. Clearly, under these conditions the measurement is less precise than in other ranges.

The range in which the smallest relative error in the results occurs is between absorbance values 0.4–0.7 or 20–60% *T*. In other words, to achieve the best precision in an absorption experiment, the concentration of the sample or the path length of the sample should be adjusted so that the output of the instrument is in absorbance

**Table 17.5** Errors in Absorbance with a
Fixed Error in % $T$

| % $T$ | $A$ | $\pm 0.005\ T$ Gives Abs. Error in $A$ | Relative Error in $A$ (%) |
|---|---|---|---|
| 2 | 1.70 | 0.088 | 5.3 |
| 10 | 1.00 | 0.021 | 2.1 |
| 40 | 0.398 | 0.0054 | 1.4 |
| 80 | 0.097 | 0.0027 | 2.8 |
| 90 | 0.045 | 0.0024 | 5.2 |

range 0.4–0.7 if possible. This conclusion is reached by considering only the relationship between the absorbance and transmittance with a fixed, random error in transmittance. In fact, the error in $T$ is not constant over its entire range. Each instrument design will possess a different set of specifications. However, the general trend described here still is applicable and should be considered when developing an assay.

### Other Limits to Photometric Precision

The random error associated with measurements of light is called the **photometric error.** In this section, contributions to the random error by the instrument are described. The instruments for spectrometry in various spectral regions all have sources of noise that limit their precision. (The details of these different sources of noise can be found in texts and reviews describing the construction details of each type of instrument.) Without going into detail, you can be assured that most of the noise will be associated with the radiation source and with the transducer. The electronic systems are now inexpensive and sophisticated enough so that even relatively inexpensive contemporary instruments possess amplifiers and other output stages that contribute little to such problems as long as they are in proper operating condition.

However, one often ignored but significant contribution to photometric error in all spectrometries is the lack of reproducibility of the condition and position of samples in the instrument. For instance, for numerous analytical methods, the sample is put in a special container to be placed in the instrument. (You may be familiar with **cuvettes** or sample cells used in spectrometry in the visible range.) If its placement is not reproducible, then some error will be introduced due simply to the slightly different geometry. Figure 17.8 illustrates one specific case which was investigated carefully. The cuvette repositioning error was the major limitation in precision. The problems with nonreproducible sample placement suggest that it is better to use a sample cell that is fixed and to replace the liquid or gas sample in it.

Differences in sample preparation can, as always, contribute to imprecision. Consider, for instance, a series of samples analyzed with an absorbance technique; each sample has a different amount of particulate matter stirred up in it. The particulate

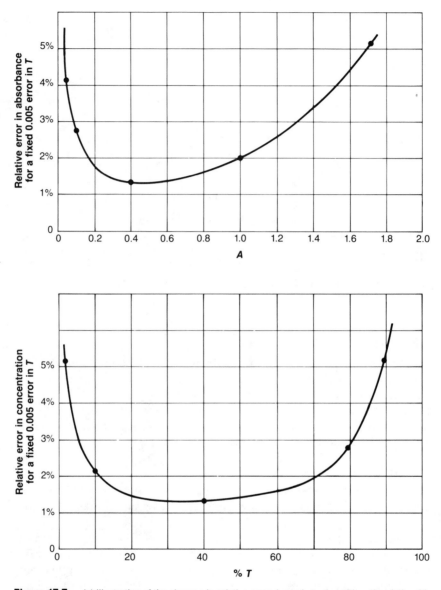

**Figure 17.7** (a) Illustration of the changes in *relative* error due only to the arithmetic relationship between absorbance and transmittance. For a fixed $\frac{1}{2}$% $T$ error, the relative error in absorbance is shown. The absorbance range from about 0.2 to 0.8 is the optimum range to minimize error in absorbances. (b) The relative error in % $T$ for a fixed $\frac{1}{2}$% $T$ error. This shows that the optimum range for determining the concentrations by transmittance measurements is between 20 and 60% $T$. Some contemporary instruments can obtain data with a precision about an order-of-magnitude better. In that case, the sample itself limits the precision more. See Figure 17.8.

**Figure 17.8** Graph of experimental results made to determine the relative error in an absorption experiment. The data were obtained in the visible range on solutions of potassium dichromate, $K_2Cr_2O_7$, in sulfuric acid. The solutions were held in standard 1.000-cm square-cross-section cells made of fused silica. The holder for the cells only allowed 15 min of angle error ($\pm 0°7.5'$), but that was sufficient for this placement to be the major source of error (squares) in the optimum absorbance range. The filled circles indicate the contribution from the next largest source of error — the inherent noise in the transducer, which was a high-quality photomultiplier (Supplement 17A). Compare this graph with that in Figure 17.7a. [Ref: Reprinted with permission from Rothman, L. D.; Crouch, S. R.; Ingle, J. D. *Anal. Chem.* **1975**, *47*, 1226–1233. Copyright 1975 American Chemical Society.]

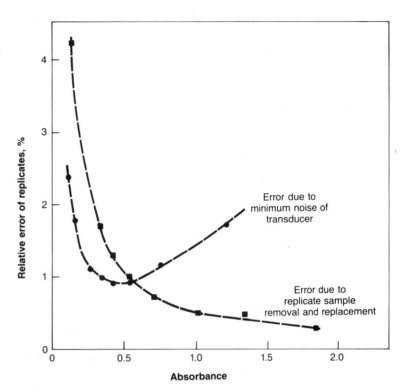

matter will block some of the light and result in a changing baseline and, possibly, ruined results. Thus, as has been mentioned before, the limitation in precision can have more to do with the care and skills of the individuals doing the analysis than with the instrument.

## 17.7 Fluorescence/Phosphorescence Spectrometry

Fluorescence spectra are obtained with the general instrumental geometry shown in Figure 17.9. As when an absorbance is measured, the sample does not emit light on its own so a light source for the appropriate energy region is required. The incident light is monochromatic after passing through the first monochromator. This light impinges on the sample, is absorbed, and then is reemitted at longer wavelengths, in all directions. This reemission of radiation is called *luminescence.* The light that comes from the sample is measured at an angle that is not in line with the source–sample axis. This angle is often, but not necessarily 90°. Figure 17.2 shows the spectra of both absorption and reemission for the same sample.

If the sample luminescence is proportional to the analyte content, it can be used to quantify the analyte. In addition, if a time-dependent spectrometric method is used, the light due to phosphorescence and that due to fluorescence may be separated. On

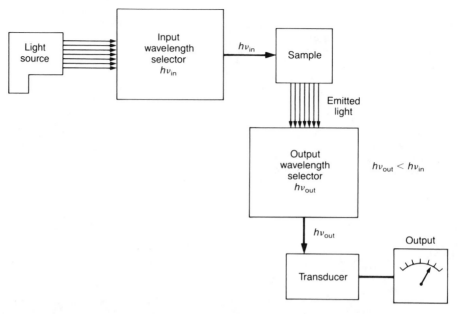

**Figure 17.9**    Generalized experimental equipment design to measure sample luminescence (fluorescence, phosphorescence) in the wavelength regions from infrared through X-ray. The source light is collected and passed through a monochromator where a specific wavelength is allowed through to impinge on the sample. The light emitted from the sample is collected and passed through a second monochromator set at a different wavelength. The light energy at that wavelength is converted to an electrical signal in the transducer. This is amplified and displayed in any of a number of ways as the instrument output. Spectra from such equipment are illustrated in Figures 17.2 and 17E.2.

The source monochromator is not needed if the light source is monochromatic (such as a specific radiofrequency or a laser light emission) with no spurious radiation. Wavelength selective filters are commonly used in place of the monochromators.

this practical basis, phosphorescent reemission generally can last a much longer time — even after the light source is no longer providing radiation to the phosphorescing sample. A more precise and correct definition of the differences between fluorescence and phosphorescence is presented in Supplement 17E. This differentiation will not concern us further here.

### Comparisons of the Methods

When does one consider using luminescence as opposed to absorption methods? Since all compounds that fluoresce also absorb light, a number of analytical samples can be analyzed either by absorption or luminescence spectrometry. If an analyte does not fluoresce strongly, the choice seems easy: Use absorption. However, the choice of method is often not straightforward since there are a number of benefits and drawbacks to consider in such a choice.

First, just as your eye finds it easier to see a light spot on a dark background than to see slight changes in brightness, so the lower limits of detection are inherently higher in luminescence spectrometry than in absorption. A problem arises as a result. Some of the incident light can be **scattered** directly from the source into the transducer from particles in the sample. It is necessary to be sure that no significant amount of light is scattered from particles in the sample (be it solid, liquid, or gaseous). This scattered light can interfere severely with the precision of measuring replicate samples. For good precision at the highest sensitivity, the samples must be carefully filtered to remove such particles. When less sensitive analyses are needed, there is a relatively easy way to separate the scattered radiation from the reemitted radiation: The wavelength of the scattered radiation will always be near that of the source, whereas the radiation that is absorbed and reemitted will be at longer wavelengths. This difference is seen in Figure 17.2.

A second basis for choice is the analyzable range of analyte concentrations. The fluorescence emission intensities can be linear with concentration for up to a factor of $10^5$ or more. A nonlinear but usable working curve can add another factor of 10 to that. In contrast, analyses depending on absorbance tend to be linear with concentration only over a factor of 10 to 100 in concentration. Luminescence has a greater linear range than absorbance spectrometry.

There is a drawback to luminescence spectrometry for more concentrated solutions. Recall that in luminescence spectrometry, the source power is absorbed by the sample and subsequently reemitted. However, the reemitted light can also be absorbed by the sample. This process is called, reasonably, **self-absorption.** See Figure

**Figure 17.10**   Hypothetical but representative fluorescence working curve. The plot is the log (power) versus log (concentration). At higher concentrations, the working curve becomes nonlinear and can even drop enough to produce a double-valued function — two different concentrations having the same output.

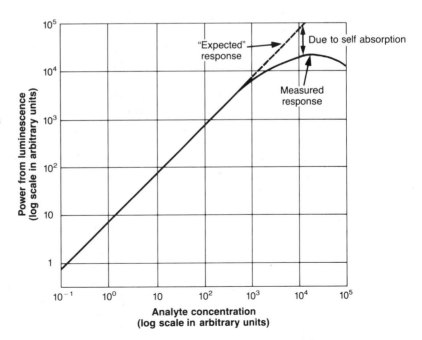

17.10. As a result of self-absorption, less light will be emitted at higher concentrations of analyte than we would expect by extrapolation of the results from lower concentrations. As illustrated, the measured intensity of the luminescence can even decrease, producing a double-valued working curve. *Double-valued* means that one intensity reading can have two different concentrations associated with it. In general, calibration curves will become nonlinear in fluorescence spectrometry at lower analyte concentrations than for absorption spectrometry. But again, the problem may simply prove an inconvenience since the sample can be diluted. On balance, luminescence is used more for trace analyses, while absorption is applied to problems with more concentrated analytes.

As opposed to fluorescence or absorption, emission spectrometry can be used only when the analyte itself emits light. There are two common cases. The first is when the sample can be heated to high temperatures (4000 – 10,000 K); a number of techniques utilizing hot samples tend to be used only for elemental analyses and are described in Chapter 18. In the second case, the origin of light emission is quite special; only a relatively few examples have been found. The energy is introduced from chemical bond rearrangements. This is **chemiluminescence.** A description of an analysis based on chemiluminescence is presented in Supplement 17F. Chemiluminescence is not a general spectrometric method but a description of the origin of the light.

## 17.8 Spectral Purity

No matter how excellently crafted a monochromator may be, or how well constructed a filter, it always passes a *range* of wavelengths. Also, atomic-emission lines have a narrow range of wavelengths, as do lasers. For all of these cases a plot of the power versus the wavelength of the light looks something like that in Figure 17.11.

A more quantitative measure of the range of wavelengths (the **spectral purity**) of light is a number called the **spectral bandwidth, spectral linewidth,** or simply **bandwidth** or **linewidth.** The measurement used, as shown in the figure, is the width of the band at a point half way between the baseline and the peak. This is called the *full width at half height* (fwhh) or the *full width at half maximum* (fwhm). Incidentally, the band envelope is not in the shape of a gaussian peak. The lineshape is closer to a shape called *lorentzian*. Compared with a gaussian, it is narrower in the middle and spreads over a wider range at the edges.

For the time being, let us consider only a monochromator that operates over a broad spectral range, for example, all visible light. For light that passes through a monochromator, with a given dispersing element, the resulting spectral bandwidth depends on the size of the **slit width.** As indicated in Figure 17.12, the slit of a monochromator is exactly that, a slit cut out of metal. Recall that the dispersion of electromagnetic radiation is accomplished by spreading the component wavelengths over a range of angles (see Supplement 17G). The exit slit, placed at some distance from the dispersing element, allows only a narrow range of wavelengths to pass through while masking the unwanted range. The slit width is usually measured in

**Figure 17.11**    Illustration of the definition of spectral bandwidth. (a) is a plot of radiant power versus wavelength. The nominal wavelength is at the peak of the power spectrum with power decreasing to both sides. (b) approximates what your eye would see as the output from a monochromator projected on a screen.

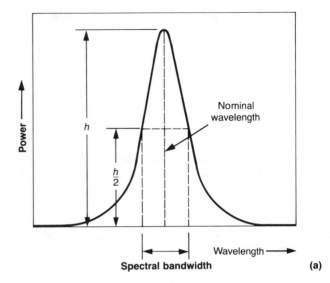

(a)

Appears to eye:

(b)

**Figure 17.12**    Illustration of the operation of a monochromator in selecting out a specific range of wavelength from a **continuum** source. The term *continuum* describes a source with light power at all wavelengths over a range. The white light enters through a slit and is dispersed by a grating. (See Supplement 17H.) A fraction of these wavelengths is selected by a second slit. The output as a plot of power versus wavelength is illustrated to the right. The spectral purity — the narrowness of the distribution of wavelength — is a function of the slits and grating and their geometry.

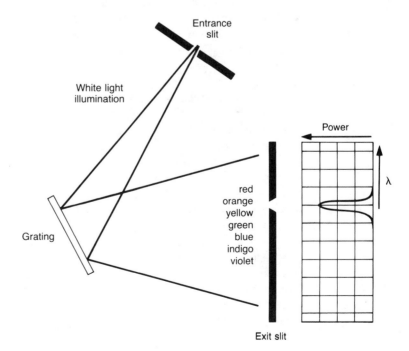

millimeters or micrometers. As described in Supplement 17H, to calculate the wavelength range requires knowing, among other quantities, the qualities of the grating (that is, the number of lines/cm), the distance from the grating to the slits, and the slit widths. To simplify things somewhat, most monochromators are calibrated by their **spectral slit width.** This is the spectral bandwidth that passes through the monochromator. For a fixed slit width, the bandwidth that passes through may depend on the wavelength of the light. Thus a statement of the spectral slit width will be similar to "0.5 nm at 350 nm."

One other point is that the amount of light — the total power — passed through the slit is greater the wider the slit. However, the spectral bandwidth is larger then as well. This trade-off always exists. When the spectral power is raised by using a wider slit, spectral purity is reduced.

## 17.9 Spectral Slit Width and the Measurement of Spectra

In spectrometry, the measurement of spectra depends intimately on both the spectral properties of the sample and the measuring properties of the instrument. These two cannot be separated, as the following illustration shows. Consider the absorption spectrum that is shown as a plot of power versus wavelength in Figure 17.13a. How is such a plot made in an actual experiment? First, there must be a radiation source that is stable in intensity. If the source is not stable, then the power measured will fluctuate rapidly, and the plot might look like Figure 17.13b. (This is true also for illumination in emission and luminescence spectroscopies.)

Assume, then, that we have managed to obtain a source with a stable intensity. As much of this light as possible is directed into a monochromator. The monochromator is designed to scan over the wavelength range of interest. The scan is generally done by rotating the grating (Supplement 17B) or prism. Light at changing wavelengths passes through the exit slit onto a transducer. (Assuming that the monochromator is scanning from longer to shorter wavelengths over time, the longer wavelengths pass through first and the power measured — followed by progressively shorter wavelengths.) If a pen moves on a piece of graph paper at a rate coordinated with the wavelength being measured at the time, the plot of Figure 17.13a will result — the graph of the spectrum.

However, we have not yet considered the spectral slit width. The spectral slit widths are indicated by pairs of parallel lines on the wavelength axis. These lines bracket the range of wavelengths that pass through the monochromator at any **nominal wavelength** setting. The nominal wavelength is at the center of the wavelength range. Let us see how the spectrum changes as the slit width is varied. We shall be especially interested in the appearance of the narrow feature.

First, consider what happens when the spectral slit width is approximately the width of the sharp feature. The spectrum will appear as shown in Figure 17.13c. Why? The reason is that the power of *all* the wavelengths passing through the monochromator is averaged; the transducer responds only to the total light power falling on it, with effectively no discrimination of wavelength. So, as the monochromator scans

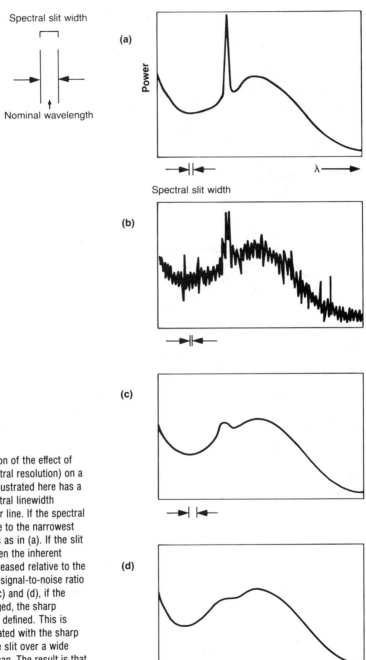

**Figure 17.13**    Illustration of the effect of spectral slit width (or spectral resolution) on a spectrum. The spectrum illustrated here has a feature with a narrow spectral linewidth superimposed on a broader line. If the spectral slit width is narrow relative to the narrowest line, the spectrum appears as in (a). If the slit is narrowed even more, then the inherent noise of the system is increased relative to the signal, as seen in (b). The signal-to-noise ratio is decreased. As seen in (c) and (d), if the spectral slit width is enlarged, the sharp feature becomes less well defined. This is because the power associated with the sharp feature passes through the slit over a wide range of the wavelength scan. The result is that sharp changes in power are smoothed out. This point is illustrated further in Figure 17.14.

through wavelength, when the *edge* of the slit reaches the spectral range of the sharp feature, the *average* power falling on the transducer rises. This is seen as a rise in the power level before the nominal wavelength reaches the base of the sharp peak. When the nominal wavelength of the sharp peak is reached, the highest point (power) of the peak appears to be reduced, since the transducer is averaging the peak power with the points of lower power on either side.

If the slit is made even wider, so that the spectral width is greater than that of the sharp feature, the sharp peak *appears* to be even broader and lower, as illustrated in Figure 17.13d. Notice that the broad peak appears relatively unchanged through all the slit-width changes.

The difference in effect on the broad and narrow features depends on the ratio of the **inherent linewidth** (or **natural linewidth**) of each spectral feature compared with the spectral slit width. As a general rule, in order to measure the spectral features precisely,

$$\text{spectral bandwidth} < (\tfrac{1}{10}) \text{ natural linewidth}$$

The reason that no effect is seen on the broad line of the emission is that this relationship is true for all three slit widths. For quantitation (as opposed to identification), the constraint can be relaxed somewhat so that

$$\frac{\text{natural linewidth}}{\text{spectral bandwidth}} \geq 5$$

As you read in the previous section, there is always a trade-off between obtaining higher power at the transducer and having a narrow spectral slit width. Although we lose power with a narrower slit, we gain resolution: Spectral features with narrow linewidths can be resolved. An example of the increased resolution achieved can be seen in the spectrum of benzene vapor with different spectral slit widths, as shown in Figure 17.14. The explanation of the observed effects are as have been related in this section. Benzene vapor is convenient and commonly used to check ultraviolet spectrophotometers for the accuracy of their wavelength scales and slit-width settings.

### Spectral Slit Width and Error in Concentration Measurements

To make an accurate spectrometric measurement of concentration, the spectral bandwidth should be small — less than 0.2 of the inherent linewidths of the features in the spectrum. The reason for this rule lies in the nonlinear relationship between absorbance and transmittance. To see this idea in a less abstract way, look at Figure 17.15. Assume that the spectral bandwidth and nominal wavelength are such that one end of the slit-width range lies on the peak of a spectrum, here 30% $T$, and the other end is at the baseline at 100% $T$. Also assume that the natural band shape is quite unnaturally linear in % $T$. Then, the average value of the power is 65% $T$, corresponding to 0.187 absorbance units. Now double the concentration of the sample. The transmittance at 100% $T$ does not change — all the light is transmitted,

**Figure 17.14** Illustration of the change in the appearance of a spectrum with changes in the spectral bandwidth. The spectra are all from the same sample of a saturated benzene vapor in the ultraviolet region 225–275 nm. This is a commonly used standard to calibrate slit widths and determine limits of resolution for analytical instruments. The slit width is fixed, and so the bandwidth changes over the wavelength range. The bandwidth values are (a) 0.5 to 1.0 nm, (b) 0.25 to 0.5 nm, and (c) 0.05 to 0.1 nm. [Ref: Reprinted from West, M. A.; Kemp, D. R. *Intl. Lab.* **1976**, May/June, 28. Copyright 1976 by International Scientific Communications, Inc.]

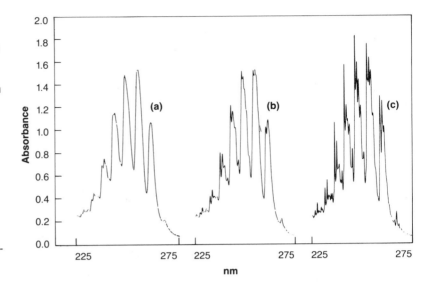

whatever the sample concentration. The transmittance at the peak becomes 9% $T$. The average transmittance that will be measured for the higher concentration is 54% $T$, corresponding to 0.263 $A$. But having measured the absorbance of 0.187, from Beer's law, we expect that doubling the concentration would double the absorbance from 0.187 to 0.374. The sample does not *appear* to follow Beer's law. However, this is because the spectrometer's spectral bandwidth is incorrectly set. A sufficiently narrow spectral bandwidth would show that at each wavelength, the spectral power versus wavelength *does* follow Beer's law. This is another example of how the sample and spectrometer are closely tied together in making an analysis. More will be said about other causes of deviations from Beer's law in Section 17.15.

## 17.10 Derivative Spectrometry

**Derivative spectrometry** refers to having the output from a spectrometer plotted as the slope of the absorbance versus wavelength (or energy). The output appears as $dA/d\lambda$ versus wavelength instead of $A$ versus wavelength. The advantage of this type of plot is that small, relatively rapid changes in the spectrum are emphasized in comparison with a background that changes more slowly over the wavelength range. Sharper features stand out from the broader background. Absorbance and derivative absorbance spectra of the same sample are shown in Figure 17.16. Note that the peak-to-peak height of the derivative spectra will be proportional to the concentration. The proof of this assertion is problem 17.10.

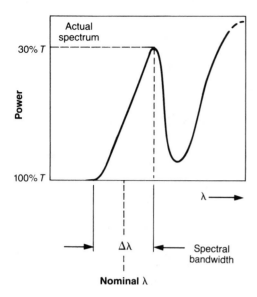

30% $T$ — Actual spectrum

Power

100% $T$

λ ——→

$\Delta\lambda$ — Spectral bandwidth

Nominal λ

**Figure 17.15** Illustration of how the choice of an instrument's spectral bandwidth can cause an apparent deviation from Beer's law. In spectrometric instruments, the transducer responds only to the *average* power over the range of wavelengths passed through a monochromator. Thus, when the transmittance changes significantly over the spectral bandwidth, the absorbance is not linear with concentration. For this reason, the slit width is kept as narrow as possible while still keeping a large S/N. Alternatively, the assay wavelength is chosen where the absorbance curve is flat with good sensitivity — usually at spectral band peaks.

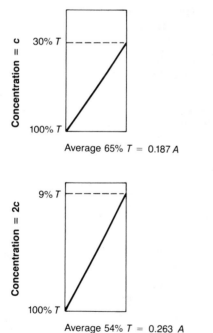

Concentration = c

30% $T$

100% $T$

Average 65% $T$ = 0.187 $A$

Concentration = 2c

9% $T$

100% $T$

Average 54% $T$ = 0.263 $A$

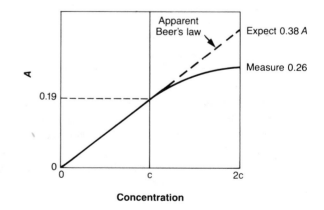

Apparent Beer's law

Expect 0.38 $A$

Measure 0.26

$A$

0.19

0

0   c   2c

Concentration

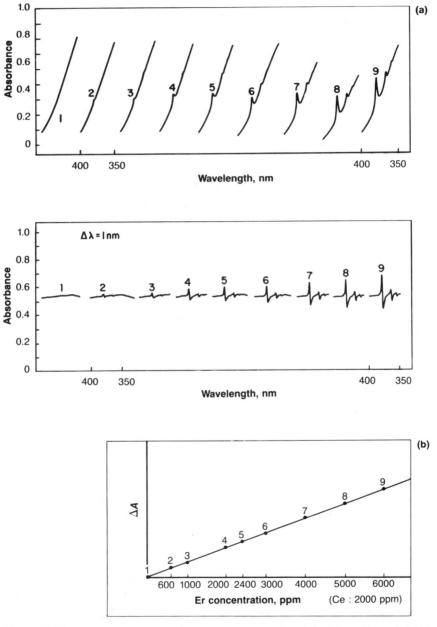

**Figure 17.16** Absorption spectrum and derivative spectrum of a solution of an erbium salt (which consists of narrow lines) in the presence of a cerium salt (which contributes a strong, broad background absorption). By taking a derivative spectrum, we suppress the broad background, so that the sharp features are accentuated. Quantitation can be done by measuring the peak-to-peak height of the derivative spectrum, as shown in the plot of $\Delta A$ versus Er concentration. The concentration of Ce is 2000 ppm and is constant. The concentrations of erbium (1)–(9) are, respectively, 0, 600, 1000, 2000, 2400, 3000, 4000, 5000, and 6000 ppm. [Ref: Reprinted with permission from Porro, T. J. *Anal. Chem.* **1972,** *44*(4), 93A–103A. Copyright 1972 American Chemical Society.]

## 17.11 Signal-to-Noise Ratios

In Section 7.2 we briefly mentioned the idea of signal-to-noise effects in measurements. Observing a dye solution is an example. There is a *lower* concentration limit, below which you cannot observe whether or not any dye is present. The actual concentration could be *any* amount less than the lowest visible concentration. Similarly, there is an *upper* concentration value, above which you cannot see through the solution. Once the concentration of a sample is above that value you cannot observe whether or not the solution is more concentrated.

These limitations are cited as examples of the problem with finding the information desired — the signal of changing concentration — in the presence of other stimuli — noise. Notice that the problem with measuring the solutions with your eyes is one of *seeing* the effects of concentration changes, not in the changes themselves. The difficulty resides as an inherent limitation of your eyes.

The same kind of problems arise in spectrometry. The noise has its origins in every part of the instrument: the illumination source, the sample, the monochromator, the transducer, and the electronic circuits for amplification. (If you ask someone who is working with a complicated instrument, you likely will hear about the phase of the moon as a source of unexplainable noise.) The noise seen on the output display is the total of the noise introduced by every part.

Look at Figures 17.13a and 17.13b. The spectra have different amounts of unwanted output fluctuations: The noise is clearly greater in spectrum b. The ratio of the output signal to the output noise is called the **signal-to-noise ratio,** often abbreviated (S/N).

One form of this measure uses the ratio of signal voltage to noise voltage. That is,

$$\frac{S}{N} \text{ (voltage units)} = \text{constant} \cdot \frac{\text{signal voltage}}{\text{noise voltage}} \qquad \textbf{(17-9)}$$

This is expressed as a simple ratio or as a number, such as, "The signal-to-noise ratio was 3" or "With a 3 : 1 signal-to-noise ratio . . . ." This is the manner in which signal-to-noise ratios are usually reported for spectra such as illustrated in Figure 17.13b.

However, defining the voltages to be placed in Eq. 17-9 is not quite as simple as naming them. Let us consider a specific case: the simple example shown in Figure 17.17. The signal, which is the difference in the voltages of the blank and sample, is easy to label. Here, as noted on the right, the signal voltage is measured between the tops of the plateaus. But how do we measure the noise voltage? Do we take the magnitude of the noise to be the differences between the highest and lowest noise voltages? This measure is called the **peak-to-peak** voltage (p – p); it is labeled in Figure 17.17a. Or do we measure the average noise voltage? In determining an average noise voltage, we deal with the random noise as if it were a sinusoidally varying potential. For a simple sine wave, the average voltage is the **root mean square** voltage. The

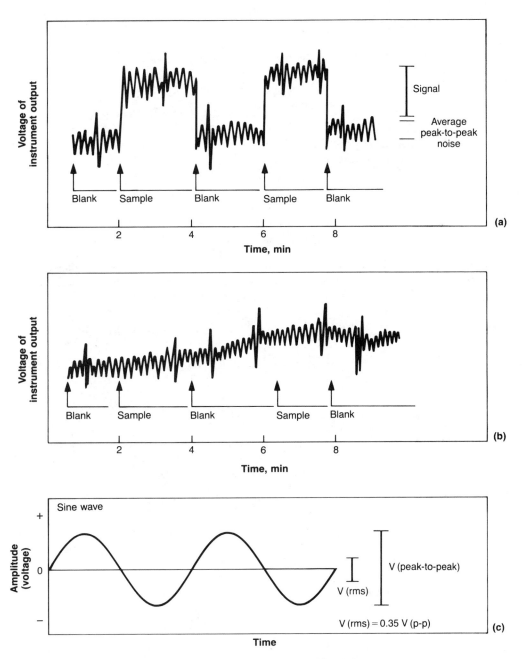

**Figure 17.17** Illustration of the concepts of signal-to-noise ratio and detection limits. Plot (a) shows an instrument output when the noisy signal changes due to a sample and blank being interchanged. The signal and average peak-to-peak levels are labeled on the right. In (b), the sample is now one tenth the concentration, and no obvious difference with the blank can be seen. There are a few spikes in the noise and some *baseline drift* present (the baseline is not strictly horizontal). The concentration of the sample is below the detection limit of the instrument under the conditions of the experiment. In (c), if the noise is treated as if it were a sine wave, then a root-mean-square (rms) noise can be defined.

numerical value of the root mean square (rms) voltage is

$$\text{rms voltage} = \frac{0.707}{2} \text{ (sine wave peak-to-peak voltage)} \qquad \text{(17-10)}$$

$$= 0.35 \ V_{\text{p-p}}$$

This relationship is illustrated in Figure 17.17c. It turns out that both rms and p–p measurements are given in the literature. Fortunately, most authors specifically state which they are using. If the signal is considered to be constant, then the two types of signal-to-noise ratios differ by the inverse of 0.707/2:

$$\text{S/N (rms)} = 2.8 \ \frac{S}{N} \text{ (p-p)} \qquad \text{(17-11)}$$

### The Power S/N Ratio

Another form of the signal-to-noise ratio uses the ratio of signal power to noise power. The unit often used to express this is the **decibel.** A decibel (abbreviated dB) is a measure of the ratio of two power measurements:

$$\text{power ratio (decibels)} = 10 \ \log_{10} \frac{P_2}{P_1} \qquad \text{(17-12)}$$

where $P_2$ and $P_1$ are two different power measurements. The S/N ratio in decibels becomes

$$\text{S/N (dB)} = 10 \ \log_{10} \left( \frac{\text{signal power}}{\text{noise power}} \right) \qquad \text{(17-13)}$$

This is related to the signal and noise voltages through the expression for electrical power,

$$P(\text{watts}) = \frac{V^2}{R}$$

The resistance $R$ for both the signal and noise is assumed to be the same, so

$$\text{S/N (dB)} = 10 \ \log_{10} \left[ \frac{(\text{signal voltage})^2}{(\text{noise voltage})^2} \right]$$

or

$$\text{S/N (dB)} = 10 \ \cdot \ 2 \ \log_{10} \left( \frac{\text{signal voltage}}{\text{noise voltage}} \right) \qquad \text{(17-14)}$$

### 17.12 The Detection Limit

Assume the sample that produced the results shown in Figure 17.17a is diluted by a factor of 10, and the same experiment is run again—alternately substituting the blank and sample in the instrument. The results that would be obtained are illustrated in Figure 17.17b. Other than by the notation along the axis, we cannot tell when the sample was measured and when the blank was measured. Clearly, that amount of material is not detectable with the level of noise in the instrument. However, we would like to have a single number that can be used to characterize our ability to detect the sample. We do this using the number called the *detection limit.* This number is either a measure of weight – such as *nanograms of mercury* — or of a concentration — such as *the detection limit is $2 \times 10^{-7}$-M cadmium.* This number is calculated as the amount of material producing a *signal* that is *2 to 3 times the rms noise level* (about equal to the p–p noise level). In general, this is an extrapolated value and usually is calculated for the best possible set of conditions with minimal interferences. Usually, the lowest limit that can actually be measured is 3 to 5 times the detection limit. The detection limit is closely related to the standard deviation of a series of measurements including blanks. The statistical relationships are described in greater detail in Supplement 18B.

## 17.13 Interferences in Quantitative Spectrometry

Noise is not the only contribution to the limit of detection or the only limit to the precision of an analysis. There are, as always to be expected, possible interferences. In spectrometry, these interferences can be classified into three types.

1. Spectral interference: The absorption or emission of other components of the matrix occurs at the same wavelength(s) used for the assay.
2. Chemical interference: The material being determined is not in the same chemical form over the determinable concentration range, or else the form varies significantly with changes in the matrix. Thus, part of the spectral power expected may not appear at the wavelength(s) chosen for the analysis.
3. Instrument interference: Excess illumination arrives at the transducer due to imperfections in the instrument.

These three problems will be discussed in the remainder of this chapter.

## 17.14 Spectral Interference

Spectral interference may be defined as absorption of radiation or luminescence at the assay wavelength(s) that interferes with the analyte quantitation. The effect results from other components in the matrix.

The interference is, somewhat arbitrarily, divided into two categories: interference from the background and interference from spectral features at or near the same wavelength. The difference between a background and a spectral feature is the rate of change of the signal magnitude with wavelength. The background varies slowly *in relation to the desired signal.* The signal from an interferent (a spectral feature) is approximately as "sharp" as the signal from the analyte. As examples, in Figure 17.13a the spike in the output might be considered a signal on the broader background; and in Figure 17.16, even though the origin of the broad absorption is known to be cerium, the signal was considered to be the peak on the sloping background. This rule for assignment of signal and background is not rigid, though. If the important part of the spectrum of Figure 17.13 is the broad part, the spike might be considered noise or just ignored. After all, if the monochromator is set at the wavelength of the maximum of the broad peak of Figure 17.13, the spike will not have any influence on the result since its contribution will be eliminated.

The origin of the spectral interferences, the problems they introduce to the assay, and the techniques used to trace and overcome them depend on the understanding of three fundamental points.

1. The selectivity and sensitivity of spectral methods depend on the isolation of a narrow wavelength band. This point was illustrated in Figure 17.13.
2. The transducer responds to the *average* of the total power of this narrow bandwidth. This point was mentioned in the discussion with reference to Figure 17.15.
3. To a good approximation, the total power detected is the sum of the illumination (or absorption) due to *all* the materials emitting (or absorbing) at the wavelength. This point is illustrated in Figure 17.18.

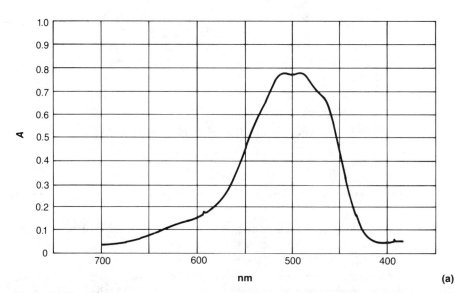

**Figure 17.18**    These are spectra of metal–edta solutions in a phosphate buffer. (a) The absorbance spectrum of a standard Co–edta solution made by diluting 20 mL of 0.150-M $CoCl_2$ to 50 mL with the edta–phosphate solution.

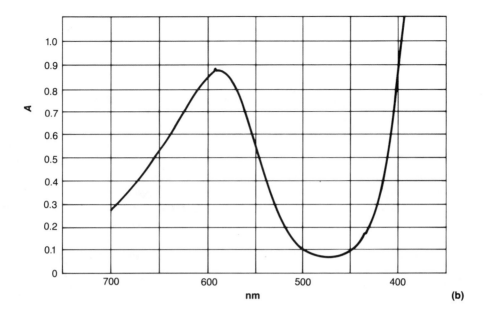

**(b)**

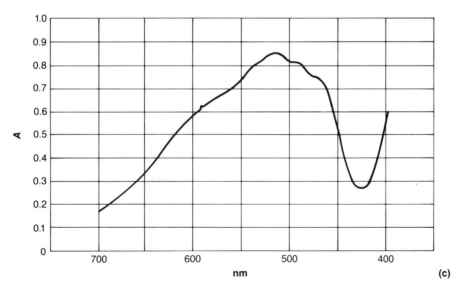

**(c)**

**Figure 17.18** *(continued)* (b) A similar solution made for nickel. For the Ni standard, 20 mL of 0.250-M NiCl₂ solution was diluted to 50 mL. (c) A mixture of the two solutions. The spectrum of the mixture is the sum of the spectra of the two components in it.

A few specific examples may clarify where these ideas fit into the elimination of spectral interferences and the development of spectrometric assays. Such techniques are applicable in all wavelength regions.

### Background Correction

The output of a spectrometer is proportional to the sum of the radiant power due to all the components at each wavelength. As you have read, the part of the signal *not* from the analyte can be considered background if it varies relatively slowly with wavelength. A spectrum of an analyte peak on a sloping background is shown in Figure 17.19. To determine the illumination due to the analyte alone, it is necessary to compensate for the background signal.

For spectrum 1, one way of compensating for the background is to take the plotted spectrum, draw a "best" baseline, and manually determine the height or area of the peak from the analyte. How could we avoid this manual work and have the instrument do it?

The instrument could be set to take three output readings and store the information electronically. The three readings would be at the wavelength labeled $B_1$, the peak position, and the wavelength $B_2$. $B_1$ and $B_2$ are equidistant from the peak in this case. The output with time would look like the plot 1 in Figure 17.19b. The instrument would store the readings at the three plateaus.

Since the background positions are equidistant from the central peak, the signal from the background *at the peak position* B is

$$B = \frac{B_1 + B_2}{2}$$

The peak signal is due to both the background B and the analyte peak output S. It is the sum of these, (B + S). Therefore, the signal due to the analyte alone is

$$S = (B + S) - B = (B + S) - \tfrac{1}{2}(B_1 + B_2)$$

Notice, in this special case, *no blank was necessary.* The instrument output due to the blank is shown in spectrum 3. In fact, note from the signal values that if the output were only measured at the peak position and corrected by the measured blank, a highly inaccurate result would have been obtained. The background during the assay is significantly higher than from the blank alone. Does this example then violate the principles presented in Chapter 6 concerning standards? Not really. As shown here, some other interferent is present in the sample that was not included in the blank. The blank was imperfect; it does not correctly account for all the components that contribute to the spectral interference.

The necessity for a separate chemical blank has been eliminated through a *spectral separation* of the contribution from the matrix components and the contribution from the analyte. Thus, by measuring the background separately, we obtain the same effect as making a correction from a correctly formulated blank.

The preceding example is a special case: There is only one isolated analyte peak on a very simple background. What if the background is not simple, so that there is a peak due to some matrix component at the wavelength chosen for $B_1$? This is illustrated in spectrum 2. The output that the instrument would measure is shown in

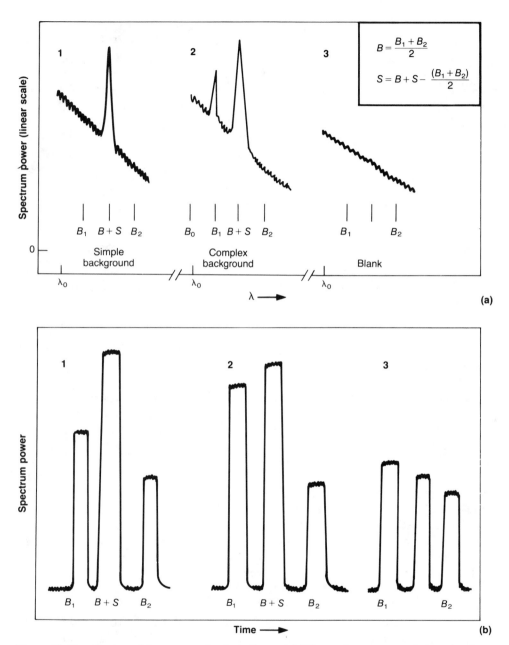

**Figure 17.19** Spectra and the compensation for background. (a) The spectra are measured at wavelengths $B_1$, $B + S$, and $B_2$. (b) The measurements as they would appear on a chart recorder. The signal from the analyte alone can be retrieved by subtracting the average background.

graph 2 of Figure 17.19b. If the instrument were programmed to do the same calculation as before, namely, that the background at the analyte peak is found to be $\frac{1}{2}(B_1 + B_2)$, the result of the analysis would be highly inaccurate. Such surprises are not uncommon. Ask any analyst, and you probably can hear a story about such problems. These are the sorts of difficulties that teach us humility in approaching a complex analysis.

The way out of this difficulty is to find two wavelengths, such as $B_0$ and $B_2$ to take the background measurements and use them to calculate B at the peak wavelength. The formula for calculating B from $B_0$ and $B_2$ differs from that for $B_1$ and $B_2$, which are equidistant from the peak. It is through such strategies that background corrections can be made so that single components can be assayed spectrometrically even in complex matrices.

### Quantify Two Species with Spectral Interference Present: Resolved Lines

The next example of correcting for spectral interference derives from an emission assay. The general technique works for absorbance and luminescence as well. Let us assume that the assay is carried out using a highly stable emission source, so that the emission intensity can be written

$$I(\lambda)_{\text{emission}} = k[\text{analyte}] \qquad (17\text{-}15)$$

The amount of light emitted $I$ at the assay wavelength $\lambda$ is linearly proportional to the amount of analyte in the samples.

We want to determine the amounts of Zn and Cu in a sample. Typical experimental results are shown in Figure 17.20. The instrument output is plotted on a linear intensity scale. Note that the intensity scales of the two wavelength regions are different. The emission lines are characterized in Table 17.6. There is clearly a spectral interference between Cu and Zn at 213.85 nm. (The emission may appear to be characterized by a relatively broad distribution of intensity. But the wavelength scale is quite spread out. For instance, compare the wavelength range of Figure 17.20 and the spectra in Figure 17.18.)

To begin the determination, a series of different concentrations of the two metal standards is run. Shown in Figure 17.21 are plots of the peak emission intensities versus the concentrations of the standards. These calibration plots are straight lines for this range of concentration. The lines are extrapolated to zero intensity at zero concentration, although the noise level will not allow precise measurements to be made in the low-concentration range.

First, let us consider how to do the determination of copper in the sample. There are two emission lines, each with a different intensity. The change in light emission of the 324.75 line with a change in copper concentration is much greater than that for the emission at 213.85. The analytical sensitivity at 324.75 is much greater. The difference in sensitivity can be found from the difference in the slopes of the calibration plots. Either of the two lines could be used to determine the concentration of copper, however. Under the conditions of the experiment,

$$I_{\text{Cu}}(324.75) = 195\, I_{\text{Cu}}(213.85) \qquad (17\text{-}16)$$

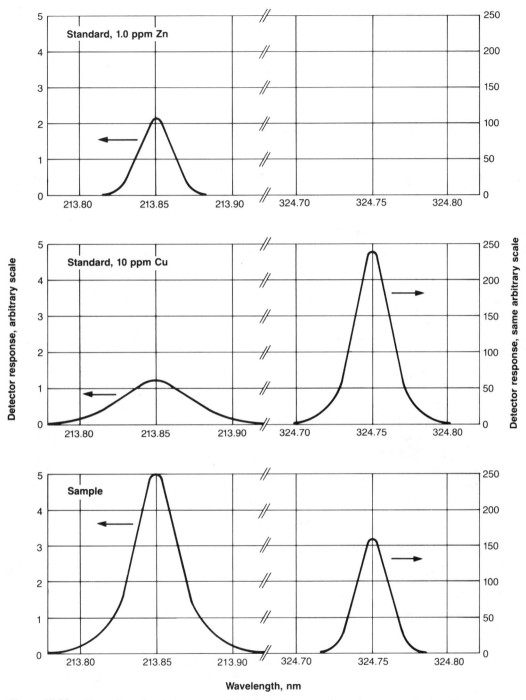

**Figure 17.20** Illustrations of emission spectra of solutions of separate zinc and copper standards and an analyte solution containing both. Note that the right and left peaks are on different scales of instrument output as indicated by the adjacent arrows. Also note that the Cu peak at 213.85 is broader then that due to zinc. The output of the instrument is linear with concentration although it is plotted on a scale of arbitrary linear magnitude. The linearity is shown in the calibration curves in Figure 17.21.

**Table 17.6**   Wavelengths Used in Zn/Cu Emission Assay

| Metal | Wavelength (intensity) | Wavelength (intensity) |
|-------|------------------------|------------------------|
| Zn | 213.85 nm (strong) | |
| Cu | 213.85 nm (weak) | 324.75 nm (strong) |

(If the experimental conditions were varied, most likely the ratio of intensities would vary.)

We choose to determine the copper content using the emission intensity at 324.75 nm for two reasons:

1. The line offers more sensitivity (195 times more).
2. There will be no interference from zinc.

Next consider the zinc determination. The calibration of the assay for zinc could be done the same way if zinc alone were present in the sample. However, since copper is also present, the assay is not as simple. There is a spectral interference from the copper: Some emission from copper appears at the same wavelength as the zinc emission, and, inevitably, the sum of the two is measured by the instrument. We know the intensity measured at 213.85 *must* be due in part to copper. How is this known? We observe emission simultaneously at the copper wavelength 324.75 nm and know that there must be some accompanying emission at the shorter 213.85-nm wavelength summing with the zinc emission. We can make a correction for this interference and thus obtain a more accurate assay of zinc. (However, the precision will not be as good as if no copper were interfering.)

The problem is to find out how much of the ultraviolet light is due to copper and how much is from zinc at the peak of the 213.85-nm line. This intensity can be expressed as the sum

$$I(213.85) = I_{Zn}(213.85) + I_{Cu}(213.85)$$

The intensity due to zinc alone must be

$$I_{Zn}(213.85) = I(213.85) - I_{Cu}(213.85) \tag{17-17}$$

To solve this equation for $I_{Zn}(213.85)$, we must find the part of the total intensity due to copper. But $I_{Cu}(213.85)$ is related to the intensity of the other, independent copper peak at 324.75 through Eq. 17-16. Thus, if we measure the intensity of the independent peak, we can calculate the intensity "underneath" the zinc peak:

$$I_{Zn}(213.85) = I(213.85) - I_{Cu}(324.75)/195 \tag{17-18}$$

**EXAMPLE**

The results of the assay for Zn and Cu in a sample are shown in Figure 17.20. What are the Cu and Zn concentrations? The calibration curves appear in Figure 17.21.

**SOLUTION**

There is some copper in the sample, since we measure emission at 324.75 nm. The peak height

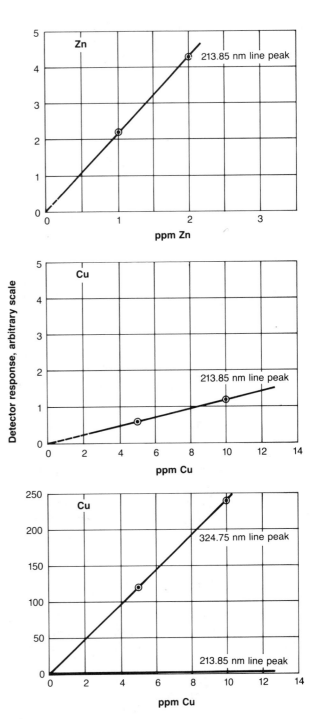

**Figure 17.21** Calibration plots of Zn and two copper peaks from the data in Figure 17.20. The calibration curve for the Cu 213.85 line is shown on both copper graphs.

is 160. On the calibration curve for copper, that peak intensity corresponds to 6.6 ppm Cu in the unknown sample.

Further, from Eq. 17-16, at 213.85 nm,

$$I_{Cu}(213.85) = \frac{I_{Cu}(324.75)}{195} = \left(\frac{160}{195}\right) = 0.82$$

Thus, from Eq. 17-17,

$$I_{Zn}(213.85) = 5.0 - 0.8 = 4.2$$

From the calibration curve for zinc in Figure 17.21, an emission intensity of 4.2 corresponds to 2.0 ppm Zn in the sample.

---

To review: To find the concentration of Zn in the presence of a Cu interference, the total emission intensities at two wavelengths were required. First, the Cu emission intensity at its primary wavelength was measured. Then, the interfering Cu intensity at the Zn wavelength could be calculated from the relative intensities at the two Cu-emission wavelengths. Finally, the Zn concentration was found from the total emission intensity at the Zn wavelength by correcting for the contribution from Cu.

For this assay for two components, Cu and Zn, the quantities that are measured are the following:

1. For the unknown sample, the instrument response at two wavelengths
2. For calibration, the relative emission at those same two wavelengths *for each component separately*

The technique used to solve for the concentrations of Cu and Zn is a general one. The generalities are discussed next.

### Spectrometric Determination of Components of Mixtures

The problem just solved is a specific example of a general analytical method. It is *not* necessary for there to be independent spectral lines for each component in a mixture. If enough information is available, all of the components of spectrally unresolved mixtures can be quantified. An example of such a mixture is shown in Figure 17.18.

The problem is mathematically equivalent to having two simultaneous equations with two unknowns. Similar calculations are done all the time in quantitative spectrometry, both in emission spectrometry as well as in absorption and fluorescence quantitation. It is important, therefore, for you to understand the manner in which such calculations are made. The algebraic calculations for absorption, emission, and luminescence methods are nearly the same.

Let us consider the calculations for absorption experiments. You will see that the molar absorptivities of absorption spectrometry are used in this calculation in the same way as the relative emission intensities for copper were used in the preceding example.

The required information is as follows:

1. For the unknown sample, the absorbance at a number of wavelengths.

The number of wavelengths must be equal to (or greater than) the number of components that interact with the light.

2. The relative molar absorptivities at those same wavelengths *for each component separately*

Let us state these requirements algebraically. For a sample containing two components, identified by subscripts 1 and 2, both of which absorb light at the same wavelength $\lambda_i$, the total absorbance is

$$A(\lambda_i) = A_1(\lambda_i) + A_2(\lambda_i) \qquad \textbf{(17-19)}$$

This says that at any wavelength the total absorbance measured is the sum of the absorbances of the two components. (This simple relationship will *not* be true of the sum of transmittances.)

We could proceed and solve this equation in the same way as above for the simultaneous emission of Zn and Cu. However, for a number of reasons, in absorbance spectrometry a different form of this equation is used: Eq. 17-19 is recast with the molar absorptivities and molar concentrations instead of with the absorbance due to each species. With slightly different labels, we use Eq. 17-8. Then, at wavelength $i$ for component number 1,

$$A_1(\lambda_i) = \epsilon_{1i}bC_1 \qquad \textbf{(17-20a)}$$

A similar expression describes the absorbance of the second component.

$$A_2(\lambda_i) = \epsilon_{2i}bC_2 \qquad \textbf{(17-20b)}$$

Substituting Eqs. 17-20 into Eq. 17-19 gives the expression for the total absorbance at wavelength $\lambda_i$:

$$A(\lambda_i) = \epsilon_{1i}bC_1 + \epsilon_{2i}bC_2 \qquad \textbf{(17-21)}$$

There is nothing different in principle between Eqs. 17-19 and 17-21.

---

**EXAMPLE**

Two components absorb at the same wavelength in a mixed solution. Two standards show that $\epsilon_{1i} = 500$ and $\epsilon_{2i} = 3000$. A mixture of the two components has $A(\lambda_i) = 0.450$. Find the concentrations of each of the components. The cell path length is 1.000 cm, and the absorptivities are in units of L $mol^{-1}cm^{-1}$.

**SOLUTION**

Substituting into Eq. 17-21 gives

$$(\text{at } \lambda_i) \qquad 0.450 = 500\, C_1 + 3000\, C_2 \qquad \textbf{(17-22)}$$

What values can the concentrations $C_1$ and $C_2$ have? It turns out that this equation can be solved with any one of an *infinite* set of $C_1$ and $C_2$ values. Thus the information given is *not* sufficient to solve the problem. We must find another equation to limit the possibilities to a unique result.

---

As you have observed, a measurement at a single wavelength will not be sufficient to find the amounts of *two* components. This is the reason that the standards and total absorption at several wavelengths must be measured. For a two-component mixture, only one more wavelength set is needed. The second equation then must relate $C_1$ to $C_2$. Experimentally, the necessary information can be obtained by redoing the same set of experiments at second wavelength $\lambda_j$ and putting the results into Eq. 17-21.

We restate the problem with sufficient information to solve it.

---

**EXAMPLE**

Two components absorb in the same wavelength range in a mixed solution. At wavelength $\lambda_i$,

$$\epsilon_{1i} = 500, \qquad \epsilon_{2i} = 3000, \qquad \text{and} \qquad A(\lambda_i) = 0.450$$

At a second wavelength $\lambda_j$,

$$\epsilon_{1j} = 2100, \qquad \epsilon_{2j} = 160, \qquad \text{and} \qquad A(\lambda_j) = 0.565$$

Find the concentration of each of the components. The cell path length is 1.000 cm, and the absorptivities are in units of L mol$^{-1}$cm$^{-1}$.

**SOLUTION**

Two components comprise the mixture. The concentration of each component can be found from the data, since the information at two wavelengths is now included.

$$\text{(at } \lambda_i) \qquad 0.450 = 500\, C_1 + 3000\, C_2 \qquad \text{(17-22)}$$
$$\text{(at } \lambda_j) \qquad 0.565 = 2100\, C_1 + 160\, C_2 \qquad \text{(17-23)}$$

There are the two simultaneous equations needed to find a unique solution for the two unknowns. Equations 17-22 and 17-23 can be solved in a straightforward way to find unique values of $C_1$ and $C_2$. The result is

$$C_1 = 2.61 \times 10^{-4} \text{ M}$$
$$C_2 = 1.07 \times 10^{-4} \text{ M}$$

---

Problems in emission and luminescence assays on mixtures are solved in a similar way. In place of the molar absorptivities and path length, the relative emission intensities (or machine-response values) for each species at each wavelength are used.

## 17.15 Chemical Interference

In absorption and luminescense spectroscopies, the wavelength is characteristic of a specific chemical species absorbing or emitting the light. The magnitude of the interaction is found to be proportional to the amount of the species in a sample, all other factors being constant. It is this property that allows us to use spectrometry to analyze a material. However, if all other factors are *not* constant, then an error could be introduced. One type of change is inadvertent variation of the concentration of the chemical species being measured. This is classified as a chemical interference. A few examples of chemical interference follow.

Consider an unexpected result in the experiment such as described in the preceding section. The molar absorptivities of two colored standard solutions are determined. However, when the two are added together to form a mixture, a chemical reaction occurs between the two, and the resulting solution becomes colorless. A spectrometric analysis of the mixture cannot be done, and the reaction is classified as a chemical interference.

Suppose that you wish to measure the concentration of a colorimetric indicator (Section 9.9) by absorption spectrometry. A solution is made by mixing the indicator with solvent, and a set of standards is made by successively diluting this solution with solvent. However, as the solution becomes diluted, the pH changes, and the indicator may change its chemical form and color. If so, Beer's law will not be followed, and a nonlinear working curve will result. Even though the indicator molecules are in solution in the light beam, the *species* in solution does not remain the same. The change in conditions has produced an error: This is a chemical interference. This problem could be overcome by making all the standards with a fixed-pH buffer with the buffer at constant concentration. It would even be unwise to start with a buffered indicator solution and then dilute it with pure solvent. Even though the pH would remain nearly the same upon dilution, the ionic strength changes.

A third example is described in more detail. This involves a chemical equilibrium in a flame. The assay is for potassium and is done by emission spectrometry. The emission is due to light given off by *atomic* potassium heated in a flame. We expect the instrument output to be linear with the concentration of the analyte.

The monochromator is set so that the light that is measured is that from atomic potassium, $K^0$. Any ionic potassium, $K^+$, in the gas phase emits at a different wavelength (and with a different intensity per potassium). The light from $K^+$ will not be passed through the monochromator to the transducer, so no evidence of $K^+$ will show up in the measurement. If $K^+$ form, the measured intensity will be less than if all the potassium were atomic.

The ionization creates no problem if the ratio $[K^+]/[K^0]$ remains constant. The light measured will still be proportional to the potassium in the sample.

$$I_{measured} = I_{K^0} \quad \text{and} \quad I_{K^0} = \text{constant} \cdot [K^0]$$

But the ratio $[K^+]/[K^0]$ is not constant with concentration. That fact is found experimentally. The changes, as measured in flame emission experiments, are shown in Figure 17.22.

The reason for the inconsistency lies in the chemistry of potassium in flames. An equilibrium exists between the ionic and atomic potassium in the gas phase,

$$K^+ + e^- \leftrightharpoons K^0 \tag{17-24}$$

Unlike an oxidation–reduction reaction in solution, a flame under constant conditions can have a finite concentration of free electrons in it. The equilibrium constant (which might look strange) is

$$K_{eq} = \frac{[K^0]}{[e^-][K^+]} \tag{17-25}$$

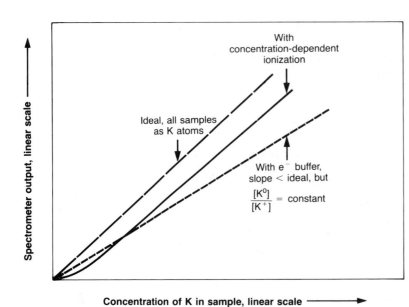

**Figure 17.22**   Representative calibration curves for an atomic potassium emission when a chemical interference changes the results. The top line illustrates "ideal" behavior, as if all the potassium of the sample were in the atomic form. The working curve below that is the type of curve that would be obtained if some of the atomic potassium were converted to the ionic form depending on the total potassium concentration. The atomic concentration apparent from the output is less than that expected in the ideal case and exhibits nonlinear behavior at the low end. The lowest curve (on the high concentration end) shows the type of behavior with an electron buffer present to keep the ratio $[K^0]/[K^+]$ constant. The calibration curve is linear but at a lower emission level than the ideal.

If potassium is the major ionizable element in the flame, then according to Eq. 17-24, the electron concentration will equal the concentration of ionic potassium. As the concentration of potassium decreases, the electron concentration decreases as well. From Eq. 17-25, you can calculate that with less potassium the equilibrium shifts to the left ($[e^-] = [K^+]$). A greater proportion of the potassium is $K^+$. As a result, there is less light emitted from atomic potassium at the lower concentrations than expected from the rest of the curve. This is classifiable as a chemical interference.

The problem can be overcome by buffering the flame with free electrons. An easily ionized material such as cesium can be added in excess and at a fixed concentration. The electron concentration is then fixed by the cesium, and the ratio $[K^+]/[K^0]$ remains constant over the entire concentration range. (Note the similarity between the use of cesium and an ionic strength buffer—Supplement 10A.)

There are a number of alternatives if there is no "chemical fix." You might run a large number of standard samples in the low-concentration range and determine a precise working curve in the nonlinear range. However, nonlinear working curves require more standards than linear ones to achieve the same precision with a consequent increase in analysis time.

You might also work only in the linear region of the working curve. The samples falling in the nonlinear, dilute region could then be concentrated. However, such a concentration step could introduce contamination and, perhaps, cause worse errors than working in the nonlinear range.

The use of internal standards might help. However, the concentrations of the spikes should be small relative to the concentration of the analyte to minimize the problem with the nonlinear response. If the analyte is already at a trace level, this approach may not work.

Another approach might be to change the assay to one using an entirely different method in which either the ions will not form or the chemical form is irrelevant. An example of an alternate method for potassium is neutron activation analysis in which the potassium is rendered radioactive through a nuclear reaction. The subsequent radioactivity is measured and related to the potassium in the sample.

## 17.16 Instrument Interference

Instrument interference can be defined as excess illumination that impinges on the transducer due to imperfections in the instrument hardware. This is also described by the phrase *stray light*. Each spectrometer design differs from others in its stray-light characteristics — even those designed for the same spectral range. Unfortunately, even the algebraic definition of stray light can differ among manufacturers.

One example of the effects of instrument interference is shown in Figure 17.23. This is a spectrum in the ultraviolet region and is a plot of absorbance. In the spectrum where the absorption of light is so great that the output should be reaching a value of $A$ far off the scale (nearly 0% $T$), the response does not reach that level. Some

**Figure 17.23**    Apparent absorption curves of saturated lithium carbonate solutions. At wavelengths less than 225 nm, the spectra suffer from the effects of stray light. Lithium carbonate solutions have a rapid rise in absorbance at around 225 nm, and this does not decrease at shorter wavelengths. The lithium solution is said to have a cutoff at 225 nm. The apparent decrease in absorbance in the region below 225 nm (to shorter wavelengths) is due to stray light. Results for two different instruments are shown. There is less stray light in instrument A in the ultraviolet region. [Ref: Reprinted from West, M. A.; Kemp, D. R. *Intl. Lab.* **1976**, May/June, 30. Copyright 1976 by International Scientific Communications, Inc.]

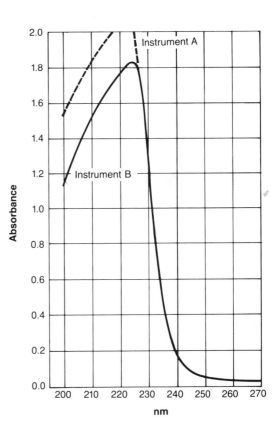

stray light that is not passing through the sample is arriving at the photomultiplier. This can be caused by dust that scatters the light or simply by light leaks in the instrument. Whatever the cause, it is independent of the transmittance of the sample.

It should be clear from the figure that the *relative* error due to the presence of stray light will be greater as less light is transmitted through the sample. This error can be calculated if the relative amount of stray light is known. Let $\zeta$ be the *ratio* of stray light to incident light at a given wavelength. Algebraically,

$$\zeta = \frac{P(\text{stray})}{P_0} \qquad \text{(17-26)}$$

Then, from the definition of transmittance,

$$T_{\text{obs}} = \frac{P + P(\text{stray})}{P_0 + P(\text{stray})} \qquad \text{(17-27)}$$

But

$$T_{\text{true}} = \frac{P}{P_0}$$

Thus,

$$T_{\text{obs}} = \frac{T_{\text{true}} + \zeta}{1 + \zeta} \qquad \text{(17-28)}$$

In order to use this equation to determine the error in concentration due to stray light, it is easiest to calculate the absorbance and, from the absorbance, the equivalent concentration error. In other words use

$$A_{\text{obs}} = -\log T_{\text{obs}} \qquad \text{and} \qquad A_{\text{true}} = -\log T_{\text{true}}$$

---

**EXAMPLE**

What is the relative error in concentration from 1% stray light for a sample with molar absorptivity $2.0 \times 10^4$ and concentration $1.20 \times 10^{-5}$ M held in a 1.000-cm cell?

**SOLUTION**

From the data given,

$$A_{\text{true}} = 2.0 \times 10^4 \cdot 1.000 \cdot 1.20 \times 10^{-5} = 0.24$$

Thus,

$$T_{\text{true}} = 10^{-0.24} = 0.57_5$$

We add in 1% (of $P_0$) stray light to find the expected $T_{\text{obs}}$ using Eq. 17.28.

$$T_{\text{obs}} = \frac{0.575 + 0.01}{1.01} = 0.58$$

Therefore,

$$A_{\text{obs}} = -\log 0.58 = 0.23_6$$

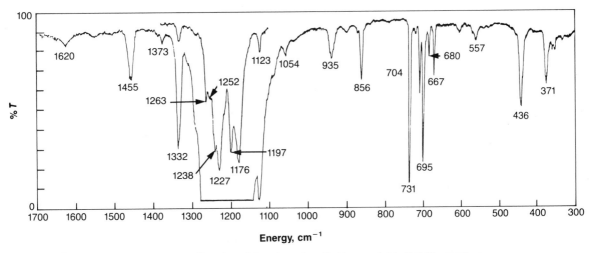

**Figure 17.24**    Illustration of stray-light effects in an infrared spectrum that is recorded in % *T*. The spectrum never reaches 0% *T* at the bottom of the plot because of infrared emission. If no emission were present, the intense band in the 1100 to 1200 cm$^{-1}$ region would coincide with the 0% *T* line where it is now squared off. The spectrum from 1100 to 1400 cm$^{-1}$ was run with two different concentrations of analyte.

which corresponds to a sample concentration of $1.18 \times 10^{-5}$ M. The relative error in concentration is

$$\frac{(1.20 - 1.18) \times 10^{-5}}{1.20 \times 10^{-5}} \times 100 = 1._7\%$$

A second example of changes in spectra with stray light is shown in Figure 17.24 in which a % *T* output was produced by a spectrometer for the infrared region. The stray light originates not from scattered light or light leaks but from infrared radiation emitted by the warm sample, sample holder, and perhaps some other parts of the instrument. (Recall we calculated that 300 K $\approx$ 200 cm$^{-1}$.) Nevertheless, conceptually and computationally the problem is treated in the same way as before.

As can be seen, the 0% *T* line of the spectrum itself does not reach the calibrated instrumental 0% *T* line. (The 0% *T* line was found by blocking off the infrared source with the sample holder absent.) This stray light can be reduced significantly by cooling the sample and the sample holder, which tend to be warmed by the incident radiation.

## Suggestions for Further Reading

Descriptions of the construction of UV–visible spectrophotometers and the limitations of the data. Electronics is of the vacuum-tube type, but this does not detract from the quality of the review.

Schilt, A. A.; Jaselskis, B. In "Treatise on Analytical Chemistry," Pt. I, Vol. 5; I. M. Kolthoff
and P. J. Elving, Eds.; Interscience: New York, 1964, Chap. 58.

This chapter describes and compares gratings and prisms for optical spectrometry.
Meehan, E. J. In "Treatise on Analytical Chemistry," Pt. I, Vol. 5; I. M. Kolthoff and P. J.
Elving, Eds.; Interscience: New York, 1964, Chap. 55.

More-advanced treatment of quantitation in spectrometry.
Bauman, R. P. "Absorption Spectroscopy"; Wiley: New York, 1962, Chap. 9.

An excellent, brief review of the effects and sources of stray light. Highly recommended.
Sharpe, M. R. "Stray Light in UV–VIS Spectrometers"; *Anal. Chem.* **1984,** *56,* 339A–356A.

A short, useful mathematical description on derivative spectrometry. Worth reading despite
its age.
Hager, R. N., Jr. "Derivative Spectroscopy with Emphasis on Trace Gas Analysis"; *Anal.
Chem.* **1973,** *45,* 1131A.

This short monograph contains details of procedures to calculate and obtain quantitative
fluorescence spectra. Some points apply to absorption as well. Outstanding and concise.
J. N. Miller, Ed. "Standards in Fluorescence Spectrometry"; Chapman and Hall: London and
New York, 1981.

A book mostly of applications of fluorescence and phosphorescence, including fluorescence
immunoassays.
S. G. Schulman, Ed. "Fluorescence and Phosphorescence Spectroscopy: Physicochemical
Principles and Practice"; Pergamon: New York, 1977.

A specialist review of fluorescence analysis written at a moderately advanced level.
Warner, I. M.; McGown, L. B. "Recent Advances in Multicomponent Fluorescence Analysis"; *CRC Crit. Rev. Anal. Chem.* **1982,** *13,* 155–222.

Two interesting review articles in the area of chemiluminescence.
Kricka, L. J.; Thorpe, H. G. "Chemiluminescent and Bioluminescent Methods in Analytical
Chemistry"; *Analyst* **1983,** *108,* 1274–1296.
Miller, J. N. "Recent Developments in Fluorescence and Chemiluminescence Analysis";
*Analyst* **1984,** *109,* 191–198.

## Problems

17.1     This exercise refers to Figure 17.5. Assume that the concentrations are in mM and the
absorbance maximum of the least concentrated solution occurs at $A = 0.120$. The compound's molecular weight is 320.4, and the cell holding the sample has a path length of
2.000 cm.
    a.   What is the value of the molar absorptivity at the band maximum?
    b.   What is the absorptivity in $\mu g^{-1}$ L cm$^{-1}$ at the band maximum?
    c.   What is the sensitivity of the assay in absorbance units mol$^{-1}$ L?

17.2     With the conditions the same as in problem 17.1, what are the values of % $T$ at the
band maxima for the four concentrations?

17.3    This exercise refers to Figure 17.10. If the output power is $1.5 \times 10^4$,
    a.  Can you tell what the concentration of the analyte is within $\pm 20\%$?
    b.  If the answer to part a is no, in what way can you decide the true value of the analyte's concentration?

17.4    What temperature is equivalent to
    a.  17,000 cm$^{-1}$?
    b.  1 eV?
    c.  1.25 Å?

17.5    From the data shown in Figure 17.18, what are the molar absorptivities of Co-edta and Ni-edta at their peak maxima?

17.6    [Figure reprinted with permission from Patton, C. J. et al., *Anal. Chem.* **1982**, *54*, 1113. Copyright 1982, American Chemical Society.]
Figure 17.6.1 shows the output from a continuous flow analyzer running at 120 samples h$^{-1}$

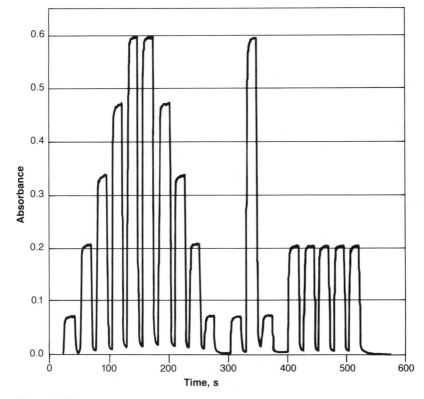

**Figure 17.6.1**

and testing for nitrite. The samples are separated by bubbles of air to prevent cross contamination. Each sample is reacted with sulfanilimide, and the nitrite present forms a colored compound which is measured as it flows through a spectrometric detector. The response for each of the test solutions rises to the measurement level and then falls off as the sample passes through.

The protocol of the determination shown here consists of standards of 2-, 6-, 10-, 14-, and 18-$\mu$M nitrite ($NO_2^-$) followed by a test of intersample contamination and then a repeatability test. What is the sensitivity of the instrument in (absorbance unit) ($\mu$g $NaNO_2$)$^{-1}$ L? Assume that the samples were 100 $\mu$L each.

17.7     On an instrument with an absorbance output, a sample has an absorbance of 0.335 at its peak when the baseline is at $A = 0.0$. Somehow, some opaque spot got onto the cuvette in the light path, and the baseline showed up at 0.120. At what absorbance reading with the peak now be for the sample?

17.8     Shown in Figure 17.8.1 are two runs of samples on an instrument with % $T$ output. The analyte is the same and the sample conditions are the same for both samples. The 0% and

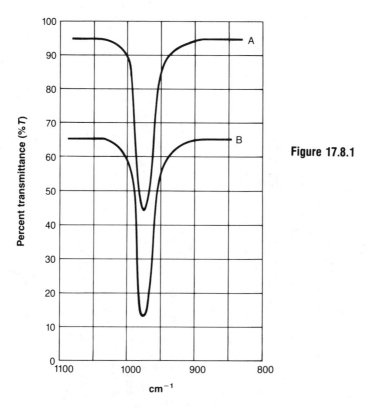

**Figure 17.8.1**

100% $T$ values were checked and found to coincide and to be correct for both runs. What are the relative concentrations of the two samples: that is, [A]/[B]?

*17.9   [This problem requires calculus.] A relative change in concentration causes a change in transmittance. This change can be written as $(1/C)(dC/dT)$, where $T$ is transmittance (not

---

* Denotes the more involved problems.

temperature). Beer's law can be written as $C = -k \ln T$. The constant $k$ includes absorptivities and cell constants. Find the value of $T$ for which the relative error $(1/C)(dC/dT)$ is a minimum.

17.10  [Requires calculus.] Show that if the absorption spectrum follows Beer's law with concentration changes, then the magnitude of the derivative spectrum will be linear with concentration as well. Assume that the spectrum can be described mathematically by a gaussian

$$A(\bar{v}) = A_0 \exp[-(\bar{v} - \bar{v}_0)^2/\sigma^2]$$

in a plot of absorbance versus energy. This is a peak centered at $\bar{v}_0$.

17.11  [Figure courtesy SLM Instruments, Inc., The American Instrument Company, Urbana, IL 61801 as in *J. Chem. Educ.* **1973**, *50*, A389.]

**Figure 17.11.1**

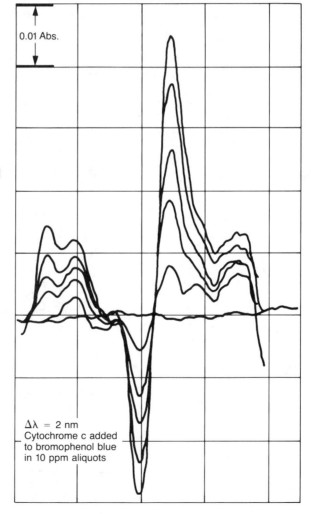

Figure 17.11.1 is a derivative absorption spectrum of the sample noted. (Peak-to-peak measures give higher precision in derivative spectrometry.)

    a.  Using the optimum wavelength positions, what is the sensitivity of the assay in absorbance units per 2 nm/ppm cytochrome C? Cytochrome C is a protein (m.w. 12,400) that acts as an electron carrier involved in photosynthesis and respiration.

    b.  Assuming that the value of $(\Delta A/\Delta\lambda)$ is linear with $\Delta\lambda$, by what factor would the sensitivity of the assay change if the value of $\Delta\lambda$ were 1 nm?

17.12   What are the concentrations, in grams of metal $L^{-1}$, of Co and Ni in the mixture that produced the spectrum of Figure 17.18c?

17.13   The following questions refer to the data shown in Figures 17.20 and 17.21.

    a.  Write out the two equations in the forms of Eqs. 17-22 and 17.23 for the data involving the quantitation of Cu and Zn that is shown in Figures 17.20 and 17.21. Use relative emission intensities, with the smallest being 1.00. Use general constants $k_i$ to account for the geometry and emissivities.

    b.  Show that the result agrees with that found by solving the problem the way illustrated in the chapter.

17.14   This exercise refers to Figure 17.19. Write the equation for the emission power at the baseline of the analyte if the background signal is sampled at the wavelengths associated with backgrounds $B_0$ and $B_2$, as shown in 2.

    Note: $[\lambda(B + S) - \lambda(B_0)] = 2[\lambda(B_2) - \lambda(B + S)]$

17.15   [Figure courtesy of United States Biochemical Corp., P.O. Box 22400, Cleveland, OH 44122.]

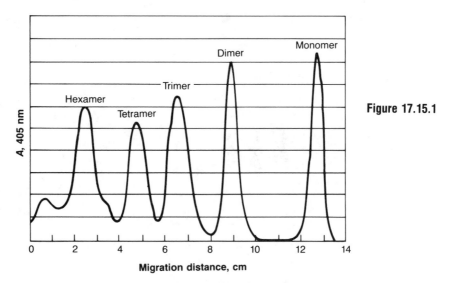

**Figure 17.15.1**

Figure 17.15.1 shows the result of a separation of cytochrome C from its di- , tri- , tetra- , and hexamers. The separation was done by placing the mixture on a gel and driving the molecules through the gel by migration in an electric field. [This is the technique called *gel electrophore-*

*sis.*] After the separation has been made, the gel is scanned at 405 nm to yield the graph of absorbance versus distance moved. The integrals under the peaks are proportional to the product of molar absorptivity and moles of substance. Upon polymerization, the light absorption properties of each monomer do not change. Thus, the molar absorptivity of the dimer is twice that of the monomer, and so on. If there is 10 $\mu$g of monomer on the gel, what mass of each of the other forms is present? Assume that the absorbance of each band would reach the baseline if it were alone. Separate the trimer and tetramer by a vertical line at the minimum absorbance.

17.16   [Ref: Coppola, E. D.; Hanna, J. G. *J. Chem. Educ.* **1976,** *53,* 322.]
Figure 17.16.1 shows the results of the preliminary data taken while developing a fluorometric

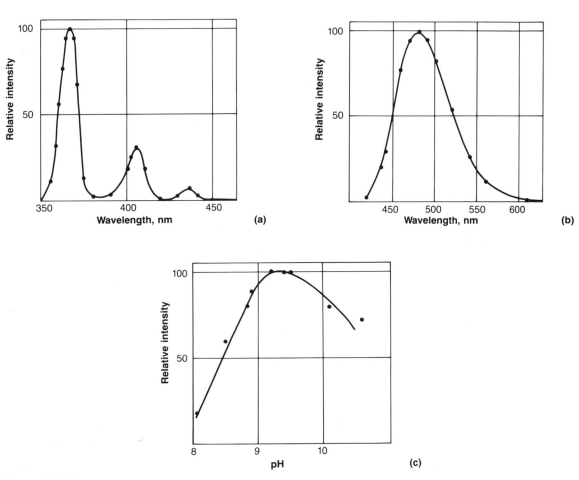

**Figure 17.16.1**

assay for the amino acid glycine. The glycine was reacted with a reagent, fluorescamine, which forms a fluorescent product with amines. The three scans shown are (a) the relative intensity of emission as the excitation wavelength was scanned—an excitation spectrum at a fixed-emission wavelength; (b) the emission spectrum found at a fixed-excitation wavelength; and (c) a plot of the relative fluorescence intensity with both excitation and emission wavelengths fixed but varying the pH. To optimize the assay's sensitivity to glycine, at what wavelengths should the excitation and emission monochromators be set, and what should the pH of the solution be?

*17.17    [Combination electrochemistry–spectrometry problem.]
[Figure reprinted with permission from W. R. Heineman et al. *Anal. Chem.* **1975**, *47*, 79. Copyright 1975, American Chemical Society.]
Spectrometry can be used to monitor the changes in redox equilibria, as shown in Figure 17.17.1. Illustrated is the spectral change in the redox protein cytochrome C as it changes from

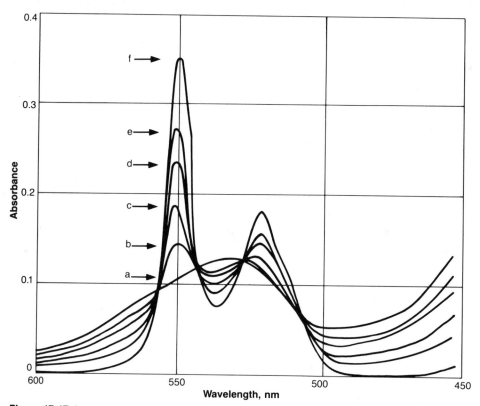

**Figure 17.17.1**

its oxidized to its reduced form. The protein is held in a thin spectral cell, and the solution potential is changed by setting the potential on partially transparent electrodes. The solution potentials are, in mV versus SCE, (a) 250.0, (b) 50.0, (c) 30.0, (d) 10.0, (e) −10.0, and (f) −250.0. The spectrum (a) has all the protein in the oxidized form. The spectrum (f) has all the protein in its reduced form. The four spectra in between are of mixtures of the two forms. The points that do not change are called **isosbestic** points. The best wavelength to probe these changes is where the changes are greatest: those near 550 nm. The absorbance of the peak is the sum of the absorbances of the two components—the oxidized and reduced forms. The two extreme peaks give you the information you need to find their spectral properties.

   a.    Let $x$ be the fraction or protein in the reduced form. Then $x = 0$ for peak (a) and

$x = 1$ for peak (f). Find the fraction of protein in the reduced form for the four middle curves.

b. From the results of part a, find the equilibrium constants for the middle four curves.

c. From the derivative of the Nernst equation $dE/d(\log K)$, the slope of a plot of $E$ versus $\log K$ will have a slope of 59 mV/$n$. Find $n$ for this reaction.

d. The point on the curve where $K = 1$ is where $E = E^{0\prime}$. Find $E^{0\prime}$ for the redox reaction.

17.18   Demonstrate that Eqs. 17-26 and 17-27 do, in fact, result in Eq. 17-28 for stray light.

17.19   Consider that a single sample was analyzed using two different spectrometric instruments. One has 1.0% stray light at the assay wavelength, the other effectively no stray light. What relative error in concentration would the 1.0% stray radiation cause if, on the better instrument, the sample had $T = 0.40$?

17.20   Under the same conditions described in problem 17.19, what is the relative error due to stray light if, on the better instrument,
   a. the sample had $T = 0.10$?
   b. The sample had $T = 0.05$?

This is another reason that, to achieve the best precision, the middle absorbance range ($A = 0.3 - 0.6$) should be used if possible.

*17.21   [Combination kinetics–spectrometry problem.]
Carboxypeptidase Y is an enzyme that aids cleavage (hydrolysis) of certain peptide bonds. The enzymatic activity can be analyzed by monitoring the production of an amino acid from a dipeptide that has the amino group at one end blocked by a bulky organic group. The amino acid that is formed produces an intense blue color when reacted with ninhydrin reagent—a reagent specific for primary amino groups. The general reactions are

$$\text{CBZ-L-phenylalanyl-L-leucine} \xrightarrow{\text{carboxypeptidase Y}} \text{CBZ-L-phenylalanine} + \text{L-leucine}$$

$$\text{L-leucine} + \text{ninhydrin} \longrightarrow \text{colored product}$$

By definition, 1 unit of carboxypeptidase Y hydrolyzes 1 $\mu$mol of benzyloxycarbonyl-L-phenylalanyl-L-leucine min$^{-1}$ at 25 °C and pH 6.5 in a phosphate buffer. In the equation above, CBZ- abbreviates the benzyloxycarbonyl group, the blocking group.
The assay is carried out as follows.

**Reagents Required**
   # 1-mM CBZ-L-phenylalanyl-L-leucine
   # 0.050-M sodium phosphate in 0.15-M NaCl solution, pH 6.5
   # Freshly made ninhydrin reagent solution consisting of an equal mixture of (a) 4% ninhydrin in 2-methoxy ethanol and (b) 0.2-M sodium citrate, pH 5.0 with 7.1-mM tin(II) chloride.
   # 50% v/v $n$-propanol in water
   # 0.05-M L-leucine
   # A 1.00-mg mL$^{-1}$ solution of the enzyme in doubly distilled water.

**Standard Curve**
Standard L-leucine solutions are made by mixing the 0.05-M L-leucine solution with the sodium phosphate buffer in the proportions listed in the table. These solutions were then diluted to exactly five times their volumes with more of the buffer solution.

| # | Buffer (mL) | 0.05-M L-leucine ($\mu$L) | Concentration ($\mu$M L-leucine)* |
|---|---|---|---|
| 1 | 4.8 | 200 | — |
| 2 | 4.9 | 100 | — |
| 3 | 4.95 | 50 | — |
| 4 | 4.98 | 20 | — |
| 5 | 5.0 | 0 | 0 |

* After dilution

### Enzyme Solution

Prepare the enzyme solution by using 10.0 $\mu$L of the 1-mg mL$^{-1}$ stock solution and bringing the volume(s) to 1.00 mL with distilled water.

### Procedure

Pipette 1.00 mL of the substrate solution into a test tube labeled A (or Assay) and another labeled B (or Blank). Pipette 1.00 mL of each of the standard L-leucine solutions into tubes labeled with their concentrations. Thermally equilibrate the tubes A and B for 10 min at 25 °C. Begin the enzyme reaction by pipetting 50 $\mu$L of the enzyme solution into A and 50 $\mu$L of water into tube B. Allow the reaction to proceed for 10 min at 15 °C. Then add 1.0 mL of the ninhydrin reagent to each of the seven tubes. Place all the tubes into a boiling water bath for 15 min. This quenches the enzyme reaction and develops the ninhydrin color. Remove the tubes from the bath and allow them to cool to below 30 °C, and then add 5.00 mL of the 50% propanol solution and mix well. Read the absorbance of all the solutions at 570 nm using a 1.00-cm-path-length cell.

The results obtained in one batch of the enzyme were as tabulated.

| Label | Absorbance |
|---|---|
| 1 | 0.488 |
| 2 | 0.238 |
| 3 | 0.117 |
| 4 | 0.051 |
| 5 (blank) | 0.010 |
| A | 0.447 |
| B (blank) | 0.026 |

Answer the following questions.

a. What are the concentrations of the standard L-leucine solutions?

b. What is the response of the assay for leucine in $\mu$M (absorbance unit)$^{-1}$? Use all the standards to determine the value.

c. What concentration of L-leucine equivalent is in the enzyme blank B?

d. What is the total number of micrograms of enzyme in the test solution A when it is reacting?

e. How many units mg$^{-1}$ are there in this enzyme preparation?

f. The L-leucine standard solutions contain 50 $\mu$L less volume than do the enzyme and enzyme blank solutions. Assuming a linear (in time) production of substrate, what error in the 10-min reaction time will be equivalent to the change in absorbance from this volume difference? Is it worth accounting for the difference?

g. What percentage of the substrate is used up in the 10-min reaction run? Is this a pseudo first-order reaction?

h. To what class of kinetic analysis does this one belong?

## SUPPLEMENT 17A

# Representative Transducers for Electromagnetic Radiation

In the radiofrequency and microwave region, the electromagnetic radiation is converted to a DC current by rectification. A number of circuits are made that will rectify the current, all of which contain diodes that have a low resistance to current flow in one direction (forward) and a high resistance to current flow in the other (reverse). If a single rectifier is used, the resulting currents appear as illustrated in Figure 17A.1.

In the far-infrared to mid-infrared regions, changes upon heating and cooling of a small mass are the basis of a number of transducers. Their operation is based on the measurement of some electrical characteristic of the transducer that varies with temperature. A **thermocouple** is a junction of two wires — each composed of a different metal or metal mixture — which produces a voltage that depends on the junction temperature. For optimum response, the thermocouple junction is blackened. When infrared radiation falls on the junction, the change in temperature will show up as a change in potential. Thermocouple detectors are one type of a general class called **bolometers** (from the Greek word meaning *power*).

Two other bolometers that respond to infrared radiation are **thermistors** and **pyroelectric crystals.** Thermistors are constructed from either a thin metal conductor or a semiconductor, both of which have a DC resistance that changes with temperature. The changes in resistance are measured and amplified.

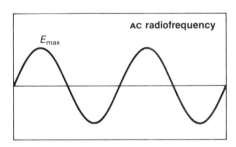

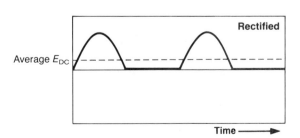

**Figure 17A.1**    Rectification converts AC current into DC in which the current is allowed to flow in one direction only. The AC current is induced in a wire by the radiofrequency radiation. In the microwave region, the radiation falls directly onto a semiconductor diode.

A number of contemporary IR instruments use pyroelectric detectors. A crystal that is pyroelectric generates an electrical charge across it when heated. One surface of the crystal becomes positive and the opposite one negative. To make a transducer, a minute capacitor is fabricated with a triglycine sulfate (TGS) crystal as the filling of the capacitor "sandwich." As the impinging IR radiation heats the TGS crystal, a voltage appears across the capacitor. The greater the heating effect, the greater the voltage.

In the near-infrared region, semiconductor detectors are used. These detectors are equivalent to light-sensitive resistors. The conductivity is low in the dark and high when exposed to light. Cadmium sulfide, lead sulfide, and indium antimonide are three materials that exhibit this effect. Semiconductors of silicon respond from the infrared into the middle of the visible region and can be used into the visible range.

However, in the visible region, the detectors of choice are **photoelectric** detectors. These detectors utilize the photoelectric effect in their operation. The photoelectric effect involves the ejection of an electron by a material when irradiated by light (one

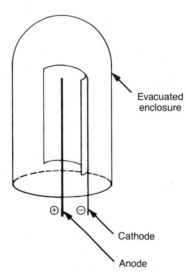

**Figure 17A.2**    Construction of a simple phototube with an opaque photocathode.

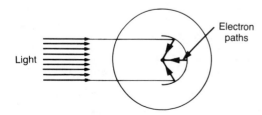

**Figure 17A.3**   Response of a representative phototube to light of various wavelengths. The output scale is logarithmic in amperes/watt. Amperes measure the electron current whereas watts measure the photon power. The label *quantum efficiency* means the average fraction (or percentage) of incoming photons that produces photoelectrons from the photocathode. [Ref: Gumm, F. In "Physical Methods of Chemistry," Vol. 1, Pt. IIIB; A. Weissberger and B. W. Rossiter, Eds.; Wiley–Interscience: New York, 1972. Copyright © 1972. Reprinted by permission of John Wiley & Sons, Inc.]

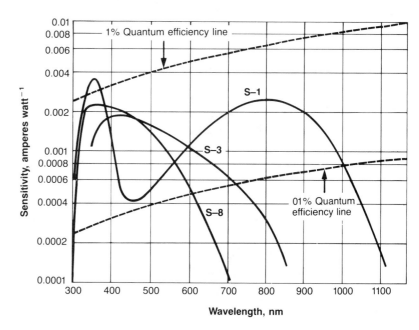

of the processes illustrated in Figure 17.1). The simplest such transducer is a **phototube.** The structure of a phototube is diagrammed in Figure 17A.2.

The apparatus is encased in an evacuated glass tube. The cathode is a half-cylinder of metal coated on the inside surface with successive layers of silver, alkali metal, and alkali metal oxide or other counter-ion. Examples are KNaCsSb and $Cs_3SbO$. The anode is a positively charged wire at the cylinder axis. The potential across the tube usually ranges around 100 V. When light hits the surface of the cathode, electrons are expelled, cross the gap, and are collected at the anode. This shows up as an electrical current. The current is in the $\mu A$ range for phototubes such as illustrated in Figure 17A.3.

Greater sensitivity can be achieved by constructing a **photomultiplier tube (PMT).** Figure 17A.4 illustrates the operation of a photomultiplier. The photocathode electrons are focused onto a second metal surface (a dynode) coated with BeO, GaP, or CsSb. The electrons, which are accelerated to around 100 eV by the potential, hit the dynode and produce more electrons (secondary electrons), which are accelerated through 100 V to a second dynode, and so on. This operation produces increases in sensitivity up to $10^8$ compared with a simple phototube. Typical responses are shown in Figure 17A.4. However, the photomultiplier also amplifies the noise. The greatest contribution to this noise is the random ejection of electrons from the photocathode —called **shot noise**— and is the limiting factor in the sensitivity of photomultipliers.

One final point is worth noting. *All* transducers have output responses that are somewhat wavelength dependent as seen in Figures 17A.3 and 17A.4.

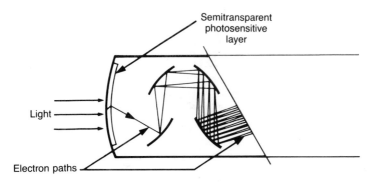

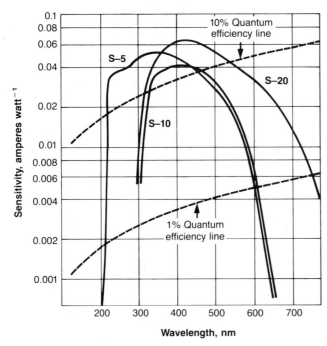

**Figure 17A.4**     One form of photomultiplier. This type has a semitransparent photocathode. The light impinges from its "back," and electrons are ejected from the "front" side. The photoelectrons are amplified by generating more electrons on each dynode. The response of representative photomultipliers is also shown with the same scales as in Figure 17A.3. [Ref: Gumm, F. In "Physical Methods of Chemistry," Vol. 1, Pt. IIIB; A. Weissberger and B. W. Rossiter, Eds.; Wiley-Interscience: New York, 1972. Copyright © 1972. Reprinted by permission of John Wiley & Sons, Inc.]

## SUPPLEMENT 17B

# A Grating Monochromator

The monochromator described in this supplement utilizes a reflection grating (Supplement 17H) as its dispersing element. It is used in the optical-wavelength region. The arrangement of the slits, mirrors, and a grating, illustrated in Figure 17B.1, is called a Czerny–Turner mounting. The light path is as follows. The light enters

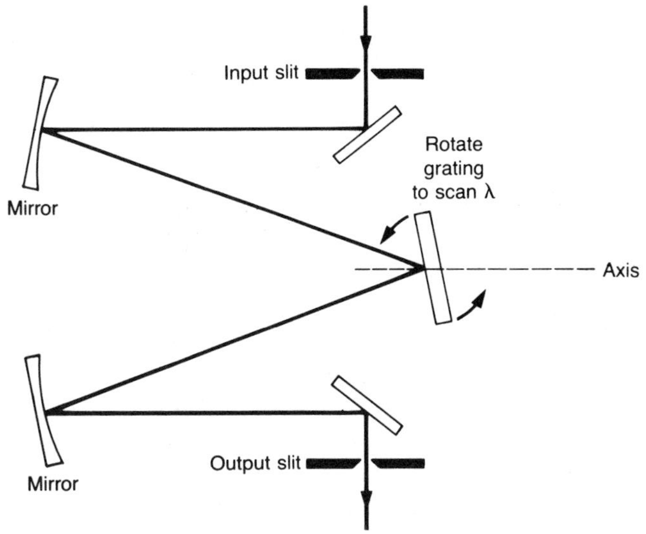

**Figure 17B.1**    General design of a Czerny–Turner monochromator. The wavelength is scanned by rotating the grating about the axis indicated.

through slit 1, reflects off a fixed mirror to a concave mirror, which collimates and focuses the light onto the flat, reflection grating. The diffracted light is collimated and focused by a second concave mirror, is reflected off a flat mirror, and leaves through slit 2. The wavelength is scanned by rotating the grating about its center as indicated.

The output spectral purity depends on the widths of both slits, the geometric relationship between the slits and the grating, the angle of the grating to the axis, and the closeness of the grating lines. Some of the best gratings for spectrometers are **holographic gratings** made with the aid of light-interference patterns; these have fewer imperfections and can be made in special shapes. The benefits are, respectively, that less stray light arises from grating imperfections and that the holographic grating can replace a number of optical elements, thus simplifying the optical design of monochromators.

Two other monochromator designs are illustrated in other figures. Figure 18.22 shows a schematic figure of an **Ebert** mounting. A single mirror replaces the two focusing mirrors of the Czerny–Turner mounting. In Figure 20B.2 (in Supplement 20B) is shown a **double monochromator,** which has two gratings. The light that is dispersed by the first grating is passed through a middle slit and then again dispersed by the second grating. The dispersion is far higher: Thus the resolution of a double monochromator is greater.

## SUPPLEMENT 17C

# Continuum Sources of Electromagnetic Radiation: 190 nm to 50 $\mu$m

The range of electromagnetic radiation used in chemical analysis ranges from the radiofrequency of nuclear magnetic resonance to the gamma-radiation of radioactivity. The source of the radiation can be the analyte itself or an outside source. The use of emission from the analyte itself is most common in atomic emission spectrometry, and such sources are described in Chapter 18. In this supplement are described the most common continuum sources of radiation in the ultraviolet through infrared regions. Continuum sources are those that emit over a broad range of wavelength. On the other hand, sources that produce radiation over only fixed narrow ranges, are called line sources.

More and more, lasers are becoming the spectrometric light sources of choice because of their high emitted power and narrow bandwidth. However, at this time, lasers cannot routinely be scanned over a broad enough wavelength range to replace continuum sources and monochromators in spectrometers. In addition, lasers still are somewhat noisier, and their power output tends to drift more than the continuum sources described here. It is entirely likely that all these problems will be overcome in the future.

### The Ultraviolet Region, 190 to 320 nm: Deuterium, Xenon, and Mercury Arcs

For absorption spectrometry, an enclosed arc lamp filled with deuterium ($^2$H)$_2$ produces a continuum spectrum with some line emissions superimposed. Its output

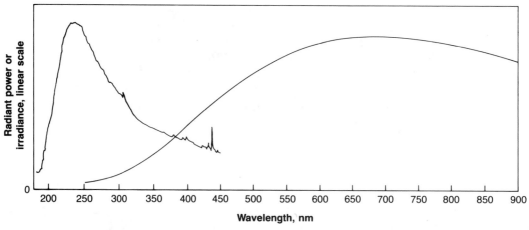

**Figure 17C.1**    Radiant-power spectrum of a deuterium arc lamp (left) and an incandescent source such as a tungsten filament lamp (right) with a temperature of 4200 K.

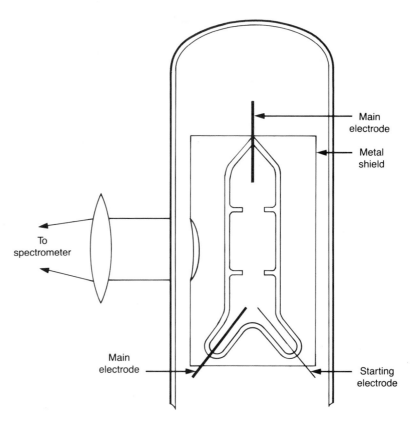

**Figure 17C.2** General design of an arc lamp. The un-ionized gases that fill the inner tube have a high resistance. An arc is struck between the closely spaced starting and main electrodes. The resistance of the resulting plasma is far lower than the un-ionized gas, and the arc is sustained with voltages between 40 and 300 V DC between the two main electrodes. The current through the low-resistance plasma is limited by an external inductor.

spectrum is illustrated in Figure 17C.1. The emission intensity is adequate to carry out absorption spectrometry in the range from 190 through about 320 nm. Intensity below 190 nm is not required in routine work since air itself absorbs in that range. The general structure of an arc lamp is illustrated in Figure 17C.2.

For fluorescence measurements, since the fluorescence emission is proportional to the source power, a higher intensity lamp can be useful. To produce this higher power, arc lamps that are filled with xenon or mercury vapor at high pressures can be used. The arc is excited by a carefully regulated steady DC current passing through it. There are two limitations to the power that can be used. One is that as the power increases, the lamp and its housing must be larger and do not fit conveniently in a benchtop spectrometer. Another is that a fluorescent sample will often be decomposed by light, and if the power is too high, the sample will decompose during the measurement time.

### The Visible Region, 320 to ~750 nm: Tungsten Filament Lamps

An incandescent lamp is the common choice source for absorption spectrometry in the visible region. Its radiation spectrum is a broad continuum ranging from the ultraviolet through the near infrared (to about 4000 cm$^{-1}$ = 2.5 μm wavelength). The power spectrum is illustrated in Figure 17C.1 for a bulb with a filament temperature of 4200 K.

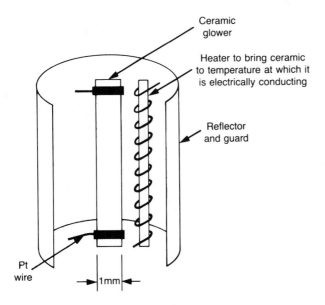

**Figure 17C.3**    (a) The general design of a Nernst glower. The ceramic is heated by the heater until it comes to a temperature that allows it to be electrically conducting. (b) A logarithmic plot of the radiant-power spectrum of a Nernst glower at 2100 K. [Ref: Olsen, E. D. "Modern Optical Methods of Analysis"; McGraw–Hill: New York, 1975. Reprinted with permission.]

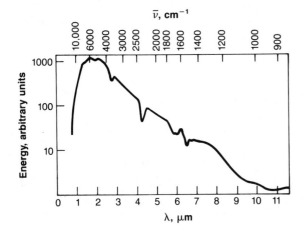

### The Infrared Region, 2.5 to 50 $\mu$m: Nernst Glower

Glass and silica are opaque to the radiation in the infrared region. As a result, glass-enclosed sources no longer can be used. The most commonly used source of radiation for infrared spectrometers is the Nernst glower, a heated rod about 1 mm in diameter and a few centimeters long. The material in the rod is called yttria-stabilized zirconia, a ceramic material, composed of an oxide of zirconium with some yttrium in it. This material is used because, at its operating temperature of around 1800 K, it is electrically conducting. Its ceramic properties make it relatively stable at high temperatures; it resists burning out in air. Its conducting properties make the ceramic easy to heat by simply passing a current through it. Because of the higher operating temperature, the wires connecting it to the current source are made of platinum (m.p. 2046 K). A diagram of the construction and the emission spectrum of a Nernst glower are shown in Figure 17C.3.

## SUPPLEMENT 17D

# Derivation of Beer's Law

Beer's law, as stated algebraically in Eq. 17-8, had its origins in experiment. Its currently accepted form was postulated in 1852, well before contemporary molecular spectroscopic theory or equipment was available. Three quantities are involved in the effect described by the equation.

1.  The amount of light incident on the sample and passed through it
2.  The thickness of the sample
3.  The amount of material in the sample: its concentration

The general form of the Beer–Lambert law (usually called Beer's law) can be rationalized as shown in Figure 17D.1a. We start with a thin slab of material of thickness $\Delta x$ with light of power $P_0$ incident on it. Some fraction of the incident light is passed through the slab. No matter what the conditions or molecular mechanism of light attenuation, the amount of light that passes through, $P_1$, can be expressed as a number, $k$, times $P_0$; for $0 \leq k \leq 1$.

Now we put two of the slabs together. If we ignore the possibility of having any light reflected from the surfaces (not completely true) then the result will be the same as if two slabs were placed apart as in Figure 17D.1b. The amount of light coming through the first slab is the incident light on the second slab. Thus,

$$P_2 = kP_1 = k \cdot kP_0 = k^2 P_0; \qquad k < 1$$

If we now think of a sample used in determination as having a path length $b$, then the number of thin samples that will fit into this length is

$$n = \frac{b}{\Delta x}$$

and

$$P_n = k^n P_0 = k^{(b/\Delta x)} P_0$$

Rearranging this equation and taking the logarithm of both sides, we find

$$\log \left( \frac{P_n}{P_0} \right) = b \frac{\log k}{\Delta x}$$

**(a)**

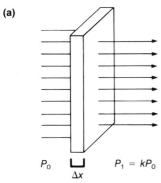

$P_0$ ⊔ $P_1 = kP_0$
$\Delta x$

**(b)**

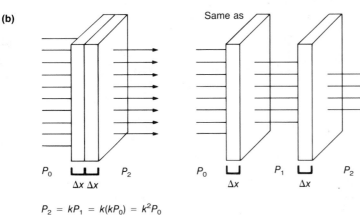

Same as

$P_0$ ⊔⊔ $P_2$        $P_0$ ⊔ $P_1$ ⊔ $P_2$
$\Delta x\ \Delta x$        $\Delta x$   $\Delta x$

$P_2 = kP_1 = k(kP_0) = k^2P_0$

**Figure 17D.1**   Illustration of a derivation of Beer's law.

**(c)**

$\Delta x$

$P_0 \rightarrow$         $\rightarrow P_n, n = \dfrac{b}{\Delta x}$        $P_n = k^nP_0 = k^{\frac{b}{\Delta x}}P_0$

$\log \dfrac{P_n}{P_0} = \dfrac{b}{\Delta x} \log k = \alpha b$

$\longmapsto$ Length $b$ $\longmapsto$        where $\alpha = \dfrac{\log k}{\Delta x} = $ constant

or, since $\log k$ and $\Delta x$ are both constant, and since $k < 1$,

$$-\log \frac{P_n}{P_0} = \alpha b$$

where $\alpha$ is a positive constant, and $b$ is the total path length. This is the reason that the absorbance is defined as $-\log(P/P_0)$. The absorbance changes linearly with the path length.

The constant $\alpha$ can be expressed as the product of a concentration of the material that absorbs the light and a new constant that is a property of the material. There are two common ways to express this.

$$\alpha = ac$$

where $a$ is a constant with the units $cm^{-1}(g/L)^{-1} = L/g \cdot cm$. The constant $a$ is called the *absorptivity*. In older literature it is also called the *absorption coefficient*. The units of the whole equation are then

$$A \text{ (unitless)} = a \text{ (L/g} \cdot cm) \, b \text{ (cm)} \, c \text{ (g/L)}$$

In chemical analysis, it is more common to use

$$\alpha = \epsilon c$$

$\epsilon$ has the units $cm^{-1}(mol/L)^{-1}$, and $c$ is in mol/L. $\epsilon$ is called the *molar absorptivity* (now preferred) or *molar extinction coefficient*. The equation for absorbance is then

$$A \text{ (unitless)} = \epsilon \text{ (L/mol} \cdot cm) \, b \text{ (cm)} \, c \text{ (mol/L)}$$

as is stated in Eq. 17-8.

## SUPPLEMENT 17E

# The Nomenclature of Molecular Absorption and Luminescence

As you may know, molecules can exist only in specific molecular energy levels. The energies are quantized in a manner similar to atomic energy quantization. However, not only are the electronic energy levels quantized, but the molecules can exist only in quantized vibrational and rotational energy states. As a result, spectra from even simple molecules exhibit large numbers of lines in the gas phase. Such a set of lines, due to spectral transitions to individual rotational levels, is shown in Figure 17E.1. The spectrum is in the visible region. Notice that the individual emission lines are about 0.03 nm apart. In a solid or solution, these individual lines broaden and become unresolved **bands.** All the spectra that we use for analysis in condensed media (liquids and solids) have broad lines due to large numbers of unresolved spectral transitions. Examples are seen in Figures 17.18 and 17E.2.

Two different conventions are used to illustrate the origins of these types of spectra. The first shows a plot of energy versus distance between atoms, as illustrated in Figure 17E.3. Each of the curves shows how the energy of the molecule changes with the interatomic distance. The individual, narrow lines indicate the allowed quantum vibrational energies. Figures such as these usually are used only for simple molecules,

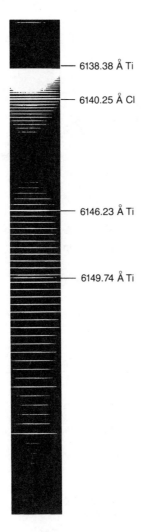

— 6138.38 Å Ti

— 6140.25 Å Cl

— 6146.23 Å Ti

— 6149.74 Å Ti

**Figure 17E.1**    One vibrational band in the emission spectrum of titanium nitride, TiN, in the gas phase. The lines are due to transitions between quantized rotational levels of the molecule. Four emission lines from atoms that were present also appear and are labeled at the top. The wavelength range of the spectrum shown here is approximately 2 nm.

because for more complicated molecules there are too many different interatomic distances to show on a page.

The second convention in diagramming the energies of atoms and molecules is exemplified in Figure 17E.4. In this case, there are two separate "stacks" separated by spin states. The states with all electron spins paired, $\downarrow\uparrow$ — the singlet states — are on the left. The states with two electrons parallel, $\uparrow\uparrow$ — the triplet states — are on the right. *Singlet* and *triplet* are two special cases of different **multiplicity.** Each stack has a different multiplicity, which is characterized by $S$, the quantum number of the total spin. Singlets have $S = 0$, and triplets have $S = 1$. When only photons are involved (absorption or emission), molecules seldom transform from a singlet to triplet or vice versa. This is called a *selection rule:* $\Delta S = 0$, generally. However, selection rules are not absolute. *Some* transitions have $\Delta S \neq 0$, as shown in Figure 17E.5.

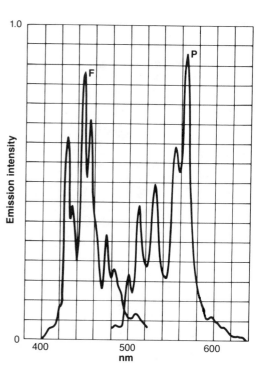

**Figure 17E.2** Emission spectrum of coronene dissolved in a polymer. *F* labels the fluorescent emission, and *P* labels the phosphorescence. Temperature, ambient; phosphorescent emission is multiplied by 5.5. [Ref: Reprinted from West, M. A.; Kemp, D. R. *Intl. Lab.* **1976,** May/June, 39. Copyright 1976 by International Scientific Communications, Inc.]

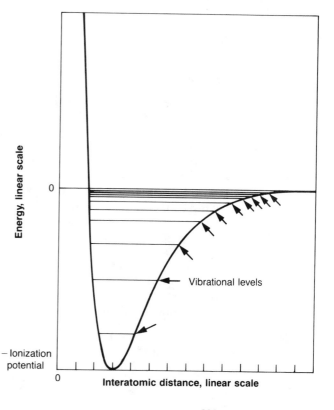

**Figure 17E.3** Calculated but representative plot of the intramolecular potential energy of a diatomic molecule as it varies with interatomic distance. The molecule can reside in any one of numerous quantized vibrational energy levels that are indicated by horizontal lines. Transitions between these vibrational levels have energies that are in the infrared region.

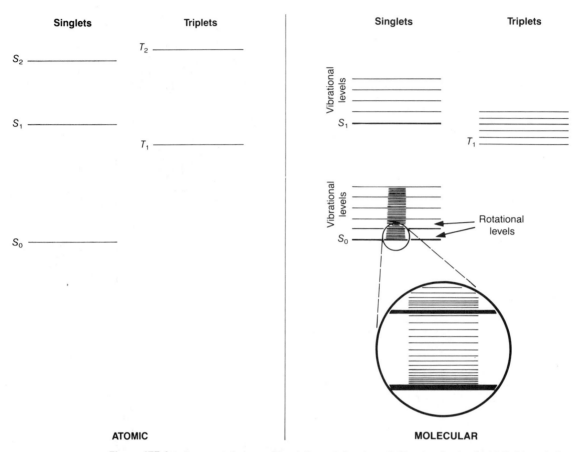

**Figure 17E.4**   Representative energy-level diagrams for atoms (left) and molecules (right) that have both singlet and triplet states (S = 0 and S = 1, respectively). The horizontal axis does not represent distance in this case.

Each thick line, which indicates an electronic state, has a stack of more closely spaced lines above it. These thinner lines indicate the vibrational energy levels associated with each electronic state, just as do the sets of lines in Figure 17E.3. The even more closely spaced rotational energy levels are only indicated to avoid having a completely blackened figure.

The previous paragraphs present the necessary prelude to correct, simple definitions of fluorescence and phosphorescence — two forms of luminescence.

> *Fluorescence* is a radiative transition in which a photon is emitted, but the multiplicity does not change. This is illustrated in Figure 17E.5.
> *Phosphorescence* is a radiative transition in which a photon is emitted, and the multiplicity changes.

In general, phosphorescence emission lasts a longer time than fluorescence. This is the reason for the original separation of luminescence into two separate phenomena.

The energy changes that occur by transitions between molecular energy levels can be either radiative (solid arrows), which means emitting a photon, or nonradiative

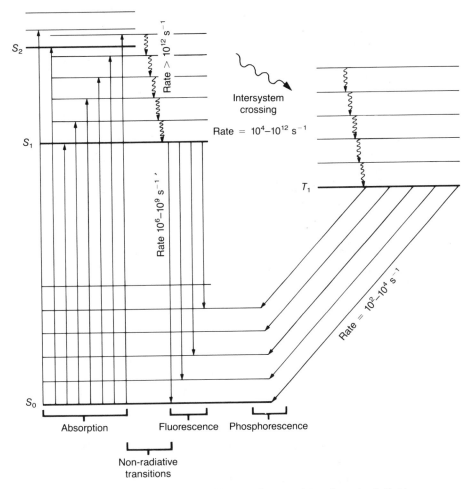

**Figure 17E.5**  Representative transitions and transition rates between states of a molecule that has both singlet and triplet states. The types of transitions are labeled with arrows pointing from the original level to the final level of the transition.

(wavy arrows), which means producing heat. Note that the arrows go between individual molecular energy levels. (If the transitions occurred only between the lowest level of each electronic state, the spectrum in Figure 17E.1 would have one line; molecular spectra would be as "simple" as atomic spectra, such as those shown in Figures 18.1 and 18.2.)

It is important to recognize that transitions of different types occur at different rates. For instance, the nonradiative transitions between vibrational levels occur in less than about $10^{-12}$ s. Other rates are illustrated in Figure 17E.5. Because the nonradiative transitions are so fast, a molecule excited to any vibrational or rotational level rapidly falls into the lowest energy level of the state. Thus, fluorescence and phosphorescence emissions tend to originate from these lowest levels, as illustrated in Figure 17E.5. (The origins of the solid arrows indicate the type of emission.)

One other general characteristic is of interest. As shown in Figure 17E.5, the lowest excited-state triplet lies lower in energy than the lowest excited-state singlet. As a result, phosphorescent transitions most often occur at lower energies (longer wavelengths) than fluorescent ones, as is true for the spectra shown in Figure 17E.2.

In addition, some molecules can be "stored" in states that cannot make a transition to the ground state without violating a selection rule. Examples are excited triplet states ($\Delta S \neq 0$) or atomic states such as a 2S state in which the 1S is a ground state (1S $\rightarrow$ 2S has $\Delta l \neq 1$; see Section 18.2). The energy is released only slowly as a spontaneous emission. However, the transition occurs much faster if the molecule is stimulated at the phosphorescence wavelength. This is **stimulated emission:** The transition to the lower state will occur more rapidly — it will be stimulated — if the particle is illuminated at the same wavelength as the slow transition. The result is light amplification by stimulated emission of radiation, which you may recognize by its initials **laser.**

Both atomic and molecular emissions are used to produce laser light. Three lasing gaseous systems are $Ar^+$, $CO_2$, and a mixture He – Ne. A solid bar of alumina doped with $Cr^{+++}$ (ruby) and a liquid solution with special dyes also can be used to produce laser light.

## SUPPLEMENT 17F

# Application Example of Chemiluminescence

There is a continuing need to monitor industrial stack emissions for regulated pollutants. The methods used to do this must be approved by the appropriate government agencies to ensure that the results are comparable with standards. One of the methods approved by the U.S. Environmental Protection Agency for the monitoring of nitrogen oxides is an optical spectroscopic method. As defined in Section 17.3, luminescence is emission of light from an atom or molecule that has been excited in some manner. If the excitation is provided by heating the sample (thermal energy in, light energy out), this is called *thermoluminescence.* If the excitation is by light, the label is *photoluminescence.*

In certain special cases, the energy can come from a chemical reaction. The name for this process is, reasonably, *chemiluminescence.* As was found in the late 1960s, the gas-phase reaction of nitrous oxide with ozone is one of these special reactions. The resulting chemiluminescence is used for measuring nitrogen oxides selectively in air. Possible interferents in the gaseous matrix are $O_2$, $SO_2$, $H_2O$, $CO_2$, and $N_2$ as well as hydrocarbons. However, the chemical reaction and subsequent light emission is so specific that only simple controls are needed to reduce errors due to changing matrix conditions. The only significant interference is from ammonia, which is not expected to be found under oxidative conditions.

**Figure 17F.1**   Plot of the emission spectrum from chemiluminescent $NO_2$. The luminescence is monitored in the region 600–900 nm. The wavelength region of the spectrometer is selected with a filter, and thus the instrument is called a nondispersive spectrometer. It has no dispersive element such as a grating in it.

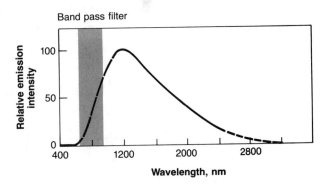

The reactions that occur are

$$NO + O_3 \rightarrow NO_2^* + O_2$$
$$NO_2^* \rightarrow NO_2 + h\nu$$

The asterisk in the equation is a commonly used convention to denote an atom or molecule in an energetically excited state. The first equation describes a reaction of ozone with nitrous oxide producing nitrogen dioxide in an excited state. The second equation indicates that the $NO_2$ in an excited state emits light, $h\nu$, and goes to its unexcited or ground state.

Figure 17F.1 shows the emission spectrum observed. The light is emitted over a range from about 500 to 3000 nm. This is from the orange range through the near-infrared region. The rest of this description is quoted from the EPA reference.

> Nitrogen dioxide ($NO_2$) does not undergo this reaction and must be reduced to NO before it can be measured by this method. Most commercial analyzers contain a converter that catalytically reduces $NO_2$ to NO.
>
> $$NO_2 \xrightarrow[\text{catalyst}]{\text{heat}} NO + \tfrac{1}{2}O_2$$
>
> The NO produced is then reacted with the ozone and the chemiluminescence measured to give a total $[NO] + [NO_2] = [NO_x]$ reading. Figure 17F.2 shows a schematic typical of this class of instruments.
>
> **Operation of a Chemiluminescence Analyzer**   Ozone is generated by the ultraviolet irradiation of oxygen in a quartz tube. The ozone is provided in excess to the reaction chamber to ensure complete reaction and to avoid quenching effects. [*Quenching* is the term used when less light than expected is emitted. This is caused by molecular collisions that remove the excitation energy as heat instead of allowing light emission.] Since the photomultiplier signal is proportional to the number of NO molecules, not to the NO concentration, the sample flowrate must be carefully controlled. . . .
>
> Molecules such as $O_2$, $N_2$, and $CO_2$ quench the light radiation of this chemiluminescent reaction . . . . The quenching problem has, however, been uniquely solved by choosing a flowrate of ozone into the sample chamber much greater than that of the sample flowrate. The resulting dilution gives a relatively constant

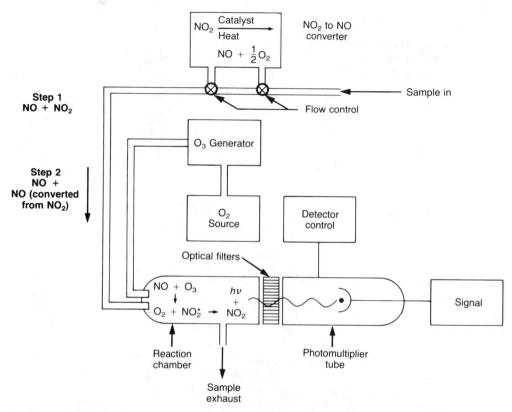

**Figure 17F.2**   Operation of a chemiluminescence analyzer. Ozone is generated by ultraviolet irradiation of oxygen in the ozone generator. $O_3$ is supplied in large stoichiometric excess into the reaction chamber. Since the chemiluminescent reaction goes to completion, any variation in flow rate of the sample will produce fluctuations in the output. As a result, the sample flow rate must be carefully controlled. The flow controls can, in addition, be used to switch the input between *step 1* and *step 2* to allow measurement of both NO and $NO_2$.

background gas composition, and the effects caused by different quenching efficiencies of different molecules are minimized. The only serious interference is ammonia, which will oxidize to NO in stainless steel converter chambers. [Converters made of] molybdenum operated at lower temperatures will not oxidize . . . ammonia. [This oxidation] is not usually a problem when the monitor is placed on a combustion source, but care should be taken in other applications. [Ref. Handbook: "Continuous Air Pollution Source Monitoring Systems." United States Environmental Protection Agency, June 1979; EPA 625/6-79-005, pp. 5–11 to 5–20.]

## SUPPLEMENT 17G

# Waves, Wavelength, and Spectra

### Waves

Waves, or the idea of waves, is one of the primary concepts of the action of the physical world. In chemical analysis, you will encounter the concepts of electromagnetic waves — radio, light, and X-rays, for example.

Waves carry energy from one location to another. They have a finite velocity. The "shape" of a wave is usually represented by a graph of height (for a water wave) or electric field strength (for light waves) or pressure (for a sound wave) versus time or versus distance. A graph of a general wave is shown in Figure 17G.1.

Much of the behavior of light can be explained by using the properties of sine waves. A graph of a sine wave is shown in Figure 17G.2. A sine wave can be described mathematically by

$$A = A_0 \sin \theta \tag{17G-1}$$

where

> $A$ is the amplitude,
> $A_0$ is the peak amplitude, and
> $\theta$ is a continuous variable, the form of which is shown below.

The peak amplitude is usually called simply the **amplitude** of the wave. The distance $\lambda$ between peaks (or valleys) of the wave is the wavelength. In terms of the variable $\theta$, the wavelength is $2\pi$. A simple, static sine wave is completely characterized by the amplitude $A_0$ and the wavelength $\lambda$.

Since the properties of sines and sine waves depend on factors of $\pi$, it is convenient to associate a sine wave with the properties of angles. Recall that a change in angle of $360° = 2\pi$ radians. This is equivalent to one complete *rotation* in a plane. If the language of angles is used for Eq. 17G-1,

> $\theta =$ an angle (in radians)
> $2\pi =$ a **cycle**

**Figure 17G.1**    Plot of a general wave. The vertical direction represents the height or amplitude of the wave. The horizontal direction can be either time or distance. This graph might represent a cross section of an ocean wave or, perhaps, the voltage output from a microphone over a short time duration.

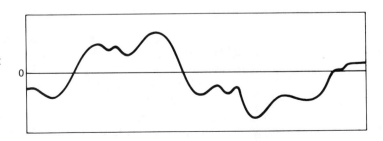

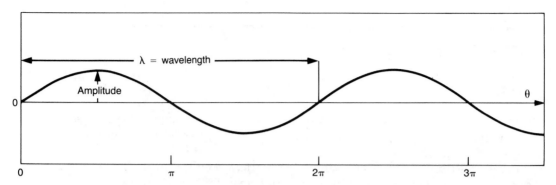

**Figure 17G.2**    Graph of a sine wave with the amplitude and wavelength annotated. This is a specific example of an harmonic wave. For our interest, it represents an electromagnetic wave. In that case, the amplitude is the magnitude of an electric or magnetic field oscillating in time.

When the angle is a function of time, we write

$\theta = \omega t$

$\omega$ = an angular velocity (radians/unit time)

When $\theta$ is substituted by the product of a constant and time, the algebraic form for a time-dependent sine wave is

$$A = A_0 \sin \omega t \qquad \textbf{(17G-2)}$$

A wavelength of this time-dependent wave is completed when $\omega t$ has changed in value by $2\pi$; one wavelength occurs over one cycle. This is also called one complete **oscillation** or one **period**. The time over which one complete cycle occurs will be

$$t_{\text{cycle}} = \frac{\omega}{2\pi}$$

The **frequency** of oscillation $v$ is, thus,

$$v = \frac{1}{t_{\text{cycle}}} = 2\pi/\omega \qquad \textbf{(17G-3)}$$

### Wavelength

Electromagnetic radiation is describable as electric and magnetic fields that oscillate sinusoidially in amplitude while traveling through space. The direction of travel is perpendicular to the direction of oscillation of both the electric and the magnetic fields. These ideas are illustrated in Figure 17G.3.

What is important to note from the figure is that we can use the properties of simple sine waves to explain the properties of the more complex electromagnetic waves. To do so, we choose a wave that describes either the oscillating electric or magnetic field moving through space.

**Figure 17G.3**   Diagram of the geometric relationship of the oscillating electric and magnetic fields of a polarized electromagnetic wave. Polarized means that the electric field oscillates in a single plane and not in a random position about the z-axis. Notice that the electric and magnetic fields are perpendicular to each other and both are perpendicular to the direction the wave is moving—along the z-axis.

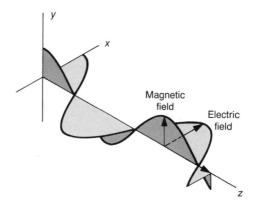

For any wave moving at a constant velocity, the frequency $v$, wavelength $\lambda$, and velocity $v$ are related to each other. This relationship is simply

$$v = v\lambda \qquad \text{(17G-4)}$$

For electromagnetic radiation, the units in this equation can be

$v$ in m s$^{-1}$
$v$ in Hertz (cycles per second or s$^{-1}$)
$\lambda$ in meters

The speed of light in a vacuum has been measured precisely and has been given the label $c$ ($c = 2.9979 \times 10^8$ m s$^{-1}$). From this value and the frequencies, the wavelengths of electromagnetic radiation *in vacuo* (in a vacuum) can be calculated. An overview of the electromagnetic spectrum is shown in Figure 17G.4. The *in vacuo* wavelengths are displayed on the right.

It is known from experiment that if electromagnetic radiation can pass into some material (if it is not totally reflected), its velocity will be slower than in a vacuum. Since the frequency is unchanged, the wavelength must be shorter. This difference is characterized by the index of refraction $n$.

$$n = \frac{c}{v} = \frac{\lambda_{(vac)}}{\lambda_{(mat)}} \qquad \text{(17G-5)}$$

In the equation, $c$ is the speed of light *in vacuo,* and $v$ is the speed of light in the material. The index of refraction is a function of both the frequency of the electromagnetic radiation and the identity of the material through which is passes.

## The Spectrum and Spectra

The word *spectrum* has two separate meanings. The first is illustrated in Figure 17G.4. This shows the range of electromagnetic radiation sorted by frequency. The frequency scale is on the left. Specific names are applied to different ranges from radiofrequencies through $\gamma$-ray frequencies. These are listed in the central column. At

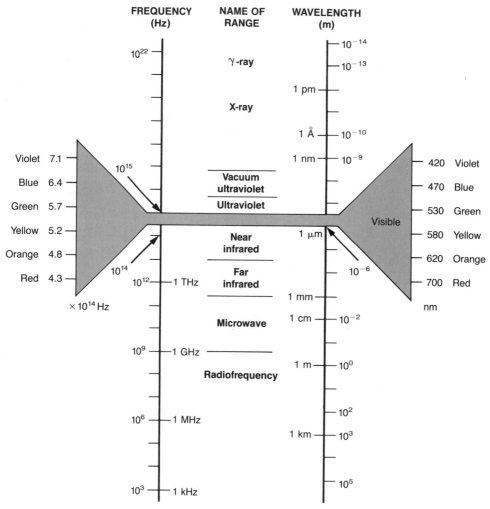

**Figure 17G.4** A diagram of the names, frequencies, and wavelengths (in a vacuum) of electromagnetic radiation. Each region of the spectrum is used in one or more analytical methods. It is probably worth remembering that visible light is centered around 500 nm.

the right is a wavelength scale, which applies exactly only for the radiation in a vacuum. As noted above, the wavelength is shorter inside any material.

The second use of the word *spectrum* (plural, *spectra*) is important in spectrometry. *Spectrum* is the name given to a figure showing wavelength (or frequency) on one axis together with some measure of the intensity of the light emitted or absorbed at the wavelength. Two types of spectra are shown in Figure 17G.5. One can record a spectrum photographically with areas of greater and lesser exposure indicating the light intensity at each wavelength. Also, a line graph can be made having the wavelength plotted on one axis together with a perpendicular axis on which is plotted the amount of light at each wavelength.

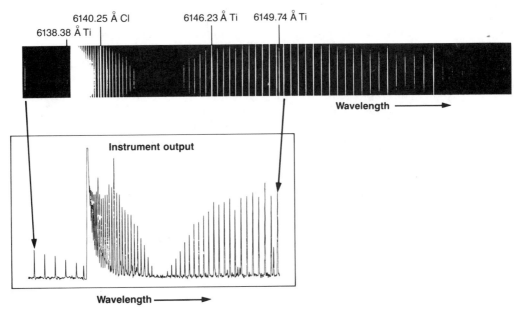

6138.38 Å Ti

6140.25 Å Cl

6146.23 Å Ti    6149.74 Å Ti

Wavelength ⟶

Instrument output

Wavelength ⟶

**Figure 17G.5**    A picture of the light emitted from the molecular species TiN, as recorded on a photograph. The emission lines are light and, if the picture were in color, would be red. On the adjacent graph are plotted a measure of the intensity of the light on the ordinate and the wavelength on the abscissa. Both the photograph and graph are examples of spectra. The graphical type is much more common. [Reproduced by permission of the National Research Council of Canada from *Canadian J. Phys.* **1970,** *48,* 1657–1663.]

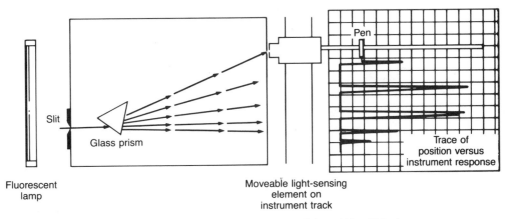

Pen

Slit

Glass prism

Trace of position versus instrument response

Fluorescent lamp

Moveable light-sensing element on instrument track

**Figure 17G.6**    Diagram of how a graphical spectrum of the output from a fluorescent light could be obtained. The light from the lamp is passed through a slit to produce a narrow band of illumination that is dispersed by passing through a prism. If a screen were placed in the now dispersed beam, a series of bands of light would show up with dark spaces in between. The bands of light are not equally intense across their whole range. To record the variation in light intensity, the screen is replaced with a device that senses the light and moves a pen to a position proportional to the amount of light falling on the sensor. If this device moved along the position of the screen, it would trace out a graph of instrument response versus wavelength. In most instruments, however, this graph is obtained with a fixed sensor and a process equivalent to rotating the prism while simultaneously moving the paper under the pen.

Figure 17G.6 illustrates a possible experiment to record the spectrum of light emitted from a common fluorescent lamp. The light from the lamp is passed through a slit, then through a prism. If we looked at the colors displayed on a screen in a dark room, we would see bands of color of varying intensity with areas of darkness in between. The photographic recording of this spectrum would reproduce the bands. The graphical form of the spectrum could be obtained by removing the screen, and replacing it with the instrument illustrated in the figure.

This instrument consists of a light-sensitive device, a sensor, with a pen anchored onto a track attached to it. The pen moves on the track, and its position depends on the intensity of the light falling on the sensor. When the light is brighter, the pen is farther to the right; when there is less light, the pen is farther to the left. To record the spectrum, the sensor is placed at the lower part of the lighted area. The sensor then is moved upward along where the screen would be. A line plot of the spectrum will be drawn. It is a graph of the intensity of the light versus the position of the light-sensitive detector. However, because the light is spread out into colors by the prism, we can label the graph *light intensity versus wavelength*. Another name for such a line graph is a **spectral curve.**

## SUPPLEMENT 17H

# Spectroscopic Gratings and Wavelength Dispersion

Here we describe the basic ideas of wave interference. The aim is to show the fundamentals of how **spectroscopic gratings** work. A spectroscopic grating is a device that can **disperse** electromagnetic radiation just as a prism disperses visible light into its component colors. Different types of gratings are usable from infrared to X-ray wavelengths.

The sine waves of electromagnetic radiation are **linear waves.** Linear is a mathematical term for *being independent of each other.* This means that the presence of one wave does not affect the properties of another wave. Mathematically, this means that we can take the expression describing one wave and add it to another. The sum will describe the actual physical result.

### Relative Positions of Two Stationary Waves

Understanding the interactions (sums, differences, and products) between two sine waves is necessary for describing a wide range of physical phenomena. The relationships must include a description of their relative positions. For instance, as illustrated in Figure 17H.1, assume that two waves are present in the same space, both with the

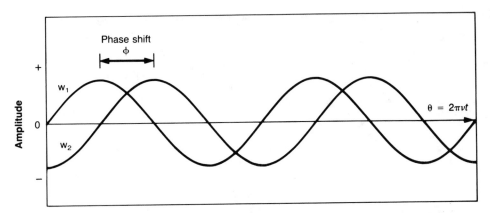

**Figure 17H.1** Plot of two sine waves of equal amplitude and wavelength (or frequency) in the same region of space. They are out of phase (or *phase shifted*) by $\lambda/4$.

same amplitude $A_0$ and frequency $v$. However, the peaks (and valleys) do not coincide. As discussed in Supplement 17G, the wave $w_1$ can be described by

$$A_1 = A_0 \sin 2\pi vt \qquad \text{(17H-1)}$$

and $w_2$ can be described by

$$A_2 = A_0 \sin\left(2\pi vt - \frac{\lambda}{4}\right) \qquad \text{(17H-2)}$$

The general equation for describing such changes in position is

$$A_2 = A_0 \sin(2\pi vt + \phi) \qquad \text{(17H-3)}$$

The value of $\phi$ is called the **phase shift.** In the example illustrated in the figure, the magnitude of the phase shift $\phi$ can have one of three labels all of which are equivalent: $\lambda/4$, 90°, or $\pi/2$.

**Figure 17H.2** Plot of two sine waves $w_1$ and $w_2$ that are in the same region of space and equal in frequency and amplitude. They are phase shifted by $\lambda/2$ [or any value of $(2n + 1)\lambda/2$]. Their sum at every point is zero. This is the explanation for destructive interference.

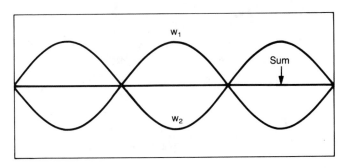

$$w_1 = \sin(\theta + \phi)$$
$$w_2 = \sin\left(\theta + \phi + \frac{\pi}{2}\right)$$
$$w_1 + w_2 = 0$$

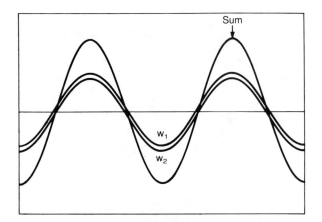

**Figure 17H.3**    Plot of two sine waves $w_1$ and $w_2$ that are in the same region of space and equal in frequency and amplitude. They are phase shifted by $\lambda$ (or any value of $n\lambda$). The sum of their amplitude at every point is twice each individual amplitude. This is the explanation for constructive interference.

## Interference

The equations representing waves can be added to represent the physical process of adding the magnitudes of electric or magnetic fields. The sums of waves are often described by comparing the resulting wave with the original ones. For example, what happens when the two waves are added together? Look at the waves drawn in Figure 17H.2. Visually, you can add the amplitudes of these waves at each point along the x-axis. Each peak adds to a trough. In fact, at each point along the x-axis, the sum is exactly zero. Mathematically,

$$A_1 = A_0 \sin(2\pi vt)$$

and[2]

$$A_2 = A_0 \sin\left(2\pi vt + \frac{\lambda}{2}\right) = -A_0 \sin(2\pi vt)$$

The sum is

$$A_1 + A_2 = 0$$

The resulting wave amplitude is zero. This is called **destructive interference** by two identical waves *out of phase* by exactly $\lambda/2$.

Another way to obtain this result is if the separation of the peaks (or valleys) of $w_1$ and $w_2$ is any *odd* multiple of $\lambda/2$. Examples of these differences are $\lambda/2$, $3\lambda/2$, $5\lambda/2$, . . . , $(2n + 1)\lambda/2$. In all such cases, the waves will exactly destructively cancel.

---

[2] This equality can be derived by recognizing that $\lambda/2$ is equivalent to $\pi$ and using the equation for the sine of a sum:

$$\sin(a + b) = \sin a \cdot \cos b + \cos a \cdot \sin b$$

Thus

$$\sin(2\pi vt + \pi) = \sin 2\pi vt \cdot \cos \pi + \cos 2\pi vt \cdot \sin \pi$$

The derivation is completed by recognizing that $\cos \pi = -1$ and $\sin \pi = 0$.

If, on the other hand, the differences between the positions of the waves is *equal* to the wavelength, then the waves will **constructively interfere** as illustrated in Figure 17H.3. Constructive interference occurs when the difference in $\phi$ is some multiple $(1, 2, \ldots, n)$ of $\lambda$.

To summarize:

Sinusoidal waves are linear waves.

Waves of equal amplitude and wavelength that are out of phase by $(2n + 1)\lambda/2$ add to zero.

Waves of equal amplitude and wavelength that are out of phase by $n\lambda$ or exactly in phase add constructively so that the sum amplitude is twice that of the original waves.

### Two-Slit Diffraction

When light is shone onto a screen through two identical, parallel slits that are close together in an opaque material, a pattern of light and dark bands is seen. This phenomenon is illustrated in Figure 17H.4. If the light were simply shining through the slits, we should see a shadow with two bright lines in it. However, this is not the pattern that is observed. Numerous bands of light appearing over a broad region shows that the light rays have been bent. This is called **diffraction** of the light.

As can be measured, the intensity of the light varies with the angle from the center of the slits, $\theta$. The spacing of the bands depends on the distance between the slits, $d$. The center of a dark band occurs at

$$2d \sin \theta = n\lambda \quad \text{or} \quad d \sin \theta = \frac{n\lambda}{2}; \quad n = 1, 2, 3, \ldots \quad \textbf{(17H-4)}$$

The center of a light (illuminated) band occurs at

$$d \sin \theta = n\lambda; \quad n = 1, 2, 3, \ldots \quad \textbf{(17H-5)}$$

The reason for this effect is that the light waves constructively interfere (to form light bands) or destructively interfere (to cause the dark areas), depending on the angle. The explanation of this effect is made in the caption to Figure 17H.4.

**Multislit Diffraction**  If, instead of having only two slits in the opaque sheet, there are many hundreds of equally spaced slits, the *centers* of the light and dark areas on the screen will still show the same angular dependence as arose from only two slits. The centers of the dark bands will still be at $n\lambda = 2d \sin \theta$, and the centers of the illuminated areas will still be at $n\lambda = d \sin \theta$. However, as illustrated in Figure 17H.5, the regions of light will have sharper edges.

We can number these bands starting from the middle one which is labeled zero. From there, bands on both sides of the middle are labeled by increasing integer values—1, 2, . . .—the further they are from the center. The bands labeled 1, 2, 3, and so on are referred to as **first order, second order, third order,** and so on. In addition, the numerical values of the label equal the value of $n$ in Eq. 17H-5.

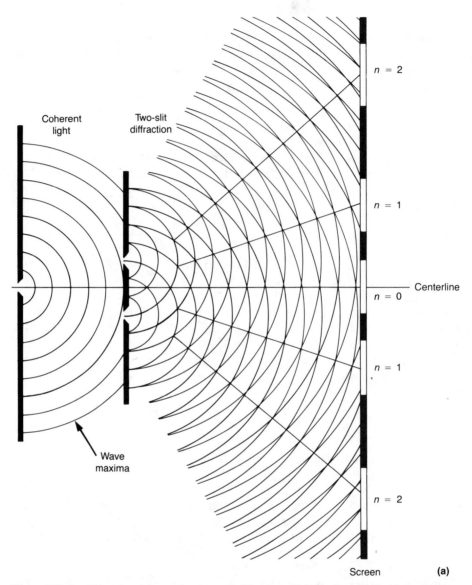

Coherent
light

Two-slit
diffraction

$n = 2$

$n = 1$

Centerline

$n = 0$

$n = 1$

Wave
maxima

$n = 2$

Screen    **(a)**

**Figure 17H.4**    Illustration of the results of two-slit diffraction of light. As illustrated in (a), the light is passed through a single slit, which causes the light to act as though all the waves are in phase *(coherent).* (A laser could be used instead.) This light then passes through two thin slits perpendicular to the page, as shown, and falls on a screen represented at the right. Each slit acts like a point source of light. The radiation has peaks and troughs of electric field like ripples in water.

Assume the light that is shown through the slits is limited to a single wavelength — a single color if the radiation is visible light. Then, each of the narrow bands of light will be the same color.

What, then, is the location of the illuminated region when light of two different

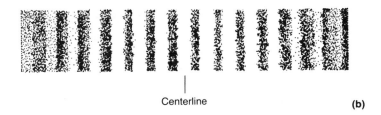

Centerline

**(b)**

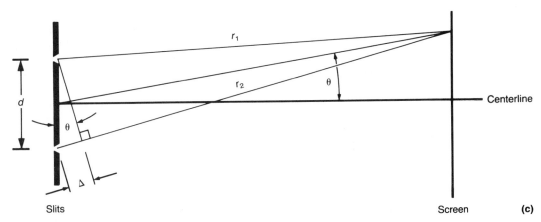

Slits                                                                    Screen    **(c)**

**Figure 17H.4** *(continued)*    As shown in (b), an image appears at the screen: a series of light and dark lines that extend on either side from the center line depending on the angle. These regions of light and dark are explained as regions of alternating constructive and destructive interference. The light regions occur where the path difference from the two slits to the screen, $r_1 - r_2$, is equivalent to a difference of $n\lambda$. The difference in length is shown by $\Delta$ in (c). This difference in distance is related to the angle at which the light band appears by $n\lambda = d \sin \theta$. The central bright line is where $n = 0$ and is called the zero-order line. The two lines on either side of this occur where $n = 1$ and are called first order. This nomenclature continues: second order, third order, and so forth. For more details see any introductory physics textbook.

wavelengths simultaneously illuminates the slits? Say the colors are blue and red: The blue light wavelength is shorter than the red.

$$\lambda_{\text{red}} > \lambda_{\text{blue}}$$

By inspection of Eq. 17H-5 you can see that for *each n*-value, each angle of $\theta$ for the blue light will be smaller than the associated angle for the red light ($\theta < 90°$).

$$\theta_{\text{red}} > \theta_{\text{blue}}$$

The two colors will be separated from each other on the screen. Thus, a multislit diffraction plate can be used to separate colors.

If white light were shining through the slits, a number of small rainbows would show up on the screen. Each of these rainbows can be labeled with an order: first

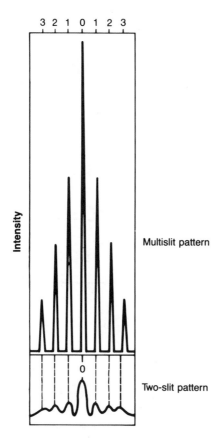

**Figure 17H.5**    The difference in light-intensity distribution between two-slit diffraction and multislit diffraction. The peaks are sharper for multislit diffraction. Therefore, if red and blue light are both in the illuminating beam, they are more clearly separated by the multislit grating than by a two-slit system. In other words, the multislit system provides better resolution of the colors.

order, second order, and so on. The white light has been dispersed into its spectrum. The plate that acts like a large number of slits is called a **diffraction grating.**

There are two general types of diffraction gratings. The light can be passed through the grating as illustrated in Figure 17H.4. This is a **transmission grating.** On the other hand, the grating could be a mirror with alternating reflecting and nonreflecting lines on it. Such a grating disperses the light that is reflected from it and is called a **reflection grating.** Reflection gratings are the most commonly used *light-dispersing elements* in most spectrometers. Instruments that include gratings in their designs are described in Supplements 17B and 20B.

# 18

# Atomic Spectrometry for Elemental Analysis

## 18.1  Atomic Spectrometry in the Visible Region

Analytical methods using luminescence and absorption spectra in the ultraviolet and visible-light regions are among the most popular and broadly used in chemistry and biochemistry, including clinical analysis. Among the reasons for this popularity is that a large number of analyses of this type were developed when there were few alternative instrumental methods. Precise and sensitive analytical instruments have been available to measure changes in light in the visible region for more than 40 years.

The elemental analyses by atomic spectrometry that are covered in this chapter are those in which we measure the absorption, emission, or fluorescence of light by gas-phase atoms (or ions): atoms or ions that are free of a condensed-phase matrix. From now on, the term *atomic* spectrum will be understood to mean the spectrum of *atoms or ions,* either of which can be used to determine the elemental content of analytes. You may read *atom* to be *atom or ion.* If the atoms are freed from a condensed-phase matrix, the analytical wavelengths in the UV–visible region then appear at predictable positions with relatively narrow bandwidths.

Not discussed here are the elemental analytical methods in which the radiation is not in the UV–visible range or where mass or electron kinetic energy is the directly measured quantity. These can be divided into methods that respond to bulk content and those that are surface-sensitive. The latter provide a signal that arises from a layer from 1–3 nm deep.[1] Also not covered are methods in which elemental content is determined by measuring spectra of molecular species in solution. Examples of such determinations are those of iron with 1, 10-phenanthroline, nickel with dimethyl-glyoxime, and molybdenum with 4-methyl-1,2-dimercaptobenzene.

The general spectrometric techniques we use to determine composition were mentioned in the last chapter. We measure either the intensity of emission (emission), the relative amount of light absorbed (absorption), or the amount of reemitted light (fluorescence). As noted before, these three general techniques are *possible* with radiation in all regions of the spectrum, but not all three can necessarily be used because of experimental difficulties. These difficulties may include

1. chemical and spectral interferences such as high background;
2. poor stability, which results in low precision;
3. poor S/N ratio, leading to bad detection limits and lowered precision;
4. low sensitivity and thus less utility for less concentrated components or smaller samples; and
5. small useful range, which may require more sample-preparation steps to bring the analyte into the range of the instrument.

As you read this chapter, you will see that all these properties pertain in some degree to quantitation with atomic spectrometry.

---

[1] Bulk elemental techniques that are used for solid and liquid samples are X-ray fluorescence (XRF), neutron activation analysis (NAA), and mass spectrometry (Chapter 16) with high-energy ionization sources such as a spark or focused laser. Two surface-sensitive techniques involve the measurement of electron kinetic energies: electron scattering for chemical analysis (ESCA) and Auger electron spectrometry (AES). A less common technique for elemental analysis of surfaces is Rutherford backscattering spectrometry (RBS), in which high-energy ions, such as $^4He^+$, bombard the surface to be analyzed. The losses in kinetic energy of the particles that "bounce off" the surface are measured and used to determine the surface composition of the solid.

Throughout this chapter, wavelengths are stated in nm. While the tendency now is toward nm units, both Å and nm are likely to be seen in the near future. Recall that 1 nm = 10 Å. Another, less widely used unit is the millimicron, $m\mu$, which equals one nanometer.

The term *sensitivity* in atomic absorption has had a specific meaning: the concentration or mass of analyte needed to produce a signal with transmittance 0.99 or its equivalent absorbance 0.0044. It is now suggested that, depending on which is applicable, the amounts needed to produce the absorbance 0.0044 be called the **characteristic concentration** or **characteristic mass.** The term *sensitivity* is to be reserved for the more broadly used definition of the slope of a working curve in the analytical concentration range.

For all the atomic spectrometric methods, the detection limit is usually defined as the amount or concentration of analyte needed to produce a signal three times the peak-to-peak noise level of a blank. The correct statistical definition of detection limits is demonstrated in Supplement 18B. The characteristic concentrations and detection limits of various atomic spectrometric techniques are included in Table 7.2.

## 18.2 Nomenclature of Atomic Spectra

The atoms from which atomic spectra arise can be in a number of different forms. For example, copper could be atomic or ionic: $Cu^0$, $Cu^+$, $Cu^{2+}$, $Cu^{3+}$, and so forth. The nomenclature of atomic spectra differs from that used to describe the chemistry of solutions and solids. The spectroscopic nomenclature we use labels an ion with the Roman numeral that is one greater than the ion's charge. The differences are shown in the chart below, with copper as an example.

| Ionic species | $Cu^0$ | $Cu^+$ | $Cu^{2+}$ | $Cu^{3+}$ |
|---|---|---|---|---|
| In atomic spectroscopy | Cu(I) | Cu(II) | Cu(III) | Cu(IV) |
| In inorganic chemistry | $\begin{cases} Cu(0) \\ Cu^0 \end{cases}$ | $\begin{matrix} Cu(I) \\ Cu^+ \end{matrix}$ | $\begin{matrix} Cu(II) \\ Cu^{2+} \end{matrix}$ | $\begin{matrix} Cu(III) \\ Cu^{3+} \end{matrix}$ |

As you are well aware from your previous chemistry studies, the energy levels of atoms and ions are quantized: Only certain, specific energies are observed. The properties of an atomic-emission spectrum, such as that of atomic hydrogen (Figure 18.1a), result from this quantization. In the range from 2000 to 100 nm (near-infrared through ultraviolet), the wavelengths of the emission lines from hydrogen that appear can be described by the simple, algebraic formula

$$\frac{1}{\lambda} = R \cdot \left( \frac{1}{n_f^2} - \frac{1}{n_i^2} \right), \qquad \text{integer values } n_i > n_f$$

The variables $n_f$ and $n_i$ are integers, and $R$ represents the Rydberg constant ($R = 109,677$ cm$^{-1}$ or $3.2898 \times 10^{15}$ Hz). For atomic hydrogen, these lines are usually divided into four sets, depending on the value of $n_f$ (between 1 and 4). Notice in the figure that as the energy levels get higher, the states are closer together.

The important point to recognize is that each line in the spectrum can be described as a difference between two algebraic terms, each having a different $n$-value. Each of these $n$-values is associated with a different quantized energy state of the atom. The $n$-value is a **quantum number** of hydrogen, and the associated states are shown in Figure 18.1b. The energies associated with the different states are the so-called **electronic energies** of the atom.

Each spectral line has associated with it a lower energy state and a higher energy state. The energy of the spectral line, and hence its wavelength, equals the difference between the two energies. The lowest energy state that an atom or ion can have is called its ground state. For hydrogen, the ground state has the electron in the $n = 1$ quantum level.

All other states — those with higher energies — are called excited states of the atom. Recall that the atomic energy levels with quantum numbers $n = 1, 2, 3, \ldots$ are also called the K, L, M, . . . shells, respectively.

The process in which an electron jumps between two states of the atom is called an atomic **spectral transition.** When the spectrum is due to changes in electron energies — as it is for hydrogen — the transition is called an **electronic transition.** The point at which the electron ceases being bound to the nucleus is the **ionization potential.** The ionization potential, written in tables in beginning chemistry texts, is the energy difference between the ground state of the atom and the state where the electron and ion are completely free of each other.

However, there are actually a number of different energies to ionize atoms. For instance, consider an atom that is in an excited state: An example is hydrogen with its electron in the $n = 2$ state. If, subsequently, just enough energy is added to allow the electron to become free, the atom is ionized. As you can understand from Figure 18.1b, all the energies required to ionize atoms in excited states are less than the ionization potential from the ground state.

Notice in Figure 18.1b that only one dimension of the figure — the vertical axis — has associated with it a scale of a physical quantity. The horizontal axis usually is reserved for labeling states of the atoms.

Atomic hydrogen produces the simplest atomic spectrum of any atom. Another relatively simple spectrum is that of lithium. Part of the emission spectrum of lithium is shown in Figure 18.2a. The energy level diagram appears in Figure 18.2b. Let us study the latter figure in detail. Again, the horizontal axis is not related to energy but to our atom energy-level labeling scheme. There are four individual stacks of energy levels that are separated by their different $l$ quantum numbers. (Recall $l = 0$, 1, 2, 3 corresponds to s, p, d, and f shells, respectively.) The arrows of such figures are used to show the directions of the observed electronic transitions. Here, the wave-

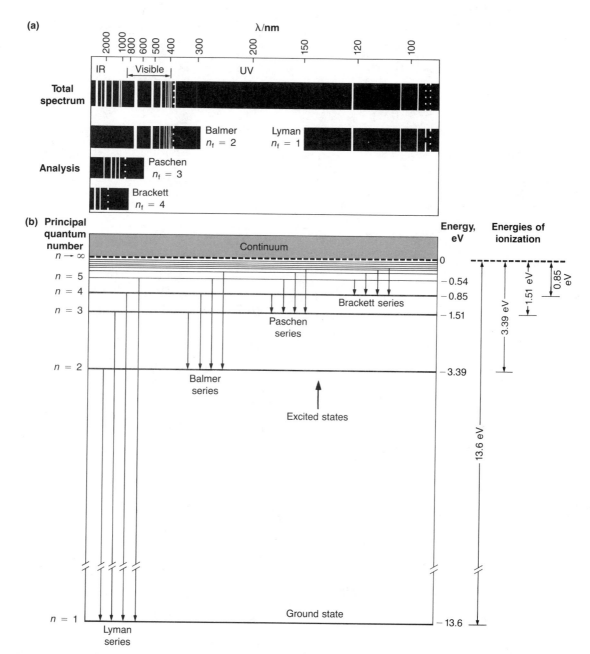

**Figure 18.1** (a) The emission spectrum of atomic hydrogen and its division into sets of lines depending on a common energy level with quantum number $n_f$. Each series of lines is named after its discoverer. [Ref: Atkins, P. W. "Physical Chemistry"; W. H. Freeman and Company: San Francisco. Copyright © 1978.] (b) The energy-level scheme that describes the spectroscopic lines of atomic hydrogen. The arrows indicate the transitions that result in light emission of the Lyman, Balmer, Paschen, and Brackett series. Four different energies of ionization are shown.

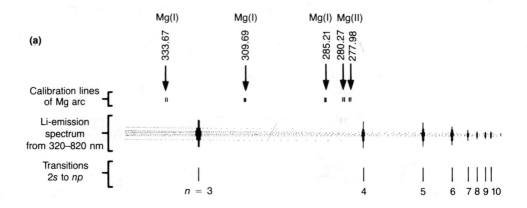

**(a)**

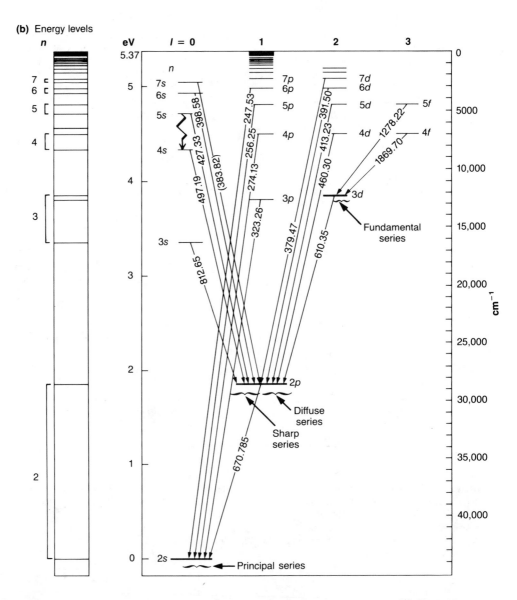

**(b)** Energy levels

lengths of the electronic transitions are listed in nm. The direction of the arrows is from a higher to lower energy state. These represent, thus, atomic-emission transitions.

Notice in the lithium energy-level diagram that none of the arrows representing observed transitions connect levels with the same value of $l$. In addition, no transitions occur with a change in $l$-value of 2. Therefore, in the emission spectrum of atomic lithium, there is a **selection rule**: $\Delta l = \pm 1$. This is a general rule describing the spectra of isolated atoms. For you right now, memorizing any specific rule is not important. However, it is worthwhile to appreciate that not all *possible* spectral lines appear — only those that "obey" selection rules do.

Let us be more precise about the atomic-emission nomenclature: If light is emitted by an atom that has no external source of illumination, the atom is said to luminesce. This process, on the atomic level, is more properly called **spontaneous emission**. A photon is spontaneously emitted from an atom in an excited state as it undergoes **relaxation** to a state of lower energy (either the ground state or a different excited state).

However, an excited atom need not emit light in the visible region to return to the ground state. The energy of a transition can also be modified by collision with another atom or ion and given off as heat. This process is called a *nonradiative transition*. A nonradiative transition is usually shown as a wavy arrow $\rightsquigarrow$ between states. A wavy arrow is drawn between the $5s$ and $4s$ levels in Figure 18.2b.

So far you have seen atomic spectral-energy diagrams of two different types: with and without $l$-quantum–number stacks. As shown in Figure 18.3, there is one more type of illustration that you might see describing atomic energy levels and transitions. Again, the horizontal axis is used for labeling the different energy levels. Here two separate sets of energy levels are drawn. This is done to show them as two classifiable sets of energy levels that exist in the same atom. In Figure 18.2, the stacks are classified depending on their $l$-quantum numbers. In Figure 18.3, the stacks differ in their spin quantum number $S$ as well. The three stacks of energy levels on the left are associated with an atom having all its electrons paired, $\uparrow\downarrow$; these energy levels are called the singlet energy levels. Singlets are indicated by a left superscript 1: $^1S$, $^1P$, and $^1D$.

The stacks of energy levels on the right exist when the atom has a set of two electrons unpaired, $\uparrow\uparrow$; these are the triplet energy levels. They are denoted by a left superscript 3: $^3S$, $^3P$, and $^3D$. More details about singlets and triplets and how they are related to phosphorescence and fluorescence are presented in Supplement 17E.

---

**Figure 18.2**    (a) The emission spectrum of atomic lithium. Calibration lines are a Mg arc with the wavelengths given in nm. The lithium spectrum was recorded on film with three different exposure times, with the longest time for the center and less time for the outer regions. This allows both weak and intense lines to appear in a sharp image on the film. [From *J. Chem. Ed.* **1974**, *51*(12), cover.] (b) The quantized energy levels of atomic lithium. The vertical scale of the figure is energy, with eV on the left and cm$^{-1}$ on the right. The horizontal scale is the $l$-quantum number assigned to the levels in the stack. Note that all observed spectral lines, as indicated by the arrows labeled with the wavelengths of the radiation, occur from transitions between adjacent stacks. This reflects the natural selection rule, $\Delta l = \pm 1$. The descriptive spectroscopic names *sharp*, *principal*, *diffuse*, and *fundamental* are the origins of the theoretical orbital names *s*, *p*, *d*, and *f*. [Ref: Walker, S.; Straw, H. "Spectroscopy"; Chapman & Hall, Ltd.: London, 1961.]

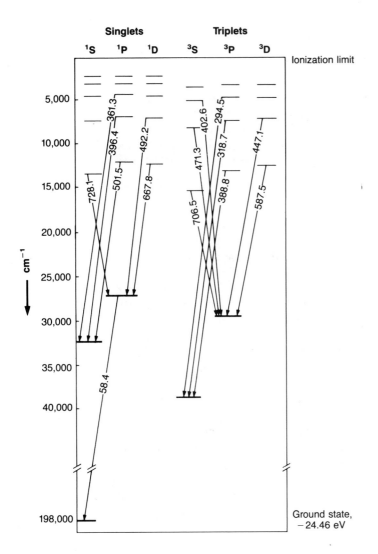

**Figure 18.3**    The quantized energy levels of atomic helium. The vertical scale in the figure is linear in energy. The straight lines indicate observed transitions, with the wavelengths in nm shown. There are two sets of three stacks. The stacks are labeled $S$, $P$, and $D$; these have $L$-quantum numbers 1, 2, and 3, respectively. The three stacks on the left indicate energy states in which the electrons are paired ↑↓. These are the singlet states and are labeled as $^1S$, $^1P$, and $^1D$. The three sets on the right have two electrons parallel ↑↑. These are the triplet states and are labeled $^3S$, $^3P$, and $^3D$. Note that transitions are observed only between stacks within a set. This reflects the operation of two natural selection rules: First, $\Delta L = \pm 1$; second, the spin state does not normally change in a spectroscopic transition, $\Delta S = 0$. Each transition reflects changes in the energies of both electrons; the states characterize the energies of helium atoms and not just one-electron orbitals. [Ref: Kuhn, H. G. "Atomic Spectra"; Academic Press: New York, 1962.]

## 18.3 Intensities and Linewidths of Atomic Spectra

The measured radiant power of each line in an atomic-emission spectrum is proportional to the algebraic product of the linewidth and the peak power. When atoms are isolated, the measured power of emission with a fixed optical geometry depends on three factors:

1. The specific spectral transition that is occurring
2. The identity of the atoms
3. The number of atoms in a given volume — the concentration

Similarly, the integrated *absorption* of an atomic spectral line depends on the same three factors when the atoms are isolated in the gas phase.

However, the *peak* emission intensities and *peak* absorptivities of the atomic spectral lines depend on the gas-phase conditions of the atoms. The magnitudes of the peaks are strongly affected by the following:

1. The temperature of the atoms
2. The pressure
3. The components of the surrounding gas (to a lesser extent)

The reason that these peak magnitudes vary is that the spectral linewidths change. The *integrated* power or absorption for each line (at a fixed analyte concentration) remains practically constant. These linewidth changes are quite important in the practice of atomic spectrometry. The effects of temperature and pressure on the linewidths are elaborated in the next few sections.

## 18.4 Linewidth Dependence on Temperature and Pressure

The general behavior of atomic-*emission* linewidths is that they increase with increasing temperature. In other words, an emission line will be wider when the atoms are hotter and narrower when the atoms are colder. In the case of atomic *absorption* of light, the quantity that changes is the width of the **absorption envelope** (see Figure 18.4). The absorption envelope is the wavelength region over which non-zero absorption would occur if light were passed through a sample.

The reason for the broadening with temperature in both emission and absorption is that the atoms move faster at higher temperatures. Recall the definition of temperature: a measure of the average speed of motion of the atoms. The random motion causes a **Doppler broadening** of the spectral lines. The basis of this broadening is a **Doppler shift** of the light frequency, which is similar to the shift in tone (sound frequency) of a horn of a moving train or car as it approaches (higher frequency) and departs (lower frequency) from you.

Consider this specific example: the change in width of one of the two yellow lines of sodium atomic emission. (Two emission lines at nearly the same wavelength cause the bright yellow color that is seen in flames when glass is heated.) One of the two sodium lines has an inherent linewidth (fwhm) of $0.33 \times 10^{-3}$ cm$^{-1}$. This is called the *natural linewidth*. (At the line's wavelength, 589.6 nm, the natural linewidth $0.33 \times 10^{-3}$ cm$^{-1}$ equals 0.00001 nm.) However, the observed width is about 0.04 cm$^{-1}$ (0.0015 nm). Of the total linewidth, the Doppler broadening comprises more than

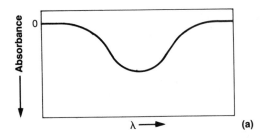

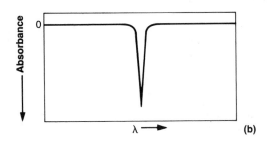

**Figure 18.4**    Illustration of the concept of absorption envelope. All three figures represent absorptions over the same spectral range. We assume that an experiment is carried out using a high-resolution spectrometer. The light from a continuum source passes through a sample and is detected. When no sample is present, the output appears as in (c). When sample A is in the beam, the output appears as in (a). When sample B is present, the output appears as in (b). Sample A has a broader absorption envelope than sample B. The absorption envelope is, in effect, a plot of absorbance versus wavelength. We think of the absorption envelope as being present even though no light source is present.

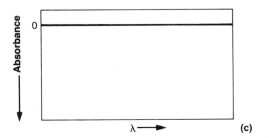

99% of the linewidth in this specific case. Thus, the Doppler shift is the major contributor to the linewidth of the atomic emission lines.

The natural linewidth and Doppler-effect linewidth have been carefully studied for the sodium emission lines. In addition, the spectral lines of most other atoms behave *approximately* similarly. The relative magnitude of natural linewidths and thermally broadened linewidths—about a factor of 100—is worth keeping in mind.

Having narrower linewidths is desirable in atomic spectrometry for two reasons: because the peak radiant power tends to be greater, and because it is less probable that atomic spectral lines overlap. As a result, the measurement sensitivity will be greater, and there will be less chance of spectral interference. From the atomic spectra seen in Figures 18.1 and 18.2a, line overlap might not seem to be a major problem. However, as illustrated in Figure 18.5, emission spectra due to more than one atomic species are much more likely to be crowded.

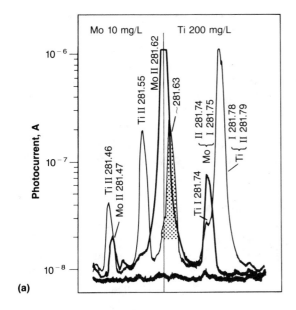

**(a)**

**Figure 18.5**   (a) Example of emission spectra: titanium atomic emission spectrum superimposed on that of molybdenum. If titanium and molybdenum were present in a single sample, the lines would overlap as shown. In the nomenclature of atomic spectroscopy, Ti(II) is Ti$^+$ and Mo(II) is Mo$^+$. Atomic titanium Ti$^0$ is written Ti(I). The stippled line is unidentified. The wavelengths are in nm. The sample concentrations are shown at the top. [Ref: Winge, R. K.; Fassel, V. A.; Peterson, V. J.; Floyd, M. A. *Appl. Spec.* **1982**, *36*, 210.] (b) Classification of three types of emission spectral interference. It is assumed that the 213.856-nm emission of Zn is being monitored. Other emissions produce (L to R) **direct overlap** of spectral lines, **wing overlap,** and a continuum background. The division between direct and wing overlap is somewhat arbitrary. These figures are *not* actual spectra but are derived from experimental spectral information. Note that the interferents are in large excess over zinc. The total spectral range shown is 0.150 nm. [Ref: Reprinted from Sobel, H. R.; Dahlquist, R. L. *Am. Lab.* **1981**, *13*, 152. Copyright 1981 by International Scientific Communications, Inc.]

**Inteferents at 1,000–10,000 > Zn Concentration**

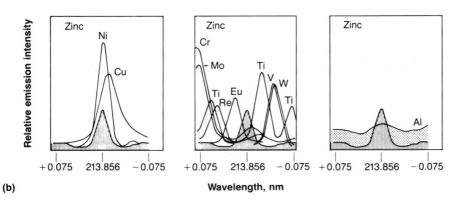

**(b)**

Higher pressures also cause broader spectral emission lines and broader absorption envelopes. The explanation of this effect is complex. However, the main idea is that collisions between atoms and ions cause small shifts in the electronic energy levels. As a result, the transitions between energy levels occur over a broader range of energies than if no collisions occurred: A broadening results. In addition, the *integrated* intensity of the line may change as well.

The line-broadening effect suggests that to avoid spectral interferences, low pressures are desirable in practice. However, this condition conflicts with having high concentrations of analyte to achieve better sensitivity. Finding an optimum range is one of the goals in analytical atomic spectrometry.

## 18.5 Simultaneous Emission and Absorption and Lineshape Changes

Atomic emission and atomic absorption are similar in a number of ways. Both are due to transitions between atomic energy states: emission from an excited state "down" and absorption from the ground state "up." Further, the widths of the emission lines or absorption envelopes depend on temperature and pressure. However, let us ignore the effects of pressure for now.

When absorption and emission of an atomic species both occur in the same region, self-absorption can occur. You read about the effects of self-absorption in fluorescence spectrometry in Section 17.7.

Now, consider what happens when both the emission lines and corresponding absorption envelopes are narrow. An interesting phenomenon occurs, one that can be both a problem and an asset. What will be the effect on an emission line if the corresponding absorptivity is relatively large? The answer is illustrated in Figure 18.6.

Assume that an experiment is done using a high-resolution spectrometer, which can scan even narrow atomic absorption lines to determine the lineshape profiles (power versus wavelength). If a single atomic line from an emission source of element $X$ is scanned, the spectrum appears as shown in Figure 18.6a.

Now, place a gaseous sample of the same element $X$ in the path of the light as it passes from the emission source to the detector. Thus, after self-absorption occurs in

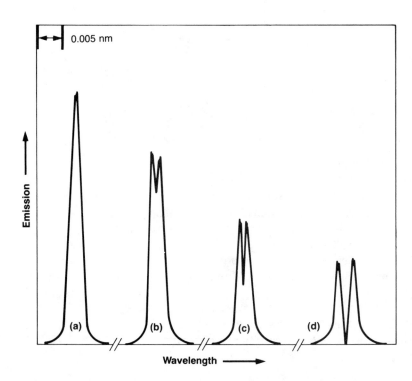

**Figure 18.6** The effects of both emission and absorption by the same element: self-absorption and self-reversal. Shown here are the spectra that would be seen from a line-source emission (a) with varying amounts of atomic vapor placed between the source and a high-resolution spectrometer (b–d). In (b), there is evidence of self-absorption. The word *self* means that the emission and absorption arise from the same elemental species. Spectrum (c) results when the atomic vapor concentration (or path length) is increased. The absorption of the middle of the line increases. At yet higher absorbing-vapor concentrations, the single line will appear to be two separate lines as shown in (d). This is called self-reversal. If the emission occurs in a hotter region, and the absorption arises from the same species in a cooler region, the absorption envelope will be narrower than the emission line. In that case, for a given concentration of absorber, the effect illustrated here is enhanced.

the gaseous sample, the spectrum appears as shown in Figure 18.6b. The energy profile of the emission line has a dip in it. The dip results from the shape of the absorption envelope: The maximum absorbance occurs at the peak of the emission.

If the absorption cell length or the concentration of the absorbing atoms is increased, more light is absorbed, and the spectrum then appears as in Figure 18.6c. Finally, in the extreme case, most of the light in the center of the emission line is absorbed; the single emission line appears as if it were two very narrow, closely spaced lines. This complete absorption of light at the center frequency of the line is called **self-reversal** and is illustrated in Figure 18.6d.

The preferential absorption of light in the center of the emission line is even more pronounced when the absorption envelope is narrower than the emission line. This situation will occur when the emitting atoms reside in a hotter region and the absorbing atoms in a cooler one. Then, the effects of self-absorption and self-reversal occur at lower absorber concentrations. This situation is quite common in flames, which usually have significant differences in temperature between the centers (usually hotter) and the outer regions (cooler).

## 18.6 Methods of Obtaining Isolated Atoms from Samples

As noted in Section 18.3, analytical atomic spectrometry depends on obtaining isolated atoms in the gas phase. Ideally, all atoms are freed of the influence of the surrounding matrix. This means that the analyte atoms are not chemically bound to

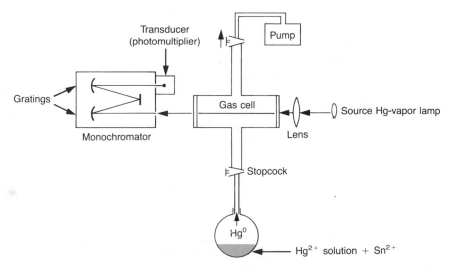

**Figure 18.7**    Diagram of the apparatus that can be used to determine the concentration of atomic mercury in a gas cell. Mercury vapor, produced by reduction of $Hg^{2+}$ to $Hg^0$ by $Sn^{2+}$, fills the previously evacuated cell. The light source is a mercury-vapor lamp. One atomic line is selected by the monochromator.

**Table 18.1** Boiling Points of
Selected Elements

| Element | Boiling Point °C |
|---------|------------------|
| Aluminum | 2057 |
| Barium | 1140 |
| Cadmium | 767 |
| Chromium | 2480 |
| Cobalt | 2900 |
| Gold | 2600 |
| Iron | 3000 |
| Lead | 1620 |
| Manganese | 1900 |
| Mercury | 356 |
| Molybdenum | 4800 |
| Nickel | 2900 |
| Platinum | 4300 |
| Potassium | 760 |
| Silver | 1950 |
| Tin | 2270 |
| Vanadium | 3000 |
| Zinc | 907 |

any other atoms: All particles, compounds, and molecules must be dissociated. The total matrix of the sample should be destroyed—**atomized** is the correct term. This ideal is not often reached. However, as long as constant conditions are obtainable, precise analyses ($\pm 1\%$) can be carried out. In effect, the recoveries and interferences are kept constant so that determinate errors do not decrease the accuracy.

If the element to be analyzed is a gas near ambient temperature, then atomization is quite simple. It is only necessary to hold the element in a cell, such as shown in Figure 18.7. For example, elemental mercury is a volatile metal and can be determined in this way with absorption, fluorescence or emission spectrometry. (Ionic mercury can be determined in a similar manner. First, the mercury can be reduced to its metallic form, which then is collected in the gas cell.)

What can be done if an element or compound is not highly volatile? This situation is far more common, as can be seen from the pure-element boiling points shown in Table 18.1. In addition, solid compounds of the metallic elements are often even more stable than the element alone. For example, iron oxide must be broken down to create free iron atoms. To achieve minimal matrix interference, high temperatures are needed.

**Table 18.2**   Sources of Free Atoms: Nonflame and Flame

| Nonflame Source | Approximate Temperature (K) | Use |
|---|---|---|
| DC arc between two carbon rods in Ar atmosphere | 6000 | Emission |
| Radiofrequency-induced plasma in argon (inductively coupled plasma, or ICP) | Up to 10,000 | Emission, fluorescence |
| Graphite furnace (sample inside a hollow graphite tube) | Up to 3000 (controllable) | Absorption |

| Flame Source | | Approximate Maximum Flame Temperature (K) | Use |
|---|---|---|---|
| *Fuel* | *Oxidant* | | |
| Acetylene | Air | 2700 | Absorption (mostly) |
| | Nitrous oxide | 3100 | Absorption (special cases) |
| | Oxygen | 3400 | Emission |
| Hydrogen | Air | 2400 | |
| | Oxygen | 3000 | |

*Source:* Massmann, H. *Angew. Chem. Int. Ed. Engl.* **1974,** *13,* 504. Mavrodineanu, R.; Boiteus, H. "Flame Spectroscopy"; Wiley: New York, 1965.

Table 18.2 lists some sources such as flames, furnaces, plasma torches, and electric arcs that are used to heat such nonvolatile samples. These atomization methods are introduced in this section.

## Flames

Flame sources of atomization have been brought to a high degree of reliability over many years of development and are the most commonly found atomization source. To obtain high sensitivities, especially in absorption, a long flame shape is desired: More absorption occurs with a long path length for the light. A long flame can be obtained by using a slot-burner, such as illustrated in Figure 18.8. A variety of different fuels and oxidants can be used. The choice depends on the desired temperature and the elements that are being determined. However, all the common flame sources remain at relatively low temperatures compared with a number of other atomizers. Flame atomizers are used for atomic emission, absorption, and fluorescence methods.

## Furnaces

Furnaces are used to atomize samples for atomic absorption. As illustrated in Figure 18.9, one type of furnace consists of a graphite tube with an inside diameter of a few millimeters. The sample placed in the furnace can be a solid or liquid. If it is a solution, the first step is to dry it by heating at a low temperature (about 100 °C). Sometimes it is beneficial to heat the furnace to a higher temperature (determined

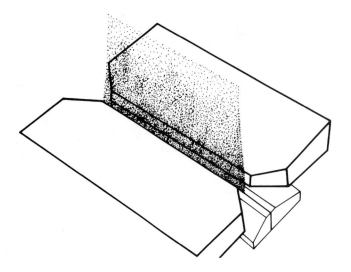

**Figure 18.8**    Illustration of a long flame produced by a slot burner. As the light path through the flame is increased, so is the absorbance of the sample.

empirically) to vaporize some of the matrix interferents. This operation is called the **char** step, and it can be helpful if the analyte is in a heat-stable form. Subsequently, the temperature is raised as rapidly as possible, and the sample is vaporized into the light path of the spectrometer. It is essential to keep the furnace in an inert atmosphere such as argon during the heating period to impede reactions with the atomized analytes. Since the furnace tube is heated by passing electrical current through the graphite (which acts as a resistor), the method is also called **electrothermal atomization.**

### Plasmas and Spark Sources

A plasma is an ionized gas. Through the addition of energy, atoms are converted to ions and the resulting free electrons, with both in the gas phase. Plasma sources are

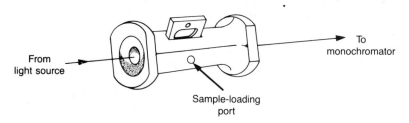

From light source

To monochromator

Sample-loading port

**Figure 18.9**    Illustration of one kind of graphite furnace. A precisely measured volume of a liquid sample can be injected through the hole into the *microboat* (here shown pulled out of its position). Other, simpler designs have the sample placed directly on the inside surface of the furnace. Electric current to heat the furnace is passed through the graphite by electrodes clamped at each end. As indicated, the light from the spectrometer passes through the annulus containing the atomized sample. [Ref: Redrawn from Sotera, J. J. *Anal. Chem.* **1983,** *55*, 204. Copyright 1983 American Chemical Society.]

quite commonly used in atomic emission. An example of a short-lived plasma is a spark that jumps across a gap with concomitant emission of light. Therefore, from the basic definition of a plasma, spark sources and plasma sources are the same. However, among other differences between plasma and spark sources in the practice of atomic spectrometry, one stands out clearly: Plasma sources are run continuously while spark sources operate intermittently.

To create the most common type of continuous plasma used in atomic emission spectrometry, the excitation energy is fed (really, "broadcast") into the gases as radiofrequency radiation. Since there is not a directly wired connection between the

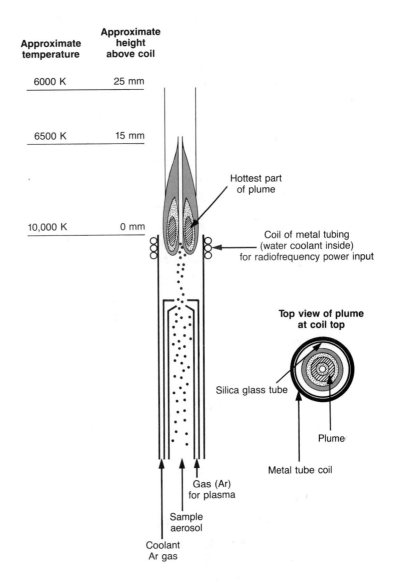

**Figure 18.10**  Schematic diagram (not to scale) of the inductively coupled plasma (ICP) torch used as an atomic-emission spectrometry source. A plasma of ionized argon is excited by radiofrequency energy "broadcast" from a coil at the base of the plume. The plasma is hollow, and the analyte is blown into the hollow region. The gases carrying the nebulized sample flow up from below together with a separate gas stream for cooling the silica glass. The plasma plume is suspended away from the walls through a combination of cooling gas flow and the shape of the radiofrequency electromagnetic field. Otherwise, the 10,000-K temperature plume would easily melt the silica.

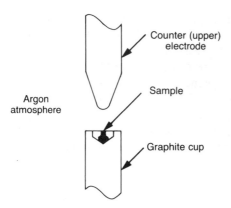

**Figure 18.11**    Illustration of one of many designs for a spark source atomizer. The electrodes are machined from graphite rods 3–8 mm in diameter. The configuration shown here consists of a cup electrode containing the solid sample and a broadly pointed counter electrode above. The electrode assembly is surrounded by an inert atmosphere enclosure. A spark is initiated by a high-voltage pulse. Once the spark plasma is initiated between the electrodes, the spark is sustainable with a low-voltage current (less than 50 V applied but up to 35 A). The electrodes heat up rapidly to around 4000 K, and the sample vaporizes. The arc temperature is generally above 5000 K, which completes the analyte atomization and excites atomic emission.

plasma and the energy source, the plasma is said to be coupled inductively to the energy source and is called an **inductively coupled plasma (ICP).** The general structure of an ICP source is shown in Figure 18.10.

Spark sources consist of two electrodes, usually made of graphite (carbon), formed as shown in Figure 18.11. The solid sample, such as a weighed steel particle, is placed on the bottom electrode. Then, a high voltage is applied across the gap. A spark — really, a high-temperature plasma — is formed, which vaporizes and atomizes the sample and excites the atoms. The excited atoms then emit their characteristic radiation, which is quantitatively measured at the characteristic wavelengths of the components being determined. Contemporary spark sources are commonly run in an inert gas atmosphere.

Two other plasma-generating methods are used: a plasma excited by a direct current voltage (a **DC plasma**) and a plasma excited by centimeter-wavelength radiation (a **microwave-induced plasma**). Neither of these two sources is as common as the two previously described. More information on these can be found in the references at the end of the chapter.

In general, spark sources are used for solids directly. The various plasma sources usually have samples fed into them in vaporized form, most commonly as a water-solution mist. However, organic solutions can be used.

## 18.7 From Samples to Atoms: Factors Affecting Atomization

### Solid Samples

The atomization of a sample is a complex process that is still not well understood in all its details, but the order of the major steps in the process has been elucidated. Atomization of a powdered solid material by rapid heating proceeds in general through the following steps.

1.   The powder heats up.
2.   Particles composed of many atoms break off from the powder surface.

3. The multiatom clusters vaporize and break up (atomize) into atoms and ions.
4. Some atoms are ionized; some ions are neutralized to atoms. The number in each form (atoms and ions) changes until equilibrium is reached. Emission and absorption can accompany these reactions.
5. Some atoms or ions may combine to form highly heat-stable molecules and compounds such as oxides, nitrides, or carbides. These will no longer absorb or emit as atoms and, as such, are effectively lost.

If the processes described in steps 1–3 do not go to completion, then there could be a problem in achieving good precision or accuracy. For instance, the powder could be composed partly of thermally stable (refractory) compounds and partly of less stable compounds. Upon being heated, the less stable compounds can fractionally boil off. This is a fractional sublimation similar in concept to a fractional distillation. So, while the less stable compounds are atomized, the process could leave the more refractory compounds intact and unatomized. As a result, part of the total elemental content would go undetected: The instrument would respond only to the atomized portion.

The process mentioned in step 5 will reduce the output response to the elements contained in the sample: This is a chemical interference. A second chemical interference, described in step 4, is less straightforward in its effect. Depending on the atomic spectrometric method chosen, one may want to monitor the spectrum of either the atomic or an ionic form of the element. The spectra of an atomic species and its ions are significantly different. *Different* means that all the spectral lines of the atom occur at different wavelengths from the lines due to its ions. If the instrument is set to detect an atomic line, it is unlikely that emission from the ions will contribute to the signal. As a result, if the atom content is being monitored, ion formation is the chemical interference. If the ionic lines are being observed, formation of atoms is the chemical interference that decreases the analytical signal. Atom–ion equilibria are discussed in Section 17.15.

Formation of compounds, in addition to signal loss, adds more problems. The stable molecules that form can contribute significant spectral interference. The reason is that absorption and emission by *molecules* that are isolated in the gas phase consist of spectral bands composed of as many as hundreds of individual lines. An illustration of such bands in emission is shown in Figure 18.12.

The errors associated with incomplete atomization and with molecule formation generally can be minimized or eliminated by using higher temperatures. The temperature should be as high as possible (10,000 K is about as high as is now routine), not only to optimize atomization of the entire sample but also to cause dissociation of molecular compounds that may form. In other words, high temperatures minimize variations in atomization that are dependent on the physical properties of the sample, such as particle size or microcrystalline form, as well as minimizing molecule formation. To undermine this simple picture somewhat, you should recognize that some stable molecules do not form *except* at high temperature.

At the same time, high temperatures increase ionization of all elements, an effect that is especially significant for the more easily ionized elements (For examples see

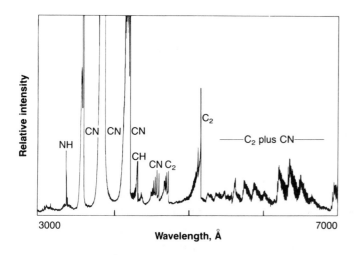

**Figure 18.12** Spectrum of molecular emissions from a nitrous-oxide/acetylene flame. Each of the simple two-atom molecular species emits large numbers of spectral lines that are seen here as bands. The spectrum shown here is low resolution and covers the entire visible range. The individual bands are composed of hundreds of lines that are not resolved. These bands can cause significant interference in atomic spectrometry. Other examples of molecular bands are shown in Figure 18.15 and Supplement Figure 17G.5. [Ref: Reprinted with permission from Christian, G. D. *Anal. Chem.* **1969**, *41*(14), 24A. Copyright 1969 American Chemical Society.]

Table 18.3.) Emissions from the alkali metals (Li–Cs) have the potential to be particularly bad interferents. Somewhat less so are those of the alkaline earths (Mg–Ba). But this is not a firm rule. Even though the spectral interference may increase in their presence, some of the chemical interferences, such as ionization and molecule formation, may be suppressed by the presence of $K^+$ and $Cs^+$ (see Section 17.15).

From the above discussion, you can observe that trade-offs must be made in choosing the conditions of atomic spectrometric determinations. The highest temperatures are best for atomization of the solids. On the other hand, higher temperatures can shift the gas-phase equilibrium toward the ionic species, which may or may not be desirable.

### Liquid Samples

When the sample is introduced into a flame or plasma source as a liquid, there are a few additional steps in the atomization process. Liquid samples are most commonly blown into the atomization source as a mist. As the droplets of this mist become heated, the liquid evaporates leaving small, solid particles that become atomized in

**Table 18.3** Proportion of Ionized Atoms of Some Elements in Two Different Flames

| Element | Ionization Potential (eV) | Proportion of Ionized Atoms (%) | |
| --- | --- | --- | --- |
| | | *Propane/Air* | *Acetylene/Nitrous oxide* |
| Cs | 3.87 | 28 | 96 |
| Li | 5.37 | 0.01 | 16 |
| Sr | 5.69 | — | 20–80 |
| Ca | 6.11 | — | 10–40 |

*Source:* Massmann, H. *Angew. Chem. Int. Ed. Engl.* **1974**, *13*, 504, refers to M. Pinta, "Spectrometrie d'absorption atomique," Masson et Cie., Paris, 1971.

the manner described in steps 1 – 5 above. Thus the possible chemical interferences remain the same.

There are numerous apparatus designs in use to convert bulk liquid samples into droplets (called *nebulizing* from the Latin *nebula* meaning *cloud*) to feed into the thermal source. The part of an atomic spectrometer (either emission or absorption) that nebulizes the liquid is called, reasonably, a **nebulizer.** One of the many types of nebulizers used in commercial instruments is illustrated in Figure 18.13.

Nebulization of a liquid sample requires a rapidly flowing stream of gas to interact with a relatively slowly flowing feed-stream of liquid. The gas flow breaks the liquid into droplets in the same way that a perfume atomizer (really, a nebulizer) or paint sprayer does. In most types of nebulizers, the droplets have a large range of sizes, from fog to fine raindrop size. After the initial nebulization, this heterogeneous mist interacts with a baffle or impinger in the gas stream, which both breaks up and removes the larger droplets by causing the gas stream to take a sinuous course. The larger droplets move more sluggishly, strike a surface, and either break up or flow out a drain. Generally less than 10% of the liquid reaches the thermal atomizer.

The amount of liquid sample and the size of the droplets that feed into the flame or plasma depend on the

1. design of the nebulizer,
2. gas flow rate,
3. viscosity of the liquid,
4. density of the liquid, and
5. surface tension of the liquid.

For a given set of determinations on the same instrument, the first two factors can be kept constant. However, the last three depend on the properties of the liquid of *each* sample and cannot be ignored for a precise and accurate analysis. Some specific details are described next.

As illustrated in Figure 18.13, the liquid sample is drawn through a small plastic tube into the nebulizer. Under a fixed set of conditions (gas flow rate, temperature,

**Figure 18.13**   Diagram of one of the many types of nebulizers that are part of atomic spectrometers with flame atomizers. The liquid sample flows through a feed tube to the region in which the liquid stream is broken up by a turbulent gas flow. The liquid droplets are carried by the gas stream toward the burner head. However, the larger droplets cannot closely follow the flow and impact on the bead in the middle. Only the smaller droplets continue along into the flame. The excess liquid, which in this design is greater than 90% of that aspirated, runs off through the drain. The brittle diaphragm protects the system (and operator) from explosion if the flame flashes back into the nebulizer.

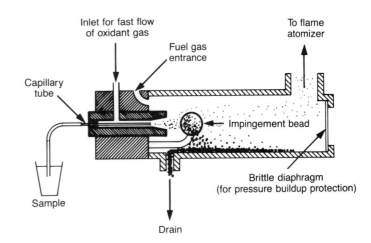

**Table 18.4**  Viscosities of Solvents and Solutions at 25 °C

| Liquid | Viscosity (centipoise) | Viscosity Relative to Water |
|---|---|---|
| Water | 0.894 | 1.00 |
| 1-M HCl | 1.01 | 1.13 |
| 1-M $H_2SO_4$ | 1.18 | 1.33 |
| 1-M $HNO_3$ | 1.03 | 1.15 |
| 1-M NaCl | 1.09 | 1.22 |
| 1-M KCl | 0.99 | 1.11 |
| 1-M $Ca(NO_3)_2$ | 1.29 | 1.44 |
| Methanol | 0.553 | 0.62 |
| Benzene | 0.602 | 0.67 |

*Source:* "International Critical Tables," Vol. 5; McGraw–Hill: New York, 1929.

and nebulizer design), the flow of liquid through the feed tube is inversely proportional to the viscosity of the liquid. This means that if the viscosity doubles in magnitude, the rate that the liquid flows through the tube is cut in half under the same pressure difference. In Table 18.4 are listed the viscosities of some liquids that might be used in atomic spectrometry.

Consider that an aqueous-solution standard is prepared. It is composed of a few ppm of some salt such as $CuCl_2$ and will be used as a standard for a Cu determination. This dilute solution flows about as easily as water. But what happens if the sample itself was digested in concentrated $HNO_3$ and diluted so that the solution is 1 M in the acid? As noted in the right-hand column of Table 18.4, the solution will flow about 15% more slowly under the same conditions. Therefore, due to the difference in viscosity, about 15% less sample will pass through the nebulizer to be atomized. The copper signal from the sample can be 15% smaller than it would be if the sample and the standards were equally viscous.

The viscosity effects in aqueous solutions can be compensated in three different ways.

1. Add an acid such as $HNO_3$ or $H_2SO_4$ to the standards so that they have viscosities equal to that of the samples. This process is called making *balanced* solutions.
2. Pump both the standards and samples into the instrument at a fixed rate. Small pumps with inert flow paths are common for this use.
3. Use internal standards.

Two other liquid properties—surface tension and density—are also changed by the addition of acids and salts. These properties are not as easy to compensate; nevertheless, because they are sources of possible error, they must be considered. Spiking is really the only way to enable us to correct for these complicated matrix effects.

Note that it would be foolish to use aqueous standards to calibrate samples that are dissolved in organic solvents. In such cases, the errors due just to viscosity could be in the range of 40% — without even considering the fact that the surface tensions, enthalpies of vaporization, and densities of the organic solvents are significantly different from those of water.

## 18.8 Background Correction in General

In common with all spectrometric methods, the light that is monitored in atomic spectrometry contains contributions from the element being assayed and some background radiation. To obtain the signal due only to the desired element alone, we must subtract the signal due to the background. (The ideas of background correction in spectrometry were presented in Section 17.14 and Figure 17.19 and are applicable in atomic spectrometry.)

As you now know, there are two general ways to ascertain the background contribution. The first is to run blanks. The second is to scan the wavelengths and record the signal of the background separately from that of the elemental analyte. Since the background in atomic spectrometry can be relatively large and variable, ascertaining the spectral background correction for each sample independently is the best choice in most cases. In the next section, some of the causes of the variability of spectral interferences are illustrated.

## 18.9 Causes of Spectral Interferences in Gases

As noted in the last chapter, spectral background consists of electromagnetic radiation from species in which we are not interested. The spectral interferences arise from other components in the original sample and from the combustion gases of the flame (or the species present in a plasma). We are concerned with background radiation only in the optical spectrometry region. (Chemical interference — the reduction of light intensity at the analytical wavelength that is caused by chemical alteration of the analyte species — does not enter into this discussion of spectral interferences.)

In this section, we shall describe how we classify different characteristics of the spectral background and explain what past research has shown to be the origins of this background. It is worth understanding the sources of background radiation so that the sampling and sample preparation can be better designed. Generally, reducing background radiation improves the precision of an atomic spectral analysis.

The spectral interference effects in atomic spectrometry are due to the following phenomena in the gas phase. The light can originate in sample emission or from an external source.

1. Light scattering from particles and droplets. As is illustrated in Figure 18.14, light scattering usually decreases the amount of light that reaches

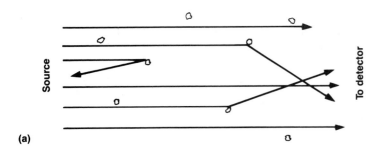

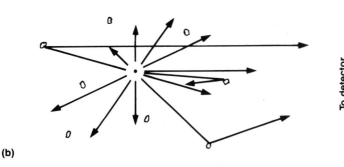

**(a)**

**(b)**

**Figure 18.14**    Illustration of light scattering from particles ($> 1\,\mu$m). (a) Scattering by particles can reduce the amount of light passing through a region. The light can be deflected or reflected away from the transducer at the right side. (b) Particles can increase the amount of light reaching the transducer from a luminous source. The light that otherwise might not reach the transducer may be reflected back toward the transducer or scattered into it. The scattering of light is more complicated than indicated here; the process depends on the sizes of the particles, their index of refraction, and the wavelength of the light.

the transducer. However, if the reflection of light is greater than the blocking, an increased signal can be detected. The particles can originate from incomplete atomization and the droplets from poor nebulization.

2.  Molecular absorption and emission, which consists of significantly more closely spaced discrete lines than do most atomic spectra. The difference between a clear baseline and one that includes a molecular species is illustrated in Figure 18.15. The type of background seen at the right of the figure, where the signal rapidly varies with wavelength, is called **structured** background. In atomic emission determinations, finding wavelengths at which the background can be sampled may be difficult. One way to overcome the difficulties of effective background subtraction is to use an atomic line in a different wavelength region where the background is not structured. Sometimes regular background-correction methods of atomic absorption function well even in the presence of structured backgrounds. These atomic absorption techniques are discussed in Section 18.11.

3.  Radiation from the continuum part of the spectra of matrix atoms. A continuum spectrum results when the separation between energy levels becomes smaller than the spectral linewidth; the spectral lines run together and are unresolved. The shift from discrete transitions to the continuum spectra of atoms is illustrated in Figure 18.16. In the contin-

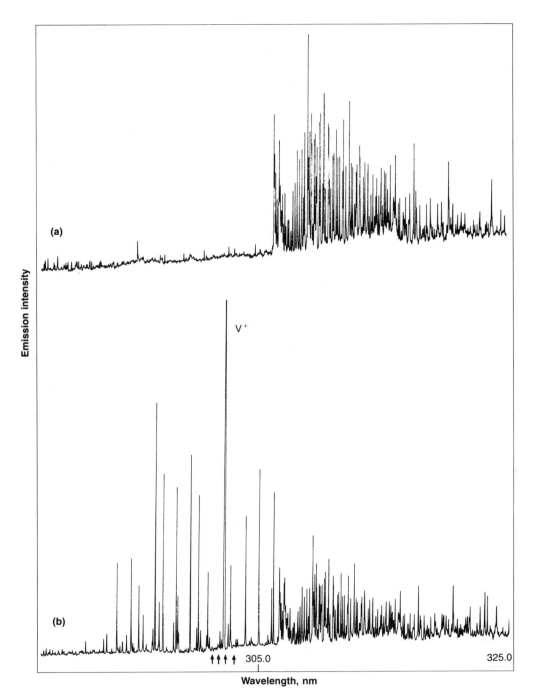

**Figure 18.15**   Spectra of molecular bands and atomic lines. The spectra here were obtained by wavelength scanning of the emission from a nebulized liquid analyte that was atomized in an ICP torch. (a) The spectrum of a water blank showing the molecular band spectrum of OH and a few atomic lines. (b) Spectrum of a water solution with vanadium present showing the atomic lines that fall outside the molecular band region. Precise background correction is easier in the region of unstructured background than in the region in which lines from the molecular band appear. The arrows in (b) indicate some of the wavelengths at which background measurements might be taken on either side of the emission line of $V^+$.

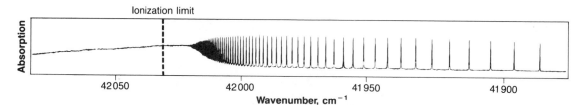

**Figure 18.16**    Absorption spectrum of barium illustrating the progressive narrowing of spectral line spacing until a continuum absorption is reached. The continuum produces an unstructured background, which can be subtracted to determine the absorption from the analyte alone. [Ref: Clark, C. W.; Lu, K. T.; Starace, A. F. In "Progress in Atomic Spectroscopy," Pt. C; H. -J. Beyer and H. Kleinpoppen, Eds.; Plenum: London, 1983, Chapter 7.]

uum region, this background is said to be **unstructured** and usually can be easily and effectively compensated.

4.    The light emitted by all heated objects—from red hot to white hot. This is called **blackbody radiation.** The temperature dependence of blackbody radiation is illustrated in Figure 18.17. As you can see from the figure, blackbody radiation is smoothly varying over a range of wavelength. It too is unstructured and can be compensated effectively. The blackbody in atomic spectrometry can be the flame or plasma as well as any material passing through, such as unatomized particles. The applicable blackbody temperatures are those noted in Table 18.2 (p. 715).

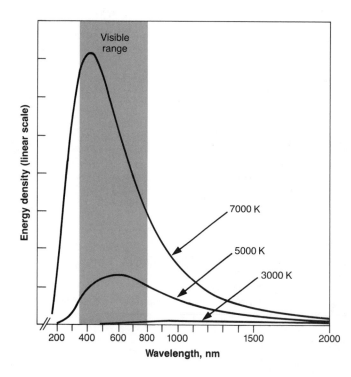

**Figure 18.17**    Illustration of the spectrum of blackbody radiation at three temperatures that are characteristic of atomizers used in analytical atomic spectrometry. The hotter the atomizer, the greater will be the contribution of blackbody radiation to the background. For flame and furnace sources around 3000 K there is little contribution. However, for spark and plasma sources, blackbody radiation can contribute a significant background emission. Blackbody emission is relatively easy to compensate.

5.  Self-reversal of spectral lines due to the absorbance of emitted radiation by cooler parts of the source. The effect of self-reversal on the line shape was shown in Figure 18.9. Self-reversal always reduces the amount of light that reaches a transducer. We can compensate for the effects of self-reversal, but the flame/plasma conditions must be highly stable to do so with high precision.

As noted earlier in this section, it is worthwhile understanding the origins of spectral interferences in atomic spectrometry in order to make intelligent choices of sample pretreatments. As you now realize, the kinds of interference effects depend on *all* the materials that go into a sample (that is, the original sample plus the chemicals used in sample preparation).

Not only does the sample content affect the background, the *manner* of atomization does too. This means that the background will differ if the sample is atomized in a flame or plasma or by heating in a furnace. In other words, the chemical species present in the gas phase depend not only on the components of the sample and the pretreatment reagents, but also on the detailed manner in which we add energy to atomize a given sample. These details include the temperatures of the various regions of the atomizer (flame, plasma, or furnace) as well as the length of time the material remains at each temperature. Thus, precise results are obtainable only when care is taken to make the analysis reproducible throughout all the steps.

The three types of atomic spectral methods—atomic emission, atomic absorption, and atomic fluorescence—are susceptible to the same spectral interferences to different degrees. However, each method is designed to overcome interferences in a different way. In Sections 18.10 through 18.12, the general manner of coping experimentally with the backgrounds is discussed for each of the three spectrometric methods.

## 18.10 Atomic-Emission Methodology

You have seen illustrations of equipment used to nebulize and atomize samples. In this and the next two sections, we shall concentrate on the light that is quantified and the general properties of the instruments used to do this.

Regardless of the source of atomic emission—DC plasma, inductively coupled plasma (ICP), spark source, or flame emission—the light emitted contains lines characteristic of all or most of the elements in the sample. It is possible, therefore, to determine a number of different elements from the same source. Emission spectrometry is intrinsically a simultaneous multielement analysis technique.

To determine the elemental composition, the radiant power from each element must be measured at the appropriate wavelength(s) and corrected for background. Two general techniques are used to carry out such multielement determinations, **simultaneous** detection and **sequential** detection. The primary parts of such instruments are illustrated in Figure 18.18. Like any other emission spectrometer, atomic-

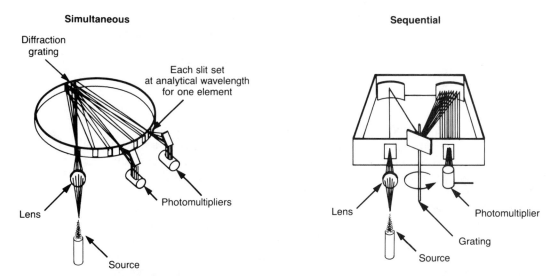

**Figure 18.18**     Diagram of the general layout of two types of spectrometers used with emission sources. The simultaneous instrument has a number of separate photomultiplier transducers (described in Supplement 18B). Each transducer is placed at the appropriate angle to detect light from one of the analytes. A sequential instrument produces an output for one element at a time. After the background and peak intensities are measured for one element, the grating is rotated to the wavelength for another analyte. Figure 18.19 further illustrates this protocol. [Redrawn from Sobel, H. R.; Dahlquist, R. L. *Am. Lab.* **1981**, *13*, 152. Copyright 1981 by International Scientific Communications, Inc.]

emission spectrometers are composed of a light source, a light-dispersing element, and a transducer, as illustrated in Figure 17.3. Whether the spectrometer is a simultaneous or sequential instrument depends on the design of the light-dispersing section.

In atomic-emission (AE) spectrometry, the background is usually corrected by using two background measurements: one each to lower and higher wavelengths. This technique is illustrated in Figure 18.19 and also was discussed in Section 17.14.

With the most sophisticated equipment, the presence or absence of 70 of the chemical elements can be determined with one experiment if enough sample is available. With contemporary instrumentation, precision is 1–3%. Atomic-emission methods can be used for elemental analysis with analyte content from 100% to ppm or ppb, depending on the element, the matrix, and the mode of thermal excitation.

However, as for any technique for multielement determinations, the optimum experimental conditions for the various elements differ. When the source and spectrometer are set to optimize the quantitation of one element, the response to another element of interest will probably not be optimized. For instance, in an *ICP torch,* the plasma chemistries will be different for different elements. The extent of chemical reactions and the rates of thermal excitation can differ among elements in the analyte. Therefore, steady-state emission intensities can differ even from different *parts* of an emission source. As shown in Figure 18.20, the relative intensities of the emissions from atomic phosphorous and singly ionized copper differ significantly at

**Figure 18.19** Background correction can be carried out by measuring the emission intensity on either side of the peak and subtracting the average value from the intensity at the peak wavelength of the analyte. In this case, the background is measured at different distances from the uranium analyte peak to avoid the other atomic peaks in the spectral region. In contemporary instruments a computer will usually control the scanning of the monochromator, coordinate the scan with the intensity measurements, and calculate the background correction. [Figure courtesy of the Perkin–Elmer Corp.: Norwalk, CT]

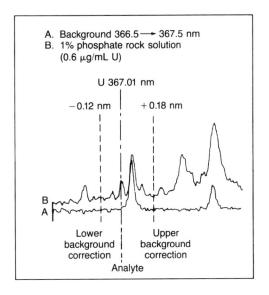

different heights in the plume. This is analogous to looking at spectra at different heights in a Bunsen burner flame. Thus, even the height of the ICP torch in the optical system is important (and must be kept constant) to obtain the most precise results. The relative intensities of the light emissions from an ICP may also differ depending on the gas flows feeding the plasma and on the radiofrequency power coupled to the plasma.

To review the characteristics of emission spectrometry: Emission spectrometry is inherently a multielement technique. The emission source must possess an extremely stable temperature to obtain the best precision. (See Supplement 18A for elaboration.) Temperature stability also usually implies that the shape of the source is constant. In addition, the temperature should be as constant as possible across the whole source to limit the effects of self-absorption. Also, as noted earlier, hotter sources usually produce more complete atomization. As a result, in emission spec-

**Figure 18.20** Three measurements of emission output from three different heights in the plume of an ICP torch. The different intensities of the peaks arise from differences in the gas-phase kinetics and equilibria of P and Cu. The emission lines are due to P(I) at 213.618 nm and Cu(II) at 213.598 nm. The nomenclature of atomic spectrometry is used. Hence, the two emitting species are atomic phosphorus and Cu⁺. [Ref: Xu, J.; Kawaguchi, H.; Mizuike, A. *Appl. Spec.* **1982**, *37*, 123.]

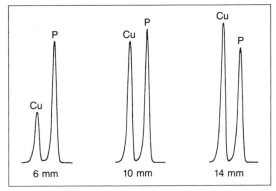

**Height above load coil**

trometry, a plasma source is preferred to a flame source because plasma sources are hotter and more homogeneous in temperature than are flame sources. The major drawback to emission spectrometry is spectral interference.

## 18.11 Atomic-Absorption Methodology

Unlike atomic-emission spectrometry, atomic absorption (AA) spectrometry is inherently a *single-element* method. The instrument is usually designed to be used to determine one element at a time. On the other hand, again unlike atomic emission, there are nearly no problems with spectral interference from other atomic or ion lines. In general, it is possible to find an atomic line that is relatively free of direct overlap. The most commonly used wavelength for each element is shown in the periodic table in Figure 18.21. Some of the few spectral interferences known are listed in Table 18.5. Precision of better than $\pm 1\%$ in an assay is possible when the sample is aspirated as a solution. As discussed in Supplement 18A, the precision of atomic absorption measurements is less dependent than emission spectrometry on the atomizer temperature. A drawback is that the absorbance is linear with analyte concentration over only about a 20-fold range of concentration.

The major parts of an atomic absorption spectrometer are shown in Figure 18.22. A flame atomizer is shown in the illustration. You will recognize the four parts of a general absorption spectrometer: the light source, absorber, monochromator, and transducer/detector. Two of the parts are unique in an atomic absorption spectrometer: the light source and a means of compensating for the background. Let us consider each of these.

### The Light Source in Atomic Absorption

As mentioned earlier in this chapter, the effects that produce self-reversal in atomic spectral lines (Figure 18.6) are used to great advantage in atomic absorption spectrometry: The light source is always a single narrow-linewidth emission of the *element being assayed.* The emission line has a linewidth around $10^{-3}$ nm.

The linewidth of the absorption envelope (at half height) in the atomizer is about $10-100$ times wider—around 0.03 nm. Thus, although the hot atomizer absorption envelopes may overlap, the narrow source lines may not be absorbed by the envelope of a possible interferent. This effect is illustrated in Figure 18.23. Also, in Table 18.5, notice how close in wavelength the interferences are. If there is extensive interference at these "first choice" wavelengths (see Figure 18.21), another one of the atomic emission/absorption wavelengths can be used.

A second reason to use a narrow-linewidth source is to enhance the linear range of the working curve compared to a wider-linewidth source; the Beer's law region is extended. This effect is presented in more detail in Section 17.9 and Figure 17.15.

A reasonable question is, Why not adjust the slits of the spectrometer so that the bandpass is narrow enough to shut out the spectral interference and extend the Beer's law range? After all, that will produce the same effect as using the narrow-linewidth

**Figure 18.21** A periodic table showing the elements analyzable by flame emission and absorption spectrometry. The first-choice wavelengths are also shown for each element. Detection limits and characteristic concentrations of various atomic spectrometric methods in comparison with other methods are shown in Table 7.2. [Courtesy of Instrumentation Laboratory, Inc. Lexington, MA.]

Legend (Sample):
- Atomic absorption wavelength, nm → 357.9
- Atomic absorption detection limit, µg/ml → 0.002
- **Cr**
- Atomic emission wavelength, nm → 425.4
- Atomic emission detection limit, µg/ml → 0.004

- Elements best analyzed by atomic absorption shown in light type
- Elements best analyzed by flame emission in lightly shaded boxes
- Elements suitable for analysis by either method shown in bold type
- Elements unsuitable for analysis by either absorption or emission in darkly shaded boxes
- Criteria for preference based on significant difference in detection limit

The elements Na, K, and Rb were determined in an air-acetylene flame. All others in a nitrous oxide-acetylene flame. Detection limits in many cases can be improved by the addition of an easily ionized substance such as sodium or potassium.

*Band emission

**Table 18.5**  Direct Spectral Overlap
Interferences in Atomic Absorption

NOTE: This list does not include absorbance
lines of nonanalytes that appear in the background region and might interfere with
background correction.

| Elements/Interferent | Wavelengths (nm) | |
|:---:|:---|:---|
| Cu/Eu | Cu | 324.7540 |
|       | Eu | 324.7530 |
| Si/V  | Si | 250.6899 |
|       | V  | 250.6905 |
| Fe/Pt | Fe | 271.9025 |
|       | Pt | 271.9038 |
| Al/V  | Al | 308.2155 |
|       | V  | 308.2111 |
| Hg/Co | Hg | 253.652 |
|       | Co | 253.649 |
| Mn/Ga | Mn | 403.3073 |
|       | Ga | 403.2882 |
| Zn/Fe | Zn | 213.856 |
|       | Fe | 213.859 |
| Fe/Cr | Fe | 302.064 |
|       | Cr | 302.067 |

*Source:* Pickett, E. E.; Koirtyohann, S. R. *Anal.
Chem.* **1969,** *41*(14), 28A. Miller-Ihli, N. J. et al.,
*Anal. Chem.* **1982,** *54*, 799.

light source. The answer is that technically this is more difficult. Also, such a narrow slit would cause a significant decrease in the S/N ratio, making background correction less precise.

An atomic-absorption light source must be quite bright—at least as bright as a blackbody radiator at the temperature of the atomizer. This idea might be difficult to envision until you consider how difficult it is to see whether a flashlight is lit when it is in direct sunlight. The sun is a blackbody radiator. You cannot perceive the relatively low-intensity light of the flashlight. On the other hand, if the light is intense enough, you can see it quite clearly. For instance, you can easily see the light from an arc welder in sunlight.

It is exactly the same in a determination using atomic absorption spectrometry. The instrument monochromator is set at the analytical wavelength. However, the instrument's detector will not register usable changes in absorption unless the power arriving from the source is at least as great as the power from background emission that is also arriving from the flame.

**Figure 18.22**    The general layout of a single-beam atomic absorption spectrometer. The hollow cathode lamp is described in more detail in Figure 18.24. More details of the slot-burner can be seen in Figure 18.8. One type of nebulizer is illustrated in more detail in Figure 18.13.

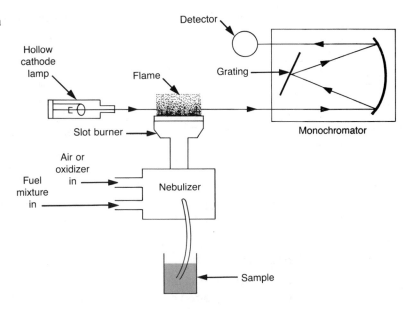

One light source that possesses these two properties — narrow linewidths and high intensity — is a **hollow cathode lamp.** The principles of operation of a hollow cathode lamp are described in Figure 18.24. A different hollow cathode lamp light source is usually used for each element, although some are manufactured for use with two or three different elements.

**Figure 18.23**    Comparison of the linewidths of the absorption envelope and the source line. The narrowness of the source line means that the effects of element Y, which would be an interferent of X in emission, causes no problem in atomic absorption. The absorption envelope has a fwhm around 0.03 nm, while the emission width is a tenth of that or less.

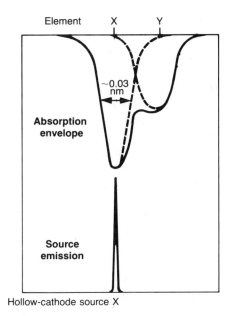

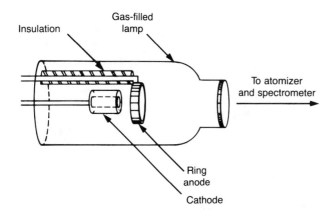

**Figure 18.24**    Hollow-cathode lamps are the most common light sources for atomic absorption. They are designed to produce intense, narrow emission lines of the element(s) that compose the cathode of the lamp. Most lamps are designed for one element, although multielement lamps are available. An electric discharge is set up in the lamp's fill gas (such as neon). The excited, ionized atoms of the fill gas are accelerated between the ring-shaped anode and the cathode and impact onto the inside surface of the cuplike cathode. The high-energy impact causes some of the cathode material to vaporize at low temperature. This process is called **sputtering**. The low-temperature vapor is excited and emits the narrow, characteristic spectral lines of the element. The lamp also emits the characteristic lines of the fill gas.

## Background Correction

The precision and accuracy of atomic absorption spectrometry depends on the quality of the background correction. (Some concepts of spectral background correction were discussed in Section 17.14.) In order to do the background correction, the absorbance must be measured at wavelengths on either side of the absorption envelope of the analyte. This is similar to measuring the emission background, as illustrated in Figure 18.19.

To measure the absorption in the background requires a light source with wavelengths outside of the absorption envelope of the assayed element. But the linewidth of the atomic-absorption light source is narrower than the absorption envelope. How can we get some light into the broader wavelength region?

There are three basic strategies to produce light in the correct wavelength region. Currently, the most common method is to use a second light source that produces a continuum spectrum. This means that it produces continuum radiation which includes the wavelengths to be used for determining the background absorbance. The method is illustrated in Figure 18.25.

The second method utilizes a special property of spectral lines in a strong magnetic field, the **Zeeman effect.** In the presence of the magnetic field, instead of one line, two or more separate lines appear at wavelengths on both sides of the original line. The effect is illustrated in Figure 18.26. Either the source or the atomizer can be in the magnetic field. In the former case, it is the split emission lines of the source that probe the surrounding background. In the latter case, what is measured is the change in absorbance of the analyte with the field on and with it off. The ratio of the absorption at the analytical wavelength and that found in the background region produces the corrected experimental signal.

The third method of obtaining light in the background region takes advantage of the self-absorption within hollow-cathode light sources. An illustration of the light output associated with the technique is shown in Figure 18.27. It has long been known that when the operating current of the lamp is raised, the emission linewidth

**Figure 18.25** (a) Atomic absorption background correction with a continuum source. The slit width of the spectrometer is broader than either the source line or absorption envelope. Therefore, the spectrometer output is a measure of the average power over the bandpass. (The bandpass is represented by the region between vertical lines.) The instrument measures alternately the output using the line source and the output using the continuum source such as a deuterium lamp (Supplement 17C). The output observed with the line source depends on the sum of background absorbance, analyte absorbance, and background luminescence and emission. The output observed with the continuum source depends on the average background only. The analyte's effect on the continuum is small since its linewidth is relatively quite narrow. By phase-sensitive detection (described in Supplement 18C), these two different responses are either subtracted or the ratio is taken to correct for background, depending on the specific instrument. The general design of one instrument is shown in (b).

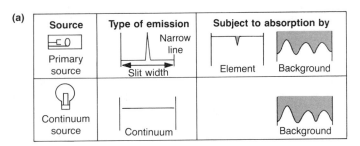

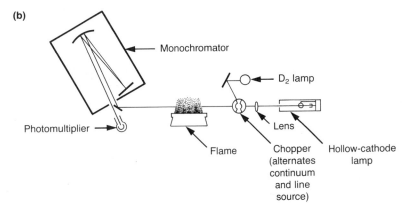

broadens. At the same time, the output at the nominal wavelength decreases due to increased self-absorption. Light emission averaged over the low-current bandwidth decreases around 80%. As a result, in comparison with the narrow-line emission under normal operating condition, the light from a pulsed lamp predominately probes the background region. Recently, this lamp characteristic has been used to enable correction of the background. When the lamp is run normally, the output is a signal proportional to (analyte + background at assay wavelength). When the lamp current is periodically increased, the output is proportional to (background + small amount of analyte). Electronic subtraction of the latter (high-current conditions) from the former (normal operating conditions) produces a signal proportional to the analyte concentration alone. Note that sensitivity to the analyte is reduced if the self-absorption does not reach the point of self-reversal.

Each of the three different background correction techniques has its own strengths and weaknesses. The details of the arguments for each can be found in more advanced texts and from the manufacturers of the instruments that use each type of background correction. However, nothing is as effective as a low-interference sample to enhance precision and accuracy.

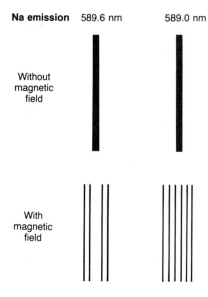

**Na emission**    589.6 nm        589.0 nm

Without magnetic field

With magnetic field

**Figure 18.26**    Atomic absorption background correction utilizing the Zeeman effect. In the presence of a magnetic field, atomic spectral lines split into two or more lines as illustrated by the splitting of two sodium spectral lines. The spectra here are recorded on photographic film. Such lines provide illumination outside the absorption envelope of the analyte. Depending on the instrument design, the magnetic field can be applied to the flame atomizer or to the hollow-cathode tube source lamp. Because the correction wavelengths are immediately adjacent to the analyte line, less error is introduced by structured background than with a continuum source. The measurements of the analyte and background are alternated. [Ref: Back, E.; Lande, A. "Zeeman-effekt u. Multiplettstrucktur"; Berlin: Springer, 1925.]

To carry out the background subtraction and enhance the S/N, AA spectrometers utilize **phase-sensitive detection,** which is explained in Supplement 18C. In essence, the signal from the analyte + background and the signal from the average background are both stored. Then, the background signal intensity is subtracted electronically, leaving the signal from the analyte alone as the instrument output. At the same time, any steady spurious emission that might be generated by the source is compensated.

In atomic absorption, the signals from the analyte and background are sampled alternately and, compared to the speed of contemporary electronic circuits, relatively slowly—in the range 30 to 100 Hz. As a result, an option to phase-sensitive detection

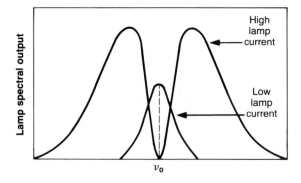

**Figure 18.27**    Background correction by changing the hollow-cathode emission linewidth. The output of the hollow-cathode lamp can be changed from a narrow emission line to a broader self-absorbed line as illustrated. This latter emission is obtained by pulsing a higher current through the lamp.

is increasingly being used. The alternating analyte and background signals are digitized in real time and processed directly by a computer.

## 18.12 Atomic-Fluorescence Methodology[2]

Atomic fluorescence spectrometry is a relative newcomer to analytical atomic spectrometry. It is, as the name states, a fluorescence method: Light absorbed by and reemitted from an analyte is measured. The intensity of the reemission is directly proportional to the concentration of the analyte and the radiant power of the source at the analytical wavelength. The radiation source intensity is kept constant so that the measured signal depends directly on the analyte concentration.

Analytical atomic fluorescence is a hybrid method; it has a combination of characteristics that are also seen in atomic absorption and atomic emission. The fluorescence method has the potential for keeping the benefits of both.

For instance, three of the practical differences between atomic absorption (AA) and atomic emission (AE) spectrometry are the following:

1. AA is a single-element technique; AE is potentially a simultaneous multielement technique.
2. AA has its optimum precision in a narrow concentration range — about a factor of 100 in concentration; AE has a far wider useful linear range — up to $10^5$, or even higher for some elements.
3. AA tends to be free of spectral interference; AE tends toward having spectral interferences.

The characteristics sought through atomic fluorescence are a capability for multi-element analyses, with a wide linear range, and freedom from interference such as is seen in AA. As yet, atomic fluorescence spectrometry is not widely accepted as sufficiently superior to AA or AE.

The major limitations to precise and sensitive atomic fluorescence has been the high, fluctuating backgrounds in furnace and flame atomizers and the availability of multielement, high-intensity light sources.

However, some of these problems have been overcome by the use of an inductively coupled plasma (ICP) torch atomizer. (This is the only atomizer discussed here.) The atomizer and optics of a commercially available instrument are shown in Figure 18.28. The light sources are hollow cathode tubes as commonly used in atomic absorption. They are chosen for the same reasons: because of narrow emission

[2] Refs: Demers, D. R.; Busch, D. A.; Allemand, C. D. *Amer. Lab.* **1982,** *14* (3), 167. Van Loon, J. C. *Anal. Chem.* **1981,** *53*, 332A.

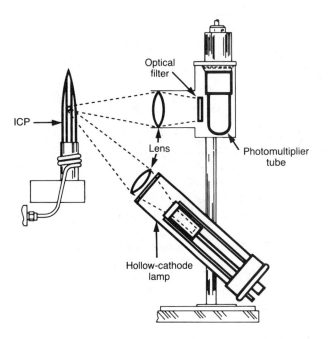

**Figure 18.28**    Representation of the instrumental layout for an atomic fluorescence spectrometer. The atomic line excitation source is focused on a space in the ICP torch. The emission is collected and is passed through a narrow bandpass optical filter and onto the photomultiplier tube. The filter allows little background light outside of the fluorescent line. Thus, there is a different filter and hollow-cathode line source for each element. [Ref: Reprinted from Demers, D. R.; Busch, D. A.; Allemand, C. D. *Amer. Lab.* **1982,** *14*(3), 167. Copyright 1982 by International Scientific Communications, Inc.]

linewidths and relatively high brightness. Each element has its own fluorescence unit with its single-element lamp, narrow bandpass filter centered at the analytical wavelength, and photomultiplier transducer. Multielement analyses of each sample are made possible by placing a number of these units around a single ICP atomizer.

Most of the relative freedom from atomic spectral interference derives from the narrow linewidth of the spectral source lamp. The spectral background interference is reduced by focusing the light and detector at the least luminous part of the ICP plume. Chemical interferences are the same as for any ICP atomizer.

For each element, the average background signal is measured. Then the hollow-cathode lamp is pulsed on to measure the *background + sample* fluorescence. Electronic subtraction yields the fluorescence signal alone.

The potential of atomic fluorescence spectrometry has not yet been reached. Since the intensity of the fluorescence depends directly on the source intensity, higher-intensity sources such as lasers could significantly increase the sensitivity at an increased cost.

## 18.13  How Good Is the Method? Figures of Merit and the Detection Limit

In Section 17.14, you saw that if spectral backgrounds varied significantly between blanks and analytical samples, running the blanks could be useless. In order to overcome the problem, a wavelength scan can be used to obtain the magnitude of the background, and from that the signal of the analyte alone.

Now, consider this question. What would be the result if there were no noise at all associated with the spectral measurements? In that case, a perfect correction for the background could be made without error no matter how small the signal. Perfectly precise results could always be obtained. (Incidentally, any method with this characteristic would soon be the only analytical method used for elemental analysis. Also, the practice of chemical analysis would be *much* simpler!) However, a noiseless experiment is an impossibility.

It must be, then, that the noise level is one of the limitations to finding a perfect method of analysis. Similarly, if the background signal is not reproducible or constant, the quality of the analysis suffers. However, the determinate error of the background itself cannot be determined if the noise level is high. Thus, the precision and accuracy of an atomic spectrometry method must depend on all three factors: the magnitude of the background signal, the magnitude of the analyte signal, and the noise level. These three factors are interdependent.

The background, analyte signal, and system noise are characterized by sets of numbers that are used to denote the quality of a specific atomic spectrometric method. Three such numbers, which have been mentioned earlier, are the signal-to-noise ratio (Section 17.11), sensitivity (Sections 18.1 and 7.2), and detection limit (introduced qualitatively in Sections 7.2 and 17.12). These numbers are called **figures of merit**. A figure of merit is a number derived from actual measurements that aids us in evaluating an instrument or analytical technique.

The figures of merit, such as sensitivity, characteristic concentration, and the S/N ratio, are useful by themselves in evaluating the quality of instruments. However, for chemical analyses, the experimentally determined detection limit is probably the best figure of merit to use as a general guide. The judgment of "best" is based on the following three criteria:

1.  The detection limit is a single number.
2.  It can be used to compare data from different laboratories.
3.  When calculated with caution, it can be used to compare various spectroscopic (and other instrumental) techniques employed in chemical analysis.

A special committee of the American Chemical Society has defined the limit of detection thus:[3]

> The limit of detection is the lowest concentration of an analyte that the analytical process can reliably detect.

A key word in this definition is *reliably,* which suggests, correctly, that the detection limit is based on a statistical calculation. In reading published reports, you should be

---

[3] ACS Committee on Environmental Improvement. *Anal. Chem.* **1980**, *52*, 2242–2249.

aware that there is still some debate about the calculation of detection limits from the experimental data.

In a useful article, the following practical points about detection limits are stated.

> . . . when scientific researchers measure a detection limit, they give themselves the *best* possible opportunity to obtain the *lowest* value for which their instrument is capable, i.e., the simplest matrix (usually a simple, high purity salt in dilute, aqueous solution) and optimum experimental conditions for the analyte under consideration.
>
> When considering detection limits, several points must be kept solidly in mind:
>
> A. Quantitative determinations cannot be made at the detection limit. Relative precision at the detection limit will be $\pm 30\%$ to $\pm 50\%$ depending on the definition used.
> B. In simple cases, concentrations of 5 to 10 times the limits of detection are required for quantitive determinations.
> C. In samples with complex matrices, factors of 20 to 100 times the limits of detection are often required.

[Ref: Parsons, M. L.; Major, S.; Forster, A. R. *Appl. Spec.* **1983**, *37*, 411–418.]

In addition, when comparing published detection limits, it is often necessary to see whether the detection limit cited refers only to the spectrometric assay or to the entire analysis. For instance, it is not uncommon for workers to include the effects of preconcentration steps in the detection limits. Without a clear statement of what part(s) of the analysis are accounted for in the calculated detection limit, direct comparison between methods and equipment is made difficult. If no details of the detection limit calculation are included, comparison is useless.

Yet another caution is needed when comparing limits of detection for trace as opposed to microanalyses. This problem relates to comparisons between concentrations and quantities of material. For instance, in an aqueous sample, the detection limit might be, for instance, 1 ng/mL. However, for a furnace atomizer, the minimum detectable quantity might be stated as 1 ng. These are not directly comparable. A more detailed treatment of the statistical basis of limits of detection is presented in Supplement 18B.

## 18.14 Furnace or Flame/Plasma: Which to Use?

You have now read about how sources of free atoms are obtained for atomic spectrometry. Let us now assume that atomic spectrometry is the first choice to solve the problem at hand.

What method of atomization should you use? If your sample is a gas and can be held in a gas cell, you have an easy task. But, should a *liquid* sample be nebulized into a flame or plasma, or injected, dried, and then atomized in a furnace? There are, in general, no hard and fast rules to make choices for you. However, a few generalizations can be made to guide such choices about sample form.

First, some questions need to be asked.

1. Is the untreated sample itself solid or a liquid? A gas will be loaded into a gas cell.
2. How large is the sample (micro or macro)?
3. Are you interested in determining many or only one of the elements present?
4. Are the elements in which you are interested major, minor, or trace?
5. Is the sensitivity of the spectrometric method high or low for the *other* matrix components at the analytical wavelength of the component(s) of interest?

Your answers to these questions will specify the type of sample, the level of analyte in it, and the types of expected interferences. Much of this was discussed before in Chapter 7. The answers are then combined with the following general guidelines.

a. *Backgrounds* from solid samples are generally greater than from liquids. In atomic absorption, the background from graphite-furnace atomization is greater than that from nebulization plus flame/plasma atomization. (Furnaces for atomic absorption generally can be used for samples up to about 50 $\mu$L that are dried before atomization.) In emission, DC arcs can be used for solids directly and tend to have greater backgrounds than ICP torches with aspirated samples.
b. With furnace atomization, better results often can be obtained for *micro*chemical samples.
c. No simple criteria are established to decide which is best for *trace* levels of specific elements.
d. If the sample contains major or minor amounts of the element(s) of interest, liquid samples usually allow more precise results. This guideline is made assuming that the sample pretreatment does not contribute significant interferences.

As usual, all the factors of sample selection and pretreatment enter into the decision of which source to use.

Unfortunately, it is impossible to separate the choice of nebulization/atomization method from the choice of spectrometric method. Furnace atomization is used for atomic absorption. Heating metal filaments or metal ribbons or a graphite furnace can atomize samples for plasmas. Flames are used for all three types: atomic absorption, atomic emission, and atomic fluorescence spectrometry. Plasmas are used for atomic emission and atomic fluorescence. DC arcs are applied only to emission.

Let us assume that a wide selection of spectrometers and atomization sources is conveniently available to you. Which of these will you choose? Some guidelines for such a choice will be presented in the next section.

## 18.15 Atomic Absorption, Emission, or Fluorescence: Which One to Use?

Taken together, the various atomic spectrometry techniques can be used to quantify nearly every element, to measure analyte concentrations from 100% to ppb, and to measure the elemental content of samples from macro to micro in size. However, no single atomization technique and no single spectrometric method is best for all determinations. How can anyone choose which of the many atomic spectrometry techniques to use?

If the sample is large, if an assay is needed for only one element at a time, and if the element is a major or minor component, atomic absorption of an aqueous solution is probably the method to use. Interferences are minimized, and the possible relative

**Table 18.6**  Summary of Atomic Spectral Methods for Trace Analysis

Abbreviations used:
FAA = Flame atomic absorption
FAE = Flame atomic emission
EAA = Electrothermal (furnace) atomic absorption       ■ Best method(s)
 ICP = Inductively coupled plasma emission
DCP = DC plasma emission                                ▨ Next best
  AF = Atomic fluorescence

At moderate levels (multi-ppm and above) many of the methods are comparable in precision. Then, the choice depends on the sample matrix and whether a single or multielement technique is the best for the problem at hand.

(Key: ■ = Best method(s); ▨ = Next best)

| Elements | FAA | FAE | EAA | ICP | DCP | AF |
|---|---|---|---|---|---|---|
| Group I-A, alkali metals | ▨ | ■ | | | | |
| Group II-A, alkaline earths | | | ■ | | ▨ | ▨ |
| Group III-A | | | ■ | ▨ | | |
| Group IV-A* | | | ■ | ▨ | | |
| Group V-A* | ■ | | ■ | | | ▨ |
| Group VI-A* | ■ | | ■ | ▨ | | ■ |
| Group VII-A + other nonmetals | | | | ■ | | |
| First-row transition metals | ▨ | | ■ | ■ | | |
| Refractory metals | | | ■ | ■ | ▨ | |
| Noble metals | | | ■ | ▨ | ▨ | |

* Excluding nonmetals.
*Source:* Parsons, M. L.; Major, S.; Forster, A. R. *Appl. Spec.* **1983**, *37*, 411–418.

precision is better than $\pm 1\%$. The sample digestion and pretreatment will require dilution to the relatively narrow working range of the instrument.

If the sample is large, and more than two or three elements are to be determined, emission methods will probably be better. However, a number of atomization methods are available, which usually means selecting one of a number of different instruments. Generally, each emission instrument is built around a specific atomization technique, for instance, ICP, DC-plasma, or spark source.

In other situations, the choice is less straightforward — assuming you have a choice of instruments. For instance, if the elements of interest are at trace levels, one spectrometric method might be inherently more sensitive and precise for a given element. One way to judge is to look at a table of detection limits. For elements of a chemical group, the choice can be restricted to a smaller number of techniques because of their similarities in numerous chemical properties. Table 18.6 shows the "best" and "next best" technique for groups of elements. These qualitative labels do not account for any interferences in the matrix.

You might assume that matrix interferences will always be minimized by using the highest temperature source: plasma and spark; but, accompanying the higher temperatures, which produce better matrix and molecule atomization, are greater background emissions. Nevertheless, for some elements, notably the heavier ones, the increase in emission intensity is much more significant at the higher temperatures. The relationship between spectral emission intensities and temperature is described further in Supplement 18A.

In addition, you may choose some atomic spectral method simply because it is available. Other possible instrumental analysis methods for the elements also can be used and may be cheaper and/or more convenient and/or more precise and accurate than the atomic spectrometry methods described here. For instance, for some elements direct competition exists from electrochemical methods such as anodic stripping and differential pulse polarography.

## Suggestions for Further Reading

This chapter contains some relatively advanced theoretical parts but excellent and quite readable descriptions of the equipment and methodology for emission spectrometry.
Sacks, R. D. In "Treatise on Analytical Chemistry," Pt. I, Vol. 7, 2nd ed.; I. M. Kolthoff and P. J. Elving, Eds.; Wiley: New York, 1981, Chap. 6.

A clearly written review with comparisons of various techniques of AA.
Robinson, J. W. In "Treatise on Analytical Chemistry," Pt. I, Vol. 7, 2nd ed.; I. M. Kolthoff and P. J. Elving, Eds.; Wiley: New York, 1981, Chap. 8.

A broad treatment of the principles, applications, and instruments of flame AA.
Christian, G. D.; Feldman, F. J. "Atomic Absorption Spectroscopy"; Wiley–Interscience: New York, 1970.

Everything you wanted to know about emission spectrometry used in chemical analysis. Many chapters at advanced levels.

Grove, E. L., Ed. "Analytical Emission Spectroscopy," Vols. 1 and 2; Marcel Dekker: New York, 1971.

Atomic analyses of organic compounds also can be done by AA. This book is an entrance to the extensive literature in the area.
Hassan, S. S. M. "Organic Analysis Using Atomic Absorption Spectrometry"; Halsted Press: Chichester, E. Horwood, and New York, 1984.

In more than four-fifths of this book are described applications of AA to specific areas of analysis.
Van Loon, J. C. "Analytical Atomic Absorption Spectroscopy: Selected Methods"; Academic Press: New York, 1980.

A listing of analytical wavelengths for each element and the possible interferences along with the interfering wavelengths.
Parsons, M. L.; Forster, A.; and Anderson, D. "An Atlas of Spectral Interferences in ICP Spectroscopy"; Plenum: New York, 1980.

A highly useful reference set. The two volumes list citations of AA analyses along with a separate author and subject index. An unparalleled entrance into the vast literature of atomic absorption.
Varma, A. "CRC Handbook of Atomic Absorption Analysis"; CRC Press: Boca Raton, FL, 1984.

Articles
Barnes, R. M. "Recent Advances in Emission Spectroscopy: Inductively Coupled Plasma Discharges for Spectrochemical Analysis"; *Crit. Rev. Anal. Chem.* **1978**, *7*, 203–296.
Brown, S. D. "Zeeman Effect-Based Background Correction in Atomic Absorption Spectrometry"; *Anal. Chem.* **1977**, *49*, 1269A–1281A.
Stephens, R. "Zeeman Modulated Atomic Absorption Spectroscopy"; *Crit. Rev. Anal. Chem.* **1980**, *9*, 167–195.
Sturgeon, R. E. "Graphite Furnace Atomic Absorption Spectrometry"; *Anal. Chem.* **1977**, *49*, 1255A–1267.

# Problems

18.1   [Table reprinted from Liddell, P. R., *Am. Lab.* **1983** (March), *15*, 111. Copyright 1983 by International Scientific Communications, Inc.]
An automated flame AA analysis was obtained for a number of samples of river and estuary waters, which were collected and then stored as a solution containing 1% v/v $HNO_3$. The data used in the determination of magnesium are shown below. The measurements were done in triplicate and recorded. The operating conditions were wavelength 202.5 nm, lamp current 10 mA, and spectral bandpass 1.0 nm.

   a.  Complete the table.
   b.  The RSD are for the three measurements for each sample. Is the RSD of the determination including sampling, etc., equal to or greater than the individual RSD? (This question is not strictly correct statistically.)
   c.  Estimate the RSD for samples 2 and 3, taking into account the error of the standards. Assume that the RSD for the standards equals the mean RSD of the three.
   d.  It turns out that samples 1, 5, and 6 were from the same river source. They were

| Solution | Concentration (mg/L) | Absorbance Readings | | | Relative Standard Deviation (%) |
|---|---|---|---|---|---|
| Blank | 0.000 | 0.001 | 0.001 | 0.001 | 0.0 |
| Standard 1 | 6.000 | 0.351 | 0.357 | 0.361 | 1.4 |
| Standard 2 | 15.000 | 0.714 | 0.711 | 0.712 | 0.1 |
| Standard 3 | 30.000 | 1.252 | 1.264 | 1.268 | 0.6 |
| Sample 1 | — | 0.558 | 0.566 | 0.563 | — |
| Sample 2 | — | 0.364 | 0.369 | 0.372 | — |
| Sample 3 | — | 0.385 | 0.388 | 0.385 | — |
| Sample 4 | — | 1.008 | 1.020 | 1.011 | — |
| Sample 5 | — | 0.555 | 0.563 | 0.565 | — |
| Sample 6 | — | 0.590 | 0.598 | 0.595 | — |

included to test the sample collection and storage errors. What is the RSD due to sample collection and storage alone?

18.2   Atomic hydrogen has one of the simplest emission spectra of all the elements. Nevertheless, since emission lines can arise from transitions between any two levels, numerous lines arise from only a few levels. The emission wavelengths for hydrogen can be calculated with the simple expression

$$\frac{1}{\lambda} = R \cdot \left( \frac{1}{m^2} - \frac{1}{n^2} \right)$$

with $R = 109{,}677.8$ cm$^{-1}$ for the spectrum in a vacuum, and with $n$ greater than $m$. The reported spectrum is in the following table.

Atomic Hydrogen Emission Spectrum Wavelengths (in nm) from 365.6 to 656.3 nm

| | | | | | | |
|---|---|---|---|---|---|---|
| 365.66 | 366.22 | 366.94 | 368.28 | 371.19 | 379.79 | 434.05 |
| 365.76 | 366.34 | 367.13 | 368.68 | 372.19 | 383.54 | 486.13 |
| 365.80 | 366.46 | 367.37 | 369.15 | 373.44 | 388.90 | 656.28 |
| 366.03 | 366.61 | 367.63 | 369.71 | 375.01 | 397.00 | — |
| 366.12 | 366.77 | 367.94 | 370.38 | 377.00 | 410.17 | — |

a.   Calculate the values of the energy levels and see how many of the lines of the reported spectrum you can assign. You may wish to write a short computer program for the calculation.

b.   The numbers you calculate may not fit too well with the values measured since the spectrum was measured in air. From the difference between the calculated values and those measured, find the index of refraction of air. The index of refraction is always greater than 1.

18.3   [Ref: Ng, K. C.; Caruso, J. A. *Anal. Chim. Acta* **1982**, *143*, 209.]
Figure 18.3.1 shows the lithium emission signals that arise from evaporating samples inside a graphite furnace and feeding the vapor into an ICP spectrometer. Graphite is somewhat porous; so to increase the quality of sample vaporization, the inside surface of the graphite tube was treated in two ways. To test the effects of the treatment, after the sample was introduced, the furnace was heated, cooled, and reheated in the original way. The signal was observed for each heating cycle. From the figure, it is clear that the behavior of the two coatings is quite

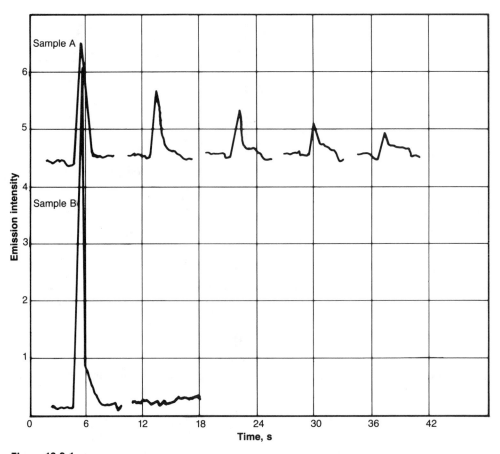

**Figure 18.3.1**

different. For sample B (1-ng Li on the furnace surface) apparently all the lithium that might come off did so in one heating; the *decontamination heating* shows no further Li signal. For sample A (10-ng Li) there was a rather large "memory" of the sample.

    a.   Assuming that the signal area of B arises from all the lithium of the sample, what fraction of sample A was vaporized and was carried into the ICP in the first heating?

    b.   After five firings of sample A, what fraction of the original Li had been vaporized and had reached the ICP?

18.4   Determination of sodium and potassium in blood or serum by low-temperature flame emission on simple instruments is complicated by the instability of the flame and the complexity of the matrix. One way to overcome the problems is to add an excess of lithium to serve as both an ionization suppressor and an internal standard. The reason it can serve as an internal standard is that the lithium emission responds similarly to the emissions of Na and K to changes in the flame conditions. The responses found for calibration solutions are listed in the table at the top of page 747.

    a.   Plot the working curves for Na and K with the Li internal standard present.

    b.   A blood sample of 10 μL is diluted in 1.00 mL of 5000-ppm Li solution. This is

| Concentration (ppm) | | | Signal (relative to Na in 1.0-ppm Na sample) | | |
| --- | --- | --- | --- | --- | --- |
| Na | K | Li | Na | K | Li |
| 0.5 | 0.5 | 500 | 0.43 | 0.57 | 31.2 |
| 1.0 | 1.0 | 500 | 1.00 | 1.25 | 36.3 |
| 5.0 | 5.0 | 500 | 4.9 | 6.4 | 36.2 |
| 10.0 | 10.0 | 500 | 8.8 | 11.7 | 31.7 |

further diluted with doubly distilled water to 10.00 mL and fed into the spectrometer. The readings with the blood were Na 2.9, K below detection, and Li 32.5. A second 10-$\mu$L sample was taken and brought to 2.00 mL with 500-ppm Li solution. The readings were Na 13.6, K 1.34, and Li 30.4. What are the serum concentrations of Na and K in mM? Assume Na and K do not interfere with each other.

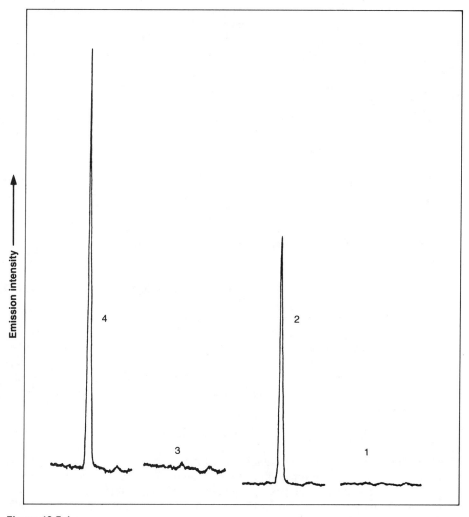

**Figure 18.5.1**

18.5   [Figure reprinted with permission from Allain, P.; Mauras, Y. *Anal. Chem.* **1979,** *51*, 2089. Copyright 1979, American Chemical Society.]

Assume that Figure 18.5.1 shows emission spectra (swept from 395.90 to 396.35) of the following solutions: (1) pure water, (2) aluminum 250 $\mu$g/L, (3) calcium 2500 $\mu$g/L, and (4) Al 250 $\mu$g/L + Ca 2500 $\mu$g/L.

a.   If you wish to determine the aluminum content of solutions containing around 2500 $\mu$g/L calcium, which peak do you use as a standard, (2) or (4)?

b.   If the calcium solution measured in spectrum (3) was made from high-quality $CaCl_2 \cdot 6\,H_2O$ (f.w. 208.49), what is the impurity of Al in this calcium reagent in ppm to within the precision possible with the data at hand?

18.6   [Data reprinted with permission from Allain, P.; Mauras, Y. *Anal. Chem.* **1979,** *51*, 2089. Copyright 1979, American Chemical Society.]

A method has been devised to determine aluminum in serum and blood by ICP spectrometry. The data in the table show the changes in the aluminum emission intensity (above baseline and corrected for any aluminum impurities in the reagents) in the presence of other ions. Anions had no effect on the emission at this wavelength: 396.15 nm. The data are recorded as aluminum emission with the intensity for a 1-mg/L solution equal to exactly 100. Normal blood serum levels center around Na 140 mM, K 6.5 mM, Ca 3.4 mM, and Mg 1.6 mM.

Al Emission with Cation Present

| g/L Cation | K | Na | Ca | Mg |
|---|---|---|---|---|
| 3 | — | 195 | — | — |
| 1 | 192 | 188 | — | — |
| 0.75 | 186 | 182 | — | — |
| 0.50 | 178 | 174 | 181 | 152 |
| 0.25 | 166 | 159 | 172 | 146 |
| 0.10 | 149 | 142 | 159 | 133 |
| 0.05 | 139 | 134 | 148 | 127 |
| 0.01 | 118 | 116 | 126 | 116 |
| 0.00 | 100 | 100 | 100 | 100 |

a.   Normal serum levels of the listed cations may vary by ±2.5%. These variations can produce errors in the measured aluminum content. Which cation will produce the largest error in the aluminum measurement if its normal serum concentration varies ±2.5%?

b.   One protocol requires that the whole blood be diluted by a factor of 10 before aspiration into the ICP. If this dilution is carried out, which cation will then produce the largest error in the aluminum measurement if its concentration varies ±2.5%?

c.   How would you minimize the problem caused by the potential variations in serum metal-cation content?

18.7   [Figure courtesy of Hitachi Corp., Tokyo, and Nissei Sangyo America, Ltd., Mountain View, CA.]

Figure 18.7.1 shows duplicate results obtained by Zeeman atomic absorption. Graphed is the absorption at 228.8 nm versus time. The analysis involved a determination of Cd in fruit juice. The juice was placed directly into the graphite furnace without pretreatment. Each plot results from graphite furnace vaporization of a 10-$\mu$L sample spiked as labeled on the figure. What is the concentration of Cd in the fruit juice?

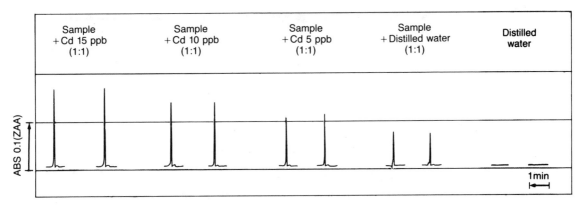

**Figure 18.7.1**

18.8   [Figure from Hwang, J. Y.; Thomas, G. P. In "Spectroscopy" Series II, Vol. II, p. 88; Intl. Scientific Communications: Fairfield, CT. Copyright 1977 by International Scientific Communications, Inc.]
In a graphite furnace, the atomization rate of analytes is kinetically controlled and depends on the composition of the matrix. Figure 18.8.1 shows the effect of calcium on the signal for aluminum when atomized at 2400 °C. The samples consisted of, respectively, 1.0 ng of Al

**Figure 18.8.1**

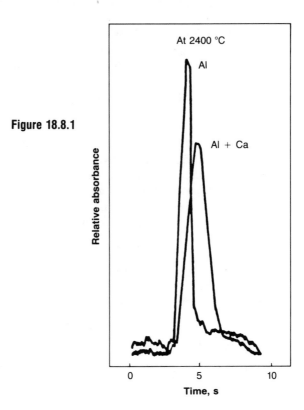

alone and a mixture of 1-ng Al + 10-ng Ca. Obviously, the peak heights are not equal and are not comparable.

    a.   Could the peak areas be used to quantify aluminum if $\pm 5\%$ precision is acceptable?

    b.   Which set of conditions provides the greatest sensitivity?

## SUPPLEMENT 18A

# Emission and Absorption Spectra of Atoms and the Boltzmann Distribution

As noted in this chapter, every emission or absorption of a photon of light is due to a transition between quantized energy states. However, we have not yet considered the following point: Even though a *group* of the same atomic (or molecular) species may be both absorbing and emitting light, an individual atom cannot simultaneously absorb and emit. We then must ask, What is the relationship between the properties of individual atoms and the property of the whole group? Perhaps surprisingly, this question has a rather direct application in choosing an atomic-spectrometry assay method.

As you already have learned, atoms that are in the ground state can only absorb light under common spectrometric conditions (not being illuminated by high-power lasers). Atoms that are in excited states only emit light.

In particular, for the emission of species $m$, the radiant power of the emission at a wavelength $\lambda$ is proportional to the number of atoms present, $\mathcal{N}_m$.

$$P_m(\lambda) = C(\lambda, m) \cdot \mathcal{N}_m \qquad \text{(in emission)} \qquad \textbf{(18A-1)}$$

The specific value of the constant $C(\lambda, m)$ depends on all the conditions of the experiment, and the radiant power can be either an integrated value over all wavelengths or the magnitude at any specific wavelength.

However, the atomic species must be in a specific excited state to produce an emission at the wavelength $\lambda$. Not all the atoms present are in that state. We can write this idea by changing Eq. 18A-1 slightly:

$$P_m(\lambda) = C'(\lambda, m) \cdot \mathcal{N}_m^* \qquad \text{(in emission)} \qquad \textbf{(18A-2)}$$

where the prime $(')$ now indicates a different value of the constant. The asterisk indicates that the number of atoms emitting is not the total of all the atoms, but only the atoms in the excited state that leads to emission at $\lambda$. In other words, we can describe the experimentally determined intensity in at least two different ways algebraically: Eqs. 18A-1 and 18A-2 are equally applicable for use in elemental analysis. Equation 18A-1 is much more directly useful, however.

Similarly, in absorption, the absorbance $A$ is given by

$$A_m(\lambda) = C''(\lambda, m) \cdot \mathcal{N}_m \qquad \text{(in absorption)} \qquad \textbf{(18A-3)}$$

One form of this equation is Beer's law. However, since only the ground-state atoms absorb, we can also write

$$A_m(\lambda) = C'''(\lambda,m) \cdot \mathcal{N}_{m,\text{ground state}} \qquad \text{(in absorption)} \qquad \textbf{(18A-4)}$$

As you will see, whereas the emission constants $C$ and $C'$ are quite different, the absorption constants $C''$ and $C'''$ are usually nearly equal.

Is there any way to calculate or estimate the fractions of atoms in the ground and excited states? It turns out that there is, and the numbers depend on the temperature. As a result, both the precision and the sensitivity of the optical atomic spectrometric methods depend on the temperature of the atoms in a sample in a second way (besides the linewidth effects). The relationship between the concentrations of ground and excited states is described by the **Boltzmann distribution.**

The Boltzmann distribution is written as a simple equation, Eq. 18A-5, which describes a property of a *large number* of particles: atoms, ions, or molecules. Specifically, we label the energy levels that the atoms can have as $E_1, E_2, E_3, \ldots, E_N$. All the atoms are at a temperature $T$ (in K). Now, call any two of the energy levels $E_i$ and $E_j$, with $i \neq j$. The ratio of the numbers of atoms in each energy level, $E_i$ and $E_j$, is

$$\mathcal{N}_i/\mathcal{N}_j = \text{factor} \cdot \exp[-(E_i - E_j)/kT] \qquad \textbf{(18A-5)}$$

$\mathcal{N}_i$ and $\mathcal{N}_j$ are the numbers in each state, and $k$ is Boltzmann's constant, $1.381 \times 10^{-23}$ J K$^{-1}$. We drop the species label $m$.

Note that a special factor is included in the right-hand side of Eq. 18A-5. The factor accounts for the quantity called the **degeneracy** of the specific energy level. The value of the degeneracy factor is, in general, an integer around 1. For the cases considered here, the value is 1, 2, or 3. For a ratio like $\mathcal{N}_i/\mathcal{N}_j$, the degeneracy factor is unimportant compared with the value of the exponential factor. Thus, let us ignore the degeneracy in all the calculations below; this clarifies the calculation with little loss of information.

If the $j$ state in Eq. 18A-5 is the ground state, then defining the ground-state energy as zero, $E_0 = 0$, gives

$$\frac{\mathcal{N}_{\text{excited state }i}}{\mathcal{N}_{\text{ground state}}} = \exp\left(\frac{-E_{\text{excited state }i}}{kT}\right) \qquad \textbf{(18A-6)}$$

An illustration of the changes in populations of the ground and excited states with temperature is shown in Figure 18A.1.

Since the exponent in the equation is a ratio of energies, the same units must be used to express $E_i$ as for the value of $kT$. Some representative temperatures and values of $kT$ in cm$^{-1}$ are shown in Table 18A.1. You should be able to convert the values of $kT$ and the energies $E_i$ to any energy unit you understand best.

Let us see how the distributions of sodium atoms and zinc atoms in their ground and first excited states change with temperature. In sodium, the first excited state lies at 17,000 cm$^{-1}$ = 2.1 eV = 589 nm above the ground-energy level. A transition between this energy level and the ground state would appear at 589 nm — in the yellow – orange region of the visible spectrum. In zinc, the first excited state is at 46,750 cm$^{-1}$ = 5.8 eV = 214 nm. This wavelength is in the ultraviolet. As you can

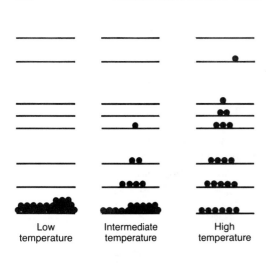

**Figure 18A.1**   Representation of the change in population of energy levels of identical atoms with variation of their temperature. Here, there are 22 atoms. A ball represents an atom in a specific quantized level. The populations are described by the Boltzmann distribution.

Low temperature     Intermediate temperature     High temperature

see from Table 18A.1, even at the hottest temperature, a great majority of both sodium and zinc atoms will still be in the ground state.[4]

One of the reasons that the concepts of the Boltzmann distribution are important relates to choosing the best of the various atomic spectrometric assay methods. For instance, consider the example of sodium and zinc in Table 18A.1. At *any* temperature, a significantly greater (but still small) proportion of the sodium atoms will be in the excited state. From the Boltzmann equation, we expect that the lower the energy of excited energy levels in the atoms, the more sensitive will an emission assay be. Thus, a simple calculation using the Boltzmann equation leads us to our first suggestion of the benefits and drawbacks of using atomic emission versus atomic absorption for atomic analysis: The sensitivity of emission spectrometry to a specific element is approximately inversely proportional to the ionization potentials of that element.

### Temperature Fluctuations and Noise

Let us now make a calculation using the Boltzmann equation to illustrate what relative concentration changes we might expect in the excited and ground states with a change in temperature of an emission source. Let us consider that the temperature varies by $\pm 10$ K. If the temperature is 2000–10,000 K, 10 K represents a small change, such as might occur from variations in the rate that gases are fed to the emission source.

As you can see from Eqs. 18A-4 and 18A-2, changes in atomic-state concentrations directly affect the light detected. Changes in population of the ground or excited states due to temperature fluctuations will be seen directly as changes in the absorption or emission, respectively. Depending on the rate of these fluctuations, the effects will be

---

[4] This result is quite intriguing. The orange–yellow emission of sodium that you see when heating a glass rod in a Bunsen burner is quite bright. However, the emission arises from only the small fraction of sodium atoms that are vaporized from the glass rod surface. And, given that the temperature of the flame is about 2000 K, it must be that the obvious, bright emission arises from only a small fraction of the atoms, those in the excited state.

**Table 18A.1** Representative Values of $kT$ in $cm^{-1}$ Together with Energy Level Populations for Na(I) and Zn(I).*
The factor counting for degeneracies is ignored.

| $T(K)$ | $kT(cm^{-1})$ | Na = 589 nm $\mathcal{N}_1/\mathcal{N}_0$ | Zn = 214 nm $\mathcal{N}_1/\mathcal{N}_0$ |
|---|---|---|---|
| 298 | 207 | $2 \times 10^{-36}$ | $8 \times 10^{-99}$ |
| 1000 | 695 | $2 \times 10^{-11}$ | $6 \times 10^{-30}$ |
| 5000 | 3476 | $8 \times 10^{-3}$ | $1 \times 10^{-6}$ |
| 10,000 | 6952 | $9 \times 10^{-2}$ | $1 \times 10^{-3}$ |

* Atomic spectroscopy nomenclature.

seen as noise (relatively fast) or drift (low-frequency noise), both of which degrade the precision and detection limits of the methods.

As an example, let us continue with the calculation for sodium shown in the preceding section and in Table 18A.1. Say that the temperature is 2700 K, a common sample temperature in flame atomic spectrometry. The wavelength of emission is 589.59 nm, which means that the excited state for this emission line (the yellow sodium line) is 16,951 $cm^{-1}$ or $3.3692 \times 10^{-19}$ J atom$^{-1}$ above the ground state. Let us make the calculation using joules. In joules, Boltzmann's constant $k$ equals $1.3806 \times 10^{-23}$ J atom$^{-1}$ K$^{-1}$. The *calculated* ratio of populations (not considering the constant degeneracy) of the excited state relative to the ground state ($E_0 = 0$) is

$$\frac{\mathcal{N}_1}{\mathcal{N}_0} = \exp\left(\frac{-E_1}{kT}\right)$$

$$= \exp\left(\frac{-3.3692 \times 10^{-19}}{1.3806 \times 10^{-23} \cdot 2700}\right)$$

$$= 1.19 \times 10^{-4}$$

And, at ten degrees higher temperature, 2710 K,

$$\frac{\mathcal{N}_1}{\mathcal{N}_0} = \exp\left(\frac{-3.3692 \times 10^{-19}}{1.3806 \times 10^{-23} \cdot 2710}\right)$$

$$= 1.23 \times 10^{-4}$$

What do these ratios mean?

Compared with the ground-state population, only around 0.01% of all Na atoms are emitting yellow light at 2700 K. However, by raising the temperature only 10 K, the population of the excited state—and intensity of this specific emission—increases by more than 3% (of the 0.01% population). The calculation is independent of degeneracy, which is a constant:

$$\frac{(1.23 - 1.19) \times 10^{-4}}{1.23 \times 10^{-4}} = 0.032$$

To find the change in the population of the ground state requires some care because the decrease in the number of atoms in the ground state equals the *sum* of gains of *all* the excited states. In this case, all excited states but one are so much higher in energy that they need not be considered in the calculation. The only exception is a state at nearly the same energy. The calculation shows that the population of the ground state decreases only about 8 ppm (2 states × 4 ppm each) with a 10-K change in temperature.[5]

What are the consequences of this behavior? From our calculation, we can conclude that for atoms that absorb and emit in the UV–visible region (250–750 nm), atomic absorption spectrometry is inherently less susceptible to temperature fluctuations than is emission spectrometry. The atomizer temperature is more critical in emission, so the factors that control the temperature must be more carefully controlled, including gas flow rates and keeping the same solvent for all samples and standards.

## SUPPLEMENT 18B

# The Limit of Detection in More Detail

A more precise definition and calculation of the detection limit is demonstrated here. As noted earlier, a statistical approach is required in which all three important factors enter: background effects, instrument sensitivity to the analyte, and signal-to-noise (S/N) ratio. A rigorous derivation is somewhat beyond the level of expertise being pursued here. A less difficult demonstration is, therefore, presented.

In the statistical treatment of the data, the distribution of error is assumed to be gaussian (a normal distribution, as described in Section 4.6). Thus, all the rules you learned in Chapter 4 about the standard deviation, expressed as $s$ (for a finite number of experiments) and $\sigma$ (for a very large number of experiments), hold here. As with any statistical measure, by running more independent experiments, you can be more

---

[5] The calculation of the relative change in ground-state population with a 10-K increase in temperature is done as follows. We calculate

$$\Delta \mathcal{N}_i / \mathcal{N}_0 = (1.23 - 1.19) \times 10^{-4} = 0.04 \times 10^{-4} = 4 \text{ ppm}$$

Only two excited states (with essentially the same $E_i$) and the ground state need be considered. We let $\mathcal{N}_0 = 1$, but since $\mathcal{N}_i \ll \mathcal{N}_0$,

$$\mathcal{N}_0 + 2\mathcal{N}_i \approx 1$$

Then

$$2 \cdot \frac{\Delta \mathcal{N}_i}{\mathcal{N}_0 + 2\mathcal{N}_i} \approx 2 \cdot 4 \text{ ppm}$$

and, with the approximation of ignoring the $\mathcal{N}_i$ in the denominator,

$$2 \cdot \Delta \mathcal{N}_i = 8 \text{ ppm} \cdot \mathcal{N}_0$$

certain of the value of the detection limit *for a specific set of experimental conditions.* Different detection limits will be found for different experimental conditions (for example, note points B and C in the quote in Section 18.13, page 740).

The following abbreviations are used in the discussion:

$S_t$ represents the signal that is obtained from the instrument output.

$S_X$ represents the part of the total signal that is due to analyte X. This is the net signal.

$S_b$ is the part of the total signal that is due to the blank (baseline).

$S_L$ is the signal due to the analyte alone at the detection limit.

$C_X$ is the concentration of the analyte.

$s_b$ is the standard deviation (finite number of runs) of the blank measurements.

$c_L$ is the concentration of analyte at the limit of detection. This is also called simply the detection limit.

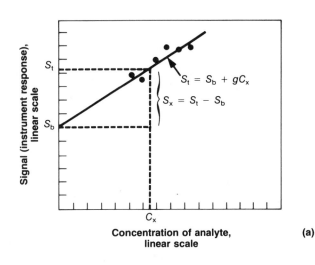

**(a)**

**Concentration of analyte, linear scale**

**Figure 18B.1**    (a) Plot of the signal output of an instrument versus the analyte concentration. The value $S_b$ is the signal due to the background alone as extrapolated from the data. A straight-line relationship is assumed. (b) The solid line is the same plot as above. The dashed lines indicate the error limits at some confidence level. The curvature is intentional and due to the fact that the ends of the straight line will move further than the middle with a change of slope in the line.

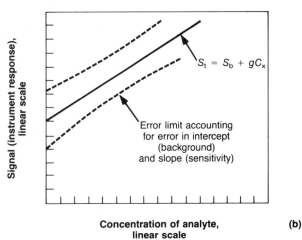

**(b)**

**Concentration of analyte, linear scale**

The experimentally determined data are assumed to be well fit by a straight line. Such a fit is illustrated in Figure 18B.1a. The linear calibration line has a slope that equals the analytical sensitivity. Let us label the slope (sensitivity) by the letter $g$. Algebraically, this is expressed in a simple equation; the sensitivity relates the signal to the analyte concentration.

$$S_X = g\, C_X \tag{18B-1}$$

Assume that the signal that is measured can be expressed as a simple sum of the background and analyte signals. Algebraically, this is

$$S_t = S_b + S_X \tag{18B-2}$$

Combining these two ideas—and two equations—we obtain an expression for the total signal,

$$S_t = S_b + g\, C_X \tag{18B-3}$$

This is the equation for the line in Figure 18B.1a. Note that this equation takes into account two of the three factors involved in determining the limit of detection—the instrument sensitivity and the background effects.

You should understand two points about Eq. 18B-3. First, it is a correct description *only* if there are no determinate errors such as incorrect standards or nonzero blanks. Second, it is true only if the background is exactly constant for all samples and standards.

It should be clear that if both of the above points were true, any amount of analyte, no matter how small, would be detectable. As you know, there are always some errors in determining both the background and the analyte signals for real instruments and samples. And that is the crucial point. The random errors can be determined and described by a confidence limit (C.L.).

Figure 18B.1b shows the same graph but with confidence limits indicated by dashed lines. This *could* be a 95% C.L., but the specific value is not important here. What is significant is that the intercept at zero analyte concentration is uncertain due to errors in the measurement. Note that this intercept determines the magnitude of the background signal. Thus, the background level is uncertain. If the errors are normally (gaussian) distributed, then the results can be shown in a different way from that in Figure 18B.1. This alternate graph appears in Figure 18B.2.

In Figure 18B.2, the distribution of the background measurements is plotted as the gaussian curve (a histogram of the results). The gaussian curve can be characterized by a mean and standard deviation, which describe the mean value and statistical error of the background signal. The left scale of the graph represents the scale of measured signal.

On the right-hand side of the figure is the scale of the signal that is due to analyte *alone*. It is zero at the *mean value* of the background signal, which is the behavior that is expected. In words, for the best-fit line, the intercept at zero concentration $S_b$ occurs at the average background signal.

The question that we now ask is, How much analyte signal do we need to be *certain*

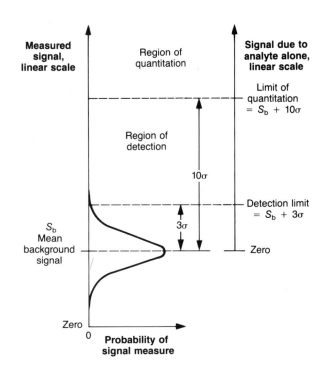

**Figure 18B.2**  Illustration of the relationship between the distribution of error in the background measurement, the detection limit, and the limit of quantitation. The definition of detection limit is more widely accepted and used than the quantitation limit. The total measured signal is plotted on the left scale, while the signal due to the analyte alone is plotted on the right.

that we can observe it? Recall from Chapter 4 that we can discuss certainty more precisely using statistics. The statistical certainty is expressed as a confidence limit. *By definition,* the detection limit of the analyte *signal* is at the 99.7% C.L. This is three times the standard deviation of the background $s_b$. The magnitude of the signal at the detection limit ($S_b + 3\sigma$) is indicated in Figure 18B.2.

The ideas in the paragraph above can be expressed using the following algebraic expression.

$$\text{signal for analyte at limit of detection} = S_L = S_t - S_b \geq 3\,s_b \quad \textbf{(18B-4)}$$

However, this expression concerns only the signal itself and not the concentration of analyte. Thus, to calculate the detection limit of concentration $c_L$, the sensitivity must be included. Thus,

$$c_L = \frac{S_t - S_b}{g} = \frac{S_L}{g} \quad \textbf{(18B-5)}$$

However, into the definition of $c_L$ (Eq. 18B-5), we can substitute for $S_L$ from Eq. 18B-4 and get

$$c_L \geq \frac{3 s_b}{g} \quad \textbf{(18B-6)}$$

for the detection limit *under the specified conditions* during the data collection.

Recall that the value of a limit of detection depends on three factors: background effects, instrument sensitivity to the analyte, and signal-to-noise (S/N) ratio. However, the signal-to-noise ratio does not seem to appear in Eq. 18B-6. What happened to it? The answer is that the effects of the noise appear in the value of $s_b$, the standard deviation. Intuitively, you might be able to see that as the noise level increases, determining the value of the background becomes more difficult (as illustrated in Figure 17.17, p. 643). As the noise level increases, the value of $s_b$ must increase. Thus, as Eq. 18B-6 states, the detection limit as determined from extrapolation of experimental values will be higher.

Under matching experimental conditions, it is more difficult to determine the concentration of an analyte than merely to detect its presence or absence. Thus, relatively recently, a new term has been suggested to delineate the lower end of the region where we can measure the quantity of analyte present. As shown in Figure 18B.2, the **region of quantitation** is defined when the signal is above the **limit of quantitation.** The limit of quantitation is where the signal due to the analyte alone is 10 times the standard deviation of the background signal.

$$S_X = 10s_b \qquad \textbf{(18B-7)}$$

It is not yet clear whether the term *quantitation limit* or its definition in Eq. 18B-7 will be widely accepted and used.

## SUPPLEMENT 18C

# Lock-in Amplifiers/Phase-sensitive Detectors

The names *lock-in amplifier* and *phase-sensitive detector (PSD)* apply to the same device. There are a number of types of these instruments, and the operation of one of the simplest is described here. The description is itself simplified. An up-to-date reference, with citations to some earlier reviews, is Meade, M. L., *J. Phys. E.* **1982,** *15,* 395.

The operation of a lock-in amplifier is illustrated in Figures 18C.1 and 18C.2. The benefits of a PSD derive from the use of *synchronous* switching between an input signal from the sample and one from a reference. At the time the sample input is connected to the signal processor, the sample signal itself is generated. Similarly, at the time the reference signal is connected, the output due to the reference is generated. Any input that is not synchronous with the signals from the sample or reference can be separated out. Such nonsynchronous inputs comprise the noise.

Phase-sensitive detection is commonly used in spectrometry, and the diagram of the apparatus in Figure 18C.1 is representative of a generalized spectrometer. In this spectrometer, the signals from both the sample and the reference are **chopped** or **modulated.** However, the chopping is done alternately so that the sample and reference light levels appear as shown in lines C and D of Figure 18C.2. This figure consists

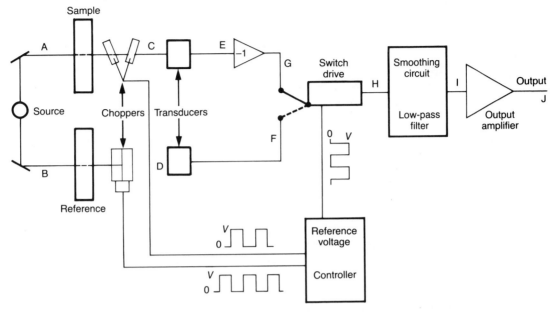

**Figure 18C.1**  Diagram of the principal components of a phase-sensitive detector. Measurements of light or voltages at the positions labeled with circled letters show the wave forms illustrated in Figure 18C.2.

of a series of graphs with the *x*-axis being time and the *y*-axis being either the amount of light present (A – D) or the voltages at various parts of the circuit. In the remainder of this supplement, the signal at each point A – J will be explained in order.

In the system illustrated here, the chopping of the light beam is done by electrically driven vanes that open and close in response to a voltage from a controller.[6] In this case, the controller puts out a voltage that changes in time as shown in Figure 18C.1. This is called a *square wave.* A second signal with an identical form also goes to a switching circuit in the signal processing circuit. The light from the sample and reference has an intensity that varies with time as graphed in plots C and D of Figure 18C.2.

The transducers convert the light into an electrical signal. Let us assume that the magnitude of the transducer output is zero when no light is present and that the output is perfectly linear. In this case, the heights of the output signals indicate that there is more light passing through the reference than through the sample. The output curves are not perfectly square because the circuit has some inductance and/or capacitance in it.

---

[6] Another way to chop the light is to use a rotating vaned wheel similar to a flat fan. Modulation may also be done by varying the wavelength or power of the electromagnetic source. Numerous other waveforms can be beneficially used in addition to the square wave illustrated here.

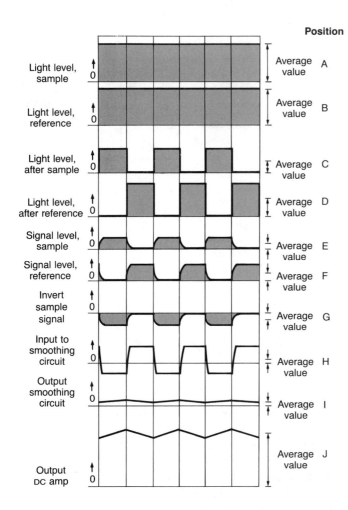

Light level, sample — Average value A

Light level, reference — Average value B

Light level, after sample — Average value C

Light level, after reference — Average value D

Signal level, sample — Average value E

Signal level, reference — Average value F

Invert sample signal — Average value G

Input to smoothing circuit — Average value H

Output smoothing circuit — Average value I

Output DC amp — Average value J

**Figure 18C.2**   Plots of the time-dependent signal as it passes through a spectrometer with phase-sensitive detection. The modulation is a square wave, although other waveforms are used as well.

The signal from the sample is inverted by an inverting amplifier (multiply by $-1$) so that the voltages at points F and G are as shown. The voltage from the sample now varies at G in the opposite direction from zero than at E.

Now the signals pass through the synchronous switch. The property of this switch is that when the modulated signal from the sample is *on*, the switch connects with the sample channel. When the signal from the reference is *on*, the switch connects with the reference channel. The signal passing to the amplifier then appears as on line H.

The sum of these two signals is averaged by a smoothing circuit. It is called a **low-pass filter.** This unit allows only low-frequency signals through — the limit of which is a constant DC current. (The average signal would be zero if the reference and the sample were the same.) This voltage is then amplified to produce the output. There is a slight fluctuation of the output voltage since the filter is not perfect.

As a final point, note that there is a trade-off for the use of a PSD. Half the light is lost, as can be seen from the average values of graphs A – D of Figure 18C.2. In effect, the signal from the analyte is *on* only half the time.

# 19

# Nuclear Magnetic Resonance Spectrometry

## 19.1 Introduction and Case Reading

Nuclear magnetic resonance spectrometry (NMR) involves measuring the absorption of radiofrequency radiation by a sample material that is placed in a strong magnetic field. The radiation used is in the 100-MHz range: the frequency of FM radio in the U.S. The magnetic fields are large; the most sophisticated instruments use some of the largest constant magnetic fields that can be generated. Supplement 19A contains further information on instrument design and operation.

The nuclei of many atoms possess a **magnetic moment.** This means that they act like small bar magnets — magnetic dipoles. The radiofrequency radiation is absorbed by the nuclear magnets which are thereby raised from their ground states to their excited states. As illustrated in Figure 19.1, the ground state is the energy level of the nucleus when the magnetic dipole is aligned along the magnetic field, and the excited state is the energy level when the magnetic dipole is aligned against the magnetic field. Nuclei, such as those of hydrogen or $^{13}$C, behave as if they can be oriented in only these two ways. The orientation of a nuclear dipole is quantized.

As in other spectrometries, the energy of the transition is determined by molecular and atomic properties; but in NMR spectroscopy, the energy also depends on the

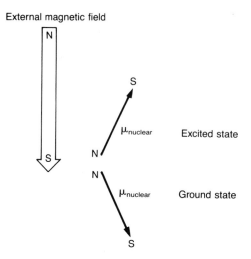

External magnetic field

**Figure 19.1**  The spectroscopic transition for nuclear magnetic resonance is described as the absorption of energy needed to reorient magnets in a magnetic field. The absorbed energy is converted to heat in a nonradiative transition. The energy between the ground and excited states is in the radiofrequency region.

$\mu_{nuclear}$     Excited state

$\mu_{nuclear}$     Ground state

magnetic field. Also, as in other spectrometries, the magnitude of the power absorbed is proportional to the concentration of the absorbing species. Thus, for a sample containing protons, if we excite the proton nuclei, the signal that is measured is proportional to the number of protons in the sample. This is the basis of the measurement described in the NMR application presented below.

The nuclei are sitting in the middle of the electron clouds of their atoms. It may seem surprising, but as far as the nuclear magnet is concerned, this is an isolated environment. Once the radiofrequency energy excites the nuclei from the ground to the excited state, they stay in the excited state for a relatively long time: up to seconds. When we measure the number of nuclei returning to the ground state in an experiment, the signal looks like a change in concentration with first-order kinetics (see Supplement 1B). Such a signal is illustrated in Figure 19.2. The characteristic time associated with this first-order rate process is called the relaxation time. For more advanced techniques in NMR, this is an extremely important quantity. However, in this chapter we shall not delve deeply into the subject of nuclear relaxation times.

The practical example that follows describes the use of an NMR method to measure the moisture content of tobacco. A common way to do such a determination without NMR is to dry a weighed sample until its weight remains constant and then measure the weight loss. The determination usually requires a number of hours of drying in an oven. The NMR method does not require any drying time, in fact, it requires no weighing of the sample, and an answer is obtained in about one minute. Sample weighing is eliminated by a clever use of an experimental fact: Protons in solids and in liquids have different nuclear relaxation times.

The relaxation times for protons in solids are much shorter (tens of microseconds) than the relaxation times for protons in liquids (as long as a few seconds). Because of this significant difference in relaxation times, the signal from protons in a liquid phase will remain detectable long after the signal from protons in the solid phase of the same sample has decayed away. Therefore, by measuring the amplitude of the

**Figure 19.2** The excitation of the nuclear magnets from ground to excited states can be achieved in a short time by a pulse of intense radiofrequency radiation. The nuclei then relax back to the ground state with a characteristic relaxation time. Part of the energy for this process is emitted as radiofrequency luminescence and can be detected as a free induction decay. This output signal follows the same time course as the plot of the number of nuclei in the excited state.

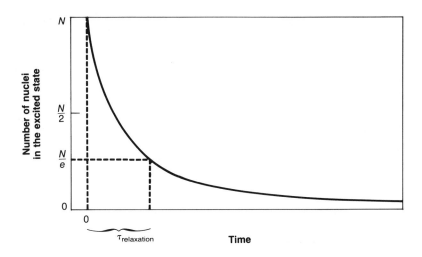

**Figure 19.3** The free induction-decay signal due to protons in a sample that has part of its protons in liquid water and the remainder in the solid material. After the excitation is finished, the first 10 $\mu$s of the output signal is not recorded. Subsequently, the signal is sampled at two times, and each point is extrapolated back to the initial time, that is, immediately after the excitation *90° pulse* was applied. The total number of protons contained in both solid and liquid give rise to the initial signal measured at 12 $\mu$s. The part of the initial signal due to the liquid-phase protons alone is found from the second point at 70 $\mu$s. The difference is due to solid-phase protons alone. The ratio is used to calculate the %(w/w) water in the sample. [Redrawn from a figure courtesy of IBM Instruments.]

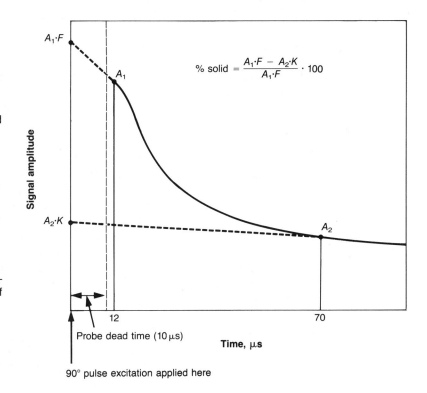

$$\% \text{ solid} = \frac{A_1 \cdot F - A_2 \cdot K}{A_1 \cdot F} \cdot 100$$

**Table 19.1**  Verification* of NMR Method for Measuring Moisture in Tobacco

| Sample | Loss on Drying (% $H_2O$ by Weight) | % $H_2O$ by NMR | Standard Deviation on 10 NMR Trials |
|---|---|---|---|
| 1 | 12.99 | 13.11 | 0.05 |
| 2 | 13.32 | 13.48 | 0.06 |
| 3 | 12.97 | 12.94 | 0.05 |
| 4 | 14.56 | 14.37 | 0.04 |
| 5 | 13.45 | 12.80 | 0.04 |
| 6 | 12.48 | 12.37 | 0.06 |
| 7 | 11.55 | 11.26 | 0.04 |
| 8 | 21.96 | 19.20 | 0.16 |
| 9 | 15.77 | 15.30 | 0.09 |
| 10 | 21.24 | 20.83 | 0.26 |

* Correlation 0.9819 for linear least squares.

signal at two times, the *ratio* of protons in the solid phase and protons in the liquid phase can be determined without weighing the sample.

The amplitude of the whole signal at an early time is determined by the sum concentration [H(solid + liquid)], while the signal at later times is proportional to [H(liquid)] only. The instrument must be calibrated to relate the signal magnitude to the concentrations of protons in each phase.

You will need to have some definitions in order to understand the following quote. A **free induction decay** is another name for the signal in this type of NMR experiment. This is abbreviated *FID*. An example of such a signal from the NMR spectrometer is shown in Figure 19.3.

The experiment is initiated (the nuclei are raised into the excited state) by a short pulse of radiofrequency radiation. The instrument delivers the maximum sensitivity when the length of the pulse (a few microseconds) and the power transmitted to the sample are adjusted to a condition called a **90° pulse.** The pulse is transmitted to the sample, which is contained in a glass tube placed inside a small **coil** of wire. The coil acts as an antenna for the radiofrequency energy (see Supplement 19A). There is a short time in which the coil cannot be used as a receiver antenna for the reemitted radiofrequency radiation. This period is called the *dead time.* Both the solid and liquid decays are extrapolated back through the dead time to the time immediately after the irradiation pulse. The proton concentrations are calculated from these extrapolated values.

**Determining Solid/Liquid Ratios**  The solid to liquid ratio [abbreviated S/L] calculation is described here. . . . Since two points on the FID are digitized [by the computer] for this measurement, the weight of the sample is not involved in the calculation . . . [Figure 19.3 illustrates] how the microprocessor . . . does the S/L calculation.

The liquid signal is measured at point $A_2$ [at 70 $\mu$s]. It is multiplied by the K-factor, a correction factor which is determined during the NMR [calibration] test. The K-factor adjusts for the slight decay of the liquid signal that has occurred by 70 $\mu$s . . . . The solid signal is measured at 12 $\mu$s after the 90° pulse. The signal amplitude cannot be measured immediately after the 90° pulse because the coil used to transmit the r.f. pulse is the same coil that is used to receive the signal from the sample. [The probe dead time is the period of] 10 $\mu$s after the r.f. energy is transmitted. The FID decays during the 10 $\mu$s dead time, so the signal measured at 12 $\mu$s is multiplied by the F-factor, a correction factor determined during calibration. The F-factor extrapolates the measured signal height back to the signal height immediately after the pulse ends [zero time].

The calculation of S/L ratios relies on the assumption that the K-factor times the [signal size at the] second point, $K \cdot A_2$, will give [a measurement equal to] the undecayed liquid signal and that the F-factor times the first point, $F \cdot A_1$, will give the total undecayed signal. Since free induction decays are not linear, the second assumption is not exactly correct, but it is close enough for most measurements . . . . [Verification of the method is illustrated in Table 19.1.]

[Ref: Manufacturer's literature from IBM Instruments, courtesy of L. D. Jones.]

## 19.2 General Principles of NMR

The type of NMR assay described in the introduction is one special use of the NMR phenomenon. The remainder of this chapter describes more customary applications: determining organic molecule structures. For such studies, spectra of radiofrequency-energy absorption are measured and not the time dependence of the signal.

The frequency at which the absorption of energy occurs in an NMR experiment depends on two factors:

1. The identity of the nucleus
2. The magnetic field strength

The relationship is written algebraically as

$$v_{\text{res}} = \left(\frac{\gamma}{2\pi}\right) \cdot H_0 \qquad \text{(19-1)}$$

where $v_{\text{res}}$ is the **resonance frequency** and $H_0$ is the magnetic field strength of the spectrometer. $\gamma$ is a constant, the value of which depends on the identity of the nucleus; it is called the **magnetogyric ratio.** In Table 19.2 are listed the magnetic field strengths and the resulting resonant frequencies of the five most common NMR nuclei. These five are, by far, the most easily investigated experimentally (although the instrument electronics is still quite sophisticated) and exhibit the most easily understood spectra. In this chapter, only the analytical use of proton, $^1$H, NMR spectra is presented in some detail.

The early developers of the NMR techniques were hoping that NMR could be used to make highly precise measurements of the magnetogyric ratios of nuclei. However, reality intervened, and they found that the frequency of resonance varied in a manner

**Table 19.2**   Field Strengths and Associated Resonance Frequencies of the Five Most Commonly Investigated Nuclei

| Isotope | Field Strength $H_0$ in Gauss | Resonance Frequency in MHz | Nuclear Spin Quantum Number I |
|---------|-------------------------------|----------------------------|-------------------------------|
| $^1$H   | 14,100 | 60.0  | $\frac{1}{2}$ |
|         | 23,500 | 100.0 | |
|         | 51,480 | 220.0 | |
|         | 70,200 | 300.0 | |
| $^2$H   | 23,500 | 15.3  | 1* |
| $^{13}$C | 23,500 | 25.1  | $\frac{1}{2}$ |
| $^{19}$F | 23,500 | 94.0  | $\frac{1}{2}$ |
| $^{31}$P | 23,500 | 40.5  | $\frac{1}{2}$ |

\* Deuterium, with a nuclear quantum number of 1, has *three* different energy levels in a magnetic field.

that depended on the chemical environment of the nuclei. This effect was named the **chemical shift.** However, this "problem" is one of the reasons that NMR is such a useful analytical technique in chemistry.

It was found experimentally that the chemical shift was proportional to the magnitude of the magnetic field. This result can be expressed algebraically by modifying Eq. 19-1 with a magnetic field **shielding parameter** $\sigma$. To express the chemical effects, we substitute $H_0(1 - \sigma)$ for $H_0$ and get

$$v_{\text{shifted res}} = \left(\frac{\gamma}{2\pi}\right) \cdot (H_0 - \sigma H_0) \qquad \textbf{(19-2)}$$

However, there is no way to know $H_0$, the field at the sample nucleus without any shielding. So Eq. 19-2 is of little use. The problem can be circumvented by using a relative frequency shift compared with a reference compound. The equation describing the chemical shift $\delta$ is

$$\delta(\text{ppm}) = \frac{H_0(\text{reference}) - H_0(\text{sample})}{H_0} \times 10^6 \qquad \textbf{(19-3)}$$

The units of the chemical shift are **parts per million** (ppm). (This should not be confused with the content measure, which is also ppm.) A reference compound is used as an internal standard of the chemical shift. The internal standard for $^1$H-NMR spectra of samples that dissolve in organic solvents is now mutually agreed upon: It is tetramethylsilane $(CH_3)_4Si$, abbreviated TMS. The reasons for this choice will become clear as the properties of NMR spectra are described. On the chemical-shift ($\delta$) scale, the tetramethylsilane $^1$H resonance signal appears at $\delta = 0.00$ ppm *by definition.* In aqueous solvents, a common standard is trimethylsilylpropanesulfonate with $\delta = 0.015$ ppm. Its formula is $(CH_3)_3SiCH_2CH_2CH_2SO_3^-$.

Because the field and resonance frequency are simply related (Eq. 19-1), an equivalent definition of the chemical shift as described by Eq. 19-3 is

$$\delta(\text{ppm}) = \frac{\nu(\text{sample}) - \nu(\text{reference})}{\nu(\text{reference})} \times 10^6 \qquad \textbf{(19-4)}$$

---

**EXAMPLE**

Assume that a spectrum is run at a frequency of 60 MHz and a field of 14,092 gauss with TMS as reference. The resonance of a proton from the sample appears at a frequency 216 Hz higher than that of the standard. What is the chemical shift?

**SOLUTION**

From the definition of the chemical shift in the form of Eq. 19-4,

$$\delta = \left(\frac{216}{60 \times 10^6}\right) \times 10^6 = \frac{216}{60} = 3.6 \text{ ppm}$$

---

NMR, like most other analytical techniques, was the product of the work of many people. During this development, two different chemical-shift scales arose: the delta ($\delta$) scale and the tau ($\tau$) scale. By definition, $\tau = 10.00$ ppm for TMS and runs *with* the field, in the opposite direction of the $\delta$ scale. The relationship between these two scales is illustrated in Figure 19.4. Most data in the chemical literature are now reported using the delta scale. Clearly, it is easy to convert the two scales using

$$\tau = 10 - \delta \qquad \textbf{(19-5)}$$

The frequency differences due to the chemical-shift effect are relatively small. For a 100-MHz resonance frequency, the range of chemical shifts found for protons is only about 1000 Hz. As a relative change, this is only 10 ppm. Experimentally, it is quite difficult to keep the irradiation frequency, detector amplifiers, and magnetic field so stable that these measurements can be made accurately and reproducibly. Nevertheless, such conditions must be maintained, because individual spectral lines can have linewidths less than 0.1 Hz. In NMR, as in other spectrometries, the instrumental bandwidth must be significantly smaller than the inherent spectral linewidth in order to obtain undistorted sample spectra. (See Section 17.9.)

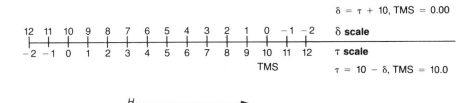

**Figure 19.4** Illustration of the relationship between the two NMR chemical shift scales. The resonance position of tetramethylsilane (TMS) is defined as $\delta = 0.0000$. The range shown here is typical for $^1$H spectra. All other NMR nuclei, for example, $^{19}$F and $^{13}$C, exhibit much wider chemical-shift ranges.

### Origin of σ, the Shielding Parameter

Equation 19-2 is an algebraic description of an experimental result: The instrument magnetic field is apparently reduced by some opposing magnetic field that arises *inside* the sample. This reduction in the magnetic field at the nucleus is called **shielding.** We explain this internal field as being due to the electron density around the nucleus: The greater the electron density, the larger the shielding. Since the field-frequency ratio is fixed by $\gamma$, if we want to observe a resonance absorption of energy, then we must *increase* the external field to compensate for the *reduction* due to the shielding effect. The greater the electron density around a proton, the more the external magnetic field must be raised to observe resonance absorption. Therefore, highly shielded protons show absorption lines that are **upfield** from less shielded **(deshielded)** protons. These relationships are illustrated in Figure 19.5. A most important general result is explained through this relationship: *The chemical shift of a proton depends on the chemical structure near it.*

### Signal Magnitude and Concentration

As in other spectrometric methods, the frequency (or wavelength) is determined by internal properties of molecules, and the magnitude of the energy absorption is determined by the concentration. In NMR spectrometry, the areas under the peaks are measured by electronic integration. An NMR spectrum and its integral are shown in Figure 19.6. The spectrum, shown as peaks arising from a baseline, was run; then a second run was made with the instrument set to integrate the area. (Computerized instruments obtain the integral by numerical integration of the stored spectrum.)

In NMR spectrometry, the integrated peak areas are only relative values, depending on the number of nuclei that contribute to the peak. This is one of the great strengths of the NMR method: *Every proton, when its absorbs energy, contributes equally to an NMR absorption spectrum.* The magnitude of the absorption is inde-

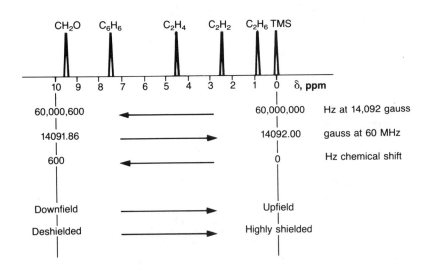

**Figure 19.5**  Some of the nomenclature and numerical values found for $^1$H-NMR spectra. The numbers are typical for the smaller, more common instruments, which operate with a radiofrequency near 60 MHz.

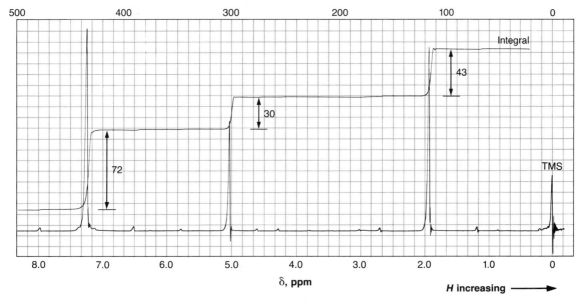

**Figure 19.6**   A ¹H-NMR spectrum with integration of the peaks. The sample is benzyl acetate with an internal standard of TMS.

pendent of the chemical shift. In effect, every proton has the same absorptivity. Obtaining *absolute* concentrations using NMR spectra is done using concentration standards.

---

**EXAMPLE**

Assign the three peaks of the benzyl acetate NMR spectrum shown in Figure 19.6.

**SOLUTION**

The structure of benzyl acetate is

$$\langle\bigcirc\rangle-CH_2-O-\overset{\overset{\displaystyle O}{\|}}{C}-CH_3$$

There are five protons on the ring, three on the methyl group, and two on the benzyl group. The area under each peak must be relatable to an integral number of protons. Let us find what area measure corresponds to *one* proton in the molecular formula. In this case, we use the heights of the integrals of the peaks. The integrated spectrum is the line with three steps of heights 72, 30, and 43 divisions on the chart paper.

Let us try dividing by the smallest height. That is

$$\frac{72 \text{ divisions}}{30 \text{ divisions}} = 2.40 \qquad \frac{30 \text{ divisions}}{30 \text{ divisions}} = 1.00 \qquad \frac{43 \text{ divisions}}{30 \text{ divisions}} = 1.43$$

The ratios here are not whole number ratios as desired. Nevertheless, note that the ratios will be almost whole numbers if we multiply each by 2. This is equivalent to having 15.0 divisions on the integral curve correspond to one proton. Multiplying the ratios by 2.0 produces

$$4.80 : 2.0 : 2.86$$

These ratios approximate

$$5 : 2 : 3$$

within the precision of the measurement of the integrals, approximately $\pm 10\%$.[1]

From this simple calculation of ratios (allowing for some error in the integral measurement) we assign the furthest downfield peak, $\delta = 7.3$, as the resonance energy absorption of the five ring protons. Similarly, the middle peak, at 5.1 $\delta$, is assigned to the two benzyl protons. And the furthest upfield peak, at 2.0 $\delta$, is assigned to the three protons of the methyl group.

## 19.3 Typical Chemical Shifts

Not only are the NMR absorption-peak areas proportional to the number of each kind of proton, another great advantage of NMR spectra is that the chemical shifts for protons in different chemical environments occur in *ranges* that are specific to that environment. Therefore, we can make a **correlation table** of the chemical shifts

---

[1] A number of integration runs could be carried out to increase the precision of the measurement. The process would be equivalent to making more runs in the determination of a single sample.

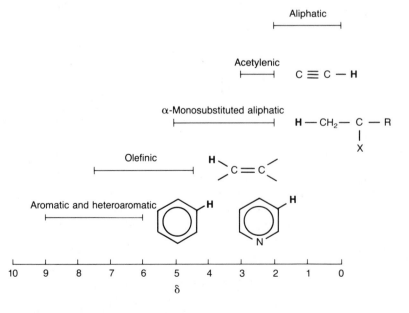

**Figure 19.7**  Typical ranges of the chemical shift of protons in organic molecules. The substituent noted by X can be a group such as O, N, Cl, Br, and $NO_2$. More detail is presented in Figure 19.8 and Table 19.3.

**Structural type**

$\delta$ and $\tau$ value and range[†]

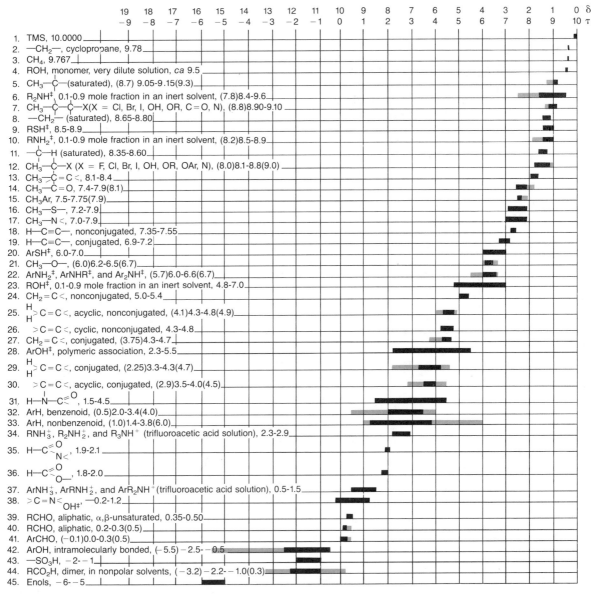

1. TMS, 10.0000
2. —$CH_2$—, cyclopropane, 9.78
3. $CH_4$, 9.767
4. ROH, monomer, very dilute solution, $ca$ 9.5
5. $CH_3$—C—(saturated), (8.7) 9.05-9.15(9.3)
6. $R_2NH^‡$, 0.1-0.9 mole fraction in an inert solvent, (7.8)8.4-9.6
7. $CH_3$—C—C—X(X = Cl, Br, I, OH, OR, C=O, N), (8.8)8.90-9.10
8. —$CH_2$—(saturated), 8.65-8.80
9. $RSH^‡$, 8.5-8.9
10. $RNH_2^‡$, 0.1-0.9 mole fraction in an inert solvent, (8.2)8.5-8.9
11. —C—H (saturated), 8.35-8.60
12. $CH_3$—C—X (X = F, Cl, Br, I, OH, OR, OAr, N), (8.0)8.1-8.8(9.0)
13. $CH_3$—C=C<, 8.1-8.4
14. $CH_3$—C=O, 7.4-7.9(8.1)
15. $CH_3Ar$, 7.5-7.75(7.9)
16. $CH_3$—S—, 7.2-7.9
17. $CH_3$—N<, 7.0-7.9
18. H—C≡C—, nonconjugated, 7.35-7.55
19. H—C≡C—, conjugated, 6.9-7.2
20. $ArSH^‡$, 6.0-7.0
21. $CH_3$—O—, (6.0)6.2-6.5(6.7)
22. $ArNH_2^‡$, $ArNHR^‡$, and $Ar_2NH^‡$, (5.7)6.0-6.6(6.7)
23. $ROH^‡$, 0.1-0.9 mole fraction in an inert solvent, 4.8-7.0
24. $CH_2$=C<, nonconjugated, 5.0-5.4
25. H(H)>C=C<, acyclic, nonconjugated, (4.1)4.3-4.8(4.9)
26. >C=C<, cyclic, nonconjugated, 4.3-4.8
27. $CH_2$=C<, conjugated, (3.75)4.3-4.7
28. $ArOH^‡$, polymeric association, 2.3-5.5
29. H(H)>C=C<, conjugated, (2.25)3.3-4.3(4.7)
30. >C=C<, acyclic, conjugated, (2.9)3.5-4.0(4.5)
31. H—N—C=O, 1.5-4.5
32. ArH, benzenoid, (0.5)2.0-3.4(4.0)
33. ArH, nonbenzenoid, (1.0)1.4-3.8(6.0)
34. $RNH_3^+$, $R_2NH_2^+$, and $R_3NH^+$ (trifluoroacetic acid solution), 2.3-2.9
35. H—C(=O)(N<), 1.9-2.1
36. H—C(=O)(O—), 1.8-2.0
37. $ArNH_3^+$, $ArRNH_2^+$, and $ArR_2NH^+$ (trifluoroacetic acid solution), 0.5-1.5
38. >C=N(OH‡), −0.2-1.2
39. RCHO, aliphatic, α,β-unsaturated, 0.35-0.50
40. RCHO, aliphatic, 0.2-0.3(0.5)
41. ArCHO, (−0.1)0.0-0.3(0.5)
42. ArOH, intramolecularly bonded, (−5.5) −2.5- −0.5
43. —$SO_3H$, −2- −1
44. $RCO_2H$, dimer, in nonpolar solvents, (−3.2) −2.2- −1.0(0.3)
45. Enols, −6- −5

† Normally, absorptions for the functional groups indicated will be found within the range shown. Occasionally, a functional group will absorb outside this range. Approximate limits for this are indicated by absorption values in parentheses and by shading in the figure.

‡ The absorption positions of these groups are concentration dependent and are shifted to higher $\pi$ values in more dilute solutions.

**Figure 19.8** NMR absorption chemical shifts for protons in various chemical environments. The lighter areas are extreme but possible regions. [Ref: Dyer, J. R. "Applications of Absorption Spectroscopy of Organic Compounds"; Prentice–Hall: Englewood Cliffs, NJ © 1965. Reproduced with permission.]

**Table 19.3** Approximate NMR Chemical Shifts of Protons (R—— = alkyl group, Ar—— = aromatic group)

### CH₃ — Protons

| Proton | δ (ppm) |
|---|---|
| CH₃—C | 0.9 |
| CH₃—C—C=C | 1.1 |
| CH₃—C—Cl | 1.4 |
| CH₃—C—O | 1.4 |
| CH₃—C≡C | 1.6 |
| CH₃—C—NO₂ | 1.6 |
| CH₃—C—Br | 1.8 |
| —C=C(CH₃)—CO | 1.8 |
| CH₃—CO—N—R | 2.0 |
| CH₃—C≡C—CO | 2.0 |
| CH₃—CO—R | 2.2 |
| CH₃—Ar | 2.3 |
| CH₃—N | 2.3 |
| CH₃—CO—O—Ar | 2.4 |
| CH₃—CO—Ar | 2.6 |
| CH₃—Br | 2.7 |
| CH₃—Cl | 3.0 |
| CH₃—O—R | 3.3 |
| CH₃—O—Ar | 3.8 |
| CH₃—O—CO—R | 3.7 |

### —CH₂— Protons

| Proton | δ (ppm) |
|---|---|
| —C—CH₂—C | 1.3 |
| —C—CH₂—C—C=C | 1.7 |
| —CH₂—C—Cl | 1.8 |
| —CH₂—C—Br | 1.8 |
| —C—CH₂—C—O | 1.9 |
| —C—CH₂—C—NO₂ | 2.1 |
| —C—CH₂—CO—O—R | 2.2 |
| —C—CH₂—C≡C | 2.3 |
| —C—CH₂—CO—R | 2.3 |
| —C—CH₂—C≡C—CO | 2.4 |
| —C=C(CH₂)—CO | 2.5 |
| —C—CH₂—N | 2.7 |
| —C—CH₂—Ar | 3.4 |
| —C—CH₂—O—R | 3.4 |
| CH₂—Cl | 3.4 |
| CH₂—Br | 3.6 |
| —C—CH₂—O—H | 4.1 |
| —C—CH₂—O—CO—R | 4.3 |
| —C—CH₂—O—Ar | 4.4 |
| —C—CH₂—NO₂ | |

### —CH— Protons

| Proton | δ (ppm) |
|---|---|
| —C—CH—C | 1.5 |
| —CH—C—Br | 1.9 |
| —CH—C—Cl | 2.0 |
| —C—CH—C—O | 2.0 |
| —C—CH—CO—R | 2.7 |
| —C—CH—N | 2.8 |
| —CH—Ar | 3.0 |
| —C—CH—O—R | 3.7 |
| —C—CH—O—H | 3.9 |
| —CH—Cl | 4.0 |
| —CH—Br | 4.1 |
| —C—CH—NO₂ | 4.7 |
| —C—CH—O—CO—R | 4.8 |

### Other Groups

| Proton | δ (ppm) |
|---|---|
| —C=CH | 5.3 |
| —C=CH— | 5.1 |
| —C=CH—(cyclic) | 5.3 |
| R—C≡CH | 3.1 |
| —C=CH—CO | 5.9 |
| —CH=C—CO | 6.8 |
| R—CHO | 9.9 |
| Ar—CHO | 9.9 |
| H—CO—O | 8.0 |
| H—CO—N | 8.0 |
| CHCl₃ | 7.25 |
| H₂O | ~5 |

*Aromatic Protons*

Benzene —H: 7.2

X-substituted benzene —H: 7.0–9.0

Toluene CH₃: 2.3, —H: 7.1

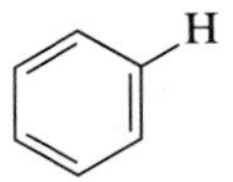

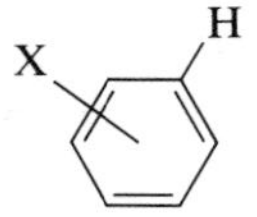

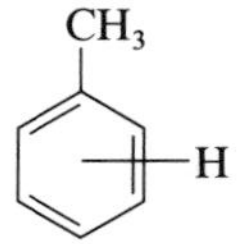

*Source:* Pavia, D. L.; Lampman, G. M.; Kriz, G. S. "Introduction to Spectroscopy: A Guide for Students of Organic Chemistry." Copyright © 1979 by W. B. Saunders Company. Reprinted by Permission of CBS College Publishing.

observed and the identity of the chemical groups in which the protons are bonded. Figure 19.7 shows some general correlations so that you can see what a few of the chemical regularities are. A more detailed figure of this type is shown in Figure 19.8. The chemical shift ranges are listed in Table 19.3 by chemical type. Because of this regularity, the assignments of the benzyl acetate resonance peaks could have been made from the chemical shifts alone without measuring the integral values. The chemical shifts in Table 19.3 show that protons of an aromatic ring with a carbon attached should be expected at around 7.1 ppm. The methyl protons of an acetate group are expected at around 2.0 ppm ($CH_3$—CO—O—Ar). The benzyl protons are the only ones left.

## 19.4 Equivalence and Inequivalence

You have seen how the three peaks occurring in the NMR spectrum of benzyl acetate can be assigned to the three different types of protons in the molecule. However, one peak appears from, for example, five ring protons. This is also true of the internal

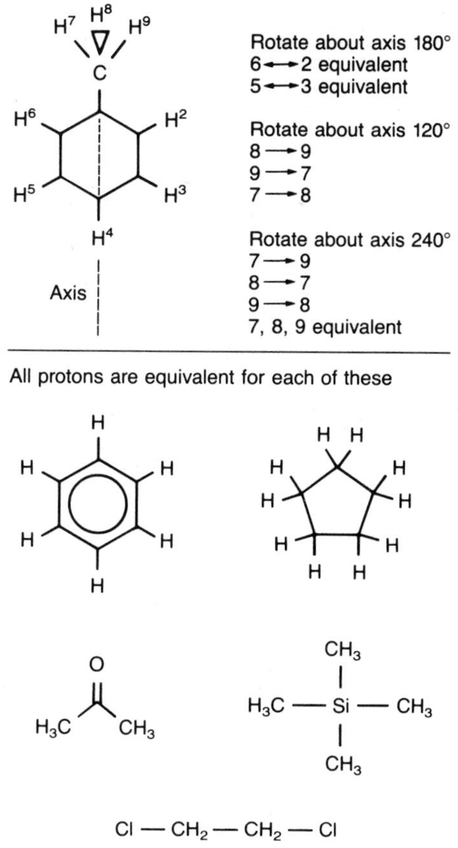

**Figure 19.9**  Illustration of the concept of equivalence of protons. Experimentally it is found that equivalent protons exhibit the same NMR spectral properties. The way we determine if two protons are equivalent (slightly oversimplified) is to see if some rotation will produce a molecule that looks the same. The protons that are interchanged are equivalent. The dotted line shows the axis about which the molecule is rotated.

Rotate about axis 180°
6 ↔ 2 equivalent
5 ↔ 3 equivalent

Rotate about axis 120°
8 → 9
9 → 7
7 → 8

Rotate about axis 240°
7 → 9
8 → 7
9 → 8
7, 8, 9 equivalent

All protons are equivalent for each of these

standard TMS. There are 12 protons in a TMS molecule, $(CH_3)_4$ Si, but only one peak appears in its NMR spectrum, as seen on the right in Figure 19.6. When the resonances of two or more protons give rise to only one absorption peak, they are said to be **chemically equivalent**: They are called **equivalent** protons. Illustrated in Figure 19.9 are some examples of molecules that show chemically equivalent peaks. A simple technique can be used to see if there is chemical equivalency between protons. If it is possible to rotate the molecule in some way so that you cannot tell which protons were interchanged, then the protons are equivalent. An example in Figure 19.9 illustrates the sorts of rotations that are used. The study of this subject is part of an area of mathematics called *group theory*. Further discussion of this topic is outside the subject of this book.

When we carry out the rotations of the ring in benzyl acetate to see which protons are equivalent, the rule given above suggests that the ring protons should fall into three groups: two *ortho,* two *meta,* and one *para.* However, only one peak is seen at 7.3 $\delta$. The reason is that the differences in the chemical shifts among these three chemically distinct groups is too small to be resolved in this spectrum. In other words, simply because chemically distinct protons *can* have different chemical shifts does not mean they *will.*

## 19.5 Nuclear Spin – Nuclear Spin Interaction

As noted above, the frequency at which the NMR peak appears is determined by the size of the external magnetic field of the instrument and by the internal magnetic field due to the electrons surrounding the nucleus. However, one other magnetic field is present: the magnetic field from nearby protons since every proton is itself a magnet. The effect of the magnetic field of other protons is to shift the NMR-absorption line. The shifts show up as **split** lines, and so the effect is called **spin – spin splitting** or **hyperfine splitting.** As you shall see, this effect is extraordinarily useful in determining chemical structures.

To explain spin–spin splitting, first consider two protons attached to adjacent carbons. The experimental spectrum of such a compound is shown in Figure 19.10. There are only two protons, but four lines are seen. What has happened is that each resonance line for a proton has been split into a **doublet.** Each line of the doublet has half the area of the original line. The separation of the two lines (between the line centers) is called the **hyperfine coupling constant.** Usually the capital letter $J$ is used to signify this value, and the magnitude is expressed in frequency units (Hz). The hyperfine coupling constant in Figure 19.10 is the spacing, in Hz, between the two lines on the left *or* the two lines on the right. The hyperfine coupling constant is the same for both doublets. This is a general result. If the splitting by proton A of the resonance line of proton B is $J_{AB}$, then the resonance line of A will be split by the same amount. Algebraically, we write $J_{AB} = J_{BA}$.

Next look at Figure 19.11, which shows NMR spectra of $CH_2CHCN$, acrylonitrile, at three different field values (obtained on three different instruments). Observe the four lines on the right-hand side. They are split by the same amount *regardless of the*

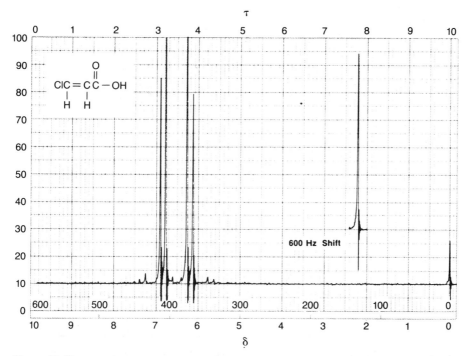

**Figure 19.10** The NMR spectrum of *cis*-3-chloroacrylic acid showing the effect on the spectrum of hyperfine splitting. The protons on either side of the double bond have different chemical shifts: one at 4.2 ppm and one at 4.85 ppm. Each of the lines is split into two by the adjacent proton. The peak for the hydroxyl proton is moved onto the scale; it appears at 12.2 ppm. [Spectrum from Pouchert, C. J.; Campbell, J. R. "The Aldrich Library of NMR Spectra." Courtesy Aldrich Chemical Co.]

*size of the magnetic field.* However, the other lines are shifted further away from these four as the field gets larger. These spectra illustrate an important general result. The absolute value of the shift due to shielding by the electrons depends on the external field; the spin–spin splitting, due to adjacent protons, does not.

Let us see how the existence and properties of the spin–spin splitting can be explained. Consider that the absorption being measured is that of proton $H^a$ in Figure 19.12. Two factors must be considered.

1. The magnetic field position of each NMR absorption is determined by the internal (molecular) field. But now there is an extra magnetic field coming from an adjacent magnet, that of the nucleus of proton $H^b$. Thus the line position will depend on *both* the chemical shift and the magnetic field of the adjacent proton.

2. The other nuclear spin, due to $H^b$, is quantized in direction either along or away from the external magnetic field (recall Figure 19.1). Thus, the magnetic field of $H^b$ can have two different values.

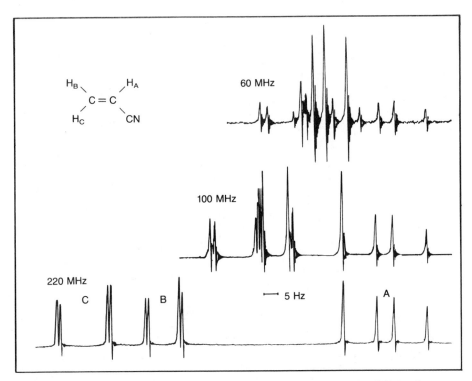

**Figure 19.11**    The hyperfine splitting remains the same with changes in magnetic field, but the multiplet positions are field dependent. Illustrated are NMR spectra of the same compound run with successively higher magnetic fields but plotted on the same scale in Hz. The size of the magnetic field is directly proportional to the frequency. The spectra are usually labeled with a frequency scale, as is done here: 60 MHz, 100 MHz, and 220 MHz. The constancy of $J$ is seen most clearly for the four peaks on the right. [Ref: Reprinted with permission from Johnson, L. F. *Anal. Chem.* **1971**, *43*(2), 29A. Copyright 1971 American Chemical Society.]

When $H^b$ is aligned along the field direction, the NMR absorption of $H^a$ will occur at a lower external field than when $H^b$ is not present. Conversely, when $H^b$ is aligned against the external field, the NMR absorption of $H^a$ will occur at a greater external field than if $H^b$ is not present. Thus the $H^a$ absorption line appears to be split. No line appears at the unsplit line position, because $H^b$ always has an effect.

Experimentally, the size of the splitting $J_{AB}$ depends on a number of factors. But basically the magnitude of $J$ decreases with distance between the protons. Unless there is a conjugated system, spin–spin coupling is rarely observed when more than three bonds separate the hydrogen nuclei.

The labeling that is used to describe the spectrum with spin–spin splitting is as follows. The chemical shift is expressed as if the spin–spin splitting did not occur. In the example of Figure 19.12, this is at the average position (midpoint) of the doublet: the doublet's center of energy. For the spectrum in Figure 19.10, the lines can be reported as

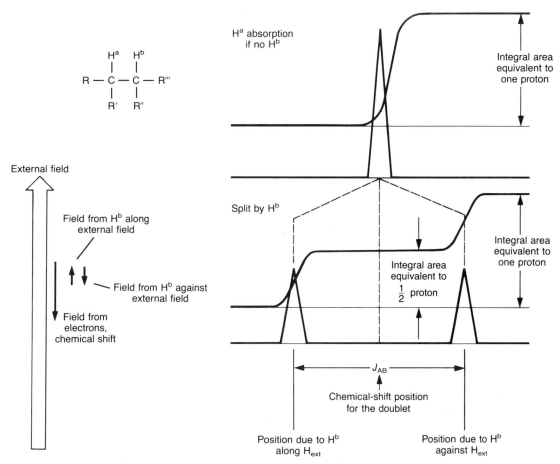

**Figure 19.12**   The origin of the intensities and positions of split NMR lines. Shown here is the situation observed experimentally in the spectrum in Figure 19.10. As illustrated, the positions of the resonance lines for proton $H^a$ is determined by three interactions: with the external magnetic field, with the field due to the electrons causing a chemical shift, and with the two different alignments of the adjacent proton $H^b$. $H^b$ is equally likely (in time) to be in either orientation, so there are two peaks equal in magnitude from $H^a$. Each of the peaks is half the size of an unsplit peak.

a doublet at 6.2 $\delta$, with $J = 6.8$ Hz,
a doublet at 6.8 $\delta$, with $J = 6.8$ Hz, and
a singlet at 12.2 $\delta$.

Note that the chemical shifts are reported only to tenths of a ppm even though the graph can be read more precisely. This is the limit of accuracy of the instrument, as can be found by running the same spectrum repeatedly.

## The Heights and Areas of Split Peaks

As illustrated in Figure 19.12, each of the peaks in the doublet has one-half the area that would be measured for a singlet (unsplit) peak from the proton. The two doublet

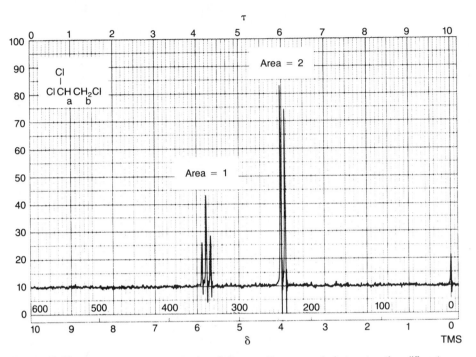

**Figure 19.13**   When a proton resonance is split by more than one equivalent proton, then different **splitting patterns** are observed. In this spectrum from 1,2,2-trichloroethane are resonances from two equivalent protons adjacent to a single proton. The pattern from the single proton (area = 1) is a typical NMR triplet. Its origin is explained in Figure 19.14. The spectrum from the two equivalent protons (area = 2) is a doublet with its splitting equal to the triplet's. [Spectrum from Pouchert, C. J.; Campbell, J. R. "The Aldrich Library of NMR Spectra." Courtesy Aldrich Chemical Co.]

peaks are equal, because the $H^b$ proton can be aligned either away from or along the external magnetic field, and both directions are, essentially, *equally probable.*[2] A simple rule can be written about the area: The integrated area per nuclear spin remains the same regardless of the number of lines into which it may be split.

This brings us to the next topic. What happens when more than one proton splits the resonance? The answer comes from an experimental spectrum, that of 1,1,2-trichloroethane ($ClH_2CCHCl_2$) as shown in Figure 19.13. The integrated areas confirm the assignment of the spectrum. The resonance of proton $H^a$ is split into three lines, a **triplet,** and the two equivalent protons $H^b$ appear as a doublet.

We explain the triplet splitting and intensities in the following way, which is illustrated in Figure 19.14. Either or both of the two equivalent protons can be

---

[2] This point can be understood with a calculation using the Boltzmann distribution equation (Section 17.2) and the energy equivalent of 100 MHz (Table 17.2).

aligned along or against the external field, so that for the two protons there are four possible sets of orientations of equal probability. These four sets produce only three different magnetic fields (Figure 19.14) because two sets produce the same (zero) magnetic field.

Since *each* of these four sets of orientations is equally probable, the areas of the lines associated with each set of orientations are equal. Therefore, we expect an unshifted line twice as large as the two shifted lines. The relative areas of the three lines will be, as observed, $1:2:1$. The total integrated area of the three lines is still that for one proton.

On the other hand, the two equivalent $H^b$ protons are split into a doublet by a single $H^a$. The area of the doublet is that expected for two protons. Note in the experimental spectrum that the splittings between the triplet lines equals the splitting between the doublet. Again, $J_{AB} = J_{BA}$.

When describing the NMR spectrum, the chemical shift of the triplet is, as before, taken at the frequency (or field) at which there is no contribution from the spin–spin interaction. When an *odd* number of lines exists in the multiplet (the general name of

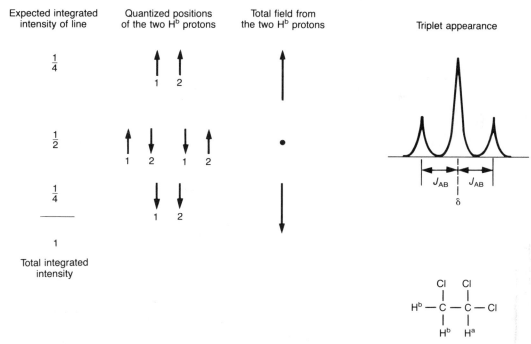

**Figure 19.14**   When two equivalent protons split the resonance line of another proton, the pattern is as shown at the right. Three lines are seen, with the middle one twice the area of the others. The area of each of the three lines is proportional to the average time the two protons are in each of three different states. These three states are shown in the second column. The four possible sets of positions are equally probable. However, there are two ways that the sum of the fields from the two protons can be zero. Thus the central line is two times as intense.

sets of lines: doublets, triplets, quartets, and so on), the chemical shift for the set is the position of the middle line.

### General Splitting Patterns

Hyperfine splitting patterns get more complicated as the number of equivalent protons increases. A set of $n$ equivalent protons splits adjacent protons into $(n + 1)$ lines. The number of lines in the splitting pattern is called the **multiplicity.** The relative integrated area of each of the lines in the multiplet (the set of lines) can be calculated from probability theory or by drawing diagrams such as in Figure 19.14. However, a far simpler way to obtain the relative areas is to use a Pascal Triangle. This is illustrated in Figure 19.15. Each number in the triangle is the sum of the two numbers above it (above left and above right), assuming zeros in the blanks. The areas of the individual lines in a multiplet are in the proportions given after the multiplicity name. The integrated areas of all the lines of a multiplet are proportional to the number of protons absorbing there.

Let us review the rules to explain splitting patterns.

1.  *The number of lines in a multiplet equals* n *+ 1, where* n *is the number of equivalent protons contribution to the splitting.*
2.  *The expected integral areas of each of the component lines in a multiplet are given by the Pascal Triangle values.*
3.  *The area of all the lines in a multiplet is proportional to the number of equivalent nuclei absorbing there.*
4.  *The splittings* $J_{AB} = J_{BA}$.
5.  *Seldom will there be splittings between protons that are more than three bonds removed.*

| Number of equivalent protons causing splitting ($n$) | Number of peaks in multiplet ($n + 1$) | Multiplicity name | Relative intensities of lines in multiplet |
|---|---|---|---|
| 0 | 1 | Singlet | 1 |
| 1 | 2 | Doublet | 1  1 |
| 2 | 3 | Triplet | 1  2  1 |
| 3 | 4 | Quartet | 1  3  3  1 |
| 4 | 5 | Quintet | 1  4  6  4  1 |
| 5 | 6 | Sextet | 1  5  10  10  5  1 |
| 6 | 7 | Septet | 1  6  15  20  15  6  1 |

**Figure 19.15**    The Pascal Triangle. This is a useful mnemonic device for obtaining the relative areas of the lines of a multiplet produced by a number of equivalent protons (or other nuclear-spin-$\frac{1}{2}$ equivalent nuclei). The number of equivalent nuclei is shown on the left. The intensities are shown in the triangle. The rule for forming the numbers in the triangle is that each value is the sum of the two immediately above it to the right and left. Zeros are assumed to be outside of the *ones* at the edges.

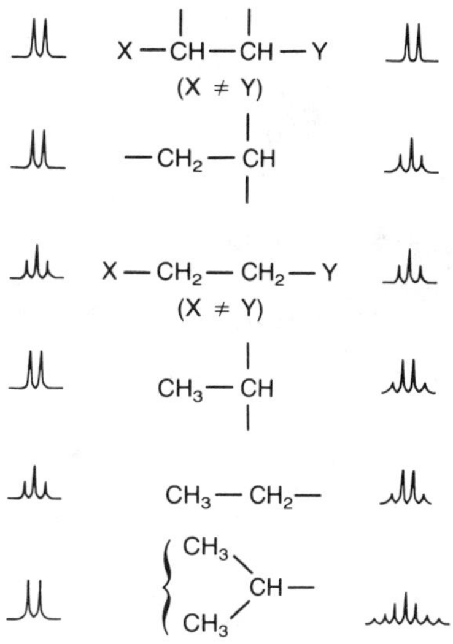

**Figure 19.16** Since hyperfine splittings are molecularly quite localized, a number of commonly occurring structures have sets of multiplets that show up together. Some are represented here. An NMR spectrum of the groups will have these *sets* of splitting patterns. The relative magnitudes of the integrated areas will also be constant. [From Pavia, D. L.; Lampman, G. M.; Kriz, G. S. "Introduction To Spectroscopy: A Guide for Students of Organic Chemistry." Copyright © 1979 by W. B. Saunders Company, Philadelphia, PA. Reprinted by Permission of CBS College Publishing.]

Figure 19.16 illustrates the splitting patterns for common organic groups. Let us consider the information contained in such sets of multiplets.

A.   *The chemical environments of the nuclei are indicated by the value of δ.*

B.   *The number of nuclei absorbing energy at a specific δ is proportional to the integrated area of each multiplet.*

C.   *The number and intensities of lines in the multiplet tell you how many protons are adjacent.*

D.   *The identity of the adjacent protons can be ascertained from* their *chemical shifts and splitting pattern.*

E.   *The spin–spin coupling values can help confirm the above assignments. (See Table 19.4.)*

[1]H-NMR is, indeed, a powerful method for determining the structures of molecules containing protons.

## Some Complications

You may have observed that a number of details in the experimental NMR spectra have been ignored so far. Some of these complications will be pointed out next.

The first problem is seen most clearly in Figure 19.11. All the scans of the spectra proceed from left to right, and on the right side of each line is a large amount of "jiggle." If a single peak were scanned in the same way, it would appear as shown in Figure 19.17. The undulation to the side of the main peak is called **ringing**. This arises from a fast scan through the spectral peak by an instrument that is working well. The better tuned the machine is, the larger the ringing effect will be. Ringing can be reduced significantly by scanning more slowly. Ringing is not a problem with Fourier transform instruments.

**Table 19.4**  Spin–Spin Coupling Constants in Some Common Systems

| System | $J_{H,H'}$, Hz Full Range | Typical | System | $J_{H,H'}$, Hz Full Range | Typical |
|---|---|---|---|---|---|
| H–C–H' (geminal) | 0–25 | 12–15 | cis >C=C< (H, H') | 0–12 | 7–10 |
| >CH–CH'< | 0–8 | ~7 | trans >C=C< (H, H') | 12–18 | 14–16 |
| CH₃–CH₂– (Free rotation) | 6–8 | ~7 | aromatic $J_{H,H'}$ | 6–10 | 8 |
| (CH₃)₂CH'– (Free rotation) | 5–7 | ~6 | aromatic $J_{H,H''}$ | 0–37 | 2 |
| >CH–C–CH'< | 0–1 | 0 | aromatic $J_{H,H'''}$ | 0–1 | 1 |
| >C=CH–CH'< | 4–10 | 5–7 | =C< (H, H') | 0–3.5 | 0–2 |
| >C=CH–CH'=C< | 6–13 | 10–13 | –CH=C–CH'< | 0–3 | 0.5–2 |
| >CH–CH'O | 0–3 | 2 | >CH–C=C–CH'< | 0–2 | 1 |
| C=CH–CH'O | 5–8 | 7 | >CH–C≡CH' | 2–3 | 2–3 |

*Source:* Courtesy Varian Associates.

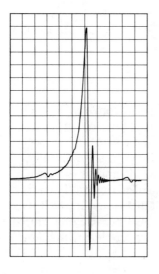

**Figure 19.17**  An NMR peak with ringing. The spectrum was scanned from left to right at a relatively fast speed. This pattern is indicative of a well-tuned spectrometer. The word *tuned* is used in the sense of tuning a radio station.

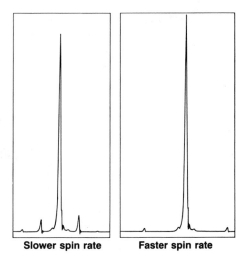

**Figure 19.18** Spinning side bands occur when the sample tube is not perfectly cylindrical or the magnetic field is not well adjusted. They appear symmetrically placed on either side of each true spectral line. Their positions move further away from the main peak as the rate of sample-tube spinning is increased.

Slower spin rate      Faster spin rate

**Glycolic acid, 99 + %**

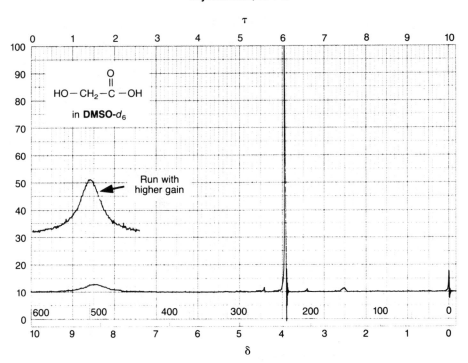

**Figure 19.19** Illustration of one kind of effect that can be seen when a proton is not covalently bound to the molecule. The two hydroxy protons are **chemically exchanging** between their sites on the molecule and on the solvent. Their resonance appears broad and shifts with pH. In addition, there is no hyperfine splitting due to either of them. [Spectrum from Pouchert, C. J.; Campbell, J. R. "The Aldrich Library of NMR Spectra." Courtesy Aldrich Chemical Co.]

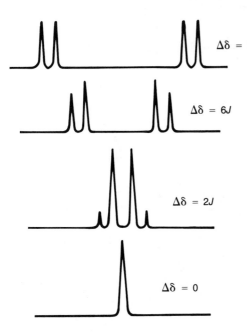

**Figure 19.20**   Simulation of the effects on the peak intensities when two protons produce a spectrum in which each proton splits the resonance of the other. The peaks become unequal in area (or height) as the difference in the chemical shift $\Delta\delta$ and hyperfine coupling $J$ become nearly the same. Note that such an effect is seen in the spectrum in Figure 19.10.

The second effect can be seen in Figures 19.6 and 19.10. Notice the small peaks on either side of the main peaks. These are not resonance peaks. They are **spinning side bands.** Briefly, inside the instrument, typically the sample is contained in the bottom of a glass tube 5 mm in diameter and about 20 cm long. The tube containing the sample is mechanically spun around its long axis. Slight imperfections in the tube (out of round) or in the instrument cause side-band peaks to show up, equidistant on each side of each absorption peak. It is easy to determine that they are not impurities from their symmetrical positions and from the fact that they change position with changes in the spinning rate. This change is shown in Figure 19.18.

The third problem is that the chemical shift of a resonance may be pH dependent. An example is shown in Figure 19.19. This dependence appears for protons that are loosely held on a molecule. The chemical shift and shape of such peaks usually depend on the solvent and on pH. Such peaks tend to be broad.

Fourth, notice that the two lines in the doublets in the spectrum of Figure 19.10 are not the same height (or area). The details of the cause of this effect are beyond the level of this text. However, the effect can be illustrated pictorially as in Figure 19.20. In addition to the changes in peak height, there is a distortion in the splitting values, which is not so obvious.

### Splitting by More Than One Set of Equivalent Nuclei

If a nuclear resonance is split by more than one set of equivalent spins, the result is a multiplet of multiplets, as illustrated in Figure 19.21 for the spectrum of 1-nitropropane. There is no guarantee that all the expected multiplet lines will show up in the complicated splitting patterns. In fact, usually fewer lines are seen than expected. A

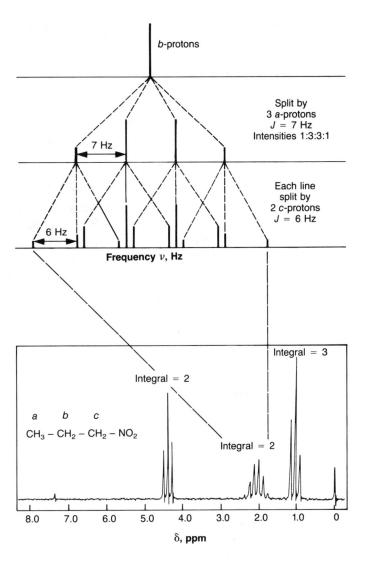

**Figure 19.21**   When a spectral line is split by more than one set of equivalent protons, the effects are additive. This is illustrated by the spectrum of 1-nitropropane. The chemical formula is shown. The stick figure at the top explains the splitting of the *b*-protons by the *a*- and *c*-protons. A total of 12 lines is expected (a quartet of triplets, or a triplet of quartets). However, only six are seen since many lines overlap due to a lack of resolution. The effect of the *b*-protons on the resonances of the *a* and *c* groups is straightforward: producing triplets for both. [NMR spectrum of 1-nitropropane courtesy of Varian Associates.]

typical case is the multiplet for the middle methylene group in 1-nitropropane. The lines are due to the protons labeled *b*. As shown in the figure, we expect a quartet of triplets: a total of 12 lines. In the experimental spectrum only six lines appear. The reason is that, because the spin–spin splitting value $J_{AB}$ and $J_{CB}$ are nearly equal, a number of peaks overlap, as illustrated in the stick diagram. Thus, even for relatively simple molecules it is necessary to be cautious and to be sure that the multiplet patterns and integrals make sense *as a whole* in order to assign structures. Notice that with four bonds between them, the *a*- and *c*-protons do not show any spin–spin splitting.

NMR spectra easily become far more complicated in appearance. The complexity especially shows up for molecules that are not highly symmetric or molecules in which more than two sets of equivalent protons contribute to spin–spin splittings. An example is shown in Figure 19.22, which shows the spectrum of the relatively simple molecule 1,4-pentadiene. The same rules about splitting apply to this spectrum, but the various contributions to each multiplet are more difficult to pick out. This becomes especially hard when $J_{AB} \approx |\delta_A - \delta_B|$. In those situations, either the spectra can be compared with standard spectra for qualitative identification or computer programs can be used to fit the spectra. Further information on the latter method can be found in the Suggested Readings.

### Spin–Spin Splitting from Nonresonant Nuclei

After your first look at the spectrum shown in Figure 19.23, you might immediately start searching for some other multiplet with an integral equivalent to three $I = \frac{1}{2}$ equivalent spins. However, you would not find it even if you searched past the normal spectral range. This is because the splitting is being caused by nuclei that are not themselves near their resonant frequency. In the case of 2-iodo-1,1,1-trifluoro-

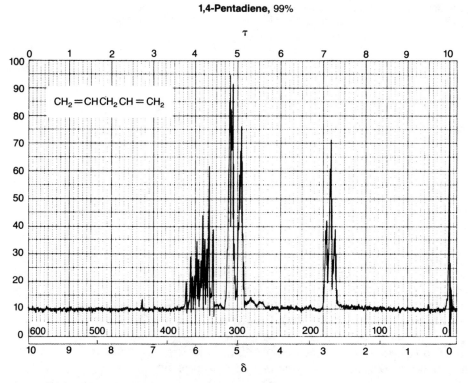

**Figure 19.22**    The NMR splitting patterns can get complicated, as illustrated in the spectrum of the apparently simple molecule 1,4-pentadiene. [Spectrum from Pouchert, C. J.; Campbell, J. R. "The Aldrich Library of NMR Spectra." Courtesy Aldrich Chemical Co.]

**2-Iodo-1,1,1-trifluoroethane,** 99% (2,2,2-trifluoroethyl iodide)

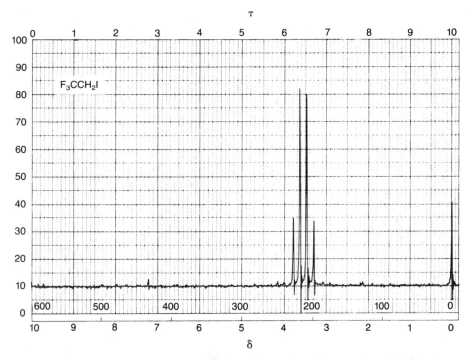

**Figure 19.23**  Resonance lines can be split by nuclei that are themselves not observable. Here, three equivalent $^{19}$F atoms ($I = \frac{1}{2}$) split the spectral line of the protons in 2-iodo-1,1,1-trifluoroethane. [Spectrum from Pouchert, C. J.; Campbell, J. R. "The Aldrich Library of NMR Spectra." Courtesy Aldrich Chemical Co.]

ethane, the splittings are from $^{19}$F. (If you were able to obtain the $^{19}$F-NMR of the compound, what kind of multiplet splitting would the fluorines exhibit?[3])

Incidently, notice that the integral of the absorption peaks conveys no useful information alone. Some other multiplet or standard is needed in order to obtain either the relative or the absolute concentrations, respectively.

## 19.6  $^{13}$C-NMR

You might ask, Why, if nuclei other than protons can split the spectral lines, don't the $^{13}$C nuclei do so in all these molecules? The answer is that the splitting does occur, except that only 1% of the carbons are the isotope $^{13}$C; so the split peaks that do show up are from only 1% of the carbons. The peaks are quite small relative to the peak left

---

[3] The fluorine multiplet would be a $1:2:1$ triplet from two equivalent protons.

788

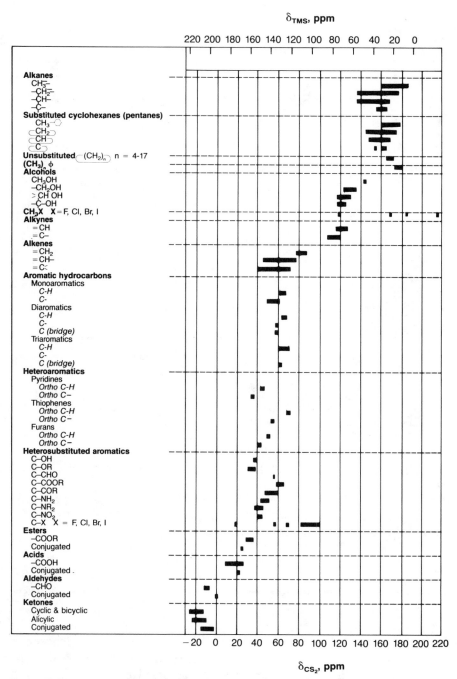

**Figure 19.24** Characteristic $^{13}$C-NMR chemical shifts in organic materials relative to a $^{13}$CS$_2$ standard (lower scale) and to $^{13}$C-TMS (upper scale) [Ref: Jensen, R. K.; Petrakis, L. *J. Magn. Reson.* **1972**, *6*, 105–106. Reproduced with permission.]

**Figure 19.25**    The nonresonant $^1$H nuclei can split $^{13}$C. Only the bound protons actually produce a splitting. This can aid in assignment of the spectrum to a specific molecule. However, the large number of lines generated often causes significant complexity in the spectra. The coupling can be removed by proper experimental conditions. This is called **decoupling** of the protons. The effect of decoupling is shown in the lower spectrum. [Ref: Moore, J. A.; Dalrymple, D. L. "Experimental Methods in Organic Chemistry"; Philadelphia: W. B. Saunders & Co., 1976.]

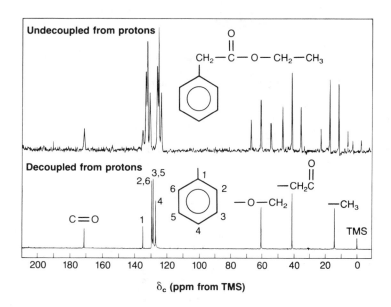

unsplit by $^{12}$C (nuclear spin $= 0$; 99% abundance). NMR spectra cannot be obtained for $^{12}$C, which has no nuclear spin. The low natural abundance of $^{13}$C makes $^{13}$C-NMR more difficult to do experimentally. Either the molecules must be enriched in $^{13}$C, or much more sample must be used to obtain a good spectrum, one with a high S/N. But $^{13}$C spectra can be measured routinely with appropriate instruments, and the resonances show patterns of chemical shifts that depend on the chemical type of carbon. The general regions are shown in Figure 19.24. Notice that the range of chemical shifts is quite large in comparison with that of $^1$H: around 200 ppm as opposed to 10 ppm. It follows that the $^{13}$C chemical shifts are far more sensitive to changes in environment than are the $^1$H spectra.

Therefore, $^{13}$C-NMR is an extremely powerful method to investigate the structures of molecules. One reason is that all the protons connected to a carbon cause spin–spin splitting. This means that by looking at the $^{13}$C-NMR spectrum, the number of protons (nonresonating) bonded to a carbon can be observed merely by noting the multiplet number of each carbon resonance. This effect can be seen in Figure 19.25 in the upper spectrum: quartets for —CH$_3$, triplets for —CH$_2$—, and a singlet for C=O. However, the spectra quickly get so complicated that even for relatively simple molecules with a few similar groups, it becomes difficult to interpret.

The problem is remedied by using the technique of **proton noise decoupling**. The details of the method are beyond the level explored here. However, the effect is that the proton–carbon spin–spin coupling can be eliminated, and the multiplets of $^{13}$C lines "collapse" into a single line for each carbon. A proton noise-decoupled spectrum is shown in the lower part of Figure 19.25. If two carbons are equivalent, they will exhibit only one absorption line. However, for various reasons, quantitation by integration of $^{13}$C spectra is quite inaccurate. You can see the effect by comparing the heights of the peaks in Figure 19.25. They are far from proportional to the number of carbons. Nor are the peak areas.

## Suggestions for Further Reading

A chapter at a slightly more advanced level than this one. The authors describe identification of organic molecule NMR spectra.
Silverstein, R. M.; Bassler, G. C.; Morrill, T. C. "Spectrometric Identification of Organic Compounds," 4th ed.; Wiley: New York, 1981; Chaps. 4 and 5.

A complete review of proton and carbon NMR in chemical analysis. Some chapters are at an advanced level. Two chapters concern analytical applications.
Leyden, D. E.; Cox, R. H. "Analytical Applications of NMR"; Wiley–Interscience: New York, 1977.

An intermediate level monograph which includes theory, instrumentation, and representative applications of proton NMR.
Kasler, F. "Quantitative Analysis by NMR Spectroscopy"; Academic Press: London, 1973.

Middle-range theory with numerous practical tips for NMR in general and examples of biochemical NMR.
Knowles, P. F.; Marsh, D.; Rattle, H. W. E. "Magnetic Resonance of Biomolecules"; Wiley–Interscience: London, 1976; Chaps. 1–5.

More or less a "classic." Relatively high-level introduction to Fourier transform NMR.
Farrar, T. C.; Becker, E. D. "Pulse and Fourier Transform NMR"; Academic Press: New York, 1971.

A short article describing the abilities of contemporary NMR in medicine and polymer studies.
Jelinski, L. W. "Modern NMR Spectroscopy"; *Chemical and Engineering News* **1984,** November 5; pp. 26–47.

An outstanding nonmathematical intermediate introduction to contemporary NMR.
Akitt, J. W. "NMR and Chemistry: An Introduction to the Fourier Transform–Multinuclear Era"; Chapman and Hall: London and New York, 1983.

This book consists of 100 pages of NMR theory and then 75 example problems of $^{13}$C-NMR spectra. Worked-out solutions to the problems follow. A unique and useful introduction.
Breitmaier, E.; Bauer, G. "$^{13}$C NMR Spectroscopy. A Working Manual with Exercises"; Harwood: Chur, Switzerland, 1984.

## Problems

19.1   A 60-MHz $^1$H-NMR spectrum was run, and six peaks with equal integral areas were found. The peaks were at 0, 83, 98, 112, 228, and 429 Hz downfield relative to TMS. In the sample are tetramethylsilane, acetone, benzene, cyclohexane, *t*-butanol, and dioxane. The hydroxy proton of the alcohol could not be seen.
    a.   Assign the six peaks to the six compounds.
    b.   If the spectrum were a 100-MHz $^1$H-NMR spectrum, at what frequencies relative to TMS would the six resonances be?
    c.   If the TMS is assigned a concentration of 10.0, what are the concentrations of the other five components of the mixture?

19.2   A spectrum of toluene (that is, methylbenzene) was run. The spectrum consists of two singlets. The singlet at 7.2 $\delta$ has an integral of 72 graph spaces. The singlet at 2.35 $\delta$ has an integral of 36 graph spaces. Is the toluene pure?

19.3    [Spectra from Pouchert, C. J.; Campbell, J. R. "The Aldrich Library of NMR Spectra."
Courtesy Aldrich Chemical Co.] Assign spectra I through V in Figure 19.3.1. The possible
compounds are

a.    1-chloropropane, $CH_3CH_2CH_2Cl$
b.    1,2-dichloropropane, $CH_3CH(Cl)CH_2Cl$
c.    1,3-dichloropropane, $Cl(CH_2)_3Cl$
d.    isopropanol, $(CH_3)_2CHOH$
e.    allyl chloride, $H_2C{=}CHCH_2Cl$
f.    1-propanol, $CH_3CH_2CH_2OH$

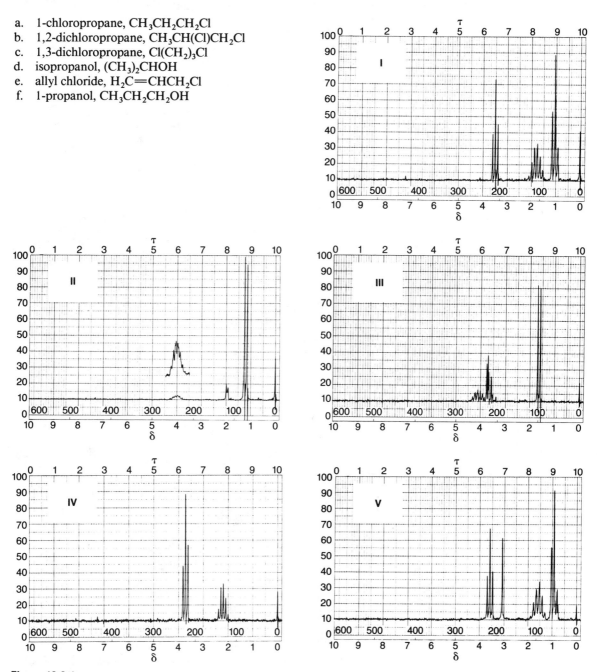

**Figure 19.3.1**

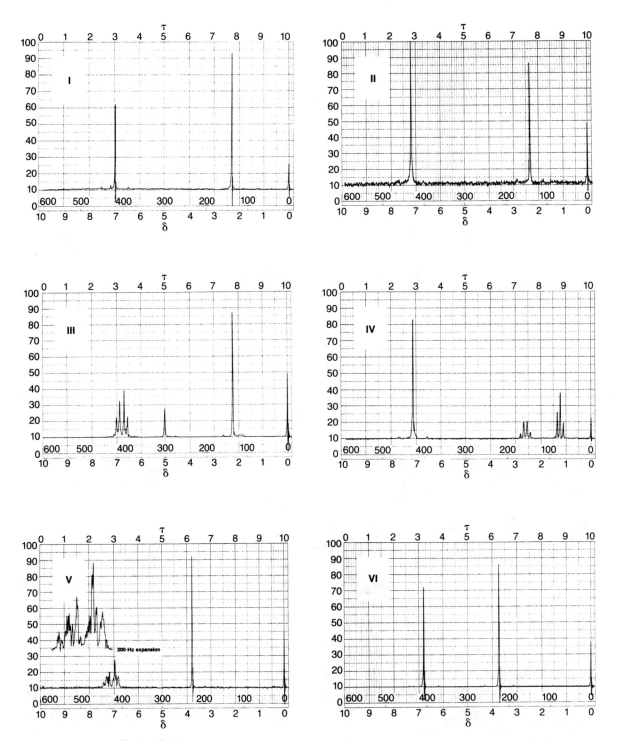

**Figure 19.4.1**

**19.4** [Spectra from Pouchert, C. J.; Campbell, J. R. The Aldrich Library of NMR Spectra. Courtesy Aldrich Chemical Co.] Assign spectra I through VI in Figure 19.4.1. The possible compounds are

    a.  ethylbenzene, $C_6H_5C_2H_5$
    b.  *p*-cresol, $CH_3C_6H_4OH$
    c.  *p*-xylene, $CH_3C_6C_4CH_3$
    d.  benzene, $C_6H_6$
    e.  toluene, $C_6H_5CH_3$
    f.  anisole, $C_6H_5{-}OCH_3$
    g.  *p*-dimethoxybenzene, $CH_3O{-}C_6H_4{-}OCH_3$

**19.5**  A new sulfonate drug was being developed that contained the following group in its structure.

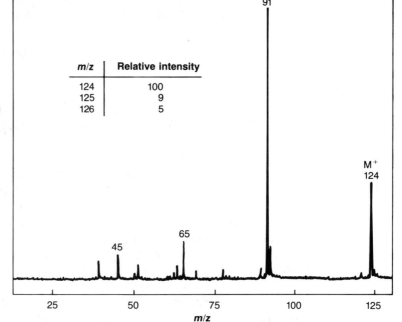

Running a conventional mass spectrum was not possible because the compound degrades when heated; so it was decided to find the molecular mass by NMR. To do this, 10.0 mg of the compound was weighed into an NMR tube. To the tube was added 5.0 mg of anhydrous sodium acetate (f.w. 82.04). Both compounds were dissolved in $D_2O$. An NMR spectrum was run. The sodium acetate peak at 1.90 $\delta$ had an integral area of 82.0 graph spaces (average of 3 runs). A singlet at 8.0 $\delta$ (no other peaks in the region) had an integral of 30.5 spaces (average of 3 runs). What is the formula weight of the analyte?

**19.6**  [Figure courtesy of Varian Associates]
Figure 19.6.1 shows the NMR and mass spectra of a compound. What is the compound?

**Figure 19.6.1**  *(Continued on next page.)*

| m/z | Relative intensity |
|-----|--------------------|
| 124 | 100 |
| 125 | 9 |
| 126 | 5 |

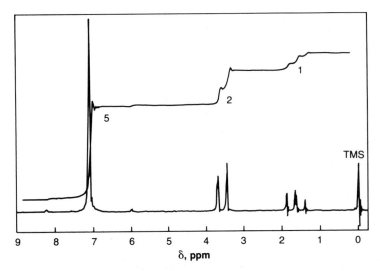

**Figure 19.6.1**     *(continued)*

19.7   Two different energy levels exist for a proton pointing along and away from the external field, and the transition between them occurs at 100 MHz at 298 K.

    a.   What fraction of the nuclei are in the upper and lower levels?

    b.   Do the same calculation for 60 MHz and 300 MHz.

    c.   You read that the intensities of the component lines of a multiplet are due to the probabilities that the protons will be pointing away from or along the magnetic field direction. Will there be any observable differences in the relative heights of the components of multiplets in NMR spectra at these different frequencies?

## SUPPLEMENT 19A

# NMR Instruments and Samples

The NMR signal from a sample has approximately the same magnitude as random noise generated by the sample itself in the instrument. Obtaining routine NMR spectra requires sophisticated electronics. The methods of construction and optimization of NMR instrument response is a topic of advanced study. Thus only the more general points of operation are described.

More sophisticated NMR instruments use pulses of radiofrequency radiation to excite the protons in the sample. Then they collect data of the type illustrated in the application in Section 19.1. The NMR spectrum is constructed from these data. The mathematical relationship between the output over time and the output displayed with frequency/field is called a Fourier transform. Instruments of this latter type are called **Fourier transform NMR spectrometers,** abbreviated FT-NMR. The type of NMR spectrometer described below is called a *continuous wave spectrometer,* abbreviated CW-NMR. CW-NMR instruments are still quite widely used.

The major features of an NMR instrument are illustrated in Figure 19A.1. The instrument consists of a radiofrequency source that is extremely stable in both frequency and power, a highly sensitive radiofrequency receiver, and a magnet that produces a steady, strong field. Of course, a method of recording the spectrum is required. This is usually a recorder, which is used to plot the energy absorbed versus the chemical shift.

The sample in a common NMR experiment consists of a relatively concentrated solution of the solid or liquid being investigated. If the sample is a liquid, it is best to use the pure liquid. The most sensitive CW instruments can obtain a spectrum in a few minutes with a mg of sample, close to the lower limit for CW $^1$H-NMR. The more commonly available instruments require 10 mg of sample for proton NMR. The sample is placed in the bottom of a precisely cylindrical tube 5 mm in diameter to a depth of 2–3 cm. Standard tubes are 20–25 cm long. (If the tube is not precise in its shape, the spinning side-bands are enhanced and may even be larger than the absorption peaks.) This tube is spun around its long axis in the sample compartment. The reason for spinning the sample is to average out some of the imperfections in the constant magnetic field. This is critical because the resolution of the spectrum depends on the quality (the homogeneity) of the magnetic field.

As noted in the main part of the chapter, the linewidth of an NMR-absorption peak can be as narrow as 0.1 Hz. This is with a resonance frequency of $10^8$ Hz (100 MHz). Since the field and frequency are directly related (Eq. 19-1) the width of the line must reflect the precision of the magnetic field in which the sample sits: a few parts in $10^9$. Making the magnetic field this homogeneous requires careful construction. Regardless of the type of magnet used (permanent magnets, electromagnets, or supercon-

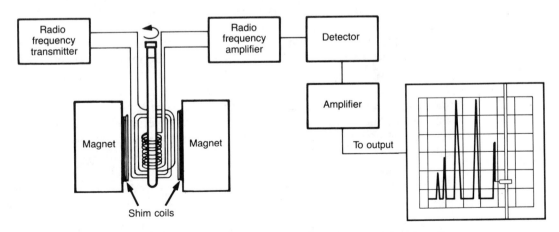

**Figure 19A.1**    Figure of a general NMR machine. The radiofrequency excitation is transmitted through a coil. A coil with its axis perpendicular to the first acts as an antenna to receive the output. The magnetic field is made homogeneous through the use of shim coils. The difference between CW-NMR and FT-NMR is in the way the radiofrequency signals are handled. The general layout is the same.

ducting magnets, depending on the quality of the instrument), the field must be made more homogeneous by using small electromagnet coils called *shims* or *shimming coils*. Further, to keep the fields constant, the magnets are thermostatted. Then, as a final step, the sample is spun.

Again, precise control (parts in $10^8$) is required in the radiofrequency source. The source frequency is controlled by the vibrations of a quartz crystal that is carefully thermostatted. Recall that we want an accurate measurement of the chemical shift; this is a ratio of the frequency of the irradiation relative to the field. The two cannot be allowed to drift relative to each other if we want a precise measurement. This problem is solved by adding a circuit to the instrument that provides a *field–frequency lock*. This is done by having, in effect, a second, simple NMR spectrometer that measures the resonance of some major species in the field and keeps the instrument correctly tuned. In the more sophisticated instruments, the major species that is used to "lock on" is the $^2H$ absorption of the solvent. In $^1H$-NMR, either solvents containing no hydrogen (for example, $CCl_4$) or deuterated solvents must be used. If they are not, the proton NMR absorption from the solvent can overwhelm that of the sample, since each proton in the sample contributes equally to the signal although at different field positions.

To obtain an NMR spectrum, either the field can be scanned at a constant radio-frequency or the frequency can be scanned with a fixed field value. In most contemporary instruments, the frequency is varied, and the output is plotted as if the field were scanned. This scanning frequency is generated by using a second oscillator that operates at about 1 kHz — in the region of audiofrequencies. The outputs from the audiofrequency oscillator and the locked-in radiofrequency oscillators are mixed to produce a sum of the two frequencies. (The details of frequency mixing can be found in more advanced references.) In this way, a very stable and precisely measurable frequency can be generated to obtain precise measurements of relative chemical shifts. The audiofrequency oscillator can be scanned over a wide enough range to obtain a spectrum. For $^1H$, this is, as you have seen, about 1000 Hz.

The radiofrequency radiant energy is transmitted to the sample through a coil surrounding the sample tube, as shown in Figure 19A.1. In essence, the energy is absorbed and reradiated by the sample. This radiation is detected and amplified by sophisticated noise-rejecting amplifiers and put out as the output signal. When an NMR absorption is reached, less power is reradiated by the sample, and the difference shows up as a peak (which is really a dip) in the spectrum output. The power is lost through a nonradiative process, as is the case in other spectral regions. For the best results, the magnet, transmitter coil, and detector coil are perpendicular to each other.

# 20

# Infrared and Raman Spectrometries: Vibrational Spectrometries

## 20.1 Introduction and Case Reading

Infrared spectrometry and Raman spectrometry both are used to probe the characteristic vibrations of molecules, crystals, and glasses, although the manner in which the spectra are obtained experimentally are significantly different.

The spectra appear with numerous characteristic bands each with fwhm about $4-7$ $cm^{-1}$ over the infrared spectral range from around 100 $cm^{-1}$ to 4000 $cm^{-1}$. Significantly narrower bands appear in the spectra of gases. Individual bands can often be associated with specific chemical groups such as carbonyl, $>C=O$, or organic chlorine, $-C-Cl$. As for all spectrometries, the wavelengths of the bands depend on properties of the molecules while the magnitudes can be used to determine concentrations.

The spectra are often complicated, and seldom can every band be assigned to its origin in specific groups in the molecule. Only in the simplest molecules can all the bands be assigned to specific atomic motions. However, an incomplete understanding of the spectra does not detract from their usefulness in both qualitative and quantitative analysis. Infrared spectrometry with mass spectrometry (and GC/MS) and nuclear magnetic resonance form the basis for contemporary organic chemical qualitative analysis: identifying the molecular structure of unknown compounds and mixtures. The practical example in this section illustrates solving a problem of this type. Among the techniques employed was infrared spectrometry. Even with all this powerful analytical equipment, the problem required a month's work by a number of analysts. In reading the example you will find that some details of mass spectrometry and NMR spectrometry are mentioned. If you have not already studied the respective chapters, pass over the details and continue. One piece of nomenclature may be unfamiliar. A *tertiary amide* is an amide with two organic groups in place of the two protons on the nitrogen of the amide group.

$$
R-\overset{\overset{\displaystyle O}{\|}}{C}\diagdown_{N}\diagup^{R'}_{\diagdown R''}
$$

*Lactose* is milk sugar. It is especially worth noting how much easier identification is when done by comparing with a standard instead of determining "from scratch" the structure of the compound being analyzed.

The example comes from a published case from the U.S. Drug Enforcement Administration Special Testing and Research Laboratory.

We at this laboratory are often asked how we go about identifying an unknown. In our college days, unknowns generally consisted of several grams of a single substance, identifiable by use of a variety of classical procedures, and for which an abundance of reference data was readily accessible. Forensic drug exhibits, however, usually consist of powders, liquids, and amorphous masses that are seldom obtained from reagent bottles, but, more likely, from scrapings of glassware in illicit laboratories or from matchbooks that have changed hands under cloak-and-dagger circumstances.

We consider ourselves fortunate if we have gram quantities to work with, and extremely fortunate if we discover that we have but a single ingredient to consider . . . .

We have to maintain complete flexibility in our approach as we gather information on the nature of the exhibit. Where a sizable amount of sample is available, for example, attempts are generally made to analyze separate portions by microscopy; thin layer chromatography; and mass, magnetic resonance, and infrared spectrometries. In many instances, the compounds of forensic interest are identified within a matter of hours, sometimes minutes, by a combination of interpretation and comparison of data with those obtained from reference substances.

With no suitable reference data, however, identification becomes increasingly difficult with increased molecular complexity . . . .

**The China White Case**   In the matter of "China White" we were confronted with an impure, complex material with which the forensic community had no prior experience. It was available, initially, in only minute concentrations, and its identification was of the greatest urgency. Among drug users word had begun to spread of a synthetic substitute for heroin. Among law enforcement people a macabre aspect became prominent; people were dying and no one knew why . . . .

Our initial examination of the powder revealed lactose and nothing more. Subsequent analysis by GC/MS of a highly concentrated extract produced a weak spectrum that was totally unfamiliar, and, unfortunately, whatever produced this spectrum was present only at very low concentration levels . . . .

A short time later, however, an exhibit was referred to us that consisted of about 200 mg of powder having a greater concentration of this same substance. It was accompanied . . . by an infrared spectrum of a hydrochloride salt prepared from the suspected active principal.

The mass spectrum resembled that obtained from the earlier exhibit. The infrared spectrum bore similarities to that of amphetamine; however, a band attributed to carbonyl was present, its location suggesting a tertiary amide [Figure 20.1] . . . .

Upon receiving the powder, we dry-extracted a substantial portion with deuterochloroform [$CDCl_3$], then filtered it to remove the lactose. Examination by $^1$H-NMR revealed the presence of at least three ethyl groups, a possible cyclic group, and aromatic protons nearly equal in number to all the others. Following extractions from the chloroform with deuterated water and backwashes with additional chloroform, it was clear that we were dealing with three major components . . . .

It was evident from the weakness of the $^1$H-NMR spectrum that . . . we had no more than about 1 mg of this compound to work with. The limitations this

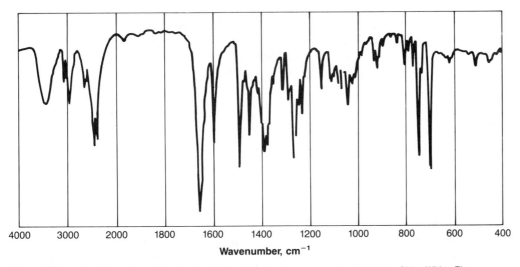

**Figure 20.1**   The infrared spectrum of a hydrochloride salt of a compound isolated from China White. The salt was crushed and mixed with dry potassium bromide powder. This mixture was placed in a hydraulic press under a pressure of several tons/cm² to form a transparent KBr pellet, which was placed in the sample beam of the infrared spectrometer. Dry KBr has no absorbance bands in this spectral region.

would place on our analytical flexibility would undoubtedly delay the desired outcome.

[One of the three major components] produced a mass spectrum that had stymied us previously . . . we spent more time in trying to find clues in the fragmentation that would allow us to assemble building blocks adding up to [its parent-ion mass of] 259 [found from electron impact mass spectrometry]. At this point it was clear that we had to make a more meaningful determination of molecular weight . . . .

GC/CIMS [gas chromatography/chemical ionization mass spectrometry] provided a major breakthrough for us. . . . [The molecule had a parent molecular mass] of 350. The loss of 91 mass units to produce the 259 fragment in the E.I. [electron impact] mass spectrum might tentatively be ascribed to loss of a benzyl group . . . .

Integration of the $^1$H-NMR spectrum told us that we probably had approximately 30 hydrogen atoms. . . .

By now we had some standards available for comparison . . . . The 3-methylfentanyl standard, following purification by HPLC, did not compare with the unknown, although some structural similarities were evident . . . we could find no reason to discount [the following structure].

All efforts were concentrated on obtaining a reference standard of this material as quickly as possible. Within days it was synthesized, purified, and subjected to spectrometric examination. It was examined as a base and as a hydrochloride salt. The outcome? A match with the sample in all respects!

[Ref: Kram, T. C.; Cooper, D. A.; Allen, A. C. *Anal. Chem.* **1981,** *53,* 1379A–1386A.]

## 20.2 Infrared Spectra

Notice in the China White work that the infrared spectrum was used to identify a specific organic functional group in the unknown molecule. This is possible because the wavelength at which energy is absorbed depends on

a. the identity of the atoms in a molecule;

b. The molecular structure (for example, $CH_3CHO$, $H_2C\overset{O}{\overbrace{\quad}}CH_2$, and $CH_2{=}CHOH$: all $C_2H_4O$ isomers with significantly different spectra); and

c. the bonding between the atoms (for example, ${>}C{=}O$ differs from $-C-O-$).

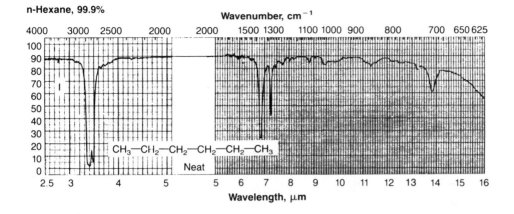

n-Hexane, 99.9%

CH₃—CH₂—CH₂—CH₂—CH₂—CH₃

Neat

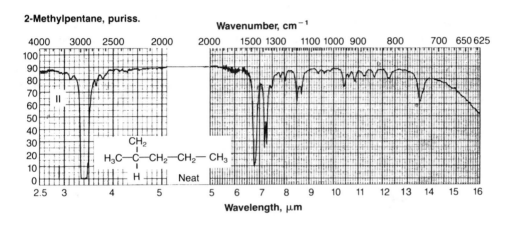

2-Methylpentane, puriss.

H₃C—C—CH₂—CH₂—CH₃

Neat

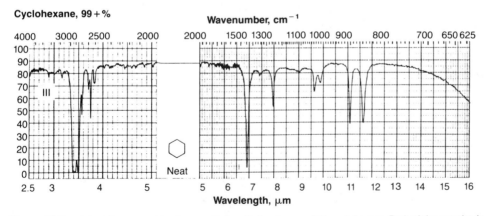

Cyclohexane, 99+%

Neat

**Figure 20.2**  Infrared spectra of I, *n*-hexane; II, 2-methylpentane; and III, cyclohexane. Each of the samples is a **neat** (meaning *pure*) liquid. Spectra I and II are of two hexane isomers; III lacks only two protons. The three spectra together illustrate the sensitivity of infrared spectra to molecular structure as well as the persistence of bands associated with specific chemical groups. No spectrum is recorded in the white space at 2000 cm⁻¹. [Spectra from Pouchert, C. J. "The Aldrich Library of Infrared Spectra." Courtesy Aldrich Chemical Co.]

The spectrum of energy absorption due to excitation of molecular vibrations is quite sensitive to differences in structure. To illustrate this sensitivity, the infrared spectra of three closely related molecules are shown in Figure 20.2.

As seen here, it has been the practice in chemical infrared spectrometry to present the spectra as a plot of $\% \ T$ versus wavenumber, $cm^{-1}$. Notice that there are two separated regions of the spectra in Figure 20.2 and that the $cm^{-1}$ scales differ in each. The longer wavelength region (lower wavenumber value) has a wider scale. Compare these scales with that in Figure 20.1; there is a break in that scale as well at 2000 $cm^{-1}$. However, the scale in Figure 20.1 is linear in $cm^{-1}$ while those in Figure 20.2 are linear in wavelength (here $\mu$m or microns). These are the two conventions used in infrared spectrometry. Some contemporary instruments produce an output of absorbance versus $cm^{-1}$. (See Figure 20.17b, page 825.)

It is worth noticing that the individual narrow absorptions in the spectra have widths at half-height (in $\% \ T$) of less than about 10 $cm^{-1}$. As can be seen in Figure 20.3, the *apparent* widths can depend significantly on the transmittance of the band. This is merely a consequence of having recorded the spectra with a $\% \ T$ scale.

Let us return to spectra I and II of Figure 20.2, which are the spectra of two different hexane isomers. Even though there are distinctive similarities in the spectra, there are some significant differences in the number of bands and the wavelengths of the absorptions. The three most obvious differences are

a.  the split of the band at 1360 $cm^{-1}$,
b.  the bands around 1150 $cm^{-1}$, and
c.  the shift of the long wavelength peak from about 725 $cm^{-1}$ to 740 $cm^{-1}$. Shifts of this magnitude in infrared spectra are significant.

Even greater differences are seen between the spectra of these two hexanes and the spectrum of cyclohexane. Especially obvious are

a.  the two bands between 850 and 900 $cm^{-1}$,
b.  the set of bands around 2700 $cm^{-1}$, and
c.  the broad band around 725–750 $cm^{-1}$, which is not present in the cyclohexane spectrum.

But not only are the differences useful, the similarities are important as well. For these three alkanes, the persistent bands are those in the ranges 2800–3000 $cm^{-1}$ and 1450–1470 $cm^{-1}$. Because of the sensitivity to bonding and structure, which you have now observed, we expect that these bands must result from vibrations of groups of atoms that occur in all three molecules: These are the C—H and C—C group vibrations.

Comparing these simple spectra in some detail has been done to illustrate that, in infrared spectra, some bands are characteristic of *specific* chemical groups in the molecules, such as C—C and C—H, while others can be use to recognize the structure of each molecule. Let us next look more closely at some of the factors that determine the frequencies of the vibrational bands.

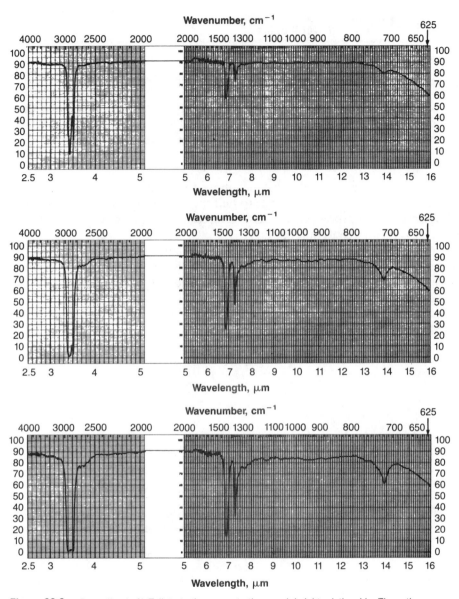

**Figure 20.3** Recording in % *T* distorts the concentration–peak height relationship. These three spectra are of the same material, mineral oil (also called *Nujol*). The liquid is held between windows made of sodium chloride. Shown are three different thicknesses of the oil layer. The slow decrease in % *T* at the right is due to NaCl absorption. [Spectra from Pouchert, C. J. "The Aldrich Library of Infrared Spectra." Courtesy Aldrich Chemical Co.]

## 20.3 Vibrations of Diatomic Molecules

We explain the origin and changes in vibrational spectra using the ideas of the following picture (or model), illustrated in Figure 20.4. Each chemical bond acts like a spring connecting two atoms with masses $M_1$ and $M_2$. The spring is a Hooke's law spring, which exerts a force that is directly proportional to the distance through which it is compressed or expanded from its resting position. Algebraically stated,

$$\text{force} = -k\,\Delta x \tag{20-1}$$

where $k$ is a constant and $\Delta x$ the distance moved from the resting position. The negative sign means that the force tends to pull the spring back to where it started: When the spring is compressed, it exerts a force to expand; when the spring is stretched, it exerts a force to pull back.

A set of masses that are connected by the spring naturally tends to settle into a specific motion when agitated. The masses vibrate with a frequency that depends on the masses and the strength of the spring. Such vibrations, which occur at *characteristic frequencies,* are what we measure in vibrational spectra.

For diatomic molecules, this vibration is described by Eq. 20-2. The variables are illustrated in Figure 20.4.

$$v = (2\pi)^{-1}\sqrt{\frac{k}{\mu}} \tag{20-2}$$

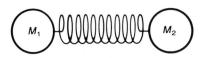

$$v = \frac{1}{2\pi}\sqrt{\frac{k}{\mu}}$$

$$\text{and } \frac{1}{\mu} = \frac{1}{M_1} + \frac{1}{M_2}$$

$$\text{or } \mu = \frac{M_1 M_2}{M_1 + M_2}$$

Force $= -k(\Delta x)$,
$\Delta x$ is the displacement from
the point where the spring is
neither compressed nor stretched.

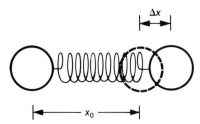

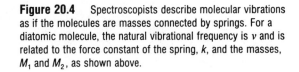

**Figure 20.4**    Spectroscopists describe molecular vibrations as if the molecules are masses connected by springs. For a diatomic molecule, the natural vibrational frequency is $v$ and is related to the force constant of the spring, $k$, and the masses, $M_1$ and $M_2$, as shown above.

**Table 20.1** Energies (cm⁻¹) of Typical
Vibrations Showing Differences with
Bond Order

| Group | Frequency Range (cm⁻¹) |
| --- | --- |
| C—C | 800–1200 |
| C=C | 1600–1680 |
| C≡C | 2100–2250 |
| C—N | 1000–1350 |
| C=N | 1640–1690 |
| C≡N | 2240–2260 |

and

$$\frac{1}{\mu} = \frac{1}{M_1} + \frac{1}{M_2}$$

or

$$\mu = \frac{M_1 M_2}{M_1 + M_2}$$

$v$ is the natural frequency of vibration of the masses.

$k$ is the **force constant** of the "spring" of the chemical bond; $k$ describes the magnitude of the force exerted when it is compressed or stretched. A large value of $k$ indicates a "strong" spring.

$\mu$ is the **reduced mass.**

Notice the way the frequency changes with $k$ and the masses $M_1$ and $M_2$. If the spring becomes stiffer, which means $k$ becomes larger, then the natural frequency rises: The vibration is faster. Conversely, if the spring becomes weaker, $k$ is reduced, and the natural frequency falls.

What happens if either of the masses is decreased and the same "spring" is retained? In that case, the value of $\mu$ decreases. Therefore, $1/\mu$ becomes larger, and, as can be seen in Eq. 20-2, the natural frequency will rise. So a decrease in mass yields the same effect as an increase in the strength of the spring, and vice versa.

These are the general rules on which we base our intuitive understanding of changes in the vibrational spectra of molecules:

A. *Stronger chemical bonds tend to cause increases in observed frequencies.*

B. *Lighter atomic masses tend to cause increases in observed frequencies.*

Rule A suggests that we might expect the frequency to increase between groups containing C—N, C=N, and C≡N bonds or C—C, C=C, and C≡C. The trend expected from this simple model is indeed observed experimentally, as seen in Table

**Methanol**

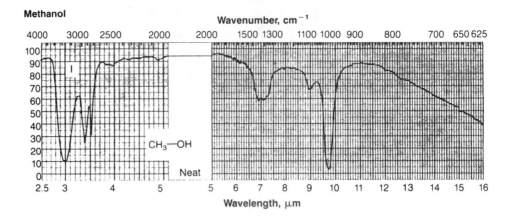

Wavenumber, cm⁻¹

CH₃—OH

Neat

Wavelength, μm

**Methyl alcohol-*d*** (methanol-*d*), **99%**

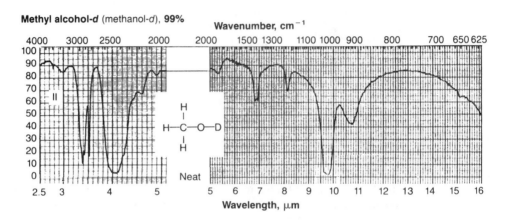

Wavenumber, cm⁻¹

H
|
H—C—O—D
|
H

Neat

Wavelength, μm

**Methyl alcohol-*d*₄** (methanol-*d*₄), **99.5%**

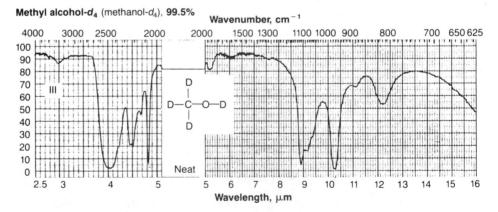

Wavenumber, cm⁻¹

D
|
D—C—O—D
|
D

Neat

Wavelength, μm

**Figure 20.5**  Assignment of vibrations to different parts of a molecule can be achieved by observing changes in the vibrational spectra when different isotopes are substituted in the molecule. In the example shown here, methanol vibrations are probed. I is the spectrum of methanol, $CH_3OH$. II has the alcoholic hydrogen substituted by deuterium giving $CH_3OD$; those bands that shift to lower energy are due to normal modes (see Section 20.4) that have a contribution from the O—H/O—D motion. III is the spectrum of $CD_3OD$; the bands that are shifted relative to II are due to normal modes that have a contribution from the motion of —$CH_3$/—$CD_3$. [Spectra from Pouchert, C. J. "The Aldrich Library of Infrared Spectra." Courtesy Aldrich Chemical Co.]

**Table 20.2**  Infrared Bands in Methanol and the Changes That Occur with H/D Substitution (from Figure 20.5)

| Band in $CH_3OH$ (cm$^{-1}$) | Shifted in $CH_3OD$ | Shifted in $CD_3OD$ | Assignment |
|---|---|---|---|
| 3340 | Yes | Yes | OH (stretching vibration) |
| 2840 and 2940 | No | Yes | $CH_3$ (stretching vibration) |
| 1029 | No | Yes | $H_3C—O$ (stretching vibration) |
| 1400 | Partly | Yes | $CH_3$ and $C—O—H$ (bending vibrations) |

20.1, a short **correlation table.** In correlation tables we record some of the regularities of descriptive vibrational spectroscopy. Molecules that contain the groups noted in the table exhibit infrared or Raman spectra with bands in the associated wavelength regions.

From rule B, we expect the frequency to change with a change in the vibrating masses. One way to see the effects of changing masses without changing the force constant appreciably is to compare vibrational spectra of molecules that are the same except for substituting one atomic isotope for another. The clearest changes occur when deuterium, $^2H$, is substituted for hydrogen, $^1H$. The deuterium mass is about twice that of hydrogen. A simple example of the spectrum changes with changes in mass is shown in Figure 20.5, which shows the infrared spectra of methanol and two different isotopically substituted forms, $CH_3OD$ and $CD_3OD$. The shifts in the bands with deuterium substitution are all to lower energy (or frequency). This conforms to rule B above.

Notice which spectral bands are shifted when each deuterium substitution is made. When the hydroxy proton is substituted, the bands between 2800 and 4000 cm$^{-1}$ show the most obvious shift. Simply from the fact that a shift occurs, we can assign the bands as being due to vibrations of specific chemical groups. The reasoning and band assignments of the **group frequencies** of methanol are outlined in Table 20.2. By following procedures like this, the information in correlation tables is verified. Table 20.3a and b present more comprehensive correlations than does Table 20.1. Another form for displaying this information appears in Figures 20.6a and 20.6b (pp. 811– 813).

## 20.4 Vibrations in Molecules That Are Not Diatomic

When a molecule has more than two atoms in it, the vibrational frequencies cannot be described with as simple an equation as Eq. 20-2. The vibrations are due to simultaneous motions of more than two atoms and are called **normal modes.** If we think only in terms of the simple model of two atoms with one bond, we may be surprised at the frequency changes of the normal modes with changes in mass or bond strengths in more complicated molecules. The details of calculating the variations in normal-mode frequencies as they change with atomic mass or bond strength are not

**Table 20.3A**   Simplified Correlation Chart of Molecular Vibrations by Type*

| | Type of Vibration | | Frequency (cm⁻¹) | Intensity |
|---|---|---|---|---|
| C—H | Alkanes | (stretch) | 3000–2850 | s |
| | —CH₃ | (bend) | 1450 and 1375 | m |
| | —CH₂— | (bend) | 1465 | m |
| | Alkenes | (stretch) | 3100–3000 | m |
| | | (out-of-plane bend) | 1000–650 | s |
| | Aromatics | (stretch) | 3150–3050 | s |
| | | (out-of-plane bend) | 900–690 | s |
| | Alkyne | (stretch) | ca. 3300 | s |
| | Aldehyde | | 2900–2800 | w |
| | | | 2800–2700 | w |
| C—C | Alkane | | not interpretatively useful | |
| C=C | Alkene | | 1680–1600 | m-w |
| | Aromatic | | 1600 and 1475 | m-w |
| C≡C | Alkyne | | 2250–2100 | m-w |
| C=O | Aldehyde | | 1740–1720 | s |
| | Ketone | | 1725–1705 | s |
| | Carboxylic acid | | 1725–1700 | s |
| | Ester | | 1750–1730 | s |
| | Amide | | 1670–1640 | s |
| | Anhydride | | 1810 and 1760 | s |
| | Acid chloride | | 1800 | s |
| C—O | Alcohols, ethers, esters, carboxylic acids, anhydrides | | 1300–1000 | s |
| O—H | Alcohols, phenols<br> Free | | 3650–3600 | m |
| |  H–bonded | | 3500–3200 | m |
| | Carboxylic acids | | 3400–2400 | m |
| N—H | Primary and secondary amines and amides    (stretch) | | 3500–3100 | m |
| | (bend) | | 1640–1550 | m-s |
| C—N | Amines | | 1350–1000 | m-s |
| C=N | Imines and oximes | | 1690–1640 | w-s |
| C≡N | Nitriles | | 2260–2240 | m |
| X=C=Y | Allenes, ketenes, isocyanates, isothiocyanates | | 2270–1950 | m-s |
| N=O | Nitro (R—NO₂) | | 1550 and 1350 | s |
| S—H | Mercaptans | | 2550 | w |

*(continued)*

**Table 20.3A**  *(continued)*

|  | Type of Vibration | Frequency (cm$^{-1}$) | Intensity |
|---|---|---|---|
| S=O | Sulfoxides | 1050 | s |
|  | Sulfones, sulfonyl chlorides | 1375–1300 and | s |
|  | Sulfates, sulfonamides | 1200–1140 | s |
| C—X | Fluoride | 1400–1000 | s |
|  | Chloride | 800–600 | s |
|  | Bromide, iodide | <667 | s |

*From Pavia, D. L.; Lampman, G. M.; Kriz, G. S. "Introduction to Spectroscopy: A Guide for Students of Organic Chemistry." Copyright © 1979 by W. B. Saunders Company. Reprinted by permission of CBS College Publishing.

needed for normal chemical analysis and will not be pursued further; however, the general characteristics of complicated vibrations can easily be illustrated pictorially.

Consider the example of a linear triatomic molecule such as $CO_2$, as illustrated in Figure 20.7 (p. 814). You might expect that there would be one vibrational frequency, that of a C=O stretch. However, each C=O bond does not vibrate separately. Instead, the molecule vibrates at two different frequencies, each one corresponding to a different normal mode. The vibrational motions are illustrated in Figure 20.7. The arrows show the direction of initial motion of each atom for one part of a vibration. The bonds are stretched, and the atomic motion stops. Then the atoms return in exactly the opposite direction until the bonds are maximally compressed; once again, the motion stops. Then bond expansion repeats.

For a $CO_2$ molecule, the motion of one normal mode consists of the two C=O bonds both moving outward and inward together while the carbon remains fixed in position; this is called the **symmetric stretch.** The normal mode in which the oxygen atoms move together in the same direction while the carbon moves in the opposite direction is called the **asymmetric stretch.**

Molecule bending is yet another normal mode of $CO_2$. Here the oxygens both move up while the carbon moves down. The result is a bond-angle **bending vibration.** Bending vibrations occur at longer wavelengths than stretching vibrations of the same group. We explain this difference in frequency in the following way. The restoring force, characterized by $k$, is smaller for molecular-bending vibrations than for molecular stretches.

You have already seen some of the effects of normal modes in the spectrum of methanol in Figure 20.5. The atomic motions associated with the methanol normal modes are shown in Figure 20.8 (p. 815). The two bands at 2840 and 2940 *both* came from the —$CH_3$ stretch. One of these is the asymmetric mode and the other the symmetric. The —$CH_3$ bending vibrations are around 1400 cm$^{-1}$.

Because the vibrations that actually occur in the molecules are normal modes, the vibrational frequencies do not depend only on the identities of the atoms and the bond strength between them. The atoms and groups of atoms nearby are also influential in determining the normal-mode frequencies. These influences help explain why

**Table 20.3B** Selected Group Frequencies in Order of Descending Wavenumber

| Frequency (cm$^{-1}$) | Group and Assignment |
| --- | --- |
| 3650–3600 | Alcohols, phenols (dilute) O—H stretch |
| 3500–3300 | Primary amines (doublet), secondary amines N—H stretch |
| 3500–3200 | Alcohols, phenols (H-bonded) O—H stretch |
| 3500–3100 | Amides N—H stretch |
| 3400–2400 | —COOH O—H stretch |
| ~3300 | Alkynes C—H stretch |
| 3150–3050 | Aromatic C—H stretch |
| 3100–3000 | Alkenes C—H stretch |
| 3000–2850 | Alkanes C—H stretch |
| 2900–2700 | Aldehydes C—H stretch |
| 2550 | Mercaptans S—H stretch |
| 2270–1950 | Allenes, ketenes, isocyanates, isothiocyanates X=C=Y stretch |
| 2260–2240 | Nitriles C≡N stretch |
| 2250–2100 | Alkynes C≡C stretch |
| 1810 and 1760 | Anhydrides C=O stretch |
| 1800 | Acid chlorides C=O stretch |
| 1750–1730 | Esters C=O stretch |
| 1740–1720 | Aldehydes C=O stretch |
| 1725–1705 | Ketones C=O stretch |
| 1725–1705 | —COOH C=O stretch |
| 1715–1650 | Amides C=O stretch (Amide I band) |
| 1690–1640 | Imines and oximes C=N stretch |
| 1680–1600 | Alkene C=C stretch |
| 1670–1640 | Amides C=O stretch (Amide II band) |
| 1640–1550 | Amides, primary and secondary amines N—H bend |
| 1600 and 1475 | Aromatic C=C stretch |
| 1465 | Alkane —CH$_2$— bend |
| 1450 and 1375 | Alkane —CH$_3$ bend |
| 1400–1000 | Fluoride C—F stretch |
| 1375–1300, 1200–1140 | Sulfones, sulfonyl chlorides, sulfates, sulfonamides S=O stretch |
| 1350–1000 | Amines C—N stretch |
| 1300–1000 | Alcohols, esters, ethers, —COOH, anhydrides C—O stretch |
| 1150 and 1350 | Nitro (R—NO$_2$) N=O stretch |
| 1050 | Sulfoxides S=O stretch |
| 1000–650 | Alkenes C—H out-of-plane bend |
| 900–690 | Aromatic C—H out-of-plane bend |
| 800–600 | Chloride C—Cl stretch |
| <667 | Iodide and bromide C—X stretch |

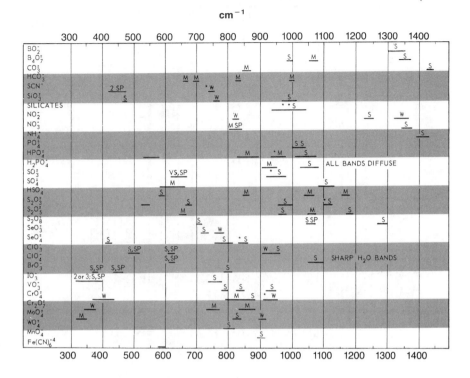

cm$^{-1}$

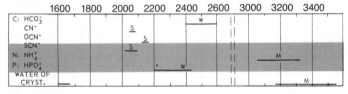

*In most, but not all, examples.
**Literature value.
S, strong; M, medium; W, weak; SP, sharp

**Figure 20.6** Charts of characteristic frequencies of various chemical groups: (a) polyatomic inorganic ions and (b, pp. 812–813) organic and some inorganic compounds. The characteristic frequencies are also called group frequencies. Notice that some of the regions are far wider than others, such as alcohols. These vibrations occur over wider ranges usually because the group can interact with other molecules in the matrix. The positions of such bands are often significantly dependent on the method of sample preparation. [Refs: (a) Ferraro, J. R. *J. Chem. Ed.* **1961,** *38,* 201–208. (b) Colthup, N. B. *J. Opt. Soc. Am.* **1950,** *40,* 397.]

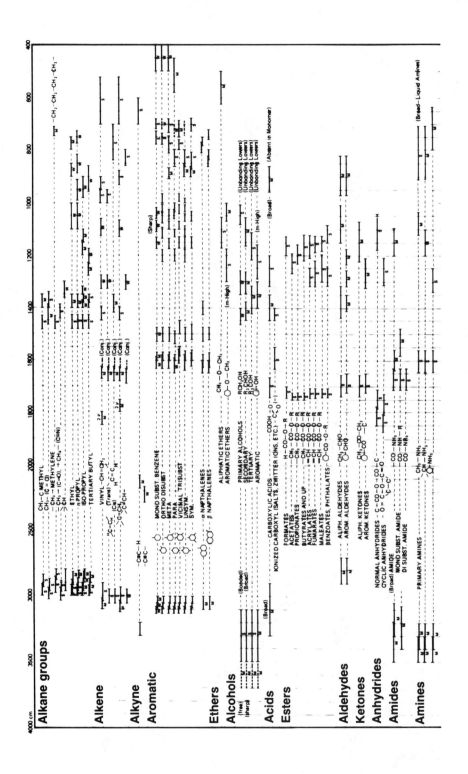

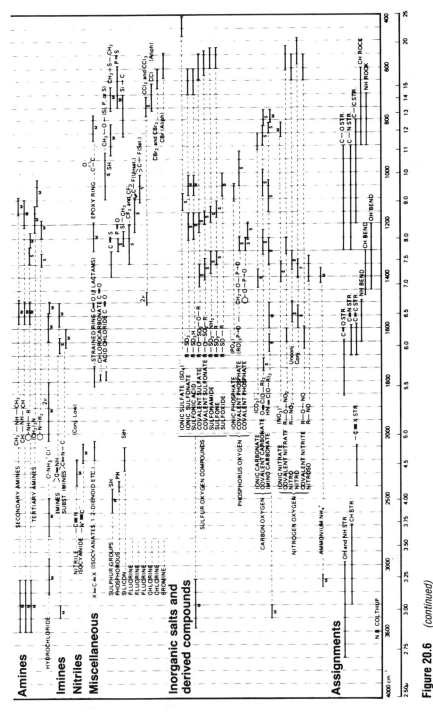

**Figure 20.6** *(continued)*

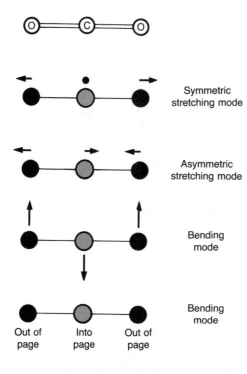

**Figure 20.7**    Normal modes of a linear triatomic molecule, here carbon dioxide, $CO_2$. There are two stretching modes, symmetric and asymmetric, as well as two bending modes. The two bending modes—one in the plane of the page and a similar one perpendicular to the plane—are said to be **degenerate modes.**

there are some bands that depend so sensitively on the differences in molecular structure, as you saw in the spectra in Figure 20.2.

## 20.5 Qualitative Information from IR Spectra

Since no two molecules have the same infrared (IR) spectrum—with every band matching in peak position (wavenumber), intensity, and bandwidth—infrared spectra can be used to identify molecular components of samples by matching the spectrum of an unknown to a library spectrum. An example of such spectrum matching is shown in Figure 20.9.

Spectra can be obtained on samples as small as a few $\mu$g. (But it is easier with mg samples.) In this manner, IR spectra are used in qualitative analysis of solid, liquids, and gases. An example of a qualitative identification of an inorganic sample by infrared spectrometry is illustrated in Figure 20.10. Also, IR spectrometry is occasionally used as a detector in chromatography. The instrumentation is usually quite sophisticated: It can obtain spectra on the order of once a second.

In addition to identification of compounds, a second *qual*itative use is in obtaining structural information, as was demonstrated in the determination of the structure of China White. Functional groups can be identified, because their absorption bands lie in relatively narrow, characteristic regions of the infrared region. Infrared spectra of organic compounds can be divided into three general regions:

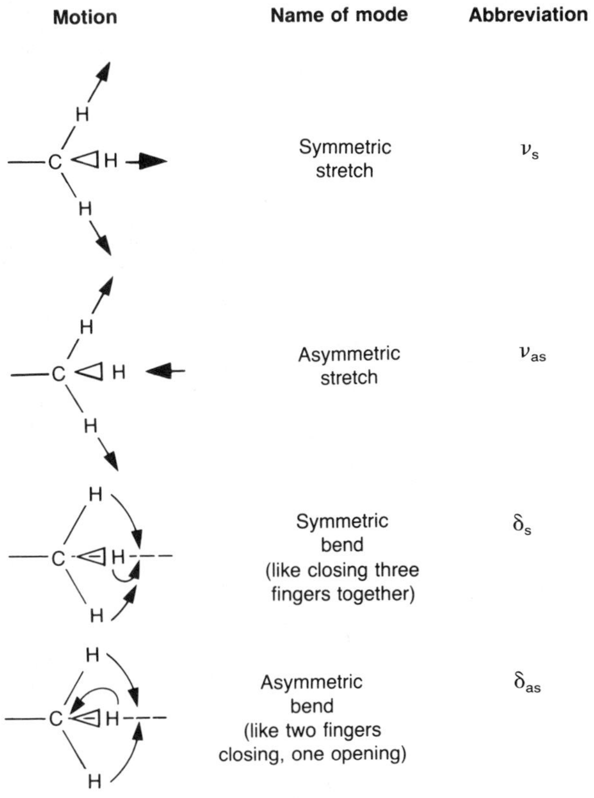

| Motion | Name of mode | Abbreviation |
|---|---|---|

|  | Symmetric stretch | $\nu_s$ |
|  | Asymmetric stretch | $\nu_{as}$ |
|  | Symmetric bend (like closing three fingers together) | $\delta_s$ |
|  | Asymmetric bend (like two fingers closing, one opening) | $\delta_{as}$ |

**Figure 20.8**    Normal modes within the methyl group. There are two stretching modes, symmetric and asymmetric, as well as two bending modes. In fact, there are three ways to draw the asymmetric stretch and asymmetric bend; the unique hydrogen could be any of the three. Thus these three asymmetric modes are degenerate, and since there are three ways to draw the identical vibration, the asymmetric stretch and asymmetric bend are said to be **triply degenerate.**

| The **functional group** region | $4000-1300$ cm$^{-1}$ |
|---|---|
| The **fingerprint** region | $1300-910$ cm$^{-1}$ |
| The **aromatic** region | $910-650$ cm$^{-1}$ |

The region between 200 and 650 cm$^{-1}$ is also useful for inorganic and organometallic compounds. The vibrations involving metals and metal ions occur in that region.

Bands in the functional group region indicate the presence of specific functional groups. These are indicated in Table 20.3. Easily discerned functional groups are carbonyl, C=O, and hydroxyl, —OH, for example.

The fingerprint region contains peaks that arise from complicated normal modes involving bending motions and are not usually assignable. But because of their origin, they are the most sensitive to differences in compound structure. So they represent a "fingerprint" for each specific compound. In addition to the hexanes of Figure 20.2, these structure-sensitive effects are illustrated in the spectra of four different diene isomers in Figure 20.11.

One or more strong peaks in the aromatic region indicates the possible presence of an aromatic compound. If there are no bands in this region, it is highly likely that there is no aromatic center. The bands arise from the motion of C—H bonds bending out of the plane of an aromatic ring. With practice, one can glean a great deal of

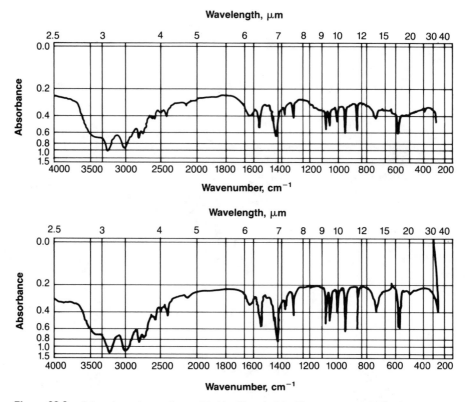

**Figure 20.9**    Infrared spectra can be used to identify materials. The upper spectrum is of some unidentified solids collected from the furnace tubes of a naphthal reboiler. The bottom spectrum is of the tars formed on heating HCl salts of a corrosion inhibitor. There is a one-to-one correspondence of all the bands in the spectra. Due to the sensitivity of IR spectra to structure, the identity of the two samples is quite certain. [Ref: Schmidt, S. A.; Gaylor, V. F. *Anal. Chem.* **1976,** *48*(12), 974A.]

structural information from this region. For instance, the number and positions of substituents on a benzene ring can be related to the spectrum. Indicative of such changes in the spectra are the examples shown in Figure 20.12. Much practice is required to interpret them, however, because of numerous deviations from the expected appearance.

## 20.6 Raman Spectra

Before delving into the causes of the differences in the absorptivities of vibrational bands, let us take a short digression and describe a typical Raman spectrum. An example is shown in Figure 20.13 (p. 821). The Raman spectrometer is operated similarly to a fluorescence spectrometer but with high-power illumination. (A laser

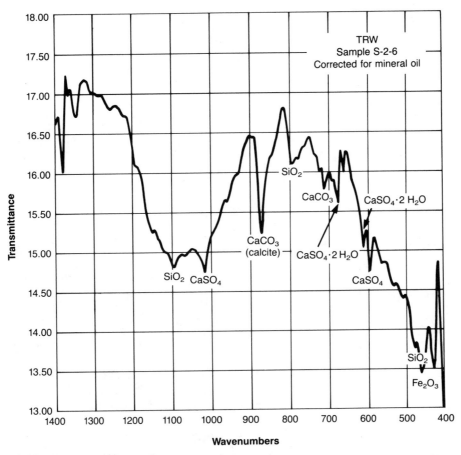

**Figure 20.10**   Infrared spectrum of the particulate matter emitted by a fluidized bed combustor. The assignments of the peaks are shown. The instrument used was a Fourier transform spectrometer (see Supplement 20B). The material was ground with mineral oil to form a paste called a mull. The paste is placed between two AgCl plates like the filling of a sandwich. This sandwich is placed in the sample beam of the spectrometer. The spectrum has had the lines due to the mineral oil subtracted out. The windows of silver chloride are infrared transparent. [Ref: Maddalone, R. F.; Ryan, L. E.; Delumyea, R. G.; Wilson, J. A. "EPA/IERL-RTP Procedures Manual: Level 2 Sampling and Analysis of Oxidized Inorganic Compounds"; EPA-600/2-200: Washington, DC, 1979.]

Raman spectrometer is described in Supplement 20B.) The spectrum is plotted as the intensity of reemitted light versus the energy in $cm^{-1}$. Note that although this is a spectrum with which we can probe molecular vibrations of infrared frequencies, the spectrum is obtained with *visible* light. The large peak in the middle, which is in the blue-green region of the visible spectrum, is due to light that is scattered (that is, absorbed and reemitted) by the molecules in the sample. This light is centered at the same wavelength as the incident light. It is called the **Rayleigh** line.

In addition to the Rayleigh line, there are a number of other spectral lines. They occur in sets, with each member of the set displaced in energy (or wavenumber)

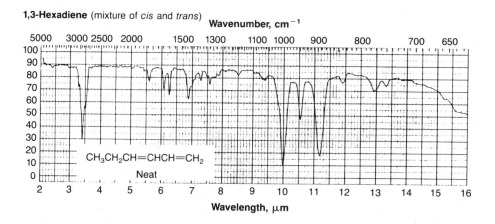

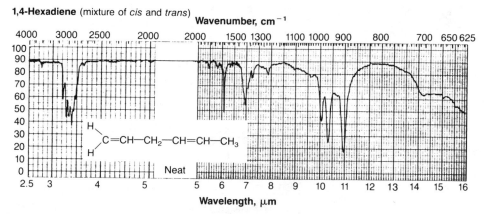

**Figure 20.11**   Illustration of the information obtainable in the functional group region (4000–1300 cm$^{-1}$) and the fingerprint region (1300–900 cm$^{-1}$) in infrared vibrational spectra. All four compounds are isomeric hexadienes. The region from 2800 to 3100 cm$^{-1}$ are the C—H stretching vibrations. In the range from 1600 to 1680 cm$^{-1}$ are the vibrations from the alkene groups, C=C. All the compounds have at least one band in that region. In addition, a band around 1430 cm$^{-1}$ appears in each spectrum. This is attributed to the bending motion of the —CH$_3$ group. However, in the range 850 to 1000 cm$^{-1}$, the bands for each compound are all significantly different in shape, intensity, and wavenumber. These are in the fingerprint region. [Spectra from Pouchert, C. J. "The Aldrich Library of Infrared Spectra." Courtesy Aldrich Chemical Co.]

equally on either side of the central line. These are the **Raman** lines. They arise as a result of the vibrations in the molecular species scattering the light. The displacements of the emissions from the Rayleigh line are equal to the vibrational frequencies of the normal modes of the molecules.

The total amount of light that is emitted as Raman lines is miniscule: about $10^{-5}$ to $10^{-6}$ of the light that is scattered by the sample (the Rayleigh line). Because of the small proportion of Raman emission, an intense source of light is needed to obtain good spectra. Therefore, a focused laser illuminates the sample. Also, a high-quality monochromator is required to minimize any instrumental interference from the

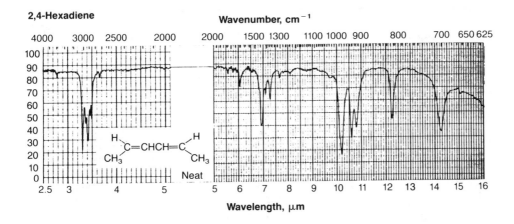

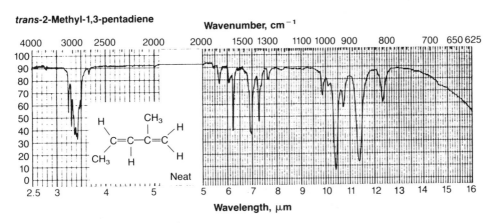

**Figure 20.11**    *(continued)*

intense Rayleigh line that enters the monochromator. Otherwise, the small Raman bands would be overwhelmed by a huge background.

## 20.7 Band Intensities of Vibrational Spectra

To catalog infrared spectra for qualitative determinations, it is sufficient to label the individual bands as being either strong (small % $T$), moderate, or weak (large % $T$)—abbreviated s, m, and w, respectively. Such abbreviations are used in Table 20.3. However, such imprecise measures are not usable when quantitative analysis is done with infrared or Raman spectrometry. Measurements of molar absorptivities (or their equivalents) are needed, and standards must be run. For the maximum sensitivity, usually the most strongly absorbing IR bands or the most strongly emit-

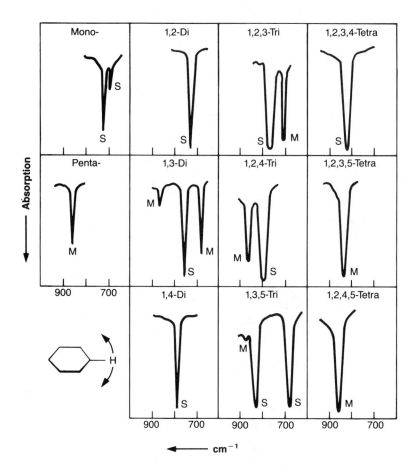

**Figure 20.12** The bands in the aromatic region are sensitive to the number and relative ring positions of substituents. The bands originate from the motion of bending of the C(ring)—H bond out of the plane of the ring (see inset). The abbreviations M and S mean that the bands are moderate or strong in magnitude. Such patterns can help to determine the substitution in the ring. They are much more useful if no other bands from other groups interfere. [Ref: Baker, A. J.; Cairns, T. "Spectroscopic Techniques in Organic Chemistry"; Heydon & Son, Ltd.: London, 1966. Copyright © 1966. Reprinted by permission of John Wiley & Sons, Ltd.]

ting Raman bands will be the best to use for the measurement. Other, less intense bands will have to be used if there is spectral interference from vibrational bands of any solvents or other components present. Consequently, the assay will be less sensitive.

It turns out that the two different spectrometries complement each other in probing molecular vibrations. It is worthwhile to look more closely at the origins of the magnitude of the IR absorption and Raman emission to decide whether to suggest one or the other for quantitation of specific substances or in specific solvents.

Observe in Figure 20.14 the Raman and infrared spectra of a single compound. Its structure is

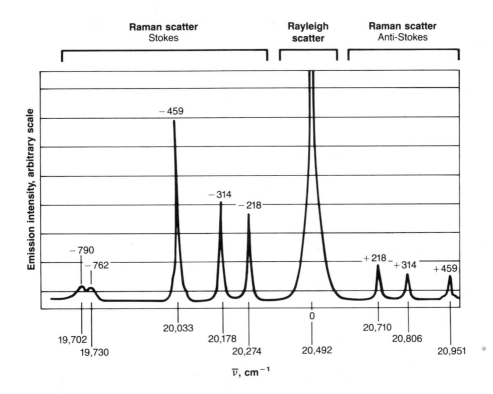

**Carbon tetrachloride,** spectrophotometric grade.

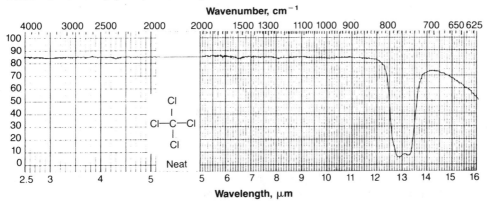

**Figure 20.13** Raman and infrared vibrational spectra for carbon tetrachloride, $CCl_4$. The Raman spectrum was obtained with laser radiation at 20,492 $cm^{-1}$ (488.0 nm, one line of an argon-ion laser). The Raman bands show up at wavelengths both positive and negative to the central (Rayleigh scattered) line. The lines to higher energy to the right are called the anti-Stokes bands and those at lower energy the Stokes bands. The Stokes bands are usually the ones measured in routine use. The *difference* in wavenumber between the central line and a Raman line equals the natural vibrational frequency of the molecules. The infrared spectrum is shown in the conventional form. [Refs: Redrawn from Strommern, D. P.; Nakamoto, K. *Am. Lab.,* Oct. 1981, pp. 70–77; Collins, D. W., et al. *Appl. Optics* **1977,** *16*, 252; and Pouchert, C. J. "The Aldrich Library of Infrared Spectra." Courtesy Aldrich Chemical Co.]

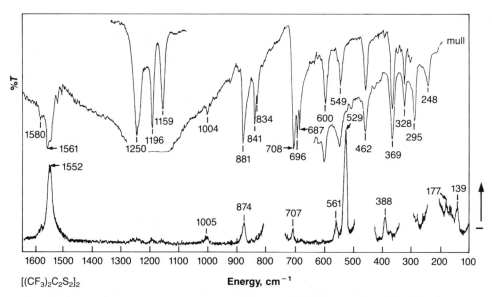

$[(CF_3)_2C_2S_2]_2$

**Figure 20.14**     Infrared (upper) and Raman (lower) spectra of $[(CF_3)_2C_2S_2]_2$, a trifluoromethyldisulfide. This illustrates how the IR and Raman spectra complement each other, since their intensities arise in different ways. The bands in the 1100–1250 range are from $CF_3$ while the band at 529 is a S—S stretch. Each is nearly invisible in one of the spectrometries. Raman spectra are more intense the more polarizable the vibrating group, while infrared spectra depend on a change in the dipole moment. C—F bonds are not very polarizable and thus do not produce intense Raman peaks. On the other hand, a S—S bond, while quite polarizable, is not ionic and does not have a dipole to change with stretching or compressing the bond between the sulfurs. The associated IR absorption is low, while the Raman intensity is large. The blank regions in the Raman spectrum are the regions of the $CCl_4$ solvent bands.

Notice that many of the vibrations that show up in the infrared spectrum do not appear in the Raman spectrum, and vice versa. Let us concentrate on sets of lines in two regions: the three lines from 1159 to 1250 and the line at 529. The former are the strongest infrared-absorption lines but are practically absent in the Raman spectrum. These are due to C—F stretching modes. The single line at 529 cm$^{-1}$ is the strongest Raman-emission line but is absent in the infrared spectrum. It is due to the S—S stretching vibration.

The following ideas are used to explain these differences in intensities. The infrared modes absorb more strongly (have a larger $\epsilon$) the larger the *polarity* of the bonds involved. For example, C—F has a highly polar bond. (This is often explained as resulting from the different electronegativities of the atoms.) On the other hand, any perfectly covalent bond, such as C—C, I—I, or S—S, is not polar at all. We do not expect vibrations of these latter groups to show up in infrared spectra.

Raman spectral intensities are proportional to the "squishiness" of the electron clouds of the atoms involved in the bonds. The correct term and concept is that the bond **polarizabilities** are large. Generally, the polarizabilities of the electron clouds increase from top to bottom of the periodic table. For the molecule here, the S—S group is quite polarizable, and its vibration gives the largest Raman intensity. C and F both are on the first full row of the periodic table and are not very polarizable. Therefore, vibrations involving the —CF$_3$ groups are very weak in the Raman spectrum.

Thus, Raman spectrometry and infrared spectrometry are complementary although both probe the normal modes of vibration in molecules. With current instrumentation, both are equally easy to run.

Not all normal modes appear as IR or Raman spectral bands. The reasons why specific vibrations do appear in infrared and Raman spectra are significantly more complicated than outlined above. To predict correctly which of the molecular vibrations appear in each type of spectrum requires the study of vibrational selection rules. Since we are interested only in understanding how these spectra can be used to solve analytical problems, we shall go no further in the details.

### Solvents for Solution Spectra

Solvents vary as to their suitability for IR and Raman samples. For instance, water is a terrible solvent to use for IR samples because it absorbs so strongly in the IR region. However, water is composed of atoms high on the periodic table—it is not very polarizable. Thus, water is excellent as a solvent for samples in Raman spectrometry. Again, Raman and IR are complementary. Similarly, CCl$_4$ is effectively transparent in most of the IR range (CCl$_4$ is not polar although the individual C—Cl bonds are) but has rather large bands in Raman, which could interfere in the region of interest. The infrared and Raman spectra of CCl$_4$ are shown in Figure 20.13.

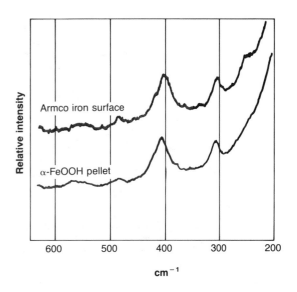

**Figure 20.15**   Raman spectrum of pure iron corroded by atmospheric exposure. The product spectrum is compared with that of α-FeOOH. Other techniques can also be used for analyses of these corrosion films. However, they do not provide information on the chemical bonding that this Raman spectrum does. In addition, this spectrum can be obtained without placing the sample in a vacuum and, if necessary, it can be run with the sample in a corrosive atmosphere. [Ref: Reprinted from Keiser, J. T., et al. *Amer. Lab.* **1982**, *14*(4), 17. Copyright 1982 by International Scientific Communications, Inc.]

## 20.8 Other Applications of Infrared and Raman Spectrometry

Both infrared and Raman spectra are used for identifying and quantifying the components in solids, liquids, and gases. The samples can be mixtures or solutions. The spectral properties can be measured in transmission on solids, liquid films, bulk solutions, or gas samples or by reflection from solids and liquids. Only two of these

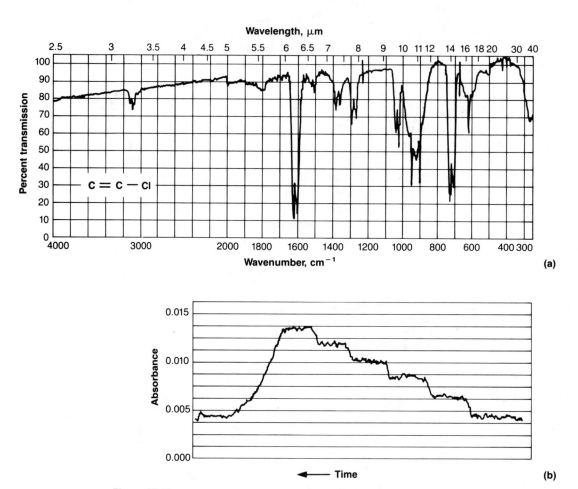

**Figure 20.16**   (a) The infrared spectrum of gaseous vinyl chloride in a multipass cell with a path length of 20.25 m. (b) The plot of instrument response (absorbance scale on the right) with changes in concentration of vinyl chloride. Each step is a successive increase of 1 ppm vinyl chloride, beginning at zero on the right. Instrument conditions: path length, 20.25 m; slit, 2 mm; time constant, 2.5 s; assay wavelength 917 cm⁻¹. [Ref: Reprinted from Lavery, D. S.; Wilks, P. A., Jr. in "Spectroscopy," Series II, Vol. II, International Scientific Communications, Inc.: Fairfield CT, 1977, pp. 127 and 128. Copyright 1977 by International Scientific Communications, Inc.]

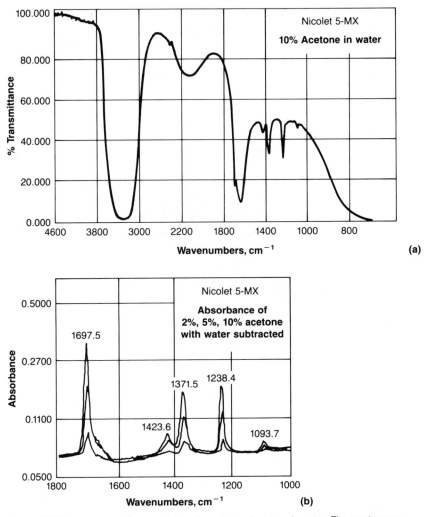

**Figure 20.17**    Internal reflectance infrared spectrum of acetone in water. The spectrum was obtained with a Fourier transform instrument (Nicolet 5-MX). A computer is an integral part of the apparatus. (a) The full-range IR spectrum of 10% acetone in water. (b) The spectrum in the wavelength range 1000 – 1800 cm$^{-1}$ after the background water spectrum was subtracted. The resulting spectrum from acetone was plotted on an absorbance scale. Three different concentrations of acetone were run: 2%, 5%, and 10% (v/v). [Analysis performed using a standard CIRCLE™ accessory, Courtesy of Barnes Analytical/Spectra-Tech, Inc.]

applications are illustrated here. Some information on the cells to hold samples for IR and Raman spectrometries is presented in Supplement 20A.

Figure 20.15 illustrates an identification of a solid by Raman spectrometry. The oxide scale on the surface of an atmospherically corroded iron surface of pure iron was found to be $\alpha$-FeOOH. There are various forms of FeOOH, and each form, $\alpha$-, $\gamma$-, $\delta$-, and amorphous FeOOH, has a characteristic Raman spectrum.

The determination of gaseous vinyl chloride by infrared spectrometry is illustrated in Figure 20.16. The whole infrared spectrum and an illustration of instrument response with concentration are shown. The absorbance measured is for a pathlength for transmission of 20.25 m. This is done by using mirrors to reflect the radiation many times through a shorter cell. Some further examples of quantitation of certain properties of polymers with IR spectrometry are presented in Chapter 21.

## 20.9 Quantitation

Quantitation of the composition of a solution with IR spectrometry is illustrated in Figure 20.17. The solution is acetone in water. The spectrum was able to be measured using a technique called **internal reflection.** The principle of internal reflection and an internal reflection cell are discussed briefly in Supplement 20A. The large background absorption due to water was subtracted, and the % $T$ signal was converted to absorbance by the instrument's computer. After calibration, the percent acetone in water could be determined in about 10 min. (Can you think of other instrumental or wet-chemical methods to do this specific determination?[1])

In general, quantitation can be done relatively easily when individual bands are completely separated. It is usually best to leave the instrument settings (0 and 100 %

---

[1] GC, UV, and redox titration, to name a few.

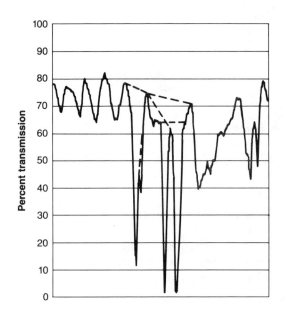

**Figure 20.18**    Infrared spectrometry and Raman spectrometry can be used for quantitative analysis, just as is UV–visible. However, sometimes there are problems choosing a baseline. Illustrated is a rather easy case as well as a few possibilities for the much more difficult choice of the correct baseline for the middle peak.

$T$) the same for all samples and calculate the absorbances from the % $T$-values of the baseline and peak, even if the baseline changes.

Some complications arise in vibrational spectrometry due to the large number of bands. When bands are close together, deciding how to find the baseline for quantitative determinations can be quite complicated as indicated in Figure 20.18. It is better to find a well-separated band that can be used.

## Suggestions for Further Reading

A clear description of the techniques of infrared analysis. The best place to go next.
Potts, W. J., Jr. "Chemical Infrared Spectroscopy"; Wiley: New York, 1963.

Comprehensive review of infrared sample-preparation techniques and problems with quantitative analysis.
Smith, A. L. In "Treatise on Analytical Chemistry," Pt. I, Vol. 6; I. M. Kolthoff and P. J. Elving, Eds.; Interscience: New York, 1965; Chap. 66.

A comprehensive review book encompassing all aspects of infrared spectrometry in analysis. The theoretical sections are advanced. The chapters on experimental techniques and applications are extensive.
Kiss–Eross, K. In "Comprehensive Analytical Chemistry," Vol. 6; G. Svehla, Ed.; Elsevier: Amsterdam, 1976.

A useful book of instrument properties and spectra properties.
Meloan, C. E. "Elementary Infrared Spectroscopy"; Macmillan: New York, 1963.

Contains an introductory chapter on assigning bands in infrared spectra of organic molecules. Good references in the area.
Silverstein, R. M.; Bassler, G. C.; Morrill, T. C. "Spectrometric Identification of Organic Compounds," 4th ed.; Wiley: New York, 1981; Chap. 3.

A relatively advanced book on infrared instrumentation. Not primarily chemical applications.
Vanzetti, R. "Practical Applications of Infrared Techniques"; Wiley–Interscience: New York, 1972.

An advanced book on its topic, FT–IR. It is a "how-to" book but relatively nonmathematical.
Griffiths, P. R. "Chemical Infrared Fourier Transform Spectroscopy"; Wiley: New York, 1975.

A short monograph showing numerous examples of the type of information Raman spectra can provide. Not a "how-to-do-it" book.
Strommen, D. P.; Nakamoto, K. "Laboratory Raman Spectroscopy"; Wiley: New York, 1984.

A description of the various parts of Raman spectrometers.
Bulkin, B. J. *J. Chem. Ed.* **1969,** *46,* A781–A800 and A859–A868.

A review article describing how to optimize infrared spectrometer measurements.
Potts, W. J., Jr.; Smith, A. L. *Applied Optics* **1967,** *6,* 257–265.

Excellent, relatively detailed, practical discussion of precision of quantitative IR measurements and how to increase it. Also, a good discussion of mixture determination.
Perry, J. A. *Appl. Spec. Rev.* **1970,** *3,* 229–262.

## Problems

**20.1**   An IR spectrum of vinyl acetate shows both a C=C bond vibration and a C=O vibration. Both bands have baselines at 80% $T$. The peaks are, respectively, at 23% $T$ and 5% $T$. What is the ratio between the absorptivity of the C=O compared with the C=C vibration?

**20.2**   Two spectroscopists are exchanging information on a compound in solution. One is using a machine with its output in absorbance. A specific band is shaped as an isoceles triangle (a good approximation in this instance) with a fwhm of 8 cm⁻¹. Both workers use the same pathlength cells for two solutions. The second worker has an instrument that records in % $T$. Two standard solutions being compared between the two laboratories have the analytical band with 10% $T$ and 1% $T$, respectively, and the baseline at 85% $T$. What is the measured fwhm for the two solutions measured on the instrument that records in % $T$?

**20.3**   [Data from Fairless, B. J., et al. *J. Chem. Ed.* **1971**, *48*, 827.]

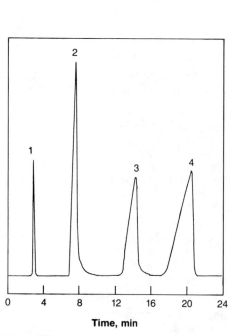

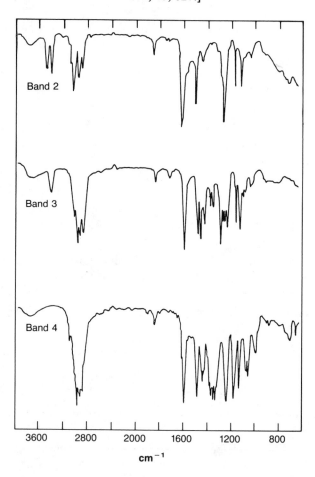

**Figure 20.3.1**

Figure 20.3.1 includes the gas chromatogram of a mixture of N,N-diethyl-*p*-toluidine, N-ethyl-*p*-toulidine, and *p*-toluidine. The structure of N-ethyl-*p*-toluidine is

The gas chromatogram was run using 25 $\mu$L of a mixture of these toluidines dissolved in ether. The peak labeled 1 is the ether. The effluent associated with each band was collected, and the IR spectra were run. The spectra are shown in the figure. What are the identities of the compounds in bands 2, 3, and 4?

20.4 [Data from the Sadtler Library of Infrared Spectra. © Sadtler Research Laboratories, Division of Bio-Rad Laboratorie., Inc., 1966.] Assign the nine infrared spectra in Figure 20.4.1. The possible compounds are the following:
  a. Decane, $CH_3(CH_2)_8CH_3$
  b. Acetophenone, $CH_3COC_6H_5$
  c. Butyronitrile, $CH_3CH_2CH_2C{\equiv}N$
  d. Benzoic acid, $HOOCC_6H_5$
  e. Nitrobenzene, $NO_2C_6H_5$
  f. *n*-Butylamine, $CH_3CH_2CH_2CH_2NH_2$

  g. Cyclopentanone, $O{=}C$

  h. Methyl methacrylate, $H_2C{=}\underset{\underset{COOCH_3}{|}}{C}{-}CH_3$

  i. Cyclohexane, $C_6H_{12}$
  j. 1-Nitropropane, $NO_2CH_2CH_2CH_3$

20.5 [Figure from Pavia, D. L.; Lampman, G. M.; Kriz, G. S. "Introduction to Spectroscopy: A Guide for Students of Organic Chemistry"; W. B. Saunders, Philadelphia, 1979. Copyright © 1979 by W. B. Saunders Co. Reprinted by permission of CBS College Publishing.]
Figure 20.5.1 shows the mass spectrum, NMR spectrum, and IR spectrum of a compound. Identify the compound.

20.6 The anti-Stokes lines (defined in Figure 20.13) of a Raman spectrum are reduced in magnitude by an amount proportional to the Boltzmann distribution. The Stokes lines remain the same since they originate in the ground state of the molecule. Thus, at very high temperatures, both Stokes and anti-Stokes lines could be equal in magnitude. If a vibrational energy level is at 459 cm$^{-1}$ and the sample temperature is 298 K, what is the expected ratio of intensities of the Stokes and anti-Stokes lines at 459 cm$^{-1}$? Does your calculation agree with the data shown in Figure 20.13?

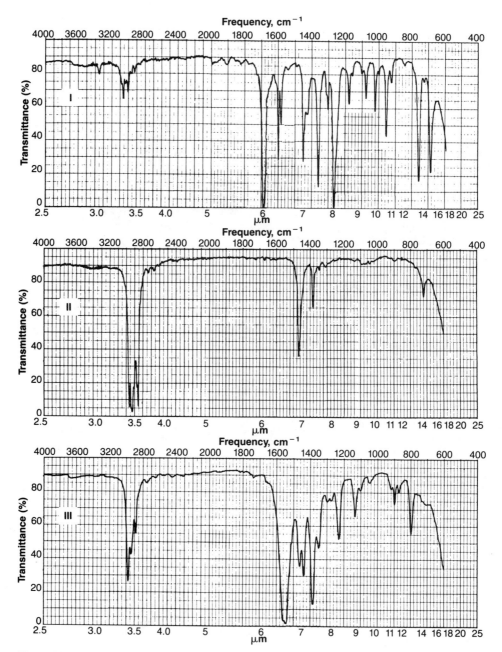

Figure 20.4.1

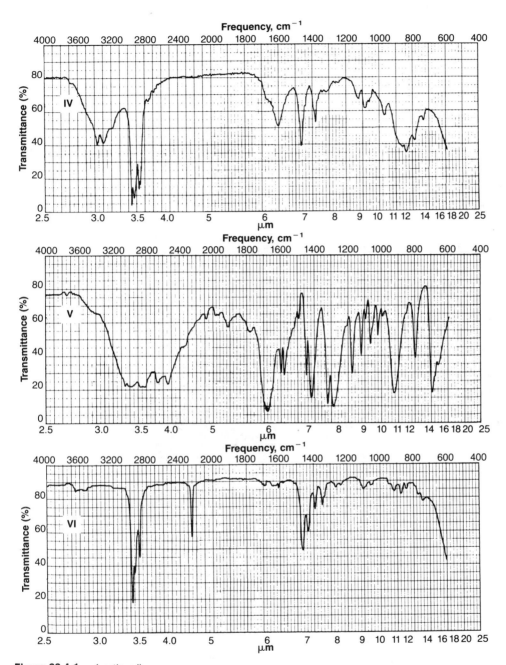

**Figure 20.4.1** *(continued)*

832

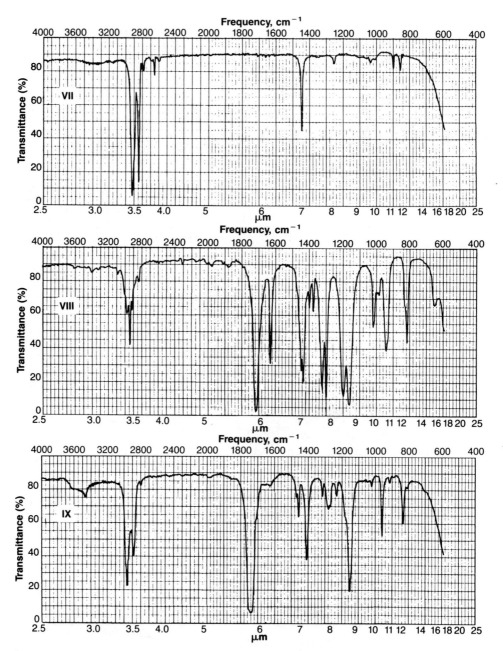

**Figure 20.4.1** *(continued)*

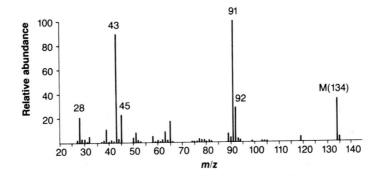

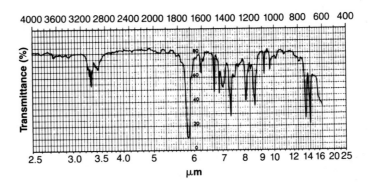

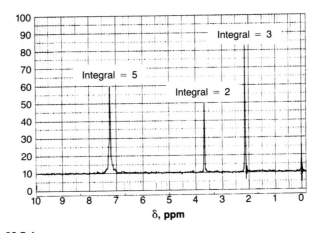

**Figure 20.5.1**

## SUPPLEMENT 20A

# Samples for Infrared and Raman Spectrometries

### Samples for Infrared Spectrometry

Some significant problems arise when trying to construct sample containers for vibrational spectrometry, because every material has *some* vibrational absorption. What is done is to use material that has a minimum interference in the regions of interest. The material of choice for IR is a solid potassium bromide plate. Such plates

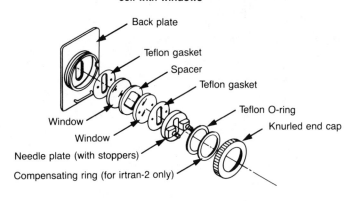

**Exploded view—demountable cell with windows**

Back plate
Teflon gasket
Spacer
Teflon gasket
Teflon O-ring
Knurled end cap
Window
Window
Needle plate (with stoppers)
Compensating ring (for irtran-2 only)

**Figure 20A.1**   Two of many designs for gas and liquid sample IR cells. The path length of the liquid-sample cell is determined by the thickness of the spacer that is inserted. These range from 0.015 to 0.5 mm. The sample is injected with a syringe through the ports on the needle plate. The ports align with holes in the front window. [Courtesy Aldrich Chemical Co., Inc.]

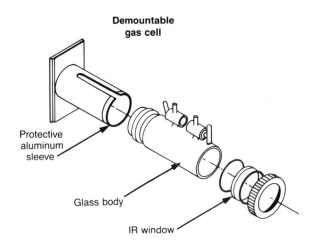

**Demountable gas cell**

Protective aluminum sleeve

Glass body

IR window

are used in a number of ways. A solid may be deposited as a film on a single KBr plate by allowing the solvent to evaporate from a solution of the sample. A liquid can be spread as a film on the plate if it is concentrated enough and will not evaporate. Otherwise, as illustrated in Figure 20A.1, the liquid can be placed in a solution cell consisting of two KBr plates that have a space between them of around 0.01 – 0.5 mm. The two plates are called **windows.** Gases are contained in longer cells.

As mentioned in Section 20.7, water is not a good solvent for IR spectrometry. However, occasionally it is necessary because no other solvent will work and a solid sample cannot be used. In that case, the window material cannot be KBr because the salt would easily dissolve in the sample solution. In this case, materials like polyethylene and polycrystalline calcium fluoride or polycrystalline zinc selenide can be used as windows.

An alternative to the usual methods for measuring IR spectra is to use the technique of total internal reflection. As illustrated in Figure 20A.2, the technique works as follows. All the light illuminating an interface between two materials is reflected from the interface if it approaches in a direction more parallel to the surface than the critical angle. This is called total internal reflection and can be seen in the visible region as a mirrorlike appearance of the surface of, for example, a fish tank seen from below the surface. However, the light that is reflected in fact *does* penetrate the surface a short distance. This light then interacts with the sample material within a few micrometers of the other side of the surface. An infrared absorption spectrum can be obtained that *appears* as if it came from a transmittance cell having a sample solution only a few micrometers thick.

Solids can be probed with infrared spectrometry even if they cannot be dissolved; they can be suspended in the infrared beam. One way to do this is to make a powder and place it on a total internal reflectance cell surface. A second way is to grind the solid into small particles and suspend them in a viscous liquid that is placed in the light beam between plates composed of KBr or NaCl. A liquid that is commonly used for this purpose is mineral oil — also called *Nujol*. It has a few strong infrared bands that might interfere, however. (See Figure 20.3.) The mixture of the solid particles suspended in the viscous liquid is called a **mull.** Another popular method to use for solids is to mix and crush the sample intimately with dry KBr. A measured amount of this finely ground powder is then pressed under several tons/cm$^2$ to form a transparent KBr disk about 1 cm in diameter and less than 1 mm thick. The KBr can, however, interact with some groups such as $-NH_2$, causing distortion of the spectrum. In addition, water in the KBr can be hard to eliminate. The water absorption interferes with $-OH$ and $-NH_2$ bands.

## Samples for Raman Spectrometry

Raman spectrometry is different from infrared in that light in the visible region is used to excite the molecules in the sample. The geometry of a Raman instrument (Supplement 20B) is similar to that of a luminescence spectrometer. The major difference is that the light source, usually a laser, is intense and focused to a small spot on the sample to enhance sensitivity. A solution can be held in a glass capillary, a solid can be placed directly in the beam, and a gas can be held in a glass container.

**(a)**

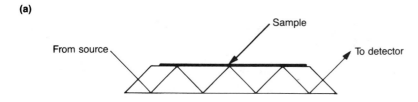

**(b)**

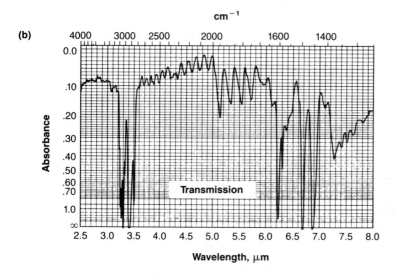

**(c)**

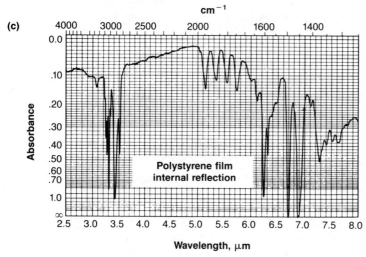

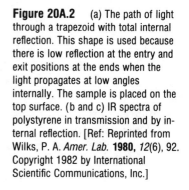

**Figure 20A.2** (a) The path of light through a trapezoid with total internal reflection. This shape is used because there is low reflection at the entry and exit positions at the ends when the light propagates at low angles internally. The sample is placed on the top surface. (b and c) IR spectra of polystyrene in transmission and by internal reflection. [Ref: Reprinted from Wilks, P. A. *Amer. Lab.* **1980,** *12*(6), 92. Copyright 1982 by International Scientific Communications, Inc.]

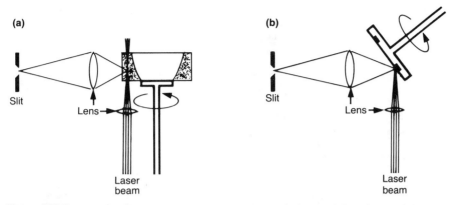

**Figure 20A.3**　To prevent overheating of colored samples by the laser radiation, the sample is constantly exchanged by rotating the sample. This spreads the heat over a larger area. (a) A liquid sample (stippled) is held in a cell which is rotated. The laser beam irradiates the sample from below, and the reemitted light is collected by the lens and focused to the monochromator. (b) Solid samples can be packed into an incised ring on a rotating wheel (which can be cooled as well).

However, a major problem arises if the intense light beam heats the sample (a problem with colored samples). Liquids will boil, solids can vaporize, and gases will either increase in pressure or expand, depending on the container. The heating problem can be overcome by distributing the light over a larger amount of sample. A few of the techniques for doing this are shown in Figure 20A.3.

## SUPPLEMENT 20B

# Infrared and Raman Spectrometers

The sources and detectors of electromagnetic radiation for the infrared region were described in Supplements 17A and 17C. Commercial infrared spectrometers are of two types: double-beam instruments and the Fourier transform type. We shall not describe the Fourier transform instruments here; details can be found in Low, M. J. D. *J. Chem. Ed.* **1970,** *47,* A163, A255, A349, and A415; and Koenig. J. L. *Appl. Spec.* **1975,** *29,* 293.

The monochromators of IR instruments usually are constructed with a combination of filters and gratings along with the usual slits to select and scan the wavelength range. A unique feature of many instruments is the manner in which the detector is used. The output from the detector is electronically connected to a mechanical **optical wedge** in the reference beam. See Figure 20B.1. An initial adjustment of 100% *T* is made with blanks in both the sample and reference beams. The 0% *T*-adjustment is made when the sample beam is blocked. When a sample and reference (blank) are

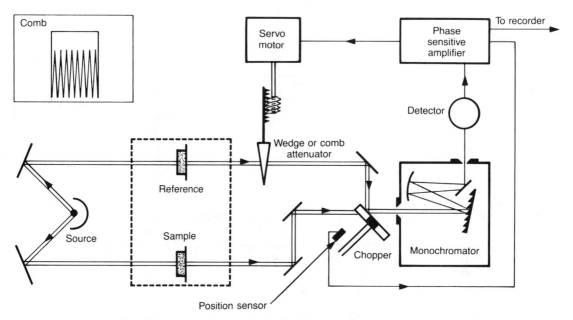

**Figure 20B.1** The general layout of a conventional double-beam infrared spectrometer. The wedge (or comb) is moved to block more or less light from passing through the blank until the blank + wedge and the sample pass the same amount of light. The position of the wedge determines the output in % $T$. Thus the linearity of the instrument depends crucially on the linearity of the attenuator element. A comb attenuator is illustrated in the inset. The signal is phase-sensitive detected, as described in Supplement 18C. Note that the path lengths of the sample and reference beam are equal to allow good compensation for background from, for instance, $CO_2$ and $H_2O$ bands in the atmosphere.

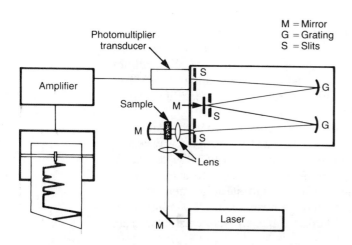

**Figure 20B.2** Layout of a Raman spectrometer. A double monochromator is needed to eliminate stray light from the Rayleigh line and other scattered light, such as that off particles in solution.

in place, as the wavelength is scanned, the electromechanical circuit moves the wedge until the power falling on the detector is the same for both sample and reference. The position of the pen on the graph paper (linear in % $T$) depends directly on the physical position of the wedge in the beam—that is, on how much the wedge attenuates the light power. This design produces a wider linear range than direct amplification of the detector signal.

A Raman instrument (Figure 20B.2) is similar to a simple single-beam fluorescence spectrometer (Figure 17.9). The primary differences are the light source, which is much more intense, and the monochromator, which must be of exceptionally high quality because the intensities of the Raman bands are so small ($10^{-5}$–$10^{-6}$ of the scattered light). If the monochromator is not high quality, then, among a number of other effects, stray light will cause a large background over a wide range of wavelengths. Commonly, a double monochromator is used. Such a monochromator is equivalent to having the light output from one monochromator enter a second one. As such, a double monochromator contains two gratings with a slit in the light path between them. The laser light illuminates the sample 90° off the axis of the monochromator. The light, being in the visible region, is detected by a sensitive, low-noise photomultiplier (Supplement 17A). The monochromator scans wavelength and is coordinated with a chart recorder to record the spectrum.

# Part VII
# POLYMER ANALYSIS

# 21

# Polymer Analysis

## 21.1 Polymers

Plastics, rubber, adhesives, paints, wood, sand, muscle, hair, silk, flour, and gelatin are all materials composed of organic, inorganic, or biological polymers. The chemical structures of a few are illustrated in Figure 21.1. In view of the pervasive influence these materials enjoy, it is not surprising that a large part of the effort and applications of chemical analysis is directed toward polymer analysis.

In this chapter, some of the primary aspects of analysis as applied to polymers are discussed. The methods used are mostly those described in earlier chapters, so only the techniques found to be especially useful for investigating macromolecules are presented in any detail. The measurement of physical properties such as electrical resistance, hardness, ductility, and other rheological properties (rheological properties include deformation and flow as a function of time, force, and temperature) are not included here.

In polymer science, the purpose of chemical analysis is to determine chemical structures and relate them to the useful properties of polymeric materials. The same

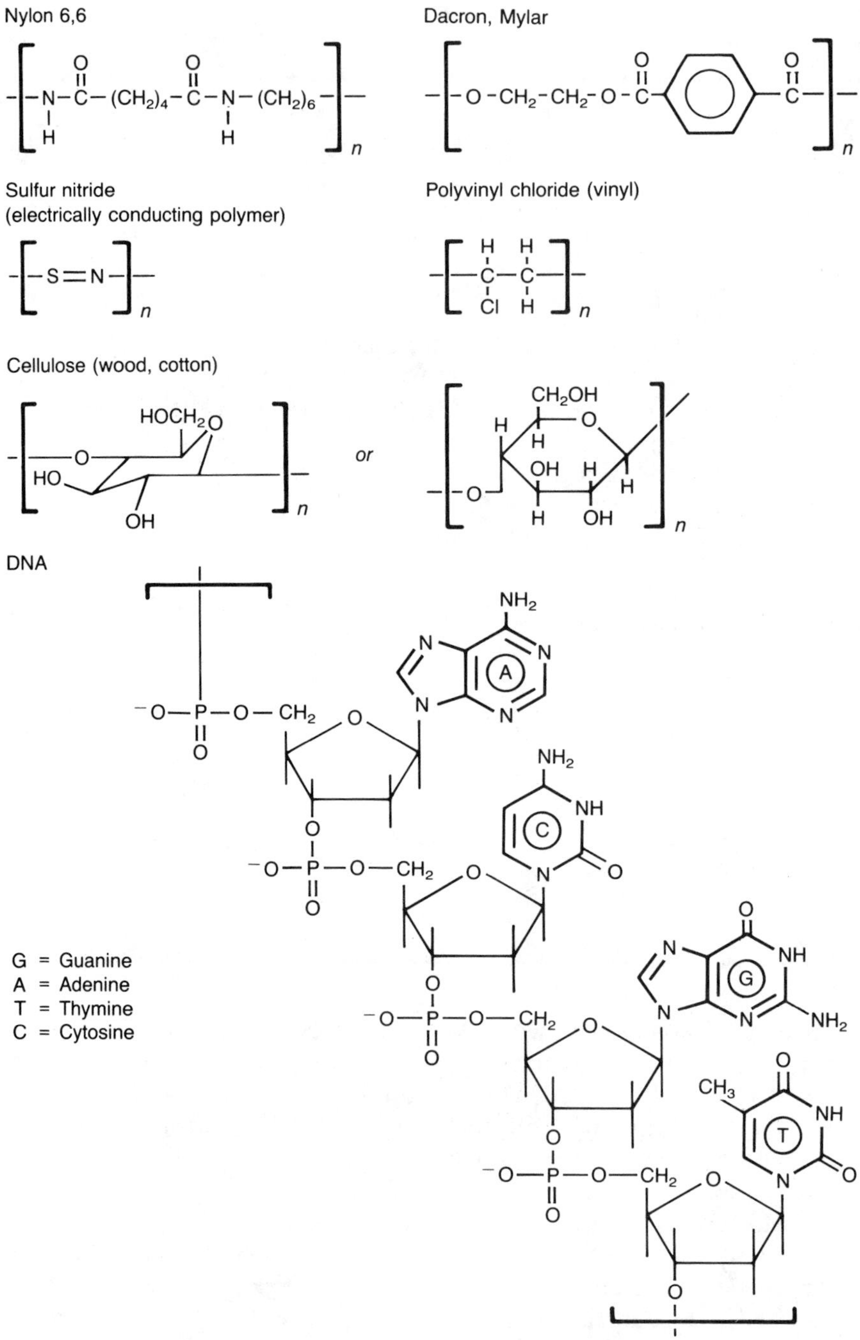

844

**Figure 21.1** Chemical structural formulas and common names of selected natural and man-made polymers.

analytical methods are also used in controlling the manufacturing processes for chemically modified naturally occurring polymer materials and for those polymers synthesized from monomer precursors.

Many of the sample preparation techniques that were outlined in Chapter 6 are not applied to polymeric materials. We do not want to digest the sample completely, unless an elemental analysis of the material is wanted. Nondegradative sample preparation methods for polymers are limited mainly to two: working the polymer into a convenient shape (film, plate, or rod) for use in a specific instrument or dissolving the polymer, if that is possible. Less drastic digestion techniques may be used in the appropriate circumstances to break down the macromolecules into the smaller chemical units composing them. Examples of the units are monomers of natural products such as the sugars of cellulose and the amino acids of proteins. Interestingly, as you will see, many of the chemical properties of the monomer units are retained in the polymers.

What are the chemical properties that are measured for polymers? The primary ones are

1. the molecular mass of polymeric molecules,
2. the relative amounts of amorphous and crystalline material in solids, and
3. the chemical functional groups and chain structures of the polymers.

Before pursuing the techniques used to elucidate these properties, first let us note some of the special nomenclature of polymer science.

## 21.2 Definitions

Polymers are composed of repeated units that are linked together into long chains. The units can be joined in a number of different ways, but three different general classifications of chain types are adequate: linear, branched, and cross-linked. These are illustrated in Figure 21.2a. The atoms that are linked together in the longest segment form the **backbone** of the polymer chain. In cross-linked polymers, the repeat units are actually linked into a three-dimensional network of macroscopic size, but this is difficult to draw clearly in two dimensions. There are usually varying lengths of backbone chain between cross-link points.

If a polymer contains two different types of monomers, then it is more correctly called a **copolymer**. There are three general classifications of copolymer structures: *random, block,* and *graft*. These are illustrated in Figure 21.2b. The block and graft copolymers generally are made in at least two steps — reacting one of the monomers to form a homopolymer, followed by a second reaction with the other monomer to produce the block or graft form.

The monomers that serve as building blocks of polymers can polymerize in a number of ways. For example, consider a polymer with a backbone formed of atoms that can form four bonds to other atoms—carbon and silicon are typical. The four bonds point toward the vertices of a tetrahedron. With chains of carbon or silicon, chemically substituted polymers can form in three different ways. These are illustrated in Figure 21.3. The shaded circle on every second carbon represents, for instance, a chlorine in poly(vinyl chloride) or a methyl group in polypropylene. If the

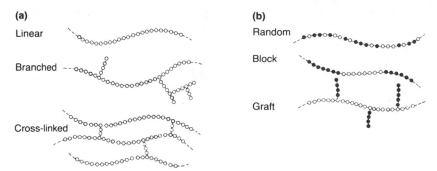

**Figure 21.2**     Modes of polymerization. (a) Illustration of three types of polymer structures: linear, branched, and cross-linked. The circles represent monomeric or repeat units. The main chain is called the backbone. The shorter chains branching off the main chain are side chains if they do not connect with another main chain. If a short chain covalently binds two main chains, it is called a cross-link. (b) Copolymers are polymers containing more than one monomer species. Three classifications of copolymers are illustrated here: random, block, and graft. Open circles represent monomers of one species and shaded circles monomers of a different one. Block copolymers have the polymer sequences in the main polymer chain. Graft copolymers have sequences of one monomer attached to a backbone of another.

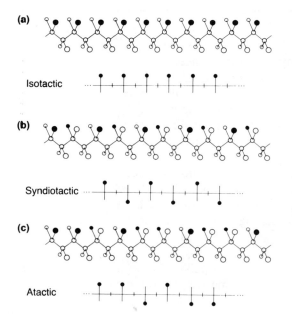

**Figure 21.3**     For tetrahedrally bonding atoms in a backbone that are substituted at one position each, there are three different stereochemical modes of polymerization. The example shown here is polyvinyl chloride, with the shaded atom being chlorine. Two different representations of each structure are shown. (a) Isotactic, all the chlorines are identical in disposition. (b) Syndiotactic, the chlorines alternate in a regular pattern. (c) Atactic, the chlorines are randomly disposed.

*cis*-1,4-polyisoprene
(natural rubber)

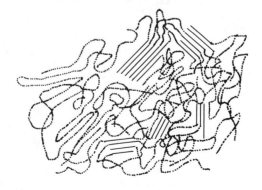

**Figure 21.4**  When double bonds exist in a backbone, then each unsaturated center can be either *cis* or *trans*. Illustrated here are the unsaturated centers of polyisoprene: *cis*-polyisoprene is natural rubber; *trans*-polyisoprene has the common name *gutta percha*.

*trans*-1,4-polyisoprene
(gutta percha)

Isoprene

$$CH_2=\overset{\overset{\displaystyle CH_3}{|}}{C}-CH=CH_2$$

substituents all project in the same direction from the carbon chain, the polymer is **isotactic.** If the substituents alternate in direction, the polymer is **syndiotactic.** And if the projection direction is random, the polymer is said to be **atactic.** These are three measures of the stereochemical regularity of the polymer. The stereochemical regularity of the chain generally affects the crystalizability and that, in turn, contributes to a number of properties of the polymer such as its brittleness.

A different kind of isomerism can occur when the backbone is composed of *un*saturated polymer chains. The polymers can then form with either *cis* or *trans* isomers or mixtures of them. These are illustrated in Figure 21.4. In the example of

**Figure 21.5**  Two-dimensional representation of the structure of a solid polymer with crystalline and amorphous regions. Only the backbone of the polymer is represented.

**Crystalline region**

**Table 21.1**   Degree of Crystallinity and Polymer Chemical Structure

| Less Crystalline | More Crystalline |
|---|---|
| Bulky side chains on monomer | No side chains on monomer |
| Long repeat distance (long distance before chain repeats) | Short repeat distance |
| Low stereochemical regularity, atactic | High stereochemical regularity, isotactic and syndiotactic |
| Highly branched | Linear polymer chain |

polyisoprene, the *trans*-1,4-polymer is a plastic while the *cis*-1,4-polymer is a rubber; natural rubber is *cis*-1,4-polyisoprene. (Vulcanization is a reaction in which some of the residual double bonds are induced to form bonds between chains to create a cross-linked network polymer.)

The chains of different polymers can pack together in different ways: as crystals, in which the chains are highly ordered, or as amorphous *(without form)* regions, in which the chains are randomly packed together. Other packing structures include the $\alpha$-helix and $\beta$-sheet structures of proteins. The packing is one characteristic of the **morphology** (Greek for *form*) of the polymer. Solid polymers tend to have mixtures of amorphous regions together with crystalline or other highly ordered regions such as $\alpha$-helices. This combination is illustrated in two dimensions in Figure 21.5. The **degree of crystallinity** (the fraction of the whole mass that is crystalline) depends on

1.  the chain structure (linear or branched),
2.  the geometric structure (isotactic, syndiotactic, or atactic),
3.  the substituents on the chain (chloride, methyl, and so on), and
4.  the preparation conditions (for instance, temperature and cooling rate).

The degree of crystallinity as a fraction of the total mass in polymeric solids varies from zero to about 0.95. The tendency toward crystallinity usually varies with the chemical structure as listed in Table 21.1.

## 21.3 Chemical Structure and Physical Properties

The molecular mass and degree of crystallinity have a major influence on the physical properties of polymers. Figure 21.6 illustrates a simplified relationship between these properties. In addition to the physical properties indicated in the figure, the chemical structure of the polymer also determines

1.  the material's resistance to chemical attack by acids or bases,
2.  how much solvent can be absorbed into the crystalline or cross-linked polymer lattice (causing it to swell),
3.  the solubility of the polymer in aqueous or organic solvents, and
4.  the temperature range over which the material is useful.

**Figure 21.6** The physical properties of a polymeric material are related to the constituent polymer molecular mass and to the relative amount of crystalline regions in its structure. These relationships are indicated in a general, qualitative way in this diagram.

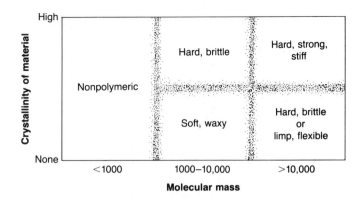

## 21.4 Molecular Mass and Molecular Mass Measurements

There are a number of different techniques that can be used to measure the *average* molecular weights of polymer molecules as long as they are not cross-linked. Cross-linked molecules form a single molecular network that cannot be separated (by dissolution) into groups of molecules. Therefore, the average molecular mass is a property that can be measured only for linear and branched polymers: those that can be dissolved in some solvent.

The average molecular mass that is measured can depend on the technique used to measure it. This kind of result probably contradicts all the training you have had in science: You might think, If the quantity measured depends on the method, then something must be wrong with at least one of the methods. But this is not the situation with polymers. The reason for this apparently strange circumstance is that synthetic polymers are not pure compounds; they are mixtures of species having different molecular weights. A range of masses exists. A typical distribution of masses is shown in Figure 21.7. The figure will be explained in more detail shortly.

When every molecule of a sample has the same molecular mass, the polymer is said to be **monodisperse.** When there is a distribution of molecular masses, the polymer is called **polydisperse.** Numerous biological macromolecular species, such as enzyme proteins, are nearly monodisperse. All man-made macromolecules are polydisperse.

### Definitions of Molecular Mass

The average molecular mass of a polymer is represented as a capital $M$ with a bar over it, $\overline{M}$. $\overline{M}$ is usually expressed in molar units, as is any other molecular weight — that is, in daltons or, equivalently, a.m.u. One way to measure the average mass is to weigh a sample and then determine the *number* of molecules in the sample by experiment. Algebraically,

$$\overline{M}_n = \frac{\text{sample mass}}{\text{number of moles present}} \qquad \textbf{(21-1)}$$

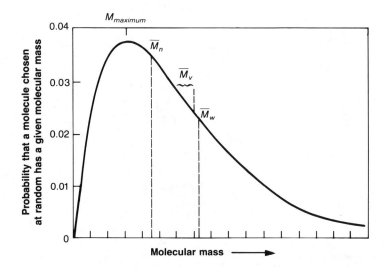

**Figure 21.7**    Many polymers are formed with a range of molecular masses. Plotted here is a distribution function (not unlike a gaussian probability function) plotted as a probability versus the molecular mass. The integral of the graph equals 1.0, that is, 100% of the material. The various average molecular masses are indicated on the figure. $\overline{M}_n$ is the number-average molecular mass determined by measurement of colligative properties. $\overline{M}_v$ indicates the range of the average molecular mass found from viscometry. $\overline{M}_w$ is the weight-average molecular mass. It can be found from light scattering. [Redrawn from Collins, E. A., et al. "Experiments in Polymer Science"; Wiley–Interscience: New York, 1981.]

$\overline{M}_n$ is called the **number-average molecular weight** or **number-average molecular mass.**

---

**EXAMPLE**

A 0.6582-g sample of a polymer contains $4.87 \times 10^{-6}$ mol of material. What is $\overline{M}_n$?

**SOLUTION**

The molecular weight is

$$\frac{0.6582 \text{ g}}{4.87 \times 10^{-6} \text{ mol}} = 1.35 \times 10^5 \text{ daltons}$$

---

To express the algebra of molecular masses concisely, we can imagine that we divide the mass distribution of Figure 21.7 into a set of narrow ranges of molecular mass. The properties of the "slice" of masses are signified by a subscript $x$. For example, there are $N_x$ moles of molecules with a molecular mass $M_x$.

The total mass (or weight) of a sample is simply the sum of the individual molecular masses. If a sample has $N_x$ moles of polymer that has molecular mass $M_x$, the total sample mass is the sum of all the fractions:

$$M_{\text{sample}} = \sum_x N_x M_x$$

In addition, the total number of moles in the sample is equal to the sum of the $N_x$-values of all the fractions.

$$N_{\text{sample}} = \sum_x N_x$$

Therefore Eq. 21-1 can be written more precisely in the form

$$\overline{M}_n = \frac{\displaystyle\sum_x N_x M_x}{\displaystyle\sum_x N_x} \tag{21-2}$$

**Table 21.2**   Representative Ranges of $\overline{M}_w/\overline{M}_n$ for Polymers

| Polymer Type and Examples | $\overline{M}_w/\overline{M}_n$ |
|---|---|
| Monodisperse polymer | 1.000 |
| Biological polymers | 1.0 |
| Polymers made by ionic polymerization<br>  a. Polystyrene with $KNH_2$ catalyst<br>  b. Poly(ethylene oxide) with $SrCO_3$ | 1–2 |
| Condensation polymers (step reaction)<br>  a. Polyesters<br>  b. Polyamides<br>  c. Polyesters | 1.5–2 |
| Addition polymers (radical chain reaction)<br>  a. Polystyrene<br>  b. Polyvinyl chloride | 2–5 |
| Polymers made by coordination polymerization<br>  a. Polypropylene with Ziegler catalysts<br>  b. Polybutadiene with Ziegler catalysts | 8–25 |
| Polymers with random branching | 20–50 |

*Sources:* Collins, E. A.; Bares, J.; Billmeyer, F. W., Jr. "Experiments in Polymer Science"; Wiley–Interscience; New York, 1973; p. 123.
   Ward, T. C. "Molecular Weight and Molecular Weight Distributions in Synthetic Polymers"; *J. Chem. Educ.* **1981**, *58*, 867–879.

The techniques used to measure $\overline{M}_n$ will be described later.

To calculate a second measure of the average molecular mass, assume that we separate from a sample all the molecules with $M_x$ and that we can do this for all possible values of $M_x$. Then assume that we can weigh each $M_x$ fraction independently. Let $W_m$ be the *total* weight of each fraction, so that it is not confused with $M_x$, the molecular mass of the fraction. Then the **weight-average molecular weight** (as it is called historically) is defined as

$$\overline{M}_w = \frac{\displaystyle\sum_x W_x M_x}{\displaystyle\sum_x W_x} \tag{21-3}$$

As shown in Figure 21.7, $\overline{M}_w$ is not equal to $\overline{M}_n$. $\overline{M}_w$ lies higher on the mass scale because it is determined more by the larger molecules in the sample. However, as the spread of the molecular masses becomes narrower, the two measures of average molecular mass become closer in value. One way to describe the spread of the distribution of molecular masses is through the ratio $\overline{M}_w/\overline{M}_n$. The ratio $\overline{M}_w/\overline{M}_n$ is often called the **polydispersity index** of the polymer. Table 21.2 lists this ratio for a number of macromolecules polymerized under different conditions.

A third measure of the average molecular mass is found from an experimental determination using the changes in viscosity of a solvent due to the presence of small amounts (less than about 1% w/w) of dissolved polymers. (The techniques used to measure the viscosity and to calculate the average molecular mass will be described in

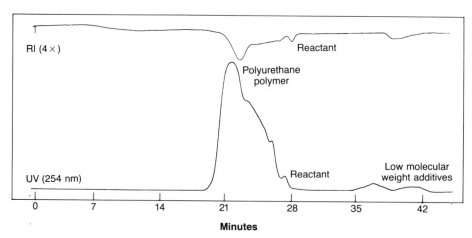

**Figure 21.8**    Gel-permeation liquid chromatogram of a commercial polyurethane coating. The effluent of the column was passed through two different detectors: (upper curve) a refractive-index detector, (lower curve) a UV detector at 254 nm. The low-molecular mass additives elute last as expected in GPC. [Reprinted from Alfredson, T. *Am. Lab.* **1981**, *13*(Aug), 44. Copyright 1981 by International Scientific Communications, Inc.]

more detail later.) The average molecular mass determined this way is called $\overline{M}_v$, and it lies between $\overline{M}_n$ and $\overline{M}_w$ but usually is closer to the value of $\overline{M}_w$, as shown in Figure 21.7. $\overline{M}_v$ lies higher on the mass distribution curve because the more massive molecules have proportionally a greater effect on increasing the solution's viscosity (slowing the solution's flow rate) than do the smaller polymers in the sample.

A significant amount of useful information is obtained by measuring the actual mass distribution and not just an average molecular mass. To determine the mass distribution, the molecules with different masses must be separated from each other. This is now usually done by gel-permeation liquid chromatography, as described in Chapter 14. The importance of this method can be most clearly seen when used with commercial polymers. These may contain various additives together with the major macromolecular species. These additives may be small molecules used as antioxidants to stabilize the organic groups against oxidation, plasticizers to change the flexibility of the polymeric material, and/or bacteriocides or fungicides to inhibit attack by those agents. An example of such a GPC result is shown in Figure 21.8. Including these small molecules in a single measurement of $\overline{M}_n$ and $\overline{M}_w$ significantly changes the values, as shown in the following example.

---

**EXAMPLE**

20% *by weight* of monodisperse molecules with mass 500 were added to a monodisperse sample of molecules with mass 100,000. Answer the following questions.

a. What is the percentage by number of the m.w.-500 molecules in the final sample?
b. What is the value of $\overline{M}_n$?
c. What is the value of $\overline{M}_w$?
d. What is the value of the polydispersity index?

**SOLUTION**

a. Assume that the mass of the 100,000-molecular-mass sample is 1 g exactly. (Any other weight would suffice.) Adding 20% by weight to this produces a sample weighing 1.20 g. From the data given,

$$N_{500} = \frac{0.2 \text{ g}}{500} = 4 \times 10^{-4} \text{ mol}$$

and

$$N_{100k} = \frac{1.0 \text{ g}}{10^5} = 10^{-5} \text{ mol}$$

Thus, the number percentage of the m.w.-500 molecules is given by

$$\frac{N_{500}}{N_{100k} + N_{500}} \times 100 = \frac{4 \times 10^{-4}}{4.1 \times 10^{-4}} \times 100 = 97.6\%$$

The sample is predominently composed of m.w. 500 *by number* even though the mixture is only 16.7% (w/w) of m.w. 500. [16.7% = (0.20/1.20) × 100.]

b. From the definition of $\overline{M}_n$ is Eq. 21-2 and assuming the same 1.20-g total polymer mixture,

$$\overline{M}_n = \frac{N_{500} \cdot 500 + N_{100k} \cdot 10^5}{N_{500} + N_{100k}}$$

$$= \frac{0.20 \text{ g} + 1.00 \text{ g}}{4.1 \times 10^{-4}} = 2900 \qquad \text{(zeros place holders only)}$$

The number-average molecular weight is changed from 100,000 to below 3000 by adding 20% by weight of the smaller molecular species.

c. The weight-average molecular weight is calculated as described by Eq. 21-3.

$$\overline{M}_w = \frac{0.20 \text{ g} \cdot 500 + 1.0 \text{ g} \cdot 100,000}{1.20 \text{ g}} = 83,400$$

$\overline{M}_w$ is relatively less sensitive than $\overline{M}_n$ to the addition of lower molecular weight species.

d. Finally, the polydispersity index is the ratio

$$\frac{\overline{M}_w}{\overline{M}_n} = \frac{83,400}{2900} = 29$$

compared with a polydispersity index of 1.0 before adding the m.w.-500 fraction.

## 21.5 Overview of the Methods for Measuring Average Molecular Masses

To measure the number-average molecular mass of a sample, somehow the number of molecules must be measured. The commonly used analytical methods all require that the sample be dissolved in a solution. Then, the **colligative properties** of the solvent–solute system can be used to determine the number of solute molecules. You will recall that the colligative properties include melting point lowering and boiling point elevation. These two properties are not often measured because both are relatively insensitive when molecular masses are above a few thousand.

Measurements relying on two other colligative properties, however, can be used; these are the measurement of osmotic pressure and of solvent vapor-pressure lowering. The osmotic pressure is the pressure that is thermodynamically equivalent to a difference in the concentrations of solute of two solutions. This pressure can be measured by **osmometry** in a number of ways. One is described below. The change in the vapor pressure of a solvent as it varies with polymer concentration is the basis for the method of **vapor-pressure osmometry.** The details of this method are described in Section 21.7.

Another method can be used for determining the average value of absolute molecular masses: **light scattering.** In essence, the scattering of light is due to the same process that causes the sky to be blue and sunsets to be red, *Rayleigh scattering.* Mass determinations by light scattering are generally more difficult to do than by the other methods mentioned here. As a result, light scattering is used more as a research tool than for daily analysis. It will be described briefly in section 21.10.

Both the osmometry and light-scattering methods are absolute methods. Osmometry produces values of number-average molecular weights $\overline{M}_n$. Light scattering provides the weight-average molecular weights $\overline{M}_w$.

Viscometry—measuring the viscosity of dilute polymer solutions—is a relative method requiring well-defined standards with which to calibrate the results. It is quicker and requires much simpler apparatus than the absolute methods. This technique is described in more detail in Section 21.8.

### Extrapolation to Zero Concentration

Macromolecules in solution have a strong tendency to interact with each other. The particles act as if they spend part of the time stuck to others, and the interactions tend to cause the measured masses to *appear* to be larger since there are fewer particles in the solution. Therefore, measuring the colligative properties of a weighed polymer sample dissolved in a known volume of solution can produce a highly imprecise result for the molecular mass. However, the amount of interaction is reduced as the solutions are made more dilute, and the polymers are, on average, separated further apart. As you might expect, the polymer–polymer interactions will tend to vanish when the concentration approaches zero. On the other hand, we cannot measure any colligative properties of solutions with zero concentration of the solute.

There is a way to resolve this problem. We can extrapolate the results to find the effect (such as osmotic pressure) at zero concentration. The extrapolation is made from measurements for a series of solutions at different concentrations. To explain this in more detail, let us use the example of osmometry. The same general extrapolation method is used for viscometry, vapor-pressure osmometry, and light scattering.

## 21.6 Osmotic Pressure

The osmotic pressure generated by a solution can be determined with an instrument such as that illustrated in Figure 21.9. A number of other types of osmometers can be used with just as much precision.

The osmotic pressure produced by an ideal solution is described by the equation

$$\pi = \frac{n}{V} RT \qquad\qquad (21\text{-}4)$$

$\pi$ is the osmotic pressure; $n$ is the number of moles of solute dissolved in $V$, the volume of solution; $R$ is the gas constant; and $T$ is the temperature in K. The units of

**Figure 21.9**   Diagram of one form of membrane osmometer. The principle of operation of this membrane osmometer is to adjust the hydrostatic pressure of the solvent to equal the osmotic pressure of the sample. When the equivalent pressure is reached, there is no solvent flow across the semipermeable membrane. The pressure difference is registered to a precision of 0.01 cm of solvent. If the solvent is water, 0.01-cm pressure equals 0.007 mm Hg or about $10^{-5}$ atm. The instrument operates in the following way. The solvent and sample compartments are filled with solvent and polymer solutions, respectively. Liquid flows into the sample compartment due to osmosis. The volume increase causes the diaphragm to move. Motion of this thin metal diaphragm is detected by an electronic circuit (a capacitance sensor). The circuit initiates a hydrostatic pressure change in the solvent compartment by moving the plunger in the tank. This adjustment continues until the volume ceases changing. The hydrostatic pressure equivalent to the osmotic pressure is then read from the counter. The system can detect less than a 1-nL volume flow. [Reprinted with permission from Rolfson, F. B., *Anal. Chem.* **1963,** *35,* 1303, and Rolfson, F. B.; Coll, H., *Anal. Chem.* **1964,** *36,* 888. Copyright 1963 and 1964 American Chemical Society.]

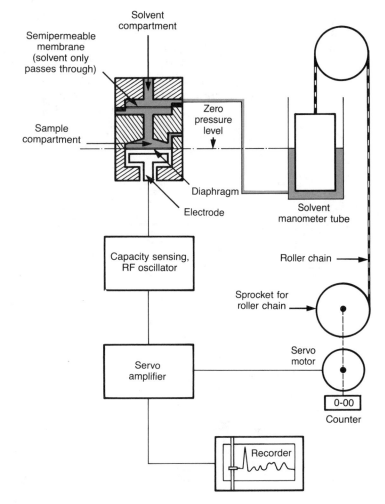

$R$ must be correct for the units chosen for $n$, $V$, and $\pi$. For instance, if $V$ is in liters, $\pi$ in atmospheres, and $n$ in moles, then one must use the value 0.0821 L atm K$^{-1}$ mol$^{-1}$ for $R$.

A few changes must be made for this equation to describe the results for polymeric solutions. Eventually, we want to relate the osmotic pressure to $\overline{M}$. Also, we must account for the nonideality of the solution. Thermodynamic arguments suggest that a term be added to the equation. This is the term that includes the parameter $A_2$ in Eq. 21-5 below. Equation 21-5 is called a **virial equation,** and $A_2$ is called the **second virial coefficient.** One other factor to take into account is that concentrations of polymers are usually expressed as g/cm$^3$ of solution. Thus, in place of $n/V$ in Eq. 21-4 we substitute the factors $\overline{M}_n$ and a concentration $c$, and include the concentration dependence of the effect, Eq. 21-4 becomes

$$\pi = \frac{c}{\overline{M}_n} \cdot RT + RTA_2c^2 \qquad \text{(valid in the low-concentration limit)} \quad \textbf{(21-5)}$$

Note that the first term is similar to Eq. 21-4. The second term, which is proportional to $c^2$, accounts for the interaction between molecules.

Clearly, as the concentration of polymer approaches zero, the osmotic pressure will also approach zero. But the osmotic pressure for an infinitely dilute solution is not itself the value we seek. We want to measure the osmotic pressure due to a certain nonzero concentration of polymer *without polymer–polymer interactions.* The way this can be done is suggested by rewriting Eq. 21-5 by dividing both sides by the factor $cRT$. We obtain

$$\frac{\pi}{cRT} = \frac{1}{\overline{M}_n} + A_2c \qquad \textbf{(21-6)}$$

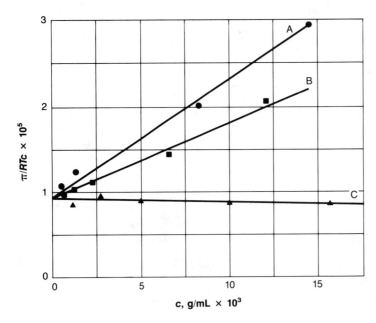

**Figure 21.10** Plots for extrapolating the data of osmotic pressure measurements to zero polymer concentration. $\pi/RTc$ versus $c$ for nitrocellulose in three solvents: Curve A, acetone; curve B, methanol; and curve C, nitrobenzene. All three sets of data extrapolate to the same value of $1/\overline{M}_n$ at zero concentration. The different slopes of the lines illustrate that the second virial coefficient, $A_2$, is different in each solvent. [Ref: Gee, G., *Trans. Farad. Soc.* **1944,** *40,* 261–266, from data of Dobry, 1935.]

If the value of $\pi/cRT$ is plotted versus $c$, a straight line is obtained. The intercept at $c = 0$ is the reciprocal of $\overline{M}_n$, and $\overline{M}_n$ is the value we seek. Such a plot is shown in Figure 21.10.

Incidentally, from plots such as that in Figure 21.10, an additional piece of information is obtained about the physical chemistry of the polymers under the specific experimental conditions. The slope of the line equals $A_2$; this quantity is related to the strength of the polymer–polymer interactions.

## 21.7 Vapor-Pressure Osmometry

Unlike a solution osmometer, a vapor-pressure osmometer requires no membrane. In regular osmometry, membrane properties are crucial and can be highly variable. Vapor-pressure osmometry thus eliminates one of the headaches (always a good reason to choose one method over another). The drawback to vapor-pressure osmometry is that its sensitivity is lower. Thus, it is more popular when $\overline{M}_n$ is less than about 15,000–20,000. But 15,000 is near the useful lower limit of regular osmometry. In the end, then, the two techniques are complementary rather than competitive.

The heart of a typical vapor-pressure osmometer is shown in Figure 21.11. The apparatus is kept in an oven with a precisely stabilized temperature ($\pm 0.001$ K or so). The atmosphere in the oven is saturated with the solvent vapor, and the temperature is below the boiling point of the solvent. In this solvent-saturated atmosphere, there are two thermistors (resistors for which the resistance is highly temperature-dependent), which are thermally insulated. The thermistors are part of a bridge circuit. Thus a signal directly proportional to their temperature difference $\delta T$ can be precisely measured. The actual temperature difference is not needed.

The temperature difference between the pure solvent and the solution arises in the following way. The vapor pressure of a solvent at a temperature $T$ is lowered when a nonvolatile solute is dissolved in it. This vapor-pressure lowering in the low-concentration range used is described by Raoult's law of a solution's vapor pressure as a function of a solute's mole fraction $X_i$.

$$p_{\text{solvent}} = X_{\text{solvent}} p^{\bullet}_{\text{solvent}} \tag{21-7}$$

$p^{\bullet}_{\text{solvent}}$ is the vapor pressure of the pure solvent.

On the molecular level, this equation means that the equilibrium between vaporizing and condensing solvent is shifted. This shift is represented in Figure 21.11 by the lengths of the arrows near the droplets on the thermistors. There is more vapor condensing on the thermistor holding the sample than on the thermistor with pure solvent. But for every mole of excess condensation, heat amounting to $-\Delta H_{\text{vaporization}}$ is added to the droplet. This raises the temperature an amount depending on the heat capacity of the solution.

The excess of condensation of the solvent is linear with solute concentration. This is so for two reasons.

A. The mole fraction of a solute is a linear function of concentration at low-solute levels.

B. The evaporation is slowed in proportion to the solute concentration.

Therefore, we can write the change in temperature as

$$\delta T = kc_{\text{solute}} \qquad\qquad \textbf{(21-8)}$$

where $k$, a constant, is determined by experiment. The value of the constant is found by calibrating with a solution of known concentration of a pure standard of known

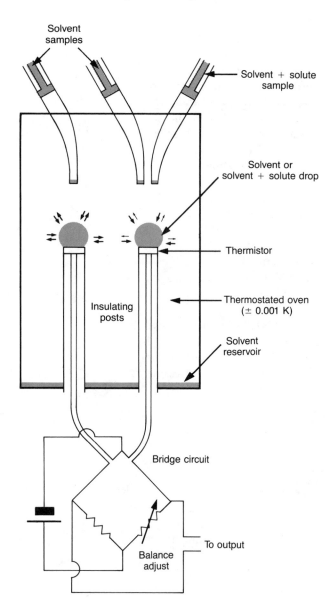

**Figure 21.11** Illustration of the operation of a vapor-pressure osmometer. Inside a highly stable oven are two thermistors mounted on insulating posts. (Thermistors are electrical resistors in which the resistance changes significantly with temperature changes.) The thermistors act as two resistors of a Wheatstone bridge circuit. The bridge is adjusted for balance after drops of identical solvent are dropped onto the two posts. After balancing, the solvent is dropped on one thermistor and the polymer solution on the other. A signal is produced that is proportional to the concentration of the polymer. The evaporation rate of the polymer solution is slower than that of the pure solvent. Thus there is an excess condensation on the polymer solution side with a concomitant heating from the extra heat of condensation there. The different lengths of the arrows around the drops indicate the relative rates of evaporation and condensation in the vapor-saturated atmosphere.

molecular mass (say 1000). The standard need not be the same compound type as the sample. With a standard, the absolute value of $\delta T$ is not needed, and the concentration can be found from the calibrated machine response alone.

In analogy with membrane osmometry determinations, we eventually seek to find the change in the signal $\delta S$ for a polymer solution that is nonideal. We thus must extrapolate a number of runs of varying concentrations to zero concentration as before. In this case,

$$\frac{\delta S}{kc} = \frac{1}{\overline{M}_n} + A_2 c \tag{21-9}$$

which should be compared with Eq. 21-6.

In operation, preheated solvent drops are injected into the oven to drop and settle on both thermistors, and the null response is set. Then a drop of the polymer solution is dropped in place of the pure solvent on one side. The calibrated response is used to determine the concentration. Together with the mass of the polymer sample and the volume of the solution, the average molecular mass $\overline{M}_n$ can be calculated.

## 21.8 $\overline{M}_v$ by Viscometry

Viscosity is a measure of resistance to flow and has the units of poise (P). At 20 °C, water has a viscosity of about 0.01 P = 1 cP (centipoise) while the value for mineral oil is about 10 P = 1000 cP. Viscosity is represented by the Greek letter $\eta$. Solutions of polymers have higher viscosities than the solvent alone at the same temperature. This change in viscosity can be used to determine the average polymer molecular weight.

For polymer solutions, three measures are commonly used in connection with viscosity. They are defined below along with their symbols. In all the equations that follow,

$c$ = concentration in g solute/100 mL solution,
$\eta$ = viscosity of the polymer solution, and
$\eta_0$ = viscosity of the solvent alone.

The three definitions are

$$\eta_r = \frac{\eta}{\eta_0} = \text{relative viscosity}$$

$$\frac{\eta - \eta_0}{\eta_0} = \text{specific viscosity} = \eta_{sp} \tag{21-10}$$

$$\lim_{c \to 0} \left( \frac{\eta_{sp}}{c} \right) = \lim_{c \to 0} \left\{ \frac{1}{c} \ln \left( \frac{\eta}{\eta_0} \right) \right\} = \text{intrinsic viscosity} = [\eta]$$

$\eta_r$ and $\eta_s$ are dimensionless ratios. The intrinsic viscosity has units of (concentration)$^{-1}$.

**Table 21.3**  Index Values for Selected Polymer Solutions

| Polymer and Conditions | $K$ | $a$ |
|---|---|---|
| Poly(methyl methacrylate) in toluene at 25 °C | $8.12 \times 10^{-5}$ | 0.71 |
| Polystyrene in toluene at 25 °C | $1.7 \times 10^{-4}$ | 0.69 |
| Poly(ethylene oxide) in water at 30 °C | $12.5 \times 10^{-5}$ | 0.78 |

It has been shown that the following relationship *describes* (not explains) the relationship between the intrinsic viscosity and the molecular weight for *linear* polymers. It is called the Mark–Houwink equation:

$$[\eta] = K\overline{M}_v^a \qquad \qquad (21\text{-}11)$$

The constants $K$ and $a$ depend on the identity of the polymer, the specific solvent, and the temperature of the solution during the experimental determination. A few examples of both $K$ and $a$ are listed in Table 21.3. As introduced in Figure 21.7, $\overline{M}_v$ is the viscometrically determined average molecular mass. The exponent is found to have values between 0.5 and 2.

The values of $K$ and $a$ are determined for each system by measuring the intrinsic viscosities (Eq. 21-10) of polymer standards. The standards consist of a series of molecular weight fractions of the polymer of interest. Each fraction has a narrow molecular weight distribution, the average of which has been determined by an absolute method such as light scattering or measurements of osmotic pressure.

As you may have already noted from Eq. 21-10, the intrinsic viscosity $[\eta]$ is found by extrapolating to zero polymer concentration. As seen from the equation, there are two different ways to carry out the extrapolation, through either the specific viscosity or the logarithm of the relative viscosity. Both can be used.

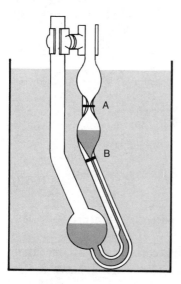

**Figure 21.12**  A Cannon–Fenske viscometer, which is about 15 cm long, in a constant-temperature bath. A fixed volume of solution is introduced into the large bulb on the lower left. The polymer solution is drawn up to a point above line A and then is allowed to flow by gravity through the capillary tube. The time required for a miniscus to pass from line A to line B etched on the glass is proportional to the viscosity of the solution. Each viscometer must be individually calibrated.

**Figure 21.13** Two different methods to extrapolate viscosity data to find the intrinsic viscosity are shown. The data are for a sample of polystyrene in benzene. The two different extrapolations help ensure precision in the determination. [Ref: Ewart, R. H. In *Adv. Coll. Sci.*, Vol II; H. Mark and G. S. Whitby, Eds.; Interscience Pub.: New York, 1946; 197–251.]

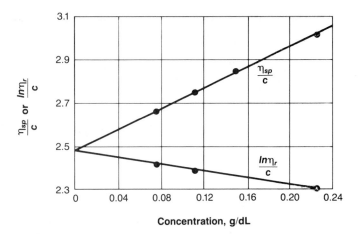

In practice, for relatively dilute solutions (<1% w/v), a viscometric assay consists of a series of measurements of the time required for a fixed volume of polymer solution to flow through a capillary tube. One type of simple viscometer that can be used is shown in Figure 21.12. At a fixed temperature, the following relationship holds for the times of flow. In the equation, $t_0$ refers to the flow time of the pure solvent.

$$\frac{\eta}{\eta_0} = \frac{t}{t_0} = \eta_r \qquad (21\text{-}12)$$

A series of runs is done for a set of solutions with different concentrations, and the value of $[\eta]$ is found through the two extrapolations in a graph versus concentration. An illustration of this procedure is shown in Figure 21.13. The extrapolated value $[\eta]$ is then used with Eq. 21-11 to determine the molecular mass $\overline{M}_v$.

## 21.9 Molecular Mass Distribution with Calibrated Gel-permeation Chromatography

As described in Section 14.2, gel-permeation liquid chromatography can be used to determine the molecular masses of eluting polymers. The separation depends on the molecular *sizes* of the polymers; on a logarithmic scale, the calibration is logarithmic in molecular mass over a useful range. (There can be exceptions to this generalization.)

The properties of a gel-permeation separation can be utilized to determine $\overline{M}_n$ and $\overline{M}_w$ of a polydisperse polymer sample. What is required is a detector that responds to the weight of polymer solute (weight/unit volume) followed by some straightforward algebra. Two types of detectors that satisfy the requirement are those that measure differential refractive index (RI) and optical absorbance. A requirement of the latter is that the absorbance due to each monomer unit does not depend on the polymer molecular mass or concentration. This property is commonly fulfilled.

To calculate the average molecular masses from the chromatographic data, the equations defining $\overline{M}_n$ and $\overline{M}_w$ can be recast in a more useful form. Let the magnitude of the detector response be $h_x$ for the fraction $x$ of the polymer. The letter $h$ is short for height above the baseline on a chromatogram chart. For a linear-response detector, $h_x$ is proportional to the mass,

$$h_x = \text{constant} \cdot W_x \tag{21-13}$$

Substituting Eq. 21-13 into Eq. 21-3, which defines $\overline{M}_w$, we obtain

$$\overline{M}_w = \frac{\sum\limits_x h_x M_x}{\sum\limits_x h_x} \tag{21-14}$$

This equation applies when the volume fractions are all equal.[1]

$M_x$ is the nominal (the median or mean) molecular mass associated with each fraction. Since the total mass in a volume fraction equals the product of the number of moles and the molecular mass,

$$W_x = N_x M_x \tag{21-15}$$

then Eqs. 21-13 and 21-15 together give

$$h_x = \text{constant} \cdot N_x M_x$$

Substitution of this expression for $h_x$ into Eq. 21-2, which defines the number-average molecular weight $\overline{M}_n$, we find

$$\overline{M}_n = \frac{\sum\limits_x h_x}{\sum\limits_x (h_x/M_x)} \tag{21-16}$$

Calculating $\overline{M}_n$ and $\overline{M}_w$ from the data then rests on delineating the equal-volume fractions of the continuous distribution of masses.

The results will be more precise as more divisions are taken, each with a narrower range of mass. For computer calculation, this is no problem, but for a hand calculation let us limit the number of divisions.

---

[1] This can be seen in the integral form of the equation. If the variables are labeled by the eluent volume as subscript $V$ ($M_V$ should not be confused with $\overline{M}_v$), then

$$\overline{M}_w = \frac{\displaystyle\int_{V_1}^{V_2} h_V M_V \, dV}{\displaystyle\int_{V_1}^{V_2} h_V \, dV}$$

The limits of integration include the entire elution volume of the polymer. The finite sum of equal volumes (Eq. 21-14) has the $\Delta V$-terms divided out.

## EXAMPLE

The following data were found from a GPC of polystyrene in tetrahydrofuran at 25 °C. The calibration was with narrow-mass-distribution polystyrene standards. The detector was a differential refractive index instrument which responds to the total mass of solute, that is, $N_x M_x$. Find $M_n$, $M_w$, and the polydispersity index.

| Retention Volume* (mL) | Height at Volume (Arbitrary Linear Scale) | m.w. at Volume |
|---|---|---|
| 70 | 0.8 | 98 000 |
| 75 | 10.0 | 55 000 |
| 80 | 42.9 | 31 000 |
| 85 | 74.3 | 18 500 |
| 90 | 66.8 | 10 000 |
| 95 | 33.6 | 5 700 |
| 100 | 9.7 | 3 250 |
| 105 | 2.0 | 1 800 |
| | Sum   240.1 | |

\* Defined in Chapter 13.

## SOLUTION

The calculation can be simplified somewhat if the value of $h_x$ is normalized, that is, $\sum_x h_x = 1.00$. As a result, Eq. 21-14 becomes

$$\overline{M}_w = \frac{\sum_x h_x M_x}{1}$$

and Eq. 21-16 becomes

$$\overline{M}_n = \frac{1}{\sum_x (h_x/M_x)}$$

The normalization of $h$ is done on the data here by dividing each of the given $h_x$ values by 240.1, the sum of the heights of the equal-volume fractions. The table can be recast as below. Terms for Eqs. 21-14 and 21-16 are at the right.

| $M_x$ | Normalized Height | $h_x M_x$ | $h_x/M_x \times 10^5$ |
|---|---|---|---|
| 1 800 | 0.0083 | 15 | 0.461 |
| 3 250 | 0.0404 | 131 | 1.243 |
| 5 700 | 0.1399 | 797 | 2.454 |
| 10 000 | 0.2782 | 2 782 | 2.782 |
| 18 500 | 0.3094 | 5 724 | 1.672 |
| 31 000 | 0.1786 | 5 537 | 0.576 |
| 55 000 | 0.0416 | 2 288 | 0.075 |
| 98 000 | 0.0033 | 323 | 0.003 |
| Sum | 0.9997 | 17 597 | 9.266 |

Thus $\overline{M}_w = 17{,}600$. $\overline{M}_n = 1/(9.266 \times 10^{-5}) = 10{,}800$.

The polydispersity index is

$$\frac{\overline{M}_w}{\overline{M}_n} = \frac{17,600}{10,800} = 1.63$$

## 21.10 Light Scattering and Gel-permeation Chromatography

Rayleigh light scattering by polymer solutions can be used to determine $\overline{M}_w$, the mass-average molecular mass of the polymers in solution. The solution must be completely free of materials other than the solvent and polymer solute—a speck of dust in the light beam can ruin the results. A series of experiments used to determine a single value of $\overline{M}_w$ are quite time consuming. However, in conjunction with gel-permeation chromatography coupled with a differential refractive-index detector, a simpler sort of light-scattering experiment can be used to determine the values of $\overline{M}_n$ and $\overline{M}_w$ without calibration runs being needed. This technique and the mathematical relationships are described in this section.

The experimental arrangement for light scattering is illustrated in Figure 21.14. It is similar to that used for fluorescence. In the usual fluorescence spectrometer, the intensity of the emitted light is measured at an angle of 90° from the illumination direction. In contrast, the Rayleigh scattering of polymers can be determined with a number of values of the scattering angle $\theta$ (defined in the figure). When the emission is measured close to the transmitted beam, this is called **low-angle light scattering.**

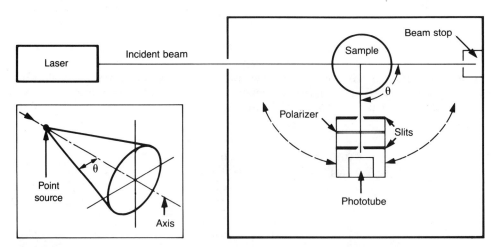

**Figure 21.14** Diagram of a simple light-scattering apparatus. The light is scattered from the sample and measured at a number of different values of the angle $\theta$ by moving the detector. The laser light is polarized. The polarizer associated with the detector is used to select polarized or depolarized scattered light. The polarized scattered light is used to determine macromolecular weights. The incident light power can be determined by setting the detector at $\theta = 0°$ with an attenuator in the beam. The enclosure and beam stop are designed to minimize stray scattered light. The cone geometry of light scattered at angle $\theta$ is illustrated in the inset.

In this case, the equations that describe the relationship of the intensity of the scattered light to $\overline{M}_w$ can be written as follows. Call the power of the incoming radiation $P(0)$. Call $P(\theta)$ the total power of the radiation emitted from the sample in a cone-shaped beam with the angle between the cone and axis being $\theta$. (The sample is assumed to be a point source.) See Figure 21.14. A number called the Rayleigh ratio is defined as

$$R(\theta) = r^2 \frac{P(\theta)}{P(0)}$$

The term $r^2$ compensates for the fact that a detector subtending a fixed *solid* angle will intercept a greater fraction of the scattered light if placed closer to the sample.

Of interest in a polymer-solution scattering experiment is the scattering due to the polymer solute alone. This is written algebraically as simply

$$\overline{R}(\theta) = R(\theta)_{\text{solution}} - R(\theta)_{\text{solvent}} \qquad \textbf{(21-17)}$$

With these definitions, you can understand the equation that describes the relationship between the low-angle light scattering and the value of $\overline{M}_w$,

$$\frac{Kc}{\overline{R}(\theta)} = \frac{1}{\overline{M}_w} + 2A_2c \qquad \textbf{(21-18)}$$

You should recognize that this equation has the same general form as those relating osmotic pressure or viscosity to the value of $\overline{M}_n$, Eqs. 21-5 and 21-9.

$K$ is a constant that depends on $n$, the solvent's index of refraction, $\lambda$, the wavelength of the light that is scattered, and $N$, Avogadro's number:

$$K = \frac{2\,\pi^2 n^2\,(dn/dc)^2}{N\lambda^4} \qquad \textbf{(21-19)}$$

$(dn/dc)$ is the change in the solution's index of refraction with concentration of the polymer. This can be measured easily using a refractometer[2]. The constant $K$ depends on the specific polymer, the solvent that is used (at a specific temperature and in the absence of concentration gradients), and $\lambda$, the wavelength of the scattered light. $K$ can be found from experiments run on the polymer solution. Incidentally, Eq. 21-19 is called the Debye equation.

In practice, the effluent from a GPC-column separation of about 1 mg of polymer is passed through both a differential refractive-index detector (output proportional to solute mass) and a low-angle light-scattering instrument. Since $K$ is known, $c$ is measured from the response of the refractive-index detector, and $\overline{R}(\theta)$ is measured by the light-scattering apparatus, a computer can take the data from both detectors and

---

[2] A refractometer is used to measure the refractive index (defined in Eq. 17G-5) of a liquid sample. The most widely used instrument is the Abbé Refractometer. Various interferometric methods can also be used with even greater accuracy. Further information may be found in Lewin, S. Z.; Bauer, N. in "Treatise on Analytical Chemistry," I. M. Kolthoff and P. J. Elving, Eds.; Interscience: New York, 1965, Part I, Vol. 6, Chap. 70.

calculate the value of the absolute molecular mass at each point of the chromatogram. The absolute molecular mass can be calculated, since for each separated monodisperse polymer fraction passing through the detectors,

$$\overline{M}_w = \overline{M}_n = M_x$$

## 21.11 Chemical Analysis

Probably the most important information about a polymer is the various average molecular masses and the mass distribution. However, the chemical composition (the monomeric units), topology (linear/branched), and stereochemistry (isomers

**Styrene**
$C_6H_5CH{:}CH_2$   M.W.   104.15

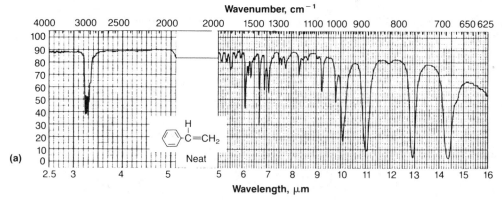

(a)

**Poly(styrene)**, secondary standard
Typical M.W. 321,000. Typical M.N. 85,000

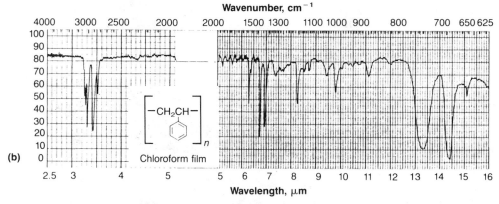

(b)

**Figure 21.15**   Infrared spectra are sensitive to structural changes upon polymerization. Contrast the spectra for (a) monomeric styrene and (b) polystyrene. One especially good way to compare these is to hold the page just below your eyes so you can peer nearly parallel to the page. Note the extra C—H-stretch band that appears and the significant changes in the intensities in the 900–1000 cm⁻¹ region as well as the shifts in positions in the lower wavenumber region.

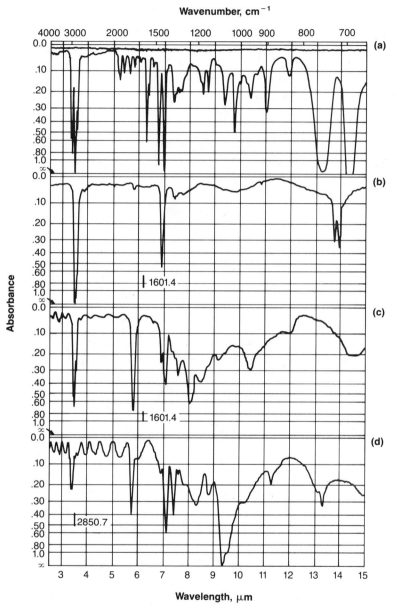

**Figure 21.16**  Polymers exhibit significantly different infrared spectra depending on their compositions. Here are spectra from four common commercial polymers. They are (with their origins) (a) baseline and polystyrene (an envelope window); (b) polyethylene (Handiwrap®); (c) polyvinyl chloride (Stretch 'n Seal®); and (d) polyvinyl chloride/polyvinylidene chloride copolymer (Saran Wrap®). [Ref: Lowry, S. R.; Gallaher, T. N.; Leary, J. J. *J. Chem. Educ.* **1975**, *52*, 684–685.]

and/or tacticity) are also useful and sometimes essential to understand the polymer properties. Determining all this information can be approached with spectrometric methods and especially with vibrational and NMR spectrometries.

## Chemical Composition

As you learned in the previous chapters on vibrational and NMR spectrometries, the spectral properties that are seen depend on the chemical structures of molecules or

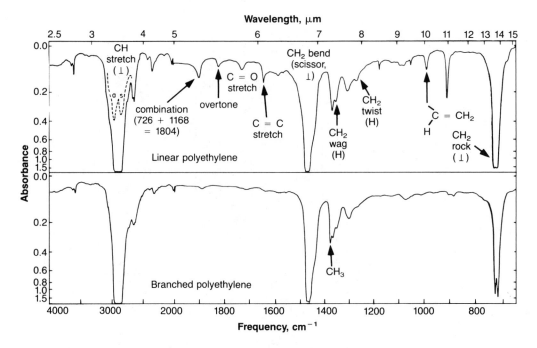

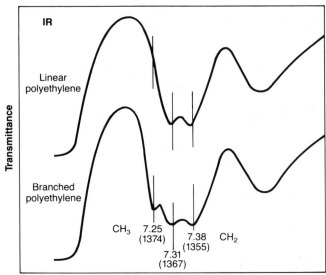

**Figure 21.17** Infrared spectra of polymers can be used to observe the amount of branching in the chain. The example here is for polyethylene. The full spectrum and an enlargement of the region around 1350 cm⁻¹ (for a melted sample at 140 °C) are shown. The relative intensities of the bands can be used to approximate the percentage of carbons with side chains attached. [Refs. Whole spectrum: Bovey, F. A.; Kwei, T. K. In "Macromolecules"; F. A. Bovey and F. H. Winslow, Eds.; Academic Press: New York, 1979; p. 214 (Fig 3.8). Partial spectrum, reprinted with permission: Bryant, W. M. D.; Voter, R. C. *J. Amer. Chem. Soc.* **1953**, *75*, 6113–6118; copyright 1953 American Chemical Society.]

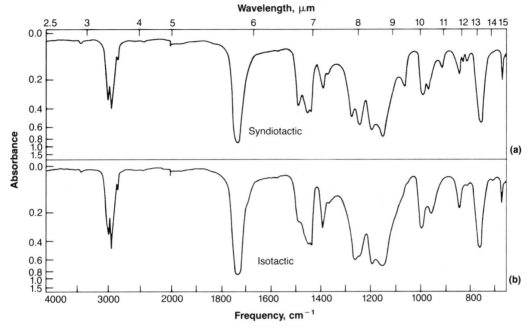

**Figure 21.18**    Infrared spectra are sensitive to polymer tacticity. Illustrated here are the infrared spectra of (a) predominantly syndiotactic and (b) predominantly isotactic films of polymethyl methacrylate. The major difference can be seen in the band between 1000 and 1100 $cm^{-1}$ (arrow). NMR spectra are more sensitive to tacticity. However, in some instances an IR spectrum is easier to obtain. [Ref: Bovey, F. A.; Kwei, T. K. In "Macromolecules"; F. A. Bovey and F. H. Winslow, Eds.; Academic Press: New York, 1979.]

chemical groups of molecules. Therefore, when monomers are polymerized, we expect the IR and NMR spectra to reflect the change in structure. But the polymer spectra can also be similar to those of the monomers when the spectral features arise from unchanged parts of the monomer. For example, as can be seen in the IR spectra of styrene and polystyrene shown in Figure 21.15, the regions that are changed and the regions that remain quite similar are obvious. Since infrared spectral "finger-prints" are sensitive to monomer structure, you might expect them to be sensitive to polymer composition. In Figure 21.16 you can see from the spectra how highly sensitive the IR spectra are. Thus, infrared spectra are useful to determine the identity of the polymer and the component monomer(s). The procedure for identification can be as simple as heating a sample of the polymer, pressing it into a thin film, running the spectrum of the film, and comparing the spectrum with a catalog of spectra.

If the polymer is new or unknown, the infrared spectrum can be used to determine if it contains esters (as in polyesters) or amides (as in polyamides), and so on. But even more information on the structure can be obtained. As an example, consider polyethylene. There are at least three types of carbon–hydrogen groups in polyethylene: methyl —$CH_3$; methylene —$CH_2$—; and tertiary carbons —CH—. The methyl groups appear at the ends of a chain and its branches, the methylenes in the middle of

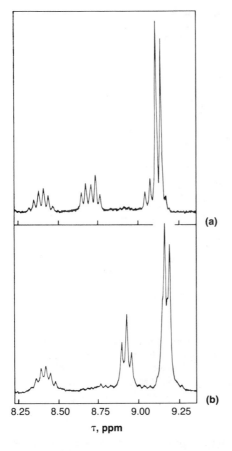

**Figure 21.19**    220-MHz $^1$H-NMR spectra of (a) isotactic and (b) syndiotactic polypropylene in *o*-dichlorobenzene at 165 °C. $^{13}$C-NMR spectra are also sensitive to tacticity, as shown in Figure 21.21. [Ref: Bovey, F. A. "Polymer Conformation and Configuration"; Academic Press: New York, 1969.]

a chain, and the tertiary carbons at the connecting points of chain branches. Each of these groups produces spectral bands that appear at different wavelengths in both vibrational spectrometries and NMR. In other words, the polymer spectra can be used to determine the microstructure of the macromolecules.

### Topology and Stereochemistry

In a perfectly linear polyethylene polymer, there are only two methyl groups, those at the ends of the long backbone chain. The ratio $—CH_2—/—CH_3$ can be many thousands to one. The ratio is intimately related to the value of $M_n$, the number-average molecular mass. If the chain is perfectly linear, and if we can quantitate the ratio $—CH_2—/—CH_3$, then we can determine the molecular mass. This is called an **end-group determination** of the molecular mass. In the best possible case, the range of the spectral-measuring methods generally limits the method to molecular masses less than 25,000 (when the middle groups are about 12,500 times as concentrated as the two end groups). The presence of any side chains will confuse the determination.

On the other hand, end-group determinations can be helpful if the average number of side chains is being determined. Figure 21.17 illustrates the kind of change that can

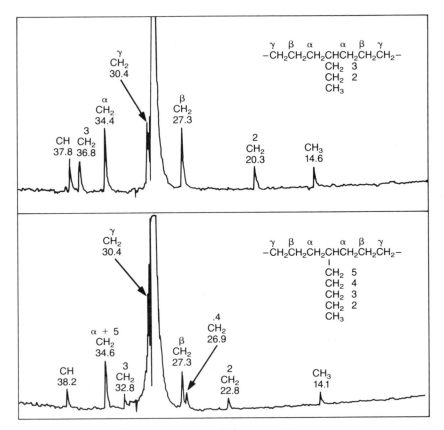

**Figure 21.20**    $^{13}$C-NMR can be used to investigate the lengths of side chains since many of the individual carbons can be resolved in the spectra. When possible, the carbons at specific positions of the backbone and side chain are assigned. Note the near overlap of the β and 4-position resonances in the lower spectrum. [Reprinted with permission from Woodward, A. E., and Bovey, F. A., Eds. "Polymer Characterization by ESR and NMR"; American Chemical Society: Washington, D.C., 1980; ACS Symposium Series #142. Copyright 1980 American Chemical Society.]

be seen in infrared spectrometry. By comparing the spectra, the differences between linear and branched polyethylene can be seen. If the baseline is accounted for, then the absorbance of the 1374-cm$^{-1}$ band reflects the methyl content of the sample. The number of side-chain ethyl groups can be determined by quantitation using bands in the 900-cm$^{-1}$ region and near 1644 cm$^{-1}$. In quality control, however, a more straightforward way to test the polymer might be simply to use the ratio of *absorbances* of two or three bands as is done for any mixture (Section 17.14). In this case, the "mixture" is composed of different carbon structures.

In addition to the chemical structure, effects of the stereochemistry can be observed in infrared and Raman spectra. Shown in Figure 21.18 is a representative set of spectra of poly(methyl methacrylate) with different tacticities. However, as shown next, NMR is normally more sensitive to these differences.

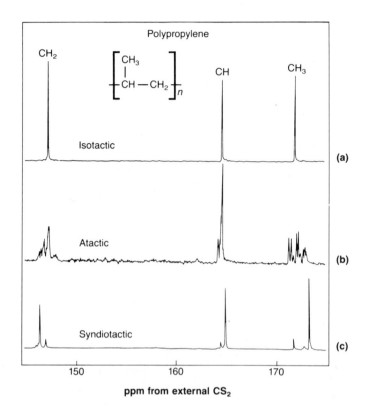

(a)

(b)

(c)

**Figure 21.21**   13C-NMR spectra can be used to investigate the tacticity of a polymer. Sample: 20% (w/v) solution of polypropylene in 1,2,4-trichloro-benzene at 140 °C. 25-MHz NMR. [Ref: Bovey, F. A.; Kwei, T. K. In "Macromolecules"; F. A. Bovey and F. H. Winslow, Eds.; Academic Press: New York, 1979.]

If the polymer can be dissolved in a solvent or melted at low enough temperatures, then NMR spectra of the type described in Chapter 19 can be measured. If the resonance positions of end groups, chains, and branch points can be resolved, then by using NMR alone, it is possible to measure average values of chain length and branching.

In addition, NMR resonance frequencies are sensitive to the structures of adjacent atomic centers. Therefore, spectra are sensitive to the tacticity. Shown in Figure 21.19 are the 1H-NMR spectra of isotactic and syndiotactic polypropylene. There are obvious differences between the spectra. Note also that each polymer has "impurities" of the other stereoisomer type.

Of even greater potential analytical use is 13C-NMR. Experimentally, carbon NMR is more difficult than proton NMR. But the spectra are significantly more sensitive to differences in the structures of the macromolecules. For example, the lengths of the chain branches can be investigated, as shown in the spectra of Figure 21.20. In addition, significant differences in the 13C-NMR spectra can depend on the tacticity, as seen in Figure 21.21. Figure 21.22 illustrates the differences that appear in 13C-NMR between *cis* and *trans* isomers in a specific example.

There is one more development of which you should be aware. Both carbon and proton NMR spectra can be obtained from solids. Thus, even high-melting materials and cross-linked polymers, which are insoluble without degradation, can be analyzed. One polymeric material that can be investigated this way is coal. The method-

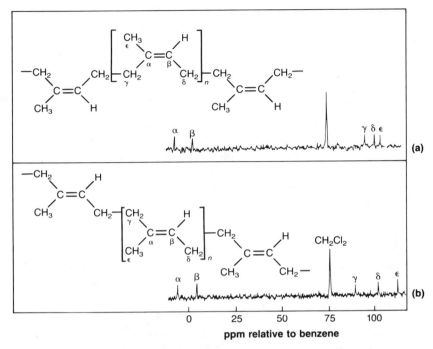

**Figure 21.22**   The differences between isomers can be investigated with ¹³C-NMR. Here can be seen the differences between *cis-* and *trans-*polyisoprene. Sample: 10% solution in CH₂Cl₂ of (a) poly(*cis*-1,4-isoprene) (natural rubber) and (b) natural poly(*trans*-1,4-isoprene)(balata). The NMR reference is benzene. [Ref: Bovey, F. A.; Kwei, T. K. In "Macromolecules"; F. A. Bovey and F. H. Winslow, Eds.; Academic Press: New York, 1979; quoting Duch and Grant, *Macromolecules* **1970**, *3*, 165.]

ology to do such experiments and interpretation of the data are well beyond the level of this text, however.

## Morphology

The way the macromolecules are packed in solid or rubbery polymers can be approached by the above spectroscopic techniques if there is some clear difference in the spectra of crystalline and amorphous regions of the polymers. However, quantitating the amount of crystalline solid in a polymer is more commonly accomplished using other methods. Among them are thermal methods such as differential scanning calorimetry, the scattering of X-rays from the solid, and measurement of small differences in the polymer density, which varies between crystalline and amorphous forms.

## Suggestions for Further Reading

A complete introduction to spectroscopy as it is used in polymer analysis. Nicely done. Klopfter, W. "Introduction to Polymer Spectroscopy"; Springer–Verlag: Berlin, 1984.

A broadly ranging reference at an intermediate level. Mostly polymer chemistry.
Billmeyer, F. W., Jr. "Textbook of Polymer Science," 3rd ed.; Wiley: New York, 1984.

A fine introduction in a paperback which is unfortunately out-of-print.
Billmeyer, F. W., Jr. "Synthetic Polymers"; Doubleday: Garden City, NY, 1972.

A laboratory coursebook with fine writing and illustrations. Introductory level.
Collins, E. A.; Bares, J.; Billmeyer, F. W., Jr. "Experiments in Polymer Science"; Wiley: New York, 1981.

A good and more advanced text on aspects of polymer molecular masses.
Billingham, N. C. "Molar Mass Measurements in Polymer Science"; Wiley: New York, 1977.

An advanced book with numerous applications cited.
Bark, L. S. and Allen, N. S., Eds. "Analysis of Polymer Systems"; Applied Science Publishers, Ltd.: London, 1982.

Mostly NMR probing of polymer structure.
Bovey, F. A. "Polymer Conformation and Configuration"; Academic Press: New York, 1969.

Intermediate level description of the thermal analysis of polymers and the information obtained.
Runt, J.; Harrison, I. R. In "Methods of Experimental Physics"; R. A. Fava, Ed.; Academic Press: New York, 1980; Chap. 9.

An advanced text on vibrational spectroscopy of polymers with numerous spectra illustrated. All aspects are covered: instrumentation, sampling, and quantitation.
Siesler, H. W.; Holland–Moritz, K. "Infrared and Raman Spectroscopy of Polymers"; Dekker: New York, 1980.

An American Chemical Society Symposium edition on GPC of polymers. The chapters are uneven in quality, with some highly specific and some more general and more useful at an introductory level.
Provder, T., Ed. "Size Exclusion Chromatography: Methodology and Characterization of Polymers and Related Materials"; American Chemical Society: Washington, D.C., 1984.

A collection of chapters by different authors on gel-permeation chromatography. They cover the details and problems of GPC. Especially recommended is Chapter 7 on precision and accuracy. This is an advanced level book.
Janca, J., Ed. "Steric Exclusion Liquid Chromatography of Polymers"; Dekker: New York, 1984.

A collection of methods and applications at an advanced level as done by specialists in thermal analysis.
Turi, E. A., Ed. "Thermal Characterization of Polymeric Materials"; Academic Press: New York, 1981.

A comprehensive handbook of experimental methods.
Rabek, J. F. "Experimental Methods in Polymer Chemistry"; Wiley: New York, 1980.

A well-known compilation of the properties of polymers including values of the variables in the Mark–Houwink equation.
Brandrup, J.; Immergut, E. H. "Polymer Handbook," 2nd ed.; Wiley: New York, 1975.

A general textbook of polymer chemistry that contains intermediate to advanced level chapters on osmometry, viscometry, and light scattering.
Hiemenz, P. C. "Polymer Chemistry"; Marcel Dekker: New York, 1984.

## Problems

21.1   The boiling point of water is raised by 0.512 °C mol$^{-1}$ (1000 g solvent)$^{-1}$, and the freezing point is lowered by 1.86 °C mol$^{-1}$ (1000 g solvent)$^{-1}$.
  a.   If 0.1 g of a 100,000 $\overline{M}$ polymer is dissolved in 100 g of water, how much will the freezing and boiling points be changed assuming an ideal solution?
  b.   Assume that a temperature change of $1 \times 10^{-4}$ °C can be detected. What would be the relative error in molecular weight determined by the freezing-point lowering described in part a?

21.2   You begin with a monodisperse sample of $M = 100,000$.
  a.   Calculate the resulting values of $\overline{M}_n$, $\overline{M}_w$, and $\overline{M}_w/\overline{M}_n$ if the sample has added to it 20% by weight $M = 10,000$ polymer.
  b.   Calculate the resulting values of $\overline{M}_n$, $\overline{M}_w$, and $\overline{M}_w/\overline{M}_n$ if the sample has added to it 20% by number $M = 1 \times 10^6$ polymer.
  c.   Calculate the resulting values of $\overline{M}_n$, $\overline{M}_w$, and $\overline{M}_w/\overline{M}_n$ if the sample has added to it 20% by weight $M = 1 \times 10^6$ polymer.

21.3   For poly(ethylene oxide) in water, the equation $[\eta] = K\overline{M}_v^a$ at 30 °C has $K = 12.5 \times 10^{-5}$ and $a = 0.78$.
The following data were recorded for an experimental viscometric determination of poly(ethylene oxide) in water at 30 °C.

| Concentration of Polymer (g/100 mL) | Triplicate Times (s) for Flow of 10 mL | | |
|---|---|---|---|
| 0 | 238.6 | 240.1 | 238.4 |
| 0.125 | 273.4 | 273.4 | 269.1 |
| 0.250 | 309.0 | 305.9 | 305.3 |
| 0.500 | 380.0 | 383.4 | 386.9 |
| 1.00 | 582.5 | 590.6 | 586.1 |

Note that the basis of the measurement is the relatively small difference between two larger numbers.
  a.   What is the value of $\overline{M}_v$? Use both extrapolation methods. (Plot the average values of the points as if they were exactly the algebraic mean-values.)
  b.   Draw the extrapolation lines leading to the highest and lowest values of $\overline{M}_v$, including points within one standard deviation of the mean-value points. Find the resulting upper and lower limit to the molecular mass value.

21.4   [Adapted from Trathnigg B.; Jorde, C. *J. Chromatog.* **1983**, *259*, 381.] You may wish to set up a short computer program to help solve this problem.
The following data points were recorded at 9.0-s intervals for a gel-permeation chromatogram of polystyrene. The eluent flow rate was 1.00 mL min$^{-1}$. Narrow molecular mass standards

produced a calibration curve that could be described by the following equation.

$$\ln M = 17.378 - 0.742\ V_e; \qquad V_e \text{ in mL, } M \text{ in a.m.u.}$$

| $t_x$ (min:s) | Output Corrected for Baseline (Arbitrary Linear Scale) |
|---|---|
| 11:28 | 2.66 |
| :37 | 7.47 |
| :46 | 17.28 |
| :55 | 31.20 |
| 12:04 | 45.00 |
| :13 | 53.22 |
| :22 | 53.13 |
| :31 | 46.34 |
| :40 | 36.66 |
| :49 | 26.37 |
| :58 | 18.49 |
| 13:07 | 12.60 |
| :16 | 8.51 |
| :25 | 5.82 |
| :34 | 3.34 |
| :43 | 2.25 |
| :52 | 1.36 |

a. What is the value of $\overline{M}_w$?
b. What is the value of $\overline{M}_n$?
c. What is the polydispersity index?

# Solubility Products

Unless noted, all values are extrapolated to ionic strength zero at 25 °C. As the ionic strength approaches 0, molar and activity $K_{sp}$-values are equal.

| Formula | $pK_{sp}$ | $K_{sp}$ | Temperature (°C) | Ionic strength (M) |
|---|---|---|---|---|
| *Bromides* | | | | |
| AgBr | 12.30 | $5.0 \times 10^{-13}$ | | |
| CuBr | 8.3 | $5 \times 10^{-9}$ | | |
| HgBr$_2$ | 18.9 | $1.3 \times 10^{-19}$ | | 0.5 |
| Hg$_2$Br$_2$* | 22.25 | $5.6 \times 10^{-23}$ | | |
| TlBr | 5.44 | $3.6 \times 10^{-6}$ | | |
| *Carbonates* | | | | |
| Ag$_2$CO$_3$ | 11.09 | $8.1 \times 10^{-12}$ | | |
| BaCO$_3$ | 8.30 | $5.0 \times 10^{-9}$ | | |
| CaCO$_3$ (calcite)† | 8.35 | $4.5 \times 10^{-9}$ | | |
| CdCO$_3$ | 13.74 | $1.8 \times 10^{-14}$ | | |
| CoCO$_3$ | 9.98 | $1.0 \times 10^{-10}$ | | |
| CuCO$_3$ | 9.63 | $2.3 \times 10^{-10}$ | | |
| FeCO$_3$ | 10.68 | $2.1 \times 10^{-11}$ | | |
| Hg$_2$CO$_3$* | 16.05 | $8.9 \times 10^{-17}$ | | |
| La$_2$(CO$_3$)$_3$ | 33.4 | $4.0 \times 10^{-34}$ | | |
| MgCO$_3$ | 7.46 | $3.5 \times 10^{-8}$ | | |
| MnCO$_3$ | 9.30 | $5.0 \times 10^{-10}$ | | |
| NiCO$_3$ | 6.87 | $1.3 \times 10^{-7}$ | | |
| PbCO$_3$ | 13.13 | $7.4 \times 10^{-14}$ | | |
| SrCO$_3$ | 9.03 | $9.3 \times 10^{-10}$ | | |
| Y$_2$(CO$_3$)$_3$ | 30.6 | $2.5 \times 10^{-31}$ | | |
| ZnCO$_3$ | 10.00 | $1.0 \times 10^{-10}$ | | |
| *Chlorides* | | | | |
| AgCl | 9.77 | $1.7 \times 10^{-10}$ | | |
| CuCl | 6.73 | $1.9 \times 10^{-7}$ | | |
| Hg$_2$Cl$_2$* | 17.91 | $1.2 \times 10^{-18}$ | | |
| PbCl$_2$ | 4.78 | $1.7 \times 10^{-5}$ | | |

* Ionizes or dissociates into Hg$_2^{2+}$.
† This is a specific solid-phase structure.

| Formula | $pK_{sp}$ | $K_{sp}$ | Temperature (°C) | Ionic strength ($M$) |
|---|---|---|---|---|
| *Chlorides (continued)* | | | | |
| TlCl | 3.74 | $1.8 \times 10^{-4}$ | | |
| *Chromates* | | | | |
| $Ag_2CrO_4$ | 11.89 | $1.3 \times 10^{-12}$ | | |
| $BaCrO_4$ | 9.67 | $2.1 \times 10^{-10}$ | | |
| $CuCrO_4$ | 5.44 | $3.6 \times 10^{-6}$ | | |
| $Hg_2CrO_4$ | 8.70 | $2.0 \times 10^{-9}$ | | |
| $Tl_2CrO_4$ | 12.01 | $9.8 \times 10^{-13}$ | | |
| *Cyanides* | | | | |
| $AgCN(2AgCN \rightleftharpoons Ag^+ + Ag(CN)_2^-)$§ | 15.66 | $2.2 \times 10^{-16}$ | | |
| $Hg_2(CN)_2$* | 39.3 | $5 \times 10^{-40}$ | | |
| $Zn(CN)_2$ | 15.5 | $3 \times 10^{-16}$ | | 3 |
| *Fluorides* | | | | |
| $BaF_2$ | 5.76 | $1.7 \times 10^{-6}$ | | |
| $CaF_2$ | 10.41 | $3.9 \times 10^{-11}$ | | |
| $MgF_2$ | 8.18 | $6.6 \times 10^{-9}$ | | |
| $SrF_2$ | 8.54 | $2.9 \times 10^{-9}$ | | |
| $PbF_2$ | 7.44 | $3.6 \times 10^{-8}$ | | |
| $ThF_4$ | 28.3 | $5 \times 10^{-29}$ | | 3 |
| *Hydroxides* | | | | |
| AgOH | 8.17 | $6.8 \times 10^{-9}$ | | 1 |
| $Ag_2O(\rightleftharpoons 2 Ag^+ + 2 OH^-)$§ | 15.42 | $3.8 \times 10^{-16}$ | | |
| $Al(OH)_3(\alpha)$† | 33.5 | $3 \times 10^{-34}$ | | |
| $Au(OH)_3$ | 5.5 | $3 \times 10^{-6}$ | | |
| $Ba(OH)_2 \cdot 8 H_2O$ | 3.6 | $3 \times 10^{-4}$ | | |
| $Ca(OH)_2$ | 5.19 | $6.5 \times 10^{-6}$ | | |
| $Cd(OH)_2(\beta)$† | 14.35 | $4.5 \times 10^{-15}$ | | |
| $Ce(OH)_3$ | 21.2 | $6 \times 10^{-22}$ | | |
| $Co(OH)_2$ | 14.9 | $1.3 \times 10^{-15}$ | | |
| $Co(OH)_3$ | 44.5 | $3 \times 10^{-45}$ | 19 | |
| $Cr(OH)_3$ | 29.8 | $1.6 \times 10^{-30}$ | | 0.1 |
| $Cu(OH)_2$ | 19.32 | $4.8 \times 10^{-20}$ | | |
| $Cu_2O(\rightleftharpoons 2 Cu^+ + 2 OH^-)$§ | 29.4 | $4 \times 10^{-30}$ | | |

\* Ionizes or dissociates into $Hg_2^{2+}$.
† This is a specific solid-phase structure.
§ For the dissolution reaction written in parentheses.

| Formula | $pK_{sp}$ | $K_{sp}$ | Temperature (°C) | Ionic strength (M) |
|---|---|---|---|---|
| *Hydroxides (continued)* | | | | |
| $Fe(OH)_2$ | 15.1 | $7.9 \times 10^{-16}$ | | |
| $Fe(OH)_3$ | 38.8 | $1.6 \times 10^{-39}$ | | |
| $HgO$ (red) ($\rightleftharpoons Hg^{2+} + 2\ OH^-$)†§ | 25.44 | $3.6 \times 10^{-26}$ | | |
| $In(OH)_3$ | 36.9 | $1.3 \times 10^{-37}$ | | |
| $La(OH)_3$ | 20.7 | $2 \times 10^{-21}$ | | |
| $Mg(OH)_2$ | 11.15 | $7.1 \times 10^{-12}$ | | |
| $Mn(OH)_2$ | 12.8 | $1.6 \times 10^{-13}$ | | |
| $Ni(OH)_2$ | 15.2 | $6 \times 10^{-16}$ | | |
| $Pd(OH)_2$ | 31 | $1 \times 10^{-31}$ | | |
| $PbO$ (red)($\rightleftharpoons Pb^{2+} + 2\ OH^-$)†§ | 15.3 | $5 \times 10^{-16}$ | | |
| $PbO$ (yellow)($\rightleftharpoons Pb^{2+} + 2\ OH^-$)†§ | 15.1 | $8 \times 10^{-16}$ | | |
| $SnO$($\rightleftharpoons Sn^{2+} + 2\ OH^-$)§ | 26.2 | $6 \times 10^{-27}$ | | |
| $UO_2$($\rightleftharpoons U^{4+} + 4\ OH^-$)§ | 56.2 | $6 \times 10^{-57}$ | | |
| $UO_2(OH)_2$($\rightleftharpoons UO_2^{2+} + 2\ OH^-$)§ | 22.4 | $4 \times 10^{-23}$ | | |
| $V(OH)_3$ | 34.4 | $4.0 \times 10^{-35}$ | | |
| $VO(OH)_2$ ($\rightleftharpoons VO^{2+} + 2\ OH^-$)§ | 23.5 | $3 \times 10^{-24}$ | | |
| $Y(OH)_3$ | 23.2 | $6 \times 10^{-24}$ | | |
| $Zn(OH)_2$ amorphous† | 15.52 | $3.0 \times 10^{-16}$ | | |
| *Iodates* | | | | |
| $AgIO_3$ | 7.51 | $3.1 \times 10^{-8}$ | | |
| $Ba(IO_3)_2$ | 8.81 | $1.5 \times 10^{-9}$ | | |
| $Ca(IO_3)_2$ | 6.15 | $7.1 \times 10^{-7}$ | | |
| $Cd(IO_3)_2$ | 7.64 | $2.3 \times 10^{-8}$ | | |
| $Ce(IO_3)_3$ | 10.86 | $1.4 \times 10^{-11}$ | | |
| $Hg_2(IO_3)_2$* | 17.89 | $1.3 \times 10^{-18}$ | | |
| $La(IO_3)_3$ | 10.92 | $1.2 \times 10^{-11}$ | | |
| $Pb(IO_3)_2$ | 12.61 | $2.5 \times 10^{-13}$ | | |
| $Sr(IO_3)_2$ | 6.48 | $3.3 \times 10^{-7}$ | | |
| $Th(IO_3)_4$ | 14.62 | $2.4 \times 10^{-15}$ | | 0.5 |
| $TlIO_3$ | 5.51 | $3.1 \times 10^{-6}$ | | |
| $UO_2(IO_3)_2$ ($\rightleftharpoons UO_2^{2+} + 2\ IO_3^-$)§ | 7.01 | $9.8 \times 10^{-8}$ | | 0.2 |
| $Y(IO_3)_3$ | 10.15 | $7.1 \times 10^{-11}$ | | |

\* Ionizes or dissociates into $Hg_2^{2+}$.
† This is a specific solid-phase structure.
§ For the dissolution reaction written in parentheses.

| Formula | $pK_{sp}$ | $K_{sp}$ | Temperature (°C) | Ionic strength (M) |
|---|---|---|---|---|
| *Iodides* | | | | |
| AgI | 16.08 | $8.3 \times 10^{-17}$ | | |
| CuI | 12.0 | $1 \times 10^{-12}$ | | |
| $Hg_2I_2$ | 27.95 | $1.1 \times 10^{-28}$ | | 0.5 |
| $PbI_2$ | 8.10 | $7.9 \times 10^{-9}$ | | |
| TlI | 7.23 | $5.9 \times 10^{-8}$ | | |
| *Oxalates* | | | | |
| $BaC_2O_4$ | 6.0 | $1 \times 10^{-6}$ | 20 | 0.1 |
| $CaC_2O_4$ | 7.9 | $1.3 \times 10^{-8}$ | 20 | 0.1 |
| $La_2(C_2O_4)_3$ | 25.0 | $1 \times 10^{-25}$ | 20 | 0.1 |
| $SrC_2O_4$ | 6.4 | $4 \times 10^{-7}$ | 20 | 0.1 |
| $Th(C_2O_4)_2$ | 21.38 | $4.2 \times 10^{-22}$ | | 1 |
| $UO_2C_2O_4 \ (\rightleftharpoons UO_2^{2+} + C_2O_4^{2-})$§ | 8.66 | $2.2 \times 10^{-9}$ | 20 | 0.1 |
| *Phosphates* | | | | |
| $Ag_3PO_4$ | 17.55 | $2.8 \times 10^{-18}$ | | |
| $BaHPO_4 \ (\rightleftharpoons Ba^{2+} + HPO_4^{2-})$§ | 7.40 | $4.0 \times 10^{-8}$ | 20 | |
| $CaHPO_4 \cdot 2\,H_2O \ (\rightleftharpoons Ca^{2+} + HPO_4^{2-})$§ | 6.58 | $2.6 \times 10^{-7}$ | | |
| $FePO_4 \cdot 2\,H_2O$ | 26.4 | $4 \times 10^{-27}$ | | |
| $Fe_3(PO_4)_2 \cdot 8\,H_2O$ | 36.0 | $1 \times 10^{-36}$ | | |
| $GaPO_4$ | 21.0 | $1 \times 10^{-21}$ | | 1 |
| $Hg_2HPO_4 \ (\rightleftharpoons Hg_2^{2+} + HPO_4^{2-})$*§ | 12.40 | $4.0 \times 10^{-13}$ | | |
| $InPO_4$ | 21.63 | $2.3 \times 10^{-22}$ | | 1 |
| $LaPO_4$ | 22.43 | $3.7 \times 10^{-23}$ | | 0.5 |
| $Pb_3(PO_4)_2$ | 43.53 | $3.0 \times 10^{-44}$ | 38 | |
| $SrHPO_4 \ (\rightleftharpoons Sr^{2+} + HPO_4^{2-})$§ | 6.92 | $1.2 \times 10^{-7}$ | 20 | |
| $(VO)_3(PO_4)_2 \ (\rightleftharpoons 3\,VO^{2+} + 2\,PO_4^{3-})$§ | 25.1 | $8 \times 10^{-26}$ | | |
| $Zn_3(PO_4)_2 \cdot 4\,H_2O$ | 35.3 | $5 \times 10^{-36}$ | | |
| *Sulfates* | | | | |
| $Ag_2SO_4$ | 4.83 | $1.5 \times 10^{-5}$ | | |
| $BaSO_4$ | 10.04 | $9.1 \times 10^{-11}$ | | |
| $CaSO_4$ | 4.62 | $2.4 \times 10^{-5}$ | | |
| $Hg_2SO_4$* | 6.13 | $7.4 \times 10^{-7}$ | | |
| $PbSO_4$ | 6.20 | $6.3 \times 10^{-7}$ | | |
| $RaSO_4$ | 10.37 | $4.3 \times 10^{-11}$ | 20 | |

* Ionizes or dissociates into $Hg_2^{2+}$.
§ For the dissolution reaction written in parentheses.

| Formula | $pK_{sp}$ | $K_{sp}$ | Temperature (°C) | Ionic strength (M) |
|---|---|---|---|---|
| *Sulfates (continued)* | | | | |
| $SrSO_4$ | 6.50 | $3.2 \times 10^{-7}$ | | |
| *Sulfides* | | | | |
| $Ag_2S$ | 50.1 | $8 \times 10^{-51}$ | | |
| $CdS$ | 27.0 | $1 \times 10^{-27}$ | | |
| $CoS(\alpha)$† | 21.3 | $5 \times 10^{-22}$ | | |
| $CoS(\beta)$† | 25.6 | $3 \times 10^{-26}$ | | |
| $CuS$ | 36.1 | $8 \times 10^{-37}$ | | |
| $Cu_2S$ | 48.5 | $3 \times 10^{-49}$ | | |
| $FeS$ | 18.1 | $8 \times 10^{-19}$ | | |
| $HgS$ (black)† | 52.7 | $2 \times 10^{-53}$ | | |
| $HgS$ (red)† | 53.3 | $5 \times 10^{-54}$ | | |
| $In_2S_3$ | 69.4 | $4 \times 10^{-70}$ | | |
| $MnS$ (green)† | 13.5 | $3 \times 10^{-14}$ | | |
| $MnS$ (pink)† | 10.5 | $3 \times 10^{-11}$ | | |
| $NiS(\alpha)$† | 19.4 | $4 \times 10^{-20}$ | | |
| $NiS(\beta)$† | 24.9 | $1.3 \times 10^{-25}$ | | |
| $NiS(\gamma)$† | 26.6 | $3 \times 10^{-27}$ | | |
| $PbS$ | 27.5 | $3 \times 10^{-28}$ | | |
| $SnS$ | 25.9 | $1.3 \times 10^{-26}$ | | |
| $Tl_2S$ | 21.2 | $6 \times 10^{-22}$ | | |
| $ZnS(\alpha)$† | 24.7 | $2 \times 10^{-25}$ | | |
| $ZnS(\beta)$† | 22.5 | $3 \times 10^{-23}$ | | |
| *Thiocyanates* | | | | |
| $AgSCN$ | 11.97 | $1.1 \times 10^{-12}$ | | |
| $CuSCN$ | 13.40 | $4.0 \times 10^{-14}$ | | 5 |
| $Hg(SCN)_2$ | 19.56 | $2.8 \times 10^{-20}$ | | 1 |
| $Hg_2(SCN)_2$* | 19.52 | $3.0 \times 10^{-20}$ | | |
| $TlSCN$ | 3.79 | $1.6 \times 10^{-4}$ | | |

* Ionizes or dissociates into $Hg_2^{2+}$.
† This is a specific solid-phase structure.

*Sources:* Martell, A. E.; Smith, R. M. "Critical Stability Constants," Vol. 4; Plenum Press: New York, 1976. Sillén, L. G.; Martell, A. E. "Stability Constants of Metal-Ion Complexes," Suppl. No. 1, Special Publication No. 25; The Chemical Society: London, 1971.

# Acid Dissociation Constants

At 25 °C and zero ionic strength unless noted otherwise. For bases $K_b = K_w/K_a$. Each acid is written in its protonated form. The acid protons are indicated by boldface.

| Name | Structure | $pK_a$ | $K_a$ |
|---|---|---|---|
| Acetic acid | $CH_3COOH$ | 4.757 | $1.75 \times 10^{-5}$ |
| Acrylic acid (propenoic acid) | $H_2C=CHCOOH$ | 4.258 | $5.52 \times 10^{-5}$ |
| Alanine | $NH_3^+$ $\mid$ $CHCH_3$ $\mid$ $COOH$ | 2.348 (COOH) 9.867 ($NH_3$) | $4.49 \times 10^{-3}$ $1.36 \times 10^{-10}$ |
| Aminobenzene (aniline) | ⬡—$NH_3^+$ | 4.601 | $2.51 \times 10^{-5}$ |
| 4-Aminobenzenesulfonic acid (sulfanilic acid) | $^-O_3S$—⬡—$\overset{+}{N}H_3$ | 3.232 | $5.86 \times 10^{-4}$ |
| 2-Aminobenzoic acid (anthranilic acid) | ⬡ with $NH_3^+$, COOH | 2.08 (COOH) 4.96 ($NH_3$) | $8.3 \times 10^{-3}$ $1.10 \times 10^{-5}$ |
| 2-Aminoethanol (ethanolamine) | $HOCH_2CH_2NH_3^+$ | 9.498 | $3.18 \times 10^{-10}$ |
| 2-Aminophenol | ⬡ with OH, $NH_3^+$ | 4.78 ($NH_3$) (20°) 9.97 (OH) (20°) | $1.66 \times 10^{-5}$ $1.05 \times 10^{-10}$ |
| Ammonia | $NH_4^+$ | 9.244 | $5.70 \times 10^{-10}$ |
| Arginine | $NH_3^+$ $\mid$ $CHCH_2CH_2CH_2NHC\overset{\displaystyle NH_2^+}{\underset{\displaystyle NH_2}{=}}$ $\mid$ $COOH$ | 1.823 (COOH) 8.991 ($NH_3$) (12.48) ($NH_2$) | $1.50 \times 10^{-2}$ $1.02 \times 10^{-9}$ $3.3 \times 10^{-13}$ |
| Arsenic acid | $O=As(OH)_3$ | 2.24 6.96 11.50 | $5.8 \times 10^{-3}$ $1.10 \times 10^{-7}$ $3.2 \times 10^{-12}$ |
| Arsenious acid | $As(OH)_3$ | 9.29 (13.5) | $5.1 \times 10^{-10}$ $3 \times 10^{-14}$ |
| Asparagine | $NH_3^+$ $O$ $\mid$ $\parallel$ $CHCHOCNH_2$ $\mid$ $COOH$ | 2.14 (COOH) ($\mu = 0.1$) 8.72 ($NH_3$) ($\mu = 0.1$) | $7.2 \times 10^{-3}$ $1.9 \times 10^{-9}$ |
| Aspartic acid | $NH_3^+$ $\mid$ $CHCH_2COOH$ $\mid$ $COOH$ | 1.990 ($\alpha$-COOH) 3.900 ($\beta$-COOH) 10.002 ($NH_3$) | $1.02 \times 10^{-2}$ $1.26 \times 10^{-4}$ $9.95 \times 10^{-11}$ |

| Name | Structure | $pK_a$ | $K_a$ |
|---|---|---|---|
| Benzoic acid | ⬡—COOH | 4.202 | $6.28 \times 10^{-5}$ |
| Benzylamine | ⬡—$CH_2NH_3^+$ | 9.35 | $4.5 \times 10^{-10}$ |
| 2,2'-Bipyridine | [bipyridine structure] $\overset{+}{N}H$ $N$ | 4.35 | $4.5 \times 10^{-5}$ |
| Boric acid | $B(OH)_3$ | 9.236 (12.74) (20°) (13.80) (20°) | $5.81 \times 10^{-10}$ $1.82 \times 10^{-13}$ $1.58 \times 10^{-14}$ |
| Butane-2,3-dione dioxime (dimethylglyoxime) | HON   NOH [structure] $CH_3$   $CH_3$ | 10.66 12.0 | $2.2 \times 10^{-11}$ $1 \times 10^{-12}$ |
| [Carbonic acid] = $[H_2CO_3] + [CO_2(aq)]$ | $\overset{O}{\overset{\|}{HO-C-OH}}$ | 6.352 10.329 | $4.45 \times 10^{-7}$ $4.69 \times 10^{-11}$ |
| Chloroacetic acid | $ClCH_2COOH$ | 2.865 | $1.36 \times 10^{-3}$ |
| Chlorous acid | $HOCl{=}O$ | 1.95 | $1.12 \times 10^{-2}$ |
| Chromic acid | $\overset{O}{\overset{\|}{HO-\underset{\underset{O}{\|}}{Cr}-OH}}$ | −0.2 (20°) 6.51 | 1.6 $3.1 \times 10^{-7}$ |
| Citric acid (2-hydroxypropane-1,2,3-tricarboxylic acid) | $\overset{COOH}{\overset{\|}{HOOCCH_2CCH_2COOH}}$ $\underset{OH}{\|}$ | 3.128 4.761 6.396 | $7.44 \times 10^{-4}$ $1.73 \times 10^{-5}$ $4.02 \times 10^{-7}$ |
| Cysteine | $\overset{NH_3^+}{\overset{\|}{CHCH_2SH}}$ $\underset{COOH}{\|}$ | (1.71) (COOH) 8.36 (SH) 10.77 ($NH_3$) | $1.95 \times 10^{-2}$ $4.4 \times 10^{-9}$ $1.70 \times 10^{-11}$ |
| Dichloroacetic acid | $Cl_2CHCOOH$ | 1.30 | $5.0 \times 10^{-2}$ |
| Diethylamine | $(CH_3CH_2)_2NH_2^+$ | 10.933 | $4.7 \times 10^{-10}$ |
| 1,2-Dihydroxybenzene (catechol) | [benzene ring]—OH —OH | 9.40 12.8 | $4.0 \times 10^{-10}$ $1.6 \times 10^{-13}$ |

| Name | Structure | $pK_a$ | $K_a$ |
|---|---|---|---|
| 1,3-Dihydroxybenzene (resorcinol) | OH / C6H4 / OH | 9.30 / 11.06 | $5.0 \times 10^{-10}$ / $8.7 \times 10^{-12}$ |
| 2,3-Dimercaptopropanol | $HOCH_2CHCH_2SH$ / $SH$ | 8.58 ($\mu = 0.1$) / 10.68 ($\mu = 0.1$) | $2.6 \times 10^{-9}$ / $2.1 \times 10^{-11}$ |
| Dimethylamine | $(CH_3)_2NH_2^+$ | 10.774 | $1.68 \times 10^{-11}$ |
| Ethane-1,2-dithiol | $HSCH_2CH_2SH$ | 8.85 ($30°, \mu = 0.1$) / 10.43 ($30°, \mu = 0.1$) | $1.4 \times 10^{-9}$ / $3.7 \times 10^{-11}$ |
| Ethylamine | $CH_3CH_2NH_3^+$ | 10.636 | $2.31 \times 10^{-11}$ |
| Ethylenediamine (1,2-diaminoethane) | $H_3\overset{+}{N}CH_2CH_2\overset{+}{N}H_3$ | 6.848 / 9.928 | $1.42 \times 10^{-7}$ / $1.18 \times 10^{-10}$ |
| Ethylenedinitrilotetra-acetic acid (edta) | $(HOOCCH_2)_2\overset{+}{N}HCH_2CH_2\overset{+}{N}H(CH_2COOH)_2$ | 0.0 (COOH) ($\mu = 1.0$) / 1.5 (COOH) ($\mu = 0.1$) / 2.0 (COOH) ($\mu = 0.1$) / 2.68 (COOH) ($\mu = 0.1$) / 6.11 (NH) ($\mu = 0.1$) / 10.17 (NH) ($\mu = 0.1$) | 1.0 / 0.032 / 0.010 / 0.0021 / $7.8 \times 10^{-7}$ / $6.8 \times 10^{-11}$ |
| Formic acid | $HCOOH$ | 3.745 | $1.80 \times 10^{-4}$ |
| Fumaric acid (*trans*-butenedioic acid) | COOH / HOOC (trans structure) | 3.053 / 4.494 | $8.85 \times 10^{-4}$ / $3.21 \times 10^{-5}$ |
| Glutamic acid | $NH_3^+$ / $CHCH_2CH_2COOH$ / $COOH$ | 2.23 ($\alpha$-COOH) / 4.42 ($\gamma$-COOH) / 9.95 (NH$_3$) | $5.9 \times 10^{-3}$ / $3.8 \times 10^{-5}$ / $1.12 \times 10^{-10}$ |
| Glutamine | $NH_3^+$ / $CHCH_2CH_2CNH_2$ (with C=O) / $COOH$ | 2.17 (COOH) ($\mu = 0.1$) / 9.01 (NH$_3$) ($\mu = 0.1$) | $6.8 \times 10^{-3}$ / $9.8 \times 10^{-10}$ |
| Glutaric acid (1,5-pentanedioic acid) | $HOOCCH_2CH_2CH_2COOH$ | 4.34 / 5.43 | $4.6 \times 10^{-5}$ / $3.7 \times 10^{-6}$ |
| Glycine | $NH_3^+$ / $CH_2$ / $COOH$ | 2.350 (COOH) / 9.778 (NH$_3$) | $4.47 \times 10^{-3}$ / $1.67 \times 10^{-10}$ |

| Name | Structure | $pK_a$ | $K_a$ |
|---|---|---|---|
| Glycolic acid | $HOCH_2COOH$ | 3.831 | $1.48 \times 10^{-4}$ |
| Guanidine | | 13.54 (27°, $\mu = 1.0$) | $2.9 \times 10^{-14}$ |
| Histidine | | 1.7 (COOH) ($\mu = 0.1$)<br>6.02 (NH) ($\mu = 0.1$)<br>9.08 ($NH_3$) ($\mu = 0.1$) | $2 \times 10^{-2}$<br>$9.5 \times 10^{-7}$<br>$8.3 \times 10^{-10}$ |
| Hydrogen azide (hydrazoic acid) | $HN=\overset{+}{N}=\overset{-}{N}$ | 4.65 | $2.2 \times 10^{-5}$ |
| Hydrogen cyanate | $HOC\equiv N$ | 3.48 | $3.3 \times 10^{-4}$ |
| Hydrogen cyanide | $HC\equiv N$ | 9.21 | $6.2 \times 10^{-10}$ |
| Hydrogen fluoride | $HF$ | 3.17 | $6.8 \times 10^{-4}$ |
| Hydrogen peroxide | $HOOH$ | 11.65 | $2.2 \times 10^{-12}$ |
| Hydrogen sulfide | $H_2S$ | 7.02<br>13.9 | $9.5 \times 10^{-8}$<br>$1.3 \times 10^{-14}$ |
| Hydrogen thiocyanate | $HSC\equiv N$ | 0.9 | 0.13 |
| Hydroxylamine | $HO\overset{+}{N}H_3$ | 5.96 | $1.10 \times 10^{-6}$ |
| 8-Hydroxyquinoline (oxine) | | 4.91 (NH)<br>9.81 (OH) | $1.23 \times 10^{-5}$<br>$1.55 \times 10^{-10}$ |
| Hypobromous acid | $HOBr$ | 8.63 | $2.3 \times 10^{-9}$ |
| Hypochlorous acid | $HOCl$ | 7.53 | $3.0 \times 10^{-8}$ |
| Hypoiodous acid | $HOI$ | 10.64 | $2.3 \times 10^{-11}$ |
| Hypophosphorous acid | | 1.23 | $5.9 \times 10^{-2}$ |
| Imidazole | | 6.993 | $1.02 \times 10^{-7}$ |
| Iodic acid | $HIO_3$ | 0.77 | 0.17 |
| Isoleucine | | 2.319 (COOH)<br>9.754 ($NH_3$) | $4.80 \times 10^{-3}$<br>$1.76 \times 10^{-10}$ |

| Name | Structure | $pK_a$ | $K_a$ |
|---|---|---|---|
| Leucine | $\overset{NH_3^+}{\underset{COOH}{\mid}}$ $CHCH_2CH(CH_3)_2$ | 2.329 (COOH)<br>9.747 (NH$_3$) | $4.69 \times 10^{-3}$<br>$1.79 \times 10^{-10}$ |
| Lysine | $\overset{NH_3^+}{\underset{COOH}{\mid}}$ $CHCH_2CH_2CH_2CH_2NH_3^+$ | 2.04 (COOH) ($\mu = 0.1$)<br>9.08 ($\alpha$-NH$_3$) ($\mu = 0.1$)<br>10.69 ($\epsilon$-NH$_3$) ($\mu = 0.1$) | $9.1 \times 10^{-3}$<br>$8.3 \times 10^{-10}$<br>$2.0 \times 10^{-11}$ |
| Maleic acid<br>(*cis*-butenedioic acid) | COOH / COOH | 1.910<br>6.332 | $1.23 \times 10^{-2}$<br>$4.66 \times 10^{-7}$ |
| Malonic acid | $HOOCCH_2COOH$ | 2.847<br>5.696 | $1.42 \times 10^{-3}$<br>$2.01 \times 10^{-6}$ |
| Mercaptoacetic acid<br>(thioglycolic acid) | $HSCH_2COOH$ | (3.60) (COOH)<br>10.55 (SH) | $2.5 \times 10^{-4}$<br>$2.82 \times 10^{-11}$ |
| 2-Mercaptoethanol | $HSCH_2CH_2OH$ | 9.72 | $1.91 \times 10^{-10}$ |
| Methionine | $\overset{NH_3^+}{\underset{COOH}{\mid}}$ $CHCH_2CH_2SCH_3$ | 2.20 ($\mu = 0.1$)<br>9.05 ($\mu = 0.1$) | $6.3 \times 10^{-3}$<br>$8.9 \times 10^{-10}$ |
| Methylamine | $CH_3\overset{+}{N}H_3$ | 10.64 | $2.3 \times 10^{-11}$ |
| 2-Methylaniline<br>(*o*-toluidine) | CH$_3$ / $\overset{+}{N}H_3$ | 4.447 | $3.57 \times 10^{-5}$ |
| 2-Methylphenol<br>(*o*-cresol) | CH$_3$ / OH | 10.09 | $8.1 \times 10^{-11}$ |
| Morpholine<br>(perhydro-1,4-oxazine) | O $\bigcirc$ $\overset{+}{N}H_2$ | 8.492 | $3.22 \times 10^{-9}$ |
| Nitrilotriacetic acid | $H\overset{+}{N}(CH_2COOH)_3$ | 1.1 (COOH) (20°, $\mu = 1.0$)<br>1.650 (COOH) (20°)<br>2.940 (COOH) (20°)<br>10.334 (NH) (20°) | $8 \times 10^{-2}$<br>$2.24 \times 10^{-2}$<br>$1.15 \times 10^{-3}$<br>$4.63 \times 10^{-11}$ |
| Nitroethane | $CH_3CH_2NO_2$ | 8.57 | $2.7 \times 10^{-9}$ |
| *N*-Nitrosophenylhydroxyl-<br>amine (cupferron) | NO / N \ OH | 4.16 ($\mu = 0.1$) | $6.9 \times 10^{-5}$ |
| Nitrous acid | $HON{=}O$ | 3.15 | $7.1 \times 10^{-4}$ |

| Name | Structure | $pK_a$ | $K_a$ |
|---|---|---|---|
| Oxalic acid | HOOCCOOH | 1.252<br>4.266 | $5.60 \times 10^{-2}$<br>$5.42 \times 10^{-5}$ |
| Oxoacetic acid<br>(glyoxylic acid) | $\overset{\text{O}}{\overset{\|}{\text{HCCOOH}}}$ | 3.46 | $3.5 \times 10^{-4}$ |
| 1,10-Phenanthroline | | 4.86 | $1.38 \times 10^{-5}$ |
| Phenol | $\langle\bigcirc\rangle$—OH | 9.98 | $1.05 \times 10^{-10}$ |
| Phenylacetic acid | $\langle\bigcirc\rangle$—CH$_2$COOH | 4.310 | $4.90 \times 10^{-5}$ |
| Phenylalanine | $\overset{\text{NH}_3^+}{\underset{\text{COOH}}{\overset{\|}{\text{CHCH}_2}}}$—$\langle\bigcirc\rangle$ | 2.20 (COOH)<br>9.31 (NH$_3$) | $6.3 \times 10^{-3}$<br>$4.9 \times 10^{-10}$ |
| Phosphoric acid | O=P(OH)$_3$ | 2.148<br>7.199<br>12.35 | $7.11 \times 10^{-3}$<br>$6.32 \times 10^{-8}$<br>$4.5 \times 10^{-13}$ |
| Phosphorous acid | $\overset{\text{O}}{\overset{\|}{\text{HP}}}$—(OH)$_2$ | 1.5<br>6.79 | $3 \times 10^{-2}$<br>$1.62 \times 10^{-7}$ |
| Phthalic acid<br>(benzene-1,2-dicarboxylic<br>acid) | COOH<br>COOH | 2.950<br>5.408 | $1.12 \times 10^{-3}$<br>$3.90 \times 10^{-6}$ |
| Picric acid | NO$_2$, OH, NO$_2$, NO$_2$ | 0.292 | $5.1 \times 10^{-1}$ |
| Piperazine<br>(perhydro-1,4-diazine) | H$_2^+$N$\langle\ \rangle$NH$_2^+$ | 5.333<br>9.731 | $4.65 \times 10^{-6}$<br>$1.86 \times 10^{-10}$ |
| Piperidine | $\langle\ \rangle$NH$_2^+$ | 11.123 | $7.53 \times 10^{-12}$ |
| Proline | COOH, N, $^+$H$_2$ | 1.952 (COOH)<br>10.640 (NH$_2$) | $1.12 \times 10^{-2}$<br>$2.29 \times 10^{-11}$ |
| Propanoic acid | CH$_3$CH$_2$COOH | 4.874 | $1.34 \times 10^{-5}$ |

| Name | Structure | $pK_a$ | $K_a$ |
|---|---|---|---|
| Pyridine (azine) | NH$^+$ | 5.229 | $5.90 \times 10^{-6}$ |
| Pyrophosphoric acid | (HO)$_2$POP(OH)$_2$ | 0.8<br>2.2<br>6.70<br>9.40 | 0.16<br>$6 \times 10^{-3}$<br>$2.0 \times 10^{-7}$<br>$4.0 \times 10^{-10}$ |
| Pyrrolidine | NH$_2^+$ | 11.305 | $4.95 \times 10^{-12}$ |
| Pyruvic acid (2-oxopropanoic acid) | CH$_3$CCOOH | 2.55 | $2.8 \times 10^{-3}$ |
| Salicylic acid (2-hydroxybenzoic acid) | COOH ... OH | 2.97 (COOH)<br>13.74 (OH) | $1.07 \times 10^{-3}$<br>$1.82 \times 10^{-14}$ |
| Serine | NH$_3^+$ \| CHCH$_2$OH \| CO$_2$H | 2.187<br>9.209 | $6.50 \times 10^{-3}$<br>$6.18 \times 10^{-10}$ |
| Succinic acid (butanedioic acid) | HOOCCH$_2$COOH | 4.207<br>5.636 | $6.21 \times 10^{-5}$<br>$2.31 \times 10^{-6}$ |
| Sulfuric acid | H$_2$SO$_4$ | 1.99 (pK$_2$) | $1.02 \times 10^{-2}$ |
| Sulfurous acid | H$_2$SO$_3$ | 1.91<br>7.18 | $1.23 \times 10^{-2}$<br>$6.6 \times 10^{-8}$ |
| L-Tartaric acid L-2,3-Dihydroxybutane-dioic acid | OH \| HOO$_2$CCHCHCOOH \| OH | 3.036<br>4.366 | $9.20 \times 10^{-4}$<br>$4.31 \times 10^{-5}$ |
| Thiosulfuric acid | O \|\| HOSOH \|\| S | 0.6<br>1.6 | 0.3<br>$3 \times 10^{-2}$ |
| Threonine | NH$_3^+$ \| CHCHOHCH$_3$ \| COOH | 2.088 (COOH)<br>9.100 (NH$_3$) | $8.17 \times 10^{-3}$<br>$7.94 \times 10^{-10}$ |
| Trichloroacetic acid | Cl$_3$CCOOH | 0.66 ($\mu = 0.1$) | 0.22 |
| Triethanolamine | (HOCH$_2$CH$_2$)$_3$NH$^+$ | 7.762 | $1.73 \times 10^{-8}$ |
| Triethylamine | (CH$_3$CH$_2$)$_3$NH$^+$ | 10.715 | $1.93 \times 10^{-11}$ |

| Name | Structure | $pK_a$ | $K_a$ |
|---|---|---|---|
| 1,2,3-Trihydroxyben-zene (pyrogallol) | | 8.94<br>11.08<br>(14) | $1.15 \times 10^{-9}$<br>$8.3 \times 10^{-12}$<br>$10^{-14}$ |
| Trimethylamine | $(CH_3)_3NH^+$ | 9.800 | $1.58 \times 10^{-10}$ |
| *tris*(hydroxymethyl)-aminomethane (TRIS or THAM) | $(HOCH_2)_3CNH_3^+$ | 8.075 | $8.41 \times 10^{-9}$ |
| Tryptophan | | 2.35 (COOH) ($\mu = 0.1$)<br>9.33 (NH$_3$) ($\mu = 0.1$) | $4.5 \times 10^{-3}$<br>$4.7 \times 10^{-10}$ |
| Tyrosine | | 2.17 (COOH) ($\mu = 0.1$)<br>9.19 (NH$_3$)<br>10.47 (OH) | $6.8 \times 10^{-3}$<br>$6.5 \times 10^{-10}$<br>$3.4 \times 10^{-11}$ |
| Valine | | 2.286 (COOH)<br>9.718 (NH$_3$) | $5.18 \times 10^{-3}$<br>$1.91 \times 10^{-10}$ |
| Water | $H_2O$ | 13.997 | $1.01 \times 10^{-14}$ |

*Source:* Martell, A. E.; Smith, R. M., "Critical Stability Constants"; Plenum Press: New York, 1974.

# Selected Standard and Formal Electrode Potentials

The formal potential is the measured potential of a half cell (versus NHE) under conditions when the logarithmic term in the Nernst equation is zero when under the following conditions: the *concentrations* of oxidized and reduced species are expressed overtly and all other species (such as $H^+$, ligands, salts) are present at the stated concentrations. Thus, for the general reaction

$$\text{Oxidized form} + e^- \rightleftharpoons \text{Reduced form}$$

when the value of $[Red]/[Ox] = 1$, the formal potential will be measured. For a reaction with stoichiometry

$$\text{Oxidized form} + e^- \rightleftharpoons 2 \text{ Reduced form}$$

when the value of $[Red]^2/[Ox] = 1$, the formal potential will be measured.

In effect, the activity coefficients are incorporated into the potential as are any chemical reactions that occur with either member of the redox couple. The formal potential will vary from medium to medium. One other convention keeps this formalism consistent. In writing the Nernst equation, solids involved in the redox equilibrium are still treated as having unit activity.

| Half-reaction | $E°$ (V) | Formal Potential (V) |
|---|---|---|
| *Aluminum* | | |
| $Al^{3+} + 3\,e^- \rightleftharpoons Al(s)$ | $-1.662$ | |
| *Antimony* | | |
| $Sb_2O_5(s) + 6\,H^+ + 4\,e^- \rightleftharpoons 2\,SbO^+ + 3\,H_2O$ | $+0.581$ | |
| *Arsenic* | | |
| $H_3AsO_4 + 2\,H^+ + 2\,e^- \rightleftharpoons H_3AsO_3 + H_2O$ | $+0.559$ | 0.577, 1-M HCl, $HClO_4$ |
| *Barium* | | |
| $Ba^{2+} + 2\,e^- \rightleftharpoons Ba(s)$ | $-2.906$ | |
| *Beryllium* | | |
| $Be^{2+} + 2\,e^- \rightleftharpoons Be(s)$ | $-1.85$ | |
| *Bismuth* | | |
| $BiO^+ + 2\,H^+ + 3\,e^- \rightleftharpoons Bi(s) + H_2O$ | $+0.320$ | |
| $BiCl_4^- + 3\,e^- \rightleftharpoons Bi(s) + 4\,Cl^-$ | $+0.16$ | |
| *Bromine* | | |
| $Br_2(l) + 2\,e^- \rightleftharpoons 2\,Br^-$ | $+1.065$ | 1.05, 4-M HCl |
| $Br_2(aq) + 2\,e^- \rightleftharpoons 2\,Br^-$ | $+1.087*$ | |
| $BrO_3^- + 6\,H^+ + 5\,e^- \rightleftharpoons \frac{1}{2}\,Br_2(l) + 3\,H_2O$ | $+1.52$ | |

* The standard potentials of these solutions should be those with 1.00-M $Br_2$ or $I_2$, respectively. However, at 25 °C, the solubilities in water of bromine and iodine are $Br_2$ (0.18 M) and $I_2$ (0.0020 M). Therefore, these are not measurable potentials. However, these extrapolated potentials can be used as any other standard potential as a starting point for calculations for runs not under standard conditions.

| Half-reaction | $E°$ (V) | Formal Potential (V) |
|---|---|---|
| *Bromine (continued)* | | |
| $BrO_3^- + 6\ H^+ + 6\ e^- \rightleftharpoons Br^- + 3\ H_2O$ | $+1.44$ | |
| *Cadmium* | | |
| $Cd^{2+} + 2\ e^- \rightleftharpoons Cd(s)$ | $-0.403$ | |
| *Calcium* | | |
| $Ca^{2+} + 2\ e^- \rightleftharpoons Ca(s)$ | $-2.866$ | |
| *Carbon* | | |
| $C_6H_4O_2$ (quinone) $+ 2\ H^+ + 2\ e^- \rightleftharpoons C_6H_4(OH)_2$ | $+0.699$ | 0.696, 1-M HCl, HClO$_4$, H$_2$SO$_4$ |
| $2\ CO_2(g) + 2\ H^+ + 2\ e^- \rightleftharpoons H_2C_2O_4$ | $-0.49$ | |
| *Cerium* | | |
| $Ce^{4+} + e^- \rightleftharpoons Ce^{3+}$ | | 1.70, 1-M HClO$_4$; 1.61, 1-M HNO$_3$; 1.44, 1-M H$_2$SO$_4$; 1.28, 1-M HCL |
| *Chlorine* | | |
| $Cl_2(g) + 2\ e^- \rightleftharpoons 2\ Cl^-$ | $+1.359$ | |
| $HClO + H^+ + e^- \rightleftharpoons \frac{1}{2}\ Cl_2(g) + H_2O$ | $+1.63$ | |
| $ClO_3^- + 6\ H^+ + 5\ e^- \rightleftharpoons \frac{1}{2}\ Cl_2(g) + 3\ H_2O$ | $+1.47$ | |
| $ClO_4^- + 2\ H^+ + 2\ e^- \rightleftharpoons ClO_3^- + H_2O$ | $+1.19$ | |
| *Chromium* | | |
| $Cr^{3+} + e^- \rightleftharpoons Cr^{2+}$ | $-0.408$ | $-0.37$, 0.1–0.5-M H$_2$SO$_4$ |
| $Cr^{3+} + 3\ e^- \rightleftharpoons Cr(s)$ | $-0.744$ | |
| $Cr_2O_7^{2-} + 14\ H^+ + 6\ e^- \rightleftharpoons 2\ Cr^{3+} + 7\ H_2O$ | $+1.33$ | $+1.00$, 1-M HCl; 1.025, 1-M HClO$_4$ |
| *Cobalt* | | |
| $Co^{2+} + 2\ e^- \rightleftharpoons Co(s)$ | $-0.277$ | |
| $Co^{3+} + e^- \rightleftharpoons Co^{2+}$ | $+1.808$ | |
| *Copper* | | |
| $Cu^{2+} + 2\ e^- \rightleftharpoons Cu(s)$ | $+0.337$ | |
| $Cu^{2+} + e^- \rightleftharpoons Cu^+$ | $+0.153$ | |
| $Cu^+ + e^- \rightleftharpoons Cu(s)$ | $+0.521$ | |
| $Cu^{2+} + I^- + e^- \rightleftharpoons CuI(s)$ | $+0.86$ | |
| $CuI(s) + e^- \rightleftharpoons Cu(s) + I^-$ | $-0.185$ | |
| *Fluorine* | | |
| $F_2 + 2\ H^+ + 2\ e^- \rightleftharpoons 2\ HF(aq)$ | $+3.06$ | |
| *Gold* | | |
| $Au^{3+} + 2\ e^- \rightleftharpoons Au^+$ | $+1.41$ | |

| Half-reaction | $E°$ (V) | Formal Potential (V) |
|---|---|---|
| *Gold (continued)* | | |
| $Au^{3+} + 3\,e^- \rightleftharpoons Au(s)$ | $+1.50$ | |
| *Hydrogen* | | |
| $2\,H^+ + 2\,e^- \rightleftharpoons H_2(g)$ (NHE) | $0.000$ | $-0.005$, 1-M HCl, HClO$_4$ |
| $2\,H_2O + 2\,e^- \rightleftharpoons H_2(g) + 2\,OH^-$ | $-0.828$ | |
| *Indium* | | |
| $In^{3+} + 3\,e^- \rightleftharpoons In(s)$ | $-0.33$ | |
| *Iodine* | | |
| $I_2(s) + 2\,e^- \rightleftharpoons 2\,I^-$ | $+0.5355$ | |
| $I_2(aq) + 2\,e^- \rightleftharpoons 2\,I^-$ | $+0.615*$ | |
| $I_3^- + 2\,e^- \rightleftharpoons 3\,I^-$ | $+0.536$ | |
| $ICl_2^- + e^- \rightleftharpoons \frac{1}{2}\,I_2(s) + 2\,Cl^-$ | $+1.056$ | |
| $IO_3^- + 6\,H^+ + 5\,e^- \rightleftharpoons \frac{1}{2}\,I_2(s) + 3\,H_2O$ | $+1.196$ | |
| $IO_3^- + 6\,H^+ + 5\,e^- \rightleftharpoons \frac{1}{2}\,I_2(aq) + 3\,H_2O$ | $+1.178*$ | |
| $IO_3^- + 2\,Cl^- + 6\,H^+ + 4\,e^- \rightleftharpoons ICl_2^- + 3\,H_2O$ | $+1.24$ | |
| $H_5IO_6 + H^+ + 2\,e^- \rightleftharpoons IO_3^- + 3\,H_2O$ | $+1.601$ | |
| *Iron* | | |
| $Fe^{2+} + 2\,e^- \rightleftharpoons Fe(s)$ | $-0.440$ | |
| $Fe^{3+} + e^- \rightleftharpoons Fe^{2+}$ | $+0.771$ | 0.700, 1-M HCl; 0.732, 1-M HClO$_4$; 0.68, 1-M H$_2$SO$_4$; 0.46, 2-M H$_3$PO$_4$ |
| $Fe(CN)_6^{3-} + e^- \rightleftharpoons Fe(CN)_6^{4-}$ | $+0.36$ | 0.71, 1-M HCl; 0.72, 1-M HClO$_4$, H$_2$SO$_4$ |
| *Lead* | | |
| $Pb^{2+} + 2\,e^- \rightleftharpoons Pb(s)$ | $-0.126$ | $-0.14$, 1-M HClO$_4$; $-0.29$, 1-M H$_2$SO$_4$ |
| $PbO_2(s) + 4\,H^+ + 2\,e^- \rightleftharpoons Pb^{2+} + 2\,H_2O$ | $+1.455$ | |
| $PbSO_4(s) + 2\,e^- \rightleftharpoons Pb(s) + SO_4^{2-}$ | $-0.350$ | |
| *Lithium* | | |
| $Li^+ + e^- \rightleftharpoons Li(s)$ | $-3.045$ | |

\* The standard potentials of these solutions should be those with 1.00-M Br$_2$ or I$_2$, respectively. However, at 25 °C, the solubilities in water of bromine and iodine are Br$_2$ (0.18 M) and I$_2$ (0.0020 M). Therefore, these are not measurable potentials. However, these extrapolated potentials can be used as any other standard potential as a starting point for calculations for runs not under standard conditions.

| Half-reaction | $E°$ (V) | Formal Potential (V) |
|---|---|---|
| *Magnesium* | | |
| $Mg^{2+} + 2\ e^- \rightleftharpoons Mg(s)$ | $-2.363$ | |
| *Manganese* | | |
| $Mn^{2+} + 2\ e^- \rightleftharpoons Mn(s)$ | $-1.180$ | |
| $Mn^{3+} + e^- \rightleftharpoons Mn^{2+}$ | | 1.51, 7.5-M $H_2SO_4$ |
| $MnO_2(s) + 4\ H^+ + 2\ e^- \rightleftharpoons Mn^{2+} + 2\ H_2O$ | $+1.23$ | 1.24, 1-M $HClO_4$ |
| $MnO_4^- + 8\ H^+ + 5\ e^- \rightleftharpoons Mn^{2+} + 4\ H_2O$ | $+1.51$ | |
| $MnO_4^- + 4\ H^+ + 3\ e^- \rightleftharpoons MnO_2(s) + 2\ H_2O$ | $+1.695$ | |
| $MnO_4^- + e^- \rightleftharpoons MnO_4^{2-}$ | $+0.564$ | |
| *Mercury* | | |
| $Hg_2^{2+} + 2\ e^- \rightleftharpoons 2\ Hg(l)$ | $+0.788$ | 0.274, 1-M HCl; 0.776, 1-M $HClO_4$; 0.674, 1-M $H_2SO_4$ |
| $2\ Hg^{2+} + 2\ e^- \rightleftharpoons Hg_2^{2+}$ | $+0.920$ | 0.907, 1-M $HClO_4$ |
| $Hg^{2+} + 2\ e^- \rightleftharpoons Hg(l)$ | $+0.854$ | |
| $Hg_2Cl_2(s) + 2\ e^- \rightleftharpoons 2\ Hg(l) + 2\ Cl^-$ (Calomel) | $+0.268$ | SCE 0.242, sat'd KCl; 0.282, 1-M KCl; 0.334, 0.1-M KCl |
| $Hg_2SO_4(s) + 2\ e^- \rightleftharpoons 2\ Hg(l) + SO_4^{2-}$ | $+0.615$ | |
| *Molybdenum* | | |
| $Mo^{6+} + e^- \rightleftharpoons Mo^{5+}$ | $+0.48$ | $+0.53$, 2-M HCl |
| $Mo^{4+} + e^- \rightleftharpoons Mo^{3+}$ | | $+0.1$, 4.5-M $H_2SO_4$ |
| *Nickel* | | |
| $Ni^{2+} + 2\ e^- \rightleftharpoons Ni(s)$ | $-0.250$ | |
| *Nitrogen* | | |
| $N_2(g) + 5\ H^+ + 4\ e^- \rightleftharpoons N_2H_5^+$ | $-0.23$ | |
| $HNO_2 + H^+ + e^- \rightleftharpoons NO(g) + H_2O$ | $+1.00$ | |
| $NO_3^- + 3\ H^+ + 2\ e^- \rightleftharpoons HNO_2 + H_2O$ | $+0.94$ | 0.92, 1-M $HNO_3$ |
| *Oxygen* | | |
| $H_2O_2 + 2\ H^+ + 2\ e^- \rightleftharpoons 2\ H_2O$ | $+1.776$ | |
| $HO_2^- + H_2O + 2\ e^- \rightleftharpoons 3\ OH^-$ | $+0.88$ | |
| $O_2(g) + 4\ H^+ + 4\ e^- \rightleftharpoons 2\ H_2O$ | $+1.229$ | |
| $O_2(g) + 2\ H^+ + 2\ e^- \rightleftharpoons H_2O_2$ | $+0.682$ | |
| $O_3(g) + 2\ H^+ + 2\ e^- \rightleftharpoons O_2(g) + H_2O$ | $+2.07$ | |
| *Palladium* | | |
| $Pd^{2+} + 2\ e^- \rightleftharpoons Pd(s)$ | | 0.987, 4-M $HClO_4$ |

| Half-reaction | $E°$ (V) | Formal Potential (V) |
|---|---|---|
| *Phosphorous* | | |
| $H_3PO_4(aq) + 2\ H^+ + 2\ e^- \rightleftharpoons H_3PO_3(aq) + H_2O$ | $-0.276$ | |
| $H_3PO_3(aq) + 2\ H^+ + 2\ e^- \rightleftharpoons H_3PO_2(aq) + H_2O$ | $-0.50$ | |
| *Platinum* | | |
| $Pt^{2+} + 2\ e^- \rightleftharpoons Pt(s)$ | $+1.2$ | |
| $PtCl_4^{2-} + 2\ e^- \rightleftharpoons Pt(s) + 4\ Cl^-$ | $+0.73$ | |
| $PtCl_6^{2-} + 2\ e^- \rightleftharpoons PtCl_4^{2-} + 2\ Cl^-$ | $+0.68$ | |
| *Potassium* | | |
| $K^+ + e^- \rightleftharpoons K(s)$ | $-2.925$ | |
| *Selenium* | | |
| $Se + 2\ H^+ + 2\ e^- \rightleftharpoons H_2Se$ | $-0.40$ | |
| $H_2SeO_3 + 4\ H^+ + 4\ e^- \rightleftharpoons Se(s) + 3\ H_2O$ | $+0.740$ | |
| $SeO_4^{2-} + 4\ H^+ + 2\ e^- \rightleftharpoons H_2SeO_3 + H_2O$ | $+1.15$ | |
| *Silver* | | |
| $Ag^+ + e^- \rightleftharpoons Ag(s)$ | $+0.799$ | 0.228, 1-M HCl; 0.792, 1-M HClO$_4$; 0.77, 1-M H$_2$SO$_4$ |
| $AgBr(s) + e^- \rightleftharpoons Ag(s) + Br^-$ | $+0.073$ | |
| $AgCl(s) + e^- \rightleftharpoons Ag(s) + Cl^-$ | $+0.222$ | 0.228, 1-M KCl |
| $Ag(CN)_2^- + e^- \rightleftharpoons Ag(s) + 2\ CN^-$ | $-0.31$ | |
| $Ag_2CrO_4(s) + 2\ e^- \rightleftharpoons 2\ Ag(s) + CrO_4^{2-}$ | $+0.446$ | |
| $AgI(s) + e^- \rightleftharpoons Ag(s) + I^-$ | $-0.151$ | |
| $Ag(S_2O_3)_2^{3-} + e^- \rightleftharpoons Ag(s) + 2\ S_2O_3^{2-}$ | $+0.017$ | |
| *Sodium* | | |
| $Na^+ + e^- \rightleftharpoons Na(s)$ | $-2.713$ | |
| *Strontium* | | |
| $Sr^{2+} + 2\ e^- \rightleftharpoons Sr(s)$ | $-2.89$ | |
| *Sulfur* | | |
| $S(s) + 2\ H^+ + 2\ e^- \rightleftharpoons H_2S(g)$ | $+0.141$ | |
| $H_2SO_3 + 4\ H^+ + 4\ e^- \rightleftharpoons S(s) + 3\ H_2O$ | $+0.450$ | |
| $S_4O_6^{2-} + 2\ e^- \rightleftharpoons 2\ S_2O_3^{2-}$ | $+0.08$ | |
| $SO_4^{2-} + 4\ H^+ + 2\ e^- \rightleftharpoons H_2SO_3 + H_2O$ | $+0.172$ | $+0.07$, 1-M H$_2$SO$_4$ |
| $S_2O_8^{2-} + 2\ e^- \rightleftharpoons 2\ SO_4^{2-}$ | $+2.01$ | |
| *Thallium* | | |
| $Tl^+ + e^- \rightleftharpoons Tl(s)$ | $-0.336$ | $-0.551$, 1-M HCl; $-0.33$, 1-M HClO$_4$, H$_2$SO$_4$ |
| $Tl^{3+} + 2\ e^- \rightleftharpoons Tl^+$ | $+1.25$ | 0.77, 1-M HCl |

| Half-reaction | $E°$ (V) | Formal Potential (V) |
|---|---|---|
| *Tin* | | |
| $Sn^{2+} + 2 e^- \rightleftharpoons Sn(s)$ | $-0.136$ | $-0.16$, 1-M $HClO_4$ |
| $Sn^{4+} + 2 e^- \rightleftharpoons Sn^{2+}$ | $+0.154$ | 0.14, 1-M HCl |
| *Titanium* | | |
| $Ti^{3+} + e^- \rightleftharpoons Ti^{2+}$ | $-0.369$ | |
| $TiO^{2+} + 2 H^+ + e^- \rightleftharpoons Ti^{3+} + H_2O$ | $+0.099$ | 0.04, 1-M $H_2SO_4$ |
| *Tungsten* | | |
| $2 WO_3(s) + 2 H^+ + 2 e^- \rightleftharpoons W_2O_5 + H_2O$ | $-0.03$ | |
| $W^{6+} + e^- \rightleftharpoons W^{5+}$ | | $+0.26$, 12-M HCl |
| $W^{5+} + e^- \rightleftharpoons W^{4+}$ | | $-0.03$, 12-M HCl |
| *Uranium* | | |
| $UO_2^{2+} + 4H^+ + 2 e^- \rightleftharpoons U^{4+} + 2 H_2O$ | $+0.334$ | |
| $U^{4+} + e^- \rightleftharpoons U^{3+}$ | $-0.61$ | $-0.64$, 1-M HCl |
| $UO_2^{2+} + e^- \rightleftharpoons UO_2^+$ | $+0.05$ | |
| *Vanadium* | | |
| $V^{3+} + e^- \rightleftharpoons V^{2+}$ | $-0.256$ | $-0.21$, 1-M $HClO_4$ |
| $VO^{2+} + 2 H^+ + e^- \rightleftharpoons V^{3+} + H_2O$ | $+0.359$ | |
| $V(OH)_4^+ + 2 H^+ + e^- \rightleftharpoons VO^{2+} + 3 H_2O$ | $+1.00$ | 1.02, 1-M HCl, $HClO_4$ |
| *Zinc* | | |
| $Zn^{2+} + 2 e^- \rightleftharpoons Zn(s)$ | $-0.763$ | |
| $Zn(NH_3)_4^{2+} + 2 e^- \rightleftharpoons Zn(s) + 4 NH_3$ | $-1.04$ | |

*Sources for E° values:* deBethune, A. J.; Loud, N. A. S., "Standard Aqueous Electrode Potentials and Temperature Coefficients at 25 °C"; Clifford A. Hampel: Skokie, IL, 1964.

Milazzo, G.; Caroli, S.; Sharma, V. K. "Tables of Standard Electrode Potentials"; Wiley: New York, 1978.

*Sources for formal potentials:* Swift, E. H.; Butler, E. A., "Quantitative Measurements and Chemical Equilibria"; W. H. Freeman: San Francisco. Copyright 1972.

Meites, L., Ed., "Handbook of Analytical Chemistry"; McGraw–Hill: New York, 1963.

# Representative Stepwise Formation Constants

For $K_a$ values of the ligands, see Appendix II. Binding sites are boldface.

## Ammonia: $NH_3$

| Cation | Conditions* | $\log K_1$ | $\log K_2$ | $\log K_3$ | $\log K_4$ | $\log K_5$ | $\log K_6$ |
|---|---|---|---|---|---|---|---|
| $Ag^+$ | a | 3.26 | 3.94 | — | — | — | — |
| $Cd^{2+}$ | b | 2.62 | 2.17 | 1.37 | 0.94 | −0.2 | — |
| $Co^{2+}$ | c | 2.10 | 1.57 | 1.11 | 0.75 | 0.2 | −0.6 |
| $Cu^{2+}$ | b | 4.12 | 3.51 | 2.88 | 2.1 | — | — |
| $Ni^{2+}$ | a | 2.8 | 2.3 | 1.8 | 1.3 | 0.8 | 0.2 |
| $Zn^{2+}$ | b | 2.32 | 2.49 | 2.30 | 2.21 | — | — |

## Cyanide: $CN^-$

| Cation | Conditions* | $\log K_1$ | $\log K_2$ | $\log K_3$ | $\log K_4$ |
|---|---|---|---|---|---|
| $Ag^+$ | b | $\log K_1 K_2 = 20.0$ | | — | — |
| $Cd^{2+}$ | d | 5.55 | 5.2 | 4.8 | 3.5 |
| $Hg^{2+}$ | e | 18.00 | 16.71 | 3.83 | 3.0 |
| $Zn^{2+}$ | d | 5.3 | 6.4 | 5.0 | 4.9 |

## Ethylenediamine: $H_2NCH_2CH_2NH_2$

| Cation | Conditions* | $\log K_1$ | $\log K_2$ | $\log K_3$ |
|---|---|---|---|---|
| $Ag^+$ | e | 4.70 | 3.00 | — |
| $Cd^{2+}$ | f | 5.47 | 4.55 | 2.07 |
| $Co^{2+}$ | b | 5.93 | 4.73 | 3.30 |
| $Cu^{2+}$ | g | 11.12 | 9.61 | — |
| $Hg^{2+}$ | h | 14.3 | 9.0 | — |
| $Mn^{2+}$ | i | 2.73 | 2.06 | 0.88 |
| $Ni^{2+}$ | h | 7.52 | 6.28 | 4.26 |
| $Zn^{2+}$ | h | 4.71 | 4.66 | 1.72 |

## Nitrilotriacetic acid (NTA)

| Cation | Conditions* | $\log K_1$ | $\log K_2$ |
|---|---|---|---|
| $Ag^+$ | a | 5.16 | — |
| $Al^{3+}$ | c | 11.37 | — |
| $Ba^{2+}$ | a | 4.82 | — |
| $Ca^{2+}$ | a | 6.41 | — |
| $Cd^{2+}$ | a | 9.83 | 3.5 |

## Nitrilotriacetic acid (NTA) — *(continued)*

| Cation | Conditions* | $\log K_1$ | $\log K_2$ |
|--------|-------------|-----------|-----------|
| $Co^{2+}$ | $a$ | 8.1 | — |
| $Cu^{2+}$ | $a$ | 13.05 | — |
| $Mg^{2+}$ | $a$ | 5.46 | — |
| $Mn^{2+}$ | $a$ | 7.36 | — |
| $Ni^{2+}$ | $a$ | 11.54 | — |
| $Pb^{2+}$ | $a$ | 11.47 | — |
| $Zn^{2+}$ | $a$ | 10.44 | — |

## Ethylenediaminetetraacetic acid: edta

| Cation | Conditions* | $\log K_1$ |
|--------|-------------|-----------|
| $Ag^+$ | $a$ | 7.11 |
| $Al^{3+}$ | $c$ | 16.01 |
| $Ca^{2+}$ | $a$ | 10.6 |
| $Cd^{2+}$ | $a$ | 16.46 |
| $Cu^{2+}$ | $a$ | 18.80 |
| $Mg^{2+}$ | $a$ | 8.69 |
| $Mn^{2+}$ | $a$ | 14.04 |
| $Ni^{2+}$ | $a$ | 18.62 |
| $Pb^{2+}$ | $a$ | 18.32 |
| $Zn^{2+}$ | $a$ | 16.26 |

* Values for °C, $\mu$ are as follows: *(a)* 20°, 0.1; *(b)* 25°, 0.1; *(c)* 25°, 0.2; *(d)* 25°, 1.0; *(e)* 25°, 2.0; *(f)* 25°, 3.0; *(g)* 30°, 0.1; *(h)* 30°, 0.5; *(i)* 30°, 1.0; *(j)* 30°, 2.0. Ionic strength provided by assumed noncoordinating ions.

*Sources:* Sillen, L. G.; Martell, A. E., "Stability Constants of Metal-Ion Complexes," Special Publication No. 17; The Chemical Society: London, 1964.

Sillen, L. G.; Martell, A. E., "Stability Constants of Metal-Ion Complexes, Supplement No. 1," Special Publication No. 17.; The Chemical Society: London, 1961.

Smith, R. M.; Martell, A. E. "Critical Stability Constants," Vol. 4; Plenum: New York and London, 1976.

# Compounds for Elemental Standard Solutions

| Element | Compound | Formula Weight (g) | 1000 ppm (g/L) | Solvent | Comments |
|---------|----------|--------------------|----------------|---------|----------|
| Aluminum | Al metal | 26.982 | 1.0000 | Hot, dil. HCl | b |
| Antimony | $KSbOC_4H_4O_6 \cdot \frac{1}{2} H_2O$ | 333.92 | 2.7427 | Water | l |
| Arsenic | $As_2O_3$ | 197.84 | 2.6406 | dil. HCl | a,c,n |
| Barium | $BaCO_3$ | 197.35 | 1.4369 | dil. HCl | |
| Beryllium | Be metal | 9.0122 | 1.0000 | HCl | c |
| Bismuth | $Bi_2O_3$ | 465.96 | 1.1148 | $HNO_3$ | |
| Boron | $H_3BO_3$ | 61.84 | 5.7200 | Water | o |
| Bromine | KBr | 119.01 | 1.4894 | Water | b |
| Cadmium | CdO | 128.40 | 1.1423 | $HNO_3$ | |
| Calcium | $CaCO_3$ | 100.09 | 2.4972 | dil. HCl | a |
| Cerium | $(NH_4)_2Ce(NO_3)_6$ | 548.23 | 3.9126 | Water | |
| Cesium | $Cs_2SO_4$ | 361.87 | 1.3614 | Water | |
| Chromium | $K_2Cr_2O_7$ | 294.19 | 2.8290 | Water | a |
| Cobalt | Co metal | 58.933 | 1.0000 | $HNO_3$ | b |
| Copper | Cu metal | 63.546 | 1.0000 | dil. $HNO_3$ | b |
| | CuO | 79.545 | 1.2517 | HCl, hot | b |
| Dysprosium | $Dy_2O_3$ | 373.00 | 1.1477 | HCl, hot | e |
| Erbium | $Er_2O_3$ | 382.56 | 1.1435 | HCl, hot | e |
| Europium | $Eu_2O_3$ | 351.92 | 1.1579 | HCl, hot | e |
| Fluorine | NaF | 41.988 | 2.2101 | Water | g |
| Gadolinium | $Gd_2O_3$ | 362.50 | 1.1526 | HCl, hot | e |
| Gallium | Ga metal | 69.72 | 1.0000 | $HNO_3$, hot | f |
| Germanium | $GeO_2$ | 104.60 | 1.4410 | 1 $m$ NaOH, hot | |
| Gold | Au metal | 196.97 | 1.0000 | Aqua Regia, hot | b |
| Hefnium | Hf metal | 178.49 | 1.0000 | Hf, fusion | i |
| Holmium | $Ho_2O_3$ | 377.86 | 1.1455 | HCl, hot | e |
| Indium | $In_2O_3$ | 277.64 | 1.2090 | HCl, hot | |
| Iodine | $KIO_3$ | 214.00 | 1.6863 | Water | a |
| Iridium | No suitable compound was found ||||| 
| Iron | Fe metal | 55.47 | 1.0000 | HCl, hot | b |
| Lanthanum | $La_2O_3$ | 325.82 | 1.1728 | HCl, hot | e |
| Lead | $Pb(NO_3)_2$ | 331.20 | 1.5985 | Water | b |
| Lithium | $Li_2CO_3$ | 73.890 | 5.3243 | HCl | b |
| Lutetium | $Lu_2O_3$ | 397.94 | 1.1372 | HCl, hot | e |
| Magnesium | MgO | 40.31 | 1.658 | HCl | |
| Manganese | $MnSO_4 \cdot H_2O$ | 169.01 | 3.0764 | Water | j |
| Mercury | $HgCl_2$ | 271.50 | 1.3535 | Water | c |
| Molybdenum | $MoO_3$ | 143.94 | 1.5003 | 1 $m$ NaOH | |
| Neodymium | $Nd_2O_3$ | 336.48 | 1.1664 | HCl | e |
| Nickel | Ni metal | 58.71 | 1.0000 | $HNO_3$, hot | b |
| Niobium | $Nb_2O_5$ | 265.81 | 1.4305 | HF, fusion | k |
| Osmium | Os metal | 190.20 | 1.0000 | $H_2SO_4$, hot | d |
| Palladium | Pd metal | 106.40 | 1.0000 | $HNO_3$, hot | |
| Phosphorus | $KH_2PO_4$ | 136.09 | 4.3937 | Water | |
| Platinum | $K_2PtCl_4$ | 415.12 | 2.1278 | Water | |
| Potassium | KCl | 74.55 | 1.9065 | Water | b |
| | $KHC_8H_4O_4$ | 204.22 | 5.2228 | Water | a,n |
| | $K_2Cr_2O_7$ | 294.19 | 3.7618 | Water | a,n |
| Praseodymium | $Pr_6O_{11}$ | 1021.43 | 1.2082 | HCl | e |
| Rhenium | Re metal | 186.2 | 1.0000 | $HNO_3$ | |

| Element | Compound | Formula Weight (g) | 1000 ppm (g/L) | Solvent | Comments |
|---|---|---|---|---|---|
|  | $KReO_4$ | 289.30 | 1.5537 | Water |  |
| Rhodium | Rh metal | 102.91 | 1.0000 | $H_2SO_4$, hot |  |
| Rubidium | $Rb_2SO_4$ | 267.00 | 1.5628 | Water |  |
| Ruthenium | $RuO_4$ | 165.07 | 1.6332 | Water |  |
| Samarium | $Sm_2O_3$ | 348.70 | 2.3193 | HCl, hot | e |
| Scandium | $Sc_2O_3$ | 137.91 | 1.5339 | HCl, hot |  |
| Selenium | Se metal | 78.96 | 1.0000 | $HNO_3$, hot |  |
| Silicon | Si metal | 28.086 | 1.0000 | NaOH, Concd. |  |
|  | $SiO_2$ | 60.08 | 2.1391 | HF |  |
| Silver | $AgNO_3$ | 169.87 | 1.5748 | Water | b,m |
| Sodium | NaCl | 58.442 | 2.5428 | Water | a |
|  | $Na_2C_2O_4$ | 134.01 | 2.9146 | Water | a,n |
| Strontium | $SrCO_3$ | 147.63 | 1.6849 | HCl | b |
| Sulfur | $K_2SO_4$ | 174.27 | 5.4351 | Water |  |
| Tantalum | $Ta_2O_5$ | 441.89 | 1.2210 | HF, fusion | k |
| Tellurium | $TeO_2$ | 159.60 | 1.2507 | HCl |  |
| Terbium | $Tb_2O_3$ | 365.85 | 1.1512 | HCl, hot | e |
| Thallium | $Tl_2CO_3$ | 468.75 | 1.1468 | Water | b,c |
| Thulium | $Tm_2O_3$ | 385.87 | 1.1421 | HCl, hot | e |
| Tin | Sn metal | 118.69 | 1.0000 | HCl |  |
|  | SnO | 134.69 | 1.1348 | HCl |  |
| Titanium | Ti metal | 47.90 | 1.0000 | $H_2SO_4$, 1:1 | b |
| Tungsten | $Na_2WO_4 \cdot 2 H_2O$ | 329.86 | 1.7942 | Water | h |
| Uranium | $UO_2$ | 270.03 | 1.1344 | $HNO_3$ |  |
|  | $U_3O_8$ | 842.09 | 1.1792 | $HNO_3$ | a,n |
| Vanadium | $V_2O_5$ | 181.88 | 1.7852 | HCl, hot |  |
| Ytterbium | $Yb_2O_3$ | 394.08 | 1.1386 | HCl, hot | e |
| Yttrium | $Y_2O_3$ | 225.81 | 1.2700 | HCl, hot | e |
| Zinc | ZnO | 81.37 | 1.2448 | HCl | b |
| Zirconium | Zr metal | 91.22 | 1.0000 | HF, fusion | i |

a = Primary standard.

b = These compounds conform very well to the criteria and approach primary standard quality.

c = Highly toxic.

d = Very highly toxic.

e = The rare earth oxides, because they absorb $CO_2$ and water vapor from the air, should be freshly ignited prior to weighing.

f = mp = 29.6 °C. The metal may be warmed and weighed as the liquid.

g = Sodium fluoride solutions will etch glass and should be freshly prepared.

h = Sodium tungstate loses both water molecules at 110 °C. After drying, f.w. = 293.83, 1000 ppm = 1.5982 g, 0.1 m = 29.383 g. The water is not rapidly regained, but the compound should be kept in a desiccator after drying and should be weighed quickly once it is removed.

i = Zirconium and hafnium compounds were not investigated in the laboratory. The following methods have been recommended for dissolution of zirconium and hafnium.

1.0000 g of the powdered metal is placed in a platinum dish with 5 – 10 mL of water and 1 – 2 mL of HF (1:5) and covered with a platinum lid or a paraffined watch glass. Once dissolved, the fluorine may be removed by adding 1 – 2 mL of sulfuric acid (cold) and evaporating to dense fumes or to dryness if required.

A fusion method may also be used. A 5 – 10 fold excess of $K_2S_2O_7$ is placed in a platinum or quartz crucible along with the sample. After melting to a homogeneous molten mass in a muffle furnace or burner, the fusion product is dissolved in 2N sulfuric acid.

A third method avoids the use of platinum ware. The sample of the metal is finely ground and placed in a small heat-resistant beaker. Two to four grams of ammonium sulfate and 3 – 6 mL of sulfuric acid are then added. A homogeneous melt is obtained on a hot plate and dissolved in 2N sulfuric acid.

j = $MnSO_4 \cdot H_2O$ may be dried at 110 °C without losing the water of hydration.

k = Niobium and tantalum pentoxides are slowly soluble in 40% HF. The addition of $H_2SO_4$ accelerates the solution process. They may also be dissolved by a fusion technique. $K_2S_2O_7$ is an often used flux. The

*(Continued on following page.)*

pentoxides are fully decomposed at 650–800 °C in the presence of an 8–10 fold amount of potassium pyrosulfate. A quartz or porcelain crucible is suitable, and the resulting melt may be dissolved in sulfuric acid. Cold $HF/H_2SO_4$ has been used successfully in plastic beakers with a 10-h solution time.

$l$ = Antimony potassium tartrate loses the $\frac{1}{2}$ $H_2O$ with drying at 110 °C. After drying, f.w. = 324.92, 1000 ppm = 2.6687 g, 0.1 $m$ = 32.492 g. The water is not rapidly regained, but the compound should be kept in a desiccator after drying and should be weighed quickly once it is removed. The dried compound is water soluble.

$m$ = When kept dry, silver nitrate crystals are not affected by light. Solutions of silver nitrate should be stored in brown bottles.

$n$ = These compounds are sold as primary standards by the National Bureau of Standards, Office of Standard Reference Materials, Washington, D.C. 20234.

$o$ = Boric acid may be weighed accurately directly from the bottle. It will loose one $H_2O$ molecule at 100 °C and a second $H_2O$ molecule at approximately 130–140 °C and is difficult to dry to a constant weight.

*Source:* Smith, B. W.; Parsons, M. L. *J. Chem. Educ.* **1973,** *50,* 679–681.

# Values of $t$ and $Q$ to Calculate the Confidence Limit at Various Levels of Probability

For the equation $\mu = \overline{X} \pm Q \cdot s.$     See Eq. 4-10 for the definition of $t$.

| | Probability Level (%) | | | | | | | |
|---|---|---|---|---|---|---|---|---|
| | 80 | | 90 | | 95 | | 99 | |
| $N$ | $t$ | $Q$ | $t$ | $Q$ | $t$ | $Q$ | $t$ | $Q$ |
| 2 | 3.08 | 2.18 | 6.31 | 4.46 | 12.71 | 8.99 | 63.7 | 4.50 |
| 3 | 1.89 | 1.09 | 2.92 | 1.69 | 4.30 | 2.48 | 9.92 | 5.73 |
| 4 | 1.64 | 0.82 | 2.35 | 1.18 | 3.18 | 1.59 | 5.84 | 2.92 |
| 5 | 1.53 | 0.68 | 2.13 | 0.95 | 2.78 | 1.24 | 4.60 | 2.06 |
| 6 | 1.48 | 0.60 | 2.02 | 0.82 | 2.57 | 1.05 | 4.03 | 1.65 |
| 7 | 1.44 | 0.54 | 1.94 | 0.73 | 2.45 | 0.94 | 3.71 | 1.40 |
| 8 | 1.42 | 0.50 | 1.90 | 0.67 | 2.36 | 0.83 | 3.50 | 1.24 |
| 9 | 1.40 | 0.47 | 1.86 | 0.62 | 2.31 | 0.77 | 3.36 | 1.12 |
| 10 | 1.38 | 0.44 | 1.83 | 0.58 | 2.26 | 0.71 | 3.25 | 1.03 |
| 20 | 1.33 | 0.30 | 1.73 | 0.39 | 2.10 | 0.47 | 2.88 | 0.64 |
| 60 | 1.30 | 0.17 | 1.67 | 0.22 | 2.00 | 0.26 | 2.66 | 0.34 |
| $\infty$ | 1.28 | — | 1.64 | — | 1.96 | — | 2.58 | — |

# Colorimetric Indicators for Neutralization Titrations

Color abbreviations: b blue, c colorless, g green, o orange, p purple, r red, and y yellow

| Chemical Name | Common Name | Approximate pH Range | Colors | | Common Stock Solutions* |
| | | | Acid Form | Base Form | |
| --- | --- | --- | --- | --- | --- |
| 2,4,6-Trinitrophenol | Picric acid | 0.6–1.3 | c | y | — |
| Thymolsulfonphthalein | Thymol blue | 1.2–2.8 | r | y | 0.04% aq. |
| 2,4-Dinitrophenol | $\alpha$-Dinitrophenol | 2.4–4.0 | c | y | 0.1% alc. |
| Tetrabromophenol-sulfonphthalein | Bromophenol blue | 3.0–4.6 | y | b | 0.04% aq. |
| Dimethylaminoazoben-zene-$p$-sulfonate | Methyl orange | 3.1–4.4 | r | o | 0.1% aq. |
| Tetrabromo-$m$-cresol-sulfonphthalein | Bromocresol green | 3.8–5.4 | y | b | 0.1% aq. |
| Dimethylaminoazoben-zene-$o$-carboxylic acid | Methyl red | 4.2–6.3 | r | y | 0.1% in 60% alc. |
| Dibromo-$o$-cresolsulfon-phthalein | Bromocresol purple | 5.2–6.8 | y | p | 0.04% aq. |
| Dibromothymolsulfon-phthalein | Bromothymol blue | 6.2–7.6 | y | b | 0.05% aq. |
| Phenolsulfonphthalein | Phenol red | 6.8–8.4 | y | r | 0.05% aq. |
| $o$-Cresolsulfonphthalein | Cresol red | 7.2–8.8 | y | r | 0.05% aq. |
| Thymolsulfonphthalein | Thymol blue | 8.0–9.6 | y | b | 0.04% aq. |
| Di-$p$-dioxydiphenyl-phthalide | Phenolphthalein | 8.3–10.0 | c | p | 0.05% in 50% alc. |
| Dithymolphthalide | Thymolphthalein | 9.3–10.5 | c | b | 0.04% in 50% alc. |
| $m$-Nitrobenzeneazo-salicylic acid | Alizarin yellow GG | 10.0–12.0 | c | y | 0.1% alc. |
| 2,4,6-Trinitrophenol-methylnitramine | Nitramine | 10.8–13.0 | c | o | 0.01% aq. |

* Solution abbreviations: aq., aqueous; alc., lower alcohols.
*Source:* Streuli, C. A.; Meites, L. In "Handbook of Analytical Chemistry", L. Meites, Ed.; McGraw-Hill: New York, 1963; Section 3.

# ANSWERS TO PROBLEMS

## CHAPTER 1

1.1   $0.57 \times 10^{-3}$ M $K^+$

1.2   a. 10% w/w   b. 90%   c. $1 \times 10^5$ ppm

1.3   a. $K^+$ 86; $Cl^-$ 2747 ppm w/v
     b. $K^+$ $85._5$; $Cl^-$ 2733 $\mu g$ $g^{-1}$
     c. Same numbers but g $kg^{-1}$

1.4   11.1%

1.5   a. 20.0 mg $Ca^{2+}$; 30 mg $CO_3^{2-}$
     b. $9.99 \times 10^{-4}$ M

1.6   94.3 mg $qt^{-1}$

1.7   5.45 mL

1.8   504 g

1.9   a. 0.0786 $Na^+$; 0.121 $Cl^-$ g $L^{-1}$ or mg $mL^{-1}$   b. All 3.42 mM

1.10   6.0 mg $kg^{-1}$ and ppm

1.11   a. $1.22 \times 10^{-4}$ M   b. and c. 40 ppm
      d. $1.22 \times 10^{-4}$ $N$

1.12   2.00

1.13   $8.91 \times 10^{-10}$ M

1.14   $4.68 \times 10^{-4}$ $N$

1.15   0.74 g $H_2SO_4$; 0.55 g HCl

1.16   a. $2.43 \times 10^{-2}$ M   b. 0.121 $N$
      c. $2.43 \times 10^{-2}$ $N$

1.17   1.240 L

1.18   18.00

1.19   47.73 mL

1.20   0.1405 $N$

1.21   1.069 M

1.22   746.7 mL

1.23   0.1108 M

1.24   a. 0.2572 $N$   b. Molarity of what species? Cations not accounted for.

S1.1   a. and b. $2.5 \times 10^2$ s

S1.2   $2.5 \times 10^{-2}$

S1.3   $1.443 \times 10^{-4}$ s

S1.4   —

S1.5   a. $-0.441$ $s^{-1}$; 0.1016 $s^{-1}$   b. 15.7 s for both   c. No

S1.6   a. $3.03 \times 10^4$ s (approximately)
      b. $3.04 \times 10^{-4}$ c. $2.29 \times 10^{-5}$ $s^{-1}$

S1.7   1.21 h

S1.8   a. $\ln([P(t) - P_\infty]/P_\infty)$
      b. $[P(t) - P_\infty]/P_\infty$ with slope multiplied by 2.303 (or equivalent)

## CHAPTER 2

2.1   The $K_p$ expressions can also be correctly written as concentration or thermodynamic equilibrium constants.   a. $P_{H_2}^2 P_{CO}/P_{CH_3OH}$
     b. $[H_2][CO_2]/([CO][H_2O])$
     c. $[NH_4^+][Co(NH_3)_5(OH)^+]/$ $[Co(NH_3)_6^{2+}]$
     d. $[H_2]^4[NO_2]^2/[N_2]$
     e. $P_{CO_2}$   f. $1/(P_{H_2S}P_{NH_3})$

2.2   a. $[\text{gluconate}^-][I^-]^3/$ $([\text{glucose}][OH^-]^3[I_3^-])$
     b. $[\text{aniline}]^2[Cr^{3+}]^4/$ $([H^+]^4[Cr^{2+}]^4[\text{azobenzene}])$

2.3   $H_2O + Bar^- \rightleftharpoons HBar + OH^-$

2.4   pH 3.41

2.5   pH 5.92

2.6   Decrease

2.7   a. Decrease   b. Decrease

2.8   a. I, 0.018; II, 0.045   b. Equal
     c. Greater   d. Less   e. Less
     f. I, 0.74; II, 0.61   g. 0.01

2.9   Ratios A (in order a,b,c): 135, $2.34 \times 10^{-2}$, $1.35 \times 10^{-4}$   Ratios B: $5.76 \times 10^3$, 1.00, $5.76 \times 10^{-3}$
     Ratios C: $2.51 \times 10^5$, 43.6, 0.251

2.10   a,b,c: same

2.11   a. pH 1.00   b. pH 6.98

2.12   $[H_2CO_3]$ $1.59 \times 10^{-5}$ M; $[HCO_3^-]$ 0.067 M; $[CO_3^{2-}]$ 0.032 M

2.13   a. pH 2.70   b. 1.5% pH; 7.9% $[H^+]$

2.14   a. pH 11.12   b. 0.1%   c. 98.7% $NH_3$; 1.3% $NH_4^+$; ammonia water

2.15   a. 11.22   b. $HPO_4^{2-}$; $H_2PO_4^-$

2.16   8.97

2.17   a. $10 e^- + 12 H^+ + 2 ClO_3^- + 2 I^- \rightarrow 2 Cl^- + I_2 + 6 H_2O$
     b. $10 H^+ + Zn(s) + 6 e^- + NO_3^- \rightarrow Zn^{2+} + NH_4^+ + 3 H_2O$ c. $2 ReO_2 + 2 H^+ + Cl_2 + 4 e^- \rightarrow 2 HReO_4 + 2 Cl^-$

2.18   a. $Al^0 + 4 H_2O + e^- \rightarrow Al(OH)_4^- + 2 H_2$ b. $18 OH^- + I_2 + Cl_2 \rightarrow 2 H_3IO_6^{2-} + 2 Cl^- + 6 H_2O + 12 e^-$

c. $3 H_2O + {}^-OH + NiO_2 + Fe^0 \rightarrow$
$Ni(OH)_2(s) + Fe(OH)_3(s) +$
$H_2O + e^-$

2.19   a. $-0.124$   b. $-0.250, 0.513$

2.20   a. 59 mV more negative
b. 118 mV more positive
c. 118 mV more negative

2.21   a. $Mg(s) + 2 Ag^+(aq) \rightarrow$
$Mg^{2+}(aq) + 2 Ag(s)$  b. $Mg(s) \rightarrow$
$Mg^{2+}(aq) + 2 e^-; Ag^+(aq) + e^- \rightarrow$
$Ag(s)$  c. 3.162 V  d. Yes

2.22   $[Ni^{2+}] = 0.217$ M;
$[Co^{2+}] = 0.178$ M

2.23   a. Zero  b. More positive: relevant
to b, d, and e, to be consistent, the
cell must be written with the cell
having reduction on the right.
Therefore, the cell (and potentials)
must be switched appropriately so
the calculations reflect the
experimental facts. Thus d and e
are the same.  c. 0.059 V  d. More
negative  e. More negative
f. 0.059 V

2.24   a. 1.54 V  b. 1.42 V  c. $H_2O_2 +$
$2 e^- = 2 OH^-$; no ($E°$ is for pH = 0
solution)

2.25   $-0.577$ V

2.26   a. 0.06 V  b. $-0.03$ V

2.27   Yes, the measured potential (and
see note in Appendix III)

2.28   $1.66 \times 10^{-10}$

2.29   a. $1.08 \times 10^{-19}$  b. 1.20 V

2.30   a. 0.67 V  b. 0.72 V  c. 0.62 V
d. 0.73 V  e. 0.73 V

S2.1   —

S2.2   a. $5.3 \times 10^{-7}$  b. $1.5 \times 10^{-5}$
c. 0.011

S2.3   0.592 unitless

S2.4   In order pH 2, pH 12
a. $3.7 \times 10^{-14}, 0.98$  b. $1 \times 10^{-3}$,
$1 \times 10^{17}$  c. $1.3 \times 10^5, 0.035$
d. About the same at 2, much worse
at 12 (optimum around pH 4.5)

# CHAPTER 3

3.1   1.4 k$\Omega$

3.2   a. 0.9 mA  b. 90 V, 90 V, 360 V
c. Same as part 2

3.3   a. $\rightarrow \infty$  b. $\rightarrow 0$  c. 2.87

3.4   a. 1 M$\Omega$  b. 0.667 M$\Omega$
c. 0.545 M$\Omega$

3.5   a. $3 \times 10^{-6}$ s  b. 1000 M$\Omega$
c. $0.333 \times 10^{-6}$ F (i.e., $\mu$F)

3.6   a. 3 s  b. 10 k$\Omega$

3.7   300

3.8   a. 10.5 $\mu$s  b. Same as a

3.9   9.75 M$\Omega$

# CHAPTER 4

4.1   10

4.2   a. $40.4 \pm 0.7\%$  b. $40.4 \pm 0.06_6$

4.3   0, the sum

4.4   Mean 21.25, std. dev. $\pm 0.04$, 95%
C.L. $\pm 0.05$

4.5   a. and b. Written as mean $\pm$ std.
dev. (relative standard deviation);
zeros are place holders. All values
are rounded. Mg $5.1 \pm 0.3$ (6%); Al
$4.66 \pm 0.07$ (1.4%); Si $17.7 \pm 0.2$
(1.2%); Ca $7.7 \pm 0.1_4$ (1.8%); Ti
$7.8 \pm 0.1_6$ (2.1%); Fe $15.4 \pm 0.4$
(2.3%); P $410 \pm 70$ (18%); S
$1800 \pm 100$ (6.5%); K $400 \pm 100$
(30%); Cr $2800 \pm 100$ (3.9%);
$2100 \pm 100$ (5.3%)  c. From Dixon
test yes  d. K and K; or excluding
outlier Mn and P

4.6   a. 0.1%–0.2%  b. 0.035 mg
rounded to 0.04 mg

4.7   a. 4.981 mL  b. 0.023 g
c. $\pm 0.0066$  d. $\pm 0.0031$

4.8   1.7%

4.9   a. $(57.76 \pm 0.23)\%$,
$(57.67 \pm 0.30)\%$  b. #1 N = 5, #2
N = 7

4.10   a. 0.0022  —; 0.0023  0.23;
0.0033  0.84; 0.0016  0.19;
0.0039  0.30  b. Constant, subtract
2.7 mg  c. Random, 825 $\mu$g
d. 0.825 g

4.11   a. Baseline = 2.8 units exactly;
$59.4 \pm 0.3$  b. Reading error $\pm 0.1$
gives std. dev. $\pm 0.2_8$ (i.e., little effect)

4.12   a. In %, $+0.5, +0.7, -0.3, +1.1$,
$-1.2, +1.1$ (The choice determines
the sign and has little effect if the
mean values are close. Signs here
are for distillation compared with
electrochemistry.)  b. $+0.3\%$ not
significant by criterion

4.13   a. A–J, respectively, $-0.134$,
$-0.126, -0.103, -0.101, -0.127$,
$-0.128, -0.116, -0.116, -0.107$,

−0.114  b. −11.7%  c. No, must use relative  d. 1.132

4.14 More precisely from the three impurity assays

4.15 a. ±0.07  b. ±0.04

4.16 −25.88 ± 0.01

4.17 $0.01 [(4/A^2) + (26/[B + C]^2) + 2.65 \times 10^6]^{1/2}$ or equivalent

4.18 $[(0.0128/D^2) + (256[A + B − C]^2/D^4)]^{1/2}$

4.19 13.0512

4.20 No; 54% total error

4.21 a. and b. Reported as mean ± std. dev. (total error in %). All numerals are kept, although this may be a questionable practice. Al 4936 ± 1572 (52%); Ca 13,784 ± 7317 (—); Cu 953 ± 66 (89%); Fe 10,626 ± 6627 (55%); Pb 516 ± 81 (59%); Ni 165 ± 27 (99%); Ag 70 ± 25 (13%); Zn 1114 ± 350 (106%)  c. Ni lower than NBS; Zn lower than NBS

4.22 $dR = dA − dB + dC$

4.23 Absolute, $dR = (1/C^2)(CBa + CAb − ABc)$; Relative, $(dR/R) = (a/A) + (b/B) − (c/C)$

4.24 $dR = (dA/A)$  b. $dR = A \exp(A) dA$  c. $dR = 0.879(dA/A)$  d. $dR = 2.193(−A) \exp(−A) dA$

S4.1 — —

S4.2 $r^2 = x^2 + y^2 + z^2$

## CHAPTER 5

5.1 a. Approximately 5.2% b. Approximately 6.0%  c. Relative error approximately 0.13

5.2 a. 120 s  b. Not in a regular manner  c. 1.43

5.3 39.12%

5.4 0.44%

5.5 a. 9.918 L STP  b. 3228 mg/m³; 1601 mg/m³ STP  c. 4035 mg/m³ dry at STP

5.6 a. Instrument 1.7%; automatic 0.82%  b. Automatic  c. Manual 10.6%; automatic 10.8% d. Sampling

5.7 a. 11.02 kg  b. 1.792 g  c. Yes, 16.7%

5.8 a. Chrysene 70; benzofluoranthenes 58; benzo(a) pyrene 63; dibenz(a,h)anthracene 64 b. Respectively, 19%, 56%, 23%, 27%  c. Benzofluoranthenes

5.9 a. 100%  b. $4.49 \times 10^{-6}$ mol NO c. $1.00 \times 10^{-4}$ L STP  d. 10.3 L e. 79 min

5.10 a. $9 \times 10^6$  b. 0.1 mg, 0.113 g, 1.13 kg  c. 1 and 10 $\mu$m

5.11 a. No change  b. Increases c. No change

5.12 1.93%

## CHAPTER 6

6.1 Mean 383 ppm, std. dev. ±3, 95%; C.L. ±4

6.2 328 ppm

6.3 251 ppm

6.4 250 ppm; may vary somewhat

6.5 $C_{sample} = C_{spike} [A_{sample}/(A_{s+s} − A_{sample})]$

6.6 a. 0.0999 ppm  b. 5

6.7 a. From the shapes of the rest, probably should ignore and extrapolate the plateau through it b. 197 ± 4 mV; may vary somewhat  c. Apparently not d. Around 320 mg dL$^{-1}$; may vary somewhat

6.8 a. 90.5%  b. 16.$_4$%  c. 16.$_4$% d. 16%  e. 15.$_6$%

6.9 0.88, 8.61, 1.26 ng/mL

6.10 49.7%

6.11 1.44%

6.12 a. Chromate, hypochlorite, both positive  b. Chromate 0.29; hypochlorite at 10 $F_c$ = 0.14, at 100 $F_c$ = 0.043. Choose the value best for the range being used or determine a different correction equation.  c. Zero

6.13 a. 1040 mg  b. 3.01%

6.14 a. 31.4%  b. 314 kg

6.15 a. All positive  b. Probably pyridine; others within 1 std. dev.  c. No

6.16 a. —  b. Linear 21 ppm; log 19.6 ppm

## CHAPTER 7

7.1 a. 2 ng  b. 6 $\mu$m

7.2 a. Fe about 3 ppb; no  b. 58 $\mu$m c. Sample Mn = 1 ng; dust 5.2 ng; no

7.3 a. 50  b. <0.01, 0.9, 0.95, 7.5, <0.01, <0.005, 0.115 ppb  c. Yes, but barely Mn

7.4 a. 12.8%  b. 4.4%

7.5 a. Hexane  b. Depending on the size of the interferences and Zn level, either hexane, chloroform, or methanol  c. Hexane  d. Methanol; no particular preference, since Na is correctable. However, high solubility reduces solvent costs and glassware size.  e. Not toluene, perhaps not methanol  f. No. The relative error for each is needed to determine this.

7.6 a. 253; 232; 241,000; 4340; 25; 183; 19,700; zero; 1060  b. Na, K, and Zn  c. No. The Na could come from the solvent.

# CHAPTER 8

8.1 a. Low  b. Correct
8.2 a. High: more AgCl forms + excess Ag  b. Less: loss of $Cl_2$
8.3 a. 0.4770  b. 0.5641  c. 0.1892 d. 0.7203  e. 0.8794
8.4 43.9 mg
8.5 a. $3.5 \times 10^{-7}$ M  b. $5.3 \times 10^{-33}$
8.6 a. 0.017 M  b. 6.10, no  c. Add two different concentrations of acid. If same result, then OK; if not, two higher concentrations.
8.7 a. Order Cu, Mn, Ca, Ba since all same stoichiometry, $K_{sp}$ enough b. No
8.8 a. Im, Un, Ev, My  b. pH = 5.78, 9.74, 10.95, 11.65
8.9 a. None  b. $99.7_4\%$
8.10 a. $[K^+] = 0.104$ M, $[CrO_4^{2-}] = 0.0187$ M, $[NO_3^-] = 0.67$ M, $[Ag^+] = 8.3 \times 10^{-6}$ M  b. Yes $[(4.15 \times 10^{-6})/0.0187 = 2 \times 10^{-4}]$ c. 0.012%
8.11 a. $[Pb^{2+}] = 4.97$ mM, $[Ba^{2+}] = 2.92 \times 10^{-5}$ M, $[SO_4^{2-}] = 3.42 \times 10^{-6}$ M  b. mmol $BaSO_4 = 0.993$, mmol $PbSO_4 = 0.017$, mole fractions = 0.983, 0.017

# CHAPTER 9

9.1 2.72
9.2 pH 5.07; 7.15 mL

9.3

| mL | pH | mL | pH |
|----|------|-----|-------|
| 0 | 2.69 | 120 | 4.10 |
| 30 | 2.90 | 145 | 4.96 |
| 50 | 3.06 | 150 | 7.75 |
| 75 | 3.50 | 155 | 10.29 |
| 90 | 3.67 | 200 | 11.22 |

9.4 $4 \times 10^{-4}$ N
9.5 a. 243 mL  b. pH 4.0, 559 mL; pH 5.0, 803 mL
9.6 a. Water throughout  b. 75.00 mL, pH 1.48  c. 100 mL, pH 7.00 d. 75.00 mL, pH 3.48; 100 mL, pH 7.00
9.7 pH 10.97, pOH 3.03, $[CO_3^{2-}]$ $8.5 \times 10^{-3}$ M, $[HCO_3^-]$ $1.5 \times 10^{-3}$ M
9.8 All three cases: 1st 7.81 mL, 2nd 21.48 mL, no
9.9 Both same: 3.90 mL, 14.64 mL, no
9.10 a. $5.9 \times 10^{-4}$ M  b. 0.005 c. 0.010  d. 0.015  e. 25  f. 0.1, 0.105, 0.11, 0.115  g. 15% (0.15)
9.11 a. $5.9 \times 10^{-4}$ M  b. 0.005  c. 0.957, 0.886  d. 1.08
9.12 a. $-0.50$ mL pH 3.3, $-0.10$ mL pH 4.0, $-0.05$ mL pH 4.3, end point 7.00, $+0.05$ mL pH 9.7, $+0.10$ mL pH 10.0, $+0.50$ mL pH 10.7 b. Phenolphthalein, methyl red, bromocresol green  c. $-0.50$ mL pH 5.3, $-0.10$ mL pH 6.0, $-0.05$ mL pH 6.3, end point 7.00, $+0.05$ mL pH 7.7, $+0.10$ mL pH 8.0, $+0.50$ mL pH 8.7  d. 0.1-M titration phenolphthalein, methyl red; 1-mM titration, none
9.13 a. 2.75 $\mu$L  b. 0.44 $\mu$L  c. All same within precision 0.1067
9.14 It is far easier to calculate the volumes from a number of given pH values. Representative values are

| pH | Volume (mL) | pH | Volume (mL) |
|------|------|------|------|
| 1.5 | 0.15 | 3.3 | 3.8 |
| 2 | 0.5 | 3.4 | 4.2 |
| 2.5 | 1.2 | 3.5 | 4.5 |
| 2.75 | 1.9 | 3.6 | 4.9 |
| 3 | 2.7 | 3.7 | 5.3 |
| 3.2 | 3.4 | 3.8 | 5.6 |

| pH | Volume (mL) | pH | Volume (mL) |
|----|----|----|----|
| 3.9 | 6.0 | 5.75 | 9.8 |
| 4 | 6.4 | 6.0 | 9.9 |
| 4.25 | 7.3 | 6.5 | 10.0 |
| 4.5 | 8.1 | after end point | |
| 4.75 | 8.8 | done in usual way | |
| 5 | 9.2 | 10.3 | 10.1 |
| 5.25 | 9.5 | 11.3 | 11.0 |
| 5.5 | 9.7 | | |

9.15 $H_4$ and $H_3$ react with $H_0$ and $H_1$ to give $5 \times H_2$ at one end point! Volume to end point = 25 mL

9.16 0.576-M $CO_3^{2-}$ (allow for end point reading errors), 0.446-M $HCO_3^-$

9.17 a. 1.659 M  b. No  c. Yes. Orange to colorless for indication of end point.

9.18 12.5 units mL$^{-1}$

# CHAPTER 10

10.1 0.0273 M

10.2 0.0674 $N$

10.3 0.0897 $N$, 0.0179$_4$ M

10.4 a. $N_{I_2} = (1/2)[V_{S_2O_3} \times N_{S_2O_3}/V_{I_2}]$
  b. $[I_2(mg/L)] = 6.345 \times 10^4$
  $[V_{S_2O_3} \times N_{S_2O_3}/V_{I_2}]$

10.5 0.255%

10.6 a. Diphenylamine sulfate
  b. Fe(III)  c. No

10.7 a. 14.05%  b. 69.42 (f.w./3)

10.8 a. and b. 0.0542  c. $1.6 \times 10^{14}$
  d. and e. $4.33 \times 10^{-9}$ M  f. 1.19 V

10.9 $[Fe^{2+}] = 0.01111$ M, $[Fe^{3+}] =$

$[Ce^{3+}] = 0.0444$ M, $[Ce^{4+}] = 1.1 \times 10^{-15}$ M; $E = 0.807$ V

10.10 a. 1.36 V  b. $Cr_2O_7^{2-}$  c. 0.036 M (see notes to Appendix III)

10.11 a. $1.393 \times 10^{-4}$ M  b. As long as reactions are reversible, the nitrate will not end up as NO uniquely since the potentials are so close. The control must then be kinetic, and perhaps iron is acting as a specific catalyst.

10.12 3.1 ppm

10.13 a. 1.21 mg/L  b. 2.4 mg/L

10.14 861 ppm

10.15 a. 21.3%  b. Fe(III) more stable, Fe(II) less stable  c. Pink is with Zr d. and e. same: direction of the titration does not affect the chemistry f. $H_5edta^+ + Zr^{4+} \rightleftharpoons Zredta + 5 H^+$; $Zr^{4+} + 2 H_2In^- \rightleftharpoons ZrIn_2^- + 4 H^+$

# CHAPTER 11

11.1 Fixed time

11.2 a. 54.0 Instrument response/ enzyme unit  b. 0.627 unit g$^{-1}$ c. 1.96%

11.3

| | Total Enzyme Activity (units) | Enzyme Activity (units/mg-protein) | n-Fold Purification | Extraction Yield (%) |
|----|----|----|----|----|
| 1 | 12,000 | 0.063 | 1 | 100 |
| 2 | 13,200 | 1.9 | 31 | 110 |
| 3 | 10,200 | 6.9 | 110 | 85 |
| 4 | 9,000 | 98 | 1550 | 75 |
| 5 | 7,560 | 250 | 4000 | 63 |

11.4 a. Slope depends on time choice. Choose time greater than 2 s.  b. — c. Not easy choice. Probably better at 5 where irregularity is less of total, assuming equally precise reading of values.  d. —  e. Fixed time probably better. Get more calibration points over wider range.  f. 1.23 ppm. Will vary with method somewhat.

11.5 —

## CHAPTER 12

12.1   0.56

12.2   a. More positive O oxidized, more negative H reduced   b. More positive $H_2$ oxidized, more negative $O_2$ reduced   c. 1.19 V vs. NHE, $-0.04$ V vs. NHE   d. 1.23 V pH independent

12.3   a. Low   b. High   c. 4.66   d. 12.46

12.4   —

12.5   0.71 M

12.6   a. 1.92 mM   b. 0.044   c. 57.3 mM

12.7   11.0 ppm

12.8   a. $C_0 = \dfrac{C_s V_s}{\dfrac{\gamma_0}{\gamma_a} \cdot 10^{\Delta E/S} \cdot (V_0 + V_s) - V_0}$

b. $(V_0 + V_s)10^{\pm(E_a - k)/S} = \gamma_a(V_0 C_0 + V_s C_s)$

c. $C_0 V_0 = C_s V_{s\text{-intercept}}$ (titration equivalent equation)

12.9   a. $K_{sp}$ or $[Ag^+][Cl^-]$   b. $a_{X^-} = 10^{En/0.059}$ or $e^{En/0.0256}$   c. 0.075 V   d. $-0.149$ V   e. 0.370 V vs. NHE   f. $+0.149$ V   g. $1.82 \times 10^{-5}$ M ($\gamma = 1$)

12.10   Hydrogen primary on Pt and cobalt primary on Hg

12.11   (Approximately 3.2 $\mu$C) 1.14 ppb

12.12   7.6 $\mu$g Fe

12.13   a. $Ag^0 \rightleftharpoons Ag^+ + e^-$; $2\,HCl \rightleftharpoons Cl_2 + 2\,H^+ + 2\,e^-$; $3\,I^- \rightleftharpoons I_3^- + 2\,e^-$; $2\,H_2O + 2\,e^- \rightleftharpoons 2\,OH^- + H_2(g)$; $H_2O \rightleftharpoons MO_2 + 2\,H^+ + 2\,e^-$; $HgNH_3edta^{2-} + NH_4^+ \rightleftharpoons Hg^0(1) + Hedta + 2\,NH_3$ [$pK_a(edta^{4-}) = 10.2$; $pK_a(NH_3) = 9.2$]; $2\,Br^- \rightleftharpoons Br_2^- + 2\,e^-$   b. Positive, positive, positive, positive, negative, positive

12.14   a. $2\,C_5H_5N \cdot HI \rightarrow C_5H_5N \cdot I_2 + 2\,H^+ + 2\,e^-$   b. $2\,e^-$

12.15   1.47 ppm

12.16   a. L to R: Cu, Pb, Cd, Zn   b. R to L: Zn, Cd, Pb, Cu

12.17   a. 6-Hydroxydopa, $L$-dopa, tyrosine   b. No

12.18   a. About 9% low   b. approximately 25% low

12.19   $1.42 \times 10^{-3}$ M

12.20   a. B   b. D

12.21   a. —   b. (Approximately 2.12 mL) 0.84 g   c. 10 units of unknown background + $HNO_2$ only

12.22   a. $Fe(CN)_6^{4-} + Ce^{4+} \rightleftharpoons$ $Fe(CN)_6^{3-} + Ce^{3+}$   b. At negative (reduction) $Fe(CN)_6^{3-} + e^- \rightleftharpoons Fe(CN)_6^{4-}$; $Ce^{4+} \rightleftharpoons Ce^{3+}$   At positive (oxidation) $Fe(CN)_6^{4-} \rightleftharpoons Fe(CN)_6^{3-} + e^-$; $Ce^{3+} \rightleftharpoons Ce^{4+} + e^-$   c. At end point, only $Ce^{3+}$ and $Fe(CN)_6^{3-}$

12.23   (Approximately) $[Cd^{2+}] = 0.12$; $[Pb^{2+}] = 0.57$; $[Cu^{2+}] = 1.7$ $\mu$g/L

12.24   a. (Approximately) $+0.002$, $-0.116$, $-0.263$, $-0.378$ V vs. SCE   b. $-0.47$ V vs. SCE   c. $-0.23$ V vs. NHE

## CHAPTER 13

13.1   ⊓⊔; height same; area increase

13.2   Increase, decrease, same

13.3   89.0 mL; 15.51 (strongly dependent on $t_0$); 13.7 mL; 7.5 mL

13.4   $\alpha_{13} = 1.69$; $\alpha_{23} = 1.15$; $R_{13} = 5.5$; $R_{23} = 0.83$

13.5   All answers band 1 then band 2 and expected to vary with data-reading errors   a. 16.4, 18.7 mL   b. $4.7_5$, $5.4_7$   c. 3.6, 4.1   d. 1.61   e. 1.17   f. $2.42 \times 10^3$   g. 0.103 mm   h. 2.3   i. As part b.

13.6   $\alpha_{1,2} = K_{D2}/K_{D1}$

13.7   Yes; no; use areas

13.8   150, 170, 225 $\mu$g

13.9   Approximately 1, 1.8, $3._3$, 3.5

13.10   Approximately $1.6 \times 10^2$

13.11   Approximately 120%

13.12   a. 0.05%   b. 0.2%

13.13   —

13.14   0.968

## CHAPTER 14

14.1   —

14.2   0.87, 0.48, 0.32

14.3   Depends on readings and data treatment $z \approx 0.4$–$0.7$   b. $y \approx 2.3$, perhaps $\bar{u}$ not constant or surfaces different

14.4   110 kdalton

14.5   trp, pro, ala, asn

14.6   a. 39 cm   b. $1.0 \times 10^3$ psi   c. $1.6 \times 10^3$ psi

14.7   a. Assume $(V_S/V_M)$ constant; $1:2.1:3.2$   b. Assume $(V_S/V_M)$ proportional to chain length;

6.2 : 1.6 : 1. Disagrees with idea that less polar solutes are "more soluble" in organic phase.

14.8 Photometric, greatest $S/N$

14.9 a. Neutral form  b. Either H-bonding or ion exchange is occurring with the cation.

# CHAPTER 15

15.1 a. 390 s  b. 300

15.2 Expect it at ~11.5; no

15.3 Major peaks at 19.6, 20.3, $20.9_5$, 21.6, $22.2_5$, 22.9, 23.5; yes

15.4 a. Hydrogen  b. Hydrogen

15.5 $n$-hexane, 0.893; cyclohexane, 0.965; benzene, 1.00; toluene, 1.0074.

15.6 a. Ethane, 0.4643  b. Propane, 0.0084  c. Isobutane  d. Propane  e. 2610 Btu/ft$^3$; answer may vary somewhat.

# CHAPTER 16

16.1 —

16.2 a. 0.06  b. 0.18  c. 0.60

16.3 $(2.2_4 \times 10^3)$ ppm

16.4 I. $C_2H_3Cl$, vinyl chloride
II. $C_6H_5NO_2$, nitrobenzene
III. $CCl_4$, carbon tetrachloride
IV. $C_2H_4Cl_2$, dichloroethane; this is 1,2-dichloroethane
V. $C_2H_4Cl_2$, dichloroethane; this is 1,1-dichloroethane.

16.5 L to R: $^{50}$Cr, $^{52}$Cr, Cr and Fe, Mn, Fe, Fe, Ni, Co, Ni, Ni, Ni, $^{64}$Ni

16.6 Top to bottom: fluoro-, chloro-, bromo-, and iodobenzene

16.7 I (a)  II (d)  III (c)

16.8 a. 194.1952, 194.2362, 194.2234, 194.1896  b. 63,369

16.9 a. 87.51  b. 267.4 ppm  c. 2.70%  d. $1.90 \times 10^9$ yr

S16.1 8000 V

# CHAPTER 17

17.1 a. 60  b. $1.87 \times 10^{-7}$  c. 120 a.u. mol$^{-1}$

17.2 92.0, 62.0, 31.9, 14.3 %$T$

17.3 a. No—a two-valued function

b. Dilute the sample and run again.

17.4 a. 24,465 K  b. 11,600 K  c. $1.15 \times 10^8$ K

17.5 Co 13.0, Ni 8.7

17.6 $4.7 \times 10^{-4}$ a.u. L $\mu$g$^{-1}$

17.7 0.455

17.8 0.50 (approximately)

17.9 $0.368 = e^{-1}$

17.10 —

17.11 a. $1.5 \times 10^{-3}$ A (2 nm)$^{-1}$ ppm$^{-1}$  b. 1/2

17.12 Co 3.4 g/L, Ni 3.0 g/L

17.13 a. —  b. —

17.14 $(P_0 + 2 P_2)/3$

17.15 (Approximate, in $\mu$g) dimer 11.3, trimer 14.2, tetramer 10.8, hexamer 12.7

17.16 Observe at 480 nm, excite at 366 nm, pH at around 9.3 (not too critical)

17.17 a. Curves b–e; 0.170, 0.33, 0.54, 0.69  b. 0.20, 0.50, 1.20, 2.2  c. $n = 1.0$  d. $E^{o\prime} = +11$ mV

17.18 —

17.19 1.8%

17.20 a. $-3.7\%$  b. $-5.8\%$

17.21 a. 400, 200, 100, 40, 0 $\mu$M  b. 830 $\mu$M a.u.$^{-1}$  c. 22 $\mu$M  d. 0.476 $\mu$g  e. $4.9 \times 10^2$ units/mg  f. 50 $\mu$L of 7.05 mL = 0.7% equivalent to $4s$  g. 37%, no  h. constant time

# CHAPTER 18

18.1 a. Concentrations: 10.87, 6.255, 6.645, 23.80, 10.85, 11.72  RSDs: 0.7%, 1.1%, 0.3%, 0.6%, 0.9%, 0.7%  b. Greater; sampling errors and errors in standards not included in table  c. #2, 1.3%; #3, 0.8%  d. 3.3%

18.2 a. All lines are from level 2 to levels 3 through 35.  b. $n = 1.00028$ approximately

18.3 a. Approximately 5–6%  b. Approximately 15%

18.4 a. —  b. K 6.1 mM; Na 135 mM

18.5 a. 4  b. $(3 \times 10^2)$ ppm

18.6 a. Mg  b. Mg  c. Add Mg, Ca, K to reduce relative variation and place in less sensitive region. (Enhances sensitivity too)

18.7  12 ppb (may vary slightly from graph-reading differences)

18.8  a. No. The areas differ by more than 5%.  b. Peak height, without Ca; peak area, with Ca

# CHAPTER 19

19.1  a. 0 = TMS, cyclohexane, *t*-butanol, acetone, dioxane, benzene  b. 0, 139, 163, 187, 380, 715  c. TMS 10.0, cyclohexane 10.0, butanol 7.5, dioxane 6.7, benzene 5.0, acetone 5.0

19.2  No. Relative to alkyl peak, aromatic should be 60 in pure toluene.

19.3  I–V, respectively, a, e, b, c, f

19.4  I–VI, respectively, c, e, b, a, f, g

19.5  Approximately 588

19.6  —CH$_2$SH

19.7  Upper 0.499996, lower 0.500003  b. 60 MHz upper 0.499997, lower 0.500002; 300 MHz upper 0.499987, lower 0.500012  c. No

# CHAPTER 20

20.1  2.22

20.2  10 %*T*, 11.5 cm$^{-1}$; 1 %*T*, 21.0 cm$^{-1}$

20.3  *2* *p*-toluidine, *3* N-ethyltoluidine, *4* N,N-diethyltoluidine

20.4  I–IX are, respectively, b, a, j, f, d, c, i, h, g

20.5  1-phenyl-2-propanone (or, equivalently, benzyl methyl ketone)

20.6  Calculate about 1 : 10, measure height difference about 1 : 8.3

# CHAPTER 21

21.1  a. Solution is $1 \times 10^{-5}$ *m*; b.p. up by $(5 \times 10^{-6})$ K; f.p. down by $(9.9 \times 10^{-5})$ K  b. Relative error of the f.p. about a factor of 5.2 (520%)

21.2  a. 40,000; 85,000; 2.1  b. 250,000; 700,000; 2.8  c. 118,000; 250,000; 2.1

21.3  $[\eta] = 1.03$ then $\overline{M}_v = 103,000$

21.4  a. Approximately 3603  b. Approximately 3238  c. Approximately 1.11

*Credits*   *(continued from p. iv)*

Table 10.3. From I. M. Kolthoff and V. A. Stenger, *Volumetric Analysis,* 2nd ed. (New York: Wiley–Interscience). Reprinted by permission of John Wiley & Sons, Inc.

Excerpts pp. 341–343, Fig. 11.2, Fig. 11.3, Table 11.1. From B. F. Quin and P. H. Woods, *Analyst* 104 (1979): 552–559. Reprinted by permission of The Royal Society of Chemistry.

Fig. 11.4.1. From M. A. Koupparis et al., *Analyst* 107 (1982): 1039. Reprinted by permission of The Royal Society of Chemistry.

Table 12.2. Adapted by permission from *The Beckman Handbook of Applied Electrochemistry.* © 1980 Beckman Instruments, Inc.

Fig. 12.33. Reprinted with permission from *Analytical Chemistry* 44 (September 1972). Copyright 1972 American Chemical Society.

Fig. 12.34. Courtesy of Bioanalytical Systems, Inc.

Fig. 12.36. From R. N. Adams et al., "The measure of dopamine and 5-hydroxytryptamine release in CNS of freely moving unanesthetised rats," *British Journal of Pharmacology* 64 (1978): 470P–471P. Reprinted by permission of the author and The Macmillan Press Ltd.

Fig. 12.17.1. From T. A. Last, *Anal. Chim. Acta* 155 (1983), Fig. 1. Reprinted by permission of Elsevier Science Publishers B.V.

Fig. 12.23.1. From S. B. Adeloju et al., *Anal. Chim. Acta* 148 (1983), Fig. 7. Reprinted by permission of Elsevier Science Publishers B.V.

Fig. 13.1. Adapted with permission from W. R. Supina and R. S. Henley, *Chemistry* 37 (1964). Copyright 1964 American Chemical Society.

Fig. 13.3. From K. Rubinson, *Biochimica et Biophysica Acta* 687 (1982): 315–320. Reprinted by permission of Elsevier Science Publishers B.V.

Fig. 13.18. Reprinted with permission from *Journal of Chemical Education* (1969). Copyright 1969 American Chemical Society.

Fig. 13.8.1. From R. K. Beerthuis et al., "Gas-Liquid Chromatographic Analysis of Higher Fatty Acids and Fatty Acid Methyl Esters," *Annals of the New York Academy of Science* 72 (1959): 620. Reprinted by permission.

Fig. 14.6. From W. Heitz, B. Bömer, and H. Ullner, *Die Makromolekulare Chemie* 121 (1969). Reprinted by permission of the publishers, Hütig & Wepf Verlag, Basel.

Fig. 14.8.1. From D. R. Jenke et al., *Anal. Chim. Acta* 155 (1983): 279. Reprinted by permission of Elsevier Science Publisher B.V.

Fig. 15.2. Courtesy of J & W Scientific, Inc., Rancho Cordova, CA.

Excerpt on pp. 540–541. From D. I. Reese, *Analyst* 90 (1965): 568–569. Reprinted by permission of The Royal Society of Chemistry.

Excerpt on pp. 541–542. From M. L. Richardson and P. L. Luton, *Analyst* 91 (1966): 520–521. Reprinted by permission of The Royal Society of Chemistry.

Fig. 15.10. From R. J. Laub and J. H. Purnell, *Journal of Chromatography* 71 (1975): 112. Reprinted by permission of Elsevier Science Publishers B.V.

Fig. 16.2b and Fig. 16.4. From F. W. McLafferty, *Introduction to Mass Spectroscopy,* 2nd ed., © 1973. Reprinted by permission of The Benjamin/Cummings Publishing Co.

Excerpt, pp. 578–579, Table 16.1. Reprinted from R. L. Walker, J. A. Carter, and D. H. Smith, *Analytical Letters* 14(A19) (1981): 1603–1612, by courtesy of Marcel Dekker, Inc.

Fig. 16.5. From C. Fenselau in T. Kuwana and T. Osa, eds., *Physical Methods in Chemical Analysis.* Copyright © 1978 by Academic Press, Inc. Reprinted by permission.

Table 16.7. From J. H. Benyon, *Mass Spectrometry and Its Applications to Organic Chemistry* (Amsterdam: Elsevier, 1960). Reprinted by permission of Elsevier Science Publishers B.V.

Fig. 16.8. From R. Ryhage et al., *Arkiv. Kemi.* 26 (1966): 305. Reprinted by permission of The Royal Swedish Academy of Sciences.

Fig. 16.4.1. From B. S. Middleditch et al., *Mass Spectrometry of Priority Pollutants* (New York: Plenum, 1981). Reprinted by permission.

Excerpt, p. 617. From L. A. Schwalbe and R. N. Rogers, *Anal. Chim. Acta* 135 (1982): 3–49. Reprinted by permission of Elsevier Science Publishers B.V.

Fig. 17E.1. Reproduced by permission of the National Research Council of Canada from the *Canadian Journal of Physics,* Vol. 48, 1970.

**Credits**   *(continued from previous page)*

Excerpt on p. 740, Table 18.16. From M. L. Parsons, S. Major, and A. R. Foster, *Applied Spectroscopy* 37 (1983): 411–418. Reprinted by permission of the Society for Applied Spectroscopy.

Fig. 18.1a. From *Physical Chemistry* by P. W. Atkins, Fig. 14.1. W. H. Freeman and Company. Copyright © 1978. Reprinted by permission.

Fig. 18.2a. Reprinted by permission from *Journal of Chemical Education,* December 1974 (cover). Copyright 1974 American Chemical Society.

Fig. 18.2b. From S. Walker and H. Straw, *Spectroscopy,* Vol. 1, 1961. Reprinted by permission of Chapman & Hall Ltd.

Fig. 18.3. From H. G. Kuhn, *Atomic Spectra* (Orlando, FL: Academic Press, 1962). Reprinted by permission.

Fig. 18.5a. From R. K. Winge, V. A. Fassel, V. J. Peterson, and M. A. Floyd, *Applied Spectroscopy* 36 (1982): 210. Reprinted by permission of the Society for Applied Spectroscopy.

Fig. 18.16. From C. W. Clark, K. T. Lu, and A. F. Starace in H.-J. Beyer and H. Kleinpoppen, eds., *Progress in Atomic Spectroscopy — Part C* (London: Plenum, 1983). Reprinted by permission.

Fig. 18.20. From J. Xu, H. Kawaguchi, and A. Mizuike, *Applied Spectroscopy* 37 (1982): 123. Reprinted by permission of the Society for Applied Spectroscopy.

Fig. 18.3.1. From K. C. Ng and J. A. Caruso, *Anal. Chim. Acta* 143 (1982): 209. Reprinted by permission of Elsevier Science Publishers B.V.

Fig. 19.8. From John R. Dyer, *Applications of Absorbtion Spectroscopy of Organic Compounds,* © 1965, pp. 84–85. Reprinted by permission of Prentice-Hall, Inc., Englewood Cliffs, N.J.

Fig. 19.25. From *Experimental Methods in Organic Chemistry,* 2nd ed., by James A. Moore and David L. Dalrymple. Copyright © 1976 by W. B. Saunders Company. Reprinted by permission of CBS College Publishing.

Table 19.1. Reproduced with permission of IBM Instruments.

Fig. 20.6a. From N. B. Colthrop, *Journal of the Optical Society of America* 40 (1950): 397. Reprinted by permission.

Fig. 20.4.1 Permission for the publication herein of Sadtler Standard Spectra® has been granted, and all rights are reserved, by Sadtler Research Laboratories, Division of Bio-Rad Laboratories, Inc.

Fig. 21.7. Redrawn from E. A. Collins, *Experiments in Polymer Science* (New York: Wiley–Interscience, 1981). By permission of John Wiley & Sons, Inc.

Fig. 21.10. From G. Gee, *Transactions of the Faraday Society* 40 (1944): 261. Reprinted by permission of The Royal Society of Chemistry.

Fig. 21.15. Courtesy Aldrich Chemical Co., Inc.

Fig. 21.17 (whole spectrum), Fig. 21.18, Fig. 21.21, and Fig. 21.22. From F. A. Bovey and T. K. Kwei, in F. A. Bovey and F. H. Winslow, eds., *Macromolecules* (Orlando, FL: Academic, 1979). Reprinted by permission.

Fig. 21.19. From F. A. Bovey, *Polymer Conformation and Configuration* (Orlando, FL: Academic, 1969). Reprinted by permission.

# Index

Page entries in italics indicate beginnings of more extensive coverage of the listed topic.

A т after a number indicates a table.

## Physical Constants

| Term | Symbol | Multiplication Factor* | |
|------|--------|-----------------------|---|
| Avogadro's number | $N$ | 6.022 045 (31) | $\times 10^{23}$ particles mol$^{-1}$ |
| Elementary charge | $e$ | 1.602 189 2 (46) | $\times 10^{-19}$ coulomb |
| | | 4.803 242 (14) | $\times 10^{-10}$ esu |
| Electron rest mass | $m_e$ | 9.109 534 (47) | $\times 10^{-31}$ kg |
| | | | $\times 10^{-28}$ g |
| Atomic mass unit† | $u$ | 1.660 566 (8) | $\times 10^{-24}$ g |
| Gas constant | $R$ | 8.314 41 (26) | J mol$^{-1}$ K$^{-1}$ |
| | | | V C mol$^{-1}$ K$^{-1}$ |
| | | | $\times 10^7$ erg mol$^{-1}$ K$^{-1}$ |
| | | 8.205 68 (26) | $\times 10^{-5}$ m$^3$ atm mol$^{-1}$ K$^{-1}$ |
| | | 8.205 68(26) | $\times 10^{-2}$ L atm mol$^{-1}$ K$^{-1}$ |
| | | 1.987 19 (6) | cal mol$^{-1}$ K$^{-1}$ |
| Faraday constant ($=Ne$) | $F$ | 9.648 456 (27) | $\times 10^4$ C mol$^{-1}$ |
| | | 2.892 534 2 (82) | $\times 10^{14}$ esu mol$^{-1}$ |
| Boltzmann's constant ($=R/N$) | $k$ | 1.380 662 (44) | $\times 10^{-23}$ J K$^{-1}$ |
| | | | $\times 10^{-16}$ erg K$^{-1}$ |
| Planck's constant | $h$ | 6.626 175 (36) | $\times 10^{-34}$ J s |
| | | | $\times 10^{-27}$ ergs |
| Proton rest mass | $m_p$ | 1.672 648 5 (86) | $\times 10^{-27}$ kg |
| | | | $\times 10^{-24}$ g |
| Bohr magneton $\left(=\dfrac{e\hbar}{2m_e c}\right)$ | $\mu_B$ | 9.274 078 (36) | $\times 10^{-24}$ J T$^{-1}$ |
| | | | $\times 10^{-21}$ erg G$^{-1}$ |
| Speed of light in vacuum | $c$ | 2.997 924 58 (1.2) | $\times 10^8$ m s$^{-1}$ |
| | | | $\times 10^{10}$ cm s$^{-1}$ |

* Numbers in parentheses are the one-standard-deviation uncertainties in the last digits.
† Calculated from $N$ and $^{12}$C $= 12\ u$ exactly.
*Source:* Cohen, E. R.; Taylor, B. N. *J. Phys. Chem. Ref. Data* **1973,** *2,* 663.